Advances in Organic Farming

Advances in Organic Farming

L.P. Awasthi

CABI

CABI is a trading name of CAB International

CABI
Nosworthy Way
Wallingford
Oxfordshire OX10 8DE
UK

Tel: +44 (0)1491 832111
E-mail: info@cabi.org
Website: www.cabi.org

CABI
200 Portland Street
Boston
MA 02114
USA

T: +1 (617)682-9015
E-mail: cabi-nao@cabi.org

A catalogue record for this book is available from the British Library, London, UK.

ISBN-13: 9781800626836 (hardback)
9781800626843 (ePDF)
9781800626850 (ePub)

DOI: 10.1079/9781800626850.0000

Commissioning Editor: Ward Cooper
Editorial Assistant: Helen Elliott
Production Editor: James Bishop

Typeset by Exeter Premedia Services Pvt Ltd, Chennai, India
Printed in the USA

Contents

Contributors

Abbyssina Mushunje, Department of Agricultural Economics & Extension, University of Fort Hare, Private Bag X1314, Alice 5700, South Africa

Abdel Jabbar Al-Rajab, Etcectera Publications, Chesterville, ON, Canada

Abdi-Khalil Edriss, Department of Agricultural and Applied Economics, Lilongwe University of Agriculture and Natural Resources, Malawi

Abdul J. Shango, World Vegetable Center, Eastern and Southern Africa, Arusha, Tanzania

Adnan Šišić, Department of Ecological Plant Protection, University of Kassel, 37213 Witzenhausen, Germany

Adrian Crew, Department of Health and Applied Sciences, University of the West of England, Bristol, BS16 1QY, UK

Alessia Iacovone, CBC Bioplanet, Cesena, Italy

Alice Caselli, Center of Plant Sciences, Sant'Anna School of Advanced Studies, Piazza Martiri della Libertà 33, 56127 Pisa, Italy

Alison Wills, Department of Animal and Agriculture, Hartpury University, Gloucester, GL19 3BE, UK

Aliyeh Salehi, Division of Organic Farming (IFÖL), University of Natural Resources and Life Sciences, Gregor-Mendel-Straße 331180 Vienna, Austria

Amjad Ahmad, Department of Tropical Plant and Soil Sciences, University of Hawaii at Manoa, USA

Andreas Surböck, Division of Organic Farming (IFÖL), University of Natural Resources and Life Sciences, Gregor-Mendel-Straße 331180 Vienna, Austria

Anne-Kristin Løes, Norwegian Centre for Organic Agriculture (NORSØK), Tingvoll, Norway

Anthony, O. Onoja, Department of Agricultural Economics and Extension, University of Port Harcourt, Nigeria.

Arnd Verschwele, Julius Kühn-Institut (JKI) - Bundesforschungsinstitut für Kulturpflanzen, Institut für Pflanzenschutz in Ackerbau und Grünland, Messeweg 11/12,38104, Braunschweig, Germany

Arup Sarkar, Department of Agronomy, Bidhan Chandra Krishi Viswavidyalaya, Mohanpur – 741252, Nadia (WB), India

Ashoka Gamage, China Sri Lanka Joint Research and Demonstration Centre for Water Technology (JRDC), Meewathura Road, Peradeniya 20400, Sri Lanka

Ben Hassine Mortadha, Department of Chemical, Pharmaceutical and Agricultural Sciences, University of Ferrara, Via Luigi Borsari 46, 44121 Ferrara, Italy

Bernhard Freyer, Division of Organic Farming (IFÖL), University of Natural Resources and Life Sciences, Gregor-Mendel-Straße 33,1180 Vienna, Austria

Bernhard Schlatter, Research Institute of Organic Agriculture FiBL, Frick, Switzerland

Bharti Suthar, International Crops Research Institute for the Semi-Arid Tropics (ICRISAT) - Patancheru, Hyderabad, India

Bina Mistry, Thermo Fisher Scientific, Overland Park, Kansas, USA

Dang Khoa Tran, Faculty of Agronomy, University of Agriculture and Forestry, Hue University, 102 Phung Hung, Hue City 49000, Vietnam

Daniele Borgatti, Department of Chemical, Pharmaceutical and Agricultural Sciences, University of Ferrara, Via Luigi Borsari 46, 44121 Ferrara, Italy

Davi Savietto, GenPhySE, Université de Toulouse, INRAE, ENVT, F-31326 Castanet Tolosan, France

David Frost, Principal, Frost Consulting. Formerly Senior Consultant, ADAS Pwllpeiran Research Centre, Cwmystwyth, Wales, UK and Marketing Strategy Coordinator, Organic Centre Wales, Aberystwyth University, Wales, UK

Day Teixeira, Department of Animal and Agriculture, Hartpury University, Gloucester, GL19 3BE, UK

Dimitrios A. Gougoulis, Clinic of Medicine, Faculty of Veterinary Medicine, School of Health Sciences, University of Thessaly, Trikalon 224, 43100, Karditsa, Greece

Ehsan Ebrahimi, Faculty of Agricultural and Environmental Sciences, Department of Agronomy (Pflanzenbau), University of Rostock, Justus-von-Liebig-Weg 6,18059 Rostock, Germany

Emanuele Radicetti, Department of Chemical, Pharmaceutical and Agricultural Sciences, University of Ferrara, Ferrara, Italy

Fabio Caporali, DAFNE (Dipartimento di Scienze Agrarie e Forestali), Via San Camillo de Lellis, 01100 Viterbo, Italy

Fikru K. Alemayehu, Inland Business School, Inland Norway University of Applied Sciences, Lillehammer, Norway

Francesco Solfanelli, Department of Agricultural, Food and Environmental Sciences, Università Politecnica delle Marche, Via Brecce Bianche, 60129 Ancona, Italy

Gabriel Adewunmi Eyinade, Department of Agricultural Economics & Extension, University of Fort Hare, Private Bag X1314, Alice 5700, South Africa

Gabriele Gollner, Division of Organic Farming (IFÖL), University of Natural Resources and Life Sciences, Gregor-Mendel-Straße 331180 Vienna, Austria

Georgios I. Papakonstantinou, Clinic of Medicine, Faculty of Veterinary Medicine, School of Health Sciences, University of Thessaly, Trikalon 224, 43100, Karditsa, Greece

Georgia Terzidou, Food Law Experts, Themistokleous 10677, Athens, Greece

Gheorghe Cristian Popescu, Department of Applied Sciences and Environmental Engineering, National University of Science and Technology POLITEHNICA Bucharest, University Center of Pitesti, 110040, Romania

Gudbrand Lien, Inland Business School, Inland Norway University of Applied Sciences, Lillehammer, Norway.

Habtamu Alem, Department of Economics Society, the Norwegian Institute of Bioeconomy Research (NIBIO), Ås, Norway

Hamdi Özaktan, Department of Field Crops, Agricultural Faculty, Erciyes University, Kayseri, Türkiye

Hector Valenzuela, Dept. of Plant and Environmental Protection Science, University of Hawaii at Manoa, Honolulu, Hawaii, USA

Helga Willer, Research Institute of Organic Agriculture FiBL, Frick, Switzerland,

Hue, N.V. Department of Tropical Plant and Soil Sciences, University of Hawaii at Manoa, USA

Ingrid Olesen, Department of Production Biology, Nofima, Ås, Norway

Jacopo Menghi, CBC Bioplanet, Cesena, Italy

Jagjeet Singh Gill, Scientist and Quality Assurance Person, Green Mountain Health Alliance, Kaleden, BC, Canada

Jan Trávníček, Czech Organics, Staré Město, Czech Republic

Janet F. Maro, Sustainable Agriculture Tanzania (SAT), Morogoro, Tanzania

Jelena Golijan Pantović, University of Belgrade, Faculty of Agriculture, Chair for Genetics, Plant Breeding and Seed Science, Nemanjina 6, 11080 Belgrade, Serbia

Jochen Mayer, Agro scope, Department of Agroecology and Environment, Reckenholzstrasse 191, CH-8046 Zürich, Switzerland

Joel Allainguillaume, Department of Health and Applied Sciences, University of the West of England, Bristol, BS16 1QY, UK

John A. Kazembe, Department of Environmental Science, Lilongwe University of Agriculture and Natural Resources, Malawi

John Paull, School of Land & Food, University of Tasmania, Hobart, Tasmania, Australia

Joshua Cabell, Norwegian Centre for Organic Agriculture (NORSØK), Tingvoll, Norway

Jude Obi, Department of Soil Science and Land Resources Management, Faculty of Agriculture, University of Uyo, uyo, Akwa Ibom, State, Nigeria

Kandiah Pakeerathan, Department of Agricultural Biology, Faculty of Agriculture, University of Jaffna, Kilinochchi, Sri Lanka

Kanu Murmu, Department of Agronomy, F/Ag, Bidhan Chandra Krishi Viswavidyalaya, P.O- Krishi Viswavidyalaya, Mohanpur – 741252, Nadia (WB), India

Lampros Fotos, Clinic of Medicine, Faculty of Veterinary Medicine, School of Health Sciences, University of Thessaly, Trikalon 224, 43100, Karditsa, Greece

Laurence Fortun-Lamothe, GenPhySE, Université de Toulouse, INRAE, ENVT, F-31326 Castanet Tolosan, France

Leonard Pollara, Biological Agriculture Specialist, Systems Specialist, Organic Sage Consulting LLC, International Organic Inspectors Association - Continuing Education Northport, Michigan, USA

Linda Chinenyenwa Familusi, Department of Agricultural Economics and Extension, University of Port Harcourt, Nigeria

Lisa Williams, Department of Animal and Agriculture, Hartpury University, Gloucester, GL19 3BE, UK

Lucas T. Manda, CARE-WWF Alliance, Iringa, Tanzania

Luciana Bignardi Brisola da Costa, College of Veterinary Medicine, The Ohio State University (Columbus, Ohio, USA

Marek Kopecký, Department of Agroecosystems, Faculty of Agriculture and Technology, University of South Bohemia in Ceske Budejovice, Branišovská 1645/31A, 370 05 České Budějovice, Czech Republic

Maria Claudia Dussi, Department of Agricultural Sciences, National University of Comahue, Rio Negro, Patagonia, Argentina

Maria Eugenia Ortiz Escobar, Soil Science Department, Federal University of Ceara, Brazil

Maria Fernanda Reyes, Stanford University Graduate School of Business, San Francisco Bay, USA

Mariam Atait, Dept. DAFNE - University of Tuscia, Via S. Camillo De Lellis 01100 VITERBO, Italy

Marie-Luise Wohlmuth, Division of Organic Farming (IFÖL), University of Natural Resources and Life Sciences, Gregor-Mendel-Straße 331180 Vienna, Austria

Mary Cole, Director • Principal Scientist • The University of Melbourne, AgPath Pty Ltd, Bunurong Country, 105 Gunn Road, Garfield VIC 3814, Australia

Masoud Latifian, Horticulture Research institute, Agricultural Research Education and Extension Organization (AREEO), Iran

Matt J. Bell, Department of Animal and Agriculture, Hartpury University, Gloucester, GL19 3BE, UK

Matthew Russell Gates, Cannabis Horticultural Association, San Diego, California, USA

Matthias Schöller, Humboldt Universität zu Berlin, Department of Life Sciences, Berlin, Germany

Mehmet Yaman, Department of Horticulture, Agricultural Faculty, Erciyes University, Kayseri, Türkiye

Mthakati, A. R. Phiri, Department of Agricultural and Applied Economics, Lilongwe University of Agriculture and Natural Resources, Malawi

Mohamed Allam, Department of Agronomy, Faculty of Agriculture, Assuit University, Assuit, Egypt

Monica Popescu, Department of Natural Sciences, University of Pitesti, 110040, Pitesti, Romania

Moulay Chrif Smaili, National Institute for Agricultural Research (INRA), Kenitra, Morocco

Muazzez Acar Cömert, Department of Animal Science, Ege University, İzmir, Türkiye

Muhammad Ahsan, Department of Horticultural Sciences, the Islamia University of Bahawalpur, Pakistan

Nagarjuna Reddy Desam, Anantha Lakshmi Institute of Technology & Sciences, Ananthapuram, India

Nyandula S. Mwaijande, Department of Crop Science and Horticulture, Sokoine University of Agriculture, Morogoro, Tanzania

Olabisi Somefun, Department of Plant Physiology and Crop Production, College of Plant Science and Crop Production, Funaab, Nigeria

Othmane Merah, Laboratoire de Chimie Agro-industrielle (LCA), Institut National de la Recherche Agronomique, Université de Toulouse, CEDEX 4, 31030 Toulouse, France

Petr Konvalina, Department of Agroecosystems, Faculty of Agriculture and Technology, University of South Bohemia in Ceske Budejovice, Branišovská 1645/31A, 370 05 České Budějovice, Czech Republic

Pete Maxfield, Department of Health and Applied Sciences, University of the West of England, Bristol, BS16 1QY, UK

Philippe Belliard, Zoology Department, University of Innsbruck, Innsbruck, Austria

Pierre Ellssel, Division of Organic Farming (IFÖL), University of Natural Resources and Life Sciences, Gregor-Mendel-Straße 331180 Vienna, Austria

Quintarelli Valentina, Department of Chemical, Pharmaceutical and Agricultural Sciences, University of Ferrara, Via Luigi Borsari 46, 44121 Ferrara, Italy

Rachel Cavallo, Agpath Pty Ltd, 105 Gunn Road, Garfield, Victoria, 3814, Australia

Rachid Mrabet, Directeur de Recherche/Senior Scientist, Institut National de la Recherche Agronomique (INRA)Avenue de la Victoire, P.O. Box 415, Rabat 10000, Morocco

Raffaele Zanoli, Faculty of Engineering & Department of Agricultural, Food and Environmental Sciences, Università Politecnica delle Marche, Via Brecce Bianche, 60129 Ancona, Italy

Rameshwari Ramaswamy, Department of Biotechnology, Cauvery College for Women, Tiruchirappalli-620018, India

Raya J. Amara, Tanzania Agricultural Research Institute (TARI), Ukiriguru Centre, Mwanza, Tanzania

Roberto Mancinelli, Agricultural and Environmental Sciences, Dept. DAFNE - University of Tuscia, Via S. Camillo De Lellis, 01100 VITERBO (Italy)

Ruggero Petacchi, Center of Plant Sciences, Sant'Anna School of Advanced Studies, Piazza Martiri della Libertà 33, 56127 Pisa, Italy

Ryoichi Komiya, Organic Shizukuishi,102-667 Ukai Rinan Takizawa City Iwate Prefecture 020-0682, Japan

Sara Hansdotter, Norwegian Centre for Organic Agriculture (NORSØK), Tingvoll, Norway

Sito-Obong Udofia, Department of Animal and Agriculture, Hartpury University, Gloucester, GL19 3BE, UK

Shehu. F.G. Yusuf, Sustainable Livelihood Supports and Development Network for Africa (SLIDEN AFRICA), Abeokuta, Nigeria

Sramana Sen Sarma, Department of Agronomy, F/Ag, Bidhan Chandra Krishi Viswavidyalaya, P.O- Krishi Viswavidyalaya, Mohanpur – 741252, Nadia (WB), India

Stefanie Christmann, Farming with Alternative Pollinators (FAP) International Center for Agricultural Research in Dry Areas, CARDA), P.O. Box: 6299, 10112 Rabat, Morocco

Taab Alireza, Department of Agronomy and Plant Breeding, Faculty of Agriculture, Ilam University, Box: 69315-516 Ilam, Iran

Tahira Batool, Department of Health and Applied Sciences, University of the West of England, Bristol, BS16 1QY, UK

Tatiana Rittl, Norwegian Centre for Organic Agriculture (NORSØK), Tingvoll, Norway

Thamali Wanasinghe, Postgraduate Institute of Agriculture, University of Peradeniya 20400, Sri Lanka

Thangapandian Dhanikodi, Aadhavan College of Arts & Science, Bharathidasan University, Tiruchirappalli, Trichy - 621315, Tamil Nadu, India

Thierry Gidenne, GenPhySE, Université de Toulouse, INRAE, ENVT, F-31326 Castanet Tolosan, France

Timo Sipiläinen, Economics and Management, University of Helsinki, Helsinki, Finland

Thi Xuan Phuong Tran, Faculty of Agronomy, University of Agriculture and Forestry, Hue University, 102 Phung Hung, Hue City 49000, Vietnam

Uma Maheswari Rajadurai, Department of Biotechnology, Cauvery College for Women, Tiruchirappalli-620018. India

Uygun Aksoy, Department of Horticulture, Ege University (formerly), Association of Ecological Agriculture Organization, İzmir, Türkiye

Vasileios G. Papatsiros, Faculty of Veterinary Medicine, School of Health Sciences, University of Thessaly, Trikalon 224, 43100 Karditsa, Greece

Victor Olowe, Crop Production Research Programme, Institute of Food Security, Environmental Resources and Agricultural Research (IFSERAR), Federal University of Agriculture, Abeokuta (FUNAAB), Nigeria

Violeta Alexandra Ion, Research Center for Studies of Food Quality and Agricultural Products, University of Agronomic Sciences and Veterinary Medicine of Bucharest, Bucharest, Romania

Wenhao Zhang, Centre for Machine Vision, School of Engineering, University of the West of England, Bristol, BS16 1QY, UK

Wasana Gunawardana, China Sri Lanka Joint Research and Demonstration Centre for Water Technology (JRDC), Meewathura Road, Peradeniya 20400, Sri Lanka

Wilson C. Wilson, Tanzania Livestock Research Institute (TALIRI), Mbeya, Tanzania

Wing K.P. Ng, Department of Animal and Agriculture, Hartpury University, Gloucester, GL19 3BE, UK

Yasavi Jayakodi, National Institute of Fundamental Studies (NIFS), Hantana Rd, Kandy 20000, Sri Lanka

Yohana C. Haule, Sustainable Agriculture Tanzania (SAT), Morogoro, Tanzania

Youssef Bencharki, International Center for Agricultural Research in Dry Areas (ICARDA), Rabat, Morocco

Yves Theoneste Murindangab, Department of Agroecosystems, Faculty of Agriculture and Technology, University of South Bohemia in Ceske Budejovice, Branišovská 1645/31A, 370 05 České Budějovice, Czech Republic

About the Editor

Prof. (Dr.) L.P. Awasthi – Dean, School of Agriculture, R.N.B. Global University, Bikaner, India; formerly Research Scientist/Professor, Amity University; Professor and Head, Department of Plant Pathology, N.D. University of Agriculture & Technology, Faizabad, India – is a distinguished agriculture scientist with a strong background in molecular plant virology and organic farming and with proven ability to undertake independent or collaborative scientific studies, as demonstrated by his successful completion of a number of mega research projects funded by various international funding organizations. His professional experience includes more than 50 years of teaching and conducting research. He has published more than 400 research papers in foreign and Indian journals of repute, has written many popular articles, and has edited/authored books and contributed chapters for edited books and laboratory manuals. He has guided many MSc (Ag.) and PhD students and has been working as an Editor/Referee on the editorial boards of many Indian and foreign journals. He has been Visiting Professor at Karl-Marx University, Leipzig; University of Berlin; Institute of Plant Pathology, Ascherslavan; and Halle University, Germany. He has been a member in the phytosanitary certificate-issuing authorities: Ministry of Agriculture; Department of Agriculture and Cooperation; Directorate of Plant Protection, Quarantine and Storage; Government of India. He is presently special invited member in the state-level Pest and Disease Surveillance Advisory Unit, Government of Uttar Pradesh, India.

Prof. Awasthi has long been associated with many professional societies and was given a Plant Pathology Leadership Award by the Indian Phytopathological Society, India; and an outstanding virologist award by the International Virological Association for outstanding contribution, scientific excellence and distinguished services in the cause of plant pathological research, education and technology dissemination, which have impacted science of plant pathology in the country. He has been commended for disseminating knowledge of organic mushroom cultivation to improve the socio-economic conditions of farmers below the poverty line and women's empowerment, and also for the popularization of organic farming.

Foreword

Dr. Amber Sciligo
Director of Science Programs
The Organic Center, Washington D.C.

The global market for organic food, fiber and natural products continues to grow at a steady pace. This growth is supported by government initiatives and monetary incentives at a global scale, as well as an interest by younger generations to slow the trajectory of climate change and pollution. Organic farming has been formally recognized as climate-smart and sustainable, and government agencies, academics, farmers, and consumers are increasingly understanding the many benefits organic farming offers to the health of the land, animals, people, and communities.

To continue the expansion of organic production that meets sustainability goals of individuals, businesses, and governments, more research and communication of that research is needed. Every crop in every region can be organic, which means organic farming faces all agronomic challenges. And those challenges are context-dependent—production challenges can vary amongst regions and climatic zones. To address these challenges, organic farmers by regulation, utilize different tools than those used for industrialized, chemical agriculture. Organic farmers must rely more heavily on ecosystem services and natural ecosystem function, which requires system-level observation, monitoring, and innovation to optimize crop production.

Compilations of case studies can help illuminate patterns across the wide variation that exists within the definition of organic and can offer guidance to farmers and their advisors. The resulting baseline of data can guide future research questions for the next generation of organic researchers and students. This textbook offers a valuable contribution to the growing body of literature that demonstrates the benefits of organic and helps guide the readers through the opportunities and challenges that exist for organic farm management.

Through case studies and literature reviews from across the world, this book offers insight into the production challenges that organic farmers face and offers important solutions to these challenges. Readers will learn how organic is defined in practice and philosophy, for written policies and legislation, in the field as a system and as a farm management style that works with and leverages natural, ecosystem services and multifunctionality. Benefits to the sustainability of our global food system are highlighted, while acknowledging the continual challenges that farmers must overcome to ensure they produce enough yield while simultaneously replenishing health to the soil, improving biodiversity, and protecting water, air and people from toxic chemical contamination.

Some of the major categories of challenges that organic farmers face includes the management of: soil fertility, pests and diseases, weeds, increased labor requirements, increased production costs, yield gaps during and after organic transition, and lack of access to critical resources like land, community, and knowledge. The research presented in this textbook and the subsequent sharing of this information can help organic farmers overcome many of these challenges and successfully "feed the world".

Looking forward, the organic industry continually needs more research. Research not only equips farmers with tools for successful production success, but it also provides clarification of the benefits of organic to businesses, consumers and policymakers. While monetary investments into organic research are increasing around the globe, more infrastructure also needs to be built. More students and faculty need training in interdisciplinary and systems sciences, and they must also have a basic understanding of the rules and markets that guide organic farmers in their decision-making.

As the world of organic research advances, topics that need more attention and are currently outside of the scope of this textbook include the impacts of organic production on nutrition, dietary and occupational exposure to pesticides, multifunctionality analyses, measurements of the externalities (costs) of industrialized chemical farming, and whole system measurements of the benefits of organic. Many textbooks and semesters of curriculum are needed to warrant all the research that currently exists and that which will continue to expand.

While summarizing organic agriculture in a single textbook is tall order, this book offers a ccomprehensive overview of benefits, opportunities, challenges, and solutions for organic production. It pulls together valuable science from around the globe, highlighting that organic farming is a global strategy and movement. Through cooperation, collaboration, and communication at all scales, organic production will continue to thrive and grow, creating a more sustainable food future for us all.

"Advances in Organic Farming", edited by Prof. L.P. Awasthi, provides a kaleidoscopic perspective on the adaption, and expansion of different aspects of Organic Farming. Much care and foresight has been taken by Prof. L.P. Awasthi in the selection of topics and their assignments to outstanding contributing authors from across the Globe. and compiling has gone into this book. I commend him for this resulting compilation of chapter contributions. I am confident that the book "Advances in Organic Farming" will serve as a valuable resource for university students and professors and as a great resource for researchers and extension professionals who specialize in or are curious about Organic Farming.

(Dr. Amber Sciligo)

Preface

Organic farming is intended to overcome chemical problems both in the field and in plants. This is determined by the use of fertilizers of organic origin such as compost, or green manures, in combination with techniques like crop rotation, etc. Many researchers have proved that organic fertilizer application in agriculture improves soil health by enhancing biogeochemical properties. Moreover, organic farming has been claimed to promote climate-smart agriculture. In tropical countries, the influence of climate is also an important aspect that must be considered. Considering ecological principles as the base for crop production, organic farming is analogous worldwide. The General Assembly of IFOAM defines "organic Agriculture as a production system that sustains the health of soils, ecosystems, and people."

It relies on ecological processes, biodiversity, and cycles adapted to local conditions, rather than the use of inputs with adverse effects. Organic agriculture combines tradition, innovation, and science to benefit the shared environment and promote fair relationships and good quality of life for all involved. Organic farming is a fast-growing sector as people become more health-conscious, the demand for organic food is skyrocketing in the local market. The fact is, organic agriculture currently occupies only 1.5% of the global total agricultural land.

Organic farming, an approach to producing crops using natural agricultural inputs, is based on different components of crop production, ecological principles, promotion of biodiversity, biological cycles, and organic matter recycling to maintain and improve soil fertility and environmental sustainability. The regulations for organic crop cultivation prohibit the use of chemo-synthetic pesticides, mineral fertilizers, growth promoters, and Genetically Modified Organisms. Indiscriminate use of these chemicals in conventional farming poses a serious threat to the quality of products as well as the environment. Concern about food safety and security and environmental sustainability is increasing among scientists, administrators, and environmentalists.

The British botanist Sir Albert Howard is often referred to as the father of modern organic agriculture. Howard, having worked in India as an agricultural researcher, gained much inspiration from the traditional and sustainable farming practices he encountered there and advocated for their adoption in the West.

The indiscriminate use of agrochemicals is believed to be causing human and animal health problems and environmental pollution. Therefore, it is felt necessary to make people more aware of the potential adverse effects of consuming food products produced by using agrochemicals. The methods and practices of organic farming have gained more interest recently. The agricultural education sector is increasingly incorporating teaching material on organic farming or more sustainable food production. There are gaps in the existing literature that are likely to be filled up to some extent by this publication.

The book "Advances in Organic Farming" describes critically reviewed key aspects of organic farming and provides a unique and timely science-based resource for students, researchers, teachers, extension workers, primary producers, and others worldwide. This book is comprised of seventeen sections. Section 1;-Introduction to Organic Farming; Principles and Practices, and current scope in global perspective, Organic farming: concept, principles, benefits and future prospects, and Role of Agroecology as a Transdisciplinary Science for Organic Farming; Section 2; - Policies, Markets and Certification for Organic produce; Section 3;- Organic Agriculture: Opportunities and Challenges; Statistical development of organic farming in Global context, Organic agriculture: opportunities and challenges and Scale appropriate farming in organic agriculture; Section 4;- Concepts and practices of organic farming- included, Cultural practices for sustainable organic farming, Organics and agroecology: Boundaries and synergies between sustainable systems and approaches, Application of SMART technologies in organic farming: Recent advances and future directions and Synergies of farming with alternative pollinators (FAP) and certified organic farming for transformative change of agriculture; Section 5;- Organic ecosystem, concepts and role of natural resources- Land use carbon changes and monitoring, and Importance of vegetation cover in organic systems; Section 6;- Nutrient management in organic farming – Organic nutrient resources and fortification: Restricted nutrient use in organic farming, The role of arbuscular mycorrhizas in organic farming and Nitrogen fixation and dynamics in organic farming in temperate climate; Section 7;- Soil management: The basic need of organic farming- Soil management: The basic need of organic farming, Soil tillage and crop rotation: Pivotal practices of the organic cropping system management, Biodegradable organic wastes - An eco-friendly approach for nutrient management in sustainable organic farming and Plant biostimulants: Essential tools for sustainable organic farming; Section 8;- An Overview of Organic Waste Management; Section 9; - Marine-derived materials as Biofertilizers in organic farming, Application of marine-derived materials as biofertilizers in organic farming and Acceptability of marine-derived fertilizers in organic farming; Section 10;- Advances in plant disease detection in organic farming systems; Section 11;- Strategies for the Management of diseases and insect pests in organic crops; Biological pest control, Cultural practices to manage insect pests and diseases, Biopesticides and Integrated Pest Management in Organic Farming; Section 12;- Weed Management in Organic Farming; Section 13;- Organic farming in the face of climate change; Section 14;-Organic Livestock Farming, principles, benefits and challenges; Section 15;- Animal welfare and health in organic farming-focused on Animal health and welfare in organic livestock production: Current status and future challenges; Section 16;- Socioeconomic impact of organic farming and Section 17;- Organic food Quality and Safety.

This book is intended to be a unique and indispensable resource that offers a diverse range of valuable information and perspectives on organic farming at a time when the world community is increasingly aware of the problems of our current agricultural practices with importance of creating sustainable agricultural systems for the long-term health of the biosphere as a whole. The book is designed considering the requirements of students, researchers, university professors and also as per the course curriculum of different universities offering courses on Organic Agriculture. It has chapters on each and every aspect related with in organic agriculture which are compiled by researchers and eminent professors at various universities across the globe.

I wish the students and various readers who relish working in the field of organic farming will find the format of book, level and quantity of information contained in the book to be appropriate for easier learning and understanding. The wide spectrum information in various chapters with the addition of the terms related to organic agriculture and concept statements are presented very concise manner.

I am confident that this publication will be useful to university students, professors, researchers, development department officials, extension workers, policy makers, and all those interested in organic farming.

Professor L.P. Awasthi

About the Book

This book is intended to be a unique and indispensable resource that offers a diverse range of valuable information, data, and perspectives on Organic Farming at a time when the world community is increasingly aware of the problems of our current agricultural practices and the importance of creating sustainable farming systems for the long-term health of humankind and the biosphere.

1 Deals with the Introduction to Organic Farming, - principles, practices, and current scope in a global perspective and Organic agriculture: opportunities and challenges.
2 Provides comprehensive essential information about policies and markets for organic produce, and overview of organic certification Process.
3 Discussed about Organic agriculture: opportunities and challenges, and Concepts and practices of organic farming, Nutrient and soil management in organic Farming.
4 Advances in plant disease detection in organic farming systems and Strategies for the Management of diseases, insect pests and weeds in organic crops.
5 Contains updated information about organic livestock farming, animal health and nutrition and socioeconomic impact of organic farming.
6 Organic farming systems and food security: identifying values and investigating perspectives and Organic Food Quality and safety.

Acknowledgements

I would first like to express grateful thanks to my mentor Professor H.N. Verma, Pro Chancellor, Jaipur National University, Jaipur, India, who has been a constant source of inspiration, encouragement and support. I also wish to acknowledge and express my sincere thanks to my family for their untiring support, cooperation, and encouragement in the preparation of this book.

Particular thanks are also due to each of the expert contributors for their cooperation in producing up to date specialist chapters on the many sectors of Organic Farming; Dr. John Paull, School of Land & Food, University of Tasmania, Hobart, Tasmania, Australia; Prof. Uygun Aksoy, Department of Horticulture, and Dr. Muazzez Cömert Acar, Department of Animal Science, Ege University, İzmir, Turkey; Prof. Fabio Caporali, DAFNE (Department of Agricultural and Forestry Sciences) Via San Camillo de Lellis, Viterbo, Italy; Dr. Ehsan Ebrahimi, Faculty of Agricultural and Environmental Sciences, University of Rostock, Rostock, Germany: Dr. Adnan Šišić, Department of Ecological Plant Protection, University of Kassel, Witzenhausen, Germany; Prof. Raffaele Zanoli, Food Marketing and Management, Faculty of Engineering & Department of Agricultural, Food and Environmental Sciences, and Dr. Francesco Solfanelli, Faculty of Engineering, Department of Agricultural, Food and Environmental Sciences, Università, Politecnica delle Marche, Italy; Dr. Ryoichi Komiya, Organic Shizukuishi, Ukai Rinan Takizawa City Iwate Prefecture, Japan: Dr. Helga Willer, Research Institute of Organic Agriculture FiBL, Frick, Switzerland, Dr. Jan Trávníček, Czech Organics, Staré Město, Czech Republic and Dr. Bernhard Schlatter, Research Institute of Organic Agriculture FiBL, Frick, Switzerland: Prof. Victor Olowe, Institute of Food Security, Environmental Resources and Agricultural Research, Federal University of Agriculture, Abeokuta, Nigeria, Dr. Jude Obi, Department of Soil Science and Land Resources Management, Faculty of Agriculture, University of Uyo, Nigeria, Dr. Olabisi Somefun, Department of Plant Physiology and Crop Production, College of Plant Science and Crop Production, Nigeria; Leonard Pollara, Biological Agriculture Specialist, Organic Sage Consulting LLC, Michigan, USA; Dr. Raya Joseph Amara Nundu, Dr. Wilson Charles Wilson, Dr. Lucas Thomas Manda, Dr. Nyandula Samwel Mwaijande, Dr. Janet F. Maro, Dr. Yohana C. Haule, and Dr. Abdul Jafari Shango, University of Agriculture, Tanzania; Dr. David Frost, Formerly ADAS Pwllpeiran Research Centre Wales, United Kingdom; Dr. Thamali Wanasinghe, Dr. Ashoka Gamage and Dr. Wasana Gunawardana, University of Peradeniya, Sri Lanka; Dr. Yasavi Jayakodi, Kandy, Sri Lanka, and Dr. Othmane Merah, University of Toulouse, France; Dr. Bernhard Freyer, University of Natural Resources and Life Sciences, Austria; Prof. Kandiah Pakeerathan, University of Jaffna, Sri Lanka; Dr. Stefanie Christmann, ICARDA, Morocco; Dr. Hamdi Özaktan, Department of Field Crops, Agricultural Faculty, Erciyes University, Kayseri, Turkey, Dr. Gheorghe Cristian Popescu, Department of Applied Sciences and Environmental Engineering, National University of Science and Technology, Romania and Dr. Mehmet Yaman, Department of Horticulture, Agricultural Faculty, Erciyes University, Kayseri, Turkey; Dr. Wing K.P. Ng, Dr. Day Teixeira, Prof. Matt J. Bell, Department of Animal and Agriculture, Hartpury University, Gloucester, GL19 3BE, U.K., Dr. Adrian Crew, and Dr. Pete Maxfield, Department of Health and Applied Sciences, University of the West of England, Bristol, BS16 1QY, UK; Dr. Sito-Obong Udofia, Dr. Lisa Williams, and Dr. Alison Wills, Department of Animal and Agriculture, Hartpury University, Gloucester, GL19 3BE, UK; Prof. Hector Valenzuela, University of Hawaii at Manoa, United States; Dr. Amjad Ahmad, and Prof. N.V. Hue University of Hawaii at Manoa, USA and Dr. Maria Eugenia Ortiz Escobar, University of Ceara, Brazil; Dr. Mary Cole, University of Melbourne, Australia and Dr. Rachel Cavallo, Victoria, Australia; Dr. Jochen Mayer, Department of Agroecology and Environment, Reckenholzstrasse, Switzerland; Dr. S. Folaranmi G. Yusuf, Sustainable Livelihood Supports and Development Network for Africa, Abeokuta, Nigeria, Dr. Gabriel Adewunmi Eyinade, and Prof. A. Mushunje, Department of Agric. Economics & Extension, University of Fort Hare, South Africa; Dr. Roberto Mancinelli and Dr. Mariam Atait, University of Tuscia, Italy, Dr. Mohamed Allam, Assuit University, Egypt and Prof. Emanuele Radicetti, University of Ferrara, Ferrara, Italy; Dr. Yves Theoneste Murindangabo, Dr. Marek Kopecký, and Dr. Petr

Konvalina, University of South Bohemia in Ceske Budejovice, Branišovská České Budějovice, Czech Republic; Dr. Thangapandian Dhanikodi, Bharathidasan University, Tiruchirappalli, Trichy, India, Dr. Uma Maheswari Rajadurai, Cauvery College for Women, Tiruchirappalli, India, and Dr. Rameshwari Ramaswamy, Cauvery College for Women, Tiruchirappalli, India; Dr. Bina Mistry, Thermo Fisher Scientific, Kansas, USA; Dr. Bharti Suthar, ICRISAT, Hyderabad, India; Dr. Monica Popescu, University of Pitesti, Romania; Dr. Dang Khoa Tran, Dr. Thi Xuan Phuong Tran, Faculty of Agronomy, University of Agriculture and Forestry, Hue University, Phung Hung, Hue City, Vietnam and Dr. Petr Konvalina, Department of Agroecosystems, Faculty of Agriculture and Technology, University of South Bohemia, Czech Republic; Dr. Anne-Kristin Løes, Dr. Joshua Cabell, Dr. Sara Hansdotter, and Dr. Tatiana Rittl, Centre for Organic Agriculture Norway; Dr. Ingrid Olesen, Department of Production Biology, Nofima, Ås, Norway, and Dr. Violeta Alexandra Ion, University of Agronomic Sciences and Veterinary Medicine of Bucharest, Romania; Dr. Kanu Murmu, Dr. Sramana Sen Sarma and Dr. Arup Sarkar, Bidhan Chandra Krishi Viswavidyalaya, India; Dr. Tahira Batool and Dr. Joel Allainguillaume, University of the West of England, Bristol, Dr. Wenhao Zhang, School of Engineering, University of the West of England, Bristol; Dr. Alessia Iacovone and Dr. Jacopo Menghi, CBC Bioplanet, Italy; Prof. Philippe Belliard, Wageningen University, The Netherlands; Dr. Alice Caselli and Dr. Ruggero Petacchi, Center of Plant Sciences, Sant'Anna School of Advanced Studies, Piazza Martiri della Libertà 33, Pisa, Italy; Dr. Schöller Matthias, Humboldt Universität zu Berlin, Department of Life Sciences, Berlin, Germany, Professor Masoud Latifian, Agricultural Research Education and Extension Organization, Iran; Dr. Nagarjuna Reddy Desam, Anantha Lakshmi Institute of Technology & Sciences, Ananthapuram, India; Dr. Abdel Jabbar Al-Rajab, Etcetera Publications Canada; Dr. Matthew Russell Gates, Cannabis Horticultural Association San Diego, California, United States; Prof. Emanuele Radicetti, Dr. Ben Hassine Mortadha, Dr. Valentina Quintarelli and Dr. Daniele Borgatti, Department of Chemical, Pharmaceutical and Agricultural Sciences, University of Ferrara, Italy, Prof. Taab Alireza, Department of Agronomy and Plant Breeding, Faculty of Agriculture, Ilam University, Iran, Dr. Ahsan Muhammad, Department of Horticultural Sciences, The Islamia University of Bahawalpur, Pakistan Dr. Mohamed Allam, University of Tuscia, Italy, and Prof. Roberto Mancinelli, Department of Agricultural, Department of Agricultural and Forestry Sciences, University of Tuscia, Italy; Dr. Arnd Verschwele, Julius Kühn-Institut, Bundesforschungsinstitut für Kulturpflanzen, Institut für Pflanzenschutz in Ackerbau und Grünland, Messeweg, Braunschweig, Germany; Dr. Jelena Golijan Pantović, University of Belgrade, Faculty of Agriculture, Serbia; Prof. Maria Claudia Dussi, Department of Agricultural Sciences, National University of Comahue, Rio Negro, Patagonia, Argentina; Dr. Thierry Gidenne, Dr. Laurence Fortun, Dr. Lamothe, Dr. Davi Savietto, Univeristy of Toulouse, INRAE, ENVT, Castanet Tolosan France; Prof. Luciana Bignardi Casimiro da Costa, College of Veterinary Medicine, The Ohio State University, USA; Dr. Georgios I. Papakonstantinou, Dr. Dimitrios A. Gougoulis, Dr. Lampros Fotos, and Prof. Vasileios G. Papatsiros, Clinic of Medicine, Faculty of Veterinary Medicine, School of Health Sciences, University of Thessaly, Greece, Dr. Georgia Terzidou, Food Law Experts, Themistokleous, Athens, Greece; Dr. Fikru K. Alemayehu, Inland Business School, Inland Norway University of Applied Sciences, Lillehammer, Norway, Dr. Habtamu Alem, Department of Economics Society, The Norwegian Institute of Bioeconomy Research (NIBIO), Ås, Norway, Prof. Gudbrand Lien, Inland Business School, Inland Norway University of Applied Sciences, Lillehammer, Norway; Dr. Linda Chinenyenwa Familusi, Department of Agricultural Economics and Agribusiness Management, University of Port Harcourt, Rivers State, Nigeria, Dr. Abdi-Khalil Edriss and Dr. Mthakati A. R. Phiri, Department of Agricultural and Applied Economics and Africa Centre of Excellence for Agriculture Policy Analysis, Lilongwe University of Agriculture and Natural Resource, Lilongwe, Malawi, Dr. John A. Kazembe, Department of Environmental Science, Lilongwe University of Agriculture and Natural Resource, Lilongwe, Malawi, Prof. Anthony O. Onoja, Department of Agriculture and Animal Health, University of South Africa, Johannesburg, South Africa, Prof. Timo Sipiläinen, University of Helsinki, Helsinki, Finland; Dr. Jagjeet Singh Gill, Scientist and Quality Assurance Person, Green Mountain Health Alliance, Kaleden, BC, Canada; Dr. Rachid Mrabet, Institut National de la Recherche Agronomique, Avenue de la Victoire, INRA, Morocco.

I am sure that the expert knowledge provided for each of the specialized sectors of Organic Farming will be an invaluable source of material for students, researchers, academics, consultants, and extension workers.

I am highly indebted to Dr. Amber Sciligo Director of Science Programs, The Organic Center Washington DC for her erudite suggestions in the organization of table of contents.

Finally, I would like to extend my sincere thanks to CABI and their publishing staff: Ward Cooper, Commissioning Editor; Lucy Pritchard, Managing Editor; James Bishop, Production Editor; Helen Elliott, Editorial Assistant and to the extended team for getting this volume published in such an excellent format.

Professor L.P. Awasthi

1 Organic Farming: Principles, Practices and Current Scope in a Global Perspective

John Paull*

School of Land & Food, University of Tasmania, Hobart, Tasmania, Australia

Abstract

Organic agriculture is the quest for pure food. The principles and practices of organic agriculture have evolved over the past century. The New Age philosopher Dr Rudolf Steiner, in 1924, called for a differentiated agriculture that relied on biology rather than synthetic chemicals. Subsequently, a biodynamic farmer in Kent, Lord Northbourne, coined the term 'organic farming' and launched his manifesto of organic agriculture in 1940, amid the chaos of World War II. The six exclusions from organic production are: synthetic fertilizers, pesticides, genetically modified organisms (GMOs), irradiation, synthetic medications, and nanotechnology. Organic agriculture is now practised in 191 countries, it accounts for 76.4 million hectares globally, and €125 billion (US$ 136 billion) sales per annum. Australia has the most organic hectares, and Germany has the most biodynamic hectares. The long-term growth trend of the organic sector is an increase of 10.15% per annum. The goal of Europe is 25% organic by 2030. Two European countries already meet this goal: Lichtenstein with 40% organic agriculture, and Austria with 27%. India's hill state of Sikkim is the world's first 100% organic state. Glyphosate is the world's most used herbicide; it has been linked to the development of multiple chronic diseases and conditions including autism, irritable bowel syndrome, and cancer. Most non-organic food contains pesticides. Most organic food does not contain pesticides. Eating organic is a sound strategy for consuming less pesticide.

Keywords: organic agriculture, biodynamics, Rudolf Steiner, Lord Northbourne, glyphosate, certification, statistics, world map

Introduction

Organic agriculture is the quest for pure food. A primary driver of the market for organic food is to avoid ingesting pesticides (Nielsen, 2005; GfK, 2017). The data support the soundness of that proposition. Most organic food (86%) does not contain pesticides. Most non-organic food (85%) contains pesticides (DEFRA, 2007; Paull, 2023) (Fig. 1.1).

A century ago a group of farmers approached the New Age philosopher Dr Rudolf Steiner about their concerns for the nutritive value of crops produced by farms using synthetic fertilizers (Fig. 1.2). Theses farmers' concerns were in the wake of the rapid proliferation of synthetic fertilizer, as the output of the explosives factories of World War I were rapidly repurposed to produce synthetic nitrogenous fertilizers (Charles, 2005; Freemantle, 2014).

To an audience of over 100 farmers and others, in the rural village of Koberwitz (then Germany, now Kobierzyce, Poland), Rudolf Steiner called for a differentiated agriculture that was focused on natural biological processes rather than on a chemical industrial approach (Steiner, 1924; Paull, 2020). Steiner's core precept was that 'the farm is an organism' (rather than, say, 'a factory' or 'an economic unit of production') (Paull, 2006). That *Agriculture Course* was held in 1924. At Koberwitz,

*Corresponding authors: j.paull@utas.edu.au; john.paull@mail.com

DOI: 10.1079/9781800626850.0001

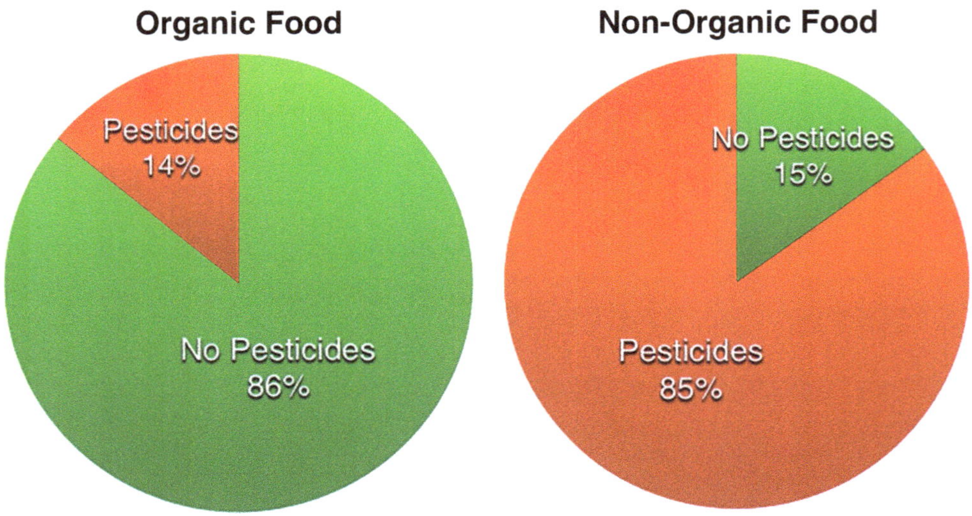

Fig. 1.1. Pesticides in food: most non-organic food has pesticides, and most organic food does not have pesticides. (From Paull, 2023)

Steiner founded the 'Experimental Circle of Farmers and Gardeners' to test and develop his ideas. Steiner called them 'hints'. The Experimental Circle evolved the ideas, and by 1938 Ehrenfried Pfeiffer published *Bio-dynamic Farming and Gardening* in which he presented principles and practices of this newly differentiated agriculture (Pfeiffer, 1938; Paull, 2011).

Organic Manifesto

In 1940, just 2 years after the public launch of Pfeiffer's *Bio-dynamic Farming*, Lord Northbourne coined the term 'organic farming' (Northbourne, 1940). It was a logical derivation from Steiner's view that 'the farm is an organism'. Launching his book in the midst of World War II, Northbourne could not have been unaware of the irony of declaring: 'We have invented or imagined a fight between ourselves and nature; so, of course the whole of nature, which includes ourselves as well as the soil, suffers .... We have tried to conquer nature by force and by intellect. It now remains for us to try the way of love' (Northbourne, 1940, pp. 191–192).

Northbourne was an Oxford University graduate in agriculture, the Governor of Wye College (aka South-Eastern Agricultural College), and a biodynamic farmer with an estate in Kent, England (Paull, 2014). He juxtaposed 'organic vs chemical farming' in his book *Look to the Land*. The book is the foundational manifesto of organic agriculture. To this day, *Look to the Land* remains salient, articulate, compelling, and is unsurpassed in making the case for farming with nature, rather than against nature, and for farming as a collaboration rather than a 'war' (Northbourne, 1940).

Organics Proliferation

The notion of organic farming quickly gained traction internationally, despite its launch in the midst of the raging catastrophe of World War II. In the USA, the entrepreneur Jerome Rodale released the first issue of his periodical *Organic Farming and Gardening* in 1942. Rodale promptly reversed the name to read *Organic Gardening and Farming*, realizing that there were lots more gardeners than farmers, that gardeners were apparently more hungry for 'know-how', and that gardeners were more likely candidates to purchase his magazine. Rodale's organic journal appeared regularly (mostly monthly) for the next 70 years (Rodale, 1965; Gross, 2008).

In Australia, in a world first, farmers founded the Organic Farming and Gardening Society in 1944 to advocate the uptake of organic farming (Paull, 2008). In the footsteps of the Australians, in the UK the Soil Association was founded in 1946 (Soil Association, 1946). In France, the International

Fig. 1.2. Rudolf Steiner first proposed a differentiated agriculture with a focus on biological management rather than chemical management. (From Kleeberg, 1928)

Federation of Organic Agriculture Movements (now Organics International) was founded in 1972 as an international umbrella group for the (by then) many and diverse organic farming groups around the world (Geier, 1998; Paull, 2010).

Since those earliest fledgling steps, organic agriculture has evolved and proliferated to be a global phenomenon (Fig. 1.3). Exclusions have evolved to six: synthetic fertilizers, synthetic pesticides, genetically modified organisms (GMOs), radiation, synthetic medications, and nanotechnology. Organic agriculture is now reported in 191 countries (Fig. 1.3), accounting for 76.4 million ha of farmland, 3.7 million farmers, and €125 billion (US$ 136 billion) of annual sales (Willer *et al.*, 2023).

Women of Organics

Women have played a key role in advancing the organics movement (Inhetveen *et al.*, 2021). At a London anthroposophy conference in London, in 1928, Marna Pease took up the challenge of advocacy. She spent the remaining two decades of her life advancing the cause of biodynamic agriculture throughout the Anglo-world, editing the earliest English-language journal (*Anthroposophical Agricultural Foundation Notes and Correspondence*), hosting the earliest conferences, maintaining an international correspondence, and fielding questions from both the curious and the converted (Paull and Harvey, 2023).

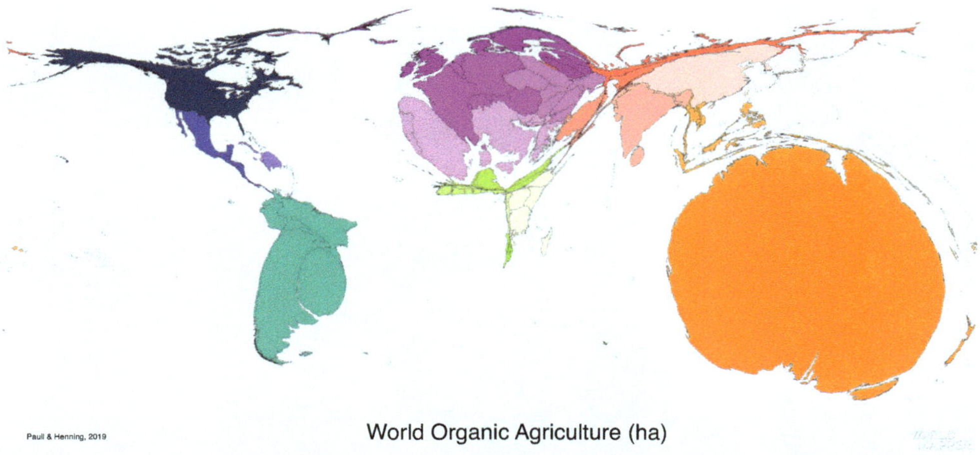

Fig. 1.3. World map of organic agriculture (countries are resized according to their tally of reported certified organic hectares). (From Paull and Hennig, 2019)

Eve Balfour founded the Soil Association in London in 1946 and was later a co-founder of the umbrella advocacy group the International Federation of Organic Agriculture Movements (IFOAM, now rebranded as Organics International) at Versailles in 1972 (Brander, 2003; Paull, 2010).

Two biodynamic farmers, Marjorie Spock and Mary T. Richards, of Long Island, New York State, sued the US government to stop the aerial spraying of DDT. Spock and Richards did not win their protracted and expensive legal proceedings, but they passed their extensive trove of research and data to Rachel Carson who used it as her primary source to write her 1962 eco-bestseller *Silent Spring*. The publication of the book precipitated the environmental movement and led to the establishment of the first Environmental Protection Authority (EPA) (in the USA); and since then, many countries have created their own EPA (Carson, 1962; Matthiessen, 1999; Paull, 2013c).

Goal Setting for Organics

The setting of goals is somewhat problematic. A goal, *per se*, is never enough; some 'goal shooting' activity is required for a goal to be realized. Bhutan is a standout fail when it comes to organics goals – there was (and perhaps still is) the goal, announced in 2013, to be 100% organic. Bhutan's Minister for Agriculture and Forestry, Dr Pema Gyamtsho, admitted that 'Going organic will take time ...We have set no deadline' (Paull, 2013a). There was, and surely remains, an uncertain number of *de facto* organic farmers in Bhutan, producing crops as they have for centuries past, and who are perhaps too isolated and/or too poor to purchase chemical inputs. Nevertheless, the goal setting was insufficient to reify the goal. Presently, Bhutan reports just 1.1% of its agricultural land as certified organic (Willer *et al.*, 2023).

The Indian north-eastern hill state of Sikkim is a standout success when it comes to organic goal setting. The Chief Minister of Sikkim, Pawan Chamling, in 2003 declared the bold decision for the state to be 100% organic. This was backed up by a four-point action plan: '(1) promotion of on-farm production of organic manures; (2) capacity building; (3) establishment of bio-fertilizer production units; and (4) establishment of soil testing laboratories' (Chief Minister's Office, 2016, p. 392). The goal was achieved in little more than a decade with a mix of push and pull (including capacity building for farmers, and banning the import of pesticides). The Indian Prime Minister, Narendra Modi, declared Sikkim as India's first fully organic state in 2016. In the light of this success, other Indian states are on the path to replicate this success (Paull, 2017).

Europe has set the goal of 25% organic by 2030: 'The Commission has set out a comprehensive organic action plan for the European Union. Through it, the Commission will aim to achieve the European Green Deal target of 25% of agricultural land under organic farming by 2030' (European Commission, 2021, p. 1). Can it be done? Yes, organic agriculture in Liechtenstein already accounts for 40% of its agriculture hectares, and in Austria the figure is 27%, so both have already met the '25% by 2030' goal. Estonia and Sweden are in 'striking distance' of the goal, at 23% and 20%, respectively. There are a further 11 European countries that are well on the way to reaching the 25% goal, with organics accounting for between 10% and 20% of their agriculture (viz. Switzerland, Italy, Czech Republic, Latvia, Finland, Slovakia, Denmark, Germany, Spain, Slovenia and Greece). Europe's 25% is a fine goal, and it looks achievable; each European country to be ≥25% organic is a finer goal in which each country 'pulls its weight' to make the fruits of the goal uniform.

Australia has achieved the largest area of certified organic agriculture of any country, accounting for nearly half of the world's certified organic hectares, and 9.9% of Australia's agricultural land (Paull, 2019; Willer *et al.*, 2023). This has been achieved without the benefit of national goal setting or government support. This success has been driven by market forces. There have been no government incentives to adopt organic practices. Australian federal and state governments have been generally indifferent to organic agriculture. Nevertheless, Australia has been an early adopter of organics with biodynamics in the country by 1928, and the world's first organics advocacy society in 1944 (Paull, 2013b).

Scope for Good News

Organics is a good-news story: 'Your food is produced without synthetic fertilizers, pesticides, GMOs, irradiation, antibiotics, and nanotechnology'. There are likely no consumers clamouring: 'I want to feed myself and my family food laced with glyphosate and insecticides' or 'I want to feed my children GMOs'.

The six exclusions for organics can be cryptic aspects of chemical food production – pesticides *et al.* are not detected by visual inpection. There is no transparency for consumers buying chemically produced food. The producer and the retailer are entirely silent as to which smorgasbord of synthetic biocides and technologies have been applied to the food in its path from field to shopping cart. Mandatory reporting of pesticides applied in production would be informative to consumers. That would add transparency to the consumer's decision to purchase (or not).

There is scope for consumer engagement by organic producers, retailers and certifiers (Fig. 1.4).

Fig. 1.4. Supermarket in-store promotion of organics, Shanghai, China. (Photo: J. Paull)

There is a congruence of interests – producers are interested in producing foodstuffs without pesticides *et al.*, and consumers are interested in consuming foodstuffs without these 'nasties'. There is room for a conversation between farms and families. There is room for a shared understanding as to why a differentiated product will and should attract a premium price, and why the price premium is 'worth it'. There is scope to discuss what is a fair and reasonable premium, from both the producer and the consumer perspectives.

The organic message in the busy marketplace is a whisper rather than a roar. There is scope to advertise organic food – its advantages for consumers, the environment, the planet, farm workers, and farm animals. There is scope for producers to get the message across that their production methods exclude synthetic chemicals and other technologies that most consumers will regard as less than desirable for their family's diet.

Avoiding Pesticides

What are the unintended consequences of chemical agriculture's addiction to pesticides and other deviations from 'natural'? Is there a price to pay for loading the food chain with synthetic fertilizers, pesticides, GMOs, nanotechnology and irradiation?

Modern maladies that are now prevalent – obesity, diabetes, autism, attention deficit hyperactivity disorder (ADHD) and various cancers – were generally unknown or little known to previous generations. What has been the role of the proliferation of a smorgasbord of pesticides in the proliferation of these maladies?

Pesticides have a four-step life cycle: discovery, enthusiasm, doubt, and ban. DDT was the 'wonder' chemical of its day. Most people carry a load of DDT and/or its metabolites despite it having been banned for half a century (EPA, 1972; CDC, 2021). Glyphosate is the present 'wonder' herbicide for chemical farming, and for weed control, by councils and householders. Glyphosate (the active ingredient in Bayer Monsanto's 'Roundup' herbicide) is cheap, available and ubiquitous in the food chain. This continues despite the WHO declaring glyphosate a carcinogen (WHO, 2018).

In a review of the toxic effects of glyphosate, the authors conclude:

> Contrary to the current widely-held misconception that glyphosate is relatively harmless to humans, the available evidence shows that glyphosate may rather be the most important factor in the development of multiple chronic diseases and conditions that have become prevalent in Westernised societies. In addition to autism, these include gastrointestinal issues such as inflammatory bowel disease, chronic diarrhea, colitis and Crohn's disease, obesity, cardiovascular disease, depression, cancer, cachexia, Alzheimer's disease, Parkinson's disease, multiple sclerosis, and ALS, among others.
>
> (Samsel and Seneff, 2013, p. 1443)

These authors continue:

> While glyphosate is obviously not the only environmental toxin to contribute to these diseases and conditions, glyphosate's ability to disrupt the gut bacteria, to impair serum transport of sulfate and phosphate, and to interfere with CYP enzymes, logically progresses to this multitude of diseased states, through well-established biological processes. And glyphosate's disruption of the body's ability to detoxify other environmental toxins leads to synergistic enhancement of toxicity. While genetics surely play a role in susceptibility, genetics may rather influence *which* of these conditions develops in the context of glyphosate exposure, rather than *whether* any of these conditions develops.
>
> (Samsel and Seneff, 2013, p. 1443)

The Australian experience of requiring GMO food to be labelled as GMO has resulted in Australian retailers offering no GMO food. That avoids its rejection at the point of sale. The zero presence of GMOs in the supermarket aisles in Australia is contrasted with the situation in USA (food packaging does not disclose the presence of GMOs) where most processed food items have GMO ingredients (often GM soy and/or GM corn) and most US soy and corn (59%) contains glyphosate (USFDA, 2022).

The best assurance for avoiding pesticides is to purchase organic.

Organic Certification

The reported data for organic farming are statistics of *certified* organic farming. A consequence is that the published statistics of organic agriculture are underestimates of the sector. There are traditional farmers who are using traditional biological practices, unaware that they are 'organic farmers'. There are other producers who are intentionally 'organic farmers' but are intentionally not certified (perhaps because of the small scale of the operation, the cost of certification, because of the intrusiveness of the certification process, because of a philosophical objection to the 'organocrats', or they cannot secure a premium for their product at their location, perhaps because a market has not been established).

Certification preserves the differential identity of organic produce as it moves through the supply chain, from producer to consumer. Certification provides oversight for producers, protection for retailers, tracking of produce, confidence for consumers that they are 'getting what they are paying for', and it secures the price premium for producers and retailers.

Carbon Sequestration

The global push for 'zero-carbon', on a planet where all life forms are carbon-based, is curious. Carbon dioxide is the essential ingredient of all food – no CO_2 means no food – means we all starve. Photosynthesis (plants creating complex carbohydrates (and oxygen) from two ingredients, CO_2, and water, in the presence of sunlight, is the superstar of biology, the underpinning of all carbon-based life and biology.

Organic farms are natural 'carbon sinks' because there is an emphasis on maintaining and enriching soil quality, considering humus, composting and the microbiome. Organic farms naturally sequester carbon. However, claiming carbon credits (directly monetizing carbon sequestration) may be an expensive and protracted process for a farm (DAFF, 2023).

Technology

Organic farmers face many of the same challenges as chemical farmers, including weeds, diseases, the price of labour, the price of other inputs, and farm gate prices (where they are often 'price takers' rather than 'price makers'). Both may be 'tech-savvy' and using, for example, drones, GPS and robots. Labour can be a greater input in organic production and this is prompting a growing uptake of robots, especially to pick and pack produce.

Organic producers eschew certain technologies such as GMOs which rely on the legal fiction that living things can be patented, based on the claim that living things can be 'invented'. Northbourne made the point succinctly: 'No chemist has ever analysed or described in chemical terms a living creature, however humble; and there is not the slightest chance that he ever will' (Northbourne, 1940, pp. 159–160). Patents on living things continues to be contested (Then and Eckhardt, 2023).

Global Organics

The global 'penetration' of organic agriculture is a niche of 1.6% as a percentage of global agricultural hectares (Willer *et al.*, 2023). Global leadership in organics is diffuse (Paull and Hennig, 2020; Willer *et al.*, 2023):

- Australia has the most certified organic agriculture (35,687,799 ha);
- India has the most organic farmers (1,599,010 producers);
- Finland has the most organic wild-culture (6,928,693 ha);
- Albania has the most organic silviculture (662,790 ha);
- Vietnam has the most organic aquaculture (40,623 ha);
- China has the most organic cereals (1,425,060 ha);
- Ethiopia has the most organic coffee (226,389 ha);
- Spain has the most organic viticulture (142,177 ha);
- Germany has the most biodynamic hectares (84,426 ha);
- Denmark has the highest share of the domestic market (13% organic);
- Switzerland has the highest per capita consumer spend (€425 p.a.; US$ 465 p.a.);
- USA has the highest retail sales of organic (€48.6 million; US$ 53.1 million).

This diffusion of leadership, coupled with many factors including the 10.15% compounding annual growth of organic hectares over the past two decades (Fig. 1.5), the growing affluence and education of consumers, the growing body of evidence linking 'bad' food to bad health outcomes, the growing consumer concern with animal welfare, the growing concern for the environment, the growing concern for climate change coupled with the aspiration for 'zero-carbon', along with declarations of aspirational goals, all augur well for the future health, growth and valorization of the organic sector.

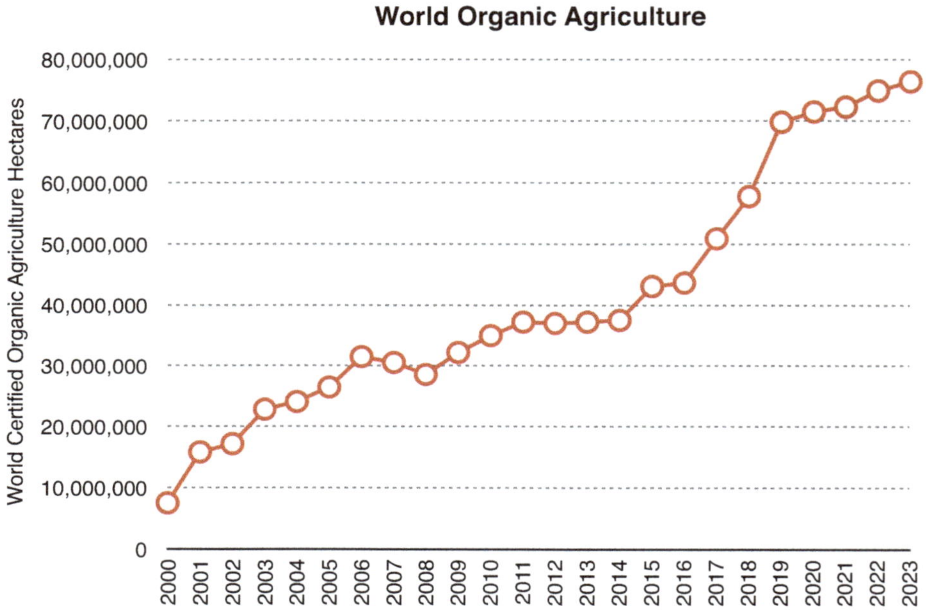

Fig. 1.5. Global growth of organic agriculture (year reported). (Data sources: annual reports to Willer *et al.*, 2023)

Conclusions

Organic agriculture has survived its first century from its Koberwitz start in 1924. It has grown at 10.15% year on year over more than the past two decades. What of the future?

The customers for organics are more affluent and better educated than average. That cohort of 'affluent and educated' consumers is expanding. Affluent countries grow in affluence and education levels. Other countries, most notably China, are rapidly shifting much of their population out of poverty and into middle-class affluence, and also despatching contingents of tertiary students to western universities, notably to the UK, Australia and the USA.

Is there a future for organic agriculture, and if so, what is it? The foundational aspiration of organic farming is the quest for pure food. That is an enduring quest, not a fad. 'Certified Organic' and 'Certified Biodynamic' are manifestations of the quest.

Organic agriculture is an important niche agriculture that has long aspired to be the mainstream. When Northbourne introduced the term 'organic farming' he was clear that this was an adversary of 'chemical agriculture'. He saw a contest of 'organic vs chemical farming'. He was under no illusions, realizing that the quest for pure food would be a task for generations: 'It is a task for generations of concentrated effort, slow and laborious, needing all available skill and resources .... A combination of cooperation and individual effort .... And those engaged will be fighting a rearguard action for many decades, perhaps for centuries' (Northbourne, 1940, p. 115).

Northbourne's contest of 'organic vs chemical farming' is an oddly asymmetrical contest. A handful of mega-multinational chemical companies (for whom the stakes are high) vs 8 billion consumers (many of whom are quite oblivious to the prevailing contest, because of, perhaps, poverty, ignorance or indifference).

The organics sector has survived a century of the 'onslaught' of technological intrusion into the food realm coupled with the market force of the multinational chemical companies who are presently consolidating their grip on the global food chain. Meanwhile there is no 'right to pure food', no right to nutrient-dense food.

Northbourne observed that 'The delusion is that cheapness leads to plenty. But of what use is plenty of rubbish?' (Northbourne, 1940, p. 66).

He declared that: 'We have plenty, actual and potential; but plenty of what? That is the real question' (Northbourne, 1940, p. 73). More than eight decades after their first expression, these sentiments still ring true, and the global epidemic of obesity testifies to their sagacity. Meanwhile, the quest for pure food continues to make headway. Perhaps 1 day the prevailing market segments can be reversed, with chemical agriculture the niche rather than the dominant segment, with perhaps chemical agriculture swapped to a niche of 2%, and organic agriculture to have grown to 98% of global agriculture? As Northbourne speculated, the contest of chemical vs organic farming may be the work for centuries.

References

Brander, M. (2003) *Eve Balfour: Founder of the Soil Association and the Voice of the Organic Movement: A Biography*. The Gleneil Press, Haddington, UK.

Carson, R. (1962) *Silent Spring* (Readers Union 1964 members' edn). Hamish Hamilton, London.

CDC (2021) Dichlorodiphenyltrichloroethane (DDT) factsheet. Centers for Disease Control and Prevention (CDC), US Department of Health and Human Services, Atlanta, Georgia.

Charles, D. (2005) *Master Mind: The Rise and Fall of Fritz Haber, the Nobel Laureate who Launched the Age of Chemical Warfare*. HarperCollins Publishers, New York.

Chief Minister's Office (2016) *Sikkim: Under the Leadership of India's Greenest Chief Minister Shri Pawan Chamling: Sustainable Development through Greening, Organic Farming and Unique Social Engineering*. Chief Minister's Office, Government of Sikkim, Gangtok, India.

DAFF (2023) Emissions reduction fund. Department of Agriculture, Fisheries and Forestry, Australian Government.

DEFRA (2007) *Characterising the Potential Risks Posed by Engineered Nanoparticles: A Second UK Government Report*. Department of Environment, Food and Rural Affairs, London.

EPA (1972) DDT ban takes effect. EPA press release. US Environmental Protection Agency.

European Commission (2021) Action plan for organic production in the EU. European Commission, Brussels.

Freemantle, M. (2014) The chemists' war: 1914–1918. Royal Society of Chemistry, Cambridge, UK.

Geier, B. (1998) *A Look at the Development of IFOAM in its First 25 Years*. IFOAM, Bonn, Germany.

GfK (2017) Decision factors on what to eat or drink: Global GfK survey. Growth from Knowledge, London.

Gross, D. (2008) *Our Roots Grow Deep: The Story of Rodale*. Rodale Inc, Emmaus, Pennsylvania.

Inhetveen, H., Schmitt, M. and Spieker, I. (2021) *Passion und Profession:* Pionierinnen des Ökologischen Landbaus. Oekom Verlag, München, Germany.

Kleeberg, L. (1928) Wege und Worte. Verlag von Rudolf Geering, Basel, Switzerland.

Matthiessen, P. (1999) Rachel Carson: Before there was an environmental movement, there was one brave woman and her very brave book. *TIME Magazine* 29 March.

Nielsen, A.C. (2005) *Consumer Attitudes towards Organic Foods: A Global Consumer Survey*. A.C. Nielsen, New York.

Northbourne, Lord (1940) *Look to the land*. Dent, London.

Paull, J. (2006) The farm as organism: The foundational idea of organic agriculture. *Journal of Bio-Dynamics Tasmania* 80, 14–18.

Paull, J. (2008) The lost history of organic farming in Australia. *Journal of Organic Systems* 3(2), 2–17.

Paull, J. (2010) From France to the world: The International Federation of Organic Agriculture Movements (IFOAM). *Journal of Social Research & Policy* 1(2), 93–102.

Paull, J. (2011) Biodynamic agriculture: The journey from Koberwitz to the world, 1924–1938. *Journal of Organic Systems* 6(1), 27–41.

Paull, J. (2013a) Bhutan's plans to go 100% organic make progress. *Organic News* 26 February.

Paull, J. (2013b) A history of the organic agriculture movement in Australia. In: Mascitelli, B. and Lobo, A. (eds) *Organics in the Global Food Chain*. Connor Court Publishing, Ballarat, Australia, pp. 37–60.

Paull, J. (2013c) The Rachel Carson letters and the making of Silent Spring. *Sage Open* 3(3), 1–12.

Paull, J. (2014) Lord Northbourne, the man who invented organic farming, a biography. *Journal of Organic Systems* 9(1), 31–53.

Paull, J. (2017) Four new strategies to grow the organic agriculture sector. *AGROFOR International Journal* 2(3), 61–70.

Paull, J. (2019) Organic agriculture in Australia: Attaining the global majority (51%). *Journal of Environment Protection and Sustainable Development* 5(2), 70–74.

Paull, J. (2020) The Koberwitzers: Those who attended Rudolf Steiner's agriculture course at Koberwitz in 1924, world's foundational organic agriculture course. *International Journal of Environmental Planning and Management* 6(2), 47–54.

Paull, J. (2023) Organic food and pesticides. *ISDRS Newsletter (International Sustainable Development Research Society)* Q1, 3.

Paull, J. and Harvey, J. (2023) Marna Pease (1866–1947): Founder of biodynamics for the English-speaking world. *Advances in Social Sciences Research Journal* 10(5), 272–301.

Paull, J. and Hennig, B. (2019) New world map of organic agriculture: Australia is 51%. *Acres Australia* 101, 35–36.

Paull, J. and Hennig, B. (2020) A world map of biodynamic agriculture. *Agricultural and Biological Sciences Journal* 6(2), 114–119.

Pfeiffer, E. (1938) . Translated by F. Heckel. *Bio-dynamic Farming and Gardening: Soil Fertility Renewal and Preservation*. Anthroposophic Press, New York.

Rodale, J.I. (1965) *Autobiography*. Rodale Press, Emmaus, Massachusetts.

Samsel, A. and Seneff, S. (2013) Glyphosate's suppression of cytochrome P450 enzymes and amino acid biosynthesis by the gut microbiome: Pathways to modern diseases. *Entropy* 15(4), 1416–1463.

Soil Association (1946) Our inaugural meeting. Mother Earth, Introductory II, 3-6.

Steiner, R. (1924) *Agriculture Course* (1929, first English language edition, George Kaufmann, trans). Goetheanum, Dornach, Switzerland.

Then, C. and Eckhardt, J. (2023) Opposition against a patent on maize with previously known traits: Patent research shows alarming new cases of patents on seeds. No Patents on Seeds (No-patents-on-seeds.org), Munich, Germany.

USFDA (2022) Questions and answers on glyphosate. US Food and Drug Administration, White Oak, Maryland.

WHO (2018) Glyphosate. *IARC Monographs* 112, 1–92.

Willer, H., Schlatter, B. and Trávníček, J. (eds) (2023) *The World of Organic Agriculture: Statistics & Emerging Trends 2023*. Research Institute of Organic Agriculture (FiBL), Bonn, Germany and IFOAM-Organics International, Frick, Switzerland.

2 Organic Farming: Concepts, Principles, Benefits and Prospects

Muazzez Acar Cömert[1] and Uygun Aksoy[2*]

[1]*Department of Animal Science, Ege University, İzmir, Türkiye;* [2]*Department of Horticulture, Ege University (formerly), Association of Ecological Agriculture Organization, İzmir, Türkiye*

Abstract

Organic agriculture practices emerged during late 19th century or onset of the 20th century by pioneers. Their focus was on nature, ecological cycles and on-farm resources aiming to maintain long-term soil health and productivity. The organic management targeted a closed system where animal and plant productions are balanced and products, by-products or waste are recycled. In the early stages, organic principles were cited as main practices or objectives. Later, in 2008, the International Federation of Organic Agriculture Movements (IFOAM) – Organics International accepted basic principles of ecology, health, fairness and care as pillars to further develop organic systems. Enlarging markets in developed countries triggered organic production in the world after the 2000s, creating global flows. Today, organic is not any more a niche but a mainstream aiming to bring solutions to the challenges faced globally. Henceforth, a closed-system approach was not achievable in all cases and stockless organic monoculture or dense animal farms appeared if standards permitted. Due to an increasing number of market standards, issues such as deviations from basic principles or organic integrity appear as a topic for discussion. The global statistics reveal that organic activities are present in almost all countries; however, nearly 60% of organic certified land is destined for animal consumption in developed countries. Thus, consumers are concerned more with how organic standards, taken as reference in certification, meet the basic principles and can address prevailing global threats. Various amendments, revisions, private organic standards or social standards, accompanying the official ones, aim to fill these gaps. Research, innovation and practitioners aim to enhance sustainable organic production worldwide taking into consideration its benefits for the global challenges faced.

Keywords: animal welfare, ecological cycles, organic integrity, sustainability

Introduction

Domestication of plants and animals and agriculture occurred quite recently in the history of mankind. The move from shifting agriculture to domesticated agriculture is attributable to the concentration of human experience with wild plants and animals and further transferring successful results of trial-and-error experimentation into practice. The shift to agricultural communities was probably gradual and triggered by one or more factors, or their combination, like change in climate, increased population or limitations in natural food resources (https://education.nationalgeographic.org/resource/development-agriculture/ [accessed 16/1/25]. Agriculture then became the major source for nutrition. The more diverse agricultural system based on producing staple crops in rotation with clovers, grasses and root crops for feeding livestock slowly diminished. During the last century, when the prevailing postwar conditions coupled with the developments in agri-science and technology, intensification occurred to obtain higher yields at low cost. However, at global scale, the process of modernization produced different types of agriculture ranging from traditional or low-input to

*Corresponding author: uygun.aksoy@gmail.com

DOI: 10.1079/9781800626850.0002

highly industrialized. Developed practices focused more on short-term productivity through off-farm inputs. A few decades later, the impact of synthetic agri-chemicals on health and environment started to unveil. Adverse effects of various synthetic inputs such as pesticides, fertilizers, growth regulators or food additives, or of methods such as irradiation or genetic engineering, were displayed, elevating food and environmental safety concerns. Monocrop and/or monoanimal enterprises destined for international markets became dominant. Introduction of genetically engineered crops resulted in extension of monoculture. Thus, sustainability became a key issue in global development, or throughout value chains or at farm level. In the last 30 years, the decrease in rural populations and in traditional farming systems has led to intensive livestock systems that also require improvements regarding animal welfare, productive efficiency, and food quality (Caroprese *et al.*, 2015). In vast monocrop or monoanimal cultures, integrating livestock into a plant production system is demonstrated to increase its economic and environmental health and diversity, thereby making an important contribution to the farm's sustainability (Coffey *et al.*, 2004). Currently, human pressures on ecosystems, especially with climate crises, and sustainability, became the core issue not only at farm level but also for human development in a well-functioning healthy and resilient environment (Rockström *et al.*, 2017). Parallel to these problems, organic concepts and sustainable practices gained global momentum in the 21st century and with the COVID-19 pandemic and climate crisis. Organic practices based on common principles, and designed accordingly, appear to bring a solution to interlinked problems in an era of crisis. In practice, increasing organically managed areas is a strategic target for sustainable development and handling current problems.

How Organic Principles are Developed and Formulized?

The individual pioneers of organic, dispersed throughout the world, later paved the way to civil society organizations. There was no common definition or guidelines/standards until the 1980s except 'biodynamic', defined in 1924 (www.demeter.net [accessed 16/1/25]). Most of the principles or practices applied were either traditional and locally tested alternatives to synthetic inputs, or based on observations derived from comparative trials. The first attempt for an international organic standard was in 1980 during the biennial IFOAM General Assembly in Brussels, based on national standards of Switzerland, France and the UK. France later developed a voluntary organic standard and a system with a special logo. In 1987, Denmark adopted the first regulatory framework on organic (Lernoud and Willer, 2019). Prior to the first official standard, certification bodies or groups carried out audits and prepared conformity documents to access distant markets based on private standards. At this stage, the focus varied and was either on major practices, banned inputs, or whole-farm management. The first set of principles was published by the Australian Organic Farming and Gardening Society (1944–1955) (Paull, 2008). The principles were later defined broadly and included many of the aspects listed below:

- Take soil fertility as the basis of production capacity and maintain soil cover to increase nutrient cycling for the long term.
- Enhance crop production through appropriate use of species, varieties, landraces or local breeds, and methods such as crop rotation and intercropping, well adapted to the prevailing conditions and in harmony with animal husbandry.
- Exclude or severely restrict synthetic fertilizers, pesticides, herbicides, growth regulators and additives, and methods such as genetic engineering and irradiation.
- Maintain and encourage diversity within the farming system and its surroundings.
- Provide attentive care including feeding strategies and stocking rates that meet both health and behavioural requirements of farm animals.
- Establish a sustainable system where the whole chain is inspected by independent bodies within an accepted regulatory framework to guarantee consumer trust, rather than the analysis of the end-product.
- Improve farmers' welfare and social standards on the farm.
- Maintain the integrity and nutritional value of organic food and processed products through each step of the process from planting to consumption.

(Aksoy, 2015)

There is still no common definition or standard of organic agriculture; however, various references define organic agriculture with the common keywords included. Standards aim to achieve set targets by putting down a framework of rules and procedures. The EC 2092 European Union regulation issued in 1991 (Council regulation (EC), 2007), first on plant production and later widened to involve livestock, created a global impact regarding preparation of standards for a major market until overtaken by the USA. The term 'organic' was a synonym for 'biological' or 'ecological' farming. This synonymous approach is accepted in many countries today even if some authors note differences, e.g. they use the term 'ecological organic' to stress the environmental component.

The EC regulation put into force in January 2009 states that:

> Organic production is an overall system of farm management and food production that combines best environmental practices, a high level of biodiversity, the preservation of natural resources, the application of high animal welfare standards and a production method in line with the preference of certain consumers for products produced using natural substances and processes. The organic production method thus plays a dual societal role, where it on one hand provides for a specific market responding to a consumer demand for organic products, and on the other hand delivers public goods contributing to the protection of the environment and animal welfare, as well as to rural development.
>
> (EC Regulation (EC) No. 834/2007)

The US Department of Agriculture has defined organic agriculture with a more technical approach focusing on some key practices:

> Organic farming is a production system which avoids or largely excludes the use of synthetically compounded fertilizers, pesticides, growth regulators and livestock feed additives. To the maximum extent feasible, organic farming systems rely on crop rotations, crop residues, animal manures, legumes, green manures, off-farm organic wastes, and aspects of biological pest control to maintain soil productivity and tilth, to supply plant nutrients and to control insects, weeds, and other pests.
>
> (Cacek, 1994)

The Codex Alimentarius prepared by the United Nations Food and Agriculture Organization (FAO) and the World Health Organization (WHO) defines organic agriculture as a 'holistic production management system which promotes and enhances agro-ecosystem health, including biodiversity, biological cycles, and soil biological activity. It emphasizes the use of management practices in preference to the use of off-farm inputs, taking into account that regional conditions require locally adapted systems' (Codex Alimentarius, 2013).

The International Federation of Organic Agriculture Movements (IFOAM), an umbrella civil society organization, worked on the definition of organic agriculture for many years and the following definition, accepted in March 2008, incorporates the four principles and states that:

> Organic agriculture is a production system that sustains the health of soils, ecosystems, and people. It relies on ecological processes, biodiversity and cycles adapted to local conditions, rather than the use of inputs with adverse effects. Organic agriculture combines tradition, innovation, and science to benefit the shared environment and promote fair relationships and a good quality of life for all involved (www.ifoam.org/organic_facts [accessed 16/1/25]).

Agriculture is one of the basic activities of mankind providing food, fibre and fuel. However, one aspect that was undermined but strongly underlined nowadays is the ecosystem services offered to the society. Organic agriculture uses precise and specific standards of production that aim at achieving optimal agro-ecosystems that are socially, ecologically and economically sustainable. Therefore, the principles of organic agriculture encompass ecological, environmental and ethical aspects to achieve a sustainable production and healthy ecosystem, and, thus, a healthy planet.

The principles are the roots from which organic agriculture grows and develops. The principles then pave the way towards the guidelines that are further elaborated into standards/regulations that govern the organic value chains, markets and the organic system (Alroe and Kristensen, 2004). Today, there are more than 100 organic national standards that are fully implemented, not yet implemented or at the stage of preparation. They are all based on common organic principles but may deviate based upon the existing national regulatory framework or the market demand (Willer *et al.*, 2024).

In the history of global organic agriculture, the foundation of IFOAM in 1972 was a turning point (www.ifoam.bio [accessed 16/1/25]). IFOAM prepares, widens and revises the basic organic standards. In 2005, to establish a common ground, IFOAM adopted a new set of basic principles of organic agriculture to serve the organic movement. These principles are based on four pillars as listed below (Anonymous, 2005):

1. Principle of Health: *Organic agriculture should sustain and enhance the health of the soil, plant, animal, human and plant as one and indivisible.* The health of the planet is closely linked with all the elements present on Earth, and health impacts all its components. A well-functioning ecosystem depends on health of individuals and their communities, which are interrelated and cannot be separated. Health is also embedded in the 'one health' phenomenon. Health is not just the state of being free from sickness, but it also means that, ecologically, physically, socially and mentally, wellness is kept for the long term.
2. Principle of Ecology: *Should be based on living ecosystems and cycles, working with them and helping sustain them.* To achieve the targeted production, where not only quantity but also quality and clean environment are main objectives, ecological processes and recycling need to be promoted rather than off-farm inputs. Since the natural cycles may vary according to site-specific conditions, organic management should be adapted to local conditions, including ecology, socio-economics, culture and scale. Due to the interlinkage, the agro-ecosystem boundaries are not limited to the farm, but its surroundings are also important in restoration, maintenance and services of the ecosystem.
3. Principle of Fairness: *Should build on relationships that ensure fairness with regard to the common environment and life opportunities.* At global level, there should be equity, respect, justice and stewardship of the shared world since all the elements of the ecosystem are interlinked and have different and complementary roles. The relationships established throughout the value chain should ensure fairness at all levels and to all parties. The rules set and the product flows must guarantee that all actors involved, e.g. farmers, labourers, processors, distributors, traders and consumers, benefit from organic agriculture in providing a better quality of life. Organic management should consider the long-term effects on all living organisms and the ecosystem at large.
4. Principle of Care: *Should be managed in a precautionary and responsible manner to protect the health and well-being of current and future generations and the environment.* Innovation and research are the tools implemented in advancing agri-food production; however, their long-term effects are important, and it requires precaution and responsibility while selecting the technology. Decisions and practices should also consider indigenous knowledge that is collected and tested by time. There are still debates ongoing for adoption of recently developed genetic engineering methods. In this respect, a precautionary approach is valid because unexpected results may occur if evaluation is based on short-term results. Science is the basis for decision making; however, it must be transparent, participatory and make long-term risk assessments. While preparing or revising standards, experts consider these aspects, and depending on scientific results obtained, standards on organic undergo frequent change.

Key Challenges in Agriculture and Benefits of Organic Agriculture

During the last decade, discussions started to focus on rethinking/redesigning of agri-food systems to address key challenges that are intertwined such as climate change, soil health, carbon emissions/sequestration, biodiversity loss, ecosystem services, authenticity, nutritional value, ageing of farming communities, adoption of smart technologies and tools, gender equality, survival of family farms, food loss and waste, and food security and safety. The impact of these challenges may vary greatly and therefore the priorities defined in each region may be different. It is often discussed if organic is a solution to all or many of these issues. Building healthy soils through enhancing organic matter is the key to organic. The organic standards prepared, and quite often amended, are based on risk assessment. While selecting the appropriate technology or inputs, organic scrutinizes thoroughly and applies a precautionary approach. Limitations imposed on use of synthetic inputs and genetic engineering aim to contribute to health in wider terms. Organic is accepted as a system composed of independently functioning but interlinked elements that come together to make the whole farm or the chain (Strassner and Kahl, 2020). The main approach is to implement holistic farm management as much as possible, like an organism within a closed system. Compared to other farming systems, organic farming has documented long-term performance. The solutions are tailored more according to specific site conditions and are therefore forced to combine the best adapted elements in a holistic approach (Köpke, 2000). The results obtained in a meta-analysis performed on data from developed countries suggest that organic practices enhanced biological communities and there was a very low variability among organic farms in respect to environmental sustainability. On the other hand, conventionally managed farms displayed a high variability regarding environmental impact. Some had higher yields obtained largely due to synthetic inputs such as pesticides and fertilizers readily available. The cost and profitability of organic and conventional agriculture were similar for crop types examined, e.g. cereals, pulses, forage crops, vegetables, oil crops (Smith *et al.*, 2019).

To achieve Sustainable Development Goals, the human, social, economic and natural resources, which are vital and interlinked, need to be secured. Organic agriculture endeavours to optimize the health and productivity of interdependent communities of soil life, plants, animals and people, which will lead to a healthy planet. At farm level, the restrictions coming up from the site or from the human resources (inclinations, values, objectives) are the main factors diverting organic farming systems from its ideal state. The standards valid in domestic or international trade provide the minimum conditions or compliance to the commonly accepted rules providing a guarantee for the distant consumer. Specific sites may force certain conditions. The housing conditions in organic livestock production differ from conventional animal husbandry in terms of form, size and stocking densities that are suited to the natural behaviour of animals. The livestock housing should be provided with sufficient clean air and daylight, and should be built to protect the animals from extreme weather conditions using friendly building materials and equipment. These aspects are essential for animal welfare and for the prevention of diseases by strengthening the immune systems; however, in some standards, these are handled either rather superficially or may undergo frequent changes based upon the pressure coming from the sector.

The role of organic agriculture goes beyond farming and includes all stages of processing, distribution and consumption. Today, organic products are distinguished by third-party certification that confirms their organic status stated on the label. As the market for organic products enlarged after the 2000s, international organizations made efforts to have a common standard. However, it failed. Therefore, today, each country where production and/or markets are developed has its own standard valid in the domestic production and imports. Third-party verification is seen as a barrier, especially for family farms and smallholders to access domestic markets due to the complexity of documentation and cost, thus hindering local organic market development. National organic standards, such as from the USA and the EU, started to allow third-party certification of groups based on internal control system (ICS). IFOAM – Organics International lobbies for promotion of a Participatory Guarantee System (PGS) as the verification scheme that relies on ICS. There are very good examples of PGSs in different regions of the world and opportunities to enhance synergies (IFOAM – Organics International, 2017, 2022). Some countries like India, Brazil, Peru and The Philippines included the Participatory Guarantee System in their national regulatory framework as a means of verification of organic.

The official organic standards embrace mostly the production phase of agri-food commodities up to the point of retail. The scope of the standards in almost all areas includes plant or animal agri-food production, wild collection and processing. Private standards try to fill the existing gaps or newly emerging areas leading to multiple certification schemes, which causes complexity. IFOAM has developed a concept of a 'Family of Standards' to promote having common approaches and similarities (www.ifoam.bio).

In organic livestock production, there is a huge demand for organic plant-origin feeds, which indicates that, especially, organic large and small ruminant production enterprises should have organic plant production (Cömert Acar *et al.*, 2019). Therefore, creating an animal–plant production cycle in organic production should preferably target a closed system (Escribano, 2016). In addition, alternating plant production unit with feed plants and supplying manure and fixing nitrogen through legumes in rotation will improve the structure and content of the soil and reduce production costs. The current statistics show that in Europe and the USA over 60% of land certified as organic is destined for animal feed, very much like shares in the conventional system (Willer *et al.*, 2024). This shows that there is a need to change not only the management but also dietary habits and to find alternative protein sources that require less land. To prioritize food for humans, 'veganic' or vegan organic, which promotes stock-free organic and avoids all forms of animal-based inputs, is gaining momentum (https://veganorganic.net/ [accessed 16/1/25]).

Stress conditions, especially those resulting from climate change, may create a disorganized influence not only on the crop, plant or animals but also on all accompanying organisms. To make real assessments of the effects of the management, long-term comparative trials of biodynamic, organic and conventional (DOK in German) are carried out and results discussed at international level (Hartmann *et al.*, 2015). General approaches may not always prove very efficient; solutions may be site-specific and require planning of combined strategies rooted in basic principles. Long-term planning based on knowledge will help achieve an environmentally and socially sound and economically viable production.

Current global challenges are matters of discussion within the organic community. Research and innovation groups such as the International Society for Organic Farming Research (ISOFAR; https://www.isofar.online [accessed 17/1/25]) or the Technology and Innovation Platform of IFOAM prioritize these aspects and promote co-operation for solutions. Further resolutions can be brought through additional measures in implementing of standards or utilizing multi-verification tools. As an example, water scarcity became an issue in some regions with global warming, and some private organic standards started to integrate water-use efficiency among their rules. Some of the private standards have articles regarding biodiversity management. To have a broader coverage for climate resilience, regenerative approaches and practices are being integrated with organic, leading to an additional label in organic (https://www.linkedin.com/company/regenerative-organic-alliance [accessed 16/1/25]).

Conclusion

Organic agriculture today is implemented in 188 countries, as of 2022. In addition to 96.4 million ha of organically cultivated area, there are 34.6 million ha managed organically and devoted to wild collection, beekeeping, aquaculture or forest areas (Willer *et al.*, 2024).

In practice, organic operators follow rules stated in different standards, having variations but all based on common principles. As the distance between the producer and the consumer lengthens and various standards come into force, multiple certifications appear more and more. These, on the one hand, help build consumer trust, and on the other trigger discussions on sustainability of the value chain due to increased cost and documentation. Globalization of the markets and consumer preferences also shape production trends and give rise to private organic standards, e.g. vegan organic, biodynamic, bio-swiss or regenerative organic.

Innovations and developments in agri-food systems that favour sustainability are affecting organic agriculture and opening new pathways to develop resilient organic systems. These developments need to be scrutinized for their long-term effects before adoption since they have technological as well as social or economic elements and may have positive or negative impacts (Rahmann *et al.*, 2021). Even if third-party verification is the widely adopted method, in many developing countries, successful PGS models are integrated into the national organic legislation paving the way for the smallholders to access export markets. Common global problems such as responding to the climate crisis, loss of biodiversity, use of plastics, yield gaps, animal welfare, gender equality, high food loss and waste rates, and lack of recycling need to be tackled in organic practices either through integrating these aspects into the organic standards or through efforts of organic farmers and processors. Overall, animal welfare standards have improved over the past 30 years; however, they still remain relatively limited (Duval *et al.*, 2020). To accomplish basic principles of health and fairness and maximizing ecosystem services, farmers, other actors and decision makers of the organic value chains need to monitor and scrutinize rapidly changing environs so as not to fall behind the basic goals.

There are other alternative certified or non-certified sustainable farming systems, agroecology and regenerative being at the top. The main difference is that organic farming has accumulated nearly a century of experience and is governed by specific national and internationally accepted standards; is accompanied by inspection and certification schemes all through the production chain; and does not allow the use of synthetic chemical inputs and methods such as genetic engineering and irradiation, whereas all the others recommend selected or reduced usage of synthetic inputs. There will be new challenges and crises in agri-food systems that need to be addressed. Organic farming, with its roots in practice, science and markets, seeks to embrace them and bring sustainable solutions.

References

Aksoy, U. (2015) General principles of organic agriculture. Mediterranean Organic Agriculture Masters' Course, academic year 2014/15. Module 1. CHIEAM/IAMB, Bari, Italy. Unpublished course lecture notes.

Alroe, H.F. and Kristensen, E.S. (2004) Basic principles for organic agriculture: Why? And what kind of principles? *Ecology and Farming* 36, 27–30.

Anonymous (2005) Principles of organic agriculture. IFOAM website. Available at: www.ifoam.org (accessed 15 June 2023).

Cacek, T. (1994) Organic farming: The other conservation farming system. *Journal of Soil and Water Conservation* 39(6), 357–360.

Caroprese, M., Albenzio, M. and Sevi, A. (2015) Sustainability of sheep and goat production systems. In: Vastola, A. (ed.) *The Sustainability of Agro-Food and Natural Resource Systems in the Mediterranean Basin*. Springer, pp. 65–75.

Codex Alimentarius (2013) CXG 32-1999. Guidelines for the production, processing, labelling and marketing of organically produced foods. Available at: https://www.fao.org/fao-who-codexalimentarius/codex-texts/guidelines/en/ (accessed 16 January 2025).

Coffey, L., Hale, M. and Wells, A. (2004) *Goats: Sustainable Production Overview*. ATTRA Publications.

Cömert Acar, M., Şayan, Y., Koşum, N. and Kandemir, Ç. (2019) The general principles of organic goat production and its role in strengthening the sustainability. In: Zencirkiran, M. (ed.) *Trends in Landscape, Agriculture, Forest and Natural Science*. Cambridge Scholars Publishing, pp. 214–227.

Council regulation (EC) No. 834/2007 of June 28, 2007, on organic production and labelling of organic products and repealing Regulation (EEC) No. 2092/91 (n.d.). https://eur-lex.europa.eu/eli/reg/2007/834/oj/eng (accessed 11 June 2025).

Duval, E., Keyserlingk, M.A.G. and Lecorps, B. (2020) Organic dairy cattle: Do European Union regulations promote animal welfare. *Animal* 10, 1786.

Escribano, A.J. (2016) Organic livestock farming: Challenges, perspectives, and strategies to increase its contribution to the Agrifood system's sustainability: A review. In: Konvalina, P. (ed.) *Organic Farming: A Promising Way of Food Production*. InTech. DOI: 10.5772/61272.

Hartmann, M., Frey, B., Mayer, J., Mader, P. and Widmer, F. (2015) Distinct soil microbial diversity under long-term organic and conventional farming. *The ISME Journal* 9, 1177–1194.

IFOAM – Organics International (2017) The full diversity of organic agriculture: What we call organic. Position Paper 4.

IFOAM – Organics International (2022) *Exploring opportunities to increase the synergy between Participatory Guarantee Systems (PGS) and Third-party Certification (TPC) in organic agriculture*. Technical Report, 26. Available at: https://www.ifoam.bio/exploring-opportunities-increase-synergy-between (accessed 16 January 2025).

Köpke, U. (2000) Organic agriculture: An organizational principle. Intensive short course on optimizing organic agriculture systems in the Mediterranean countries (unpublished course notes).

Lernoud, J. and Willer, H. (2019) Organic agriculture worldwide: Key results from the FiBL survey on organic agriculture worldwide 2019. Part 3: Organic agriculture in the regions 2017. Available at: https://orgprints.org/id/eprint/35383/3/FiBL-2019-Regions-2017.pdf (accessed 16 January 2025).

Paull, J. (2008) The lost history of organic farming in Australia. *Journal of Organic Systems* 3(2), 2–17. Available at: https://orgprints.org/id/eprint/15089/1/15089new.pdf (accessed 16 January 2025).

Rahmann, G., Rey, F., Ardakani, R., Azim, K., Chable, V. *et al.* (2021) Book of abstracts of the science forum 'From Its Roots, Organic Inspires Science, and Vice Versa. Thünen-Report 88, Trenthorst/Rennes.

Rockström, J., Williams, J., Daily, G., Noble, A., Matthews, N. *et al.* (2017) Sustainable intensification of agriculture for human prosperity and global sustainability. The Royal Swedish Academy of Sciences. *Ambio* 46, 4–17. DOI: 10.1007/s13280-016-0793-6.

Smith, O.M., Cohen, A.L., Rieser, C.J., Davis, A.G., Taylor, J.M. *et al.* (2019) Organic farming provides reliable environmental benefits but increases variability in crop yields: A global meta-analysis. *Frontiers in Sustainable Food Systems* 3. DOI: 10.3389/fsufs.2019.00082.

Strassner, C. and Kahl, J. (2020) Understanding a food systems approach. In: Auerbach, R. (ed.) *Organic Food Systems: Meeting the Needs of South Africa*. CAB International, Wallingford, UK, pp. 51–59.

Willer, H., Trávníček, J. and Schlatter, B. (eds) (2024) *The World of Organic Agriculture. Statistics and Emerging Trends 2024*. Research Institute of Organic Agriculture FiBL, Frick, and IFOAM – Organics International, Bonn. Printed by Druckerei Hachenburg PMS GmbH, Saynstraße 18, 57627, Hachenburg, Germany.

3 The Role of Agroecology as a Transdisciplinary Science for Organic Farming

Fabio Caporali*

DAFNE (Dipartimento di Scienze Agrarie e Forestali), Viterbo, Italy

Abstract

Agroecology and organic farming are two emergences in the current context of the human activity system qualified as agriculture. The former, in the field of science, and the latter, in the field of practice, oppose the dominant paradigm of conventional, industrialized agriculture and both operate in favour of a change of paradigm known as sustainable agriculture. This chapter aims at showing how this change can occur through an agroecological transition, wherein the new epistemology and ontology of systems thinking and systems practice apply to agriculture. New knowledge about the structure and functioning of nature and new potential benefits for humans and the biological community as a whole can originate from a multi-functional agriculture grounded on the application of agroecological principles and the provision of ecosystem services by organic farming.

Keywords: systems thinking, systems practice, sustainable agriculture

Introduction

The future of agriculture is not input-intensive but knowledge-intensive. We need the integrated approach that agroecology can offer. FAO Director-General José Graziano da Silva (2018) states:

As any other human activity system, agriculture requires knowledge to be successful. Knowledge stems from experience (practice) and intentional research (experimental enquiry) aimed at producing science, i.e. the knowledge of the context of life. Ecology is the more inclusive science that man has until now produced to deal with his/her context of life. Ecology is a science of relationships and Agroecology is the science of Ecology applied to the practice of agriculture.

The above reported citation from an FAO report (FAO, 2018b) is meaningful in showing the relation between three key words: agriculture, knowledge and agroecology, where 'the integrated approach' is exactly the acknowledged link connecting agriculture and knowledge, i.e. the epistemological relation defined as 'agroecology'. Integration is a property of any system, where the system's components or constituent parts show interconnections according to the system's peculiar organization. Both ecology and agroecology have a peculiar epistemological tool – the ecosystem concept (Tansley, 1935; Golley, 1993) and the agroecosystem concept (Spedding, 1971; Caporali, 1991; Caporali, 2015a), respectively – a model for representing reality as an object of study or management. Both ecology and agroecology share the epistemology of systems thinking and the ethics of systems practice (Checkland, 1993; Caporali, 2010, 2021), showing both philosophical and scientific foundations, and their nature of fields of transdisciplinary enquiry for a more sustainable agriculture (Fig. 3.1).

*Corresponding author: caporali@unitus.it

DOI: 10.1079/9781800626850.0003

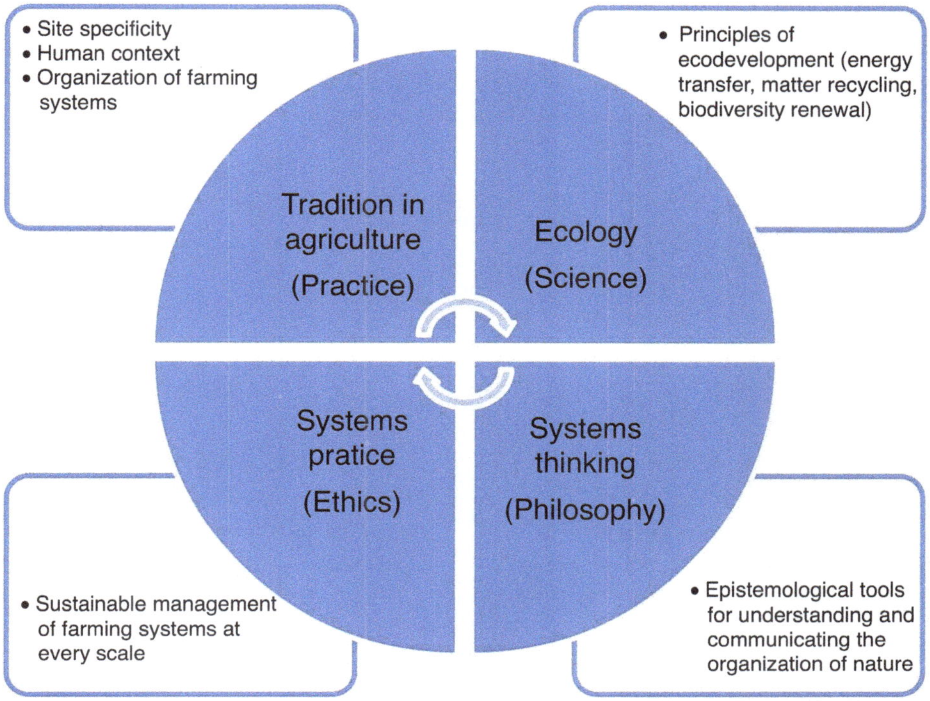

Fig. 3.1. Agroecology as a science integrating theory and practice for a more sustainable agriculture.

The Systems Paradigm as the Basis of Knowledge for Both Ecology and Agroecology

The systems paradigm is the shared epistemological background of both ecology and agroecology. The roots of this knowledge method date back to the ancient Greek philosophy as exposed by Aristotle, who describes the essence of the knowledge process as an integration of four causes: efficient, formal, material and final. The systems paradigm is also referred to as process philosophy (Barbour, 1997), where each thing, either material or immaterial, is a process, i.e. an event that happens in a spatial-temporal scale, and the becoming is the essence of reality. Each thing, or whole, continuously exchanges energy, matter and information within its context, causing reciprocal influences. In systems analysis, such a pattern of behaviour operates according to an input/output model that allows evaluation of systems performances in terms of sustainability through suitable indicators. The first decades of the 20th century were crucial in the provision of philosophical contributions to the systems paradigm, especially by Whitehead (1927) and Smuts (1927). According to the accurate enquiry of Golley (1993), the ecosystem concept is a scientific outcome of philosophical questions. Indeed, Tansley (1935) coined the term ecosystem as the result of a dialectical confrontation with the South African ecologist J. Phillips, who based its idealist foundation for ecological research on the holistic philosophy of Smuts (Caporali, 2008).

The ecosystem concept expresses an integrative idea of nature, where man is included as a major agent of change. As pictured in Fig. 3.2, it unveils the astonishing quality of the Earth as a photosynthetic planet that sustains biodiversity renewal and evolution at every scale of organization, through biochemical transformation of solar energy and continuous recycling of matter.

The ecosystem concept is a broad and powerful one revealing different meanings (Caporali, 2021):

1. a model of representation of reality or epistemological tool, i.e. a product of human mind that accounts for nature self-consciousness;

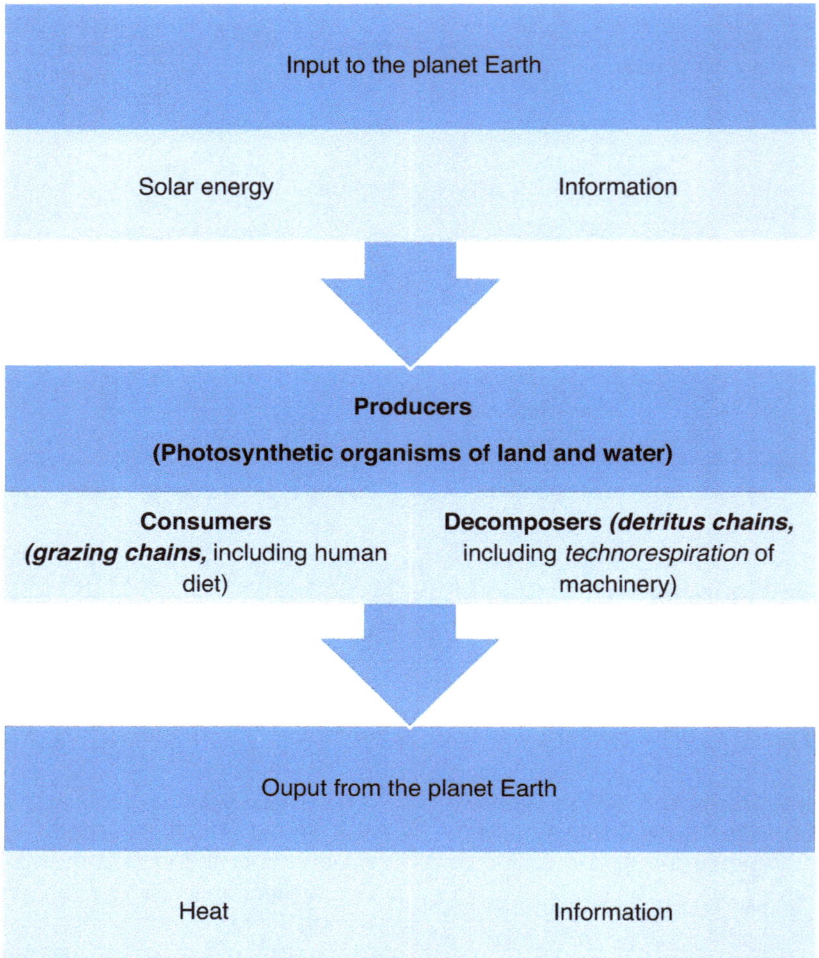

Fig. 3.2. Main components and organization of the Earth at global spatiotemporal scale according to the ecosystem model.

2. a systems model or process model or relational model, representing reality as a process of continuous organization of different components or parts in a spatial-temporal dimension;
3. an integration model that accounts for a nested or hierarchical integration in space and time of different components in a sequence of substantial change (for example, geosphere ->biosphere ->noosphere, at global level);
4. a structural-functional model that identifies: (i) discrete inorganic components and biological forms as entities linked in basic functional relationships (autotrophic for producers and etherotrophic for consumers and decomposers) performing grazing and detritus trophic chains and webs; (ii) specific functional processes as transfer of energy and information, nutrient cycling and biodiversity creation as the basis for continuous development or sustainability;
5. an aesthetic model, showing nature as a non-human marvellous construct of forms and relationships;
6. a heuristic model, accounting for both natural creative evolution and the process of increasing complexity of nature towards more widespread and purposeful self-consciousness; and

7. an ethical model, revealing that co-operation among components is the key for the sustainability of nature, which is to be intended more as a community that a commodity. It provides the principles of ecodevelopment in terms of integral use of solar radiation, material cycling and biodiversity promotion.

The utility of the ecosystem concept is in connecting components and processes into a coherent unity of organization for understanding nature. Components and processes take their name in relation to their functional role and their dynamics that yield general patterns of both biodiversity and sustainability. The ecosystem concept, better than any other concept, confirms the intelligibility of nature as an ordered construction easily grasped by human rationality. Most of scientific disciplines contribute their part of knowledge to both formulating and validating the ecosystem concept, and grounding its transdisciplinarity.

Ecosystems function through processes of self-organization referred to as ecological integrity. Karr (1991) defines integrity as an ecological quality that refers to the state of being complete and undivided having a minimal demand for external support. Ecosystem integrity represents the degree of self-organization a system has gone through (Muller *et al.*, 2000).

As to the application of the ecosystem concept to agriculture, the general concept of 'agricultural ecosystems' gained widespread acceptance in scientific circles since the 1970s, as documented by an influential paper of Spedding (1971). In 1974, this acceptance culminated with the publication by Elsevier of the scientific journal *Agro-ecosystems*, thus initiating the collection of scientific contributions grounding the knowledge body of the science of agroecology.

Spedding (1971) was prophetically able to outline the benefits of the 'systems thinking' application to agriculture, advancing the following considerations:

1. The use of the term *agricultural ecosystems* implies that this systems approach is relevant and that there are advantages in thinking about agriculture in this way.
2. The broad *system approach* is the same as an application of the scientific method itself: its value in this sense is thus evident.
3. Much agricultural research has consisted of the application of scientific method to *components* of agricultural systems.
4. A *systems approach* encourages a constant recognition that [parts] are only components and that, at some point, the agricultural systems themselves must be the subject of study.
5. Man himself (and not merely as *labour* or a source of energy), his intellect, and his tools, are all important components of agricultural systems.
6. Agricultural systems are *subsystems of something larger* [...] and agriculture itself is one subsystem of the total activity of Man.

Table 3.1. Main components and organization of an agroecosystem according to an input/output model.

Input	Agroecosystem Components	Output
Energy	Planned Associated	Heat
Matter	External (mostly Solar from fossil fuels, (gratuitous,	Flows of air, water and biomass
Information	costly, polluting) clean)	Crop yield and animal products;
Money	Machinery, Physical	natural flows of water, air and
	agrochemicals, environment feedstuff	biomass
	Producers	Cultural, social and environmental
	Selected Vegetation crops (weeds)	impacts
	Consumers	Market income
	Selected Animals livestock (insects)	
	Decomposers	
	Soil fauna	
	(micro-, meso-)	
	Technology for	
	crops, livestock and Genetic land management	
	Capital	

Following what until now has been discussed, Table 3.1 reports a general description of the agro-ecosystem model that can be adopted at every level of spatiotemporal scale (field, farm or territorial).

In any enquiry, either study or management, the first thing to do is establish the limits of the agroecosystem to design, monitor or manage. Once limits are fixed, the agroecosystem and its external environment are automatically determined, and exchanges of input and output of physical, chemical, biological or cultural nature can be assessed through suitable indicators (Tellarini and Caporali, 2000; Caporali *et al.*, 2003). Due to the activity of continual exchange (input/output) between the agroecosystem and its environment, the agroecosystem is 'open', i.e. in constant communication with its environmental context. An agroecosystem being a 'whole' in a larger one (its environmental context), it will be always constrained by the last one, for instance by the climate, soil type, topography etc., and by cultural, social and economic conditions. It follows that the choices of the farmer, the landscape planner or the politician concerning the agroecosystem composition and organization depend on the local specificity or even on the global tendency, such as the global industrialization of agriculture in the last decades, worldwide.

To give an idea of the complexity of agriculture relations according to a systems thinking approach, it is meaningful to consider it as a *human activity system*:

> In a systems framework, agriculture is regarded as a human activity system derived from the integration of natural and anthropogenic systems, both abstract and physical. The system is the product of networks of social connections linking all the important nodal points. Agriculture is clearly recognised as a multi-perspective system that provides food and other products as well as employment and income, ecosystem services, and a vital human connection to the natural environment. Sustainability emerges as a complex property of agroecosystems, embraces different aspects of the multi-functionality of human activity systems and indicates the ideal model that should be followed, that is a sustainable or co-evolutionary development.
>
> (Caporali *et al.*, 2007, pp. 23–24)

The sustainability of agricultural activity needs recognition not only from a biophysical point of view but also from a socio-economic one. Systems thinking, enlightening the hierarchical organization of reality, better helps perceive the interactive relationships between the farm and the institutions constituting its socio-economic and cultural context (political and administrative institutions, research, industry, credit and market institutions, individual and associated consumers). Agriculture pervades society as a whole and, following the market expansion in the whole planet through the process of globalization, it spreads from the local to the international level. The perception of the agricultural system implies the rediscovery of the profound linkage that binds man to the land, awakening an *ecological conscience* (Leopold, 1949; Callicott, 2013) for sustainable land governance, from the local to the planet level of organization.

As to the agroecosystem structure at the farm level, the distinction between *planned* and *associated* composition is appropriate in showing how the agroecosystem components depend on human choices resulting from a planning process. Once introduced, the planned components associate to and change a pre-existent autonomous community of biophysical components, where the underlying spontaneous organization continues to function. Although disturbed, the natural processes, such as bio-geochemical metabolism and biological diversification and control, persist inevitably. This is why the associated biological soil community is definitely important in determining the optimal conditions for maintaining soil fertility, crop productivity and the biological balance needed to preserve crops and livestock from pest and disease infestations. These conditions need sustenance through the integration of above-ground and below-ground components and processes (Ohgushi *et al.*, 2018). As to this goal, it is crucial to be able to govern successfully the interplay between the planned biodiversity of cropping systems and the change in the associated community that they determine. This is a matter of true synthesis among a vast range of disciplinary competences spanning agronomy, genetics, microbiology, entomology and pathology, but involving also aspects of economic, social, political and cultural development.

The operation of the agroecosystem has its basis in crop photosynthesis that integrates above-ground (leaves) and below-ground (roots) plant parts, accumulating biomass on cultivating fields during the growing season. CO_2 from the atmosphere, and H_2O plus nutrients from the soil, are the ingredients of the process. While water is normally supplied by rainfall in rainfed conditions, nutrients derive from mineralization by soil decomposers on crop residues or livestock-derived residues brought onto the soil during the previous crop cycles. In this way, a first cycle of integration occurs between crops and soil decomposers that synergically support themselves in a positive feedback relation where crops provide the soil with litter and decomposers supply the crop with nutrients. In adequate conditions, decomposers also provide the soil with complex and more stable organic matter through the process of humification, improving structure, porosity and water-holding capacity for the benefit of the crops. The process of humification is greatly promoted by conditions of no-tillage, especially associated with poly-annual forage crops, or by the incorporation into the soil of composted organic matter, such as the traditional farmyard manure in mixed-farming systems. Livestock rearing permits the enlargement of integration between producers, consumers and decomposers within the farming system, ensuring more ecological sustainability.

In order to protect the crops from infestations (weeds, pest and diseases), all the agroecosystem components within and between the cultivated fields need to be considered. A great deal of literature is mounting, showing how the enhancement of plant biodiversity (both planned and associated) within and between the cultivated fields is appropriate for favouring biological control in the agroecosystem trophic chains and webs (Altieri *et al.*, 2010; Kaplan and Ohgushi, 2018; van der Werf and Bianchi, 2022; Cappelli *et al.*, 2022; Deguine *et al.*, 2023). The ecological reason that justifies the use of intercropping and cover cropping within the cultivated fields, and the use of hedgerows and strips of natural or selected vegetation between the cultivated fields, lies in the positive effect of *ecotones* – areas of vegetational discontinuity between two adjacent *ecotopes* (spatiotemporal units of vegetational homogeneity) – for biological control. Crop rotation is another consolidated cropping system practice for grounding biological discontinuity within field in a spatiotemporal dimension, due to different crops alternating in the same field with different and compensatory effects on soil fertility and prevention of infestation.

The farm level of organization is the crossroads where both private and public interests and trade-offs meet and find their final outcome in the farmer's decision-making process. The farm is the appropriate level to verify if the expected multifunctionality of agriculture (productivity, preservation of environmental and social resources, food system security and safety, human and livestock health) is operating.

Systems thinking has had the potential for updating the meaning of both agriculture and the science of sustainable agriculture defined as agroecology. For a convenient comparison, the authoritative IAASTD (2009)[1] definitions of both agriculture and agroecology in the glossary of its Global Report *Agriculture at a Crossroads* follow:

> *Agriculture*: A linked, dynamic social-ecological system based on the extraction of biological products and services from an ecosystem, innovated and managed by people [...] It encompasses all stages of production, processing, distribution, marketing, retail, consumption and waste disposal.
>
> *Agroecology*: The science of applying ecological concepts and principles to the design and management of sustainable agroecosystems. It includes the study of the ecological processes in farming systems and processes such as: nutrient cycling, carbon cycling/sequestration, water cycling, food chains within and between trophic groups (microbes to top predators), lifecycles, herbivore/predator/ prey/host interactions, pollination, etc. Agroecological functions are generally maximized when there is high species diversity/perennial forest-like habitats.

According to FAO (2018a), the science of agroecology has values and potential for guiding countries 'to transform their food and agricultural systems to mainstream sustainable agriculture'. FAO has proposed the ten elements of agroecology emanated from the FAO's regional seminars[2] on agroecology as key words for guiding the process worldwide. The ten elements are interlinked and interdependent, and consonant with the systems thinking approach of agroecology (*diversity, synergies, efficiency, resilience, recycling*), the tradition in agriculture as a context feature (*human and social values, culture and food tradition*) and the systems practice approach as enabling environment (*responsible governance, circular and solidarity economy, co-creation and sharing of knowledge*).

Within the UN framework of Sustainable Development Goals (SDGs), FAO's engagement is adding strength to agroecology by the promotion of four key messages, as reported in Table 3.2.

The Systems Paradigm as the Basis for Practice in Organic Farming

Within the 'intellectual landscape of agroecology' (Tomich *et al.*, 2011), agriculture has been

Table 3.2. Key messages from FAO's Work on Agroecology. (Synthesis from FAO, 2018b)

1. Agroecology can contribute to accelerating the achievement of the 2030 Agenda
Agroecological approaches address root causes of hunger, poverty and inequality helping to transform food systems [...] through a holistic, integrated way that balances the three dimensions of sustainability – social, economic and environmental.

2. Agroecology: win-win-win for people, planet and livelihoods
By marrying scientific evidence with local wisdom, and by shortening the market chain and bringing producers and consumers closer together, agroecology can contribute to satisfying our present and future food needs.

3. Agroecology – supported by an enabling policy environment – can help transition to sustainable food systems
FAO continues to work with countries, playing a key role to bring stakeholders together to develop policies and learn from experiences in promoting and scaling up agroecology.

4. Agroecology is a living concept – together, we can help realize its full potential
Investing in knowledge and innovation is key [...] It is important to foster knowledge exchange and share practices, and to create opportunities for collaboration and innovation.

re-conceptualized according to a systems paradigm and reframed as a coupled system ('agroecosystem') involving a wide range of social and natural components and processes. After World War II, technological development massively influenced agroecosystem composition and organization following the dominant tendency of global industrialization and marketing, with inevitable negative consequences on the front of environmental impact and rural economy. To contrast these drawbacks, the practice of organic farming has progressively expanded, gaining consensus and success.

Concerning origin and meaning of both organic farming and agroecology, it is interesting to compare the following two positions, expressed in a seminal book on organic farming (Lampkin, 1994) and in a recent FAO report on agroecology (FAO, 2018a), respectively:

> The ideas behind organic farming have been around since the 1920s. They have evolved considerably and continue to evolve as new scientific research becomes available, while retaining the fundamental philosophical perspective of working with, not dominating, natural systems and respect for the environment which sustains us.
>
> (Lampkin, 1994, p. 10)

> Agroecology is not a new invention. It can be identified in scientific literature since the 1920s, and has found expression in family farmers' practices, in grassroots social movements for sustainability and the public policies of various countries around the world.
>
> (FAO, 2018a, p. 1)

Looking at them deeply, these two positions seem curiously interchangeable, with the first that sounds good also for agroecology and the second for organic farming as well. This complementarity is reasonable in that agroecology and organic farming are two faces of the same coin – the systems paradigm, constituted of scientific, philosophical and experiential components. Therefore, if you consider the practice of organic farming, you may recognize agroecology as its inspiring science (Lampkin's position), and if you consider the science of agroecology, you may recognize organic farming as the practice that confirms it (FAO's position).

Likely, each country has found its way to organic farming historically, agriculture being *de facto* 'organic' before the *conventionalization* process caused by industrialization. In Europe, due to the rich documentation on agriculture left in the Latin language by writers of Roman times (Cato, Varro, Virgil, Pliny the Elder, to cite the most famous of them), it is possible to recognize their own way of systems thinking, farm organization and best practices as follows:

> We can hardly contest that these points are still today relevant for a good modern farming carried out on a family-based organization, even if labour as human slaves has been replaced by machine-driven labour. Indeed, agricultural tradition has kept these general recommendations valid for most countries in the world because their success has been assured by the experiential learning cumulated along continuing generations of farmers. Agroecology in general recognizes the value of tradition in agriculture, establishes the scientific reasons of traditional agriculture's best practices and recognizes their values as foundation for a sustainable agriculture.
>
> (Caporali, 2010, p. 9)

That ancient tradition has never been lost in Italy, although progressively set apart under the dominant regime of industrialization which neglects systems thinking and professes the dominion of the parts on the whole, subverting and fragmenting the unitary order of the agricultural enterprise into single business activities more and more dependent on off-farm resources, which are costly and risky. Sometimes, the feeling emerges that agriculture is losing its identity and the farmer no more exists as such, but only as a driven operator (or a 'new slave'?) in the business machine of a money-addicted global society.

The systems doctrine of the farm as a 'body', and the cropping systems and the livestock rearing as its main 'organs', was the content focus of the first academically recognized curriculum in 'Agriculture and Animal Husbandry' that was established in 1844 at the University of Pisa, then belonging to the Great-Duchy of Tuscany. Cosimo Ridolfi and Pietro Cuppari were the curriculum developers, and the last one was the main advocate of the systems principle underlying the whole curriculum organization, i.e. that of considering the farm as a 'living body' made up of interacting parts to be organized harmonically under physical, biological, technological and economic constraints (Cuppari, 1862; Caporali, 2015b). In that curriculum, experiential learning occurred in two pilot farms with different soils and students benefiting from an education for analysis and synthesis of both the parts and the farm as a whole. The general rationale of the curriculum was considering the performance of the whole farm as the main purpose of agriculture teaching and research. About a century later, Draghetti (1948) recognized the relevance of Cuppari's systems approach to agriculture research and teaching and published a seminal book with a meaningful self-explanatory title 'Fisiologia dell'Azienda Agraria' (*Farm Physiology Principles*) (Table 3.3). In that book, Draghetti uses the term *agricoltura biologica* – the Italian version of 'organic agriculture' – for identifying the way a farm should be harmonically organized and managed as a whole composed of many parts in order to achieve complementary functioning and sustainability.

In 1972, the international movement for supporting organic agriculture (IFOAM)[3] was established with the mission of 'leading, uniting and assisting the organic movement in its full diversity' and with the goal of promoting 'worldwide adoption of ecologically, socially and economically sound systems that are based on the principles of organic agriculture' (IFOAM - Organics International , 2021). These principles 'are composed as ethical principles to inspire action' (Fig. 3.3) and were officially adopted by IFOAM in 2005[4]. Since then, they actually have assisted the processes of both *standards* making and *certification* for legislation on organic agriculture worldwide.

In Europe, legislation for organic agriculture dates back to 1991. The current law (Regulation 2018/848), adopted in 2018, defines organic farming as follows:

Organic production is an overall system of farm management and food production that combines best environmental and climate action practices, a high level of biodiversity, the preservation of natural resources and the application of high animal welfare standards and high production standards in line with the demand of a growing number of consumers for products produced using natural substances and processes. Organic production thus plays a dual societal role, where, on the one hand, it provides for a specific market responding to consumer demand for organic products and, on the other hand, it delivers publicly available goods that

Table 3.3. Draghetti's systems principles and pillars for supporting both agroecology and organic farming. (Elaboration from Caporali, 2010)

- A farm is viewed as a functional unity (living body) where all parts (organs) are in connection through a 'physiology' (organization), provided by the farmer's design and management, that allows 'circulation' and 'recycling' of materials in a synergistic framework between complementary components, according to their functional roles of 'organs'.
- Soil fertility maintenance is the main 'physiological' objective for ensuring long-term productivity or 'agroecosystem health', while crop rotations and mixed-farming with farmyard manure are the main 'organs' for supplying organic matter to soil.
- Pietro Cuppari and Alfonso Draghetti are two of the most prominent Italian contributors to agriculture systems thinking and practice; these scholars can be fairly recognized as proto-agroecologists in the evolving framework of agroecology and organic farming.

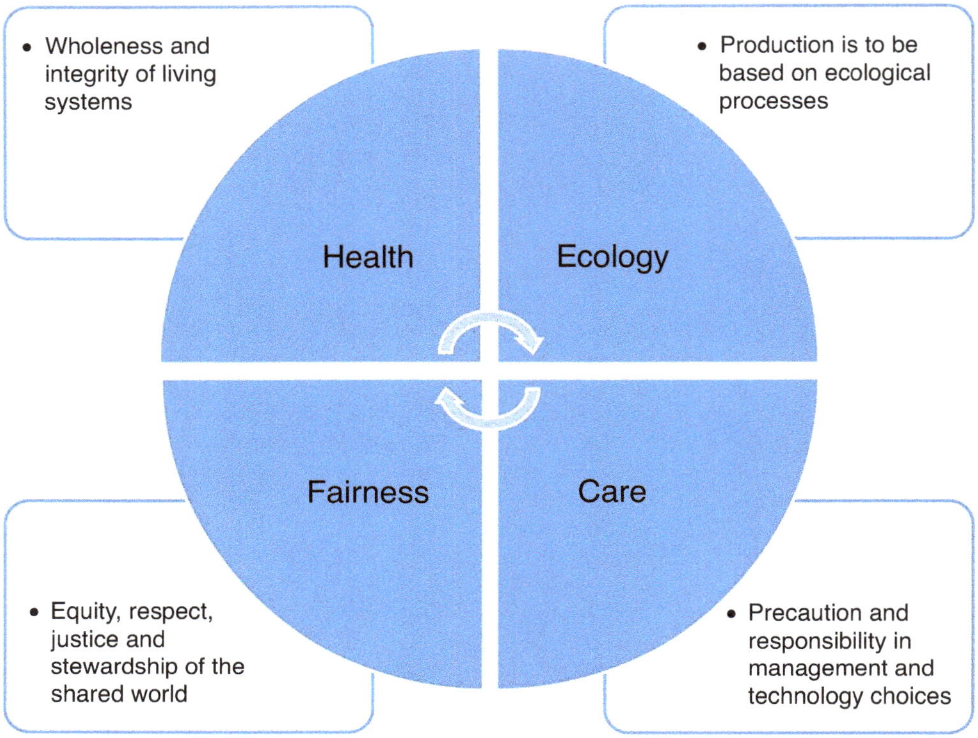

Fig. 3.3. Principles of organic agriculture. (Synthetic elaboration from IFOAM - Organics International, 2021)

contribute to the protection of the environment and animal welfare, as well as to rural development.

Conclusions

The term 'system' is recurrent in definitions and explanations concerning both agroecology and organic farming. Organic agriculture has adopted the principle of ecology as scientific framework for action, and therefore shares with agroecology the paradigmatic basis of both systems thinking (epistemology) and systems practice (ethics). Transformative redesign of farming systems based on agroecological principles is strategic against the competitive narrative of incremental efficiency in conventional agriculture. Since transformative systems such as organic farming have proven sustainability benefits, coherent policies and governments should only support agriculture and food systems that deliver on the SDGs (Eyhorn *et al.*, 2019; FAO and IFAD, 2019).

If scholars of agroecology and stakeholders of organic agriculture will be able to keep faith to the shared systems paradigm, their contribution to the development of both a sustainable agriculture and a sustainable planetary community will be fundamental.

Notes

[1] A panel of experts for international assessment of agricultural knowledge, science and technology for development.

[2] The ten elements of agroecology are based on scientific documentation (Altieri, 1995; Gliesman, 2015) and discussion during FAO's multi-actor regional meetings on agroecology from 2015 to 2017 (FAO, 2018a).

[3] IFOAM (International Federation of Organic Agriculture Movements).

[4] The General Assembly of IFOAM approved these principles on 25 September 2005.

References

Altieri, M.A. (1995) *Agroecology: The Science of Sustainable Agriculture*. CRC Press, Boca Raton, Florida.

Altieri, M.A., Nicholls, C.I., Wilson, H. and Miles, A. (2010) *Habitat Management in Vineyards: A Grower's Manual for Enhancing Natural Enemies of Pests*. Laboratory of Agroecology, College of Natural Resources, University of California. Available at: http://agroecology.berkeley.edu (accessed 17 January 2025).

Barbour, I.G. (1997) *Religion and science*. Harper, San Francisco, California.

Callicott, J.B. (2013) *Thinking Like a Planet: The Land Ethic and the Earth Ethic*. Oxford University Press, New York.

Caporali, F. (1991) *Ecologia per l'Agricoltura*. Utet-Libreria, Torino, pp. 415–429.

Caporali, F. (2008) Ecological agriculture: Human and social context. In: Clini, C., Gullino, M.L., Musu, I. and Xialing, Y. (eds) *Sustainable Development and Environmental Management*. Springer, pp. 415–429.

Caporali, F. (2010) Agroecology as a transdisciplinary science for a sustainable agriculture. In: Lichtfouse, E. (ed.) *Biodiversity, Biofuels, Agroforestry and Conservation Agriculture*. Springer, pp. 1–71.

Caporali, F. (2015a) History and development of agroecology and theory of agroecosystems. In: Monteduro, M., Buongiorna, P., Benedetto, S. and Isoni, A. (eds) *Law and Agroecology: A Transdisciplinary Dialogue*. Springer, pp. 3–29.

Caporali, F. (2015b) *Pietro Cuppari precursore dell'agroecologia e del governo sostenibile del territorio*. Edizioni ETS, Pisa.

Caporali, F. (2021) *Ethics and Sustainable Agriculture: Bridging the Ecological Gaps*. Springer.

Caporali, F., Mancin, R. and Campiglia, E. (2003) Indicators of cropping system diversity in organic and conventional farms in Central Italy. *International Journal of Agricultural Sustainability* 1(1), 67–72.

Caporali, F., Lieblein, G. and Francis, C. (eds) (2007) Integration of research and education in agroecology and organic farming. In: *Proceedings of ENOAT Workshop, Pieve Tesino*, Plant Production, University of Tuscany.

Cappelli, S.L., Domeignoz-Horta, L.A., Loaiza, V. and Laine, A.-L. (2022) Plant biodiversity promotes sustainable agriculture directly and via belowground effects. *Trends in Plant Science* 27(7), 674–687.

Checkland, P.B. (1993) *Systems Thinking, Systems Practice*. John Wiley and Sons, London.

Cuppari, P. (1862) *Saggio di ordinamento dell'azienda rurale*. Cellini, Firenze.

Deguine, J.-P., Aub, J.-N., Bellon, S., Côte, F., Lauri, P.-E. *et al*. (2023) Agroecological crop protection for sustainable agriculture. *Advances in Agronomy* 178, 1–59.

Draghetti, A. (1948) *Principi di Fisiologia dell'azienda agraria*. Istituto Edizioni Agricole, Bologna.

Eyhorn, F., Muller, A., Reganold, J.P., Frison, E., Herren, H.R. *et al*. (2019) Sustainability in global agriculture driven by organic farming. *Nature Sustainability* 2(4), 253–255.

FAO (2018a) The 10 elements of agroecology: Guiding the transition to sustainable food and agricultural systems. Available at: www.fao.org (accessed 17 January 2025).

FAO (2018b) *FAO's Work on Agroecology: A Pathway to Achieving the SDGs*. Food and Agricultural Organization, Rome.

FAO and IFAD (2019) *United Nations Decade of Family Farming 2019–2028*. Licence: CC BY-NC-SA 3.0 IGO. Global Action Plan, Rome.

Gliesman, S.R. (2015) *Agroecology: The Ecology of Sustainable Food Systems*. CRC Press, Boca Raton, Florida.

Golley, F.B. (1993) *A History of the Ecosystem Concept in Ecology*. Yale University Press, New Haven, Connecticut.

IAASTD (2009) *Agriculture at a Crossroads*. Global report of the International Assessment of Agriculture Knowledge, Science and Technology for Development. Island Press, Washington, DC.

IFOAM - Organics International (2021) Principles of organic agriculture. Available at: https://www.ifoam.bio/why-organic/principles-organic-agriculture/principle-health (accessed 18 August 2022).

Kaplan, I. and Ohgushi, T. (2018) Application and theory of plant-soil feedbacks on aboveground herbivores. In: Wurst, S. and Johnson, S.N. (eds) *Aboveground-Belowground Community Ecology*. Springer, pp. 319–343.

Karr, J.R. (1991) Biological integrity: A long-neglected aspect of water resource management. *Ecological Applications* 1(1), 66–84.

Lampkin, N. (1994) *Organic Farming*. Farming Press, Ipswich, UK.

Leopold, A. (1949) *A Sand Country Almanac and Sketches Here and There*. Oxford University Press, New York.

Muller, F., Hoffmann-Kroll, R. and Wiggering, H. (2000) Indicating ecosystem integrity: Theoretical concepts and environmental requirements. *Ecological Model* 130, 18–23.

Ohgushi, T., Wurst, S. and Johnson, S.N. (2018) *Aboveground-Belowground Community Ecology*. Springer.

Smuts, J.C. (1927) *Holism and Evolution*. Macmillan, London.
Spedding, C.R.W. (1971) Agricultural ecosystems. *Outlook on Agriculture* 6(6), 242–247.
Tansley, A.G. (1935) The use and abuse of vegetational concepts and terms. *Ecology* 16(3), 284–307.
Tellarini, V. and Caporali, F. (2000) An input/output methodology to evaluate farms as sustainable agro-ecosystems: An application of indicators to farms in Central Italy. *Agriculture, Ecosystems & Environment* 77(1–2), 111–123.
Tomich, T.P., Brodt, S., Ferris, H., Galt, R., Horwath, W.R. *et al*. (2011) Agroecology: A review from a global-change perspective. *Annual Review of Environment and Resources* 36, 193–222.
van der Werf, W. and Bianchi, F. (2022) Options for diversifying agricultural systems to reduce pesticide use: Can we learn from nature? *Outlook on Agriculture* 51(1), 105–113.
Whitehead, A.N. (1927) *Process and Reality: An Essay in Cosmology*. The Free Press, New York.

4 Organic Farming Guidelines and Alternative Food Production Systems in Comparison with Conventional Farming Practices

Ehsan Ebrahimi[1*] and Adnan Šišić[2]

[1]*Faculty of Agricultural and Environmental Sciences, Department of Agronomy (Pflanzenbau), University of Rostock, Germany;* [2]*Department of Ecological Plant Protection, University of Kassel, Germany*

Abstract

The certification and labelling of organic farming play a crucial role in ensuring the authenticity of organic products. However, the complex regulatory systems associated with organic farming pose challenges, particularly for small-scale producers in developing countries. Issues such as bureaucratic regulations, documentation requirements, access to organic seeds, and mislabelling of products hinder the growth of organic agriculture. In response, alternative food production systems, including agroecology, regenerative agriculture, permaculture and biodynamic farming, have emerged. These systems focus on sustainable and environmentally friendly practices, biodiversity conservation, and the promotion of natural cycles. While they share common values with organic farming, these alternative systems offer flexible and innovative approaches to agriculture. Agroecology emphasizes ecological and social aspects, regenerative agriculture prioritizes the regeneration of natural resources, permaculture adopts a design-oriented approach, and biodynamic farming incorporates spiritual perspectives and specific preparations to enhance soil health and plant growth.

Keywords: alternative food production systems, agroecology, regenerative agriculture, permaculture, biodynamic

Introduction

The historical development of the term 'organic' highlights the significance of safeguarding and maintaining the principles of the organic movement. While it is evident that establishing robust standards and control mechanisms is important for ensuring the integrity of organic practices, it is necessary to recognize associated trade-offs with organic farming regulations. As we navigate the certification process and the significance of labelling through the first part of this chapter, we also address some of the main challenges associated with organic farming certification regulations.

Agricultural producers and policymakers face the task of simultaneously mitigating the adverse environmental impacts of conventional food production systems and meeting the growing demand for food. At the same time, there is a growing consumer demand for environmentally friendly food options. Many alternative farming systems are supported by a wide array of scientific and empirical evidence, and have demonstrated their high productivity and ecological and economic viability. These are discussed in the second part of this chapter, which also highlights the key differences among

*Corresponding author: eh.ebrahimi@uni-rostock.de

DOI: 10.1079/9781800626850.0004

these alternative farming systems and the organic farming guidelines.

Organic Farming Guidelines

Organic farming certification and labelling

Organic farming certification and labelling ensure the integrity and transparency of organic products (Zezza *et al.*, 2020). While certification entails a verification process that includes maintaining records, conducting inspections and ensuring compliance with organic guidelines (FAO, 2012), labelling also allows consumers to easily identify organic products and make informed choices based on their preferences and values (Smith *et al.*, 2019; Miśniakiewicz *et al.*, 2021). Simultaneously, the regulatory frameworks that are designed to preserve the authenticity of organic agriculture can inadvertently create trade obstacles that disproportionately affect small-scale producers in developing countries, who are the core of the organic movement (Neuhoff *et al.*, 2014; Jouzi *et al.*, 2017).

Organic farming certification

Organic farming certification involves meticulous assessment procedures that ensure farms adhere to the necessary criteria of organic production. Accredited certification bodies play a pivotal role in assessing farmers' adherence to organic guidelines (Miśniakiewicz *et al.*, 2021). Certifiers responsible for evaluating agricultural and livestock products and foods are required to follow ISO 17065 and ISO 19011 standards. These standards outline the scope of activities for conformity assessment bodies (Brito *et al.*, 2022). The certification process encompasses several key aspects (ISO, 2012):

- documentation and record-keeping: farmers are required to maintain meticulous records of their practices, which encompasses the inputs utilized, production processes, and methods employed for handling of produced goods (Zezza *et al.*, 2020);
- on-site inspections: certification bodies conduct regular inspections to verify that organic practices are being implemented correctly and consistently on the farm;
- testing and residue analysis: samples of soil, water and products may be collected and analyzed to verify their conformity to organic standards and to ensure the absence of any prohibited substances; and
- adhering to specific guidelines: depending on the region or country, there may be specific guidelines or regulations that farmers must follow to obtain organic certification.

Significance of labelling

Labelling organic products enables consumers to distinguish and identify them. Organic logos and seals, bestowed by certification bodies, act as recognizable symbols of trust (Brito *et al.*, 2022). Furthermore, they provide transparency and assurance, allowing consumers to make informed choices aligned with their values and preferences.

Key challenges and limitations of organic farming regulations

Organic agriculture is a fundamentally straightforward and simple concept. However, as organic agriculture evolved, transitioning from niche to mainstream markets globally, governmental intervention in organic regulations also increased. Consequently, several challenges emerged in relation to organic certification and labelling:

- The growing bureaucratization of organic agriculture has led to the development of complex and fragmented regulatory systems. These regulatory systems can vary greatly across different national governments, posing significant challenges for organic producers, particularly for those in developing countries. While organic agriculture may offer promising market opportunities for these producers, they often find themselves obligated to adhere to foreign organic regulations that do not adequately consider the unique agricultural conditions specific to their regions. This can even result in the exclusion of smallholder farmers, who are the essence of the organic movement. Obtaining organic certification is also a significant financial burden in developing countries, where governmental support is often lacking (Reganold and Wachter, 2016; Jouzi *et al.*, 2017).
- The inability to fulfil the requirements of documentation and record keeping can pose a considerable challenge, particularly in regions

with a high prevalence of illiteracy among smallholder producers. This issue is exemplified by a case in Mexico, where several small-scale organic coffee farmers were de-certified due to their inability to effectively handle the paperwork required for organic certification, even though they were in compliance with organic standards (Mutersbaugh, 2002).

- Securing access to organic seeds and vegetative propagating materials is essential for organic farming. In the European Union, member states are obligated to maintain national seed databases that provide electronic access to information on organic seed availability. However, there is a significant disparity in the range of seed varieties among member states. For example, the seed database of Croatia is completely empty; Bulgaria has few entries, while the German and UK databases offer an extensive selection of organic seed varieties for various crops (https://rb.gy/njdq3 [accessed 18/1/25]).
- Complying with rigorous seed health requirements, particularly in organic vegetable production due to the necessity of ensuring the complete absence of certain seed-borne pathogens, which are challenging to eliminate entirely without the use of synthetic pesticides (Migliorini and Wezel, 2017).
- The problem of mislabelling or fraudulent utilization of organic labels occurs when products labelled as organic fail to meet the necessary standards or are deliberately mislabelled to deceive consumers. This can lead to loss of consumer trust and creation of difficulties for authentic organic producers (Manning and Kowalska, 2021).

Alternative Food Production Systems

Modern conventional agriculture promotes genetic uniformity and monocultures across extensive land areas. These food production systems are heavily reliant on synthetic pesticides and mineral fertilizers to maintain high yields. Today, agriculture is estimated to contribute directly and indirectly approximately 25–30% of the total global greenhouse gas (GHG) emissions (Tubiello *et al.*, 2013; Bennetzen *et al.*, 2016). Agriculture has been recognized as the primary non-point source of water pollution, with pesticides and nitrates from fertilizers regularly found in groundwater in most agricultural areas. Loss of biodiversity, soil degradation, deforestation, and other adverse environmental effects are hallmarks of the prevailing chemically intensive food production systems (Altieri and Nicholls, 2012; McLennon *et al.*, 2021).

The distinguishing feature of an alternative farming approach lies not in the rejection of conventional practices, but rather in the inclusion of innovative practices. In contrast to conventional systems, alternative agriculture is a system approach to farming, focusing on harmonizing with natural cycles and biological interactions. Farmers who adopt alternative practices strive for both profitability and the implementation of ecologically sustainable methods that optimize the use of the physical, chemical and biological potential of their farms. To attain these goals, they opt to diversify their operations and make efficient use of the available on-farm resources (Altieri and Nicholls, 2012). Nonetheless, alternative farming practices typically demand a greater amount of knowledge, trained labour, time, and management skills per unit of production when compared to conventional farming methods (Piñeiro *et al.*, 2020).

Alternative agriculture encompasses a diverse range of farming practices and various approaches including organic farming or biodynamic agriculture, which completely avoids synthetic chemical inputs, but also farming methods that make minimal use of mineral fertilizers, pesticides or antibiotics to address specific issues (Altieri and Nicholls, 2012). These include, but are not limited to, farming systems known as agroecology, regenerative agriculture and permaculture. All alternative farming systems share some common practices (Altieri and Nicholls, 2012; Reganold and Wachter, 2016; McLennon *et al.*, 2021), including: (i) minimizing or completely excluding the use of external inputs that negatively impact the environment and consumer health; (ii) implementing crop rotation, with a focus on nitrogen-fixing crops to enhance soil nitrogen levels and contribute to weed, pest and disease control; (iii) adopting conservation tillage practices to protect and enhance soil health; and (iv) use of resistant cultivars and integrated pest management strategies to minimize reliance on synthetic pesticides, and preventive disease management strategies in animal systems.

Agroecology

Agroecology is a holistic farming approach that emphasizes the ecological and social aspects of agriculture. It acknowledges the interconnections between soil, plants, animals and humans within the agricultural system (Altieri *et al.*, 2015; FAO, 2023). By prioritizing sustainability and resilience, agroecology advocates for the adoption of ecological principles and practices.

Biodiversity is a fundamental principle of agroecology. Agroecological farms seek to enhance biodiversity by cultivating a broad variety of crops, integrating natural habitats, and fostering beneficial interactions between different species. The presence of diverse species in agroecological farms not only contributes to the overall health of the ecosystem but also helps to mitigate pest and disease outbreaks (Migliorini and Wezel, 2017; De Marchi *et al.*, 2022). To enhance soil fertility, minimize environmental pollution and reduce reliance on external inputs, agroecological approaches employ techniques such as crop rotation, cover cropping and the utilization of organic fertilizers (Altieri *et al.*, 2015; Tittonell *et al.*, 2022).

Agroecology acknowledges the importance of farmer and community involvement. It values traditional knowledge and local wisdom. This approach empowers farmers to actively participate in decision-making processes and fosters innovation and adaptability. Moreover, agroecology acknowledges the socio-economic aspects of agriculture. It is committed to promoting social justice, ensuring equitable access to resources, and advocating for fair trade practices (Tittonell *et al.*, 2022). Agroecological farmers often adopt direct marketing and community-supported agriculture models, facilitating a direct and meaningful connection between producers and consumers (Bezner Kerr *et al.*, 2021).

While agroecology shares many principles and practices with organic farming, there are some differences. Organic farming tends to be a more technical and normative practice, tightly governed by certification programmes. On the other hand, agroecology is a broader concept and framework that offers flexibility for implementation in various environments and different contexts (De Marchi *et al.*, 2022).

Regenerative

Regenerative agriculture signifies a fundamental change in agricultural methods, placing emphasis on the conservation and rehabilitation of natural resources. It centres on the regeneration of topsoil, the promotion of biodiversity, the enhancement of ecosystem services, the improvement of resilience to climate change, and the overall health and vitality of ecosystems (Rhodes, 2017; Newton *et al.*, 2020). Regenerative agriculture is increasingly acknowledged as a promising farming system that gives priority to the interconnectedness of various elements within a farming system (Rhodes, 2017). This includes considering the well-being of the farmer and their family, understanding the significance of numerous biological balances within the system, and striving to enhance favourable biological interactions while reducing the use of materials and practices that disturb those interactions (Giller *et al.*, 2021).

Regenerative practices in agriculture not only result in cost savings for farmers and reduced environmental impacts but also contribute to carbon sequestration by storing atmospheric carbon dioxide in the soil (Newton *et al.*, 2020; Giller *et al.*, 2021). Moreover, through its emphasis on diverse crop rotations, rotational grazing and agroforestry, regenerative agriculture fosters the establishment of habitats, provides support for pollinators and beneficial insects, and enhances the overall health of ecosystems (Giller *et al.*, 2021; Tittonell *et al.*, 2022).

While regenerative agriculture and organic farming share similar objectives, there are notable differences in their approaches and emphases. Organic farming primarily focuses on avoiding the use of synthetic inputs and adhering to strict organic standards. In contrast, regenerative agriculture goes beyond organic practices by actively seeking to regenerate ecosystem functions and constantly enhance the overall health of the farm's natural resources (Tittonell *et al.*, 2022). In 2017, a certification programme called Regenerative Organic Certified was created in the USA. This certification is based on three core pillars: soil health, animal welfare and social fairness. It is proposed that each of these pillars can be evaluated using existing certification standards. The motivation behind establishing this new standard stems from the belief that there is a requirement to go beyond the criteria

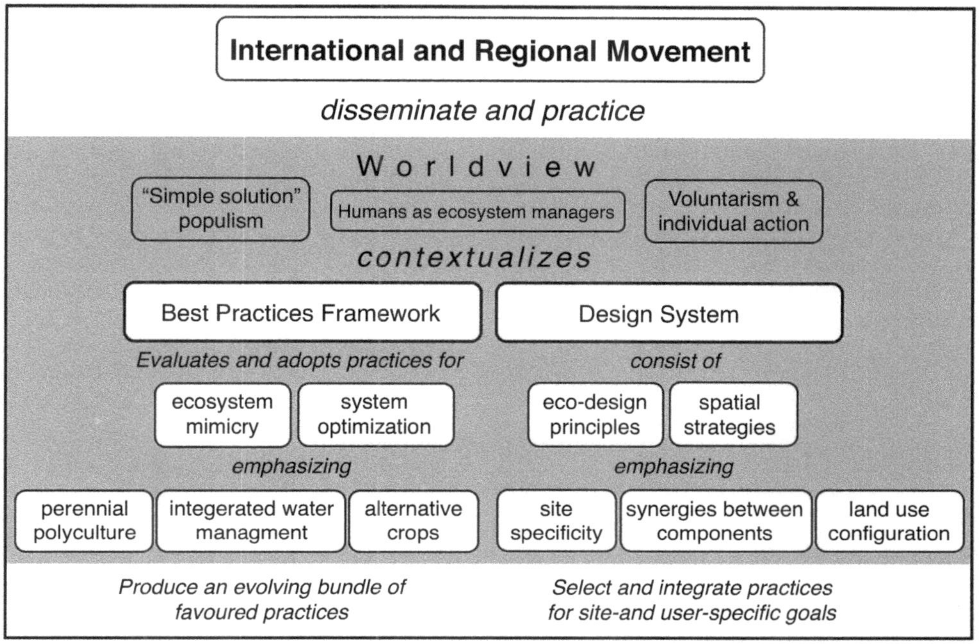

Fig. 4.1. . Stratified definition of permaculture. An international and regional movement that disseminates and practises a design system and and a best practice framework. They are contextualized by the worldview that is carried by the movement. (Adapted from Ferguson and Lovell, 2014)

outlined by the USDA Organic Certification Scheme (Regenerative Organic Alliance, 2020).

Permaculture

Permaculture is a sustainable agro-ecosystem design approach aimed at creating harmonious human settlements that imitate the patterns and functions of natural ecosystems to establish self-sustaining and regenerative systems (Fig. 4.1). This production system encompasses a comprehensive understanding of the interplay between humans and the environment, along with a framework for achieving social change (Holmgren, 2002). The core principles of permaculture embrace efficient resource utilization, biodiversity promotion, and leveraging natural cycles and patterns (Hirschfeld and Van Acker, 2021). Additionally, permaculture embodies a three-part ethical framework: (i) nurturing the environment; (ii) supporting human well-being; and (iii) establishing boundaries for consumption and reproduction, and redistribute surplus. The third aspect is often summarized as 'fair shares' (Fiebrig *et al.*, 2020). Fair shares can be seen as a component of governance that mediates between the two aforementioned principles, maintaining equilibrium between human utilization of natural resources for their well-being and the preservation of nature's inherent needs.

Evaluation of best practices in permaculture is based on two primary criteria: ecosystem mimicry and system optimization. Ecosystem mimicry involves taking cues from the structure and functioning of unmanaged ecosystems as a model to create highly productive systems. System optimization focuses on identifying strategic leverage points where minimal intervention can enhance the performance of desired functions (Morel *et al.*, 2019; Fiebrig *et al.*, 2020; Hirschfeld and Van Acker, 2021).

Permaculture and organic farming share common values like ecological sustainability and the avoidance of synthetic chemicals, although permaculture distinguishes itself by adopting a more comprehensive and design-oriented approach (Fiebrig *et al.*, 2020; McLennon *et al.*, 2021). The design idea in permaculture is derived from an examination of the current state of a specific area of land, known as the 'sector', along with a systematic arrangement known as 'zoning' that provides an organized layout for the productive system's design (Hirschfeld and Van Acker, 2021). Sectors are indicative of the movement of resources from outside the boundaries, passing through the land and influencing productivity through factors such as wind direction or water flows. The implementation of 'zoning' is essential in order to optimize the distances for transportation, resulting in reduced energy and time requirements once the system becomes operational (Holmgren, 2002; Fiebrig *et al.*, 2020).

Biodynamic

The fundamental principles and practices of biodynamic agriculture involve the utilization of specific preparations derived from medicinal plants and minerals to promote soil health and foster optimal plant growth and is nowadays regarded as a variant of organic agriculture (Santoni *et al.*, 2022; Chaurasiya *et al.*, 2023).

Biodynamic farmers adhere to the lunar calendar for planting and harvesting, implement integrated crop-livestock systems to optimize resource utilization, and prioritize biodiversity conservation along with the preservation of traditional knowledge and techniques (Muhie, 2023).

In addition, biodynamic farming is an alternative agricultural system that tends to incorporate spiritual perspectives into agricultural practices and acknowledges the cosmic forces of nature in the farming process. As a major principle of biodynamic, farmers perceive the farm as a dynamic living entity and actively work to enhance its vitality and capacity for self-regulation and biodiversity. Biodynamic farming shares a common foundation with organic farming in terms of avoiding synthetic pesticides and fertilizers, having multi-annual crop rotations and the utilization of diverse plant species (Muhie, 2023). However, biodynamic farming places strong focus on the use of biodynamic preparations (Table 4.1; Fig. 4.2), which are specially formulated substances made from natural materials aiming to enrich soil fertility and stimulate plant growth (Santoni *et al.*, 2022; Chaurasiya *et al.*, 2023). Another important aspect of biodynamic agriculture is mandatory integration of ruminant animals into the production system (pigs and poultry are not considered to fulfil the biodynamic regulations).

The scientific investigation into biodynamic agriculture is still in its early stages of development, making it difficult to draw comprehensive conclusions regarding its effectiveness as a production technique.

Conclusion

The certification of organic farming guarantees the authenticity and transparency of organic products while the labelling of organic products

Table 4.1. List of biodynamic preparations.

Number	Name of preparation	Way of use	Function
No. 500	Horn cow manure	Spray in field	Improves soil biological activity
No. 501	Horn silica	Spray in field	Resilience in plants
No. 502	Yellow blossoms (*Achillea millefolium*)	Prepare as a compost	S & K procedures
No. 503	Chamomile blossoms (*Chamomilla officinalis*)	Prepare as a compost	Ca & S procedures
No. 504	Stinging nettle (*Urtica dioica*)	Prepare as a compost	Nitrogen management
No. 505	Oak bark (*Quercus robus*)	Prepare as a compost	Calcium processes
No. 506	Dandelion flowers (*Taraxacum officinale*)	Prepare as a compost	Silica management
No. 507	Valerian flowers (*Valeriana officinalis*)	Prepare as a compost/ spray in field	P & warmth reactions
No. 508	Horse tail plant (*Equisetum arvense*)	Spray on plants	Plant tolerance

Fig. 4.2. Using cow-horn for biodynamic preparation No. 500. (Photo: Adnan Šišić, Germany, 2023)

empowers consumers by providing them with valuable information, enabling them to make informed choices and distinguish organic options from other alternatives. Accredited certification bodies rigorously evaluate farms to ensure their adherence to organic guidelines. Organic farming guidelines play a vital role in promoting transparency, traceability and accountability across the supply chain. They support the growth of the organic industry, allowing consumers to align their preferences and values with choices that encourage the adoption of sustainable agricultural practices.

The transition towards sustainable agriculture necessitates collective effort. Consumers play a vital role by supporting organic and sustainable food choices, generating demand, and providing incentives for farmers to embrace sustainable practices. Farmers play a crucial role in driving the adoption of sustainable agriculture. Through approaches such as organic farming, agroecology and regenerative agriculture, they foster environmental stewardship, enhance soil health, safeguard biodiversity and decrease dependence on synthetic inputs.

To remove obstacles related to the development and adoption of organic and other sustainable farming systems, it is important to accept variations in standards based on clear principles. Collaboration between organic, agroecological and allied movements, sectors and industries is pivotal in accomplishing this goal. This would provide farmers and stakeholders with a regulatory framework that is transparent yet flexible, allowing innovations and a sense of self-responsibility.

Policymakers bear the responsibility of establishing a supportive framework that enables sustainable agriculture. To achieve this, they can enact and enforce regulations that uphold the principles of organic farming, provide financial incentives to encourage the adoption of sustainable practices, and allocate resources for research and development. By giving priority to sustainable agriculture on national and international agendas, policymakers demonstrate their recognition of its critical role in ensuring food security, promoting environmental sustainability and fostering rural development. By collaborating through certification, labelling, consumer choices, farmer actions and policymaker support, we can create a resilient, eco-friendly and socially equitable food production system.

References

Altieri, M.A. and Nicholls, C.I. (2012) Agroecology: Scaling up for food sovereignty and resiliency. *Sustainable Agriculture Reviews* 11, 1–29.

Altieri, M.A., Nicholls, C.I., Henao, A. and Lana, M.A. (2015) Agroecology and the design of climate change-resilient farming systems. *Agronomy for Sustainable Development* 35(3), 869–890.

Bennetzen, E.H., Smith, P. and Porter, J.R. (2016) Decoupling of greenhouse gas emissions from global agricultural production: 1970–2050. *Global Change Biology* 22(2), 763–781.

Bezner Kerr, R., Madsen, S., Stüber, M., Liebert, J., Enloe, S. *et al.* (2021) Can agroecology improve food security and nutrition? A review. *Global Food Security* 29, 100540.

Brito, T.P., de Souza-Esquerdo, V.F. and Borsatto, R.S. (2022) State of the art on research about organic certification: A systematic literature review. *Organic Agriculture* 12(2), 177–190.

Chaurasiya, A., Dutta, S.K., Prasad, S., Singh, G., Kumar, S. *et al.* (2023) Maintenance of sustainability of agriculture and soil organic carbon through biodynamic agriculture practices. In: Meena, S.K., Oliviera Ferreira, A., Meena, S.V., Rakshit, A. and Shrestha, R.P. (eds) *Agricultural Soil Sustainability*

and Carbon Management. Academic Press, London, pp. 327–340.
De Marchi, M., Pappalardo, S.E. and Diantini, A. (2022) Agroecology and sustainable food systems: Inquiring technological approaches. In: De Marchi, M., Diantini, A. and Pappalardo, E. (eds) *Drones and Geographical Information Technologies in Agroecology and Organic Farming Contributions to Technological Sovereignty*. CRC Press, Boca Raton, Florida, pp. 1–21.
FAO (2012) Organic agriculture and the law. Available at: https://www.fao.org/3/i2718e/i2718e.pdf (accessed 18 January 2025).
FAO (2023) Agroecology knowledge hub. Available at: https://www.fao.org/agroecology/overview/en (accessed 18 January 2025).
Ferguson, R.S. and Lovell, S.T. (2014) Permaculture for agroecology: Design, movement, practice, and worldview. *Agronomy for Sustainable Development* 34(2), 251–274.
Fiebrig, I., Zikeli, S., Bach, S. and Gruber, S. (2020) Perspectives on permaculture for commercial farming: Aspirations and realities. *Organic Agriculture* 10(3), 379–394.
Giller, K.E., Hijbeek, R., Andersson, J.A. and Sumberg, J. (2021) Regenerative agriculture: An agronomic perspective. *Outlook on Agriculture* 50(1), 13–25.
Hirschfeld, S. and Van Acker, R. (2021) Review: Ecosystem services in permaculture systems. *Agroecology and Sustainable Food Systems* 45(6), 794–816.
Holmgren, D. (2002) *Permaculture: Principles and Pathways beyond Sustainability*. Holmgren Design Services.
ISO (International Organization for Standardization) (2012) *Conformity Assessment: Requirements for Bodies Certifying Products, Processes and Services* (ISO/IEC 17065:2012). Available at: https://www.iso.org/standard/46568.html (accessed 18 January 2025).
Jouzi, Z., Azadi, H., Taheri, F., Zarafshani, K., Gebrehiwot, K. *et al*. (2017) Organic farming and small-scale farmers: Main opportunities and challenges. *Ecological Economics* 132, 144–154.
Manning, L. and Kowalska, A. (2021) Considering fraud vulnerability associated with credence-based products such as organic food. *Foods* 10(8), 1879.
McLennon, E., Dari, B., Jha, G., Sihi, D. and Kankarla, V. (2021) Regenerative agriculture and integrative permaculture for sustainable and technology driven global food production and security. *Agronomy Journal* 113(6), 4541–4559.
Migliorini, P. and Wezel, A. (2017) Converging and diverging principles and practices of organic agriculture regulations and agroecology. *Agronomy for Sustainable Development* 37(6), 1–18.
Miśniakiewicz, M., Łuczak, J. and Maruszewska, N. (2021) Improvement of organic farm assessment procedures on the example of organic farming in Poland: Recommendations for organic farming in Poland. *Agronomy* 11(8), 1560.
Morel, K., Léger, F. and Ferguson, R.S. (2019) Permaculture. In: *Encyclopedia of Ecology*, 2nd edn. Elsevier, Amsterdam.
Muhie, S.H. (2023) Concepts, principles, and application of biodynamic farming: A review. *Circular Economy and Sustainability* 3(1), 291–304.
Mutersbaugh, T. (2002) The number is the beast: A political economy of organic-coffee certification and producer unionism. *Environment and Planning A* 34(7), 1165–1184.
Neuhoff, D., Tashi, S., Rahmann, G. and Denich, M. (2014) Organic agriculture in Bhutan: Potential and challenges. *Organic Agriculture* 4(3), 209–221.
Newton, P., Civita, N., Frankel-Goldwater, L., Bartel, K. and Johns, C. (2020) What is regenerative agriculture? A review of scholar and practitioner definitions based on processes and outcomes. *Frontiers in Sustainable Food Systems* 194.
Piñeiro, V., Arias, J., Dürr, J., Elverdin, P., Ibáñez, A.M. *et al*. (2020) A scoping review on incentives for adoption of sustainable agricultural practices and their outcomes. *Nature Sustainability* 3(10), 809–820.
Reganold, J.P. and Wachter, J.M. (2016) Organic agriculture in the twenty-first century. *Nature Plants* 2(2), 15221.
Regenerative Organic Alliance (2020) Regenerative Organic Certified. Available at: https://regenorganic.org (accessed June 2023).
Rhodes, C.J. (2017) The imperative for regenerative agriculture. *Science Progress* 100(1), 80–129.
Santoni, M., Ferretti, L., Migliorini, P., Vazzana, C. and Pacini, G.C. (2022) A review of scientific research on biodynamic agriculture. *Organic Agriculture* 12(3), 373–396.
Smith, O.M., Cohen, A.L., Rieser, C.J., Davis, A.G., Taylor, J.M. *et al*. (2019) Organic farming provides reliable environmental benefits but increases variability in crop yields: A global meta-analysis. *Frontiers in Sustainable Food Systems* 3, 82.
Tittonell, P., El Mujtar, V., Felix, G., Kebede, Y., Laborda, L. *et al*. (2022) Regenerative agriculture: Agroecology without politics? *Frontiers in Sustainable Food Systems* 6.
Tubiello, F.N., Salvatore, M., Rossi, S., Ferrara, A., Fitton, N. *et al*. (2013) The FAOSTAT database of greenhouse gas emissions from agriculture. *Environmental Research Letters* 8(1), 015009.
Zezza, A., Demaria, F., Laureti, T. and Secondi, L. (2020) Supervising third-party control bodies for certification: The case of organic farming in Italy. *Agricultural and Food Economics* 8(1), 1–14.

5 Policies for Organic Farming and Markets

RAFFAELE ZANOLI[1*] AND FRANCESCO SOLFANELLI[2]

[1]Full Professor of Food Marketing & Management, Faculty of Engineering & Department of Agricultural, Food and Environmental Sciences, Università Politecnica delle Marche Via Brecce Bianche, Italy; [2]Associate Professor of Agricultural Economics, Department of Agricultural, Food and Environmental Sciences, Università Politecnica delle Marche Via Brecce Bianche, Italy

Abstract

This chapter examines the rise of organic farming in response to market failures inherent in conventional agriculture. It explores the concept of Pareto efficiency and how externalities lead to inefficiencies in resource allocation. Organic farming, grounded in the principles of health, ecology, fairness and care, generates positive externalities and public goods, necessitating government intervention. Policies worldwide have evolved to support organic farming through legal frameworks, economic incentives, and communicative measures. These policies aim to bridge the gap between higher production costs and market prices, stimulate demand, and ensure transparency and integrity in the organic supply chain. Effective implementation and collaboration among stakeholders are crucial for policy success. By addressing supply and demand factors, governments can promote sustainability, environmental stewardship and public health while fostering the growth of the organic sector.

Introduction: The Rise of Organic Farming as a Response to the Market Failure of Conventional Farming

With the term 'Pareto efficiency', named after the economist Vilfredo Pareto, who first emphasized the concept, economists refer to a state of equilibrium in the use of resources in which '*one individual's utility could only increase at the expense of someone else's*' (Ventura *et al.*, 2016). Pareto efficiency is, however, related to the concept of 'competitive' market equilibrium, which is no longer possible when there are public goods and externalities. According to Pigou (1920), external effects that are not properly considered in the production costs would not be reflected in the prices, thus causing a market failure in achieving a Pareto efficiency condition. Environmentally friendly farming practices, such as those implemented in organic farming, are typical examples of production systems that deliver positive externalities and lower external societal costs. Researchers from diverse fields have engaged in intense discussions concerning the role of organic farming in the provision of public goods (see, among others, Gabriel *et al.*, 2013; Reganold and Wachter, 2016; Smith *et al.*, 2019; Eyhorn *et al.*, 2019). However, the positive environmental impact of organic farming is nowadays recognized by numerous scholars (see, among others, Schader *et al.*, 2012; Seufert and Ramankutty, 2017; van der Werf *et al.*, 2020). Furthermore, in many industrialized countries, the government views organic farming systems as an important instrument for providing several other ecosystem services, such as healthy food for citizens and job opportunities for the younger generation (EC, 2020). According to IFOAM (2005), organic farming is an agricultural

*Corresponding author: zanoli@agrecon.univpm.it

 Advances in Organic Farming (ed. L. P. Awasthi)
DOI: 10.1079/9781800626850.0005

method grounded on the principles of health, ecology, fairness and care. These four principles represent the fundamental values guiding the birth and development of organic agriculture. They are the root of all positive environmental and societal impacts that the organic sector can bring, promoting fair relationships among actors in the food supply chain and ultimately leading to a higher quality of life for all citizens (IFOAM, 2005). Since organic agriculture is a production method that provides societal benefits in the form of reduced external costs (negative externalities) or increased external benefits (positive externalities), it is politically relevant that farmers get compensated for the higher (internal) production costs. Organic farmers' agricultural output consists of not only marketable (private) goods but also public goods in the form of positive externalities, such as cleaner air, healthier soil and purer water. The issue of market failure was and is generally dismissed by mainstream free-market economists, who believe that the market itself should automatically lead to efficient resource allocation. In 1960, the Nobel Prize laureate R.H. Coase, in his article 'The Problem of Social Cost', suggested that negotiation among the actors involved in whatever form of externalities would still lead to a socially efficient outcome (Coase, 1960).

Nevertheless, the crucial aspect of all positive externalities, such as those generated by organic farming, is that they fall into the category of public goods, for which neither property rights nor market prices are currently available (Turner *et al.*, 1993). Most of the ecosystem services originating from organic farming practices are public goods characterized by two important features: non-rivalry and non-exclusiveness in consumption. As a consequence, the market allocation mechanism 'fails' and becomes unfair since it is not efficient in fully compensating the goods and services offered by organic farming.

The mere existence of externalities in organic farming production is a sufficient reason for advocating governmental policies to stimulate and develop the sector. Other reasons to justify policies for organic farming are related to social objectives such as a fairer distribution of income between farmers, rural development, and avoidance of land abandonment or better market transparency. The latter objective is the basis for all organic labelling regulations, such as EC Regulation 2091/92 (now replaced by EU Regulation 2018/848) or the US Organic Foods Production Act (OFPA) of 1990 (USDA, 2005).

Policies for the Organic Sector

Over the past four decades, agricultural policies in several western countries, such as those in the European Union and the USA, have developed in a way that strongly influenced the development of organic farming. According to Willer *et al.* (2023), organic agricultural land worldwide has increased about five times compared to the year 2000, increasing from about 15 million ha to more than 76 million ha in 2021 (Fig. 5.1). The growth of organic farming is unevenly spread across the five continents. Most of the organic land is in Oceania (about 36 million ha), followed by Europe with about 18 million ha, Latin America with nearly 10 million ha, Asia with more than 6.5 million ha, Northern America with more than 3.5 million ha and Africa with nearly 2.7 million ha (Willer *et al.*, 2023).

The organic market development at the country level is the joint result of specific governmental policies to support the sector and the dynamism of the private enterprises and movements operating in the market. However, in most cases, the latter factor is strongly influenced by government policies and programmes, which have been instrumental in driving the growth of the organic sector in several countries. In the EU, during the 1980s and 1990s, regulatory policies and various forms of public support for organic farming were established thanks to the impulse of environmental movements and green parties. In contrast to the USA, where policies mainly concentrated on initiatives to expand the domestic organic demand, in several EU member states, policies for the organic sector focused particularly on supporting the provision of ecosystem services associated with the supply of organic products. At the EU level, with EC Regulation 2072/92, farmers were provided with subsidies and other incentives to support conversion to and maintenance of organic practices. According to Zanoli *et al.* (2015), a favourable policy environment for organic farming can be generated by integrating the supply-side support measures with measures from the demand side, thus shifting part of the support policy burden from taxpayers – i.e. almost all citizens – to organic consumers – i.e. a group of concerned citizens that voluntarily choose to purchase organic products.

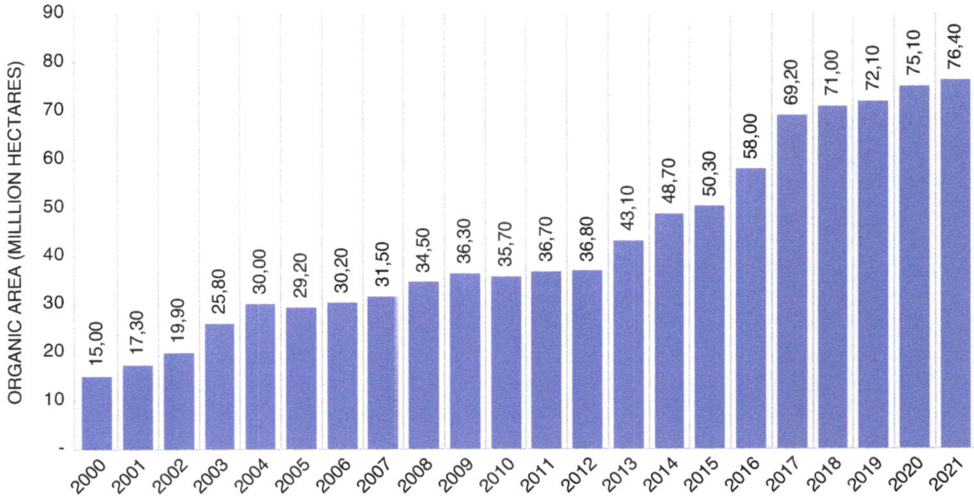

Fig. 5.1. The development of organic agricultural land worldwide. (From Willer *et al.*, 2023)

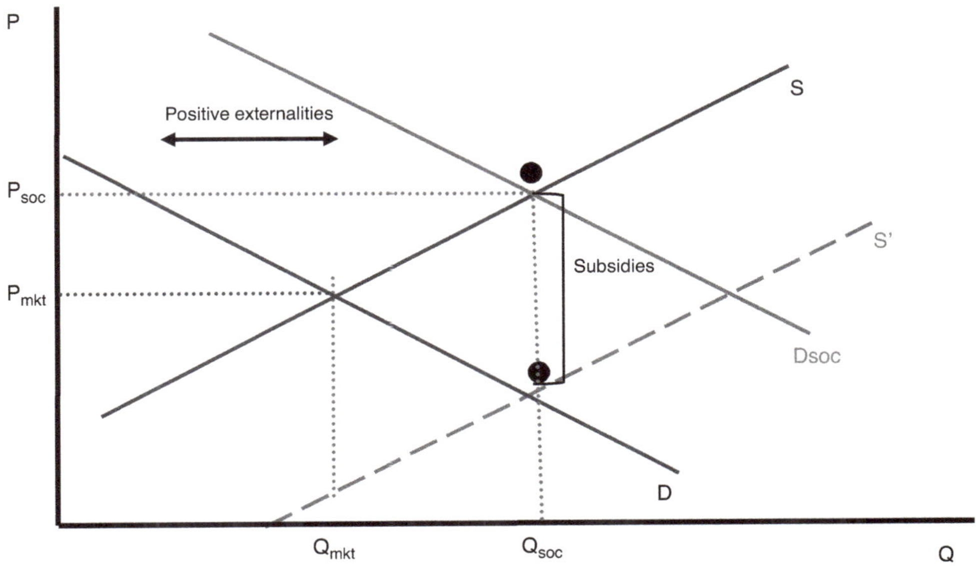

Fig. 5.2. Supply support policy. (Adapted from Turner *et al.*, 1993)

Different Types of Policy Support

Organic farmers exhibit higher production costs than conventional farmers since they voluntarily experience lower yields and production losses related to the avoidance of synthetic fertilizers and plant protection products while generating positive externalities and public goods. These externalities may be related to all the organic farming principles, not only to health and ecology but also to care and fairness, in the form of environmental, social and cultural externalities.

The situation without public support is illustrated by the supply curve S in Fig. 5.2. At this level of costs (P_{mkt}), the private demand (D) for organic products represents only a part of the total potential and latent social demand (D_{soc}). The non-exclusiveness of public goods (e.g. the ecosystem services provided by the increased biodiversity produced by organic farming practices) implies a free-riding behaviour: all consumers, including those not buying organic products, would benefit from the generated positive externalities. Therefore, since some of the consumer benefits of purchasing organic products are freely available to all, consumers are generally unwilling to pay for the full production cost. This explains why the private demand for organic products (D) is less than socially optimal (D_{soc}). Public subsidies to support organic supply aim to compensate for the higher production costs. Suppose the financial support offsets the costs associated with the production of public goods (positive externalities). Therefore, when the ecosystem services provided by organic farmers are subsidized, the farmer's production costs are reduced, and the supply curve moves from S to S'.

In that case, the quantity demanded in the market (D) equals the quantity that would be socially optimal (D_{soc}). This justifies public intervention to support the supply of organic products and related positive externalities.

Looking at the issue from another perspective (see Fig. 5.3), it is interesting to notice that the demand-side interventions can substantially reduce the public subsidies to support organic supply and, therefore, the overall burden on taxpayers. Interventions that can shift the demand for organic products in the upper right, moving the demand for organic products from D to D', can reduce the overall cost of public support policy.

Integrating supply-side and demand-side support measures may redistribute a portion of the support policy burden from taxpayers, theoretically representing all citizens of a country, to consumers of organic products, a group that voluntarily buys organic products on the market. In this context, the positive externalities provided by organic farming can be compensated through both the market, in the form of premium prices offered by the consumers, and the targeted subsidy programmes made available by the government. As we will discuss further on, public demand support can take various forms. These range from interventions made to stimulate private consumption (e.g. awareness and promotional campaigns for organic food, common logo for organic products, and related communication policies) to interventions made to stimulate public demand (e.g. organic public procurement, such as organic public canteens). Governments have a wide range of policy instruments that can be employed to foster the organic sector. They can develop a specific

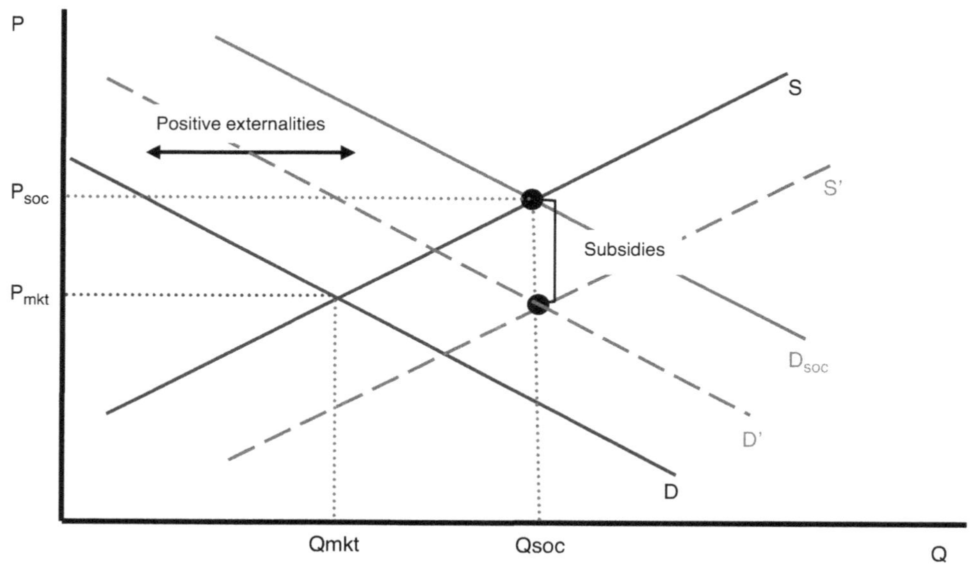

Fig. 5.3. Supply support policy combined with demand support policy. (Adapted from Turner *et al.*, 1993)

set of policy instruments adapted to specific regional or national conditions and needs. According to Mccalla and Josling (1985), policies to support the supply and demand of agricultural products can be implemented at five different levels. The first is at the level of resource availability (e.g., availability of land dedicated to organic farming); the second regards the level of inputs that can be purchased by farmers (e.g. technical inputs, such as pesticides and fertilizers); the third is related to the final product (e.g. subsidies or taxes on specific products); the fourth at the level of consumption (e.g. promotional campaign, public procurement); and the last at the level of relationship with neighbouring states (e.g. tariff rates and quota applied on exports or imports).

The policy instruments generally used to support the supply and the demand of organic products and, indirectly, the public goods which are produced through the adoption of organic farming systems may be grouped into three main categories (Stolze and Lampkin, 2009): legal and regulatory instruments, economic instruments, and informational instruments. While the first category addresses both supply and demand, establishing the legal framework for sector development, the last two categories mainly refer to measures on the supply and demand side, respectively. Nevertheless, some measures fall into the category of economic incentives that also affect the demand side and, at the same time, some measures are included in the category of communication tools that can also be used to stimulate the supply of organic products. Regardless of the goodness of the specific instruments, it is important to notice that the success of policies supporting organic farming depends on its effective implementation and ongoing evaluation, as well as on the consensus and collaboration among government agencies, farmers, industry stakeholders and the broader community. A strong public commitment to the development of organic agriculture, together with a well-co-ordinated and robust organic movement operating in the country, is one of the major positive factors influencing the development of the organic sector, as it creates confidence among stakeholders and, therefore, the basis for a positive market environment (Dabbert *et al.*, 2004).

Legal and regulatory instruments

In environmental economics, legal and regulatory instruments aimed at supporting environmental protection and sustainability are usually considered 'command and control' (C&C) measures, where government authorities enforce specific practices for alleviating the pressure of certain human activities on the environment, such as reducing pollution. In agriculture, typical examples of C&C measures are the mandatory establishment of buffer zones along the rivers to limit pesticide contamination or the restriction of time and rate of manure application on the field to prevent nitrate leaching. This approach is usually complemented with economic interventions, such as incentives, which will be discussed further. In organic farming, establishing regulations became necessary to provide a standardized framework that ensures the integrity of the relative products along the supply chain and fair competition among farmers. In several countries organic farming has been legally defined by government authorities since the beginning of the 1990s. In Europe, for instance, organic production and labelling have been legally defined since 1992, with the coming into force of the Council Regulation (EEC) No. 2092/91 (EEC, 1991), while in USA, the first national standards for organic products have been established by the Organic Food Production Act in 1990 (USDA, 2005; Mosier and Thilmany, 2016). Similar regulations have been implemented in other countries, including Japan, which established the Japan Agricultural Standards (JAS) in 2000, and Canada, which established the Canadian Organic Regime (COR) in 2009. Before government intervention in the regulation of organic farming, several private organic standards were established by private entities, mostly managed by groups of pioneers actively operating in Europe and USA in the 1970s and 1980s. This was the case of the International Federation of Organic Agriculture Movements (IFOAM), established in 1972 by several European and US organic and biodynamic associations, which published the first international standard in 1980. This standard largely inspires the original content of both the EU (EEC, 1991) and US (USDA, 2005) original organic regulations. In the same years, the Food and Agriculture Organization of the United Nations (FAO), together with the World Health Organisation (WHO), started working on the development of the so-called Codex Alimentarius Organic Guidelines, a document which could serve to initiate the process of harmonization and equivalence recognition among the organic regulations approved by the various governments worldwide, but also to provide first guidelines to governments that had not yet developed a regulation, such as those in the developing countries (Huber and Schmid, 2009). With the introduction of organic regulations, the power to define and control organic farming has directly moved from private to government authorities (Dabbert *et al.*, 2004). The regulatory introduction of standards has significantly enhanced both market transparency and trade opportunities in the producing countries, along with the mitigation of fraudulent activities of the operators. With the introduction of organic farming regulations, certification become one of the most important elements of organic farming systems, as it ensures that only products that met the standards can be labelled as organic and sold as such in the market. The main aim of the certification systems is to assure all actors of the organic food supply chain, with specific concern for the consumers, that the product is produced according to standards specified in the legal framework (Dabbert *et al.*, 2008). As organic product is defined by the production methods, rather than on the intrinsic characteristics of the product, the certification process is even more important for the supply chain actors, including the consumers. The presence of credence attributes regarding organic food products, which include specific production methods adopted along the supply chain (e.g. use of organic pesticides, animal care, biodiversity promotion) and that are difficult to assess by the consumers, together with the presence of a considerable price premium compared to the conventional counterparts, give specific relevance to the concept of fraud (Jahn *et al.*, 2005; Nuttavuthisit and Thøgersen, 2017; Manning and Kowalska, 2021). The legal framework outlines the fundamental principles and regulations for production, processing, labelling and marketing of organic products, along with the control and certification system act for its enforcement. The structure and functioning of the organic certification system may involve several different actors, depending on the specific country regulation. However, in the national standards mentioned above, with specific concern to organic regulations established in the EU and US, there are at least three main actors involved: the designated public competent authority; the private or public control bodies or certification agent; and the accreditation bodies. The competent authority has the aim to ensure proper implementation of the organic regulation in force in the respective country. In the EU, the responsible competent authorities are the single member states, which usually delegate the department of agriculture or the department of public health (EC, 2018). In the USA, the competent authority is the Department of Agriculture (USDA), which is administered by the National Organic Program (NOP) (USDA, 2005). The competent authorities usually approve control bodies and inspectors and ensure a harmonized implementation of the legal framework. Competent authority usually delegates the control task to external private control bodies, which must ensure their competence by following the ISO guide 17,020 accreditation criteria. Accreditation bodies, usually designated by the competent authority, assesses the competency of control bodies. All operators involved in producing, processing, storing, importing or selling organic products are subject to control activities. At least one annual inspection is required for each organic operator, while additional controls and samples activities are based on risk analysis.

Economic instruments

In literature, these can be classified as positive economic incentives to support the public goods provided by the organic farming practice to society or negative economic incentives to discourage conventional farming practices that cause harm to the environment (e.g. taxes and duties related to the use of pesticides and synthetic fertilizers). Among positive economic incentives, income support addressed to organic farmers are undoubtedly the most frequently used financial instrument adopted by the governments. This support measure has the aim to bridge the gap between the production costs of conventional agriculture and those of organic farming. The higher costs are associated with the production of public goods, which in the absence of public support would generate lower farm incomes and/or higher consumer prices. As previously introduced, providing subsidies to organic farming is justified by the various socio-economic and environmental benefits of this production system. Since the early 1990s, the EU and USA conversion to and maintenance of organic farming practices are financed through specific 'agri-environmental support measures'. In the EU member states, the monetary compensation addressed to farmers who voluntary uptake and implement organic practices in their farms are financed under the Common Agricultural Policy (CAP) pillar II (Lampkin and Sanders, 2022), while in the USA these initiatives are included in the Environmental Quality Incentives Program (EQIP) under the US Farm Bill (Johnson, 2023). Payments for conversion to and maintenance of organic farming are usually determined at the state and regional levels and, in most cases, are annual and area-based, differentiated by land use. For animal farming, some subsidies are paid per head. The calculation considers the income foregone and all additional and transactional costs due to implementing organic agricultural practices. The voluntary agri-environmental measures aim to encourage the adoption of organic agriculture among farmers and particularly to help the conversion phase, which is the most economically critical phase in organic farming. During the conversion period, the farmers cannot sell the product as organic, thus gaining the relative premium price. Therefore, subsidies received by the farmers during this phase are generally higher than those received for maintaining organic practices (Gemmill-Herren *et al.*, 2021).

As the support measures available are addressed only to certified organic farmers, the certification cost to organic farming regulation (e.g. USDA organic, EU organic regulation) is usually considered a transaction cost. In some countries, these costs are integrated into the annual area-based calculation (e.g. in Germany). However, it is much more frequent that the subsidies for certification services costs are managed in separate programmes. This is the case in the USA, where the National Organic Certification Cost Share Program (NOCCSP) provides reimbursement of up to 75% of the certification costs (IFOAM-International, 2017). Similar approaches are adopted in many EU countries, where the RDP covers reimbursement of certification fees through specific measured addressed to farmers participating in approved 'food quality schemes', including the organic standard (Sanders *et al.*, 2011). In Denmark, where the government entirely manages the organic certification system through designated public inspection authority, certification is free for both farmers and companies along the supply chain (EU CAP Network, 2023). In countries where group certification (GC) in organic farming has long been applied (i.e. mainly countries in Africa and Latin America), the cost of internal control systems (ICS) for GC is often supported by specific public-funded projects (IFOAM-International, 2017; Meinshausen *et al.*, 2019).

Economic instruments also include fiscal policy interventions, mainly related to the government's exemption or imposition of specific taxes. Some examples are listed below:

1. **Value Added Tax (VAT) tariff of 0% or reduced rates of VAT for organic products.** The measure has the advantage of making organic foods more affordable for consumers, stimulating private demand and supporting the growth of the organic market. The preferential VAT rate for organic products is a promising initiative. However, implementing this measure is encountering some resistance, mainly due to the perceived complexity in implementation by actors in the organic supply chain. Besides, there is a risk that the VAT reduction will be absorbed by price increases at the retail level, negating the effect of the measure.
2. **Taxes related to the use of pesticides and synthetic fertilizers.** In some EU countries, taxes on synthetic fertilizers and pesticides have been implemented since the 1980s as a potential instrument to support reducing their use, helping to mitigate GHG emissions, and decreasing water pollution. The direct effect of these measures on the environment is questionable (Meyer-Aurich *et al.*, 2020). However, they can have several indirect effects on the development of organic farming. First, they can reduce the relative economic advantage of conventional farming in comparison to organic farming (Verschuur and Well, 2001). Furthermore, they can gradually stimulate farmers to the use of alternative/organic weed and pest control, contributing to change the farmer's attitude towards organic methods and therefore helping the conversion phase. Tax revenue from pesticides and/or fertilizer could be efficiently used to finance other economic policy instruments, such as income support measure or support to research and extension. This is the case with Italy, where revenue raised through the pesticides tax is reinvested to finance public research projects exclusively dedicated to organic farming and processing.

In addition to income support and tax credits, there are several other supply-side support measures which can directly or indirectly support organic farming. These may include initiatives specifically structured to support organic supply chain, such as funding to research and extension activities, as well as initiatives which include prioritization for intervention measures such as support to organic farms'/processors' capital investment or support to product development and/or marketing initiatives (Lampkin and Sanders, 2022). Despite the recent growth, organic sector is still a niche market and, in many cases, has not yet achieved the maturity and scale to efficiently compete with the conventional sector. Public support to the organic supply chain, including all actors, such as producers, processors and retailers, is therefore essential to assist the sector and to sustain the competitive pressure in the long run (Jankowska *et al.*, 2021). According to Melitz (2005), the theory underling this concept, which takes the name of 'infant industry theory', was first proposed by the economist John Stuart Mill and later taken over by Charles Francis Bastable.

Following this theory, for a sector at an early stage of development, such as the organic for specific countries, public support in the form of subsidies, tariffs or quotas, can be instrumental to increase the probability of success in the medium to long term.

However, due to the existence of the so called 'learning by doing effect', the sector should gradually achieve a proper economy of scale and succeed in competing in the market (Kemp, 1960; Melitz, 2005; Jankowska *et al.*, 2021). Accordingly, once the emerging sector achieves the maturity to compete, the public support should be removed.

Communication instruments

In many countries, the current demand of organic food represents only a part of the potential social demand. As discussed in the introduction chapter, insufficient information regarding the externalities resulting from the organic production process is the primary factor contributing to this market inefficiency. If more consumers decide to buy organic products, supporting the production of the relative public goods, the need for supply-side incentives could be substantially diminished, with several advantages in terms of market efficiency and resource allocation. As mentioned earlier, supply subsidies are very important for developing the organic sector in all countries; however, in some cases they could generate problems of oversupply and price distortion. An excess of subsidies could lead to lower market prices than would be determined in a competitive market, leading to a lack of efficiency along the entire organic supply chain. Furthermore, an excessive supply-support-oriented policy may promote a dependence on subsidies among farmers. As the temporary lower prices of organic products might not accurately reflect the actual cost of production, farmers may struggle to compete in the market once the subsidies are reduced or eliminated.

Several factors hamper the full realization of the potential demand for organic products. The lack of awareness about the advantages of organic products for health and environment, the difficulty in identifying organic products in the market, and the consumers' trust in the organic certification systems are among the most important challenges. In this context, communication policies, such as public information campaigns (PICs) targeting organic consumers, particularly the younger generation, can be crucial in promoting awareness and enhancing demand for organic food. According to Weiss and Tschirhart (1994), PICs can be defined as a policy instrument that 'the government officials deliberately attempt to shape public attitudes, values, or behaviour in the hope of reaching some desirable social outcome'. PICs are a set of organized communication activities, usually addressed and administered to a broad number of citizens for a certain period. PICs aiming at increasing attitudes and awareness towards organic products have been experimented with with success in many countries worldwide (IFOAM-International, 2017; Bougherara *et al.*, 2022).

Nevertheless, the effectiveness of public information campaigns addressed to the organic sector is not granted. The relationship between the socially desirable outcomes, such as the increase in the demand for organic products, and the target groups is of primary importance. Apart from the messages that need to be adapted at the context in which targeted citizens receive the information and education initiatives, for the campaign to be successful, organic products should be available in adequate quantities and through various distribution channels. PICs targeting organic consumers should therefore be accompanied and co-ordinated with measures to ensure the effectiveness of organic supply chains in producing, processing and distributing the products.

Further, to increase the demand of organic products, consumers must be able to easily distinguish between organic and conventional products at the point of sale (*Dabbert et al., 2004; Schmid et al., 2004*). Logos on product packages are the most common signals used to communicate to consumers that a product is certified according to specific organic standards. This is especially relevant for new or occasional consumers of organic food, since a logo not only helps in identifying the organic product but also increases consumer trust (Schmid *et al.*, 2004). A familiar and easily distinguished national organic logo is, therefore, a necessary precondition for developing the domestic organic markets. As organic products contain several credence attributes, which are difficult for the consumers to detect and verify, the trust in the organic certification systems and in the relative logos significantly contributes to the consumer's decision to buy organic food (Zanoli *et al.*, 2015).

Conclusion

In conclusion, the rise of organic farming has been fuelled by the recognition of market failures in conventional farming systems, where externalities and public goods are not adequately priced or compensated. The principles of organic farming, grounded in health, ecology, fairness and care, underscore its role in delivering positive externalities and societal benefits. However, the market mechanism fails to fully compensate organic farmers for the public goods they produce, leading to the necessity for government intervention and policy support.

Policies for the organic sector have evolved over the past decades, with governments worldwide implementing various measures to stimulate organic farming. These policies encompass a range of instruments, including legal and regulatory frameworks, economic incentives, and communication measures. The objective is to address both supply and demand factors, ensuring the sustainability and growth of the organic sector while mitigating market inefficiencies.

Legal and regulatory instruments provide the foundation for organic farming standards and certification, enhancing market transparency and consumer trust. Economic instruments, such as subsidies and tax incentives, bridge the gap between the higher production costs of organic farming and conventional methods, thereby supporting farmers and incentivizing conversion. Communication policies, including public information campaigns and organic labelling, aim to raise awareness and stimulate demand for organic products, crucial for market expansion and consumer choice.

Effective implementation and ongoing evaluation of these policies are essential for their success, requiring collaboration among government agencies, farmers, industry stakeholders and the broader community. By aligning supply and demand interventions, governments can foster a conducive environment for the organic sector's development, promoting sustainability, environmental stewardship and public health.

References

Bougherara, D., Ropars-Collet, C. and Saint-Gilles, J. (2022) Impact of private labels and information campaigns on organic and fair trade food demand. *Journal of Agricultural & Food Industrial Organization* 20(1), 39–59.

Coase, R.H. (1960) The problem of social cost. *The Journal Of Law (III).*

Dabbert, S., Häring, A.M. and Zanoli, R. (2004) *Organic Farming: Policies and Prospects*. Zed Books Ltd, London.

Dabbert, S., Lippert, C., Schulz, T. and Zorn, A. (2008) CERTCOST – economic analysis of certification systems for organic food and farming at EU level. In: Neuhoff, D., Halberg, N., Alföldi, T., Lockeretz, W. and Thommen, A. (eds) *Cultivating the Future Based on Science. Volume 1: Organic Crop Production*. Proceedings of the Second Scientific Conference of the International Society of Organic Agriculture Research (ISOFAR), held at the 16th IFOAM Organic World Congress in Cooperation with the International Federation of Organic Agriculture Movements (IFOAM) and the Consorzio ModenaBio, 18–20 June 2008, Modena, Italy.

EC (2018) Regulation (EU) 2018/848 of the European Parliament and of the Council of 30 May 2018 on organic production and labelling of organic products and repealing Council regulation (EC) no 834/2007. *Official Journal of the European Union* L150, 1–92. Available at: https://eur-lex.europa.eu/legal-content/EN/TXT/?uri=uriserv:OJ.L_.2018.150.01.0001.01.ENG&toc=OJ:L:2018:150:TOC (accessed 19 January 2025).

EC (2020) Farm to fork strategy for a fair, healthy and environmentally-friendly food system. European Commission.

EEC (1991) Council regulation (EEC) no 2092/91 on organic production of agricultural products and indications referring there to on agricultural products and food stuffs. *Offical Journal of the European Communities* 19(L), 1–15. Available at: http://eur-lex.europa.eu/legal-content/EN/TXT/PDF/?uri=OJ:C:1992:191:FULL&from=EN (accessed 19 January 2025).

EU CAP Network (2023) Organics case study Denmark.

Eyhorn, F., Muller, A., Reganold, J.P., Frison, E., Herren, H.R. *et al.* (2019) Sustainability in global agriculture driven by organic farming. *Nature Sustainability* 2(4), 253–255.

Gabriel, D., Sait, S.M., Kunin, W.E. and Benton, T.G. (2013) Food production vs. biodiversity: Comparing organic and conventional agriculture. *Journal of Applied Ecology* 50(2), 355–364.

Gemmill-Herren, B., Baker, L.E. and Daniels, P.A. (2021) True cost accounting for food: Balancing the scale. In: Gemmill-Herren, B., Baker, L.E. and Daniels, P.A. (eds) *True Cost Accounting for Food: Balancing the Scale*. Routledge, Abingdon, UK.

Huber, B. and Schmid, O. (2009) Standards and regulations. In: Willer, L., Trávníček, J. and Schlatter, B. (eds) *The World of Organic Agriculture – Statistics and Emerging Trends*. FiBL and IFOFM.

IFOAM (2005) Four principles of organic agriculture. IFOAM. Available at: http://www.ifoam.bio/en/organic-landmarks/principles-organic-agriculture (accessed 19 January 2025).

IFOAM-International (2017) Guidelines for public support to organic agriculture.

Jahn, G., Schramm, M. and Spiller, A. (2005) The reliability of certification: Quality labels as a consumer policy tool. *Journal of Consumer Policy* 28, 53–73.

Jankowska, B., Staliński, A. and Trąpczyński, P. (2021) Public policy support and the competitiveness of the renewable energy sector – the case of Poland. *Renewable and Sustainable Energy Reviews* 149, 111235.

Johnson, R. (2023) *U.S. Farm Policy: USDA-Certified Organic Production*. Congressional Research Service, pp. 1–3. Available at: https://crsreports.congress.gov/product/pdf/IF/IF10278 (accessed 19 January 2025).

Kemp, M.C. (1960) The Mill-Bastable infant-industry dogma. *Journal of Political Economy* 68(1), 65–67.

Lampkin, N. and Sanders, J. (2022) Policy support for organic farming in the European Union 2010–2020. Thünen Working Paper.

Manning, L. and Kowalska, A. (2021) Considering fraud vulnerability associated with credence-based products such as organic food. *Foods* 10(8), 1–19.

Mccalla, A.F. and Josling, T.E. (1985) *Agricultural Policies and World Markets*. Macmillan, New York.

Meinshausen, F., Richter, T., Blockeel, J. and Huber, B. (2019) Issue 200. *Group Certification – Internal Control Systems in Organic Agriculture: Significance, Opportunities and Challenges*. Research Institute of Organic Agriculture, FiBL.

Melitz, M.J. (2005) When and how should infant industries be protected? *Journal of International Economics* 66(1), 177–196.

Meyer-Aurich, A., Karatay, Y.N., Nausediene, A. and Kirschke, D. (2020) Effectivity and cost efficiency of a tax on nitrogen fertilizer to reduce GHG emissions from agriculture. *Atmosphere* 11(6), 607.

Mosier, S.L. and Thilmany, D. (2016) Diffusion of food policy in the U.S.: The case of organic certification. *Food Policy* 61, 80–91.

Nuttavuthisit, K. and Thøgersen, J. (2017) The importance of consumer trust for the emergence of a market for green products: The case of organic food. *Journal of Business Ethics* 140(2), 323–337.

Pigou, A.C. (1920) *The Economics of Welfare*. Macmillan, London.

Reganold, J.P. and Wachter, J.M. (2016) Organic agriculture in the twenty-first century. *Nature Plants* 2, 15221.

Sanders, J., Stolze, M. and Padel, S. (2011) Use and efficiency of public support measures addressing organic farming. Institute of Farm Economics.

Schader, C., Stolze, M. and Gattinger, A. (2012) Environmental performance of organic farming BT – green technologies. In: Boye, J.I. and Arcand, Y. (eds) *Food Production and Processing*. Springer, pp. 183–210.

Schmid, O., Sanders, J. and Midmore, P. (2004) *Organic Marketing Initiatives and Rural Development*. Aberystwyth. School of Management and Business, University of Wales, UK.

Seufert, V. and Ramankutty, N. (2017) Many shades of gray: The context-dependent performance of organic agriculture. *Science Advances* 3(3), e1602638.

Smith, O.M., Cohen, A.L., Rieser, C.J., Davis, A.G., Taylor, J.M. *et al*. (2019) Organic farming provides reliable environmental benefits but increases variability in crop yields: A global meta-analysis. *Frontiers in Sustainable Food Systems* 3(September), 1–10.

Stolze, M. and Lampkin, N. (2009) Policy for organic farming: Rationale and concepts. *Food Policy* 34(3), 237–244.

Turner, R.K., Pearce, D.W. and Bateman, I. (1993) *Environmental Economics: An Elementary Introduction*. Johns Hopkins University Press, Baltimore, Maryland.

USDA (2005) *Organic foods production act of 1990* 1–21.

van der Werf, H.M.G., Knudsen, M.T. and Cederberg, C. (2020) Towards better representation of organic agriculture in life cycle assessment. *Nature Sustainability* 3(6), 419–425.

Verschuur, G.W. and Well, E.A.P. (2001) *Stimulating Organic Farming in the EU with Economic and Fiscal Instruments*. Centre for Agriculture and Environment.

Ventura, A., Cafiero, C. and Montibeller, M. (2016) Pareto efficiency, the Coase theorem, and externalities: A critical view. *Journal of Economic Issues* 50(3), 872–895.

Weiss, J.A. and Tschirhart, M. (1994) Public information campaigns as policy instruments. *Journal of Policy Analysis and Management* 13(1), 82.

Willer, H., Schlatter, B. and Trávníček, J. (2023) *The World of Organic Agriculture Statistics and Emerging Trends 2023*. FiBL and IFOFM.

Zanoli, R., Naspetti, S., Janssen, M. and Hamm, U. (2015) Mediation and moderation in food-choice models: A study on the effects of consumer trust in logo on choice. *Wageningen Journal of Life Sciences* 72–73(1), 41–48.

6 Overview of Organic Certification

Ryoichi Komiya*

Organic Shizukuishi, Takizawa City, Japan

The definition of organic agriculture is a system that illustrates potential SDG impacts without using synthetic inputs such as fertilizers and pesticides, herbicides, as well as gene-engineered seeds (United Nations, 2015). The International Federation of Organic Movements (IFOAM) summarizes organic generations as Organic 1.0, Organic 2.0 and Organic 3.0 (Arbenz *et al.*, 2016).

The generation is illustrated in Fig. 6.1.

Organic 1.0 – Pioneers from around the world

Organic 1.0 was started by numerous pioneers, who observed the problems with the direction agriculture was taking at the end of the 19th century and beginning of the 20th century. They saw the need for a radical change.

Organic 2.0 – Norming and performing

Organic 2.0 started in the 1970s when the writings and agricultural systems developed by our pioneers were codified into standards and then later into legally mandated regulatory systems.

Despite increasing success, certified organic agriculture has only reached 1.6% of global agricultural land (Willer et al., 2023). At the same time there is increasing awareness that organic can be a solution to global challenges such as soil contamination, loss of biodiversity and climate change.

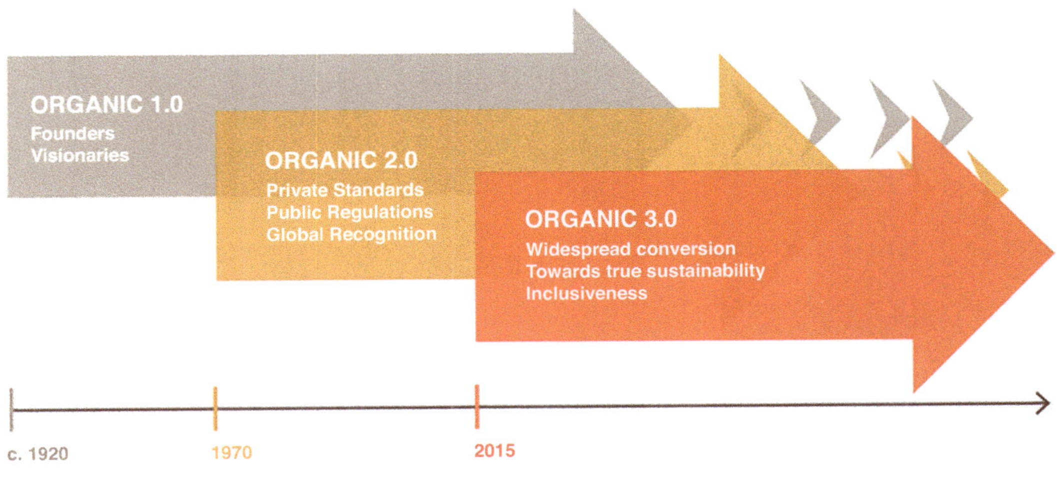

Fig. 6.1. Three generations of organic agriculture.

*Corresponding author: komiya@navy.plala.or.jp

DOI: 10.1079/9781800626850.0006

Organic 3.0 – Broad uptake of truly sustainable systems

Organic 3.0 is intended to support organic farmers by a wide variety of consumers in the districts to maintain healthy soils and to achieve agriculture sustainability. We are now in the era of Organic 3.0, but before the sale of organic produce, certification is required.

Organic agriculture is a fast-growing sector in the world, creating jobs and promoting economic growth and opportunities. The numbers of organic farmers continue to grow as consumer demand for organic grows. In 2021, there was approximately 76.4 million ha of organic land in the world (Willer *et al.*, 2023).

This chapter focuses on the Participatory Guarantee Systems (PGS) for produce certification. They have more functions to contribute in spreading organic farmers around the world.

Now, PGS is expressed as Organic+ (Plus) because it has more functions than those of the third-party certification (FAO, 2020).

Introduction

The development and professionalization of the organic sector, accompanied by increased international trade has called for third-party certification to become the norm in most developed organic markets; however, PGS have never ceased to exist and serve organic producers and consumers eager to maintain local economies and direct, transparent relationships between local people (Standards and Certification, 2023).

Thousands of organic producers and consumers are now verified through PGS initiatives around the world. Although details of methodology and processes vary, the six key elements and ten features remain consistent worldwide.

Thanks to the efforts of networks such as MAELA (the Latin American Agroecology Movement) and IFOAM, the PGS concept has gained recognition and is now viewed by many as one of the most promising tools to develop local organic markets (Altieri and Nicholls, 2017). The entire view of the PGS is well illustrated in the PGS Guidelines published by IFOAM in 2019 (May *et al.*, 2019). This chapter follows these guidelines.

PGS Specifics

PGS do not simply consist of certification systems. PGS strive for a collective commitment to a set of principles reflected through actions that demonstrate measurable compliance to the organic principles (Shaping Agriculture, 2023).

Farmers are encouraged to demonstrate their capacity to follow production standards and improve their practices by actively engaging in the management of assurance systems. At the same time, flexibility in how farmers demonstrate compliance is essential to ensure the consideration of relevant cultural and social aspects. For example, PGS can apply local social control and culturally appropriate mechanisms to demonstrate their commitment to PGS rules.

These dynamic engagement opportunities and possibilities to demonstrate compliance allow each PGS initiative to develop in its own unique way. All PGS initiatives contain six key elements and ten features, which are consistent throughout the world.

PGS six key elements

1. A shared vision

A shared vision is the starting point of a PGS. It is where key stakeholders (producers, consumers, NGOs, traders, religious institutions, governments, and others) collectively identify and agree to support the principles guiding the objectives and goals of the PGS. The shared vision can embrace organic production goals, objectives for standards, social justice, fair trade, respect for ecosystems, the autonomy of local communities, cultural differences, and more. These principles serve as a reference both for production standards to be adopted and rules on how the PGS will operate.

2. Participation

Participation is reflected through active engagement of stakeholders who share a common vision. This includes producers and consumers, retailers and traders, as well as NGOs and others that are active in PGS operations. All stages of planning, from PGS development to day-to-day processes, are facilitated by these stakeholders in various capacities. The concept of participation embodies the principle of collective responsibility to ensure the organic integrity of the PGS.

3. Transparency

Transparency is created by educating all stakeholders, including producers and consumers, on how the guarantee system works. This includes standards, norms (the organic guarantee process), and decision making processes. This does not mean that every detail is known by everyone but rather that others have at least a basic understanding of how the system functions or have a way to obtain this information.

4. Trust

The integrity-based approach which PGS relies on is rooted in the idea that producers can be trusted and that the organic guarantee system can be an expression and verification of this trust. Trust is built by key stakeholders through the collective development of a shared vision. It is maintained through continued effort to collectively shape and reinforce this vision.

This idea of trust assumes that the individual producer has committed to the shared vision of protecting nature and ensuring consumer health through organic production.

5. Horizontality

PGS initiatives are intended to be non-hierarchical. This is reflected in the overall democratic structure and through the collective responsibility taken up by those involved.

6. Learning process

Through the exchange of ideas and experiences, a learning process unfolds and becomes an ongoing dynamic of PGS. This includes technical aspects, for example, that build knowledge on organic standards and organic practices.

A social learning process happens when different stakeholders get to know each other and gain awareness of each other's situations through participation in the PGS. The knowledge obtained is fundamental for the PGS initiative and contributes to the design of components. Moreover, it helps develop trust between stakeholders and in the PGS process itself.

PGS ten features

1. Grassroots organization

In PGS, the term 'grassroots' relates to those stakeholders that are most involved in producing and consuming PGS products. As far as possible, a PGS will be built on local initiatives with and for the people it is designed to serve. This does not mean that a government body or external organization cannot be involved or even initially take the lead, but rather that the grassroots movement will be integral to the PGS development process and its operations.

2. Suitable for smallholder agriculture

A smallholder is one of the millions of producers worldwide who generate a variety of products in smaller quantities (FAO, 2013). To be suitable for smallholder agriculture means that a PGS will be designed to be affordable for smallholder family farmers and culturally appropriate in terms of paperwork, procedures and applied processes.

3. Principles and values that enhance livelihoods

PGS are characterized by clearly defined and documented principles and values. These may be expressed through a charter, standards/norms, operations manuals, public meetings, and via the farmer's pledge. The principles and values focus on enhancing livelihoods and the well-being of farming families, fair relations with consumers, and the promotion of organic agriculture. They may cover fair trade, social justice, environmental protection and cultural appropriateness.

4. Mechanisms to support farmers

There are several ways in which a PGS can provide support to producers such as facilitation of market access. The Global Comparative Study on the Interaction between Social Processes and PGS found that PGS is an important platform for the development of parallel social processes such as collective seed management, collective work, or small-scale savings systems. These processes have an important function in supporting farmers.

5. Norms conceived by all stakeholders

A reference point is needed to measure the organic integrity of a product. For organic agriculture, this reference point should be a generally recognized set of organic production rules. In many cases, for third-party certification bodies as well as for PGS initiatives, these rules are inspired by the IFOAM standards. In other cases, the national organic standards are used as a reference (e.g. the USDA Organic Standards for CNG, the NSOP for PGS-India, JAS for Organic Shizukuishi PGS). There are also examples of PGS initiatives that adapted private standards of a recognized third-party certifier (OFNZ).

6. Documented management systems and procedures

For an organic guarantee system to be transparent and deliver on a consistent and equitable basis, systems and procedures need to be documented. The depth and complexity of this documentation will vary and can evolve over time. It is important to have a system in place that can be measured in an objective and consistent manner. As with any quality assurance system, there is an expectation of ongoing improvement as lessons are learned and the organization gains experience.

The extent to which producers are expected to complete documentation and engage in record keeping can vary and might include a simple filing system (a folder) where receipts are filed to a written log cross-referenced against receipts. In some cases, a local group leader or resource person might manage the filing process on behalf of other group members.

7. Mechanisms to verify producer compliance

As previously explained, specific to PGS is the way in which PGS members verify producer compliance and manage the certification of their products without delegating this step to a third-party certification body. Mechanisms to verify producer compliance with established norms should actively stimulate participation and encourage stakeholders to embrace co-operation and provide learning opportunities. Mechanisms can include: description of the farm and farming activities (farm management plan); and peer reviews (farm inspections). The output of this process is a peer review visit record (dated and signed) that contains details of what was seen. It is important that all stakeholder representatives have a chance to join these teams and that all participants are familiar with their roles and functions. Rede Ecovida (Brazil) requires at least three individuals in the PGS take part in the review process. OFNZ (New Zealand) requires all local group members to be present at one another's review. This equals four to eight producers, depending on the size of the local group. Organic Shizukuishi (Japan), composed of six organic farmers and about 10–13 stakeholders including invited other stakeholders from other prefectures/districts, joined together for the annual farm peer reviews to proceed the participatory guarantees of the produce. Five farmers have been cultivating dry fields growing a wide variety of vegetables, fruits and herbs while one paddy field farmer has been growing Japanese rice 'Akita Komachi'.

Therefore, it is good for all organic farmers in the Organic Shizukuishi to learn a lot from its stakeholders and invited peer reviewers from different notes, comments and redress actions.

Furthermore, consumers in the locality can enjoy different produce tastes from different farms on-site during the farm peer reviews of Organic Shizukuishi. For its annual visits, Nature & Progrès (France) relies on a team composed of one producer and one consumer. This will be changed for each producer visited. These individuals are members of the PGS and are appointed by the local certification commission, which plans the team rotations to avoid conflict of interests and reciprocal review between producers. Some PGS initiatives include unexpected visits for a certain percentage of producers every year in their verification system.

8. Farmer's pledge

Producers joining a PGS agree to established norms or standards via a record or document. Depending on what is culturally and socially acceptable for a given context, this can be a signed declaration or video recording.

The pledge can be made as a self-declaration. Ideally, it is signed, witnessed by another producer, or carried out publicly via a group process.

9. Seals and labels

A seal (official stamp) is used by a PGS initiative to provide a formal endorsement of key documents, such as producer certificates. Such certificates are usually renewed each year and the period of validity is often noted on the same document. The use of the seal is controlled and managed by nominated individuals or bodies, depending on the structure of the PGS initiative (e.g. the PGS management committee). Labels have a different purpose. They are attached to a final product, and their intention is to provide the consumer with a guarantee that the product is organic. The label usually features the PGS logo and a producer number or code. Examples of PGS Labels (IFOAM Recognized PGS Initiatives) are shown in Fig. 6.2.

Fig. 6.2. Logo marks of the IFOAM-recognized PGS initiatives (as of 2024).

10. **Clear and predefined consequences for non-compliance**

It is important to keep in mind that, alongside PGS objectives, there should also be a focus on continuous improvement of practices. The different approaches that a PGS initiative adopts to verify compliance are not only a tool for controlling producer adherence to defined criteria, but also a tool for supporting producer's improvement. If the vision behind a PGS is broadly shared amongst members, and procedures are clear and well-implemented, cases in which standards and rules are abused or disrespected might be lowered.

Yet, there will always be circumstances where producers are unable to or fail to comply with their PGS's standards and norms. It is evident that the consequences for non-compliance will be graded to fit the seriousness of non-compliance. Non-compliance sanction examples are shown in Table 6.1.

Organic Shizukuishi Documents and Relevance to the PGS Six Key Elements and Ten Features

Organic Shizukuishi has made seven documents by following the PGS six key elements and ten key features and these are illustrated in Table 6.2 (Hauptfleisch, 2010). The table also illustrates the relevance to the PGS six key elements and ten features.

How to Obtain the IFOAM-recognized PGS Initiatives

Through the official IFOAM PGS recognition programme, IFOAM evaluates whether a PGS initiative operates in accordance with the six key elements and ten features and verifies the integrity of a PGS *vis-à-vis* the Principles of Organic Agriculture (IFOAM PGS Recognition, 2023).

The evaluation process is entirely free of charge, carried out with support from the IFOAM PGS Committee.

PGS initiatives that have received IFOAM PGS recognition are granted access to the IFOAM PGS logo, which can be used in communication materials, such as websites and brochures, but not on products. The PGS initiatives that are officially recognized by IFOAM – Organics International are highlighted by different-colour balloons on the IFOAM Global PGS Initiative Map (PGS worldwide, 2024).

Any PGS may apply for IFOAM PGS recognition if they are affiliates of IFOAM – Organics International and if their production standard is listed in the IFOAM Family of Standards (That's Organic, 2020; Become an Affiliate, 2023).

To initiate the evaluation process, they have to read the instructions for applicants and send their

Table 6.1. Non-compliance and sanction examples.

Non-compliance	Consequences
Missing attendance at a required field day or training	Reminder
Minor violation of the standards (i.e. farmer has no receipts for purchased inputs)	From reminder to warning
Repeated minor violations	From warning to suspension of certification for a fixed period, until farmer(s) take corrective actions
Clear violation of the standards (e.g. farmer uses prohibited inputs)	Removed from PGS, cancelled
Obvious fraud	Removed from PGS, withdraw certification, producer banned from PGS membership either permanently or for a period of time

Table 6.2. Seven documents of Organic Shizukuishi PGS initiative.

#	Organic Shizukuishi document name	Major document contents	Relevance to PGS six key elements and ten features
1	PGS Management	1. Members' annual fees 2. Organization operation 3. Procedures to get PGS certificate 4. Compliance	A shared vision, transparency, grassroots organization, clear and pre-defined consequences for non-compliance
2	PGS Basic Production Principles	a. JAS b. Soil management c. Plant management d. Water management e. Management of ecosystem f. Pest and disease management g. Pollution and contaminant control h. Harvesting and packaging i. Cleaning, disinfecting and sanitation j. Social equity and justice k. Documentation and records	Transparency, learning process, norms conceived by all stakeholders
3	PGS Grower's Pledge	Farmer's pledge declaration and his/her signature	Transparency, trust, farmer's pledge
4	PGS Application Form	Items from (b) to (k) plus field location map, field structure and vegetables/fruits/herbs fields figures, seeding/seedling start time for each vegetable/fruit/herb intended produce for the year	Transparency, learning process
5	PGS Farm Assessment Guidelines and Report	Assessment for items from (b) to (k) and notes, comments, redress actions including names of inspectors, signature of representative inspector	Transparency, horizontality, learning process, mechanisms to verify producer compliance
6	PGS Certificate	Name of farmer, PGS certified produce, signature of Organic Shizukuishi President	Shared vision, trust, Transparency
7	Use of IFOAM PGS and Logos	Background, procedures, terms and conditions to use respective logos	Transparency, seals and labels

completed self-evaluation form to IFOAM PGS committee attached to the e-mail (SEF, 2023).

Application for IFOAM-Organics International Membership

All required information is listed in Membership FAQs (Membership FQA, 2023). They fill in the 'Apply for Membership' (Membership Directory, 2023) and send it to IFOAM directly, online.

Application for IFOAM-recognized PGS

They fill in the self-evaluation form (SEF, 2023). All questionnaires in the SEF are relevant to the PGS's six key elements and ten key features.

They send it to the IFOAM PGS Committee attached to the e-mail pgs@ifoam.bio (IFOAM PGS Committee, 2021).

In the case of Organic Shizukuishi (Fig. 6.3), after the transmission of the SEF, three questions came up from the IFOAM PGS Committee.

Table 6.3. Questions placed by IFOAM PGS Committee and replies from Organic Shizukuishi (OS).

Questions (at end of 2016)	Replies (at beginning of 2017)
Q1: PGS young age (launched in 2015, first farmers certified in summer 2016) (i) Even though procedures are in place they have not been really tested.	For (i): We have really inspected producers' field based on the OS Field Inspection Form since 2016. Therefore, all producers' fields have been really tested fully.
(ii) Minimum period of 1 year (better 2 years) of full operation should be completed before the procedures can be considered as well tested.	For (ii): We have already done the full operations on our PGS in 2016 and 2017 based on our seven documents as indicated in our website. They are: [1] PGS Operation Principles [2] PGS Plant Growing Principles [3] PGS Application Form [4] PGS Producer's Pledge [5] PGS Field Inspection Form [6] PGS Member Certificate [7] PGS Logo Usage
(iii) Even if the farmers are well-experienced with organic agriculture, they may need a 'conversion process' to get familiar with PGS processes. This takes time.	For (iii): We have already got familiar with PGS processes completely. We do not understand why IFOAM PGS Committee came up with this conclusion based on our evidence. All producers do not need the 'conversion process from conventional to organic' because all of them started as organic farmers from the beginning.
Q2: PGS small size (four certified) (i) The small size *per se* is no problem.	For (i): Agreed
(ii) Shizukuishi PGS wants to grow in the future and it is not yet clear how it will be administered once more farmers join. Will farmers be organized in one big group? Will they be divided into several small groups? How will this upscaled PGS be managed? Which additional processes will be implemented, e.g. to allow decision making involving more stakeholders?	For (ii): OS operation policy: More farmers could be accepted as far as all PGS operations are workable. Key factors to make them workable are: Geographical distribution of stakeholders should be within about 20 km diameter from the PGS Office because of the easy car access to our regular PGS meetings venue and to farmers' fields for annual inspections. Our intended time durations to inspect each producer's field: 60 min including 20 min transit time to move from one producer's field to another. Currently, we have inspected four farmers per day, so the typical field inspection time schedule is as below. Field inspection: Four hours in total Lunch break: One hour Wrap-up meeting: One hour Total: Six hours
(iii) It is recommended that the PGS first grows into the envisioned structure before it can be approved.	For (iii): From the data above, for at least two more years we are going to maintain the current organization structure. There is no need to divide our PGS into two or more. In future, if total stakeholders' population increases more than twenty, it might be necessary for us to divide the PGS group into two.

Continued

Table 6.3. Continued

Questions (at end of 2016)	Replies (at beginning of 2017)
Q3: Stakeholder involvement (i) Currently the group consists mostly of farmers and there is little involvement of other stakeholders.	For (i): We have now three more farmers who are intending to apply to OS PGS. The six stakeholders other than farmers are already involved. They are a B&B owner, a nutritionist, a life consultant, a university lecturer, a CEO of an insurance company and housewives.
(ii) You should involve more people before applying again. Lacking stakeholder involvement may also be a sign that the PGS is not mature enough.	For (ii): Member breakdown of current OS is: Organic producer: 9 (4 OSPGS certified) Other stakeholders: 6 Our members are not simply categorized as 'organic producers' or 'consumers'. Almost all of them have their own expertise, as in (i).
(iii) It is recommended to first deepen the relationship with consumers and other stakeholders before you seek recognition again.	For (iii): It has been achieved at OS regular monthly meetings and field inspections, these stakeholders get together to improve our organizations and our produce quality. And besides, we have been very active to promote the relationship between producers and other stakeholders by organizing our own events, participating in the Organic Festas in Iwate several times and exchanging ideas to build organic society with local people heading for the Organic 3.0. Therefore, we are very proud of our OS PGS which has been following the IFOAM philosophy to realize Organic 3.0

Fig. 6.3. Group photo of Organic Shizukuishi stakeholders at Sachiko's farm inspection in 2020.

Table 6.3 summarizes the list of questions placed by the Committee and the replies from Organic Shizukuishi.

After the Q&R session, we have to send two recommendation letters from people who have been involved in the organic agriculture in our locality to the committee. All these communications are the packages before obtaining the IFOAM Recognized PGS Initiatives.

On 4 December 2018, Organic Shizukuishi obtained the IFOAM Recognized PGS Initiatives (Flávia Moura, 2018). It took about two-and-a-half years to complete our efforts.

Fig. 6.4. Miso making at Nishine daycare.

PGS Activities at Organic Shizukuishi

Recent activities are illustrated in the blog of the IFOAM – Organics International (Digging Deeper, 2022)

Organic Shizukuishi has tried to bridge the gap between farmers and the local community while contributing to sustainable organic farming and educating the general public to understand the advantages of a safe and secure source of organic produce.

Organic Shizukuishi's major PGS activities are farm peer reviews and promoting education of good eating habits for children at a daycare (Fig. 6.4).

Other activities have been:

- creation of jobs in the local communities through organic farming (Fig. 6.5);
- use of organic produce for their members' bed & breakfast meals (Fig. 6.6); and
- revitalization of deserted arable lands by organic no-till farming (Fig. 6.7).

Organic Shizukuishi set up the Allied PGS meeting in 2021 to get prospective IFOAM-recognized PGS initiatives in Fukushima, Chiba, Nagano, Aichi, Gifu and Hiroshima prefectures. This creates a promising perspective on the future because PGS could contribute to the increase of the share of organic land as well as raise consumers' organic awareness anywhere around the world.

They can learn a wide variety of organic agricultural techniques and know the joys of farm life.

The handicapped friends can be involved in works contributing to society.

Thus, they can help extend organic agriculture in this country.

Fig. 6.5. Participation of the handicapped in the organic farming.

Fig. 6.6. Organic meals at a member's bed & breakfast.

Fig. 6.7. Anyone can enjoy no-till farming.

IFOAM-recognized PGS Initiatives around the World

There are now eight IFOAM-recognized PGS initiatives in the world. Table 6.4 shows the basic data of relevant PGS. It is understood that the IFOAM PGS Committee is fair because the committee carefully investigates PGS candidates from the viewpoint of the execution of PGS basic concepts, six key elements and ten features, regardless of the scale of the PGS.

Conclusion

The PGS is not a simple certification system; more importantly it means achieving Organic 3.0 to accelerate the conversion from conventional to organic, the inclusiveness to support organic farmers together with local people, and the realization of Earth's sustainability. The PGS has already been greatly contributing to the organic districts construction where organic producers, public consumers, hotels, B&Bs, restaurants, hospitals and school meal providers are happy to grow and eat safe and secured foods (GAOD, 2020).

Table 6.4. IFOAM-recognized PGS initiatives in the world as of March 2024. (PGS worldwide, 2024)

#	PGS name	Description	Number of producers involved	Number of producers certified	Total area of PGS certified land (ha)	Organic standard applied
1	PGS Hòa Bình	Established in 2016 by Seed to Table, strengthened in 2018 under MOAP project of VOAA. Facilitated by Tan Lac Extension Center and Hoa Binh Farmer Union	25	25	20	PGS Standard and Vietnam national organic stds CVN11041
2	Good Market PGS	The Good Market Organic Participatory Guarantee System was started in February 2013 by producers and consumers that want to make organic more accessible and affordable in Sri Lanka. It is based on the IFOAM norms for organic production and processing. Good Market Organic Participatory Guarantee System is recognized by IFOAM Organics International.	607	408	224	It is based on the IFOAM Norms for Organic Production and Processing.
3	Bio Calédonia PGS	Created in 2009, the Bio Calédonia Association organizes the delivery of the Bio Pasifika organic farming label thanks to its Participatory Guarantee System. Bio Calédonia has about ten local groups spread over the entire territory of New Caledonia.	240	155	1676	Norme Océanienne d'Agriculture Biiologique or Pasifika Organic Standard
4	Organic Farm New Zealand	PGS for smallholders mostly in pods. Operating since 2003, currently negotiating with government for official recognition as an independent PGS certifying body.	130	130	n.a.	Biogro NZ
5	Nature et Progrès	n.a.	1621	928	n.a.	N&P Cahier des Charges (16)
6	SPG Bio Fetia	Founded in 2011, Bio Fetia is the only PGS in French Polynesia. It is an association which federates producers and consumers and delivers the BIO PASIFIKA guarantee in strict compliance with the Pacific Organic Standard.	115	67	323	PACIFIC ORGANIC STANDARD
7	Certified Naturally Grown	n.a.	750	650	8440	CNG Standards for Produce and Livestock, based on the USDA Organic standards.
8	Organic Shizukuishi PGS	Organic Shizukuishi is located in a small Shizukuishi town (population: 17,000) in Iwate Prefecture, Japan. We are an organization to connect organic farmers and other stakeholders intending to create future organic society.	8	6	2	JAS
9	MASIPAG PGS	MASIPAG is a farmer-led network of people's organizations, NGOs and scientists working towards the sustainable use and management of biodiversity through farmers' control of genetic and biological resources, agricultural production and associated knowledge.	30,000	n.a.	n.a.	n.a.

References

Altieri, M.A. and Nicholls, C.I. (2017) Agroecology: A brief account of its origins and currents of thought in Latin America. *Agroecology and Sustainable Food Systems* 41(3–4), 231–237.

Arbenz, M., Gould, D. and Stopes, C. (2016) Organic 3.0 – for truly sustainable farming and consumption. In: *IFOAM – Organics International*, 2nd edn. pp. 5–23.

Become an Affiliate (2023) Join the organic movement today. IFOAM – Organics International. Available at: https://www.ifoam.bio/about-us/our-network/become-affiliate (accessed 21 January 2025).

Digging Deeper. (2022) How PGS have been contributing to the local communities in Japan. 16 May. *Organic Without Boundaries*. Available at: https://www.organicwithoutboundaries.bio/2022/05/16/how-pgs-have-been-contributing-to-the-local-communities-in-japan/ (accessed 21 January 2025).

FAO (2013) Smallholders and family farmers. Available at: https://www.fao.org/3/ar588e/ar588e.pdf (accessed 21 January 2025).

FAO (2020) Organic-PLUS. Food and Agriculture Organization of United Nations. Available at: https://www.fao.org/family-farming/detail/en/c/1443997/ (accessed 21 January 2025).

Flávia Moura, E.C. (2018) The last issue of the Global PGS newsletter, IFOAM – Organics International, November/December 6(8), 7. Available at: https://www.ifoam.bio/sites/default/files/2020-03/pgs_newsletter_nov_dec_2018.pdf (accessed 21 January 2025).

GAOD (Global Alliance for Organic Districts) (2020) IFOAM– Organics International, IFOAM Asia, IFOAM EU. The Baltic Foundation of Lithuania and the Organic Food SystemProgram for future cooperation. Available at: https://gaod.online/about-us/ (accessed 21 January 2025).

Hauptfleisch, K. (2010) *PGS in southern Africa: draft documentation for PGS South Africa (PGSSA). June.*

IFOAM PGS Committee (2021) A newterm for the IFOAM PGS Committee. Available at: https://www.ifoam.bio/news/new-term-ifoam-pgs-committee (accessed 21 January 2025).

IFOAM PGS Recognition (2023) Official PGS Recognition Program. IFOAM – Organics International. Available at: https://www.ifoam.bio/our-work/how/standards-certification/participatory-guarantee-systems/pgs-recognition-program (accessed 21 January 2025).

May, C., Cornelia, K., Moura, F. and Flavia Moura, E.C. (2019) PGS Guidelines: How to develop and manage participatory guarantee systems for organic agriculture. *IFOAM – Organics International*. Available at: https://www.ifoam.bio/sites/default/files/2020-05/pgs_guidelines_en.pdf (accessed 21 January 2025).

Membership Directory (2023) Membership directory, apply for membership. *IFOAM Organics International*. Available at: https://directory.ifoam.bio/join/organizations (accessed 21 January 2025).

Membership FQA (2023) Membership directory. IFOAM – Organics International. Available at: https://directory.ifoam.bio/pages/faq (accessed 21 January 2025).

PGS worldwide (2024) Participatory Guarantee Systems Worldwide. IFOAM – Organics International. Available at: https://pgs.ifoam.bio/ (accessed 21 January 2025).

SEF (2023) Self Evaluation Form. IFOAM – Organics International. Available at: https://view.officeapps.live.com/op/view.aspx?src=https%3A%2F%2Fwww.ifoam.bio%2Fsites%2Fdefault%2Ffiles%2F2020-04%2Fpgs_self-evaluation-form_final_2016.docx&wdOrigin=BROWSELINK (accessed 21 January 2025).

Shaping Agriculture (2023) Four principles of organic agriculture. IFOAM – Organics International. Available at: https://www.ifoam.bio/why-organic/shaping-agriculture/four-principles-organic#:~:text=The%20Principles%20of%20Organic%20Agriculture%20are%20Health%2C%20Ecology,to%20improve%20all%20agriculture%20in%20a%20global%20context (accessed 21 January 2025).

Standards and Certification (2023) The Organic Policy & Guarantee Department coordinates our work supporting the improvement of organic standards and verification systems. IFOAM. Available at: https://www.ifoam.bio/our-work/how/standards-certification (accessed 21 January 2025).

That's Organic (2020) Worldwide, Family Standards Frame (2020) IFOAM. Available at: https://www.ifoam.bio/sites/default/files/2022-06/FamilyFrame_2020_June%202022.pdf (accessed 18 May 2020).

United Nations (2015) What are the Sustainable Development Goals? United Nations Development Program. Available at: https://www.undp.org/sustainable-development-goals (accessed 21 January 2025).

Willer, H., Schlatter, B. and Trávníček, J. (2023) The world of organic agriculture statistics and emerging trends 2023. Corrigenda and supplementary material. Research Institute of Organic Agriculture FiBL IFOAM – Organics International. Available at: https://www.fibl.org/fileadmin/documents/shop/1254-organic-world-2023.pdf (accessed 21 January 2025).

7 Statistical Development of Organic Farming in Global Context

HELGA WILLER[1]*, JAN TRÁVNÍČEK[2] AND BERNHARD SCHLATTER[1]

[1]Research Institute of Organic Agriculture FiBL, Frick, Switzerland; [2]Czech Organics, Staré Město Czech Republic

Abstract

The most recent dataset on certified organic agriculture across the globe (data as of end 2021) reveals remarkable trends: in 2021, almost 77 million ha of agricultural land, accounting for 1.6% of the total farmland, were under organic management. Impressively, the number of organic producers soared to 3.7 million. In tandem, the global organic food market surged to a value of nearly €125 billion.

The data, gathered from 191 countries, showcase an unceasing upward trajectory in organic agriculture. Organic farmland, the number of organic producers, as well as retail sales and exports, all reached all-time highs.

These insights stem from the annual survey on global organic agriculture, which has been carried out since 1999. This survey engages a diverse array of data sources, including contributions from the private sector, government agencies, esteemed research institutions, and certified organic entities.

Keywords: organic agriculture worldwide, development of organic farmland, organic retail sales and international trade

Introduction

The organic agriculture sector has emerged as a significant player in the effort to address current issues such as environmental sustainability, health and ethical consumption. The year 2021 witnessed remarkable developments in the realm of organic farming, reaffirming its position as an important component of the world's agricultural landscape. This chapter provides an overview of the state of organic agriculture worldwide, offering insights into key statistics, trends and challenges.

The latest data from 2021 paint a picture of the rise of certified organic farming worldwide. A total of almost 77 million ha of agricultural land, representing 1.6% of the world's farmland, embraced organic management practices. The development towards sustainable agriculture was driven by 3.7 million organic producers who dedicated themselves to cultivating organic produce. The economic impact of this movement was equally impressive, with global organic food sales soaring to nearly €125 billion. (For a breakdown of the data by region and country, see Table A7.2 and A7.3.)

However, challenges such as supply chain disruptions and consumer price concerns have impacted growth in recent years, and the resilience and adaptability of the organic agriculture industry will be key to meeting the growing demand for organic products worldwide.

This chapter is based on the findings of the statistical yearbook *The World of Organic Agriculture*, made possible by the dedicated efforts of the Research Institute of Organic Agriculture (FiBL) and partners.[1] FiBL has been at the forefront of compiling and disseminating data on organic agriculture since 2000.

*Corresponding author: helga.willer@fibl.org

 Advances in Organic Farming (ed. L. P. Awasthi)
DOI: 10.1079/9781800626850.0007

Table 7.1. Organic agriculture worldwide: key indicators and top countries. (FiBL survey 2023, based on national data sources, data from certifiers and IFOAM – Organics International (Willer *et al.*, 2023; with some modifications/updates))

Indicator	World	Top countries
Countries with organic activities[2]	2021: 191 countries	
Organic agricultural land	2021: 76.8 million ha (1999: 11 million ha)	Australia (35.7 million ha) Argentina (4.1 million ha) France (2.8 million ha)
Organic share of total agricultural land	2021: 1.6%	Liechtenstein (40.2%) Samoa (29.1%) Austria (26.5%)
Increase of organic agricultural land 2020/21	1.3 million ha; +1.7 %	Portugal: 450,000 ha (+140%) China: 319,000 ha (+13 %) France: 228,000 ha (+9%)
Wild collection and further non-agricultural areas	2021: 29.7 million ha (1999: 4.1 million ha)	Finland (6.9 million ha) Zambia (2.5 million ha) Namibia (2.3 million ha)
Producers	2021: 3.7 million producers (1999: 200,000 producers)	India (1,599,010) Uganda (404,246) Ethiopia (218,175)
Exports to the European Union and USA	2021: 4.7 million metric tonnes (MT)	Ecuador (583,589 MT), Mexico (425,414 MT) Peru (344,483 MT)
Organic retail sales	2021: €124.8 billion (2000: €15.1 billion)	US (€48.6 billion) Germany (€15.9 billion) France (€12.7 billion)
Per capita consumption	2021: €15.7	Switzerland (€425) Denmark (€384) Luxembourg (€313)
Number of countries/territories with organic regulations	2022: 74 (fully implemented)	

With contributions from more than 190 countries, these data collection initiatives have brought transparency to the global organic agriculture landscape, aiding policymakers, researchers and industry stakeholders in informed decision making (Urrutia, 2022). Table 7.1 shows the latest key data found in the survey.

Global Organic Statistics

Statistics on organic area

More than 76.8 million ha of organic farmland

In 2021, 76.8 million ha of organic agricultural land were recorded, including areas in the process of conversion. The regions boasting the largest organic agricultural land areas were Oceania, with 36.0 million ha (almost half of the world's organic agricultural land or 47%); and Europe, with 18.2 million ha (constituting 24%). Latin America followed with 9.9 million ha (13%), succeeded by Asia with 6.5 million ha (8.5%), North America with 3.5 million ha (4.6%), and Africa with 2.7 million ha (3.5%).

The countries leading in terms of organic agricultural land were Australia with 35.7 million ha, Argentina with 4.1 million ha, and France with 2.8 million ha (Fig. 7.1).

Globally, 1.6% of the farmland is organic

In 2021, organic farming covered 1.6% of the world's agricultural land. This was particularly pronounced in certain regions, with Oceania leading the way at 9.7% and Europe following closely behind at 3.7% (with the EU boasting an impressive 9.9% organic share).

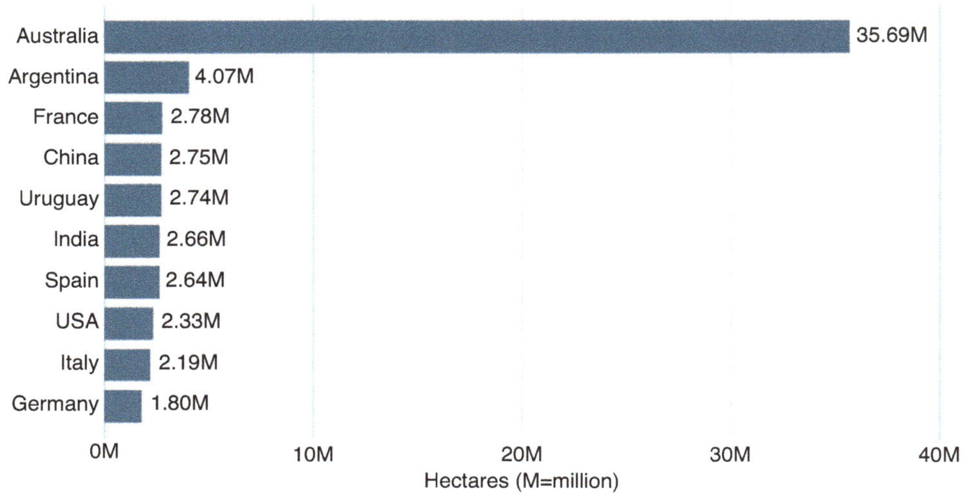

Fig. 7.1. World: the ten countries with the largest areas of organic agricultural land in 2021. (FiBL survey 2023, based on information from the private sector, certifiers and governments)

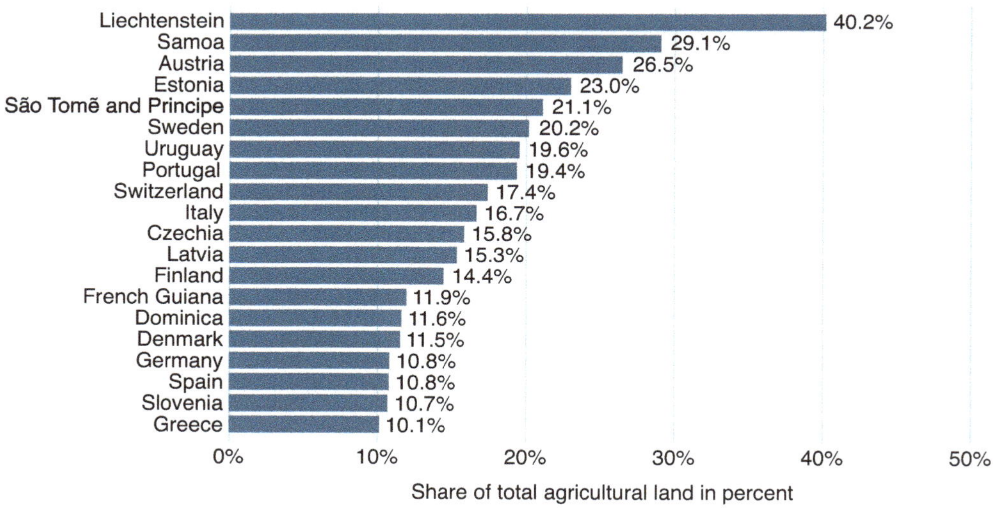

Fig. 7.2. World: countries with an organic share of the total agricultural land of at least 10% (2021). (FiBL survey 2023, based on information from the private sector, certifiers and governments)

However, the most striking revelation came from individual countries that surpassed the global average considerably. Liechtenstein stood out with a remarkable 40.2% of its agricultural land under organic management, followed by Samoa at 29.1%, and Austria at 26.5%. An even more significant milestone was achieved as 20 countries recorded organic shares of 10% or more of their total agricultural land, setting a new record in the global organic farming landscape (Fig. 7.2).

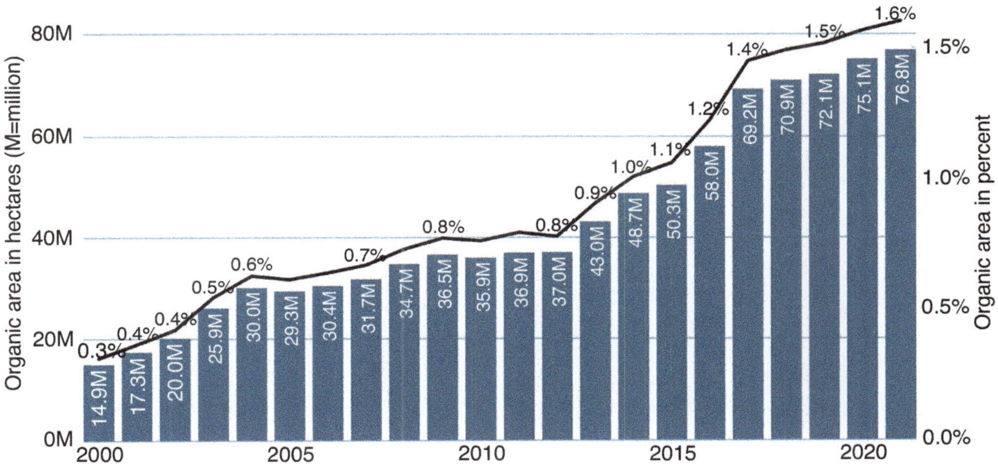

Fig. 7.3. World: growth of the organic agricultural land and organic share in 2000–2021. (FiBL-IFOAM-SOEL surveys 2001–2023)

Growth in organic farmland – increase of 1.7 million ha

Organic farmland increased by 1.3 million ha (2.3%) in 2021 (Fig. 7.3). Many countries reported a significant increase. In absolute terms, the biggest increases were in Portugal, China and France: in Portugal, organic farmland increased by almost 0.5 million ha or an exceptional 140%, mainly due to the fact that in 2021, support under the Portugese Rural Development became available for the first time in many years. In China, organic farmland increased by almost 320,000 ha (+13.1%) and in France by nearly 228,000 ha (+8.9%). However, some countries also reported decreases. The most notable decrease occurred in Argentina, which reported almost 0.38 million ha less (mainly grazing areas).

Regionally, the expansion of organic agricultural land made significant strides in 2021, with notable growth observed in Africa, Asia, Europe and Oceania. Leading the charge was Europe, which experienced the most substantial increase at 6.7%, adding a remarkable 1.2 million ha to its organic farmland. Africa followed closely, boasting an impressive growth rate of 17.3%, equivalent to an expansion of 0.39 million ha. In Asia, organic agriculture continued its upward trajectory with a 5.8% growth, resulting in an additional 0.36 million ha coming under organic management.

However, a different trend emerged in Latin America and North America, where both regions reported a decrease in organic farmland during the same period.

Growth in most major crop groups

Across most major crop groups, organic agriculture demonstrated substantial growth in 2021. Detailed information on land use and crop allocation was available for over 92% of the organic agricultural land. It is noteworthy that certain countries, including Brazil and India, with extensive organic acreage, lacked comprehensive data on their land use practices.

Approximately two-thirds of the organic agricultural land, encompassing nearly 50 million ha, was grassland/grazing areas. In 2021, this category experienced a modest decrease of 2.5% (as depicted in Fig. 7.4). Arable land, covering almost 14.8 million ha, constituted 19% of the organic agricultural land and saw a remarkable 11.4% increase since the previous year. The primary use of this category of land was for the cultivation of cereals, including rice, followed by green fodder from arable land, oilseeds, textile crops and dry pulses.

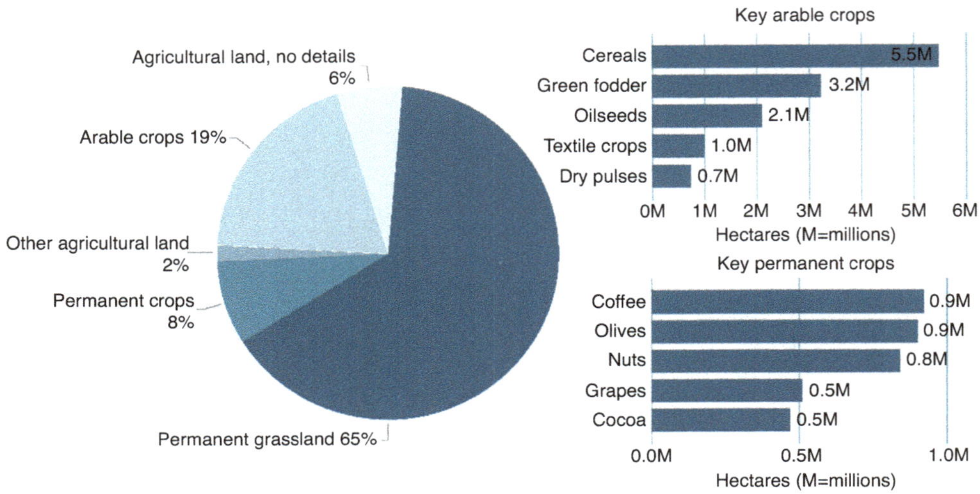

Fig. 7.4. World: distribution of main land-use types and key crop categories in 2021. (FiBL survey 2023, based on information from the private sector, certifiers and governments (Willer *et al.*, 2023))

Permanent crops, spanning 8.1% of the organic agricultural land and totalling over 6.2 million ha, witnessed substantial growth as well. Compared to the previous survey, this category reported an impressive increase of almost 829,000 ha, equivalent to a remarkable 15.4% rise. Notable crops within this category included coffee, olives, nuts, grapes and cocoa (Fig. 7.4).

Organic Operators

Organic producers on the rise – 3.7 million producers in 2021

In 2021, the global landscape of organic farming featured a remarkable presence of at least 3.7 million organic producers. This expansive community of producers was spread across the continents, with Asia leading the way, hosting 49% of the world's organic producers, followed by Africa at 31%, Europe at 12% and Latin America at 8%.

The countries that stood out included India, with a staggering 1,599,010 producers, Uganda with 404,246, and Ethiopia with 218,175 (as illustrated in Fig. 7.5). In comparison to 2020, there was a noteworthy increase of more than 170,000 producers, reflecting a growth rate of 4.9%.

It is essential to note that some countries reported numbers based on companies, projects or grower groups, each potentially comprising multiple individual producers. Therefore, it is reasonable to assume that the total number of organic producers exceeds the figures reported here.

Further operators

Additionally, the data revealed the presence of 118,747 processors, 8592 importers and 9292 exporters within the organic industry. However, it is important to acknowledge that due to some data providers not reporting figures for these operator types, it is highly likely that their actual numbers far exceed those recorded.

Organic imports and exports

International trade data for organic products are becoming increasingly comprehensive, with information available for more countries. These data can be presented in terms of metric tonnes or monetary values. Since 2018, the EU has been collecting import data on organic products (export volumes in MT), while for the USA data (export volume and value) have been available since 2014 (accessible on the USDA website). Both trade blocs provide detailed breakdowns of exports by product, product group and country of origin, measured in MT.

However, when analyzing the US organic import data, it is important to note that these do not cover all types of commodities:

- In 2021, a substantial 4.7 million MT of organic products found their way into the EU and the USA. Of this total, the EU imported 2.9 million MT, while the US imported 1.8 million MT. Importantly, organic imports into both the EU and USA increased by 0.5%, which translates to nearly 23,000 MT in 2021 (Fig. 7.6).
- Ecuador emerged as the leading exporter, with an impressive 583,589 MT of organic products shipped to the EU and USA in 2021. Mexico

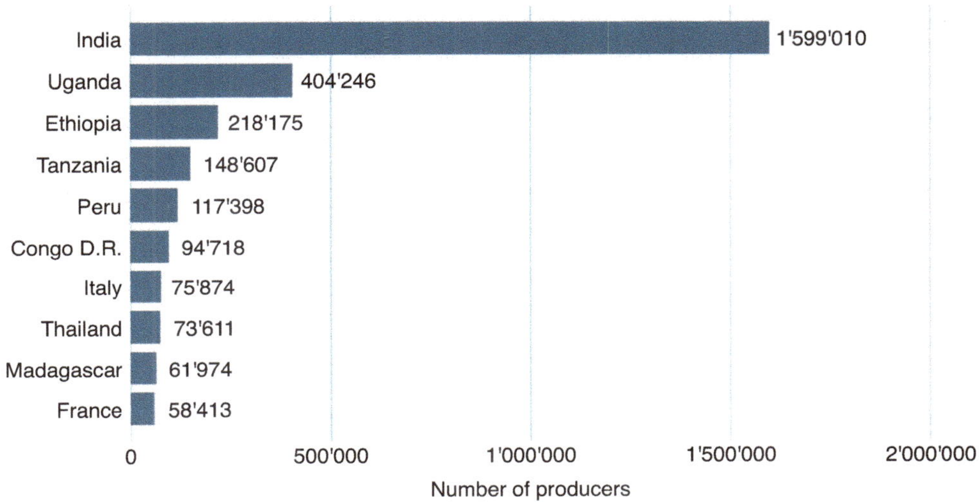

Fig. 7.5. World: the ten countries with the largest numbers of organic producers in 2021. (FiBL survey 2023, based on information from the private sector, certifiers and governments (Willer *et al.*, 2023))

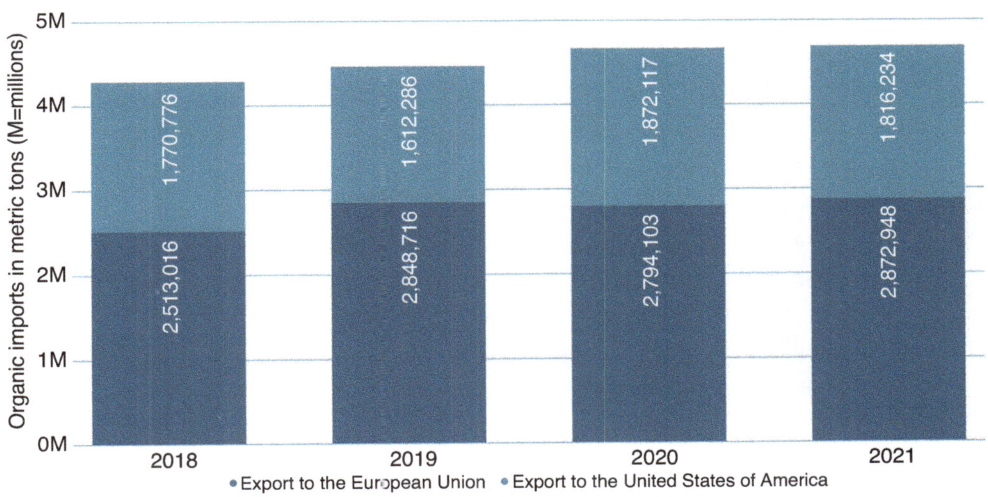

Fig. 7.6. World: development of EU and US organic imports, 2018–2021. (TRACES and USDA, 2022 (Willer *et al.*, 2023). NB: for the USA not all products are covered)

and Peru also made significant contributions, with 425,414 MT and 344,483 MT, respectively (Fig. 7.7).

- Large increase in export volume from Mexico, Canada and Colombia: Mexico (+59,880 MT, +16.4%); Canada (+38,202 MT, +29.3%); and Colombia (+27,758 MT, +11.6%) increased their exports to the EU and USA the most.
- Conversely, there were significant declines in imports from China (–80,272 MT, –35%), the Russian Federation (–55,297 MT, –47.2%), and Ukraine (–34,286 MT, –12.5%) between 2020 and 2021.
- The three most imported organic products in 2021 were banana, sugar and soybean, collectively constituting 44% of total organic imports. Banana topped the list at 1,275,723 MT, followed by sugar at 433,962 MT and oilseed at 353,808 MT (Fig. 7.8).
- Among specific product categories, tropical fruits saw a notable increase, particularly banana (+67,699 MT, 5.6%) and mango (19,804 MT, 55.1%). In contrast, cereals and soybean experienced decreases, with the most significant declines recorded for wheat (–64,463 MT, –49.7%), maize (–60,947 MT, –20.2%) and soybean (–53,695 MT, –13.2%).
- The USA, The Netherlands and Germany emerged as the primary importers, collectively handling almost 70% of organic imports into the EU and USA The USA accounted for 1.8 million MT (39% of all EU/US organic imports), followed by The Netherlands at 0.9 million MT (a notable 20% increase), and Germany with 0.5 million MT (an 11% increase). Importantly, a substantial portion of The Netherlands' imports is re-exported to other EU countries.

Global market reached nearly €125 billion

In 2021, organic food and beverage sales soared to an impressive nearly €125 billion[3]. This growth was underpinned by robust organic markets across the globe. The leading countries were the USA, with sales totalling €48.6 billion, Germany at €15.9 billion, and France with €12.7 billion in organic sales (Figs 7.9 and 7.10).

The USA emerged as the single largest market, capturing 39% of the global organic market share. Following closely behind was the EU, contributing €46.7 billion, accounting for 37% of the market, and China with €11.3 billion or 9.1% of the global market. Switzerland exhibited the highest per-capita consumption in 2021, boasting €425 per person. Notably, Denmark, Austria and Luxembourg recorded the highest organic market shares, reaching 13%, 11.6 and 11%, respectively.

However, despite the remarkable growth experienced in previous years, the global organic food and beverage market experienced a slowdown in

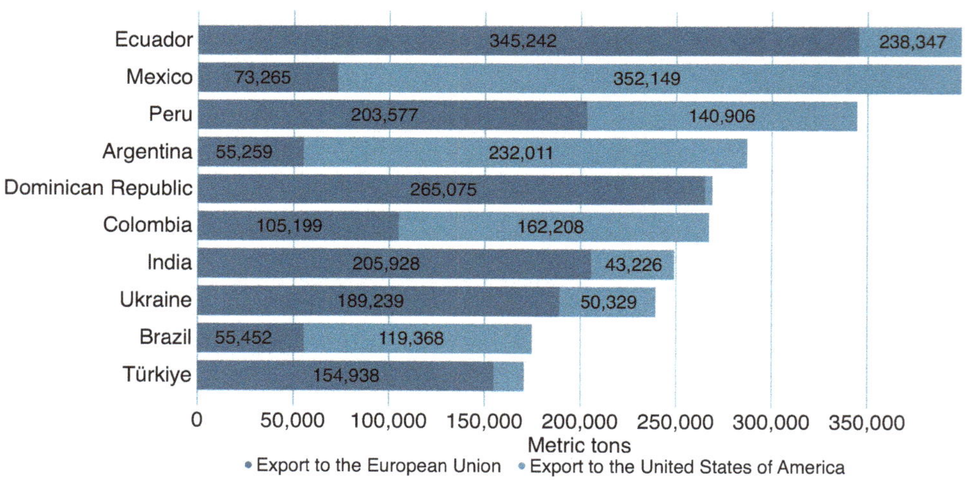

Fig. 7.7. World: key countries with organic exports to the EU and USA in 2021. (TRACES and USDA, 2022 (Willer *et al.*, 2023). NB: for the USA not all products are covered)

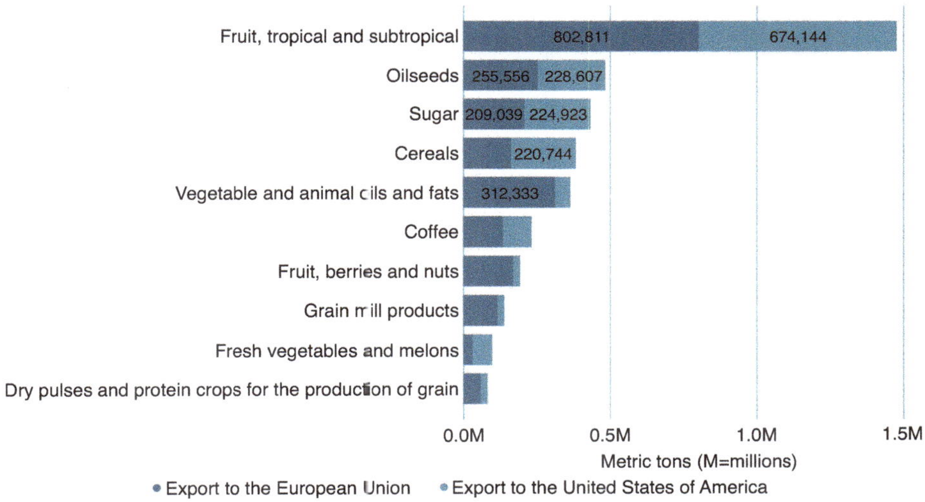

Fig. 7.8. World: key organic export product groups in 2021 (exports to EU and USA). TRACES and USDA, 2022 (Willer *et al.*, 2023). NB: for the USA not all products are covered)

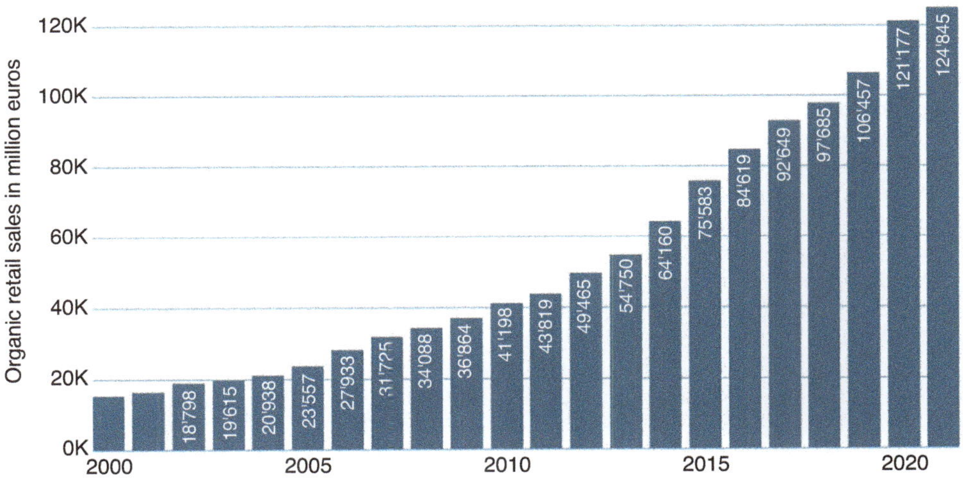

Fig. 7.9. World: organic retail sales growth. (FiBL survey, 2000–2023)

2021. This deceleration was attributed to various factors, including geopolitical conflicts and surging food prices, which cast a shadow over the organic market. After achieving record-breaking sales in 2020, the market's growth rate moderated in 2021.

Early data for 2022 paint a picture of retail sales declining in many countries, driven by weakened security that has impacted the demand for organic products. The European market, in particular, has borne the brunt of this downturn. Nevertheless, some countries, such as Austria and The Netherlands, continue to buck the trend, showing resilience, with ongoing growth in organic retail sales.

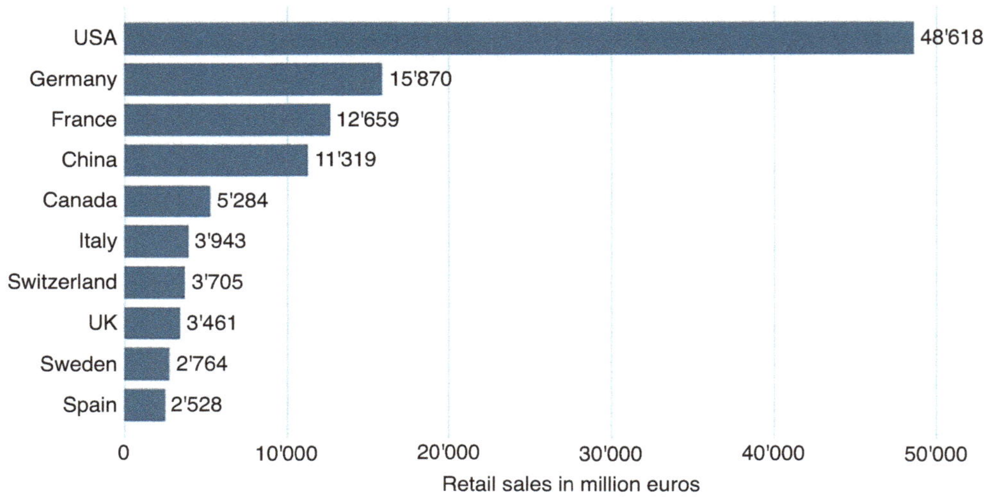

Fig. 7.10. World: top ten countries with the largest organic markets (retail sales). (FiBL survey, 2023 (Willer *et al*., 2023))

Data collection on organic agriculture worldwide: relevance and background

Since the early 2000s, FiBL has played a crucial role in compiling and publishing data on organic agriculture through its *The World of Organic Agriculture Yearbook*. In recent years, FiBL has embraced technology, introducing interactive online databases and visual elements to enhance the accessibility of this valuable information. This comprehensive annual report covers a wide range of essential aspects, including land under organic production, land use, production figures, retail sales data and international trade statistics. Remarkably, over 190 countries actively contribute their data and insights.

These data collection endeavours are made possible through the dedicated support of numerous partners from around the world. The data collection was initiated by the organizers of the BIOFACH international organic fair and initially driven by the German Foundation Ecology and Agriculture (SOEL). Today, data collection is carried out by FiBL. Notably, FiBL receives valuable support from the Swiss State Secretariat for Economic Affairs (SECO), the Coop Sustainability Fund (Switzerland), IFOAM – Organics International and NürnbergMesse (organizers of BIOFACH). This backing enabled FiBL to establish a professional database and expand data collection across various indicators.

Most of the data collection process involves personal contacts, although for certain indicators, data is sourced from the internet. For instance, Eurostat, the European Union's statistical office, offers data on its website covering areas, livestock and operators. Nevertheless, numerous countries do not publish their data, prompting FiBL to send questionnaires to gather the necessary information. It is essential to recognize that a significant disparity exists between countries with established organic legislation and those without. While most countries with organic legislation undertake data collection, many in Africa and Asia lack a national data collection system. In such cases, FiBL gathers data from international certifiers.

Apart from area and production data, market and international trade data for most countries originate from various sources. For instance, in the USA, data on organic area, production and livestock come from the National Agricultural Statistics Service (NASS/USDA), while retail sales data are provided by the Organic Trade Association (OTA). International organic trade data are sourced from the Foreign Agricultural Service (FAS/USDA, 2023). In total, FiBL collaborates with at least 200 different data sources per survey.

One significant challenge is managing the increasing volume of available data, including the recent inclusion of EU organic imports. Harmonizing this data within the FiBL system, while streamlining the data import process, demands substantial effort and time. Accessibility of data through digital tools is another challenge. While FiBL has developed interactive databases and infographics on its website (https://statistics.fibl.org [accessed 21/1/25]), further development is needed, including the addition of more indicators and improved user-friendliness. Additionally, the FiBL classification system for crops and livestock requires an overhaul to allow data extraction based on different classifications. Multilingual accessibility for interactive tools would also enhance their utility.

FiBL also aims to enhance automated production estimations, especially when production data are unavailable. This is critical for assessing the significance of organic feed imports into the EU and potentially helping improve fraud detection efforts, even though this will require special attention and resources.

Policymakers and donors rely on FiBL's data to inform their decisions and gauge the success of their initiatives. For instance, the European Commission employs FiBL's retail sales data to illustrate the increasing value European consumers place on organic products. Scientists engaged in organic agriculture research routinely cite FiBL's data in their papers and theses, underpinning their hypotheses. Thus, data reliability is paramount.

Conclusion

The latest available data on certified organic agriculture worldwide, as of the end of 2021, present significant findings. In 2021, a noteworthy 76.8 million ha of agricultural land, equivalent to 1.6% of the total farmland, were dedicated to organic farming. Additionally, the number of organic producers saw a remarkable increase, reaching 3.7 million. Concurrently, the global organic food market experienced substantial growth, reaching a total value of nearly €125 billion.

The positive development of the organic sector, which has seen continuous growth of the organic market and land under organic management, reflects the dynamic and innovative nature of organic food and farming in response to the expectations of policymakers and the demands of consumers for high-quality food. Nevertheless, despite the impressive expansion observed in former years, the worldwide organic food and beverage market encountered a reduction in momentum during 2021. This could be ascribed to a range of factors, including geopolitical conflicts and rising food prices, which cast uncertainties over the organic market. Following a record-breaking performance in sales during 2020, the market's growth rate showed signs of moderation in 2021.

In conclusion, the data unequivocally demonstrate that organic agriculture has experienced sustained growth over recent decades, despite the challenges posed by current crises affecting the organic market. Looking forward, it is important that more countries develop their own robust organic data collection systems, complete with quality checks and transparent publication procedures. Such efforts will significantly enhance the availability and accessibility of organic agriculture data, benefiting all stakeholders in the organic sector.

Notes

[1] Please note that some of the data shared in this document differ from those in the 2023 edition of *The World of Organic Agriculture* (Willer *et al.*, 2023) as some data revisions were carried out.

[2] Where the designation 'country' appears in this chapter, it covers countries and territories (see UNSTAT website: https://unstats.un.org/unsd/methodology/m49/ [accessed 21/1/25]).

[3] In 2021, 1.1827 US$ corresponded to €1.

Bibliography

European Commission (2022) EU imports of organic agri-food products. Key developments in 2021. EU Agricultural Market Briefs No. 19, September. Available at: https://ec.europa.eu/info/food-farming-fisheries/farming/facts-and-figures/performance-agricultural-policy/studies-and-reports/market-analyses-and-briefs_en#marketbriefs (accessed 21 January 2025).

United States Department of Agriculture (USDA)/ Foreign Agricultural Service (FAS) (2023) SDA Foreign Agricultural Service's global agricultural trade system. Available at: https://apps.fas.usda.gov/GATS/ (accessed 21 January 2025).

Urrutia, O. (2022) Our data are essential for policymakers. The BioEcoActual website. Available at: https://www.bioecoactual.com/en/2022/11/10/helga-willer-fibl-interview/ (accessed 10 November 2022).

Willer, H., Schlatter, B. and Trávníček, J. (eds) (2023) *The World of Organic Agriculture: Statistics and Emerging Trends 2023*. Research Institute of Organic Agriculture FiBL and IFOAM – Organics International, Frick and Bonn, Germany.

Annex: Key Data by Region and Country

Table 7.2. Key indicators by region, 2021. (FiBL survey, 2023; Willer *et al.*, 2023)

Region	Organic farmland (ha)	Organic share of total farmland (%)	Organic producers (no.)	Organic retail sales (Million €)	Export to EU and USA combined (MT)
Africa	2,658,639	0.2%	1,120,861	13	457'085
Asia	6,504,211	0.4%	1,782,133	13,747	671'233
Europe	18,258,903	3.7%	446,709	54,239	671'832
Latin America	9,870,887	1.4%	280,436	778	2'657'518
Northern America	3,542,140	0.8%	23,392	53,901	180'150
Oceania	35,985,809	9.7%	18,479	1866	34'170
World[a]	76,812,483	1.6%	3,671,242	124,545	4'671'987

Notes: Agricultural land includes in-conversion areas and excludes wild collection, aquaculture, forest and non-agricultural grazing areas. There have been some revisions compared to the data published in February 2023 by Willer *et al.*).

[a]Includes correction value for French overseas departments.

Table 7.3. Key indicators by country, 2021. (FiBL survey, 2023; Willer *et al.*, 2023)

Country	Organic area (ha)	Share of total farmland (ha)	Producers (no.)	Retail sales (million €)	Organic exports to EU and US (MT)
Afghanistan	98	0.0			0
Albania	1094	0.1	130		1970
Algeria	772	0.0	64		2919
Andorra	2	0.0			
Angola		0.0			9
Antigua and Barbuda		0.0			
Argentina	4,074,804	2.7	1336		287,269
Armenia	583	0.0	27		132
Aruba		0.0			
Australia	35,687,799	9.9	1829	1694	1895
Austria	679,119	26.5	23,961	2397	415
Azerbaijan	38,080	0.8	446	3	1228
Bahamas	49	0.3			
Bahrain		0.0			
Bangladesh	504	0.0			55
Barbados		0.0			
Belarus	6725	0.1	19		2459
Belgium	101,828	7.4	2638	978	
Belize	676	0.4	385		35
Benin	48,898	1.2	9046		7708
Bermuda		0.0			
Bhutan	5608	1.1	1998	0	
Bolivia (Plurinational State of)	179,425	0.5	14,161		21,125
Bosnia and Herzegovina	2495	0.1	90	0	2788
Botswana		0.0	1		0
Brazil	1,482,406	0.6	28,667	778	174,820
British Virgin Islands	26	0.4			
Brunei Darussalam		0.0			
Bulgaria	86,310	1.7	4352	33	15
Burkina Faso	79,285	0.7	27,021		17,556
Burundi	549	0.0	668		73
Cambodia	21,112	0.4	4135		7880
Cameroon	1594	0.0	632		288
Canada	1,215,589	2.1	6916	5284	168,305
Cape Verde		0.0			
Cayman Islands		0.0			
Central African Republic		0.0			
Chad		0.0	2960		610
Channel Islands	180	2.0			
Chile	152,773	1.0	3288		70,883
China	2,753,700	0.5	14,847	11,319	152,477

Continued

Table 7.3. Continued

Country	Organic area (ha)	Share of total farmland (ha)	Producers (no.)	Retail sales (million €)	Organic exports to EU and US (MT)
Colombia	100,874	0.2	171		267,407
Comoros	937	0.7	637		42
Congo, Republic of		0.0			
Cook Islands	15	1.0	58		
Costa Rica	10,300	0.6	47		33,647
Côte d'Ivoire	78,783	0.4	2990		35,481
Croatia	121,924	8.1	6024	99	27
Cuba	2129	0.0	8		2444
Cyprus	7738	5.7	1292		
Czech Republic	558,124	15.8	4797	226	
Democratic Republic of the Congo	89,486	0.3	94,718		12,608
Denmark	303,093	11.5	4186	2240	0
Djibouti		0.0			18
Dominica	2907	11.6	258		
Dominican Republic	117,312	4.8	15,563		268,937
Ecuador	52,185	1.0	8398		580,336
Egypt	116,000	3.0	970		52,073
El Salvador	2565	0.2	374		317
Equatorial Guinea		0.0			
Eritrea		0.0			
Estonia	226,605	23.0	2043	93	5182
Eswatini	3539	0.3	3		
Ethiopia	332,519	0.9	218,175	13	17,036
European Union undefined		#WERT!			76
Falkland Islands (Malvinas)	31,937	2.8	3		
Faroe Islands	251	8.4	1		
Fiji	30,194	7.1	13		43
Finland	327,736	14.4	5007	407	
France	2,776,554	9.6	58,413	12,659	1874
French Guiana (France)	3886	11.9	69		
French Polynesia	1592	3.5	19		79
Gabon		0.0			
Gambia		0.0			
Georgia	4278	0.2	729		730
Germany	1,802,231	10.8	36,307	15,870	245
Ghana	32,916	0.2	1039		30,398
Greece	534,629	10.1	35,813	66	1846
Greenland		0.0			
Grenada	195	2.4	5		4
Guadeloupe (France)	1300	2.5	186		
Guam		0.0			

Continued

Table 7.3. Continued

Country	Organic area (ha)	Share of total farmland (ha)	Produers (no.)	Retail sales (million €)	Organic exports to EU and US (MT)
Guatemala	71,654	1.9	141		13,201
Guinea		0.0			44
Guinea-Bissau		0.0			426
Guyana		0.0			433
Haiti	2739	0.1	4631		2174
Honduras	66,179	2.0	15,046		68,581
Hong Kong		0.0			91
Hungary	293,597	5.9	5129	30	
Iceland	6440	0.4	30		
India	2,657,889	1.5	1,599,010	186	248,273
Indonesia	83,362	0.1	17,836		16,740
Iran (Islamic Republic of)	7053	0.0	346		1641
Iraq	63	0.0			
Ireland	86,868	1.9	1914	235	16
Israel	5778	0.9	345		24,629
Italy	2,186,159	16.7	75,874	3643	27,450
Jamaica	11	0.0	2		6
Japan	11,992	0.3	3678	1419	5138
Jordan	1446	0.1	23		147
Kazakhstan	113,247	0.1	281		35,110
Kenya	128,018	0.5	44,565		9749
Kiribati		0.0			
Kosovo	1990	0.5	56		561
Kuwait	32	0.0			
Kyrgyzstan	30,259	0.3	1144		645
Lao People's Democratic Republic	7993	0.3	1519		11,692
Latvia	302,177	15.3	4171	51	
Lebanon	1671	0.3	124		100
Lesotho		0.0	10		793
Liberia	2791	0.1			
Libya		0.0			
Liechtenstein	1423	40.2	47	9	
Lithuania	261,782	8.9	2529	51	3
Luxembourg	6893	5.2	123	197	
Madagascar	95,083	0.2	61,974		5331
Malawi	324	0.0	3		3566
Malaysia	1276	0.0	30		42
Maldives		0.0			
Mali	17,840	0.0	11,608		4736
Malta	66	0.6	25		
Marshall Islands		0.0			

Continued

Table 7.3. Continued

Country	Organic area (ha)	Share of total farmland (ha)	Produers (no.)	Retail sales (million €)	Organic exports to EU and US (MT)
Martinique (France)	706	2.3	77		
Mauritania		0.0			
Mauritius	13	0.0	1		296
Mayotte	114	0.6	9		
Mexico	238,075	0.2	52,274		425,393
Micronesia (Federated States of)		0.0			
Moldova	28,368	1.3	151		19,376
Mongolia	241	0.0	314	1	
Montenegro	4404	1.7	422		17
Montserrat		0.0			
Morocco	11,452	0.0	423		21,246
Mozambique	41,048	0.1	1358		5311
Myanmar	10,143	0.1	68		95
Namibia	384	0.0	15		14
Nauru		0.0			
Nepal	2448	0.1	1048		240
Netherlands	76,375	4.2	1985	1374	669
New Caledonia	800	0.4			0
New Zealand	79,347	0.8	685	172	29,587
Nicaragua	37,357	0.7	10,912		12,584
Niger		0.0			700
Nigeria	58,028	0.1	2308		906
Niue	117	2.3	2		
North Macedonia	7794	0.6	887		662
Northern Mariana Islands		0.0			
Norway	45,112	4.6	1778	442	
Oman	7	0.0	1		
Pakistan	69,850	0.2	934		45,625
Palau		0.0			
Palestine	5517	1.2	1632		861
Panama	5929	0.3	18		1270
Papua New Guinea	88,014	7.4	12,827		2494
Paraguay	113,147	0.5	5591		78,790
Peru	374,926	1.6	117,398		344,160
Philippines	216,334	1.7	13,069		26,136
Poland	509,286	3.5	18,598	314	2
Portugal	768,800	19.4	13,263	21	3228
Puerto Rico		0.0			
Qatar		0.0			
Republic of Korea	40,663	2.5	25,362	486	331
Réunion (France)	2100	4.4	427		

Continued

Table 7.3. Continued

Country	Organic area (ha)	Share of total farmland (ha)	Produers (no.)	Retail sales (million €)	Organic exports to EU and US (MT)
Romania	578,718	4.3	11,562	41	9
Russian Federation	655,457	0.3	66	183	61,865
Rwanda	4696	0.3	8930		730
Saint Helena, Ascension and Tristan da Cunha		0.0			
Saint Kitts and Nevis		0.0			
Saint Lucia	25	0.2	1		
Saint Pierre and Miquelon		0.0			
Saint Vincent/Grenadines		0.0			
Samoa	82,299	29.1	1993		51
San Marino		0.0			
Sao Tome and Principe	9291	21.1	4201		4888
Saudi Arabia	27,110	0.0	400	306	277
Senegal	3262	0.0	18,372		2891
Serbia	23,527	0.7	458		19,391
Seychelles		0.0			24
Sierra Leone	193,954	4.9	5507		16,236
Singapore	15	2.2		16	877
Slovakia	162,565	8.5	716		
Slovenia	51,826	10.7	3718	49	
Solomon Islands	8231	7.0	898		
Somalia		0.0	4		59
South Africa	97,359	0.1	1307		28,450
South Sudan		0.0			
Spain	2,635,442	10.8	52,861	2528	10,082
Sri Lanka	66,623	2.4	1940		44,723
Sudan	960	0.0	1		4997
Suriname	52	0.1	1		88
Sweden	606,669	20.2	5360	2764	45
Switzerland	181,444	17.4	7670	3705	32
Syrian Arab Republic		0.0			6
Taiwan	11,765	1.5	4436		115
Tajikistan	22,292	0.5	166		
Tanzania, United Republic of	286,627	0.7	148,607		4788
Thailand	167,985	0.8	73,611	12	28,058
Timor-Leste	32,311	8.5	3		724
Togo	130,858	3.4	19,709		72,731
Tokelau		0.0			
Tonga	1119	3.2	81		
Trinidad and Tobago		0.0			
Tunisia	279,389	2.9	7101		68,079

Continued

Table 7.3. Continued

Country	Organic area (ha)	Share of total farmland (ha)	Produers (no.)	Retail sales (million €)	Organic exports to EU and US (MT)
Türkiye	327,583	0.9	48,244	46	170,705
Turkmenistan		0.0			
Turks and Caicos Islands		0.0			
Tuvalu		0.0			
Uganda	505,308	3.5	404,246		22,866
Ukraine	422,299	1.0	418	28	239,498
United Arab Emirates	5419	1.4	152		319
UK	489,200	2.8	3581	3461	101,326
USA	2,326,551	0.6	16,476	48,618	11,845
USA Virgin Islands		0.0			
Uruguay	2,741,845	19.6	1417		3612
Uzbekistan	4925	0.0	26		1257
Vanuatu	6281	3.4	74		21
Venezuela (Bolivarian Republic of)	2496	0.0	8		
Viet Nam	74,540	0.6	12,453		14,828
Wallis and Futuna Islands		0.0			
Western Sahara		0.0			
Yemen		0.0			8
Zambia	3376	0.0	10,872		93
Zimbabwe	1085	0.0	10,379		314
Total[a]	76,812,483	1.6	3,671,242	124,545	4,671,987

Note: Agricultural land includes in-conversion areas and excludes wild collection, aquaculture, forest and non-agricultural grazing areas.
[a]Includes correction value for French overseas departments.

8 Organic Agriculture: Opportunities and Challenges

Victor Olowe[1]*, Jude Obi[2] and Olabisi Somefun[3]

[1]*Crop Production Research Programme, Institute of Food Security, Environmental Resources and Agricultural Research (IFSERAR), Federal University of Agriculture, Abeokuta (FUNAAB), Nigeria;* [2]*Department of Soil Science and Land Resources Management, Faculty of Agriculture, University of Uyo, Uyo, Akwa Ibom, Nigeria;* [3]*Department of Plant Physiology and Crop Production, College of Plant Science and Crop Production, FUNAAB, Nigeria*

Abstract

Organic agriculture is presently the fastest-growing sector of agriculture in the world occupying 76.4 million ha of land in 191 countries. Despite its phenomenal market growth from €15.1 billion in 2000 to €124.8 billion in 2021, and an increase in the number of producers from 200,000 in 1999 to 3.7 million in 2021, the sector is still facing mutifarious challenges globally. This chapter tries to situate organic agriculture within the global perspective, elucidating the opportunities and challenges facing the sector, and the roles of research and extension in advancing the sector.

Keywords: challenges, extension, organic agriculture, opportunities, research

Introduction

The universally accepted definition of organic agriculture is that of the Codex Alimentarius Commission of the FAO/WHO (1999): 'Organic Agriculture is a holistic production management system which promotes and enhances agroecosystem health, including biodiversity, biological cycles and soil biological activity', as reported by Rahmann *et al.* (2017). It emphasizes the use of management practices in preference to off-farm inputs, taking into account that regional conditions require locally adapted systems. Globally, organic agriculture remains a niche sector since it still occupies only 1.6% of the world's agricultural land (Willer *et al.*, 2023). At present, a total of 76.4 million ha of land are being managed by over 3.7 million producers as against 11 million ha managed by 200,000 producers in 1999. In the last two decades, the organic market has increased by 87.9% from €15.1 billion in 2000 to €124.8 billion in 2021 (Willer *et al.*, 2023), thus making organic agriculture a rapidly developing sector of agriculture.

It has been estimated that the world's population will most likely increase to 8.3 billion in 2023 and 9.3 billion in 2025, and the immediate concern is how to provide food for the burgeoning population because the demand for food, fibre and fuel continues to increase (Snyder *et al.*, 2009). Consequently, organic agriculture has a crucial role to play. For example, Nigeria, which is the most populous African country, will have a predicted population of 800 million in 2100 while the projected land area available per person for farming in 2100, under modest assumption, is only 504 m^2 as against 2100 m^2 in 2019 (Olowe, 2021). Therefore, organic agriculture practice should be encouraged in the developing world in particular, in order to stem the

*Corresponding author: olowevio@funaab.edu.ng

 Advances in Organic Farming (ed. L. P. Awasthi)
DOI: 10.1079/9781800626850.0008

impending global crisis as the human population continues to increase in the developing world.

Globally, this sector is driven by several factors such as economic benefits, reduced input costs, premium process of products, ecological protection factors, health awareness, improved soil health (i.e. organic matter, biological activity, nutrient cycles, reduced erosion risks), food security, and improved rural livelihoods (Kopke, 2018). Nevertheless, the sector is faced with multifarious challenges that are militating against its development.

In order to properly situate organic agriculture in global perspective, this chapter attempts to discuss the opportunities and challenges it presents and the crucial roles research and extension can play in enhancing its development and adoption.

Opportunities of Organic Agriculture

At present, organic agriculture is being practised in two strands (certified organic and non-certified organic) in the world. Certified organic agriculture provides alternative products in the market, thereby increasing the choices for consumers, and these certified products command premium prices (Halberg and Alrøe, 2006). In the last decade (2011–2021), the land that is under organic agriculture has increased tremendously in Africa (149.0%), Asia (79.6%), Europe (69.2%), Latin America (41.7%), North America (17.3%), and Oceania (215.1%), and 108.4% on average in the world, as reported by Willer *et al.* (2023). Similarly, producers have grown significantly in number across regions, except Latin America, in the last decade: 112.8% in Africa, 119.6% in Asia, 52.7% in Europe, –8.8% in Latin America, 40.9% in North America, 30.3% in Oceania, and 107.7% on average in the world, based on the report of Willer *et al.* (2023). Organic agriculture retail in the last two decades also increased from €15.1 billion in 2000 to €124.8 billion in 2021 (Willer *et al.*, 2023).

The benefits deriveable from organic agriculture are multi-dimensional and include the following: improved soil health through the support from soil maintenance and erosion prevention; and improved nutrition because organic products are more nutrient-dense, since they are produced from healthy soil relative to produce from conventional plots. Other benefits include: environmental friendliness since no synthetic chemicals that can harm the ecosystem are used; decreased exposure of the workforce to chemicals; higher standards of animal welfare; and protection of plants from dangerous substances. Dynamic and approved regulations help to check any excesses by practitioners, and the potential for co-operative rural and regional development (Kristiansen *et al.*, 2006). However, critical evaluation of large bodies of data have shown that organic products also contain one-third of the pesticide residues of conventionally grown foods (Baker *et al.*, 2002).

Organic agriculture will likely continue to gain prominence in the world because of the acceptance by consumers, despite the higher cost of the products (Bello, 2008). Most consumers will also continue to patronize organic products because of their opposition to genetically modified (GMO) foods in the market. Biodiversity is better preserved under organic production systems because they encourage a wider diversity of crops and many different natural predators that control pests, thereby preventing the spread of diseases on the farm (Parrot and Marsden, 2002). Organic agriculture practice also results in the addition of soil organic matter (SOM) to the soil through the use of organic manures over time and incorporation of plant residues as nutrient sources (Laird *et al.*, 2001).

Challenges of Organic Agriculture

In a global context, organic agriculture is facing several challenges that differ from one region to another. Kutama *et al.* (2013) enumerated some challenges, most especially in developing countries, as follows: lack of awareness, output marketing problems, shortage of biomass, and lack of supporting infrastructure. Other challenges include high input costs, lack of adequate farm inputs, lack of appropriate agricultural policies, lack of financial support, low production output, inability to meet export demand, lack of quality standards for biomanures, as well as political and social factors. Unfortunately, these problems still exist in organic agriculture practice. The high cost of certification remains a major challenge for prospective organic farmers because many of them cannot afford the cost. However, the Participatory Guarantee System (PGS) is much cheaper and is being adopted to certify organic produce for local, and more recently national and regional, use. Pest control is also critical because in organic agriculture it has to be done without synthetic chemicals and, as such, farmers have to resort to the use of biopesticides, which they produce locally, and the production could be quite difficult, unlike ready-made synthetic chemicals. Consequently, organic production systems record lesser yields compared with conventional systems. It has been reported that the yield gap between organic and conventional farming systems ranges between 9 and 25% depending on the type of commodity being considered (Seufert *et al.*, 2012; Wilbois and Schmidt, 2019). It has been reported that yield performance of agricultural commodities is location- and management-specific and depends on prevailing agroecological conditions of the study area (Martini *et al.*, 2004). Therefore, caution must be expressed while drawing conclusions on yield gap between organic and conventional production systems

There is the need to domesticate the organic principles in a wide range of diverse systems with different conditions, as suggested by Halberg and Alrøe (2006). This will facilitate the integration of the principles into local community-level approaches. Organic products face many challenges in the global market, which is characterized by the strong role being played by key players in transport, handling, distribution, marketing and sales (Alrøe *et al.*, 2006). Often, organic products face such challenges, which are counter-productive to the basic principles of organic agriculture such as transparency and traceability, which are inherent in certified trade. In order to resolve this problem, it has been suggested that the organic movement should create a system that accommodates the interests of the farming community and those of all stakeholders (Alrøe *et al.*, 2006). Other challenges facing the organic sector include underlying inadequacies in regulatory and marketing structures (e.g. labelling, preserving food quality while increasing productivity, harmonization of international and regional standards), maintaining the integrity of organic foods, developing local solutions to major production constraints such as weeds, animal health, and soil fertility (Kristiensen *et al.*, 2006).

Role of Research and Extension in Organic Agriculture

Research and extension are the backbone of development of any nation. The development of certified organic agriculture in Africa has been partly impeded by limited sound, science-based information to tackle the multifarious challenges facing the subsector along the value chains (Olowe *et al.*, 2022). It has been advocated that research in organic agriculture should address concrete challenges on organic farms (Nicolay and Baker, 2012). Furthermore, these authors identified key areas where research is most urgently needed as follows: weed management, plant breeding, soil fertility restoration, biological pest and disease control, and marketing and policy research. Weeds can reduce yield of crops from 60% to 86% (Adigun *et al.*, 2014). The major weed control strategies in organic agriculture are the no-tillage system, cover cropping and use of organic herbicides, which are gradually emerging in organic circles. According to Kumar *et al.* (2011), organic plant breeding is a system that develops varieties of crops that are high-yielding under organic conditions, maintaining the health of the environment and human beings. The outcome of any breeding programme should be the development of highly resilient crop and animal breeds that can survive under harsh conditions and still produce reasonable outputs. Ready access to local, national and international organic markets should drive the efforts of researchers in order to assist farmers in optimizing their investments. Consumer acceptance studies should also be encouraged in order to help sharpen the focus or direction farmers need to concentrate their efforts.

One of the major problems confronting organic agriculture research is poor funding (Wynen, 2003). Limited access to research grants has been reported as one of the least available instruments to aid development of organic agriculture in Africa (UNCTAD, 2016). Execution of research-based projects on a continental or regional basis can also help in advancing the adoption of organic agriculture as a production system. A good example is the ongoing continental project on Ecological Organic Agriculture Initiative (EOA-I) whose Phase II will end in 2023. Its relevance to organic agriculture development in the continent of Africa had been highlighted by Olowe (2018). It is suggested that for organic agriculture research to significantly advance globally, scientists should engage in cross-disciplinary research collaboration under different auspices such as International Society of Organic Agriculture Research (ISOFAR) based in Bonn, Germany (www.isofar.org), Vision for an Organic Food and Farming Research Agenda to 2025 (Niggli *et al.*, 2008), Strategic Research Agenda for Organic Food and Farming (Schmid *et al.*,

2009), Network for Organic Agriculture Research in Africa (NOARA), and Organic Agriculture Research Agenda for Africa (OARAA 2032) (www.noara.bio [accessed 22/1/25]). Furthermore, in a recent review, it was advocated that future research endeavour should address the largely neglected areas of organic agriculture such as organic aquaculture, policy issues, health and safety of organic products, and standards and certification (Olowe and Somefun, 2020).

Organic agriculture is a farmer-centred and grass roots movement, and, as such, should strongly focus on educating the major stakeholders (farmers) and developing their capacities. The indigenous knowledge of the farmers is very crucial in fostering organic agriculture adoption in developing countries. Networking, knowledge dissemination, advocacy and international collaboration have been recommended as veritable tools for advancing organic agriculture (Nicolay and Baker, 2012). A recent project that is based on knowledge validation and sharing was recently commissioned in Africa and is tagged 'Knowledge Centre on Organic Agriculture in Africa' (KCOA). It is expected that the project will go a long way in linking the primary producers (farmers) to the various stakeholders in the supply chain. The project will also harvest endogenous knowledge and share among the various stakeholders in organic agriculture.

Conclusions

Organic agriculture is an eco-friendly production system that is still operating as a niche within the agriculture sector in the world. The highlighted challenges facing the sector are surmountable with research and extension services playing key roles. At present, organic agriculture is playing a crucial role in addressing the vagaries of climate change, ensuring food and nutrition security, sustaining biodiversity and championing sustainable consumption.

References

Adigun, J., Osipitan, A.O., Lagoke, S.T., Adeyemi, R.O. and Afolami, S.O. (2014) Growth and yield performance of cowpea (*Vigna unguiculata* (L.) Walp) as influenced by row-spacing and period of weed interference in South-West Nigeria. *Journal of Agricultural Science* 6(4), 188–198.

Alrøe, F.H., Bryne, J. and Glover, L. (2006) Organic agriculture and ecological justice: Ethics and practices. In: Halberg, N., Alrøe, H.F., Knudsen, M.T. and Kristensen, E.S. (eds) *Global Development of Organic Agriculture: Challenges and Prospects*. CAB International, Wallingford, UK, pp. 75–109.

Baker, B.P., Benbrook, C.M., Groth, E. and Lutz Benbrook, K. (2002) Pesticide residues in conventional, integrated pest management (IPM)-grown and organic foods: Insights from three US data sets. *Food Additives and Contaminants* 19(5), 427–446.

Bello, W.B. (2008) Problems and prospect of organic farming in developing countries. *Ethiopian Journal of Environmental Studies and Management* 1(1), 1–8.

Halberg, N. and Alrøe, F.H. (2006) Kristensen ES synthesis: Prospects for organic agriculture in a global context. In: Halberg, N., Alrøe, H.F., Knudsen, M.T. and Kristensen, E.S. (eds) *Global Development of Organic Agriculture: Challenges and Prospect*. CAB International, Wallingford, UK, pp. 344–364.

Kopke, U. (2018) *Improving Organic Crop Cultivation*, Burleigh Dodds Series in Agricultural Science, Vol. 47. Burleigh Dodds Science Publishing, Sawston, UK.

Kristiansen, P., Taji, A. and Reganold, J. (2006) Organic agriculture: Opportunities and challenges. In: Kristiansen, P., Taji, A. and Reganold, J. (eds) *Organic Agriculture: A Global Perspective*. CSIRO Publishing, Collingwood, Australia, pp. 421–441.

Kumar, A., Kumar, P., Singh, L. and Kumar, S. (2011) Biomolecule reports. Available at: https://www.researchgate.net/profile/Anand-Kumar-9/publication/339999009_Organic_Plant_Breeding_Opportunities_and_Challenges/links/5e71d8274585152cdbfa8dd8/Organic-Plant-Breeding-Opportunities-and-Challenges.pdf (accessed 3 May 2022).

Kutama, A.S., Abdullahi, M.A., Umar, S., Binta, U.B. and Ahmad, M.K. (2013) Organic farming in Nigeria: Problems and future. *Global Advanced Research Journal of Agricultural Science* 2(10), 256–262.

Laird, D.A., Martens, D.A. and Kingery, W.L. (2001) Nature of clay-humic complexes in an agricultural soil: I. Chemical, biological, and spectroscopic analyses. *Soil Science Society of America Journal* 65, 1413–1418.

Martini, E.A., Buyer, J.S., Bryant, D.C., Hartz, T.K. and Denison, R.F. (2004) Yield increases during the organic transition: Improving soil quality or increasing experience? *Field Crops Research* 86(2–3), 255–266.

Nicolay, G.L. and Baker, B. (2012) Challenges and opportunities for organic research and extension. *Rural* 21 3, 32–34.

Niggli, U., Slabe, A., Halberg, N., Schmid, O. and Schlüter, M. (2008) *Vision for an Organic Food and Farming Research Agenda to 2025*. IFOAM – Organics International, EU Group and FiBL, Brussels.

Olowe, V. (2018) Improving organic agriculture in the developing world: Africa. In: Kopke, U. (ed.) *Improving Organic Crop Cultivation*. Burleigh Dodds Science Publishing, Cambridge, UK.

Olowe, V. (2021) Africa 2100: How to feed Nigeria in 2100 with 800 million inhabitants. *Organic Agriculture* 11(2), 199–208. DOI: 10.1007/s13165-020.

Olowe, V.I. and Somefun, O.T. (2020) A synthesis of the 2nd, 3rd, 4th, and 5th ISOFAR scientific conferences: Pointers to future frontiers of knowledge. *Organic Agriculture* 10(S1), 139–147.

Olowe, V., Azim, K., Atoma, C., Odueme, P. and Somefun, O. (2022) Advancing organic agriculture research in Africa: A case study of the science track of the 1st, 2nd, 3rd and 4th African organic conferences (AOCs). *Sustainability* 14(18), 11416.

Parrot, N. and Marsden, T. (2002) The real green revolution: Organic and agroecological farming in London: Green peace research in Africa: A case study of the science track of the 1st, 2nd, 3rd and 4th African organic conferences (AOCs). *Sustainability* 14, 11416.

Rahmann, G., Reza Ardakani, M., Bàrberi, P., Boehm, H., Canali, S. *et al*. (2017) Organic agriculture 3.0 is innovation with research. *Organic Agriculture* 7(3), 169–197.

Schmid, O., Padel, S., Halberg, N., Huber, M., Darnhofer, I. *et al*. (2009) *Strategic Research Agenda for Organic Food and Farming: Technology Platform Organics*. IFOAM EU Group, Brussels.

Seufert, V., Ramankutty, N. and Foley, J.A. (2012) Comparing the yields of organic and conventional agriculture. *Nature* 485(7397), 229–232.

Snyder, C.S., Bruulsema, T.W., Jensen, T.L. and Fixen, P.E. (2009) Review of greenhouse gas emissions from crop production systems and fertilizer management effects. *Agriculture, Ecosystems & Environment* 133(3–4), 247–266.

UNCTAD (United Nations Conference on Trade and Development) (2016) Financing organic agriculture in Africa: Mapping the issues. UNCTAD/WEB/DITC/TED/2016/6. Available at: https://unctad.org/system/files/official-document/webditcted2016d6_en.pdf (accessed 22 January 2025).

Wilbois, K.-P. and Schmidt, J.E. (2019) Reframing the debate surrounding the yield gap between organic and conventional farming. *Agronomy* 9(2), 82. DOI: 10.3390/agronomy9020082.

Willer, H., Schlatter, B. and Travnicek, J. (eds) (2023) *The World of Organic Agriculture: Statistics and Emerging Trends 2023*. Research Institute of Organic Agriculture FiBL / IFOAM - Organics International.

Wynen, E. (2003) *Organic Agriculture in Australia: Levies and Expenditure*. Rural Industries Research and Development Corporation, Barton, Australia.

9 Scale-appropriate Farming in Organic Agriculture

LEONARD POLLARA*

Organic and Non-GMO Certification Compliance; Biological Agriculture and Systems Specialist; Inspector and Organic Sage Consultant, LLC; International Organic Inspectors Association, Northport, Michigan, USA

Abstract

Scale, in terms of how organic agriculture is practised, has physical, operational, financial, political and philosophical drivers and impacts. Throughout agriculture, over the past 85 years, technological advances have tended to support aggregation and consolidation of operations driven by government policy decisions pushing a 'get big or get out' mindset, each of which has a significant and varied effect on the success of individual operations. Worldwide, less-affluent, less-developed and more-remote regions tend to have smallholdings and more operators representing a higher percentage of the local population engaged in agricultural practices. As the organic product sector has risen to a significant financial segment of agricultural sales worldwide, governments have moved to regulate and license 'organic' products entering the stream of commerce. The promulgation of laws and rules requiring certification have been effected to codify what began, to a great extent, as a grass roots 'counterculture' movement against chemical-input-based industrializing agricultural practices. With growing consumer pressure, we see wider adoption of 'organic practices' in an effort to meet market demand. Also, a growing dichotomy between larger industrialized agricultural operations and smaller, less-mechanized operations raises ethical and philosophical questions around the concept and validity of 'certified organic' agriculture. Much of the market space is being filled by industrialized producers meeting the various process- and procedure-based standards sufficiently to place materials into the stream of commerce as certified organic. Many smaller operators are seemingly marginalized without access to larger market spaces or relegated to local regional collaboration and exchange. Value propositions not regularly considered in industrialized agriculture reveal layers of impact that have often been discounted and ignored in studies considering 'conventional' farming enterprises. This discussion explores the range and impact of scale in organic agricultural production while asking 'Is there such as thing as "scale-appropriate" organic farming?'

Keywords: organic agriculture, scale, value propositions

Introduction

Whether it is a smallholder in rural Kenya with 400 sq. m or the biggest cattle station in the world at 23,677 sq. km, a Del Monte pineapple plantation in the Philippines (40,000 ha) or a two-acre tangerine-under-plastic operation in South Korea, a 70,000-acre wheat farm in Saskatchewan, Canada, or a 2 ha cocoa plantation in Ecuador, these are all considered agricultural operations. They share much of the characteristics that qualify them as farms despite stark differences. Differences in scope and scale are obvious metrics of comparison. However, no organization or government, no law or regulation, uses operation size as a discriminating criterion for determining eligibility as a certified 'organic' producer.

Public Perception

Public perception of farming varies greatly depending on marketplace, individual and societal experiences. In the USA, much of the perception of what constitutes farming is driven by media and

*Corresponding author: lepollara@gmail.com

DOI: 10.1079/9781800626850.0009

advertising, playing upon old tropes and stereotypes, romanticized and wishful notions, and a lack of personal experiences, now that American farm families comprise only 2% of the population (Fb.org, 2018).

This perception gap is rather easy to reach in developed industrialized countries. In other regions where populations' direct involvement with agriculture exceeds that of the USA (e.g. Bolivia – 27%, India – 43%, Zambia – 57%, China – 23%, Ukraine – 28%), the majority of their populations have a more direct relationship with the practices of agriculture (FAO, 2022).

Why this matters is that demand for organic agricultural products has been driving the unprecedented growth in organic production and, consequently, promulgation of regulatory involvement. The US market for organic products was reported at $52 billion in 2021 (Skorbiansky, 2023) and reported to be $169 billion worldwide in 2023 (ReportLinker, 2023). This is an increase from $26.9 billion in the USA and $57.2 billion (food and beverage only) worldwide in 2010 (MarketsandMarkets, 2011) What this has to do with scale becomes evident upon understanding that major corporate entities are expanding market share in organic production, such as the Bayer AG/Monsanto merger, the China Chemical Company/Syngenta merger (Shields, 2017), and the Dow Chemical Company/DuPont merger. The list of companies in organic agriculture is long and when looking closer one finds that there has been considerable consolidation through acquisition. Taking one example, White Wave purchased Earthbound Farm (2014), and Danone purchased White Wave (2017) along with Horizon Organic. Subsequently, Earthbound Farm was sold off (to Taylor Farms in 2019), and so on. Rapid entry into the organic sector was accomplished through comparatively recent acquisition of first- and second-generation organic enterprises conceived and begun in the 1970s and 1990s. These massive business entities (Sullivan, 2017) are also enmeshed in non-organic agricultural production and moving into the organic arena is a business strategy to expand their portfolios and gain access to a market that has been growing steadily, in some years by double-digit percentages since 2000. Organic agriculture has been the fastest-growing sector in American agriculture for the past 23 years, far outpacing all other agriculture sectors (National Institute of Food and Agriculture, n.d.). Big players have big expectations and bring along the 'bigger is better' mindset that drives much of the current food and fibre production in organic agriculture, thereby having an impact on the scale of supply. Part-and-parcel of industrialized business expectations is the concept of production efficiency and efficiencies of scale.

Scale of Retail Consumer Supply

When considering the scale of retail consumer supply, one must at least acknowledge the fascination with quantity over quality that seems inherent in conventional farming today. To offer a fair assessment, a brief historical consideration seems in order. It has only been three or so generations that famine has been a reality, has been experienced by populations in developing countries, a present and recurring situation, a 'routine fact of life'. To a great degree, this is not the case in developed countries presently. And, to a significant degree, widespread famine is still known in less-developed countries. As is commonly understood, the 'green revolution' (Encyclopedia Britannica, 2009), from the beginning of the use of chemical fertilizers and more highly toxic pesticides, ushered in an unprecedented degree of production abundance. Hindsight reveals that this was due to a combination of the then-new concentrated chemicals and the existing biological activity then present in soils. Moving forward 90 years, we can see that the over-reliance on these chemical-leaning practices has severely damaged soils (Montgomery, 2017), created toxic background chemical levels, and depleted the biological resilience previously existing in many environments where farming had been practised better aligned with the naturally occurring biosphere. With this focus on maximizing production and more volume of sales, much less concern was focused on product quality and the broad impacts of conventional (read as 'rescue chemistry') agriculture. Because of the fertility-and-pest-control-for-purchase promoted by the agri-chemical industry, more, new and more-potent chemical inputs were accepted as a matter of course. Higher yields, immediate pest and weed control and a more-abundant food supply swept aside the consideration of negative criteria and impacts. The idea of rescue chemistry being available to ensure an abundant and reliable food supply has worked very well in developing the agri-chemical industrial complex while permitting further technological movement away from natural systems towards a more in-line-with-manufacturing-principles industry. The resulting evolution of the 'pesticide treadmill' (Van den Bosch, 1978) may have been an unintended consequence, although it has been a very effective driver of profitability for the agri-chemical industrial complex. Also, the indoctrination of the idea that poison residues are acceptable on and in food has been a necessary companion to support acceptance of commercial agriculture as it is generally practised today. This is, regardless of the altruistic-seeming rhetoric of 'feeding the world', clearly focused on profitability, which focus has resulted in the concentration and control of much of the world's food supply. Also, the fracturing and disenfranchising of the agricultural community has been a collateral phenomenon (Berry, 1978). When you have a handful of internationally operating corporations controlling seed, fertilizers, pesticides and herbicides, genetic provenance and land, the concentration of wealth and influence is obvious.

Market demand for organic goods may be met by aggregation of smallholder output through grower groups, brokers and distributors, and direct sales from very large operators. Of course, this is in reference to primary supply chain elements supporting processing. So, in working backwards from the market demand to the processors, brokers and distributors, and finally to the producers, we can see that, similar to present-day non-organic agriculture, possibilities exist for differing scales of operation at the producer level. And yet, consistent with the rise in total value of the organic sector, we are seeing more consolidation of ownership of business and land. The result is often larger mega-farms that offer a global supply to processors. Globalization, which may appear advantageous in regard to addressing production risk and cost, also has the practical impact of driving smaller commercial producers out of business while driving remaining operators closer to the 'conventional' industrialized farm models. This apart from the reality that fresh food global transportation amounts to a translocation of fresh water (on average, 80–96% of fruits and vegetables are water) (Fao.org, 1995) from poorer and often arid countries to wealthier countries with more abundant sources of fresh water.

Consider some of the elements that are observed in organic agriculture in the context of scale-appropriate methodologies with reference to conventional practices as well. Including 'conventional' agriculture is part of the discussion, as it has driven the manner in which the greatest volume of food and fibre production, and processing and distribution mechanisms, have developed. Conventional agriculture is keeping the attendant support infrastructures in place: equipment suppliers, material suppliers, utilities and services support, distributors, processors, retail market outlets and the like. All the while the farm gate/farmer's market persists in every culture. Many retailers are applying the imagery of 'local farm' products to greenwash customer perception and trade on romantic and nostalgic inclinations.

The phrase 'economy of scale', when discussing agricultural pursuits, most often is applied to agriculture practised over vast acreage using massive equipment and the least possible human labour. The most common inference or assertion is that bigger is better, and therefore mechanized is better, both in terms of acreage and equipment used. In the USA, the single most-expensive-over-time input cost is labour. This does not hold true in less-developed countries where labour is so cheap that products can be shipped around the world and still arrive in a developed country's marketplace below the local cost of production.

The fundamental premise in assertions of the benefits of considering 'economy of scale' is that the focal 'economy' is the financial economy. The inference is that financial return is the paramount issue. 'Cost-effective' is the term of convenience applied as the arbiter determining whether or not a practice is a preferred method. 'Return' is usually based upon the equation that considers immediate cost of operations and materials to the farmer against the return in dollars to the farmer; i.e. return on investment. The same economic model is applied to manufacturing and processing where inputs are purchased, labour and equipment via process alter the materials into marketable goods, and mass-distribution schemes deliver the products to the consumer. Revenue from sales minus costs (the cost of inputs, production and operations) equals the 'return'. None of which addresses the economy of sustainability on this planet; soil economy, biological economy, water economy, societal economy, energy economy or any of the other economies impacted directly and indirectly by the manner in which agriculture is practised. Neither is the relationship of economy to the ethos of an organic philosophy considered. Rural and farm community economy also suffers considerably and has contributed to the hollowing out of 'main street' USA over the past 40 years. And yet, cost and return fits hand-in-glove with the focus on maximizing production volume and the 'more is better'/'bigger is better' economy-of-scale paradigms currently existing in much of agriculture across the world. Even so, the loss of those 'knowing the land' and how to farm continues apace; the 'culture' part of agriculture, if you will.

Experience and observation during over 50 years in farming teaches that when you superimpose a manufacturing model over living systems, the living system will inevitably collapse without some type of rescue action that requires an infusion of external (usually externally purchased) resources. Most often these are seen as pesticides, herbicides, chemical fertilizers, antibiotics, growth supplements, metabolic stimulants, cheap labour and the like. Also, yield and nutritional quality are always impacted, generally negatively.

Production models involving monoculture agriculture hew to the line of a manufacturing mindset that focuses on and reduces the variables to the smallest set attendant to the production of one item as the most economically efficient practice. Such models transfer readily into production methods for livestock such as confined-animal feeding operations, where the focus is on maximizing production. Meeting the physiological needs of livestock and fowl are minimally addressed and are based upon what animals can tolerate rather than providing an environment in which they can thrive, all the while increasing toxic loads on soil, air and the food produced as a result of production; completely ignoring that nature and natural systems are underlying reality. This 'manufacture-over-living system' approach clearly disrupts any notion of a practical production equation reaching for biological balance, such as is inherent in the application of 'organic' ideology and ethos to guide production methodologies. Even so, the manufacturing model is the foundation template of many conventional agricultural production models. The rise of these production models is somewhat easily understood when one considers the fascination of maximizing production rather than optimizing production. When maximum-production models are overlaid on organic farm production models, doing so disrupts the opportunity to derive the value inherent in biologically well-balanced, largely self-sufficient and intensively managed operations, regardless of scale.

Witness the development of a marketplace that considers agricultural goods as commodities to be traded. I submit that putting the marketing of crude oil, gold, natural gas and copper on a par with agricultural products allows for facile equivalencies to be made. Consider that those commodities are generally the output of mining and are entirely extractive. Agriculture is the creation of new wealth by dint of managing a living system that derives from the sun and living ecologies, resulting in goods of value to man, which is broadly productive. Commodities are generally considered to be interchangeable goods of the same type. The trading of commodities engages in how financial value can be derived by those who are not involved in the production, processing or distribution of the commodities. The concentration of wealth has also impacted policy decision making driven by those concerned with profit and not by those engaged in production. This distancing of financial decision making from functional decision making further drives agriculture to grow operations to unsustainable sizes. And so, given the practical aspects of what is needed to trade in commodities, for example, it is not possible for a small coffee producer on one hectare in Honduras to gain, other than through direct sale of his beans. Yet coffee is a very lucrative commodity for those trading the aggregated work product of hundreds of thousands of smallholding growers (Rau, 2019).

Superficially, the economic face of the business of farming may look similar to other production and manufacturing. The differences are profound and not often considered when positing equivalency with non-agricultural production models. This lack of consideration and oversimplification has allowed policymakers to elect to support larger 'more-efficient' operations over smaller operations. Adding in the shift in focus away from individualized custom production to an assembly-line mindset constitutes a further driving force that has moved farm production away from quality towards quantity; all in the stated service of 'feeding the world'.

This raises several questions:

- Is cost of production an acceptable reason alone to drive decision making around how organic agriculture is practised? This question was answered resoundingly after the publication of organic standards in the USA when challenged by special interest groups to permit conventional feed in organic poultry production. The answer is 'No' (CFR, n.d.).

- Is simply substituting 'organic' inputs and ingredients into a standard commercial production model that meets the *letter* of the rule enough to make a product 'organic', even if it only marginally considers, or does not consider, the motivating aspects of the organic movement? The answer is 'Yes' – there is no enabling requirement that a certified organic entity agrees to or supports a philosophical basis for meeting organic standards; and 'No' – considering the foundation principles[1] of organic agriculture when adhering to those foundational tenets of the organic movement. Considering the four principles of organic agriculture as articulated, it is not clear that simply meeting the letter of the rules addresses the principles of health, ecology, fairness and care.
- Is it OK to produce goods as 'organic' just because the regulatory requirements are met, even when doing so in an industrial agricultural model? This is, to some degree, the question that is driving the emergence of the 'regenerative organic agriculture' movement[2]. As distinct from the existing manufacturing, industrial and agricultural input-based models, more holistic models consciously consider each of the principles stated above as guidance for 'organic' compliance to the letter and in the spirit as conceived.

Optimizing production seeks to provide for a best-case scenario of productivity over time considering all of the inputs (labour, materials, land, water etc.) against all of the return values and outcomes (nutrition, yield, environmental quality, water quality, quality of life, elevating above poverty etc.) and factoring in all costs.

Maximizing productivity looks to produce the greatest marketable volume at the lowest immediate financial cost.

Where Does Scale Enter the Discussion?

If we are speaking only of the financial economy, in the short term, scale-appropriate in the USA is presumed to mean size (acreage and revenue) only. In 1920, the US government began the national census, at which time 30.2% of the population were considered 'farm people'. The earliest data that seem reliable are from 1820, which reported 72% of the population engaged in 'farm occupations'. By 1850, that number had shrunk to 64%. Today, based on the most recent US census data available, 10.5% of US *employment* (my emphasis), including full- and part-time, were related to 'agricultural and food' sectors. Also, farm and ranch 'families' comprise less than 2% of the population (in 2022).

Clearly, in the USA, there is and has been an emigration from the farm as production technologies have been adopted and farm character has shifted. Notable agriculture-related catastrophes have added impetus to the movement off-farm[3]. The political lens reveals a policy shift that has further contributed to driving the emigration. In 1973, the then Secretary of Agriculture Earl Butz called on American farmers to plant 'fencerow to fencerow' and uttered the now infamous 'get big or get out' statement. In 1940, the average farm size in the USA was 174 acres. In 1973 there were 2.823 million farms in the USA averaging 385 acres. In 2022 there were 1,172,770 farms reported with an average of 446 acres. This is down from a peak of 6.8 million farms in 1935. In addition, total farm acreage was less in 2022, at 893 million acres, down from 915 million acres in 2012 (Usda.gov, 2017).

In Minnesota and Wisconsin, the average organic row crop farm was 325 acres (131.6 ha) based on 2021/22 reports (University of Minnesota College of Food, Agriculture and Natural Resource Sciences, n.d.).

In the European Union, the average organic farm is 41 ha (101.3 acres) based on 2020 reports (European Commission, 2023). In South Africa the average organic farm is 100 ha (247 acres), while a separate study identified that there are two general models: those who are smallholder subsistence farmers selling surplus organic goods locally, and larger third-party certified farms selling into export and large retail markets.

In Thailand, in 2019, the average size of a licensed organic farm was 1.9 ha (4.7 acres) with smallholder farms comprising 85,000 ha.

These disparate examples help to show the breadth of scale in organic farming. This is not unique to organic farms. Based on data available between 1960 and 2000, smallholding farms (< 2 ha/5 acres) and family farms comprise 87% of the world's farm area (Lowder *et al.*, 2016).

With a world population anticipated to reach 9.7 billion by 2050 and an urban population comprising 56.9% of the world's population in 2022, how those peoples will be fed will also drive the development and scale of agricultural operations.

Also, who has control of the food supply will have a profound impact on all of human society.

Continuing for a moment to look through a financial lens, that average farm 100 years ago evolved to the farm of today to fit a market logistical need that can be served by larger operations selling to brokers or processors seeking large, consistent supplies of comparatively inexpensive raw materials. This production/distribution model developed to support large urban populations distant from the farms that produce the food which sustains them.

In only the last 50 years, the corner grocer or produce stand, the local butcher and baker, or the miller has gone by the wayside, supplanted by large aggregators that buy in bulk and rely upon a distribution network to deliver farm products, most often in the form of processed goods, to local sales outlets.

Typically, such operations are supplied by producers of grain and feed materials, dairy and livestock raw materials, vegetables and fruits, all of which are supplying commodity levels of production. Few 'average'-sized farms are close enough to population centres for the goods they produce to be directly consumed as food or fibre by end-users; farmer-direct markets being an exception. And, due to the inexpensive commoditized farm models, these local farms are in direct financial competition with the larger industrialized operations. All the while, any local expertise and capacity for small processing of agricultural raw materials into finished food and goods is largely expiring as older generations with practical knowledge pass away with no one to pass on the knowledge to, and as younger generations strive to function in the industrialized world.

Speaking of locally owned farms/smallholdings, we must acknowledge that there are massive corporate plantations across the world, albeit mostly in tropical regions. These have in some cases also moved into organic production and broken through the boundaries often intrinsic in local regional agriculture.

Many commodity-level producers sell to brokers that ship internationally (coffee, tropical nuts, corn, soy, wheat, fruit, juice concentrates, sugar etc.). This impacts the scale of logistics in material movement. Logistics considerations in these cases have been assessed in 'food miles'. Logistics constitute a 'hidden to the consumer/real for the planet' cost when considering scale of operations. Even so, distribution has driven much of the consolidation as consumer buying preferences are opportunistically manipulated by clever marketing. This raises another question; for example: Is it possible for a grape grown on an organic farm in Chile to be shipped to a North American market (+/– 6000 miles) and really meet the core values of the organic movement? Is it possible for a head of lettuce to be produced on a farm in the Central Valley in California and be shipped to the east coast (+/– 3000 miles) and really meet the core values of the organic movement? What about organic coconut oil produced in Fiji shipped to Australia and processed into a personal care product that is sold in Europe?

Large-scale 'organic' farms that are copying the conventional-commodity-farm large-scale models by simply substituting allowed and restricted materials and necessary organic practices to maintain certification compliance share many of the same issues that do not entirely fit within the philosophy and ethos of the organic movement yet meet the letter of the law and regulations, or cleverly skirt along the fringes.

When considering plants: spacing, nutritional needs, water needs, sun requirements, length of growth period (annual and/or perennial), planting-cultivation-harvest labour, soil condition-health-biological activity, soil retention, soil building and inputs applied to sustain and enhance productive resilience comprise elements of the environmental economy of an operation. The foundational aspect of nutrient/resource recycling in organic systems is absent in much of 'conventional' agriculture and considered a focus in organic systems.

In the context of animals, in addition to those foci listed above: species-appropriate spacing, multi-species or individual species operations, cost of young stock, impact of socialization, feed inputs, time to market size, (broadly) physiological needs, land needs, impact on soil-water-air quality, and manure management need be considered. Also, logistics of live shipping, processing and product distribution need to be considered. And again, in all such instances, optimum versus maximum is a driving focus. Purchased inputs on large farms are allowed and yet do not fully meet the philosophy of developing and expanding on-farm sustainability.

Background ecological impacts are less-often considered. Operator and staff health and wellness are rarely considered. Local community health is rarely considered. Overall farm ecology, overall community ecology, and whole-community economy come into play as well, and are not considered in the regulations defining 'organic' production. Applying the above considerations in each of the noted farm models below may prove thought-provoking.

Among traditional methods, differing iterations of 'farming' today may include:

- an individual working and planting with minimal powered equipment (one person operation);
- several people working and planting with minimal powered equipment;
- individuals or groups using draft animals for power;
- individuals or groups using animal traction with some auxiliary powered equipment;
- individuals managing operations using modern traction equipment and tool sets;
- individuals managing large operations and staff using modern apparatus and a full tool suite;
- operations over vast acreage using state-of-the-art technology.

In each case, the scale of operations must match market demand to support financial success. And in each case the market must support the operations production scale.

Yet, considering simple size-to-size comparisons, the head lands (ends of fields where equipment turns around) alone on a large industrial-style organic farm may comprise the equivalent of a smallholding farm's entire acreage. Direct observation over 37 years reveals that the most comprehensively efficient agriculture is practised at a small and human scale supported with animal traction, powered equipment, or some combination of both. This style of farming may best approach the organic ideal, and it is often impractical in many areas to consider relying upon such farms to feed large cities. Not that the possibility of sheer productive volume would be insufficient; rather that the logistics and mechanics of getting the food to people in a form that suits the preparation capacities and skills available in population centres is impractical as an immediate replacement model for the current food distribution system.

Conclusions

Scale is one lens with which to consider the evolution of modern-day 'organic' agriculture. We can acknowledge that not all expressions of organic agriculture meet the ideal. It seems prudent to not let the ideal be the enemy of good. And yet, if organic is a standard to which many producers aspire, is it not valid for that standard to acknowledge and address the founding tenets of the organic movement? And better yet, if not working towards an ideal, is it not best to be clear about the ideal?

It is unlikely, given the mechanisms that have developed over the last 50 years, to feed large populations; and given the proliferation of regulations over the past 30 years, that all organic operations will approach the ideal.

I see no justifiable reason not to work to that goal. Farming teaches patience, persistence and the merits of long-sighted effort.

The ultimate expression of hope is to put a seed in soil and expect it to grow.

Notes

[1] The precautionary principle of 'first do no harm' and, as articulated by IFOAM, 'The Principles of Health, Ecology, Fairness, and Care are the roots from which organic agriculture grows and develops. They express the contribution that organic agriculture can make to the world, and a vision to improve all agriculture in a global context' (https://www.ifoam.bio/why-organic/shaping-agriculture/four-principles-organic#:~:text=The%20Principles%20of%20Health%2C%20Ecology,agriculture%20in%20a%20global%20context [accessed 23/1/25]). The maintenance of the fertility of the soil is the first condition of any permanent system of agriculture. In the ordinary processes of crop production, fertility is steadily lost: its continuous restoration by means of manuring and soil management is therefore imperative (Sir Albert Howard, 'An agricultural testament'. Available at: https://journeytoforever.org/farm_library/howardAT/AT1.html (accessed 23/1/25).

[2] Regenerative agriculture refers to the recycling of resources on farms and has already been taken up by many conventional marketers. The Rodale Institute has begun an add-on certification that addresses many of the basic tenets of the organic movement. Available at: https://realorganicproject.org/real-organic-farming/real-organic-difference/ (accessed 23/1/25).

[3] Apart from droughts, floods, storms and other natural disasters, man-made agricultural disasters, most spectacularly the Dustbowl, the ecological disaster from DDT, etc. Available at: https://drought.unl.edu/dustbowl/; https://www1.udel.edu/chem/C465/senior/fall97/insecticide/environ.html (accessed 23/1/25).

References

Berry, W. (1978) *The Unsettling of America*. Random House, New York.

CFR (n.d.) Allowed and prohibited substances, methods, and ingredients in organic production and handling. Code of Federal Regulations. Available at: https://www.ecfr.gov/current/title-7/subtitle-B/chapter-I/subchapter-M/part-205?toc=1 (accessed 23 January 2025).

Encyclopedia Britannica (2009) Green revolution. Available at: https://www.britannica.com/event/green-revolution (accessed 23 January 2025).

European Commission (2023) Organic farming: A decade of organic growth. Agricultural Market Brief. European Commission. Available at: https://agriculture.ec.europa.eu/system/files/2023-04/agri-market-brief-20-organic-farming-eu_en.pdf (accessed 23 January 2025).

FAO (2022) UN report: Global hunger numbers rose to as many as 828 million in 2021. Available at: https://www.fao.org/newsroom/detail/un-report-global-hunger-SOFI-2022-FAO/en (accessed 24 January 2025).

Fao.org (1995) Fruit and vegetable processing – contents. Available at: https://www.fao.org/4/V5030E/V5030E00.htm (accessed 23 January 2025).

Fb.org (2018) Fast facts about agriculture – the voice of agriculture. American Farm Bureau Federation. Available at: https://www.fb.org/newsroom/fast-facts (accessed 23 January 2025).

Lowder, S.K., Skoet, J. and Raney, T. (2016) The number, size, and distribution of farms, smallholder farms, and family farms worldwide. *World Development* 87(1), 16–29. Available at: https://www.sciencedirect.com/science/article/pii/S0305750X15002703

MarketsandMarkets (2011) MarketsandMarkets: Global organic food and beverages market worth $104.5 billion by 2015. Prnewswire.Com. Available at: https://www.prnewswire.com/news-releases/marketsandmarkets-global-organic-food-and-beverages-market-worth--10450-billion-by-2015-116804058.html (accessed 4 November 2024).

Montgomery, D.R. (2017) *Growing a Revolution: Bringing Our Soil Back to Life*. W.W. Norton & Company, New York.

National Institute of Food and Agriculture (n.d.) Organic agriculture program. Available at: https://www.nifa.usda.gov/grants/programs/organic-agriculture-program (accessed 23 January 2025).

Rau, B. (2019) Colombian coffee production stabilizes around 14 million bags. USDA Foreign Agricultural Service. Available at: https://apps.fas.usda.gov/newgainapi/api/report/downloadreportbyfilename?filename=Coffee%20Annual_Bogota_Colombia_5-15-2019.pdf (accessed 23 January 2025).

ReportLinker (2023) Organic farming global market report 2023. GlobeNewswire Newsroom. Available at: https://www.globenewswire.com/news-release/2023/02/24/2615424/0/en/Organic-Farming-Global-Market-Report-2023.html (accessed 23 January 2025).

Shields, M. (2017) ChemChina clinches landmark $43 billion takeover of Syngenta. Reuters. Available at: https://www.reuters.com/article/idUSKBN1810CM (accessed 23 January 2025).

Skorbiansky, S. (2023) USDAERS – Organic agriculture. Available at: https://www.ers.usda.gov/topics/natural-resources-environment/organic-agriculture (accessed 23 January 2025).

Sullivan, B.D. (2017) Mega-mergers in agribusiness raise concerns about food costs, biodiversity. USA Today. Available at: https://www.usatoday.com/story/news/politics/2017/08/29/mega-mergers-agribusiness-raise-concerns-food-costs-biodiversity/558980001/ (accessed 23 January 2025).

University of Minnesota College of Food, Agriculture and Natural Resource Sciences (n.d.) Organic production proves profitable despite smaller sizes and inflation effects in Upper Midwest. Available at: https://cfans.umn.edu/news/organic-production-profitable (accessed 23 January 2025).

Usda.gov (2017) Number of U.S. farms continues to decline, but farm size grows slightly. Available at: https://www.ers.usda.gov/data-products/chart-gallery/gallery/chart-detail/?chartId=106385 (accessed 23 January 2025).

Van den Bosch, R. (1978) *The Pesticide Conspiracy*. Doubleday Books, New York.

10 Cultural Practices for Sustainable Organic Farming

Raya J. Amara[1]*, Wilson C. Wilson[2], Lucas T. Manda[3], Nyandula S. Mwaijande[4], Janet F. Maro[5], Yohana C. Haule[5] and Abdul J. Shango[6]

[1]Tanzania Agricultural Research Institute (TARI), Ukiriguru Centre, Mwanza, Tanzania; [2]Tanzania Livestock Research Institute (TALIRI), Mbeya, Tanzania; [3]CARE-WWF Alliance, Iringa, Tanzania; [4]Department of Crop Science and Horticulture, Sokoine University of Agriculture, Morogoro, Tanzania; [5]Sustainable Agriculture Tanzania (SAT), Morogoro, Tanzania; [6]World Vegetable Center, Eastern and Southern Africa, Arusha, Tanzania

Abstract

Organic farming is famously known as a type of farming that discourages the use of chemical pesticides and fertilizers for sustainable well-being. It has been accompanied by strict principles, certifications and standardization, especially for exportation. The basis of certification of organic farming is its cultural practices or best practices as per the adopted standards. Many studies have assessed the level of adoption and performance of specific cultural practices against abiotic and biotic stresses. The current study assessed the sustainability of organic cultural practices from the farmers' perspective. The emphasis was on getting farmers' views on how they understand and apply organic cultural practices for sustainability and resilience to any changes in climate on their farms.

Introduction

Organic agriculture is defined by IFOAM – Organics International as a production system that sustains soil health, ecosystems and people. It relies on ecological processes, biodiversity, and cycles adapted to local conditions, rather than the use of inputs with adverse effects. Organic agriculture combines tradition, innovation and science to benefit the shared environment and promote fair relationships and good quality of life for all involved (IFOAM, 2014). Globally, organic agriculture is currently practised in 190 countries with coverage of 76.4 million ha grown by about 3.4 million farmers. The leading three countries in terms of organic agricultural land are Australia (35.7 million ha), Argentina (3.63 million ha) and Uruguay (2.7 million ha). Likewise, India has the highest number of producers (1,599,010) followed by Ethiopia and Tanzania with 219,566 and 148,607, respectively (Willer *et al.*, 2022). Besides, organic food and drink sales reached more than €120 billion in 2020 with the USA (€49.5 billion), Germany (€15.0 billion) and France (€12.7 billion) being the largest organic markets (Willer *et al.*, 2020).

Tanzania has a total area of 198,226 ha for organic agriculture distributed as follows: cotton (154,495 ha), coffee (24,785 ha), cocoa (12,630 ha), oilseeds (3116 ha), tropical and subtropical fruits (1336 ha), vegetables (1080 ha) and cereals (100 ha). Tanzania is the largest organic cotton producer in Africa in terms of total cultivated hectares followed by Uganda with almost 40,000 hectares. The country takes the second lead on the number

*Corresponding author: rayanundu8@gmail.com

DOI: 10.1079/9781800626850.0010

of organic producers in Africa after Ethiopia with almost 220,000 producers (Willer *et al.*, 2022). Organic agriculture in Tanzania is well structured, especially for exportation of the certified organic produce exported by national, regional and international organizations or companies (Bakewell-Stone *et al.*, 2008). All national organic firms or organizations are under the umbrella of the Tanzania Organic Agriculture Movement (TOAM) which oversees all organic movements in the country. TOAM partners with African Organic Network (AfrONet) and IFOAM – Organics International to implement various projects. Some of the projects include Knowledge Centre for Organic Agriculture in Africa (KCOA) and Innovative Institutions for Ecological Organic Agriculture in Africa (IIABA). KCOA provides support for organic agriculture via knowledge management, dissemination and capacity building, and market systems development while ILABA implement institutional innovations that will scale up and out ecological and organic agriculture.

Despite the significant growth and achievements of organic agriculture in Tanzania and the entire world, there are various limitations associated with cultural organic practices. The term 'cultural practices' has been referred to as 'good agricultural practices', 'good agronomic practices', 'best practices' and, recently, 'ecological organic agriculture (EOA) technologies', which is the basis for standardization in organic farming (Soaan, 2013; Shango *et al.*, 2022). Organic standards must be in line with IFOAM Standards Requirements or Common Objectives and Requirements of Organic Standards (COROS), which replaced the IFOAM Basic Standards (IBS), IFOAM Standard for Organic Production and Processing, and IFOAM Accreditation Requirements for Bodies Certifying Organic Production and Processing (formerly IFOAM Accreditation Criteria). All the above-mentioned three standard documents form the IFOAM Norms, which is the basis of the Organic Guarantee System of IFOAM – Organics International for approval of certification bodies under the IFOAM Accreditation Program. The IFOAM Norms are embedded in the Principles of Organic Agriculture developed by IFOAM – Organics International, which include health, ecology, fairness and care (IFOAM, 2019). Most studies have reported on the research and validation of various cultural practices for organic production. Some of them include mulching, biocontrol, organic fertilizers, cropping sequence, storage facility, soil fertility and water management in fruit vegetables (Mbega *et al.*, 2012; Mtui *et al.*, 2015; Shango *et al.*, 2022).

The sustainability of agricultural systems has been given high priority especially to enhance climate resilience. Many approaches have emerged including organic agriculture, agroecology, sustainable agriculture, climate-smart agriculture and, recently, regenerative agriculture. All the above approaches are identified by the composition of their cultural practices. Several studies have focused generally on the entire approaches (Mason, 2003; Lipper *et al.*, 2014; Meng *et al.*, 2017; Rosset and Miguel, 2017; Giller *et al.*, 2021), but few studies have been conducted to address the cultural practices. The current study aims at exploring the cultural practices for sustainable organic farming. The approach was to study how smallholder farmers understand and practise sustainable organic cultural practices. In this study, a total of 84 organic farmers organized under the umbrella of Sustainable Agriculture Tanzania (SAT) were interviewed. The questions focused on cultural practices under organic crop production and animal husbandry.

Materials and Methods

Study sites

Questionnaire surveys were conducted in Morogoro and Mvomero districts of the Morogoro Region, Tanzania, in October 2022. A total of nine villages (five villages in three wards of Morogoro district and four villages in four wards of Mvomero district) were purposively selected based on their importance in organic farming (see Appendix).

Sample size and selection

A total of 84 organic farmers organized under the umbrella of Sustainable Agriculture Tanzania (SAT) were purposively selected for the survey based on their current farming practices, experience and conversion to organic farming.

Survey methods and analysis

Data from the survey of farmers was collected via Android devices using structured questionnaires that were subsequently coded in KoboCollect kit. There was a wide range of themes including socio-economics, crop production, animal husbandry, biodiversity management, marketing, post-harvest handling, marketing, processing and climate resilience. The final data collected were transmitted to an online cloud server. This computer-aided personal interviewing (CAPI) was selected over a paper interview format in order to improve data quality and reduce the time spent in the post-collection data processing. R statistical software, version 4.1.1 was used to process the range of basic descriptive statistics (R Core Team, 2022).

Results and Discussion

Socio-economics

The findings (Table 10.1) revealed that the majority of interviewed farmers (90%) practise organic farming and were predominantly female farmers (60%) in the age ranging 30–60 (89%). Most of them were married (79%). The majority of the interviewed farmers (61%) have knowledge of agriculture and farming, and have their own land ranging from less than 1 ha to 10 ha (82%), while the portion of their land allocated for organic farming ranges from 1 ha to 3 ha (56%). Furthermore, many (67%) organic farmers shifted from conventional to organic farming in between one to five years.

The farmers' socio-economic status is an important aspect for the assessment of the sustainability of cultural practices. Activity which involves more women has a high chance of being sustainable because of their higher rate of adaptation and adoption compared to men. This is confirmed by studies that revealed a higher level of adaptation and adoption in female-headed households than in male-headed households (Ndambiri *et al.*, 2013; Kurgat *et al.*, 2020). On the other hand, the low

Table 10.1a. Socio-economic particulars of interviewed farmers in the Morogoro region.

Variable	Frequency	Percentage
Gender		
Female	50	60
Male	34	40
Age range		
≤ 30 years	12	14
31–40 years	22	26
41–50 years	23	27
51–60 years	18	21
> 60 years	9	11
Marital status		
Divorced	1	1
Married	66	79
Separated	3	4
Single	6	7
Widowed	8	10
Courses in agriculture		
Short course	41	49
No	33	39
SAT training	10	12
Land ownership		
All borrowed	1	1
All hired	2	2
Own land	69	82
Part is hired and part is borrowed	2	2
Part is owned and part is hired	6	7
Other (family land, inherited)	4	5

Table 10.1b. Farming practices in the Morogoro region.

Variable	Frequency	Percentage
Type of farming		
Organic	82	90.10
Conventional	6	6.60
Conservation	3	3.30
Conversion to organic		
≤ 1 hectare	4	5
≤ 5 hectares	56	67
≤ 10 hectares	13	15
> 10 hectares	8	10
None	3	4
Organic farming land		
1–3 hectares	47	56
3–5 hectares	6	7
5–10 hectares	8	10
Role in farming		
Farmer (joint owner)	5	6
Farmer (sole owner)	61	73
Farmer (spouse)	15	18
Other (family farm, group chairperson, only livestock keeping)	3	4

involvement of youths under age 30 in organic farming activities is a sign of unsustainability. The existence of a large number of farmers who have undergone training in agriculture and own their own land is another indicator of sustainability (Ndambiri *et al.*, 2013). This is because the trained farmers will succeed in their organic farming and influence others to adopt after noticing the success of their peers, and it is easy to decide to grow organic for who owns his/her land. This current study also revealed that organic farming can be conducted amply by small-scale farmers who normally own 1–3 ha of land (Sarris *et al.*, 2006). Hence, farmers who have less than one hectare of land can hire other land for organic farming, which indicates that organic farming is preferred by many farmers.

Organic Crop Production and Animal Husbandry

The results (Table 10.2) indicate that organic crop production is practised by more (57.8%) of the interviewed farmers than organic animal husbandry (39.7%). Under organic crop production, most of the interviewed farmers grow horticultural crops (49.5%) followed by other crops (cereal crops – maize and paddy, banana, grass, spices, seasonal crops) (21.2%). On the other hand, animal husbandry was dominated more by poultry, sheep/goats and dairy cattle (92.45%) than other products such as beef, pigs and fish. The majority of farmers use conservation tillage (72.1%), few (10.3%) practise no-tillage and conventional tillage (17.6%). There are many ways employed by organic farmers to improve soil fertility; some of these practices includes use of compost, green manure, crop residues, and on-farm and off-farm manure. Other techniques used by the minority to improve soil fertility include intercropping, crop rotation, soil amendments, inorganic fertilizers and other practices like fencing, terraces, mulching and agroforestry trees. Of the interviewed farmers, 39% of them reported their crops to be affected by grass weeds, broad leaves and parasitic weeds. Manual weeding and botanical herbicides are the most frequently used methods to control weeds. The most rarely practised methods of weed control are crop rotation/intercropping, burning of plant residues, grazing through animals and chemical herbicides.

The practice of either sole organic crop production or sole animal husbandry, instead of integrating

Table 10.2. List of cultural activities under crop and animal production.

Variable	Frequency	Response (%)	Cases (%)
Kind of organic farming practised			
Crop production	67	57.8	79.8
Animal husbandry	46	39.7	54.8
Aquaculture	3	2.6	3.6
Type of crops grown			
Legumes	10	10.1	11.9
Roots and tubers (cassava, potato, yam)	7	7.1	8.3
Horticulture (vegetables, fruits, spices, flowers)	49	49.5	58.3
Annual cash crops (cotton, sunflower)	9	9.1	10.7
Perennial cash crops (coffee, tea, cashew nuts)	3	3.0	3.6
Other (cereal crops (maize, paddy), fruits (banana), grass, spices, seasonal crops)	21	21.2	25.0
Type of animals raised			
Dairy	21	22.8	25.0
Beef	4	4.3	4.8
Poultry	33	35.9	39.3
Pigs	5	5.4	6.0
Sheep/goats	26	28.3	31.0
Fish	0	0.0	0.0
Other	3	3.3	3.6
Type of tillage methods			
No-tillage	7	10.3	8.3
Reduced tillage/non-inverted tillage (conservation tillage)	49	72.1	58.3
Inverted tillage/moldboard tillage (conventional tillage)	12	17.6	14.3
Soil fertility improvement techniques			
Crop rotation	7	3.8	8.3
On-farm manure	23	12.5	27.4
Off-farm manure	19	10.3	22.6
Green manure	36	19.6	42.9
Incorporation of crop residues	29	15.8	34.5
Compost	44	23.9	52.4
Soil amendments	5	2.7	6.0
Intercropping	12	6.5	14.3
Chemical/inorganic fertilizers	2	1.1	2.4
Others (fencing, terraces, mulching & planting agroforestry trees)	7	3.8	8.3
Crops affected by weeds			
No	51	61	
Yes	33	39	
Total	84	100	
Type of weeds			
Broad leaves	8	21.1	9.5
Grasses	27	71.1	32.1
Parasitic weeds	3	7.9	3.6
Others	0	0.0	0.0

Continued

Table 10.2. Continued

Variable	Frequency	Response (%)	Cases (%)
Total		100	45
Weed control methods			
Mechanical weeding, tillage, mowing, manual	22	61.1	26.2
Crop rotation intercropping	1	2.8	1.2
Burning of plant residues	1	2.8	1.2
Grazing through animals	1	2.8	1.2
Botanical herbicides	8	22.2	9.5
Chemical herbicides	2	5.6	2.4
Others	1	2.8	1.2

Table 10.3. Harvesting, post-harvest handling, and marketing of organic produce.

Variable	Frequency	Responses (%)	Cases (%)
Changes in yield after conversion			
Decreased	9	11	
Increased	47	56	
No change	28	33	
Immediate action to produce			
Processed	2	2	2
Stored	31	34	37
Immediate sale	58	64	69
Why produce is immediately sold			
Prices are good	6	9	7
Expensive processing facilities	6	9	7
Expensive storage facilities	17	25	20
Pests	7	10	8
Space	2	3	2
Others	31	45	37

both, confirms low level of sustainability of the system. This is because both crops and animals depend on each other. For instance, organic crop production depends on animal manure; likewise, animals feed on crop residues. Moreover, diversification of both crops and animals is one of the coping strategies for climate change and thereby sustainability of the system (Nambiar, 2001; Valet and Ozier-Lafontaine, 2014). The involvement of many farmers in organic horticulture could be due to the economic viability of the sector compared to other crops. On the other hand, it is a potential opportunity for other crops that do not require more intensive care and resources. The rearing of poultry, sheep/goats and dairy cattle indicates how the animal sector is diverse and promotes climate resilience through the provision of various products (Nambiar, 2001).

Harvesting, Handling and Marketing

The results (Table 10.3) show that most (56%) of the interviewed farmers reported an increase in their produce yields after converting from conventional to organic farming. Some (33%)

Table 10.4. Climate shocks and their effects.

Variable	Frequency	Responses (%)	Cases (%)
Climate shock experienced in the past 5 years			
No	8	10	
Yes	76	90	
Type(s) of climate shock(s)			
Drought	69	82	
Erratic rainfall pattern	6	7	
Flood	1	1	
None	8	10	
Specify others (decline in production, mostly the affected crops are legumes)			
Effects of climate shock(s)			
Loss of entire crop	18	9.9	21.4
Food price increase	15	8.2	17.9
Loss of income	32	17.6	38.1
Food shortage insecurity	41	22.5	48.8
Decline in crop yield	57	31.3	67.9
Death of animals	16	8.8	19.0
Others	3	1.6	3.6

farmers experienced no change while others (11%) reported a decrease in yield. A large amount of produce (64%) is sold immediately after harvesting; some (34%) is stored and a smaller amount (2%) is processed. The interviewed farmers mentioned that expensive storage facilities (25%), pests (10%), good prices (9%), expensive processing facilities (9%) and space (3%) are among the possible reasons why a large amount of produce is sold immediately after harvesting. Despite the fact that organic farming has proved to be beneficial due to high yield compared to conventional farming, it is still dominated by selling unprocessed produce due to expensive processing facilities. Improvement of post-harvest and processing units in organic farms may increase the supply of products during off-season, add value to the products and reduce the post-harvest losses due to pests (Sarma *et al.*, 2022).

Climate Shocks and Climate-Resilience Practices

The findings (Table 10.4) indicate that a large proportion (90%) of the interviewed farmers reported having experienced climate shocks in the last two years. Most (82%) reported being affected by drought; few farmers experienced a decline in production (10%), erratic rainfall (7%) and floods (1%). The effects of the above-mentioned climate shocks on organic farming and household food security included a decline in crop yield (31.3%), food shortage (22.5%) and loss of income (17.6%). Other effects of climate shocks mentioned by farmers included the loss of entire crops (9.9%), death of animals (8.8%) and increase in food prices (8.2%). On the other hand, farmers reported various climate-resilience practices used as adaptation and coping strategies including: harvesting water for irrigation (23.1%), others (19.7%) – listed in Table 10.5 – boreholes and wells (13.7%) and insurance (11.1%). On the other hand, rarely used strategies include growing drought-tolerant varieties (4.3%), diversifying both crops and animals (3.4%), changing from crops to animal husbandry (3.4%), increasing land under crops (0.9%) and changing planting dates (0.9%).

Climate shocks and their effects on organic crop production and animal husbandry reported in two years are in agreement with Lotter (2015). Most of the adaptation and coping strategies mentioned by

Table 10.5. Adaptation and coping strategies for climate-resilience organic farming.

Adaptation and coping strategies	Frequency	Responses (%)	Cases (%)
No challenge	7	6.0	8.3
Grow drought-tolerant variety	5	4.3	6.0
Resistant/tolerant animal breeds	2	1.7	2.4
Changed to other crops	4	3.4	4.8
Changed from crops to animal husbandry	4	3.4	4.8
Changed from livestock to crops	0	0.0	0.0
Diversification (both crops and animals)	4	3.4	4.8
Buy insurance	13	11.1	15.5
Water harvesting to irrigate	27	23.1	32.1
Soil conservation	10	8.5	11.9
Use bore holes and wells	16	13.7	19.0
Increase land under crops	1	0.9	1.2
Change planting dates	1	0.9	1.2
Others (buying water from another village for livestock use, buying water from water vendors, change from farming to business, cultivate more, add more energy, simply wait, irrigation, shading some crops to avoid extreme sunlight, shading the crops by banana leaves)	23	19.7	27.4

Table 10.6. Challenges affecting organic farming.

Challenges	Frequency	Responses (%)	Cases (%)
None	6	5.3	7.1
High cost of production	14	12.3	16.7
Low yield	5	4.4	6.0
High weed pressure	3	2.6	3.6
Insect pests and diseases	41	36.0	48.8
Lack of suitable market	17	14.9	20.2
Standards and regulations of organic products	7	6.1	8.3
Others (botanicals preparation is time consuming and needs more physical utilization, climate change, insufficient supply of water, drought, lack of organic inputs, lack of transport facilities to the markets to avoid contamination, diseases, lack of organic seeds, treatment using botanicals takes long time, labour, insufficient animal feed, organic farming education, transportation costs to the market, lack of organic fertilizers and pesticides, lack of organic inputs, lack of equipment for manufacturing botanicals	21	18.4	25.0

farmers are cultural practices that do not require much cost and, when integrated, reduce the magnitude of climate change effects (Bakewell-Stone *et al.*, 2008). These practices, therefore, reveal the sustainability and resilience of organic farming for sustainable development.

Challenges Affecting Organic Farming

The results (Table 10.6) indicate that, although some farmers (5.3%) reported no challenges, there is a long list of challenges of organic farming dominated by insect pests and diseases (36%), others – long list in Table 10.6 (18.4%) – lack of suitable market (14.9%) and high cost of production (12.3%). Organic farming will be more sustainable and resilient to climate threats once mutual efforts of farmers, stakeholders, governmental and non-governmental organizations including policymakers will be incorporated to combat the above-mentioned challenges. Likewise, continued applications of the reported organic cultural practices and coping strategies will increase productivity and sustainability of organic farming (Lotter, 2015).

Conclusions

More efforts should be made to promote the adoption and adaptation of organic farming and climate shocks, respectively, to emerging youths in agriculture. Model of creating more awareness to youths in primary schools and secondary schools, and to street youths, may be appropriate. Achievable plans of combating the identified challenges are vital. More research on the proper ways of addressing the challenges and better coping strategies for climate threats will increase sustainability of organic farming.

References

Bakewell-Stone, P., Lieblein, G. and Francis, C. (2008) Potentials for organic agriculture to sustain livelihoods in Tanzania. *International Journal of Agricultural Sustainability* 6(1), 22–36.

Giller, K.E., Hijbeek, R., Andersson, J.A. and Sumberg, J. (2021) Regenerative agriculture: An agronomic perspective. *Outlook on Agriculture* 50(1), 13–25.

IFOAM (2014) The IFOAM Norms for organic production and processing. IFOAM, Germany. Available at: www.ifoam.org (accessed October 2022).

IFOAM (2019) Plant reproductive material in the new organic regulation (EU) 2018/848. Available at: https://www.organicseurope.bio/content/uploads/2020/06/ifoam_eu_policy_seed_positionpaper_20190529.pdf?dd (accessed 27 January 2025).

Kurgat, B.K., Lamanna, C., Kimaro, A., Namoi, N., Manda, L. *et al.* (2020) Adoption of climate-smart agriculture technologies in Tanzania. *Frontiers in Sustainable Food Systems* 4, 55.

Lipper, L., Thornton, P., Campbell, B.M., Baedeker, T., Braimoh, A. *et al.* (2014) Climate-smart agriculture for food security. *Nature Climate Change* 4(12), 1068–1072.

Lotter, D. (2015) Facing food insecurity in Africa: Why, after 30 years of work in organic agriculture, I am promoting the use of synthetic fertilizers and herbicides in small-scale staple crop production. *Agriculture and Human Values* 32(1), 111–118.

Mason, J. (2003) *Sustainable Agriculture*. Landlinks Press.

Mbega, E.R., Mortensen, C.N., Mabagala, R.B. and Wulff, E.G. (2012) The effect of plant extracts as seed treatments to control bacterial leaf spot of tomato in Tanzania. *Journal of General Plant Pathology* 78(4), 277–286.

Meng, F., Qiao, Y., Wu, W., Smith, P. and Scott, S. (2017) Environmental impacts and production performances of organic agriculture in China: A monetary valuation. *Journal of Environmental Management* 188, 49–57.

Mtui, H.D., Maerere, A.P., Sibuga, K.P. and Bennett, M.A. (2015) Effect of mulch and different fungicide spray regimes on yield of tomato (*Solanum lycopersicum* L.) in Tanzania. *African Journal of Food, Agriculture, Nutrition and Development* 15(68), 9607–9619.

Nambiar, S. (2001) Knowledge, technology transfer and multinational corporations. In: Ghosh, B.N. (ed.) *Contemporary Issues in Development Economics*. Routledge, Abingdon, UK, pp. 156–173.

Ndambiri, H.K., Ritho, C.R. and Mbogoh, S.G. (2013) An evaluation of farmers' perceptions of and adaptation to the effects of climate change in Kenya. *International Journal of Food and Agricultural Economics* 1(1), 1–22.

R Core Team (2022) *A language and environment for statistical computing*. R foundation for statistical computing, Vienna. Available at: https://www.R-project.org/ (accessed 20 December 2022).

Rosset, P.M. and Miguel, A.A. (2017) *Agroecology: Science and Politics*. Practical Action Publishing, Rugby, UK.

Sarma, A., Binod, G., Bezbaruah, M.P., Das, A.K. and Goswami, B. (eds) (2022) Challenges of post-harvest management of farm output. In: *Agricultural Factor Markets and India's Small Farmers*. Sage, London.

Sarris, A., Savastano, S. and Christiaensen, L. (2006) The role of agriculture in reducing poverty in Tanzania: A household perspective from rural Kilimanjaro and Ruvuma. FAO Commodity and Trade Policy Research Working Paper 19. Available at: https://openknowledge.fao.org/server/api/core/bitstreams/04918604-58cb-46e7-bbd0-2d965d4fa7d0/content (accessed 24 January 2025).

Shango, A.J., Maswi, P.B., Malya, R.B., Maro, J.F., Mwaipopo, R.E. *et al.* (2022) Advances and trends in ecological organic agriculture (EOA) technologies and research on fruit vegetables produced in Tanzania.

CABI Reviews. Available at: https://www.cabidigitallibrary.org/doi/10.1079/cabireviews202217023 (accessed 11 June 2025).

Soaan, S. (2013) Best practice guidelines for agriculture and value chains. Sustainable Organic Agriculture Action Network (SOAAN). Available at: https://openknowledge.fao.org/server/api/core/bitstreams/12e41479-ea70-4ed7-a351-c290a92b898c/content (accessed 24 January 2025).

Valet, S. and Ozier-Lafontaine, H. (2014) *Ecosystem services of multispecific and multistratified cropping systems. Sustainable Agriculture Reviews 14: Agroecology and Global Change*. Springer International Publishing, Cham, Switzerland, pp. 185–268.

Willer, H., Bernhard, S., Trávníček, J., Kemper, L. and Lernoud, J. (eds) (2020) The world of organic agriculture: Statistics and emerging trends 2020. Research Institute of Organic Agriculture,(FiBL), Frick, and IFOAM – Organics International. Available at: www.organic-world.net/yearbook/yearbook-2020.html (accessed 24 January 2025).

Willer, H., Trávníček, J., Meier, C. and Schlatter, B. (eds) (2022) The world of organic agriculture: Statistics and emerging trends 2022. Research Institute of Organic Agriculture FiBL, Frick, and IFOAM. Organics International. Available at: https://www.fibl.org/en/shop-en/1344-organic-world-2022 (accessed 24 January 2025).

Appendix: Survey Areas

District	Ward	Village	Total
Morogoro Dc	Kinole	Tandai	10
	Mkuyuni	Kibwaya	7
		Mfumbwe	7
		Mkuyuni	7
	Tawa	Kisarawe	9
Mvomero Dc	Dakawa	Wami Sokoine	16
	Langali	Langali	14
	Lubungo	Mingo	8
	Mgeta	Kibaoni	6
Total			84

11 Organic Farming: A Review of Trends and Boundaries

David Frost*

Formerly Senior Consultant, ADAS Pwllpeiran Research Centre, Cwmystwyth, Wales, UK, and Marketing Strategy Coordinator, Organic Centre Wales, Aberystwyth University, Wales, UK

Abstract

Organic farming is reaching a 100-year milestone. This chapter questions whether it has been a century of continuity or of uneven growth. It suggests that a disjuncture contributed to new directions in philosophy and practice and a move away from the traditional ideal of an organic farm. This period has also seen the global expansion of organic farming with opportunities for international marketing requiring worldwide systems of organic certification. Procedural and cost barriers have made existing certification schemes inaccessible for small-scale producers resulting in the development of alternatives. Meanwhile movements concerned with transforming agriculture have proliferated. Boundary theory could help to clarify the position of organic farming, with the possibility of adopting 'transformation of the global food system' as an agreed boundary object.

Keywords: history, trends, marketing, globalization, certification, boundary theory

Introduction

What is organic farming? When Nicolas Lampkin asked this question in the first textbook on the subject in 1990, he wrote that there were, 'about 16 different names throughout the world for what we call organic farming' (Lampkin, 1990, p. 4). Since then, the number has grown, but the definition of organic agriculture adopted by the General Assembly of the International Federation of Organic Agriculture Movements (IFOAM) in 2008 reflects the themes that have come to dominate organic farming – soil health, sustainability, biodiversity, tradition, innovation, science, rejection of genetically modified organisms (GMOs) in food and farming, environmentalism, climate change and socio-economic equality (IFOAM Definition of Organic Agriculture, n.d).

In 2016, IFOAM proposed an organic timeline that identified three phases: Organic-1 (1920s to 1970s) when organic pioneers established the concept of organic farming and made connections between food and health; Organic-2 (1980s to 2020) – when organic became regulated by production and processing standards and certification schemes were introduced around the world; and Organic-3 (starting in the 2020s) when organic food and farming becomes more ecologically sound, economically viable, socially just, culturally diverse and transparently accountable. For the current Organic-3 phase, IFOAM proposed two priorities: to overcome the low rate of farmers' conversion globally; and to increase nutritional revenues from fields while maintaining ecosystem vigour (Arbenz *et al.*, 2016).

Organic Farming – Current Trends

The main data source to monitor the rate of conversion to organic farming is the annual reports published by the Research Institute of Organic

*Corresponding author: davidfrost@tyn-yr-helyg.com

 Advances in Organic Farming (ed. L. P. Awasthi)
DOI: 10.1079/9781800626850.0011

Agriculture (FiBL) based in Frick, Switzerland, and IFOAM – Organics International based in Bonn, Germany. The 24th edition of these annual reports, *The World of Organic Agriculture* (Willer *et al.*, 2023) showed that the annual trend of global growth in the past decade continued. There were almost 3.7 million organic producers worldwide in 2021[1] compared to 200,000 producers in 1999.

Organic farmland grew to a global total of 76.4 million ha in 2021 from a total of 11 million ha in 1999. Globally, 1.6% of all agricultural land was organic in 2021. The biggest increases were in China (+13.1%), France (+8.9%) and Spain (+8.1%). Thirty-seven countries reported a decrease in organic farmland (Willer *et al.*, 2023). Behind these headline figures there are many local and regional differences, and as with most social movements, organic farming's path to development is uneven. IFOAM's organic timeline could be read as 100 years of continual development, but has the development of organic farming been continuous or have there been disruptions? Have there been departures from the original concepts established by the pioneers? And in the current situation, how can organic farming define its boundaries and maintain relationships when numerous apparently similar movements are jostling for position?

Organic Farming – A Case of Continuous Development?

The narrative of historical continuity in the organic movement (Conford, 2001, 2011) is disputed by those arguing that a second organic wave emerged as part of a wider environmental movement in the 1960s and 1970s (Clunies-Ross, 1999). This second organic wave was linked to an international counter-urbanization trend in the 1960s–1980s which saw a worldwide phenomenon of ex-urban migrants setting up organic enterprises in rural areas (Frost, 2002a, b; Frost and Wacher, 2003; Guthman, 2004). One feature of this new wave was the enthusiasm for new ideas and new methods. New farmers and growers brought with them metropolitan values such as vegetarianism and veganism and prioritized concerns such as rights, the loss of wildlife and habitats, the exploitation of Third World food producers; and the need for fair trade. For more traditional organic farmers, the ideal had been a mixed agricultural holding where livestock husbandry was integrated with crop production. The farm produced crops to feed livestock, and the livestock produced manure to fertilize the land. Indeed, for proponents of biodynamic farming such as Ehrenfried Pfieffer, an organic farm without cattle was inconceivable. As he wrote, *c.*1947, 'Farms without cattle cannot be developed under the biodynamic system. In gardening, where there can be no cattle, the purchase of stable manure or other organic manure is the only way' (Pfeiffer, 1983, p. 72). Many of the newer post-1970 entrants to organic farming included a livestock enterprise on their holdings but others set out to find new ways of producing organic food. A specialized version of organic farming developed – vegetable growing on holdings primarily committed to crop production. In the UK, for example, growers broke from the leading organic certifying body, the Soil Association (founded in 1946), and set up their own producer's association, the Organic Growers Association (OGA), in 1981 (Frost and Wacher, 2003). Some members of the OGA rejected what they regarded as an almost universal myth that 'it is not possible to build up soil health and grow organically without animal manure' (Hall and Tolhurst, 2006, p. 2). These growers advocated the additions of plant-based compost and chipped branch wood as well as green manures to replenish organic matter lost through cropping and to encourage high levels of microbial activity. Using these techniques, growers could overcome the apparent dilemma of how to sustain soil health with neither animal manures nor artificial fertilizers.

Stock-free organic farming

Organic standards were first agreed for farms without livestock in 2004. These provided two additional methods for maintaining soil fertility in organic systems: green manuring and adding organic amendments (Hall and Tolhurst, 2006). The Stockfree Organic Standards (Stockfree Organic Standards, n.d.) were developed by the Vegan Organic Network in 2005–2007 and since then have been used in the UK by the Soil Association to certify farmers and growers who 'want to abide by a farming system which is free from any animal inputs' (Schiopu, 2022a, p. 7).

Hirth (2021) has argued that stock-free organic agriculture is an example of how the focus of veganism as a consumer identity, where being vegan is defined as eating solely a plant-based diet, is shifting towards vegan food practices (VFP), where no animal by-products are used in crop production. Such by-products include farmyard manure, composts made with animal manures, and by-products from abattoirs like dried blood, hoof and horn, and bonemeal. Hirth claims that this is an emerging paradigm, challenging both animal-dependent organic agriculture and mainstream vegan culinary culture.

The trajectory of the movement towards specialist organic methods of vegetable growing and the development of standards for stockless organic agriculture are traceable to the disjuncture from traditional organic farming ideas and methods that began in the 1970s.

Organic Farming – Market Growth and Certification

A second issue in the uneven development of organic farming concerns the criticism that it has 'lost its way'. A major intervention into this debate occurred with the publication of Julie Guthman's Agrarian Dreams (2004). Writing about the situation in the USA, with a focus on California, Guthman argued that in the 1980s, following food scares around the use of the pesticide Aldicarb and the growth regulator Alar, there were significant changes in attitudes to diet and health. This contributed to increased support for organic farming and encouraged conventional growers to convert to organic production. Guthman's study compiled data on 1533 registered organic growers in California and found that rather than traditional agri-business taking over organic agriculture, the organic movement had developed its own version of agri-business. This led to her claim that organic farming was becoming conventionalized.

Significantly, whereas IFOAM highlights the development of production and processing standards and the introduction of certification schemes in its Organic-2 phase of development, 1980s to 2020, for Guthman, production and processing standards in the USA/California were a main contributing factor in the conventionalization of organic farming. Standards have been accused of contributing to the conventionalization of organic farming for a number of reasons (Dufeu *et al.*, 2020). They have been accused of becoming limited and permissive when written into public standards (a particular problem in the case of the USA's Organic Rule). Further, the dilution of certification schemes allowed them to be easily adopted by industrial organic agriculture, while the rising cost of certification has become an obstacle for smaller-scale producers. Thus, the adoption of organic standards, like so many innovations in business, are more easily accommodated by economically and technically strong operations.[2]

Globalization

Is the globalization of the organic market evidence of conventionalization? As the marketing of organic became dominated by the multiple retailer sector, suppliers were required to commit to continuity of supply in sufficient volume and to conform to multiple retailers' product specifications. The seemingly irresistible trend of mergers and acquisitions is also evident in organic food marketing as successful smaller businesses have been acquired by multi-national companies.

In 2021, the global organic market reached almost €125 billion – an increase of nearly €4 billion or approximately 3% compared to 2020 (Willer *et al.*, 2023). With organic sales equivalent to €48.6 billion, the USA continued to be the world leader followed by Germany (€15.9 billion) and France (€12.7 billion). These figures on their own provide no information on the nature of the marketing firms nor of their suppliers, though some indications could be derived from a study of, for example, BIOFACH, the World Trade Fair for organic products.

In 1990, concerned at the lack of marketing opportunities for organic producers in Germany and their virtual exclusion from existing trade fairs, the founders of BIOFACH organized an exhibition in the Ludwigshafen Stadthalle (town hall), for organic producers. They named the exhibition 'BIOFACH: the 1st European Trade Fair for Organic Food and Natural Products'. It attracted 197 exhibitors and 2500 visitors and grew annually thereafter, increasingly attracting exhibitors from around the world. The trade fair located permanently to Nuremberg in 1999. That year it attracted 1276 exhibitors and more than 15,384 visitors. In 2001, the original founders transferred the organization of BIOFACH to NurnbergMesse (Nurnberg Trade Fair). BIOFACH 2023 represented its 34th successive year attracting *c.*36,000 trade visitors from 135 countries during the four-day event. All the organic food exhibited at BIOFACH is certified to the EU organic regulation or the accreditation

Fresh food	Quasani Societa Agricola a.r.l. (I): MANDORINO PISTACHIO, MANDORINO TURMERIC & PEPPER, MANDORINO WALNUT
Frozen food	followfood GmbH(D):BioPizzaBiancavegan
Grocery products, cooking & baking	SONNENTOR Kräuterhandels GmbH (A): Chakalaka , Harissa, 5-Spice-Powder, Greek Salad Seasoning, Umami Spice Preparation organic package
Grocery products, snacks & sweets	Nungesser AG (CH): Kenari nuts
Other grocery products	Uhhmami Aps (DK): Uhhmami – BACON'ish, BEEF'ish, BLUE CHEESE'ish, CHEESE'ish, CHICKEN'ish, OCEAN'ish, TRUFFLE'ish, VEGETABLES
Drinks	primoza GmbH (D): The Growable Tea
Non-food	Koriolis GmbH (D): AKINOD cutlery set 12H34

Fig. 11.1. Winners of the Best New Product Award at BIOFACH World Trade Fair 2023, by category. (Available at: https://www.biofach.de/en [accessed 24 April 2023]).

directives of IFOAM. It is an irony, however, that whereas the original founders intended to create an accessible trade fair for organic companies, BIOFACH has evolved into a world trade fair where the increasingly high costs of exhibiting are prohibitive for smaller companies and BIOFACH increasingly showcases novel and processed organic foods (see Fig. 11.1).

Although organic food companies have set up international supply chains for their raw materials, during the COVID-19 pandemic the most successful businesses were those with local supply. Companies more reliant on organic supplies from distant geographic regions were negatively affected. As Willer *et al.* (2023) said: 'The way forward appears to have more local – if not regional – supply chains. In fact, a widespread growth of direct retailing and short supply chains for organic food have mirrored the growth of global chains. In the RAFAEL EU Interreg project, for example, partners from the peripheral areas[3] in Europe came together with a shared concern for the effects of globalization on their communities. Their aim was to promote the growth of what the partners labelled 'authentic food' defined as local, people-focused, ethical and, wherever possible, certified to organic standards. The RAFAEL project found that some producers did not become certified organic because it felt important to distinguish themselves from large-scale growers supplying the multiple retail sector. These producers argued that the organic symbol does not reflect the fact that they are ethical businesses committed to many values of sustainability, local marketing, environmental protection, local communities etc. They objected that 'their' organic symbol can be used by large producers who have set up organic enterprises to market to the multiples, but who have no other commitment to the ideals of the organic/sustainable/authentic movement. Nor could these producers use the term 'local' to distinguish their operations, as 'local' is ill-defined and could include a local intensive system with no sustainable characteristics (Frost *et al.* 2007; Frost and Daniel, 2007).

Participatory certification

Participatory certification is an alternative for small-scale producers. According to IFOAM's 2008 definition, participatory certification schemes (PGS) certify that producers are based on active participation of stakeholders and are built on a foundation of trust, social networks and knowledge exchange. PGS systems are recognized internationally and by national governments. They enable producers to label their products as organic at a lower financial cost; however, producers incur opportunity costs through the work involved in monitoring, reviewing and recording the work of other farmers and growers involved in the schemes (Schiopu, 2022b).

Setting up a PGS requires at least two years before producers can be fully certified (Anselmi and e Castro, 2022; IFOAM – Organics International, n.d.). IFOAM – Organics International has recorded 242 PGS initiatives in 78 countries, with at least 1,244,239 producers involved and 1,205,050 producers certified, managing an estimated 915,997 ha of land. The global trend saw an increase from 496,104 PGS certified producers in 2019 to 1,205,050 in 2021. IFOAM reported, however, that Europe and North America had no major change in numbers of PGS-certified producers; the growth came from outside these trading blocs.

As Sahota (2022) has pointed out, consumer demand for organic products has escalated since the COVID-19 pandemic, particularly in countries like the USA and the UK where increases in the total area of organic cultivated land have been comparatively small, leaving them reliant on imports and the global organic market. Sahota warns that if production fails to meet demand, shortages and higher prices will follow with the possibility of a rising number of fraud incidents and conventional foods illegally labelled as organic. This underlines the importance of internationally agreed certification schemes. Furthermore, speakers at the 2023 BIOFACH Congress on 'Organic, Food, Sovereignty, True Prices' and declarations by La Via Campesina[4], the international organization of peasants, small-scale farmers and agricultural workers unions around the world, both express concern about food sovereignty. The extension of PGS could play a valuable role here.

Organic Farming and Its Boundaries

Organic farming is a system defined as embracing sustainability, biodiversity, tradition, innovation, science, environmentalism and socio-economic equality. The ownership of these concepts is not restricted to the organic farming movement alone, however; organic farming operates in a territory with sometimes competing, sometimes co-operating alternative movements. Many of these have been identified by the International Panel of Experts on Sustainable Food Systems (IPES) as 'terms used to describe food system transformations in global policy and development spaces (IPES-Food, 2022). They list regenerative agriculture, agroecology, nature-based solutions, sustainable agriculture, ecosystem-based approaches, conservation agriculture, circular systems and agriculture that supports biodiversity. In this crowded field, does organic farming claim to be transformative? There are already those who doubt its credentials (Lyons, 2022). The following review considers three movements whose boundaries may at times encroach upon or overlap with organic farming: permaculture, agroecology and regenerative farming.

Permaculture

The concept of permaculture envisages a sustainable agricultural system modelled on the structure of the perennial rainforest with layers of trees, shrubs, herbaceous plants, fungi and root systems[5], which is described in *Permaculture One: A Perennial Agriculture for Human Settlements* (Mollison and Holgren, 1978). In 1979, Mollison established the first Permaculture Institute to teach the practical applications of permaculture design, based at Tagari Farm in New South Wales, Australia[6]. What is the difference between permaculture and organic farming? Like organic farming, permaculture pays great attention to soil fertility. The key difference is that permaculture is an overall design framework. Permaculture has a transformative objective; it originated to support self-sufficiency initiatives in preparation for a post-petrol world (Whitefield, 2011) and criticizes the use of fossil fuels in the diesel-powered tractors and implements used by most organic farmers (Mollison, 1988).

Agroecology

Agroecology has been described as the application of ecology in agriculture, but it also seeks social and political change, 'It challenges the prevalent perception of agriculture as production of commodities and the commodification of nature and aims at reconciling agriculture with nature' (Bröckelmann-Simon's foreword to Chappell and Bernhart, 2018). The IPES noted that agroecology originated in indigenous peoples' food systems and with the emergence of environmental movements in the 1960s; it became part of a broader conversation on finding alternatives to industrial, chemical-intensive production systems. Compared to agroecology, there are already those who question organic farming's transformative credentials. For example, in a conference paper exploring how the UK wine industry and agroecology have, respectively, moved from 'positions of ridicule into a space where they are publicly respected', Lyons claims that 'agroecology needs to be brought to scale to transform the food system … because at this point previous initiatives, such as organics and sustainable intensification, have been stripped of transformative potential' (Lyons, 2022).

Regenerative farming

The term 'regenerative' was first adopted in the USA by Robert Rodale, son of the founder Jerome Rodale, founder of the Rodale Institute; and in 1983, Richard Harwood, an agronomist and the Director of the Rodale Institute, enumerated three fundamental principles of regenerative agriculture: (i) the interrelatedness of all parts of a farming system, including the farmer and farm family; (ii) biological balances in the system; and (iii) the need to maximize desired biological interactions while minimizing the use of materials and practices that disrupt those relationships (Harwood, 1983, cited in IPES-Food, 2022). Regeneration International was established in 2017 as a non-profit organization with a mission to 'promote, facilitate and accelerate the global transition to regenerative food, farming, and land management for the purpose of restoring climate stability, ending world hunger, and rebuilding deteriorated social, ecological and economic systems' (Regeneration International, cited in IPES-Food, 2022).

Sarah Compson, Director of Standards Innovation at the UK Soil Association and IFOAM – Organics International world board member, while describing IFOAM – Organics International, as 'the umbrella organization for the worldwide organic movement', notes that, some of the academic studies of regenerative farming have been carried out on farms that are already certified organic. She writes, 'If you read the definition of organic agriculture, it's so closely aligned with the best of regenerative that they're virtually synonymous' (Compson, 2022).

Rather more critically, the UK's Organic Trade Board has claimed that organic farmers already use regenerative methods throughout the year; and, anyway, organic was the first and the most legally defined and truly regenerative food and farming method.

Boundary theory

In the crowded and contested field illustrated by these three examples, how is it possible to situate organic farming? Perhaps one way is to use boundary theory. Star and Griesemer defined 'boundary' as a shared but ill-defined and ambiguous space and they introduced the term 'boundary object': an object (symbolic or material) that is flexible enough to be adapted to the needs and constraints of all those using it, but strong enough to maintain a common identity (Star and Griesemer, 1989, p. 393).

The concept of boundary object has been used in several studies relevant to organic farming (Borgen, 2021). For example, Favilli *et al.* (2015) analyzed the innovation potential of a local food network, where different actors co-operated to build a local organic food production-provision system. Their study focused on organic food and farming as boundary objects, a connecting element that helped the involved actors develop common visions, languages and goals and organize their joint activities. Hirth (2021) used boundary object theory to explore variation in the performance of veganism; and Carr and Wilkinson (2005) explored how groups can use participatory natural resource management techniques to combine farmers and scientists at the boundary between science and farming, thereby creating a powerful force for improving the management of natural resources on farms.

Could boundary theory be used to establish the borders of organic farming, regenerative farming, agroecology, permaculture etc. and provide a means to further co-operation, opportunities for collaboration, and avoidance of competition? This would depend on whether agreement could be reached on a common boundary object.

Conclusions

If, as most historians of the organic agriculture movement agree, organic farming originated in the 1920s, it is about to reach its centenary. Although commentators point to differing stages of development, the predominant narrative is a century of continuity. This narrative is supported by statistics showing sustained growth of the movement in terms both of numbers of organically certified producers globally and in the size of the market for organic produce. Increasingly, however, there have been developments in the philosophy and practice of organic agriculture, in practices of organic certification and in the number of bodies working to transform agriculture. These suggest at least a degree of disruption that challenges the concept of continuity.

The emergence of stock-free organic farming and the vegan network, for example, challenges the original ideal of a mixed farm where livestock husbandry and crop cultivation were fully integrated. Specialization is appearing, demarcating organic food produced using inputs derived from livestock production and that which has had none of these at all.

Contemporaneously, many international organizations are promoting transformative and sustainable agriculture in an increasingly competitive space. The movement represented by these bodies could work towards a common framework, a common boundary object – one uniting the call for sustainability, biodiversity, respect for cultural traditions, innovation, science, environmentalism and socio-economic equality. In short, a boundary object stressing the urgent need for comprehensive transformation of the global food system.

Notes

[1] Some countries report only the numbers of companies, projects or grower groups but these may each comprise a number of individual producers, so it may be assumed that the total number of organic producers is higher than that reported here (Willer *et al*., 2023, p. 22).

[2] Importantly, however, Dufeu *et al*.'s five-year participatory research in an organic fruit and vegetable producers' organization in western France, *Bio Loire Océan (BLO),* suggested that a weakening of the rules is not inevitable and that standardization does not necessarily weaken what is required to be organic.

[3] Rafael partners represented Alentejo, Andalucía, Brittany, Galicia, Northern Ireland, South-west England, Trás-os-Montes e Alto Douro, Wales, and Zamora.

[4] See https://viacampesina.org/en/17-april-faced-with-global-crises-we-are-building-food-sovereignty-to-ensure-a-future-for-humanity/ (accessed 24 April 2023).

[5] Modelling the ecological principles of forest ecosystems in agriculture had also been identified by earlier proponents of biodynamic farming. As Ehrenfreid Pfeiffer wrote in the 1940s: 'The basic laws of forestry – for example, the constant ground cover, the mutual help of various plants to one another, protective planting strips and a compost fertilizing as natural as possible – are laws which apply in a similar way to gardening' (Pfeiffer, 1983).

[6] Name subsequently changed to the Permaculture Research Institute (PRI), located on a permaculture demonstration site in The Channon, NSW, Australia.

References

Anselmi, S. and e Castro, F.M. (2022) Participatory guarantee systems in 2021. In: Willer, H., Trávníček, J., Meier, C. and Schlatter, B. (eds) *The World of Organic Agriculture Statistics & Emerging Trends 2022*. Fibl & IFOAM – Organics International. Available at: https://www.fibl.org/fileadmin/documents/shop/1344-organic-world-2022.pdf (accessed 13 April 2023).

Arbenz, M., Gould, D. and Stopes, C. (2016) Organic 3.0 – for truly sustainable farming and consumption. IFOAM – Organics International, Bonn, and SOAAN. Available at: https://www.ifoam.bio/sites/default/files/2020-05/Organic3.0_v.2_web.pdf (accessed 25 January 2025).

Borgen, S.O. (2021) Standard quality as a boundary object: A conceptual study. *Ecological Solutions and Evidence* 61(4), 704–722. DOI: 10.1111/soru.12345.

Carr, A. and Wilkinson, R. (2005) Beyond participation: boundary organizations as a new space for farmers and scientists to interact. *Society & Natural Resources* 18(3), 255–265. Available at: (accessed 25 April 2023).

Chappell, M.J. and Bernhart, A. (2018) *Agroecology as a pathway towards sustainable food systems*. Bischöfliches Hilfswerk MISEREOR e.V, Aachen. Available at: https://www.researchgate.net/publication/328812927 (accessed 16 March 2023).

Clunies-Ross, T. (1999) Agricultural change and the politics of organic farming. PhD thesis, University of Bath, UK.

Compson, S. (2022) When it comes to regenerative agriculture, enthusiasm and confusion seem to be increasing in equal measure. Available at: https://www.soilassociation.org/blogs/2022/october/24/can-organic-and-regenerative-co-exist/?count=13 (accessed 25 April 2023).

Conford, P. (2001) *The Origins of the Organic Movement*. Floris Books, Edinburgh, UK.

Conford, P. (2011) *The Development of the Organic Network*. Floris Books, Edinburgh, UK.

Dufeu, I., Le Velly, R., Bréchet, J. and Loconto, A. (2020) Can standards save organic farming from conventionalisation? Dynamics of collective projects and rules in a French organic producers' organisation. *Sociologia Ruralis* 60(3), 621–638.

Favilli, E., Rossi, A. and Brunori, G. (2015) Food networks: Collective action and local development.The role of organic farming as boundary object. *Organic Agriculture* 5(3), 235–243. 10.1 007/s13165-015-0118-2 (accessed 25 April 2023).

Frost, D. (2002a) Deconstructing the organic movement: UK organic research 2002. In: *Proceedings of the COR Conference*, Organic Centre Wales, University of Aberystwyth, March 26–28, pp. 113–114.

Frost, D. (2002b) The early organicists – a review of 'The origins of the organic movement' by Philip Conford. *Planet No. 150, December 2001/January 2002* 2002, 103–104.

Frost, D. and Wacher, C. (2003) A new incarnation – the role of the organic growers association in changing the production and marketing of organic produce. In: *European Society for Rural Sociology*, 20th Biennial Conference, Sligo, August 19–22.

Frost, D. and Daniel, A. (2007) RAFAEL – preserving regional cultures by promoting authentic food. In: Sorensen, N. (ed.) *Proceedings of the 1st IFOAM International Conference on Marketing of Organic and Regional Values*, Schwäbisch Hall, Germany 26–28 August. IFOAM, Bonn, pp. 177–182.

Frost, D., Sherwood, A.M. and Diepen, P. (2007) Opportunities for authentic food marketing in nine regions of the EU Atlantic area. In: *Report to the Interreg III B 'Atlantic Area' Renaissance of Atlantic Food Authenticity and Economy Links (RAFAEL) Partnership*. ADAS, Helsby, UK.

Guthman, J. (2004) *Agrarian Dreams: The Paradox of Organic Farming in California*. University of California Press, Berkeley, California.

Hall, J. and Tolhurst, I. (2006) *Growing Green: Organic Techniques for a Sustainable Future*. The Vegan Organic Network, Altrincham, UK.

Hirth, S. (2021) Food that matters: Boundary work and the case for vegan food practices. *Sociologia Ruralis* 61(1), 234–254.

IFOAM Definition of Organic Agriculture (n.d) Definition of Organic Agriculture. Available at: https://www.ifoam.bio/why-organic/organic-landmarks/definition-organic (accessed 10 May 2023).

IFOAM – Organics International (n.d.) Participatory guarantee systems. Available at: https://www.ifoam.bio/our-work/how/standards-certification/participatory-guarantee-systems (accessed 28 April 2023).

IPES-Food (2022) Smoke and mirrors: Examining competing framings of food system sustainability: agroecology, regenerative agriculture, and nature-based solutions. Available at: https://ipes-food.org/_img/upload/files/SmokeAndMirrors.pdf (accessed 23 April 2023).

Lampkin, N. (1990) *Organic Farming*. Farming Press Books, Ipswich UK.

Lyons, C. (2022) The impacts of the UK wine industry on people, place, and climate. In: Dublin Gastronomy Symposium – Food and Movement.

Mollison, B. (1988) *Permaculture: A Designer's Manual*. Tagari Publications, Tyalgum, Australia.

Mollison, B. and Holgren, D. (1978) *Permaculture One: A Perennial Agriculture for Human Settlements*. Tagari Publications, Tyalgum, Australia.

Pfeiffer, E. (1983) *Soil Fertility, Renewal and Preservation*. The Lanthorn Press, East Grinstead, UK.

Sahota, A. (2022) The global market for organic food & drink. In: Willer, H., Trávníček, J., Meier, C., Schlatter, B., Willer, H. *et al*. (eds) *The World of Organic Agriculture Statistics & Emerging Trends 2022*. Fibl & IFOAM – Organics International. Available at: https://www.fibl.org/fileadmin/documents/shop/1344-organic-world-2022.pdf (accessed 13 April 2023).

Schiopu, T. (2022a) Participatory certification. *The Organic Grower* 61, 13.

Schiopu, T. (2022b) What's new in stockfree organic? *The Organic Grower* 61, 7.

Star, S.L. and Griesemer, J.R. (1989) Institutional ecology, "translations" and boundary objects: Amateurs and professionals in: *Berkeley's museum of vertebrate zoology, 1907–39*. *Social Studies of Science* 19(3), 387–420.

Stockfree Organic Standards (n.d.) The Stockfree Organic Standard. Available at: https://stockfreeorganic.net/stockfree-organic-services#standard (accessed 24 April 2023).

Whitefield, P. (2011) *The Earth Care Manual: A Permaculture Handbook for Britain Other Temperate Climates*, Revised edn. Permanent Publications, East Meon, Uk. Available at: https://derdejan.files.wordpress.com/2020/06/the-earth-care-manual_-a-permaculture-handbook-for-britain-and-other-temperate-climates-pdfdrive.com-.pdf (accessed 25 January 2025).

Willer, H., Schlatter, B. and Trávníček, J. (eds) (2023) *The World of Organic Agriculture: Statistics and Emerging Trends 2023*. Research Institute of Organic Agriculture, FiBL, Frick, and IFOAM – Organics International, Bonn. Available at: https://www.fibl.org/fileadmin/documents/shop/1254-organic-world-2023.pdf (accessed 11 April 2023).

12 Organics and Agroecology: Boundaries and Synergies Between Sustainable Systems and Approaches

Thamali Wanasinghe[1], Ashoka Gamage[2,3]*, Yasavi Jayakodi[4], Wasana Gunawardana[2] and Othmane Merah[5,6]

[1]Postgraduate Institute of Agriculture, University of Peradeniya, Sri Lanka; [2]China Sri Lanka Joint Research and Demonstration Centre for Water Technology (JRDC), Sri Lanka; [3]Chemical and Process Engineering, Faculty of Engineering, University of Peradeniya, Sri Lanka; [4]National Institute of Fundamental Studies (NIFS), Kandy, Sri Lanka; [5]Laboratoire de Chimie Agro-industrielle (LCA), Institut National de la Recherche Agronomique, Université de Toulouse, France; [6]Département Génie Biologique, IUT A, Université Paul Sabatier, Auch, France

Abstract

Organic farming is a sustainable approach to agriculture that prioritizes soil health, biodiversity and the avoidance of synthetic chemicals. It promotes long-term sustainability through natural inputs, crop rotations and ecological balance. The global shift towards organic farming is driven by environmental concerns, health implications and climate change. Practices like crop rotation, cover cropping and composting minimize environmental impact and promote natural pest control. Organic farming can lead to long-term economic viability due to growing consumer demand and premium market prices. Asian farmers are mainly embracing organic farming to balance agricultural productivity with environmental conservation and socio-economic well-being. It combines tradition, innovation and science to benefit the environment, promote fair relationships and improve quality of life. Organic products are rich in nutrients and free of pesticide residues and additives.

Keywords: biofertilizers, biopesticides, organic farming, soil health, vermicompost

Introduction

Organic farming is a sustainable, environmentally conscious approach to agriculture, focusing on soil health, biodiversity and avoidance of synthetic chemicals. It promotes long-term sustainability through natural inputs, crop rotations and ecological balance (Singh, 2021). The global shift towards organic farming is driven by environmental concerns, health implications and climate change. This introduction explores the principles, practices, challenges and benefits of organic farming, highlighting its potential for balancing agricultural productivity and human well-being.

Organic farming is a sustainable approach to agriculture that prioritizes environmental, social and economic aspects. Practices like crop rotation, cover cropping, composting and the avoidance of

*Corresponding author: ashogamage@gmail.com

DOI: 10.1079/9781800626850.0012

synthetic pesticides and fertilizers promote healthy soil structure and nutrient cycling, and reduce soil erosion, contributing to agricultural ecosystem sustainability (Pinto *et al.*, 2023). Organic farming also promotes natural pest control. It also promotes biodiversity in agricultural landscapes, providing habitats for beneficial insects and wildlife and preserving genetic diversity in crops. It emphasizes resource efficiency by minimizing external inputs and promoting closed nutrient cycles, reducing the demand for non-renewable resources and contributing to climate change mitigation efforts. Transitioning to organic farming may present initial challenges but can lead to long-term economic viability due to growing consumer demand and premium market prices. It also promotes community well-being by fostering local food systems, supporting small-scale farmers, and encouraging fair labour practices.

Asian farmers are mainly embracing organic farming and sustainability to balance agricultural productivity with environmental conservation and socio-economic well-being. The continent's unique agricultural landscape, rooted in traditional practices, offers an ideal environment for exploring the synergies between these practices.

The economic viability of organic farming is crucial in Sri Lanka, as the demand for organic products rises globally, offering farmers access to premium markets (Anderberg, 2020). However, challenges such as transition periods, market access and certification processes need to be addressed for sustained economic success. Social dimensions of sustainability are also relevant, as organic farming can empower local communities through fair trade practices, preserving traditional agricultural knowledge and contributing to food security. Government initiatives support organic farming, but challenges like pests, diseases, education and market establishment remain.

What Is Organic Farming?

Organic farming is a sustainable farming system that avoids artificial inputs like fertilizers, pesticides, hormones, and feed additives. It relies on crop rotations, residues, animal manures, off-farm organic waste, mineral-grade rock additives, and biological systems of nutrient mobilization for optimal plant protection (Gamage *et al.*, 2023). Organic agriculture sustains soil, ecosystems and people by relying on ecological processes, biodiversity and cycles adapted to local conditions. It combines tradition, innovation and science to benefit the environment, promote fair relationships and improve the quality of life for all involved.

Organic farming enhances resilience against climate change by creating environmentally sound systems that are resistant to temperature fluctuations and drought, preventing soil erosion. It promotes sustainable management, conservation practices and restoration activities. Compared to modern agriculture, organic farming has a low financial requirement and helps farmers and communities adapt to climate change. It fulfils many requirements for successful adaptation strategies, making it a valuable option for farmers and communities.

Organic farming addresses contemporary issues in agriculture and food production by focusing on principles of health, ecology, fairness and care. Organic products are rich in nutrients and free of pesticide residues and additives. Organic livestock practices produce animal products without antibiotics and other drugs, slowing the spread of antibiotic-resistant bacteria. Organic farmers promote biodiversity, increase natural enemy populations, improve natural resources and support pollinators, which are essential for maintaining a healthy environment and producing healthy foods. Organic products have a lower level of pesticides and nitrates.

IFOAM's 2005 principles of organic production are health, ecology, fairness and care. Health focuses on sustaining and enhancing the health of the soil, plants, animals, humans and the planet as a single entity (Migliorini and Wezel, 2017). Ecology is based on living ecological systems and cycles, while fairness ensures fair relationships with the common environment and life opportunities. Care is about managing organic agriculture responsibly to protect the health and well-being of current and future generations and the environment.

Components of Organic Farming

Organic farming focuses on improving soil structure, managing yield, composting farm waste for organic manure, using non-chemical weed management and using biofertilizers instead of chemical fertilizers for better soil management.

Fig. 12.1. Soil and crop management practices. (Photo: L. Rankoth)

Crop and soil management

Soil organic matter is crucial for soil fertility and can be improved through effective farming methods. Fertile soil should have good water holding capacity and cation exchange, and reduce soil erosion. Organic farming uses green manure to enhance soil biological activity, while crop rotation and intercropping control weeds and manage soil chemical and physical properties (Angon *et al.*, 2023). Mixed cropping, using livestock, farm residues and straw, helps prevent nutrient leaching from surface soil and reduces soil erosion. Fig. 12.1 illustrates how soil and crop management practices interact to enhance agricultural sustainability.

Crop rotation and intercropping

Organic agriculture relies on soil biology and health, and practices like crop rotation, mixed cropping and intercropping (Fig. 12.2) can enhance soil life and biological activities. Crop rotation balances soil building and cash crops, allowing for bare fall periods to break weed cycles and incorporate plant matter into the soil (Soni *et al.*, 2022). This method keeps weed growth at bay and helps in nutrient recycling in the ecosystem. Farmers like Jean-Paul Courtens emphasize the importance of these practices in organic farming (O'Reilley, 2014). Intercropping is a crucial method for balancing ecosystems in intensive agriculture programmes. It allows simultaneous growth of different crops in the same field, allowing for seeds to be sown from two different plants or with a gap.

Fig. 12.2. Intercropping between maize and leguminous plants. (Photo: L. Rankoth)

Crop residues

India generates tons of crop residues annually, which are a significant source of nutrient recycling in soil. These residues, inoculated with fungal hyphae and spores, improve soil health and aid organic farming. Examples include straws, stalks, cobs of maize, beans, peas and potatoes, as well as threshing sheds like oil cakes, rice husks, peanut shells, Indian millet and pearl millet.

Organic manure

Organic fertilizers or manures are crucial for sustainable soil management, improving soil quality without compromising ecosystem health. Composting biological sources like plant or animal residues enhances soil biological activity, increasing inorganic nutrient availability and humus for good crop yield. Organic manures are categorized into bulky and concentrated forms (Lotter, 2003).

Bulky organic manure, made from decomposed animal excreta like dungs, urine and farm residues, is commonly used in developing countries like India. It includes compost, a humus-like material produced from organic waste due to microbial activity in anaerobic conditions; and green manures, which are crops grown for the benefit of soil. Green manure increases soil fertility by adding surplus inorganic nutrients, organic matter, microbial growth and humus, while preventing soil erosion and leaching of nutrients, and controlling weed growth (Kumar *et al.*, 2020a, b). They are used for sustainable annual cropping systems and are part of the crop rotation method. Legumes are essential green manures as they fix atmospheric nitrogen into the available form of nitrogen in the soil with the help of nitrogen-fixing bacteria present in their root nodules. Non-legume plants, such as *Jatropha*, *Fagopyrum esculentum*, Neem, *Hibiscus viscosa*, *Vitex negundu*, *Calotropis gigantean*, *Delonix regia*, and *Derris indica*, are also used as green manures. In India, there are two ways of green manuring: green leaf manuring, where green leaves and twigs are collected from plants grown in fields or wastelands; and on-farm green manuring, where legume plants grow as green manure in the same field under crop rotation. Although bulky organic manure has lesser nutrients than concentrated organic manures, it is widely used in developing countries like India. Concentrated organic manure is a type of organic manure made from plant and animal materials with higher nutrient content, such as meat and bone meal, blood meal, fish fertilizer, horn fertilizer and hoof meal.

Biofertilizers

Excessive use of chemical fertilizers harms soil quality and human health, negatively impacting microflora and fauna. To benefit the ecosystem, alternative methods like microbes are essential. Microbes increase crop yield by increasing micronutrient availability in soil, leading to soil productivity. Biofertilizers like bacteria, fungi and algae are used to supplement these methods, enhancing soil productivity and crop yield. Symbiosis is a biotic interaction that helps plants fix atmospheric nitrogen into various forms in the soil. The relationship between legume plants and Rhizobium bacteria is particularly beneficial, as Rhizobium bacteria reside in the root nodule of leguminous plants, making atmospheric nitrogen accessible to the plants (Kumar *et al.*, 2020a). Blue-green algae, *Mycorrhizae*, bacteria like *Azospirillum* and *Azotobacter*, and *Azolla*, a small aquatic plant, can decompose soil organic matter and convert nitrogen into nitrates, nitrites and ammonia without forming a symbiotic relationship with other organisms.

On-farm waste recycling

Organic waste recycling is crucial for sustainable farming as it reduces the use of harmful chemical fertilizers (Durán-Lara *et al.*, 2020). Farm and household waste, including pruned branches, straw, and discarded fruits and vegetables, undergoes composting, anaerobic digestion and thermochemical treatments, leading to maximum recycling and reduced use of conventional fertilizers and energy sources. Industrial and household wastes also contribute to organic waste recycling.

Weed management

Organic agriculture employs sustainable weed management methods to reduce the use of herbicides and chemicals. These methods include cultural practices like crop rotation, cultivar choice, seeding rates, row spacing, mechanical and physical weed control, tillage, solar radiation, intercropping, mulching, hand weeding, biological weed control, and bioherbicides. These methods aim to minimize the environmental impact of agriculture.

Biopesticides

Management of pests is vital in sustainable production in agriculture. Synthetic pesticides are widely used to control pests in agriculture. However, there is increasing concern over synthetic pesticide usage due to its adverse long-term effects on the public health of both users and consumers, the environment and the ecosystem of natural pests. On the other hand, biopesticides are highly selective to the target pests, do not harm the other organisms in the field, are easily biodegradable, and generate less harmful pesticide residue. Therefore, biopesticides can be used in sustainable agriculture to increase productivity (Rajapakse *et al.*, 2016; Yadav *et al.*, 2022).

Biopesticides are biological materials such as animals, plants and certain minerals that produce harmful toxins against pests in agriculture, and biopesticides produce secondary metabolites like alkaloids, phenolics and terpenoids as active biopesticides. Currently, plant powders, non-volatile and volatile oils, and plant crude extracts are commercially available and have been successfully utilized in both developed and developing countries for the management of pests; for example, Neem seeds and leaves, cow urine extract and garlic extract. Further, several bacterial and fungal biopesticides have shown promising results for efficiently managing plant pathogens in agriculture.

Tropical countries are rich in biodiversity and the majority of indigenous plants contain unique chemical properties, which can be used to produce biopesticides. Also, some countries like Sri Lanka and India have traditional knowledge of the sustainable utilization of biopesticides. Using this traditional knowledge, small-scale industries have produced biopesticides. However, research and technology development are needed to produce new products and develop currently available products for the long-term sustainability of agriculture, as well as the protection of community and public health. Biopesticides have several advantages over chemical pesticides and are expected to occupy a large share of the market in the coming period (Kumar *et al.*, 2021).

Vermicompost

Vermicomposting is a sustainable and environmentally friendly method of composting that involves different species of worms to digest organic waste materials. Usually, red wigglers, white worms and other earthworms are involved in breaking down organic waste materials, such as food scraps and yard trimmings, into nutrient-rich compost. This process requires moderate environmental conditions and microorganisms.

This process not only reduces the amount of waste sent to landfill or dumpsites but also produces a high-quality soil amendment that can be used to improve soil fertility and structure. The earthworms consume the organic matter, digest it in their gut, and then excrete nutrient-rich castings that are full of beneficial microorganisms in the soil. Vermicompost is rich in water-soluble nutrients like nitrogen, phosphorus and potassium, which are released slowly over time, making them beneficial for plant growth with a steady source of nutrition (Chatterjee *et al.*, 2020). Vermicomposting is increasingly being adopted by professional gardeners, farmers and environmentalists as a sustainable solution for managing organic waste and improving soil health. Therefore, vermicompost is widely used in organic farming (Lim *et al.*, 2014). Vermicomposting is an eco-friendly and economically favourable biotechnological process, and due to its great potential in dealing with waste accumulation, several studies have been done to examine the nutrient profile in vermicompost and its benefits, and to understand the knowledge gaps in the vermicomposting process (Vuković *et al.*, 2021).

Advantages of Organic Farming

Organic farming offers numerous benefits across ecological, agricultural and health dimensions. It prioritizes soil health by avoiding synthetic pesticides and fertilizers, focusing on natural alternatives that promote microbial diversity and ecological balance (Lehmann *et al.*, 2020). This approach not only protects the environment from harmful chemical runoff but also preserves biodiversity, promoting a more resilient and sustainable ecosystem. Organic practices like crop rotation, intercropping and companion planting enhance soil structure and fertility, reducing the need for external inputs. The prohibition of genetically modified organisms

(GMOs) aligns with consumer preferences for traditional, natural food production.

Moreover, organic farming reduces the risk of chemical residues in food by eliminating synthetic chemicals. Organic crops contain higher levels of nutrients and antioxidants, contributing to food nutritional value. The prohibition of antibiotics and growth hormones in organic animal farming ensures the production of meat and dairy products free from these additives.

Ethical animal treatment is also a key aspect of organic farming, requiring humane living conditions and outdoor spaces for livestock. This approach resonates with consumers concerned about animal welfare in agriculture.

Disadvantages of Organic Farming

Organic farming has several disadvantages, including increased pest and disease vulnerability, labour-intensive farming, and higher costs of organic feed compared to non-organic alternatives. These costs are passed on to consumers, making organic food more expensive than conventionally produced food. Despite this, many people are willing to pay more for organic food; but during times of hardship or recession, they may be less likely to buy organic food (Rümmele, 2023).

Challenges and Drawbacks of Organic Farming

The global demand for organic products and farming has increased, but despite its benefits such as natural resource preservation, environmental protection, and improved food quality, the shift to organic farming remains a significant challenge in developing countries like Sri Lanka. Other constraints include lack of good marketing, proper agricultural policies and proper certification processes (Kumar *et al.*, 2023).

Lack of awareness, scarcity of biomass, poor quality of manure and seeds, dearth of livestock, and difficulty in soil nutrients management also pose challenges. Organic pesticides are less effective than chemical pesticides, and the labour-intensive process leads to high input costs. Insufficient infrastructure and lack of finances or credit also hinder the growth of organic produce.

Organic farming has a disadvantage in producing low yields, which is balanced with the high price premium and low input cost involved in producing organic products. To support organic farming in India, the government needs to invest in providing financial incentives to farmers, collaborate with non-government agencies, and provide special training to enhance their knowledge and skills for production, processing and marketing strategies. Extension programmes should also be introduced to help farmers use their own input resources instead of purchasing and relying on external resources. Encouraging farmers to form associations and trade unions can also improve their marketing efficiency.

Organic Farming in Global Agriculture

Organic farming is a growing trend in 187 countries, with 3.1 million farmers operating 72.3 million ha of farmland organically. In 2019, organic food and drink sales reached over €106 billion, indicating a growing interest in healthy and environmentally friendly eating habits (Gamage *et al.*, 2023). Over 75% of the world's agricultural production is produced through certified organic farming, making it a cost-effective and prolific system. India was the world's largest producer of organic foods in 2019, with 1.4 million producers (Singh *et al.*, 2023). Organic farming is considered more sustainable than conventional farming but has its own set of challenges. It is less polluting and produces lower yields on average. However, widespread upscaling could lead to further loss of natural habitats and higher output prices. Organic farming is not necessarily the best option for long-term agriculture and food security, but a judicious mix of organic and conventional practices could boost global agriculture output. The principles of health, ecology, fairness and caring are the foundation of organic agriculture, highlighting its positive impact on the globe and the worldwide ambition to improve all agriculture. Organic food is becoming increasingly popular due to consumer concerns about the harmful effects of conventional agriculture on human health and the environment.

Contribution of Organic Farming to Sustainable Development

Organic farming offers several advantages, including reducing greenhouse gases, preserving culture and agriculture, and providing healthier food sources. Research from the University of California, found that organic food contains up to 58% more polyphenolics than conventionally grown food, which plants use as a natural defence system against pests (Reeve *et al.*, 2016). Organic farming also reduces the use of synthetic fertilizers, which require fossil fuels during manufacturing, and poses no environmental risks. Farmers use nutrient-rich composts and minerals like rock phosphate and greensand, and practise crop rotation to maintain soil health. Organic farming also reduces groundwater contamination, as no chemicals are used. Additionally, organic farming provides ample opportunities for rural jobs and fewer external costs for unnecessary pesticides and fertilizers. Overall, organic farming offers numerous benefits for both consumers and the environment.

Sustainable Development Goals (SDGs) aim to end world hunger, achieve food security and promote sustainable agriculture. Agriculture contributes to all 17 SDGs, either directly or indirectly (Lile *et al.*, 2023). Food security is crucial for reducing poverty, and good health starts with nutritious food. Gender equity can boost agricultural productivity, and sustainable agriculture can address water security. Organic agriculture helps overcome dependence on fossil fuels, affecting economic growth in agriculture-based countries. Land reforms can provide fairer access to rural land, and promoting sustainable agriculture can lead to physical, mental and economic health.

Ending hunger can contribute to establishing peace, justice and strong institutions. Fish provide 20% of daily animal protein and forests contain over 80% of the world's terrestrial biodiversity. However, organic farming carries higher costs and is time consuming, making it less sustainable and profitable. Studies have found that organic yields are 25% lower than conventional farming, and the cost of organic feed is higher than non-organic feed.

Sustainable agriculture adopts agroforestry practices, including integrated pest management, aquaponics and hydroponic farming. These practices use biological and mechanical methods to keep unwanted animals and insects from crops and can be used as natural insecticides. Aquaponics and hydroponic farming use fertigated water, while hydroponic farming uses materials like clay balls, coconut hair and fabric.

Sustainable farmers minimize the use of fertilizers and pesticides, reducing soil erosion and water pollution. They can adopt drip irrigation, cover crops for manure handling and graze animals. Organic farming aligns with the SDGs, promoting sustainable development and reducing soil erosion. This approach helps farmers manage manure and graze animals effectively (Lal *et al.*, 2021).

Future Prospects for Organic Farming

Organic farming is on the rise due to increasing global awareness and demand for sustainable products. Concerns about environmental sustainability, soil health and synthetic chemicals in agriculture have led to a shift towards eco-friendly and ethical practices. Countries and agricultural organizations are investing in organic farming practices. Advancements in technology and agricultural techniques are facilitating scalability and efficiency, making it economically viable for farmers. The expansion of organic farming is expected to lead to greater market access for small and medium-sized farmers, improved soil fertility, reduced environmental impact and enhanced biodiversity. The future of organic farming looks promising for both consumers and the environment.

Conclusions

Organic farming is a key solution to the global demand for environmentally friendly practices, promoting soil health and mitigating the negative effects of chemical inputs on ecosystems. It preserves biodiversity, protects pollinators and maintains a healthy balance in agricultural landscapes. Traditional, locally adapted farming methods foster resilience against climate change impacts. Organic farming empowers local farmers by reducing dependency on external inputs and promoting self-sufficiency. Consequently, embracing organic farming aligns with a nation's commitment to sustainable development, emphasizing environmental conservation and citizen well-being. Despite challenges, the long-term benefits of ecological health, economic viability and social well-being make it a vital component in shaping a sustainable future for the agricultural sector.

References

Anderberg, S. (2020) *The contribution of organic agriculture to poverty reduction*. In: Breiling, M. and Anbumohzi, V. (eds) *Vulnerability of Agricultural Production Networks and Global Food Value Chains Due to Natural Disasters*. Economic Research

Institute for ASEAN and East Asia, Jakarta, Indonesia, pp. 42–72.

Angon, P.B., Anjum, N., Akter, M.M., Shreeejana, K.C., Suma, R.P. *et al.* (2023) An overview of the impact of tillage and cropping systems on soil health in agricultural practices. *Advances in Agriculture* 1–14.

Chatterjee, R., Debnath, A. and Mishra, S. (2020) Vermicompost and soil health. In: Giri, B. and Varma, A. (eds) *Soil Health, Soil Biology*, Vol. 59. Springer, Cham, Switzerland.

Durán-Lara, E.F., Valderrama, A. and Marican, A. (2020) Natural organic compounds for application in organic farming. *Agriculture* 10(2), 41.

Gamage, A., Gangahagedara, R., Gamage, J., Jayasinghe, N., Kodikara, N. *et al.* (2023) Role of organic farming for achieving sustainability in agriculture. *Farming System* 1(1), 100005.

Kumar, D., Menon, M.K., Balasubramani, N., Sadalaxmi, A. and Banerjee, M. (2023) Certification of organic agriculture and trade issues. In: Kumar, D., Ravisankar, N. and Panghal, A. (eds) *Transforming Organic Agri-Produce into Processed Food Products: Post-COVID-19 Challenges and Opportunities*. CRC Press, Boca Raton, Florida.

Kumar, J., Ramlal, A., Mallick, D. and Mishra, V. (2021) An overview of some biopesticides and their importance in plant protection for commercial acceptance. *Plants* 10(6), 1185.

Kumar, N., Srivastava, P., Vishwakarma, K., Kumar, R., Kuppala, H. *et al.* (2020a) The rhizobium–plant symbiosis: State of the art. *Plant Microbe Symbiosis* 1–20.

Kumar, S., Samiksha, A and Sukul, P. (2020b) Green manuring and its role in soil health management. In: Giri, B. and Varma, A. (eds) *Soil Health*. Springer, pp. 219–241.

Lal, R., Bouma, J., Brevik, E., Dawson, L., Field, D.J. *et al.* (2021) Soils and sustainable development goals of the United Nations: An international union of soil sciences perspective. *Geoderma Regional* 25, e00398.

Lehmann, J., Bossio, D.A., Kögel-Knabner, I. and Rillig, M.C. (2020) The concept and future prospects of soil health. *Nature Reviews Earth & Environment* 1(10), 544–553.

Lile, R., Ocnean, M. and Balan, I.M. (2023) Challenges for zero hunger (SDG 2): Links with other SDGs. *Zero Hunger* 9.

Lim, S.L., Wu, T.Y., Lim, P.N. and Shak, K.P.Y. (2014) The use of vermicompost in organic farming: Overview, effects on soil and economics. *Journal of the Science of Food and Agriculture* 95(6), 1143–1156.

Lotter, D.W. (2003) Organic agriculture. *Journal of Sustainable Agriculture* 21(4), 59–128.

Migliorini, P. and Wezel, A. (2017) Converging and diverging principles and practices of organic agriculture regulations and agroecology: A review. *Agronomy for Sustainable Development* 37(6), 1–18.

O'Reilley, M.R. (2014) *The Barn at the End of the World: The Apprenticeship of a Quaker, Buddhist Shepherd*. Milkweed Editions, Minneapolis, Minnesota.

Pinto, A.P., Faria, J.M., Dordio, A.V. and Carvalho, A.P. (2023) Organic farming: a sustainable option to reduce soil degradation. In: Prasad, M.N.V. and Kumar, C. (eds) *Agroecological Approaches for Sustainable Soil Management*. Wiley, Chichester, UK, pp. 83–143.

Rajapakse, R.H.S., Ratnasekera, D. and Abeysinghe, S. (2016) Biopesticides research: current status and future trends in sri lanka. In: Singh, H.B., Sarma, B.K. and Keswami, C. (eds) *Agriculturally Important Microorganisms*. Springer, pp. 219–234.

Reeve, J.R., Hoagland, L.A., Villalba, J.J., Carr, P.M., Atucha, A. *et al.* (2016) Organic farming, soil health, and food quality: considering possible links. *Advances in Agronomy* 137, 319–367.

Rümmele, Y. (2023) Advantages and disadvantages of mulching in market garden systems. Doctoral dissertation, Hochschule Rhein-Waal.

Singh, M. (2021) Organic farming practices and sustainable agriculture. *Indian Journal of Organic Farming* 1(1).

Singh, R., Avasthi, R.K. and Babu, S. (2023) Organic farming: Issues and challenges in india. In: Singh, D.P., Prakash, H.G., Swapna, M. and Solomon, S. (eds) *Organic Crop Production Management: Focus on India, with Global Implications*. CRC Press, Boca Raton, Florida.

Soni, R., Gupta, R., Agarwal, P. and Mishra, R. (2022) Organic farming: A sustainable agricultural practice. *Journal of Thematic Analysis* 3(1), 21–44.

Vuković, A., Velki, M., Ečimović, S., Vuković, R., Štolfa Čamagajevac, I. *et al.* (2021) Vermicomposting: Facts, benefits and knowledge gaps. *Agronomy* 11(10), 1952.

Yadav, R., Singh, S. and Singh, A.N. (2022) Biopesticides: Current status and future prospects. *Proceedings of the International Academy of Ecology and Environmental Sciences* 12(3), 211–233.

13 Choice of Crops and Cropping Systems for Organic Farming in the Region: Concepts and Prevalent Practices

Bernhard Freyer[1]*, Pierre Ellssel[2], Gabriele Gollner[2], Aliyeh Salehi[2], Andreas Surböck[2], and Marie-Luise Wohlmuth[2]

[1]*Faculty of Business and Globalisation, University for Continuing Education Krems, Dr.-Karl-Dorrek-Straße 30 3500 Krems, Austria;* [2]*Division of Organic Farming (IFÖL), University of Natural Resources and Life Sciences, Gregor-Mendel-Straße 331180 Vienna, Austria*

Abstract

Organic farming relies on maintaining soil fertility and productivity through a comprehensive implementation of various practices, where the crop rotation plays an outstanding role. The crop rotation system in organic farming encompasses the following components: allocating 20–25% forage legumes, dedicating over 25% to catch crops and following a set of rules to ensure a continuous delivery of organic matter into the soil, utilizing rhizobia fixation for nitrogen supply, preventing erosion, boosting water retention capacity, suppressing weeds, fostering crop diversity and variety to mitigate risks from pests and diseases, while simultaneously enhancing the presence of beneficial organisms.

Additional practices integral to the sustainable and resilient nature of organic farming include liming, recycling organic residues, judicious use of specific inorganic fertilizers, as well as employing mechanical methods and liquid agents for pest and disease management. A cohesive tillage system underpins the effectiveness of all other practices. Organic farming operates not as a sum of individual practices, but rather as a fully integrated approach, where a majority of practices must be executed in harmony.

Keywords: agroecology, catch crops, crop rotation, forage legumes, intercropping, organic farming

What We Should Know Before Planning a Crop Rotation

Frequently overlooked, effective crop rotation planning and adjustment necessitate the fusion of fundamental knowledge and hands-on expertise in soil-plant-climate dynamics. The latter is crucial as numerous choices hinge on location-specific observations. All planning is based on an array of data points that can impact the overall crop rotation structure as well as the precise choices regarding crop selection and its associated practices (such as soil tillage, liming and fertilizer application).

In the present day, a multitude of tools are available for planning and analyzing crop rotations (Fenz *et al.*, 2023). Within this section, we introduce foundational principles essential for proficiently navigating organic farming crop rotations, which can also serve as a basis for modelling and evaluating such rotations.

Geology, soil and climate

Geology, soil and climate, collectively, shape the choice of crops, including the selection of particular crop varieties. To differentiate between them, we categorize their attributes into those that remain unaltered by farming practices and those that can be modified, adjusted or enhanced to achieve

*Corresponding author: bernhard.freyer@boku.ac.at

 Advances in Organic Farming (ed. L. P. Awasthi)
DOI: 10.1079/9781800626850.0013

increased crop yields and improved crop quality under specific circumstances.

Geology is a fixed factor that profoundly shapes the underlying soil attributes. A pivotal consideration pertaining to soil characteristics, which can be strategically manipulated to enhance growth conditions for a diverse range of crops, includes adjusting pH levels through liming, and optimizing nutrient content and accessibility as well as bulk density through organic matter, bolstering nitrogen availability via legumes and on-site organic fertilizers, and utilizing both internal organic sources and external organic and inorganic fertilizers. Additionally, soil carbon level and water holding capacity can be moderately enhanced through organic matter accumulation, crop biomass incorporation, and the application of organic manure.

In order to minimize the influence of slope gradient and topography to soil and water runoff, diverse crop sequences with cover crops and conservation tillage practices are necessary. The use of perennial crops and winter cover crops is important on sloping farmland to prevent soil erosion.

The length of the vegetation period primarily stems from climatic conditions. Nevertheless, through the application of relay cropping (where a second crop is sowed into a mature one, such as cowpea into sorghum), a subsequent crop can utilize residual soil moisture, at the very least serving as a catch crop for biomass production and nitrogen fixation. Certain crops offer various varieties with differing requirements for vegetation period length, thereby enhancing the flexibility of crop rotations. Nevertheless, the establishment of catch crops/underseeds depends on water availability during the vegetation period, and the water usage of intermediate crops can diminish the water availability for subsequent main crops.

The natural occurrence of annual or biannual rainy seasons is affected by climate change, leading to shifts in the commencement and conclusion of each season, including both short and long rainy periods. Consequently, unforeseen patterns of rain distribution ensue, influencing the duration of growing seasons. Under such circumstances, irrigation systems become essential for mitigating irregular rainfall distribution. Regarding irrigation, optimizing conditions for vegetables can be accomplished through the utilization of drip irrigation techniques. Larger farms have the capability to enhance water supply by employing sprinkler systems. In contrast, smallholder farmers are limited to trench irrigation and other technologies, contingent upon the accessibility of water resources.

Direct and indirect pre-crop effects

For designing crop rotations, we distinguish the direct and indirect pre-crop value, which provides us orientation for the follow-up of crops. While the first is of very high impact on the immediately proceeding crop, the indirect pre-crop informs about the accumulated impact of a crop on the current crop. The soil serves as a storage medium where various plant-specific effects are accumulated over a certain period. This assessment also includes catch crops in its consideration.

Each crop type can be characterized via its impact on the development of soil fertility, weeds, and pests and diseases that finally impact the growth conditions of the subsequent crop (refer to Table 13.1 for details; adapted from Freyer, 2019):

1. *Soil fertility development:* Shoot and root biomass with predetermined mineral content, and particularly the chemical element nitrogen and C/N ratio, along with organic matter maturity. Root attributes encompass two aspects: (i) biochemical root characteristics involving nutrient mineralization, exudates, and allelopathic effects; and (ii) biophysical root characteristics such as bulk density, biopore volume, aggregate stability and soil air and water retention.
2. *Weed development:* Capacity to compete with weeds, quantity and dispersion of weed diaspores and rhizomes in the soil, as well as the uptake and storage of (micro-)nutrients by weeds.
3. *Pest and disease development:* The pre-crop's role as a host plant for particular pests and diseases, but also beneficials, the volume and quality of organic matter, and the soil's antiphytopathogenic potential, characterized by the presence of microorganisms capable of mitigating diseases.

It is important to bear in mind that the wide range of genotypes contributes to a varied spectrum of crop-specific pre-crop values. Pertinent characteristics include: required vegetation period, yield capacity, ability to foster/reduce pests and diseases, potential for weed suppression, harvest index, shoot and root biomass, nitrogen fixation capability, nitrogen distribution between marketable produce and shoot/root residues/biomass, and root attributes, encompassing exudates.

Table 13.1. Characteristics of crops for arable use. (From Freyer *et al.*, 2020, extended)

Crop group/ family	Below-ground biomass*	Above-ground biomass*	Nitrogen fixation *	Weed suppression*	Accumulated pre-crop value*	Preferable subsequent crops/ other plant families**
Forage legumes						
Alfalfa	+++	+++	+++	+++	+++	Maize, wheat, potato, sweet potato
Clover (red)	+++	+++	+++	+++	+++	Maize, wheat, potato, sweet potato
Clover (white)	+++	+++	+++	+++	+++	Maize, wheat, potato, sweet potato
Desmodium	+++	+++	+++	+++	+++	Maize, wheat, potato, sweet potato
Forage/grain legumes						
Cowpea	++	+ to ++	++	++	++	Potato, sweet potato, teff, cereals
Lablab	++	++	++	++	+ + to +++	Maize, wheat, potato, sweet potato
Mucuna	+++	+++	+++	+++	+++	Maize, wheat, potato, sweet potato
Grain legumes						
Bean spp.	+	+	+	+	++	All cereals with higher N-demand
Chickpea	+	+	+	+	+	All cereals with higher N-demand
Faba bean	++	++	++	++	++	All cereals with higher N-demand
French bean	+	+	+	+	++	All cereals with higher N-demand
Garden pea	+	+	+	+	++	All cereals with higher N-demand
Lentil	+	+	+	+	+	All cereals with low N-demand
Lupin spp.	++	+ to ++	++	++	++	All cereals with higher N-demand
Peas	+	+	+	+	++	All cereals
Soy beans	+	+	+ to ++	+	++	All cereals
Cereals						
Barley	+	+	-	+	+	Oat, all legumes, oil crops
Maize	+	++	-	+	+	Teff, oat, millet, legumes, oil crops
Millet	+	+	-	+	+	Legumes, oil crops
Oat	+	+	-	+ to ++	+	Legumes,
Sorghum	+	+++	-	+ to ++	+ to ++	Legumes
Teff	+	+	-	+	+	Legumes
Wheat	+	+	-	+ to ++	+	Barley, teff, legumes
Root crops						

Continued

Table 13.1. Continued

Crop group/ family	Below-ground biomass*	Above-ground biomass*	Nitrogen fixation *	Weed suppression*	Accumulated pre-crop value*	Preferable subsequent crops/ other plant families**
Potato	+	+	-	+ to ++	+	All cereals, all legumes
Sweet potato	+	++	-	++	+ to ++	All cereals, all legumes
Cassava	+	+ to ++	-	+ to ++	+	All cereals, all legumes
Taro	+	+ to ++	-	+ to ++	+	All cereals, all legumes
Oil crops						
Flax	+	+	-	+	+	All crops except oil crops
Linum	+	+	-	+	+	All crops except oil crops
Hemp	++	++	-	++	++	All crops except oil crops
Rape	++	+to++	-	++	++	All crops except oil crops
Sunflower	++	+to++	-	++	+	All crops except oil and grain legume crops
Grasses						
Napier grass	++	+++	-	++	++	Legumes
Brachiaria	++	++	-	+-++	+-++	Legumes
Ray grass	++	++	-	+-++	+-++	Legumes, oat, potato
Catch crops						
Leg. mixtures	+ to ++	+ to ++	+ to ++	++	+ + to +++	Maize, wheat, potato, sweet potato
Non-leg. mixture	+ to ++	+ to ++	-	++	+ to ++	Maize, wheat, potato, sweet potato

[a]+ = low; ++ = medium; +++ = high; - = no nitrogen fixation
[b]Ranking: the first crop profits the most from the pre-crop

Each crop group possesses distinct attributes that qualify them as potential pre-crops for other crops within the group or even for themselves (refer to Table 13.1). Based on these traits, they assume designated roles and proportions within a crop rotation scheme (Kolbe, 2006).

The significance of nitrogen storage in crop residues becomes particularly pronounced when evaluating the direct pre-crop value (refer to Table 13.2). The accessibility of these residues hinges on factors such as the C/N ratio, soil biological conditions, and soil and climate characteristics.

The combined influence of crop characteristics and associated agronomic practices deposited in the soil establishes the groundwork for the indirect pre-crop effect. Together, these interventions contribute to shaping soil fertility within a cropping system. Typically, the impact of the indirect pre-crop effect on the current crop is less significant compared to that of the direct pre-crop effect.

Share of crops and interruption of years

A pivotal guideline for evaluating indirect pre-crop effects is provided by the proportion of a specific crop or crop group (e.g. cereals) within a crop rotation (Freyer, 2003). As the share of a particular crop increases, along with the prominence of a crop

Table 13.2. Nitrogen storage in root residues and nitrogen demand of a following crop. (From Freyer, 2019, modified)

Crop	Crop yield	Nitrogen demand	Root biomass	Root nitrogen	Root nitrogen
Unit	DM t ha^{-1}	Tot. N kg ha^{-1}	DM t ha^{-1}	kg 1000 kg^{-1} ha^{-1}	N kg ha^{-1}
Lucerne (1 year)[a]	10	–	5	22	110
Potato (FM)	200	100	0.5	90	45
Rye	4	100	2	12	24
Oat	3	60	2	12	24
Field bean	3	–	1.8	18	32
Winter wheat	5	125	1.5	12	18
Spelt	3	60	1.5	12	18

[a]Mineralization over 1–3 years; further 130–220 kg N ha^{-1} can be expected by the above-ground biomass

Table 13.3. Share of different crop groups in crop rotations. (From Freyer *et al*., 2020)

Crop families	% in the rotation	Crop examples
Forage legumes	20–25	Alfalfa, desmodium, clover/clover-grass mucuna, mung bean, lablab, cowpea, *Vicia* spp.
Grain legumes	10–20	Bean, pea, chickpea, cowpea, lablab, lentil, lupin, soy bean, vetch
Cereals	up to 60	Barley, maize, millet, oat, rice, rye, sorghum, teff, wheat
Oil crops	up to 20	Rape, sunflower, pumkin
Root crops	up to 20	Beet root, cassava, potato, sweet potato
Intercropping	10–25	Cereals with grain legumes
Catch crops	10–33	All kind of species rich mixtures

group (i.e. family), the risk increases for cumulative adverse factors to potentially reduce the yield of the subsequent crop and/or impact the entire crop rotation.

In the case of each crop, or other crops sharing similar characteristics from the same or different crop families, it is important not to exceed a designated maximum share (refer to Table 13.3). A substantial presence of a particular crop can result in issues like nutrient imbalances, humus reduction or the emergence of soil-borne pests and diseases that could adversely impact at least the subsequent crop (Fenz *et al.*, 2023).

Regarding cereals, we further refine the allocation within the cropping system based on the specific soil characteristics (refer to Table 13.4).

Soil-borne pests and diseases are inherently tied to specific crops. They endure for a defined span within the soil, crop residues, or organic manure. To prevent infections over a certain period, it is advisable to abstain from cultivating crops that act as hosts (refer to Table 13.5)

Catch crops

In temperate climate conditions, there exists a window of opportunity between the main crop's harvest and the subsequent main crop's sowing, during which catch crops can be established. These catch crops can serve as green manure within the same plot, be utilized via cut-and-carry methods for another plot, be used for forage through cutting or grazing, or in some cases even as food crops with a brief growth period.

The above-ground dry-matter yield of catch crops typically ranges from approximately 1–5 tons per ha, while below-ground root biomass falls within the range of 0.3–2.5 tons per ha. Nitrogen fixation by catch crops varies from 30 to 80 kg per ha.

Catch crops offer a multitude of ecological functions. They serve as a protective soil cover against intense sunlight and heavy rainfall, augment water infiltration, and contribute to soil aeration, often alongside technical interventions. Additionally, they act as nutrient reservoirs, preventing surplus

Table 13.4. Maximum share of cereal crops and crop groups in crop rotations. (From Freyer, 2019, modified)

Crop, crop groups	High soil quality	Low soil quality
	Crop %	
Wheat	33	25
Winter barley, triticale	40	33
Rye	50	33
Total winter cereals (rotations without fodder, barley/oat)	66	50
Oat	25	20
Spring barley	33	25
Total spring cereals (rotations without winter cereals)	50	33

nutrient storage from the preceding crop and averting nutrient leaching and soil erosion, especially during heavy spring rains. Furthermore, they aid in carbon sequestration and, in the case of legumes, provide nitrogen enrichment. Catch crops contribute to weed suppression and create favourable habitats for beneficial organisms, while simultaneously diminishing pest populations.

The pre-crop value of catch crops significantly varies based on distinct growth conditions and crop types. They can enhance or, in certain cases, hinder growth conditions for subsequent crops. In optimal circumstances, they can compensate for a weak pre-crop value of the prior main crop.

Species selection hinges on factors such as germination speed and soil cover, nitrogen contribution potential for the subsequent crop, weed pressure resultant from the previous crop, and the requirement for minimal weed risk for the ensuing crop. While adhering to crop rotation guidelines, catch crops must also conform to established rules. However, there is room for flexibility – after a legume, a legume catch crop can be sown, as long as the subsequent crop comes from a different non-leguminous crop family.

Varied vegetation periods among main crops present diverse opportunities and associated functions for establishing catch crops. The primary distinction lies between spring and winter catch crops. Spring catch crops succumb to frost and snow, whereas winter catch crops endure these conditions.

- *Short vegetation periods:* A minimum of approximately 40 days is required for a catch crop. The harvest of the preceding crop is at the latest in August, with the catch crop being harvested approximately two weeks before sowing a winter crop. The earliest sowing of a winter crop is approximately the beginning of September; the latest sowing of a winter crop typically occurs at the beginning to middle of November. Fast-growing crops or crop mixtures are utilized in this scenario.
- *Extended vegetation periods:* Two types of extended periods exist. The first occurs when the main crop is harvested as early as June (e.g. barley, peas), with the subsequent main crop scheduled for sowing in October. The second occurs with spring crops sown from February to May. In the first scenario, the catch crop is typically an annual, while in the second scenario, it can include biennial crops like clover or alfalfa.

An alternative approach involves undersowing forage legumes beneath a main crop such as wheat or maize, cultivating crops specifically for green manure or forage purposes. Unlike the previous system where catch crops are grown to bridge the gap between two growing seasons, this alternative system serves multiple roles. It contributes to weed suppression, diminishes the risk of pests and diseases, collects nitrogen, lowers the erosion risk, enhances aggregate stability, and improves both aeration and biological activity. Once the main crop is harvested, the forage legume continues to grow and, if biennial, can be utilized multiple times before eventually transitioning into a main crop.

Catch crop residues are typically integrated into the soil and left to decompose approximately two weeks prior to sowing the subsequent crop. In the case of green manure, it is advisable to plough to a depth of no less than 15 cm. This practice prevents nutrients from being buried too deeply in soil layers that plants cannot access. Deeper burial restricts

Table 13.5. Cultivation break in organic crop rotations in number of years to avoid seed and/or soil-borne pests and diseases (mainly temperate regions). (From Kolbe, 2006; Freyer, 2019; Landwirtschaftskammer Österreich, 2016)

Crop	Years	Pest and disease risk
Cereals		
Maize	1–2	Stemborer; root and stem rot, nematodes
Millet	2–3	Fungal pathogens, in particular foot and root diseases
Oat	3–5	Oat cyst nematodes
Rye	1–2	*Urocystis occulta, Claviceps purpurea*
Spring barley	2–3	Cereal cyst nematodes, powdery mildew
Triticale	2–4	*Septoria tritici, Claviceps purpurea*
Winter barley	2–4	*Drechslera teres, Helminthosporium teres* (also spring barley), *Pyrenophora teres*; *Ramularia collo-cygni* (also spring barley); *Cochliobolus sativus, Drechslera sorokiniana, Helminthosporium sativum* (also other cereals)
Winter wheat	2–4	*Fusarium graminearum, F. culmorum a. o. Fusarium-species* (see also maize, soybean, peas; also other cereals), *Septoria tritici, Septoria nodorum or Stagonospora nodorum* (also spring wheat, rarely barley or rye), *Drechslera tritici repentis, Gaeumannomyces graminis* (also barley and rye), *Helminthosporium tritici repentis, Tilletia caries*
Root crops		
Cabbages	3–6	Clubroot, sugar beet cyst nematodes
Potato	3–4	Potato cyst nematodes
Pumpkin	3–5	*Ascochyta cucumis*
Sugar beet, forage beet	3–5	Sugar beet cyst nematodes; *Aphanomyces Root Rot, Cercospora Leaf Spot, Fusarium Root Rot, Pythium Root Rot, Rhizoctonia Root Rot, Verticillium* wilt
Oil crops		
Flax	3–4	See sugar beet; *Sclerotinia* stem rot; cutworms, nematodes, wireworms,
Linum	3–6	*Pythium, Rhizoctonia, Thielaviopsis basicola, Olpidium, Fusarium oxysporum f. sp. lini*
Hemp	1–2	*Fusarium wilt*, various fungi damaging the roots
Rape	4–5	Various fungi damaging the roots, sugar beet cyst nematodes, clubroot
Sunflower	7–8	*Sclerotinia,* aphids, sunflower (head) moth, cutworms, thrips, downy mildew, powdery mildew, rust, *Verticillium* wilt
Grain legumes		
Faba bean	3–5	Viruses, fungal and animal pests
Field bean, Garden bean, Lupine, *Pelusium* spp., *Vicia* spp.	3–5	Incompatibility relationships, blight disease, foot and root diseases, stem nematodes, lupine wilt
Peas	4–14[a]v	See field beans
Soybeans	3–4	Seed corn maggot, wireworms, white grub, *Pythium*, *Phytophthora*
Forage legumes		
Crowtoes (*Lotus corniculatus*), Egyptian, Persian, and yellow clover	3–4	Like other clover species, white clover is largely self-compatible
Italian clover (*Trifolium incarnatum*), lucerne, red-, swedish- (*Trifolium hybridum*)	4–8	Clover cancer, *Fusarium* wilt, leaf spot disease, and others
White clover	-	-

Continued

Table 13.5. Continued

Crop	Years	Pest and disease risk
Sainfoin (*Onobrychis vicifolia*)	5–6	*Fusarium* spp., leaf spot, ring spot, leaf and stem spot, rust, chocolate spot, powdery mildew, Sitona weevils, sainfoin bruchid (*Bruchidis unicolor*)

Leitlinie für den Integrierten Feldbau (guideline for integrated agriculture); Österreichische Arbeitsgemeinschaft für integrierten Pflanzenschutz (Austrian working group for Integrated Plant Protection), and others
[a]in case soils are infected, long-term interruption is necessary or fungal pathogens, in particular foot diseases; Cereal cystella, *Thyphula*, Powdery mildew

biomass decomposition due to insufficient oxygen supply.

In certain situations, the duration of catch crop vegetation may be curtailed to accommodate the need for rigorous soil tillage aimed at controlling challenging weeds. Conversely, in scenarios where a substantial number of seeds have been dispersed from the prior harvest and can serve as the catch crop (e.g. following pea harvest), it is feasible to establish the catch crop without resorting to tillage (minimum or zero tillage). This approach not only conserves soil water but also minimizes soil disturbance.

Catch crops face limitations when dealing with inadequate rainfall, potentially jeopardizing the water supply for the subsequent main crop. If the germination rate of catch crop seeds is compromised due to poor seedbed quality, dry conditions during germination, or disruption caused by heavy rainfall, or if straw cover impedes regular germination instead of aiding weed germination, the consideration of resowing becomes necessary. In instances of elevated weed pressure, and to prevent catch crops from becoming weeds themselves, measures such as mulching or pre-germination cutting for use as forage should be implemented prior to weed (i.e. catch crop) germination.

In subtropical and tropical regions characterized by biannual rainfall patterns, the brief rainy periods, known as short rains, offer an opportunity for implementing catch crops, as an alternative to the conventional practice of leaving fields fallow for weed control. In regions with insufficient food production, these short rainy seasons are also utilized for cultivating food crops. Changing rainfall patterns pose challenges for smallholder farmers as they strive to adhere to crop rotation rules and effectively manage their crops during the short rain season.

Within the context of organic farming, legumes play a pivotal role as catch crops. Table 13.6 offers a selection of examples under subtropical and tropical climatic conditions.

Intercropping

To use vegetation periods and resources as best as possible, intercropping is a classical cropping approach of organic farming systems. The temporal and spatial distribution of a high diversity of crops and species in an organized manner offers several advantages:

1. optimizing use of the vegetation period via maximizing the use of natural resources such as water, nutrients, light and temperature;
2. reduction of pest and disease pressure and of weed pressure;
3. high availability of nutrients for the following crop;
4. optimal germinating and growing conditions for the following crop;
5. optimal provision of biomass for humus production;
6. optimal provision of nitrogen via legumes;
7. increase of biodiversity and hosts for predators (antagonists against pests);
8. contribution to increased crop yield, food, feed or green manure;
9. optimal long-term conditions for the whole crop rotation (positive indirect pre-crop effect: accumulated long-term effects of one crop on soil fertility).

Intercropping involves the simultaneous cultivation of two or more crops on the same field. Its objective is to enhance yield by maximizing resource utilization and leveraging ecological processes that would not manifest when planting a solitary crop.

Table 13.6. Suitable crops for green manure, feed and partly as food under drought, tropical and subtropical climates. (From Meitzner and Price, 1996; CIMMYT, 2015); https://www.feedipedia.org [accessed 27/1/25]; enhanced and modified)

Crop	Description	Usage
Apple-ring acacia *Faidherbia albida*	Deep rooting legume tree up to 40 m; inverse phenology: it sheds its leaves during the wet season, while the leaves mature during the dry season and the pods ripen at the end; native to arid or semi-arid areas; 30–120 kg seeds per tree per year.	Intercropped with maize, sorghum and pearl millet; used as fodder and famine food.
Cowpea *Vigna unguiculata*	Annual herbaceous legume; tolerance for sandy soil and low rainfall in semi-arid regions; shade-tolerant, intercropped with maize or sorghum; hay yield between 0.5 and 4t DM/ha, and 0.5 and 3 t/ha grains; N-fixation from 20 kg to 140 kg N/ha; strong root system.	High pre-crop value; grains for food; whole plant for feeding.
Hairy vetch *Vicia villosa*	Annual/biannual legume; medium tolerance to drought; better tolerance in acidic soils than any other legumes; up to 12t DM/ha.	Fodder use, cover and green manure crop, intercropped with small seed crops; fodder and can be grazed, or cut for hay and silage, but can be poisonous; grazing-sensitive.
Jack bean *Canavalia ensiformis*	Climbing perennial legume commonly cultivated as an annual. Up to 2m high; strong, deep root system, withstanding dry periods. Tolerant to waterlogging and salinity.	Pods and seeds are edible and used for food, the young pods being cooked as a vegetable. Plant, pods and seeds are also used as feed. Used as green manure.
Lablab *Lablab purpureus*	Fast-growing (on fertile soils) annual legume; pioneer crop; 3–9t DM/ha; N-fixation between 20 kg and 150 kg/ha; deep taproot; drought-resistant; shade-tolerant; grows well through the dry season if soils are deep; high temperature, soil type and pH tolerance; high-pre-crop value.	Fodder (pasturing, hay, silage), green manure, can be cultivated at the end of the growing season to provide dry-season pasture for livestock; intercropping with maize or sorghum; edible, but value is low.
Leucaena leucocephala	Fast-growing, evergreen legume tree, up to 5m and more. N-fixing: 150–300 kg/ha; deep taproot; erosion control, land reclamation; improving water penetration and reducing surface runoff; prevents saline subsoil water from reaching the surface.	Wood, good-quality charcoal, small furniture and paper pulp. Young shoots, young leaves and seeds used as vegetable; lightly grazed in the first year after seedling and heavily grazed after the second year. Yield 3–30t DM/ha/year; shelterbelt (preventing wind damages) to cocoa, coffee and tea and climbing crops (vanilla, pepper, passion fruit); green manure in alley cropping systems; leaves decompose quickly.
Lucerne *Medicago sativa*	Deep-rooting legume up to 9 m; drought-tolerant; can be used up to 3 years; biomass yield 6–20t DM/ha; N-fixation: 80–300 kg N/ha and year.	High pre-crop value; high forage production, specifically if irrigation is possible, protein-rich. Intercropped with maize and cassava. Best incorporated into the soil when leaves are fresh and green, approximately 16 weeks after planting or to continue and use as a forage crop.

Continued

Table 13.6. Continued

Crop	Description	Usage
Velvet bean *Mucuna pruriens*	Mainly annual legume; supplies high amount of nitrogen; covers the soil completely then climbs high; heavy nodulation in infertile and sandy soils; drought-tolerant once established; 10–35t green material/ha (3–10t DM/ha) and 250–3300 kg seeds/ha; N-fixation 80–330 kg/ha.	Fodder and feed legume; pre-crop before high-nutrient demanding main crop; also as an intercrop with maize; cover and green manure crop.
Pigeon pea *Cajanus cajan*	Under optimal conditions, it can grow up to 5 m, with a poisonous taproot reaching a length of 2 m, developing rapidly.	Providing green biomass for feed and green manure and grains for food and feed at the end of dry seasons.
Sesbania sesban	Fast-growing, perennial legume tree, up to 8 m, with a shallow root system. pH and low P level-tolerant, temperature-tolerant.	Forage for cutting or grazing, green manure, firewood and fibre for cordage, medicinal purpose against tsetse flies.
Silverleaf desmodium *Desmodium uncinatum*	Rich shallow-rooted perennial legume; biomass yield 5–10t DM/ha and year; N-fixation: 70–110 kg N/ha.	High pre-crop value; high forage production specifically, if irrigation is possible; part of push-pull system; best incorporated into the soil when leaves are fresh and green, approximately 16 weeks after planting or to continue and use as a forage crop.
Sunhemp *Crotalaria juncea*	Vigorous upright legume; good effects of suppressing weeds pest and diseases; drought-resistant; yield: 18–25 t/ha fresh biomass; N-fixation: 50–60 kg N/ha within 60–90 days of cultivation.	Good before low-demanding crops; especially suitable for weed control; can also be used as intercrop with banana, citrus and coconut, and others; use as green manure and feed.

Nutrients and water can be efficiently harnessed by blending species with distinct characteristics, such as varying rooting depths. The available space can also be judiciously employed by selecting crops with unique micro-environments that can be shared by other crops. For instance, when a tall crop is planted, space can be optimized by introducing shade-tolerant crops beneath it. The amalgamation of crops also yields positive effects in terms of weed suppression and pest control, facilitated by the synergistic behaviour of plants and the resultant soil cover. Various intercropping systems exist:

1. *Mixed intercropping:* This involves planting two crops as a mixture, such as forage legume-grass-herb mixes or catch crops with different species.
2. *Row intercropping:* In this approach, two or three crops are sown in distinct rows, like oat-faba beans, maize-beans-squash, or wheat with undersown clover.
3. *Strip intercropping:* This strategy combines crops planted in multiple rows, as seen in examples like sorghum-beans-turnip.
4. *Relay intercropping:* This entails undersowing a crop in the main crop before its maturity, such as cowpea or lablab under maize.
5. *Multi-storied intercropping:* This method combines field crops with alley crops. For instance, rows of *Leucaena leucocephala* are interspersed with a field that is 8m wide, planted with potato or maize.

The challenge lies in adhering to crop rotation rules. On one hand, crops are considered when calculating the total share within a crop rotation. On the other hand, since crops are planted in combinations, exhaustive calculations might not always be essential. The same applies to the intervals between crops.

Agronomic practices

Pre-crop effects inherently encompass all accompanying agronomic activities associated with crops (refer to Table 13.7). The influence of practices like varied tillage methods, organic solid or liquid

Table 13.7. Circumstances under which the diverse agronomic practices should be applied.

Agronomic practice	To apply, if:
Tillage	
Zero tillage 0 cm	… expected weed pressure is low, the soil is loosened, the crop residues do not hinder the germination of seeds and the period between harvest and sowing of the following crop is short. In case of relay cropping, surplus of nutrients and water can be used efficiently.
Rotor tiller, hand hoe – 8 cm	… weed pressure is low, or a second application of a harrow to reduce germinated weeds is planned; to mix higher amounts of crop residues into the soil.
Shallow ploughing –15 cm	… subsoils aren't compacted, to mix higher amounts of crop residues into the soil.
Deep ploughing >15 cm (plough turns max. 120 degrees)	… subsoils are compacted and weed pressure high.
Disc plough – 8 cm	… share of root weeds and rhizomes is very low.
Fertilizer	
Fresh animal manure	… soil structure is compacted and crops like cereals that accept top dressing are sown.
Mature animal manure	… soil structure is compacted, and root crops are planted.
Compost from alley branches and other green residues	… soil structure is compacted, humus content and soil fertility are low, but also to maintain the humus level.
Bioslurry	… nutrient reserves in soils are low, and nutrient demand specifically for nitrogen is high.
Liming	… soil structure is damaged, pH is low, nutrient availability is low due to leaching processes.
Phosphorus fertilizer	… availability and soil stock of the element is low, demand is high, e.g. for forage legumes.
Potassium fertilizer	… availability and soil stock of the element is low, demand is high, e.g. for root crops.
Sulphur fertilizer	… availability and soil stock of the element is low, demand is high, e.g. for cereals.
Off-farm compost	… the overall humus level is low.
Off-farm organic liquid fertilizer	… the nutrient demand is high, e.g. from vegetables and in case of nitrogen of relevance for wheat and maize specifically.

manure application, liming or mineral fertilization on soil attributes and crop growth conditions must be comprehensively understood and tactically integrated within the framework of crop rotation.

When it comes to different tillage 'philosophies', the following aspects need to be considered: under organic farming conditions, long-term zero-tillage systems application increases weed pressure and accumulates humus in the upper layer but less in the lower layer. Additionally, it can elevate bulk density and soil compaction. Given these considerations, it is advisable to employ various tillage systems and technologies flexibly. This should be based on the performance of the pre-crop, soil and weed status, crop and other residues, and the requirements of the subsequent crop.

Designing Organic Crop Rotations

Planning organic cropping systems entails an ongoing adjustment to shifting soil, climate and weather conditions, pre-crop performance, and market dynamics, including the economic value of crops. Crop rotations are not fixed entities; rather, they offer a framework that serves as a guiding reference.

Framing the cropping system

What distinguishes cropping systems in organic farming is their multi-faceted role, which becomes essential due to the limitations on fertilizers and the exclusion of mainstream pesticides and herbicides. Given this perspective, an effective crop rotation should be framed of crops possessing characteristics like crops with abundant biomass and minimized open soils to mitigate soil erosion; crops such as forage legumes, renowned for their substantial root biomass that enhances soil aeration, nutrient transfer from lower to upper soil layers, nitrogen fixation capability, weed competition ability, carbon contribution to soil, augmentation of microorganism activity and population, and the promotion of antiphytopathogenic potential in soils, along with prolific above-ground biomass; elevated crop diversity to diminish the threat of pests and diseases, establish habitats for beneficial organisms and other fauna, and enrich the soil with a broad array of organic matter that sustains the biodiversity of soil microorganisms.

Building upon this initial guidance, crop rotation rules incorporate the following functions:

1. Ensure 20–25% inclusion of forage legumes in the system to cover nitrogen and carbon demand.
2. Alternate between crops that require significant humus input and those that contribute to humus production.
3. Alternate between nitrogen-fixing and nitrogen-requiring crops to balance between nutrient supply and demand.
4. Alternate between crops with potential soil compaction risk during harvest and those with lower risk.
5. Alternate between spring and winter crops to reduce weed pressure.
6. Integrate catch crops, accounting for more than 25% of the rotation to reduce risk of soil erosion.
7. Adhere to the maximum crop share within the rotation to increase biodiversity and to reduce pest and disease, and weed pressure risk.

Table 13.8 provides a comprehensive framework outlining the various functions of crop rotations.

By adhering to this approach, the risk of crop failure remains minimal. In instances where

Table 13.8. Investment into crop rotation functions (temperate regions).

	Investment into crop rotation functions via crop rotation management		
Year	High	Medium	Low
1	Forage legumes	Forage legumes	Grain legumes
2	Forage legumes	Winter cereal/maize	Winter cereal/maize
		Catch crop	*Catch crop*
3	Winter cereal/maize	Winter cereal	Oil crop
	Catch crop	*Catch crop*	*Catch crop*
4	Spring cereal/oil crop	Spring cereal/oil crop	Spring cereal
	Catch crop	*Catch crop*	*Catch crop*
5	Root crop/cereal	Root crop/cereal	Grain legume
	Catch crop	*Catch crop*	*Catch crop*
6	Grain legume	Grain legume	*Root crop*
	Catch crop	*Catch crop*	*Catch crop*
7	Winter cereal	Winter cereal	Winter cereal
	Catch crop		
8	Spring cereal		

Catch crop with a high share of legumes; integration: minimum six weeks, according to site-specific conditions, pre-crop performance and demand of the succeeding crop; catch crop: can be also an undersown crop; the lower the share of catch crops and the shorter the vegetation period and catch crop yield, the lower the investment into soil fertility.

circumstances compromise the performance of a pre-crop, particularly catch crops (as detailed below) or forage legumes, temporary extensions may be considered. Another alternative involves concentrating (organic) fertilizers on specific fields.

A pivotal consideration when designing crop rotations under organic farming conditions is meeting the feed demand of animals, which should ideally be met through on-farm production. Therefore, it is advisable to maintain a surplus in forage production. The inclusion of 20–25% forage legumes, along with catch crops, assumes paramount importance for the benefit of both soil organisms and animal husbandry. In stockless farming scenarios, forage legumes directly contribute as nitrogen fertilizers.

History of a field

When a farmer decides to transition to organic farming, a thorough analysis of the previous crop rotation and all agronomic practices employed becomes essential. Conducting an assessment of the cumulative nutrient content of the soil and examining nutrient balance spanning the past decade proves advantageous for gaining insights into the overall nutrient reservoir. While the comprehensive nutrient availability might initially be of concern, the inclusion of 20–25% forage legumes and over 25% catch crops will naturally enhance nutrient availability, given proper pH levels and the consistent application of suitable lime varieties (it is important to note that 'lime is not lime', e.g. calcium carbonate, dolomitic lime, basalt meal, shell lime and others).

If the cropping system has been heavily skewed towards just two or a maximum of three crops, it is strongly advisable to consider a temporary increase in the proportion of forage legumes. In the long run, this investment will yield dividends. Otherwise, there is a risk of elevated weed pressure over the years, subsequently diminishing crop yields.

Economy of a crop and market demand

Naturally, the crop rotation must also align with the farm's economics and market requirements. To achieve both ecological and economic success, farms necessitate a diverse array of crops. Often, farmers must adjust their crop rotation to accommodate changing market demands and the evolving economic value of specific crops. Therefore, the ability to diversify the crop spectrum and maintain flexibility in crop selection from year to year is crucial for farmers. Exceptions where crop rotation rules must be amended are rare. In such instances, bolstering catch crops becomes an option to rectify deviations from established crop rotation rules or to enhance the application of organic manure.

Crop rotation examples

For annual rainfall patterns, whether in temperate, arid or select humid regions, it is advisable to consistently devise sequences comprising three or four crops (Table 13.9). However, it is imperative to maintain a minimum share of forage legumes exceeding 20% within the overall crop rotation. Examples of three- or four-crop sequences with cereals are as follows (numbers indicate the year):

A. 1. forage legumes; 2. winter cereal – catch crop; 3. spring crop
B. 1. forage legumes; 2. potato; 3. winter cereal
C. 1.–2. forage legumes; 3. winter cereal – catch crop; 4. spring crop – catch crop
D. 1. forage legumes; 2. winter cereal – catch crop; 3. grain legumes – catch crop; 4. winter cereal – catch crop

Crop rotations can span up to 12 years, particularly in biodynamic farming. In organic farming, typical rotation durations range from five to eight years, resulting in rotation patterns as outlined below:

1. Five-year rotation: 1. forage legumes; 2. winter cereal – catch crop; 3. spring cereal – catch crop; 4. root crop – catch crop; 5. vegetables
2. Six-year rotation: 1.–2. forage legumes; 3. winter cereal; 3. catch crop; 4. spring crop – catch crop; 5. root crop – catch crop; 6. vegetables
3. Seven-year rotation: 1.–2. forage legumes; 3. winter cereal – catch crop; 4. spring crop – catch crop; 5. root crop – catch crop; 6. grain legumes – catch crop; 7. winter cereal
4. Eight-year rotation: 1.–2. forage legumes; 3. winter cereal – catch crop ; 4. spring crop – catch crop; 5. root crop – catch crop; 6. grain legumes – catch crop; 7. winter cereal – catch crop; 8. spring cereal

Table 13.9. Example of a crop rotation for annual and biannual rainfall patterns with catch-, inter- and relay cropping. (adapted from Freyer *et al.*, 2020)

Year	Season	Annual rainfall/one season	Example 1: Biannual rainfall/two seasons	Example 2: Biannual rainfall/ two seasons
1	Long	Red clover	Forage legumes (e.g. Alexandrine or Persian clover)	Maize with undersown desmodium
	Short	-	-	Desmodium
2	Long	Maize-IC (white clover)	Maize-IC (grain legumes)	Wheat
	Short	White clover (the whole season with irrigation)	RC cowpea	Beans
3	Long	Potato	Potato	Potato
	Short	-	Lablab	Lablab
4	Long	Rye with undersown white clover	Wheat-IC (white clover)	Vegetables and/or herbs
	Short	-	Finger millet or oil crop	Cowpea
5	Long	Spelt with undersown red clover		

Short season: in case of annual rainfall in temperate climate it is the season for catch crops;
CC = catch crop; IC = intercropping; RC = relay cropping

Instead of cereals, oil crops such as hemp, flax, rapeseed or squash can also be integrated with a maximum share of 20% (see also spring crop).

Conclusion

Organic crop rotations, characterized by high crop diversity and rooted in science-based principles, function as a strategic framework. They hold ecological significance, exhibit climate resilience, mitigate the risk of weed pressure, pests and diseases, and contribute to a resilient and sustainable agricultural output. At the heart of these rotations, forage legumes assume a multi-faceted role, while catch crops bolster primary crop production. Despite stringent regulations, the rotations offer considerable flexibility, provided the entirety of the crop spectrum is harnessed.

To enhance their competitiveness in terms of crop yield, optimizing crop genotypes, refining tillage methodologies, optimizing pest and disease management, and particularly refining the application of liquid organic manure with cutting-edge technology to optimize nutrient management and to reduce yield gap compared to conventional farming is imperative.

References

CIMMYT (2015) Green manures help Zambian and Malawian farmers feed crops and livestock. Available at: https://www.cimmyt.org/news/green-manures-help-zambian-and-malawian-farmers-feed-crops-and-livestock/ (accessed 27 January 2025).

Fenz, S., Neubauer, T., Heurix, J., Friedel, J.K. and Wohlmuth, M.-L. (2023) AI- and data-driven pre-crop values and crop rotation matrices. *European Journal of Agronomy* 150, 126949. DOI: 10.1016/j.eja.2023.126949.

Freyer, B. (2003) *Fruchtfolgen [konventionell, integriert, biologisch]; 116*. Verlag Eugen Ulme, Tabellen.

Freyer, B. (2019) The role of the crop rotation in organic farming. In: Köpke, U. (ed.) *Improving Organic Crop Cultivation*. Burleigh Dodds Science Publishing Limited, Cambridge, Massachusetts, pp. 547–568.

Freyer, B., Fahringer, A., Ellssel, P. and Schoeber, M. (2020) Step-by-step to organic and climate change adapted agriculture in Mozambique: Practitioner's guide. Austrian Development Agency (ADA)/OeAD – APPEAR.

Kolbe, H. (2006) Fruchtfolgegestaltung im ökologischen und extensiven landbau: Bewertung von vorfruchtwirkungen. *Pflanzenbauwissenschaften* 10(2), 82–89.

Meitzner, L.S. and Price, M.L. (1996) *Amaranth to Zai Holes: Ideas for Growing Food under Difficult Conditions*. Echo, North Forth Myers, Florida.

14 Application of SMART Technologies in Organic Farming: Recent Advances and Future Directions

KANDIAH PAKEERATHAN*

Department of Agricultural Biology, Faculty of Agriculture, University of Jaffna, Sri Lanka

Abstract

Smart farming is the use of information technology and sensors to collect and analyze data about crops, soil and weather conditions to make informed decisions about crop management, such as when to irrigate, apply fertilizer or control pests and diseases. The application of SMART technologies in organic farming is a relatively new field, but there have been some recent advances that have the potential to improve the sustainability and efficiency of organic production. The future of SMART technologies in organic farming is promising. As these technologies continue to develop, they will provide organic farmers with new tools to improve their yields, reduce their environmental impact, and make their operations more sustainable. This chapter reports current developments and application of SMART technologies such as sensor technology, biotechnology, nanotechnology and information communication technology, etc. in organic agriculture and blooming technologies for scientists and students who are working in the field to use this chapter as reference material.

Keywords: nano-biotechnology, organic farming, plant protection, sensors, SMART technologies

Introduction

Organic farming is a holistic system of agriculture that strives to optimize the productivity and fitness of diverse communities within the agro-ecosystem, including soil organisms, plants, livestock and people with the expectation of economic, environmental and social sustainability (Stockdale *et al.*, 2001). There are many different approaches like crop rotation, cover cropping, composting, integrated pest management (IPM) and animal welfare that are critically considered to optimize farm production while minimizing costs. Therefore, organic farming is a complex system that requires a deep understanding of the principles of ecology and sustainable agriculture (Altieri, 2004; Francis and Porter, 2011). However, the rewards of organic farming can be improved soil health, reduced environmental impact, secure nutritious food and better animal welfare.

Apart from the benefits, there are uncertainities in organic farming such as lower yields, higher costs, more labour-intensive, more difficult to scale, less predictable yields due to unpredictable weather changes and pests outbreaks (Malhi *et al.*, 2021). SMART technologies are a set of technologies that use sensors for precision irrigation, fertilization and weed control, biotechnology to improve agricultural productivity, nanotechnology to improve the effectiveness of biopesticides, and artificial intelligence (AI)-based pest monitoring, which are really indispensable to overcome the bottlenecks in organic agriculture (Javaid *et al.*, 2022; Ali *et al.*, 2023). Smart technologies can be

*Corresponding author: pakeerathank@univ.jfn.ac.lk

DOI: 10.1079/9781800626850.0014

used in a variety of ways in organic farming, as detailed below.

Sensors for Precision Irrigation, Fertilization and Weed Control

There are a number of sensors that can be used for precision irrigation, fertilization and weed control. Some of the most common sensors include:

- soil moisture sensors: these sensors measure the amount of water in the soil, which can be used to determine how much water is needed to irrigate each field;
- leaf wetness sensors: these sensors measure the amount of water on the leaves of plants, which can be used to determine if the plants need to be irrigated (Millán *et al.*, 2019; Sánchez Millán *et al.*, 2023);
- Nitrogen sensors: these sensors measure the amount of nitrogen in the soil, which can be used to determine how much fertilizer to apply to each field (Liu *et al.*, 2023);
- weed sensors: these sensors detect the presence of weeds, which can be used to determine if herbicides need to be applied (Andújar *et al.*, 2012).

These sensors can be used to collect data using a variety of methods such as coupled plasma (ICP) spectroscopy, near-infrared spectroscopy (NIRS), X-ray fluorescence (XRF), capacitance, resistance, optical sensors, infrared sensors, and thermal sensors (Kandpal *et al.*, 2022). These data can then be used to create a management plan for each field. For example, if the soil moisture sensor indicates that a field is dry, the irrigation system can be activated. If the leaf wetness sensor indicates that the plants are not getting enough water, the irrigation system can be adjusted to provide more water. If the nitrogen sensor indicates that the soil is low in nitrogen, fertilizer can be applied. If the weed sensor indicates that there are weeds present, herbicides can be applied (Javadi *et al.*, 2021; Greenberg *et al.*, 2023).

By using sensors to collect data and create management plans, farmers can improve the efficiency of their irrigation, fertilization, and weed control practices. This can lead to increased crop yields, reduced input costs and improved environmental sustainability (Millán *et al.*, 2019; Sánchez Millán *et al.*, 2023; Liu *et al.*, 2023).

Biotechnology to Improve Agricultural Productivity

Biotechnology has been applied to breeding resistance cultivars to improve the agricultural productivity in a number of ways particularly important to organic agriculture.

Gene transfer

Genes from other organisms can be transferred into crop plants to confer resistance to diseases. For example, in the early 2000s, a group of scientists at Syngenta developed a transgenic corn by transferring genes from the bacterium *Bacillus thuringiensis* (*Bt*) into corn and cotton to provide resistance to insects, such as European corn borers, and help to protect the corn from insect damage. This technology has helped to reduce use of synthetic pesticides several-fold (Buiatti *et al.*, 2013; Romeis *et al.*, 2019).

RNA interference (RNAi)

RNAi is a natural process by which cells silence genes that are essential for the pathogen's survival. This can be used to develop crops that are resistant to diseases by silencing genes. There are a number of examples of how RNAi has been used to protect crops from disease (Halder *et al.*, 2022). For example, tomato yellow leaf curl virus (TYLCV) is a serious disease that can cause significant yield losses in tomato crops. RNAi-resistant tomato plants are produced by introducing a gene into the plant that encodes for a short interfering RNA (siRNA) that targets the TYLCV genome. When the siRNA binds to the TYLCV genome, it triggers the RNAi pathway, which leads to the destruction of the TYLCV RNA and the prevention of virus replication (Koeppe *et al.*, 2023).

The same principles used to develop many disease-resistant crops like resistant rice plants to the rice tungro virus (RTV), tobacco plants to the tobacco mosaic virus (TMV), potato plants to the potato virus Y (PVY) and soybean to the soybean cyst nematode (Koeppe *et al.*, 2023).

RNAi is a powerful new tool that has the potential to revolutionize the way that we protect crops from disease. As our understanding of RNAi improves, we can expect to see even more innovative and effective ways to use RNAi to protect crops from disease (Liu *et al.*, 2020; Koeppe *et al.*, 2023).

Genetic engineering and genome editing

Genetic engineering techniques can be used to modify the genes of crop plants to make them more resistant to diseases. For example, in the 1990s, a group of scientists at Monsanto developed a transgenic soybean plant that was resistant to the herbicide Roundup. This technology, known as Roundup Ready soybeans, has revolutionized soybean production and has helped to reduce the use of chemical herbicides (Homrich *et al.*, 2012).

In recent years, there has been a growing interest in using targeted genome editing using transcription activator-like effector nucleases (TALENs), zinc finger nucleases (ZFNs) and CRISPR/Cas9, to develop plants that are resistant to diseases. Genome editing allows scientists to make precise changes to the DNA of plants, which can be used to alter genes that are involved in disease resistance. CRISPR/Cas9 technique applied to tomato plants resistant to the tomato yellow leaf curl virus (TYLCV), potato plants resistant to the potato virus Y (PVY) and leaf hoppers, rice plants resistant to the rice tungro virus (RTV), salt-tolerant, brownplant hoppers have been successfully developed and are widely being used. As the technology continues to develop, we can expect to see even more innovative and effective ways to use genome editing to protect crops from diseases (Li *et al.*, 2020; Pramanik *et al.*, 2021).

These technologies have the potential to reduce the use of pesticides and herbicides in organic agriculture. However, there are also concerns about the safety of genetically modified crops and the potential for these crops to harm the environment (Bawa and Anilakumar, 2013).

Image processing for pest and disease detection

Image processing techniques can be utilized for the early detection of crop pests and diseases. By analyzing images of crops, it becomes possible to identify visual signs and symptoms associated with specific pests or diseases (Ngugi *et al.*, 2020).

This model is applied to unseen images to detect potential pests or diseases. The model will classify each image based on the learned patterns and provide a diagnosis. When a potential pest or disease is detected, the detection system with an alert mechanism notifies farmers or agricultural experts. The alerts can be sent through mobile applications, email notifications, or other communication channels (Li *et al.*, 2021).

As we know, image processing is a smart technology for early detection of crop pests and diseases. It can be used to identify and classify pests and diseases based on their visual characteristics, such as colour, shape and texture. This can be done by using a variety of image processing techniques, such as image segmentation, feature extraction and classification (Shahoveisi *et al.*, 2023).

Image processing for crop pest and disease detection has a number of advantages over traditional methods, such as visual inspection and laboratory testing. These advantages include: (i) speed: image processing can be used to quickly scan large areas of crops for signs of pests and diseases. This can help to identify problems early, when they are easier to control; (ii) accuracy: image processing can be very accurate in identifying pests and diseases. This is because it can be used to extract features that are not visible to the naked eye; (iii) cost-effectiveness: image processing is a relatively inexpensive method of pest and disease detection. This is because it does not require the use of specialized equipment or trained personnel. As a result of these advantages, image processing is becoming increasingly popular for crop pest and disease detection. It is being used by farmers, agricultural researchers and government agencies around the world (Shoaib *et al.*, 2023; Ngugi *et al.*, 2021; Shahoveisi *et al.*, 2023).

Here are some examples of how image processing is being used to detect crop pests and diseases:

1. The PlantVillage project is a collaborative effort between the University of Florida and Google. The project has developed an image processing system that can identify 26 different plant diseases. The system has been used by farmers in over 100 countries to help them detect and control plant diseases.
2. The Cornell University Plant Disease Detection System is another image processing system that is used to detect plant diseases. The system has been used to identify over 100 different plant diseases.
3. The European Union's e-Phyto system is a web-based system that uses image processing to detect plant pests and diseases. The system is used by EU member states to inspect imported plant material.

As the technology continues to develop, it is likely that image processing will become even more widely used in agriculture.

It is important to note that the effectiveness of image processing for early pest and disease detection depends on the availability of high-quality training data, accurate labelling, and continuous refinement of the detection model based on real-world observations. Regular updates to the model using newly acquired data can improve its accuracy over time and combining image processing with other data sources, such as weather information or soil analysis, can further enhance the detection capabilities and provide more comprehensive insights for crop management (Ngugi *et al.*, 2020; Shoaib *et al.*, 2023).

Overall, image processing is a promising SMART technology to protect crops and ensure food security by early detection of crop pests and diseases in organic agriculture (Dhanaraju *et al.*, 2022).

Nanotechnology in organic agriculture

Nanotechnology has the potential to revolutionize various industries, including agriculture. In the field of organic agriculture, nanotechnology can offer innovative solutions for plant protection (Deshpande, 2019). Here are a few potential applications of nanobiotechnology in plant protection in organic agriculture:

Nano-botanical-pesticides. Nanoparticles can be utilized to develop targeted and controlled release formulations of pesticides. By encapsulating active ingredients in nanoparticles, their delivery and efficacy can be enhanced while minimizing their negative impacts on the environment. These nanopesticides can be designed to release the pesticide gradually, providing long-lasting protection against pests and diseases (Chaud *et al.*, 2021).

Nano-organic-fertilizers. Nanotechnology can also contribute to the development of nano-fertilizers that improve nutrient uptake and utilization by plants. Nanostructured organic fertilizers can be designed to release nutrients slowly, preventing nutrient leaching and optimizing plant nutrition. Additionally, nanoscale nutrient carriers can enhance the efficiency of nutrient delivery to plants, reducing the amount of fertilizer needed (Beig *et al.*, 2022). Research on nano-organic fertilizers is an emerging field that explores the potential benefits of nanotechnology in improving the efficiency and effectiveness of organic fertilizers (Vasseghian *et al.*, 2022). Recent researches focus on (i) nanoencapsulation of organic nutrients: nanoencapsulation involves the encapsulation of organic nutrients within nanoscale materials to protect them from degradation, improve their solubility, and enhance their controlled release (Ayyaril *et al.*, 2023). Researchers are investigating the use of biocompatible nanomaterials, such as polymers or lipids, to encapsulate organic nutrients like compost, manure or plant extracts. This approach aims to improve nutrient availability and reduce nutrient losses in organic farming systems; (ii) nanostructured delivery systems: nanostructured delivery systems provide a platform for the controlled release of organic nutrients. Nanoemulsions, nanogels or nanoparticles can be designed to encapsulate and slowly release organic fertilizers, ensuring a sustained nutrient supply to plants (Jiménez-Arias *et al.*, 2022). These delivery systems can improve nutrient uptake efficiency, reduce nutrient leaching, and minimize the risk of nutrient loss to the environment (Ayyaril *et al.*, 2023); (iii) nutrient uptake enhancement: nanotechnology offers the potential to enhance nutrient uptake by plants. Researchers are exploring the use of nanoparticles to improve the solubility and mobility of organic nutrients in the soil, making them more available for plant uptake. Nanoparticles can also interact with plant roots to enhance nutrient absorption processes, leading to improved nutrient use efficiency (Yadav *et al.*, 2023).

Nano-organic fertilizers are being investigated for their impact on soil health and fertility. Nanomaterials can modify soil properties, such as water holding capacity, nutrient retention and microbial activity. Researchers are studying the effects of nano-organic fertilizers on soil microbial communities, nutrient cycling processes and overall soil fertility to assess their long-term impact on soil health in organic farming systems. Further, it is crucial to assess the potential environmental impacts of nano-organic fertilizers. Advanced researches are being conducted to evaluate the fate and behaviour of nanoparticles in the environment, their potential toxicity to soil organisms, and the risk of nanoparticle accumulation in plants and ecosystems (Ahmed *et al.*, 2023). Understanding the environmental implications is essential for the safe and sustainable use of nano-organic fertilizers. It is important to note that research in this area is still in its early stages, and more studies are needed to fully understand the benefits, risks and practical applications of nano-organic fertilizers in organic agriculture. Regulatory frameworks and guidelines are also required to ensure the safe and responsible use of these technologies (Basavegowda and Baek, 2021).

Nanosensors. Nanotechnology-based sensors can be employed for early detection of pests, diseases and environmental stresses. These sensors can monitor various parameters such as moisture levels, nutrient status and the presence of pathogens. By providing real-time data on plant health, nanosensors enable farmers to take timely and precise actions to prevent crop damage (Shaw and Honeychurch, 2022).

Nanobiosensors. Nanotechnology can be utilized to develop biosensors that can detect specific pathogens or pests. These biosensors can identify the presence of harmful organisms in plants or the surrounding environment, enabling farmers to implement targeted control measures. Nanobiosensors offer high sensitivity, specificity and speed, allowing for rapid and accurate disease diagnosis (Mondal *et al.*, 2022).

Nanocarriers for biocontrol agents. Beneficial microorganisms, such as biocontrol agents, can be encapsulated within nanoparticles to enhance their stability and efficacy. Nanocarriers protect these agents from degradation, improve their adhesion to plant surfaces and promote controlled release. This approach ensures the targeted delivery of biocontrol agents, enhancing their ability to suppress pests and diseases (Saberi Riseh *et al.*, 2022).

Regulations and guidelines should be established to evaluate and manage the potential risks associated with the use of nanomaterials in organic agricultural practices.

Conclusions

The use of SMART technologies in organic farming is a novel strategy and a promising area of research and development. These technologies have the potential to improve the sustainability, profitability and efficiency of organic farms. As these technologies continue to develop, it is likely that we will see even more widespread adoption of SMART technologies in organic farming in future. Technologies are costly and may be difficult to handle in real field scenarios with unpredicatable climate changes; therefore, appropriate choices of amenable sustainable technology when predicting future technology would have the potential to significantly improve the sustainability and efficiency of organic production.

References

Ahmed, A., He, P., He, P., Wu, Y., He, Y. *et al.* (2023) Environmental effect of agriculture-related manufactured nano-objects on soil microbial communities. *Environment International* 173, 107819.

Ali, A., Hussain, T., Tantashutikun, N., Hussain, N. and Cocetta, G. (2023) Application of smart techniques, internet of things and data mining for resource use efficient and sustainable crop production. *Agriculture* 13(2), 397.

Altieri, M.A. (2004) Linking ecologists and traditional farmers in the search for sustainable agriculture. *Frontiers in Ecology and the Environment* 2(1), 35–42.

Andújar, D., Weis, M. and Gerhards, R. (2012) An ultrasonic system for weed detection in cereal crops. *Sensors* 12(12), 17343–17357.

Ayyaril, S.S., Shanableh, A., Bhattacharjee, S., Rawas-Qalaji, M., Cagliani, R. *et al.* (2023) Recent progress in micro and nano-encapsulation techniques for environmental applications: A review. *Results in Engineering* 18, 101094.

Basavegowda, N. and Baek, K.H. (2021) Current and future perspectives on the use of nanofertilizers for sustainable agriculture: The case of phosphorus nanofertilizer. *3 Biotech* 11(7), 357.

Bawa, A.S. and Anilakumar, K.R. (2013) Genetically modified foods: Safety, risks and public concerns–a review. *Journal of Food Science and Technology* 50(6), 1035–1046.

Beig, B., Niazi, M.B.K., Sher, F., Jahan, Z., Malik, U.S. *et al.* (2022) Nanotechnology-based controlled release of sustainable fertilizers: A review. *Environmental Chemistry Letters* 50(6), 1035–1046.

Buiatti, M., Christou, P. and Pastore, G. (2013) The application of GMOs in agriculture and in food production for a better nutrition: Two different scientific points of view. *Genes & Nutrition* 8(3), 255–270.

Chaud, M., Souto, E.B., Zielinska, A., Severino, P., Batain, F. *et al.* (2021) Nanopesticides in agriculture: Benefits and challenge in agricultural productivity, toxicological risks to human health and environment. *Toxics* 9(6).

Deshpande, M.V. (2019) Chapter 3: Nanobiopesticide perspectives for protection and nutrition of plants. In: Koul, O. (ed.) *Nano-Biopesticides Today and Future Perspectives*. Academic Press, London, pp. 47–68.

Dhanaraju, M., Chenniappan, P., Ramalingam, K., Pazhanivelan, S. and Kaliaperumal, R. (2022) Smart farming: Internet of things (IoT)-based sustainable agriculture. *Agriculture* 12(10), 1745.

Francis, C.A. and Porter, P. (2011) Ecology in sustainable agriculture practices and systems. *Critical Reviews in Plant Sciences* 30(1–2), 64–73.

Greenberg, I., Vohland, M., Seidel, M., Hutengs, C., Bezard, R. *et al.* (2023) Evaluation of mid-infrared and X-ray fluorescence data fusion approaches for prediction of soil properties at the field scale. *Sensors (Basel, Switzerland)* 23(2), 662.

Halder, K., Chaudhuri, A., Abdin, M.Z., Majee, M. and Datta, A. (2022) RNA interference for improving disease resistance in plants and its relevance in this clustered regularly interspaced short palindromic repeats-dominated era in terms of dsRNA-based biopesticides. *Frontiers in Plant Science* 13, 885128.

Homrich, M.S., Wiebke-Strohm, B., Weber, R.L.M. and Bodanese-Zanettini, M.H. (2012) Soybean genetic transformation: A valuable tool for the functional study of genes and the production of agronomically improved plants. *Genetics and Molecular Biology* 35(4 (suppl)), 998–1010.

Javadi, S.H., Munnaf, M.A. and Mouazen, A.M. (2021) Fusion of Vis-NIR and XRF spectra for estimation of key soil attributes. *Geoderma* 385, 114851.

Javaid, M., Haleem, A., Singh, R.P. and Suman, R. (2022) Enhancing smart farming through the applications of agriculture 4.0 technologies. *International Journal of Intelligent Networks* 3, 150–164.

Jíménez-Arias, D., Morales-Sierra, S., Silva, P., Carrêlo, H., Gonçalves, A. *et al.* (2022) Encapsulation with natural polymers to improve the properties of biostimulants in agriculture. *Plants* 12(1), 55

Kandpal, L.M., Munnaf, M.A., Cruz, C. and Mouazen, A.M. (2022) Spectra fusion of mid-infrared (MIR) and X-ray fluorescence (XRF) spectroscopy for estimation of selected soil fertility attributes. *Sensors* 22(9), 3459.

Koeppe, S., Kawchuk, L. and Kalischuk, M. (2023) RNA interference past and future applications in plants. *International Journal of Molecular Sciences* 24(11), 9755.

Li, H., Yang, Y., Hong, W., Huang, M., Wu, M. *et al.* (2020) Applications of genome editing technology in the targeted therapy of human diseases: Mechanisms, advances and prospects. *Signal Transduction and Targeted Therapy* 5(1), 1.

Li, W., Zheng, T., Yang, Z., Li, M., Sun, C. *et al.* (2021) Classification and detection of insects from field images using deep learning for smart pest management: A systematic review. *Ecological Informatics* 66, 101460.

Liu, J., Cai, H., Chen, S., Pi, J. and Zhao, L. (2023) A review on soil nitrogen sensing technologies: Challenges, progress and perspectives. *Agriculture* 13(4), 743.

Liu, S., Jaouannet, M., Dempsey, D.A., Imani, J., Coustau, C. *et al.* (2020) RNA-based technologies for insect control in plant production. *Biotechnology Advances* 39, 107463.

Malhi, G.S., Kaur, M. and Kaushik, P. (2021) Impact of climate change on agriculture and its mitigation strategies: A review. *Sustainability* 13(3), 1318.

Millán, S., Casadesús, J., Campillo, C., Moñino, M.J. and Prieto, M.H. (2019) Using soil moisture sensors for automated irrigation scheduling in a plum crop. *Water* 11, 10.

Mondal, R., Dam, P., Chakraborty, J., Paret, M.L., Katı, A. *et al.* (2022) Potential of nanobiosensor in sustainable agriculture: The state-of-art. *Heliyon* 8(12), e12207.

Ngugi, L.C., Abelwahab, M. and Abo-Zahhad, M. (2021) Recent advances in image processing techniques for automated leaf pest and disease recognition – a review. *Information Processing in Agriculture* 8(1), 27–51.

Pramanik, D., Shelake, R.M., Park, J., Kim, M.J., Hwang, I. *et al.* (2021) CRISPR/Cas9-mediated generation of pathogen-resistant tomato against tomato yellow leaf curl virus and powdery mildew. *International Journal of Molecular Sciences* 22(4), 1878.

Romeis, J., Naranjo, S.E., Meissle, M. and Shelton, A.M. (2019) Genetically engineered crops help support conservation biological control. *Biological Control* 130, 136–154.

Saberi Riseh, R., Hassanisaadi, M., Vatankhah, M., Soroush, F. and Varma, R.S. (2022) Nano/microencapsulation of plant biocontrol agents by chitosan, alginate, and other important biopolymers as a novel strategy for alleviating plant biotic stresses. *International Journal of Biological Macromolecules* 222, 1589–1604.

Sánchez Millán, F., Ortiz, F.J., Mestre Ortuño, T.C., Frutos, A. and Martínez, V. (2023) Development of smart irrigation equipment for soilless crops based on the current most representative water-demand sensors. *Sensors* 23(6), 3177.

Shahoveisi, F., Taheri Gorji, H., Shahabi, S., Hosseinirad, S., Markell, S. *et al.* (2023) Application of image processing and transfer learning for the detection of rust disease. *Scientific Reports* 13(1), 5133.

Shaw, D.S. and Honeychurch, K.C. (2022) Nanosensor applications in plant science. *Biosensors* 12(9), 675.

Shoaib, M., Shah, B., Ei-Sappagh, S., Ali, A., Ullah, A. *et al.* (2023) An advanced deep learning models-based plant disease detection: A review of recent research. *Frontiers in Plant Science* 14, 1158933.

Stockdale, E.A., Lampkin, N.H., Hovi, M., Keatinge, R., Lennartsson, E.K.M. *et al.* (2001) Agronomic and environmental implications of organic farming systems. *Advances in Agronomy* 70.

Vasseghian, Y., Arunkumar, P., Joo, S.-W., Gnanasekaran, L., Kamyab, H. *et al.* (2022) Metal-organic framework-enabled pesticides are an emerging tool for sustainable cleaner production and environmental hazard reduction. *Journal of Cleaner Production* 373, 133966.

Yadav, A., Yadav, K. and Abd-Elsalam, K.A. (2023) Nanofertilizers: Types, delivery and advantages in agricultural sustainability. *Agrochemicals* 2(2), 296–336.

15 Synergies of Farming with Alternative Pollinators (FAP) and Certified Organic Farming for Transformative Change of Agriculture

STEFANIE CHRISTMANN[1]*, MOULAY CHRIF SMAILI[2] AND YOUSSEF BENCHARKI[1]

[1]*International Center for Agricultural Research in Dry Areas (ICARDA), Rabat, Morocco;* [2]*National Institute for Agricultural Research (INRA), Kenitra, Morocco*

Abstract

Organic farming is a holistic and ambitious way of crop production challenging farmers starting in depleted soils. During the certification process for organic farming, farmers refrain already from using multiple chemicals to boost production, but often they cannot yet gain the full payment for certified organic production. Assumed income loss can limit readiness for organic production. An approach for pollinator protection in low- and middle-income countries, farming with alternative pollinators (FAP) can fill the gap, as the habitat-enhancement measures for pollinators substantially and significantly pay off for farmers in the first year. Planting marketable habitat-enhancement plants attracts higher diversity and abundance of wild pollinators, increasing productivity. The additional plants also attract higher abundance and diversity of natural enemies during a prolonged flowering time. In a large project in smallholder fields in Morocco, they reduced pests on average by 64.8% in main crops. The total net income was, on average, 121% higher than from monocultural control fields. Combination of certified organic agriculture and FAP might accelerate transformative change of agriculture in several aspects.

Keywords: flowering time, marketable habitat-enhancement plants, natural enemies, pollinators

Introduction

Pollinator decline is observed globally, but better monitored and more often addressed in high-income countries (Potts *et al.*, 2016; Powney *et al.*, 2019; Zattara and Aizen, 2021). Eighty-seven per cent of all flowering plants depend on pollinators (Ollerton *et al.*, 2011); without cross-pollination their chances to adapt to climate change are very low, which would have tremendous impact on diversity of species and habitats but also on all ecosystem services, namely those provided by these pollinator-dependent plants (Christmann, 2019a). Therefore, pollinator loss can cause interlinked degradation spirals, first environmental, then economical and social, and finally endanger peace (Christmann, 2019a). Any agricultural approach might therefore be revisited concerning its current and potential contribution to avoid such worst-case scenario by supporting pollinator protection, and *vice versa*.

Organic farming is not, *per se*, pollinator-friendly, but in diverse landscapes organic farming can attract more pollinators and benefit from their services than from conventional agriculture (Klein *et al.*, 2012). Higher plant diversity, e.g. by flowering weeds, can enhance pollinator diets and

*Corresponding authors: s.christmann@fap-research.de; lomantang@shohat.com

 Advances in Organic Farming (ed. L. P. Awasthi)
DOI: 10.1079/9781800626850.0015

health (Rotchés-Ribalta *et al.*, 2023). According to Duque-Trujillo *et al.* (2023), in organic fields, pollinator diversity and abundance are more impacted by higher diversity and abundance of flowering weeds than by reduced pesticides and fertilizer. Smaller field size and more frequent proximity to natural or semi-natural sites have also positive effects on pollination services and thus benefit organic farmers, yet differences between both farm types are statistically not relevant (Duque-Trujillo *et al.*, 2023). The combination of organic farming and flower strips has been identified as a good option to sustain pollinator diversity (Szitár *et al.*, 2022).

However, farmers often dislike or even reject wildflower strips (Christmann *et al.*, 2017, 2021a, b; Kleijn *et al.*, 2019) because they incur opportunity costs; farmers have to invest funds in seeds, fuel and labour without revenue but face the risk of weeds – as farmers call wildflowers – spreading into crops (Christmann *et al.*, 2017). Therefore, wildflower strips often require agroecological schemes to pay farmers (Batáry *et al.*, 2015), but even based on payments they remain unpopular among farmers (Kleijn *et al.*, 2019). Low- and middle-income countries cannot even afford paid wildflower strips (Christmann, 2020a).

Globally, only 1.5% of agricultural land is used for organic agriculture (IFOAM, 2021); it is still a niche approach. Farms are business entities (Christmann *et al.*, 2017). Before starting to shift to certified organic agriculture, farmers might consider the financial robustness of their farming enterprise and calculate the risks of potential income losses during and after the certification process. Achieving high yields on depleted soils without chemicals requires increased involvement of beneficial species, e.g. soil organisms, wild pollinators and natural enemies. To revitalize soil for the return of soil organisms takes years.

Farming with Alternative Pollinators (FAP) Focuses on Wild Pollinators and Natural Enemies and Pays Off Already in the First Year

FAP has been developed as an economically self-sustaining approach to protect pollinators notably in low- and middle-income countries; the main target groups are farmers and policymakers (Christmann and Aw-Hassan, 2012; Christmann *et al.*, 2017; Christmann, 2019b; Christmann, 2020a; Christmann *et al.*, 2021a; b). In field trials, the impacts of habitat enhancement (higher plant diversity, nesting and water support out of local or waste materials) are measured concerning diversity and abundance of flower visitors, natural enemies and pests; and net income by comparing FAP and monocultural control fields.

Instead of wildflowers, only marketable habitat-enhancement plants (MHEP) are used to avoid opportunity costs, for instance oil seeds, spices, vegetables, fruits, and medicinal or forage plants. Farmers have income from these plants or consume them on-farm. Tscharntke *et al.* (2016) highlighted that crops can sustain natural enemies better than wild plants; we had expected similar impacts on flower visitors. Usually, four to eight MHEP are selected with participation of farmers, as they know about availability of seeds, flowering times and marketing options of specific fruits, seeds or plants. MHEP should have different flower types and different colour petals to attract high diversity of insects (Christmann *et al.*, 2017, 2021a). High diversity of pollinators enhances fruit sets (Garibaldi *et al.*, 2013). Therefore, about two-thirds of MHEP should flower simultaneously with the main crop to enhance productivity by attracting higher diversity of flower visitors; one-third should flower before or after to prolong the flowering time, which benefits the conservation of pollinators and prolongs natural pest control. In smallholder fields, these annual or perennial MHEP occupy 25% of FAP fields (Fig. 15.1) while the central part is used for the main crop (Christmann *et al.*, 2017).

In large-scale fields the habitat-enhancement zone consists of strips every 15m in pollinator-dependent

Fig. 15.1. A typical smallholder FAP field with eggplant as the main crop, and sunflower, coriander, melon, zucchini, anise and green pepper as marketable habitat-enhancement plants near Kenitra (Moroccan region with adequate rainfall).

Fig. 15.2. A large-scale FAP faba bean field using canola and coriander as marketable habitat-enhancement plants (Settat, semi-arid region in Morocco).

crops using only 2–3 different MHEP (Fig. 15.2) because large-scale farmers prefer to limit increase of labour costs whereas smallholders prefer the maximum revenue from their limited field areas. In large-scale fields with pollinator-dependent main crops, the habitat-enhancement zone encompasses only 7% of the field (Bencharki *et al.*, 2024). In large-scale orchards, MHEP grow between the rows of trees (Christmann *et al.*, 2017).

The methodology of trials with field and orchard crops has been described in detail (Christmann *et al.*, 2017, 2021a). The FAP approach (Christmann and Aw-Hassan, 2012) has been tested by a proof-of-concept project in Uzbekistan (2013–14), conducted in small-scale cucumber fields and large-scale cherry orchards (Christmann *et al.*, 2017). FAP has been intensively tested with ten main crops in Morocco since 2015 (Christmann *et al.*, 2021a, b). Since 2020, trials have also been conducted in Turkey, Algeria, Jordan and Palestine. In 2022, we implemented various trials in large-scale farming systems in different agro-ecosystems of Morocco.

In 2018/19, we carried out 31 trials using four very different Moroccan agro-ecosystems (mountainous, oasis, semi-arid, adequate rainfall), using seven main crops, 27 MHEP and more than 230 smallholder fields. All farmers applied dung instead of chemical fertilizer. FAP induced an average net income gain of 121% (Christmann *et al.*, 2021b), which is an indicator of an economically reliable approach to increase farmers' incomes by pollinator-friendly production (Christmann *et al.*, 2021b). FAP paid off already in the first year even in the semi-arid region, which is dominated by large monocultures of cereals and thus very degraded regarding pollinator habitats. The income gap between FAP and control was highest in the most degraded region (Christmann *et al.*, 2021b).

FAP Impacts on Diversity and Abundance of Insects

FAP increases diversity of flower visitors and natural enemies and decreases pest diversity (Christmann *et al.*, 2021a, b). Impacts on diversity of flower visitors have been assessed on species level and in different aspects (Sentil *et al.*, 2021, 2022; El Abdouni *et al.*, 2022; Bencharki *et al.*, 2022). Sentil *et al.* (2022) demonstrated that FAP fields with MHEP attract much higher diversity of flower visitors than weeds around.

FAP impacts on natural enemies and pests are substantial and significant (Christmann *et al.*, 2021b). As MHEP prolonged the flowering time substantially (on average from 63 days in control fields to 93 days in the above-mentioned large project with seven main crops in Morocco (Christmann *et al.*, 2021b), diversity and abundance of natural enemies remained high over a prolonged period ensuring better pest control also after flowering of the main crop, as late insect samplings showed. Plant protection was enhanced also at the time of fruit development, while control fields rarely hosted natural enemies at that time. We made two main observations concerning the behaviour of pests and natural enemies in FAP fields: (i) pests detected in main crops were highly attracted by MHEP (notably by coriander) and stayed there; (ii) natural enemies attracted by MHEP fed themselves on aphids and other pests in the main crop while in control fields pests remained and their abundance increased in main crops (Fig. 15.3).

In total, pest abundance in main crops was reduced by 64.8% on average in above-mentioned trials from 2018 to 2019 (Christmann *et al.*, 2021b). Most FAP farmers declared that the remaining pest abundance in the main crop would be tolerable; they would not invest money and time in chemicals in such fields. High pest infestation in

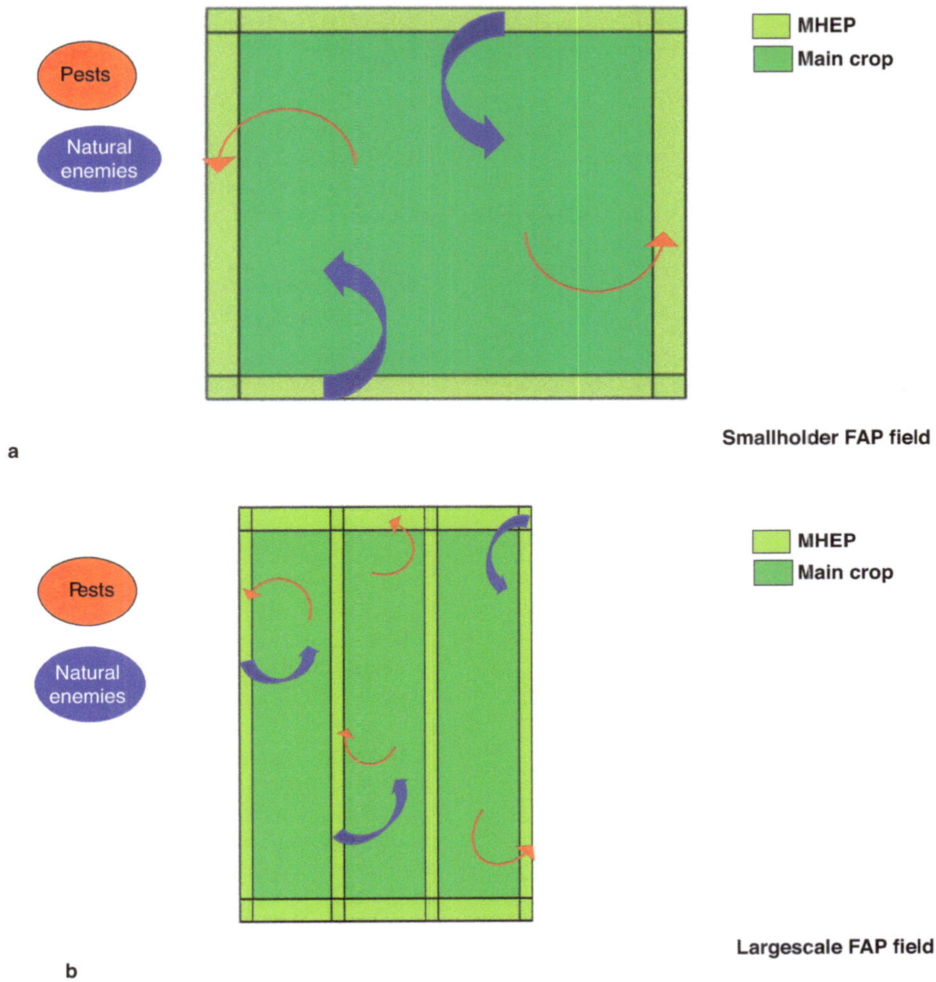

Fig. 15.3. Movements of pests and natural enemies in FAP fields; (a) movements of pests and natural enemies in FAP smallholder fields; (b) movements of pests and natural enemies in FAP large-scale fields.

MHEP was not relevant for farmers, as the increase of net income derived mainly from main crops due to higher quantity and partly also higher quality of crops (Christmann *et al.*, 2021b).

Potential Synergies between FAP and Organic Agriculture

We suggest the combination of FAP and organic agriculture due to several potential synergies. Organic farming might benefit from compensation of income loss, effective low-skill pest control as a preventive measure, and better protection of soil organisms, while pollinator protection might benefit concerning nesting success and stronger populations of ground-nesting pollinators, the most important pollinators in fields (Everaars, 2012).

Reduce or compensate income loss during certification process

The average income gain in 31 trials in Morocco is much higher and earlier than by using wildflower strips for pollinator protection (Blaauw and Isaacs, 2014; Garibaldi *et al.*, 2014; Christmann *et al.*, 2017, 2021a, b; Kleijn *et al.*, 2019; Albrecht *et al.*,

2020). High income gain is confirmed by FAP trials in Uzbekistan (Christmann *et al.*, 2017) and ongoing FAP trials in Algeria, Turkey, Palestine and Jordan. Szitár *et al.* (2022) suggested the combination of organic farming and wildflower strips; however, MHEP and the high income gain might be more attractive for farmers.

In cases where farmers produce mainly pollinator-independent crops like potato or cereals, corridors or field borders seeded with MHEP might serve pest control and support pollinators, but farmers would benefit only to a limited extent. Therefore, these farmers might ponder at least a partial shift to pollinator-dependent crops during the certification process or in the long term, to benefit economically as much as possible from FAP. Globally, but notably in dry areas, farmers have shifted to pollinator-dependent crops for decades (Potts *et al.*, 2016) and this process might be further accelerated. The growing global population, the planetary boundaries and the need to reduce, e.g., greenhouse gas emissions from livestock production might slow down the trend to animal-based food and towards balanced diets based on a mix of nuts, oil seeds, faba bean, fruits and vegetables (Springmann *et al.*, 2018; Christmann, 2020b).

Low-cost and easy-to-use preventive pest control

Integrated Pest Management, biocontrol, and intercropping have been common instruments for decades and trigger discussions on pros and cons (Gurr *et al.*, 2016; Deguine *et al.*, 2021). Some of these measures require long-term experience of farmers, training, experts, or supply centres for biocontrol agents close-by. In comparison, seeding a mix of 4–8 regionally available MHEP is easy to apply as a low-cost preventive measure for enhanced pest control. All plants provide income or are used on-farm. MHEP might work as a basic system for pest control; biocontrol agents can be added in case MHEP strong infestation by a specific pest appears.

We recommend capacity building for farmers by visual materials and social media concerning, e.g. diversity of pollinators and natural enemies and attractiveness of certain MHEP. Farmers' knowledge on insects is limited in many countries (Hanes *et al.*, 2015; Elisante *et al.*, 2019; Hevia *et al.*, 2021; Christmann *et al.*, 2021a). Pollinators are often disregarded as pests (Christmann *et al.*, 2021a), though among about six million insect species, only 20–30 are important pests for major crops (García-Lara and Serna Saldivar, 2016). Such capacity building does not require high investment, as various kinds of materials are available in several languages and also for illiterate farmers (e.g. https://www.canr.msu.edu/ipm/biocontrol/; https://mel.cgiar.org/projects/iki-pollinators [accessed 28/1/25]).

Support restoration of soil biodiversity and nesting of ground-nesting pollinators

Soil organisms are mobile to a very limited extent; soil bacteria and protists, for example, can move actively around 0.000001m only, nematodes 0.01m and arbuscular mycorrhizal fungi 0.005m per day (Orgiazzi *et al.*, 2016). As 87% of flowering plants need pollinators (Ollerton *et al.*, 2011) soil organisms highly depend on pollinators sustaining or enriching local vegetation (Christmann, 2022). Therefore, pollinator protection is important for biodiverse productive soil and organic farmers.

Vice versa and probably even more, pollinator protection might benefit from respectful and knowledgeable soil treatment in organic farming. Seventy per cent of wild bees (e.g. *Amegilla, Andrena, Anthophora, Halictus, Eucera, Lasioglossum* and *Melitta*), most solitary wasps and many syrphids nest in the ground and are threatened by late or deep tillage, and soil compactment by heavy machinery, both destroying tunnels and nest cells (Shuler *et al.*, 2005; Ullmann *et al.*, 2016; Christmann, 2022). Offspring of ground-nesting pollinators are also reduced by chemicals: pesticide risk assessments are mostly conducted on honeybees (Uhl and Brühl, 2019), though honeybees are more mobile than wild pollinators and enjoy substantial support by beekeepers (hives, transport), whereas females of ground-nesting pollinators are exposed to chemicals while digging nests. Pesticide risk assessments also do not consider pesticide accumulation and side effects over time (Christmann, 2022). Notably, neonicotinoids can reduce offspring by killing the female pollinator while digging nests (Woodcock *et al.*, 2017; Willis Chan *et al.*, 2019; Willis Chan and Raine, 2021). These specific risks for ground-nesting pollinators are lower in organic agriculture, because organic agriculture highly depends on diverse and abundant soil biota and therefore reduces pesticides, tillage and soil compaction. Organic agriculture using 4–8 MHEP and limited tillage might serve as a cradle for ground-nesting pollinators. However, for organic farmers, visual materials on natural nests might be helpful, as farmers' knowledge is low concerning nests. In Morocco, for instance, 76% of interviewed farmers never recognized a nest in or around their fields; 65% in Benin; and 38% in Turkey (Christmann *et al.*, 2021a).

Conclusion

For decades we have used more and more semi-natural land for agriculture, notably for pollinator-dependent crops (Aizen *et al.*, 2019). As FAP increases production of pollinator-dependent crops substantially and significantly, in the long term and with wide adoption it has potential to promote transformative change of agriculture towards intensified pollinator-friendly production and bend the curve of land-use change (Christmann *et al.*, 2021b). However, as the proof-of-concept project has been conducted in 2013/14, the adoption of this relatively new approach is even more limited than global use of organic agriculture. A combination of organic farming and FAP might promote the adoption of both. We suggest standard setters of organic labels to take potential synergies, notably concerning biodiversity protection, adoption and long-term transformative change of agriculture, into consideration.

Acknowledgement

This work is funded by the German Federal Ministry for the Environment, Nature Protection and Nuclear Safety (BMU) within the International Climate Initiative (IKI; 17_IV_065).

References

Aizen, M.A., Aguiar, S., Biesmeijer, J.C., Garibaldi, L.A., Inouye, D.W. *et al.* (2019) Global agricultural productivity is threatened by increasing pollinator dependence without a parallel increase in crop diversification. *Global Change Biology* 25(10), 3516–3527.

Albrecht, M., Kleijn, D., Williams, N.M., Tschumi, M., Blaauw, B.R. *et al.* (2020) Global synthesis of the effectiveness of flower strips and hedgerows on pest control, pollination services and crop yield. *Authorea* 6, April.

Batáry, P., Dicks, L.V., Kleijn, D. and Sutherland, W.J. (2015) The role of agri-environment schemes in conservation and environmental management. *Conservation Biology* 29(4), 1006–1016.

Bencharki, Y., Christmann, S., Lhomme, P., Ihsane, O., Sentil, A. *et al.* (2022) Farming with alternative pollinators' approach supports diverse and abundant pollinator community in melon fields in a semi-arid landscape. *Renewable Agriculture and Food Systems* 38, e6. DOI: 10.1017/S1742170522000394.

Bencharki, Y., Michez, D., Ihsane, O., Reverté, S., Aw-Hassan, A. *et al.* (2024) Farming with alternative pollinators' provides benefits also in large-scale fields. *Acta Oecologica* 122, 103978. DOI: 10.1016/j.actao.2023.103978.

Blaauw, B.R. and Isaacs, R. (2014) Flower plantings increase wild bee abundance and the pollination services provided to a pollination-dependent crop. *Journal of Applied Ecology* 51(4), 890–898.

Christmann, S. (2019a) Do we realize the full impact of pollinator loss on other ecosystem services and the challenges for any restoration in terrestrial areas? *Restoration Ecology* 27(4), 720–725.

Christmann, S. (2019b) Under which conditions would a wide support be likely for a multilateral environmental agreement for pollinator protection? *Environmental Science & Policy* 91, 1–5.

Christmann, S. (2020a) Pollinator protection strategies must be feasible for all nations. *Nature Ecology & Evolution* 4(7), 896–897.

Christmann, S. (2020b) Climate change enforces to look beyond the plant – the example of pollinators. *Current Opinion in Plant Biology* 56, 162–167.

Christmann, S. (2022) Regard and protect ground-nesting pollinators as part of soil biodiversity. *Ecological Applications* 32(3), e2564.

Christmann, S. and Aw-Hassan, A.A. (2012) Farming with alternative pollinators (FAP): An overlooked win-win strategy for climate change adaptation. *Agriculture, Ecosystems & Environment* 161, 161–164.

Christmann, S., Aw-Hassan, A., Rajabov, T., Khamraev, A.S. and Tsivelikas, A. (2017) Farming with alternative pollinators increases yields and incomes of cucumber and sour cherry. *Agronomy for Sustainable Development* 37(4), 1–8.

Christmann, S., Aw-Hassan, A., Güler, Y., Sarisu, H.C., Bernard, M. *et al.* (2021a) Two enabling factors for farmer-driven pollinator protection in low- and middle-income countries. *International Journal of Agricultural Sustainability* 20(1), 54–67.

Christmann, S., Bencharki, Y., Anougmar, S., Rasmont, P., Smaili, M.C. *et al.* (2021b) Farming with alternative pollinators benefits pollinators, natural enemies, and yields, and offers transformative change to agriculture. *Scientific Reports* 11(1), 18206.

Deguine, J.-P., Aubertot, J.-N., Flor, R.J., Lescourret, F., Wyckhuys, K.A.G. *et al.* (2021) Integrated pest management: Good intentions, hard realities: A review. *Agronomy for Sustainable Development* 41(3), 41.

Duque-Trujillo, D., Hincapié, C.A., Osorio, M. and Zartha-Sossa, J.W. (2023) Strategies for the attraction and conservation of natural pollinators in agroecosystems: a systematic review. *International Journal of Environmental Science and Technology* 20(4), 4499–4512.

El Abdouni, I., Lhomme, P., Christmann, S., Dorchin, A., Sentil, A. *et al.* (2022) Diversity and relative abundance of insect pollinators in Moroccan agroecosystems. *Frontiers in Ecology and Evolution* 10, 866581.

Elisante, F., Ndakidemi, P.A., Arnold, S.E.J., Belmain, S.R., Gurr, G.M. *et al.* (2019) Enhancing knowledge among smallholders on pollinators and supporting field margins for sustainable food security. *Journal of Rural Studies* 70, 75–86.

Everaars, J. (2012) The response of solitary bees to landscape configuration with focus on body size and nest-site preference. PhD thesis, Martin-Luther-Universität Halle-Wittenberg (Germany).

García-Lara, S. and Serna Saldivar, S.O. (2016) Insect pests. *Encyclopedia of Food and Health* 432–436.

Garibaldi, L.A., Steffan-Dewenter, I., Winfree, R., Aizen, M.A., Bommarco, R. *et al.* (2013) Wild pollinators enhance fruit set of crops regardless of honey bee abundance. *Science* 339(6127), 1608–1611.

Garibaldi, L.A., Carvalheiro, L.G., Leonhardt, S.D., Aizen, M.A., Blaauw, B.R. *et al.* (2014) From research to action: Enhancing crop yield through wild pollinators. *Frontiers in Ecology and the Environment* 12(8), 439–447.

Gurr, G.M., Lu, Z., Zheng, X., Xu, H., Zhu, P. *et al.* (2016) Multi-country evidence that crop diversification promotes ecological intensification of agriculture. *Nature Plants* 2, 16014.

Hanes, S.P., Collum, K.K., Hoshide, A.K. and Asare, E. (2015) Grower perceptions of native pollinators and pollination strategies in the lowbush blueberry industry. *Renewable Agriculture and Food Systems* 30(2), 124–131.

Hevia, V., García-Llorente, M., Martínez-Sastre, R., Palomo, S., García, D. *et al.* (2021) Do farmers care about pollinators? A cross-site comparison of farmers' perceptions, knowledge, and management practices for pollinator-dependent crops. *International Journal of Agricultural Sustainability* 19(1), 1–15.

IFOAM (2021) The world of organic agriculture. Available at: https://www.ifoam.bio/world-organic-agriculture-2021 (accessed 2 June 2023).

Kleijn, D., Bommarco, R., Fijen, T.P.M., Garibaldi, L.A., Potts, S.G. *et al.* (2019) Ecological intensification: Bridging the gap between science and practice. *Trends in Ecology & Evolution* 34(2), 154–166.

Klein, A.M., Brittain, C., Hendrix, S.D., Thorp, R., Williams, N. *et al.* (2012) Wild pollination services to California almond rely on semi-natural habitat. *Journal of Applied Ecology* 49(3), 723–732.

Ollerton, J., Winfree, R. and Tarrant, S. (2011) How many flowering plants are pollinated by animals? *Oikos* 120(3), 321–326.

Orgiazzi, A., Bardgett, R.D. and Barrios, E. (2016) Global soil biodiversity atlas. *European Commission, Luxemburg*. Available at: https://esdac.jrc.ec.europa.eu/content/global-soil-biodiversity-atlas (accessed June 2021).

Potts, S.G., Imperatriz-Fonseca, V., Ngo, H.T., Aizen, M.A., Biesmeijer, J.C. *et al.* (2016) Safeguarding pollinators and their values to human well-being. *Nature* 540(7632), 220–229.

Powney, G.D., Carvell, C., Edwards, M., Morris, R.K.A., Roy, H.E. *et al.* (2019) Widespread losses of pollinating insects in Britain. *Nature Communications* 10(1), 1018.

Rotchés-Ribalta, R., Marull, J. and Pino, J. (2023) Organic farming increases functional diversity and ecosystem service provision of spontaneous vegetation in Mediterranean vineyards. *Ecological Indicators* 147, 110023.

Sentil, A., Lhomme, P., Michez, D., Reverté, S., Rasmont, P. *et al.* (2021) Farming with alternative pollinators' approach increases pollinator abundance and diversity in faba bean fields. *Journal of Insect Conservation* 26(3), 401–414.

Sentil, A., Reverté, S., Lhomme, P., Bencharki, Y., Rasmont, P. *et al.* (2022) Wild vegetation and 'farming with alternative pollinators' approach support pollinator diversity in farmland. *Journal of Applied Entomology* 146(9), 1155–1168.

Shuler, R.E., Roulston, T.H. and Farris, G.E. (2005) Farming practices influence wild pollinator populations on squash and pumpkin. *Journal of Economic Entomology* 98(3), 790–795.

Springmann, M., Clark, M., Mason-D'Croz, D., Wiebe, K., Bodirsky, B.L. *et al.* (2018) Options for keeping the food system within environmental limits. *Nature* 562(7728), 519–525.

Szitár, K., Deák, B., Halassy, M., Steffen, C. and Batáry, P. (2022) Combination of organic farming and flower strips in agricultural landscapes – a feasible method to maximise functional diversity of plant traits related to pollination. *Global Ecology and Conservation* 38, e02229.

Tscharntke, T., Karp, D.S., Chaplin-Kramer, R., Batáry, P., DeClerck, F. *et al.* (2016) When natural habitat fails to enhance biological pest control – five hypotheses. *Biological Conservation* 204(B), 449–458.

Uhl, P. and Brühl, C.A. (2019) The impact of pesticides on flower-visiting insects: A review with regard to European risk assessment. *Environmental Toxicology and Chemistry* 38(11), 2355–2370.

Ullmann, K.S., Meisner, M.H. and Williams, N.M. (2016) Impact of tillage on the crop pollinating, ground-nesting bee, *Peponapis pruinose*, in California. *Agriculture, Ecosystems & Environment* 232, 240–246.

Willis Chan, D.S. and Raine, N.E. (2021) Population decline in a ground-nesting solitary squash bee (*Eucera pruinosa*) following exposure to a neonicotinoid insecticide treated crop (*Cucurbita pepo*). *Scientific Reports* 11(1), 4241.

Willis Chan, D.S., Prosser, R.S., Rodríguez-Gil, J.L. and Raine, N.E. (2019) Assessment of risk to hoary squash bees (*Peponapis pruinosa*) and other ground-nesting bees from systemic insecticides in agricultural soil. *Scientific Reports* 9(1), 11870.

Woodcock, B.A., Bullock, J.M., Shore, R.F., Heard, M.S., Pereira, M.G. *et al*. (2017) Country-specific effects of neonicotinoid pesticides on honey bees and wild bees. *Science* 356(6345), 1393–1395.

Zattara, E.E. and Aizen, M.A. (2021) Worldwide occurrence records suggest a global decline in bee species richness. *One Earth* 4(1), 114–123.

16 Organic Ecosystems: Concepts and the Role of Natural Resources

Hamdi Özaktan[1]*, Gheorghe Cristian Popescu[2] and Mehmet Yaman[3]

[1]Department of Field Crops, Agricultural Faculty, Erciyes University, Kayseri, Türkiye; [2]Department of Applied Sciences and Environmental Engineering, National University of Science and Technology Politehnica Bucharest, University Center of Pitesti, Romania; [3]Department of Horticulture, Agricultural Faculty, Erciyes University, Kayseri, Türkiye

Abstract

'Organic ecosystem' is a term and concept that refers to the complex interactions between living organisms and their environment in an organic farming system. It is characterized by the flow of energy and nutrients, as well as the continuous cycling of materials. The concept of an organic ecosystem emphasizes the interconnectedness and interdependence of all living and non-living components within a given area. Natural resources such as soil, water, air, minerals, biodiversity, but also the application of agricultural technologies play a crucial role in organic ecosystems. They provide the necessary elements for life to thrive and function within an ecosystem. For example, air provides oxygen for respiration, water supports various biological processes, soil acts as a medium for plant growth, and minerals serve as essential nutrients for organisms. The role of natural resources in an organic ecosystem can be categorized into three main functions: provisioning, regulating, and supporting. Provisioning services refer to the direct benefits that humans derive from natural resources, such as food, water, and timber. Regulating services involve the regulation of ecosystem processes, such as climate regulation, water purification, and pest control. Supporting services include the underlying processes that maintain the structure and function of an ecosystem, such as nutrient cycling and soil formation. However, the sustainable management of natural resources is crucial for the long-term stability and health of organic ecosystems. Human activities, such as deforestation, intensive and industrial agriculture, pollution, and overexploitation, can disrupt the delicate balance within an ecosystem and lead to its degradation. Therefore, it is essential to adopt sustainable practices that minimize the negative impact on natural resources and promote their conservation. In conclusion, an organic ecosystem is a complex network of organisms and their physical environment. Natural resources play a vital role in supporting the functioning and sustainability of these ecosystems. Understanding the concepts and roles of natural resources in organic ecosystems is crucial for their conservation and the overall well-being of the planet.

Keywords: organic ecosystems, natural resources, organic concepts, benefits, sustainability, farming

Introduction

Organic ecosystems are dynamic complexes in which plant, animal and microorganism communities and their inanimate environments interact with each other (Odum *et al.*, 2013). In these ecosystems, natural resources such as water, soil, air and biodiversity play an important role in supporting the functioning and sustainability of ecosystems (Costanza *et al.*, 1997). Natural resources play a crucial role in organic ecosystems, providing essential elements for the survival and functioning of living organisms (Feledyn-Szewczyk *et al.*, 2016; Suman *et al.*, 2022). The significance of these resources lies in

*Corresponding author: ozaktan_03@hotmail.com

 Advances in Organic Farming (ed. L. P. Awasthi)
DOI: 10.1079/9781800626850.0016

their ability to sustain life, maintain ecological balance and support various ecosystem services (Arora, 2018).

Organic farming, also known as biological or ecological farming, follows a well-defined set of regulations and the way of production is controlled by a certification body. While it does not allow any chemical inputs in crop production, there are restrictions on the use of antibiotics and hormones in animal production. In addition, the use of organic fertilizers and natural pesticides in crop production, long crop rotations, organic matter in the soil and the maintenance of microbial life are among the main objectives. The transition period from conventional to organic agriculture varies between 2 and 3 years.

General Consideration Related to Organic Ecosystems

The organic ecosystem is part of agricultural ecosystems. Ecosystems and natural resources like soil fertility, biodiversity and water are negatively impacted by industrial farming. The organic ecosystem is mainly that type of ecosystem promoted and built by organic agriculture. This type of ecosystem is an environmentally friendly solution to achieve food safety and security as well as environmental sustainability. The main stages of organic farming are presented in Fig. 16.1.

Organic farming is a form of agriculture that relies on natural processes and inputs, such as compost, manure, crop rotation and biological pest control to enhance soil fertility, biodiversity and food quality, while avoiding the use of synthetic fertilizers, pesticides, genetically modified organisms and other external inputs that may harm the environment or human health (IFOAM, 2021).

Natural resources refer to the materials or substances that exist naturally in the environment and are utilized by living organisms (Skinner and Castleden's, 2014). These resources can be classified into two main categories: renewable and non-renewable resources. Renewable resources, such as water, sunlight and wind, are continuously replenished by natural processes, while non-renewable resources, such as fossil fuels and minerals, are finite and cannot be replaced once depleted (Lee, 2019). In addition to supporting ecological processes, natural resources also contribute to the economic and social well-being of human communities. For instance, forests provide timber, fuelwood and non-timber forest products, which are important for livelihoods and economic development (Groth, 2007; Mair and Asada, 2019). Wetlands, on the other hand, offer various ecosystem services,

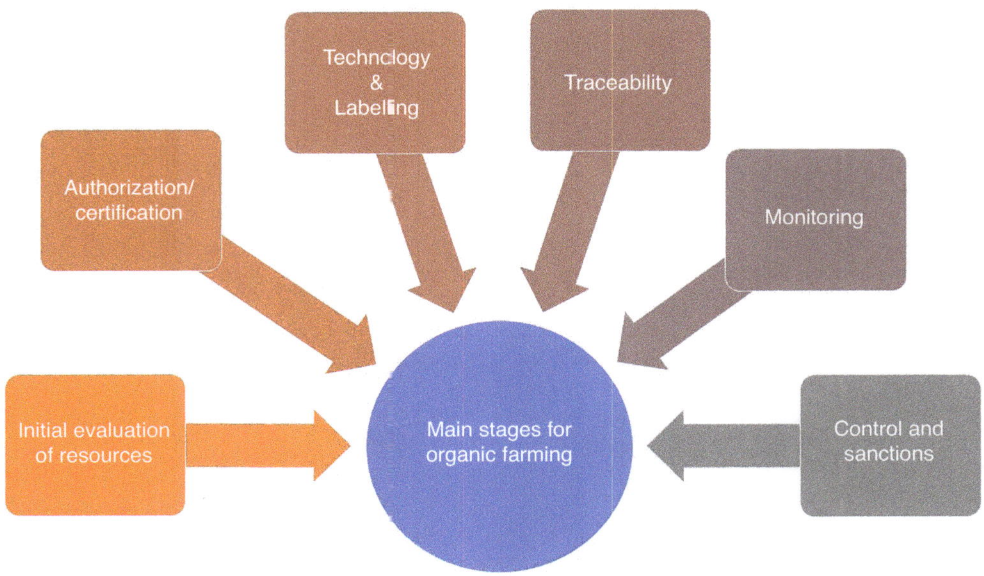

Fig. 16.1. Main stages for organic farming.

such as water purification, flood regulation and recreational opportunities, benefiting both local communities and society at large (Horwitz and Finlayson, 2011.).

One of the main environmental benefits of organic ecosystem is its contribution to soil health and carbon sequestration. Soil is a vital resource for food production and ecosystem services, but it is also under threat from erosion, degradation and loss of organic matter due to intensive agricultural practices. Organic farming practices, such as adding organic matter, reducing tillage, diversifying crops and integrating livestock, can improve soil structure, water retention, nutrient cycling and biological activity, and thus enhance soil quality and productivity (Pimentel *et al.*, 2005; Gattinger *et al.*, 2012). Moreover, organic farming can increase the amount of carbon stored in the soil, by reducing the emissions of carbon dioxide and nitrous oxide from synthetic fertilizers and pesticides, and by increasing the input and stabilization of organic matter in the soil (Lal, 2004; Muller *et al.*, 2017). According to a meta-analysis by Gattinger *et al.* (2012), organic farming can sequester 0.45–1.66 Mg C ha^{-1} $year^{-1}$ more than conventional farming, depending on the soil type, climate and management system. Organic farming also relies on natural sources of plant nutrients, such as compost, manure, crop residues, green manures and legumes, to supply the crops with the essential elements for their growth and development, while avoiding the use of synthetic fertilizers that can cause environmental problems, such as eutrophication, acidification and salinization (Gomiero *et al.*, 2011).

Another environmental benefit of organic ecosystem is its positive impact on biodiversity and wildlife. Biodiversity is the variety of life on Earth, and it is essential for the functioning and resilience of ecosystems, as well as for the provision of goods and services for human well-being. However, biodiversity is also under pressure from habitat loss, fragmentation, degradation, invasion, overexploitation and climate change, largely driven by human activities. Agriculture is one of the major drivers of biodiversity loss as it occupies about 38% of the global land area and affects 70% of all threatened bird species and 49% of all plant species (FAO, 2003). Organic farming can help conserve and enhance biodiversity, by avoiding the use of synthetic chemicals that can harm non-target organisms, by maintaining and restoring natural habitats and landscape features, such as hedgerows, grasslands and wetlands, by promoting crop and genetic diversity, and by supporting beneficial organisms such as pollinators, predators and decomposers (Bengtsson *et al.*, 2005; Hole *et al.*, 2005; Tuck *et al.*, 2014). According to a meta-analysis by Tuck *et al.* (2014), organic farming can increase the abundance and richness of species by 30% and 12%, respectively, compared to conventional farming, with the largest effects for plants, insects and birds.

Despite the environmental benefits of organic ecosystem, there are also some challenges and trade-offs that need to be addressed and balanced. One of the main challenges is the lower yield and land-use efficiency of organic farming compared to conventional farming. Yield is the amount of crop or animal product obtained per unit of land area, and it is an important indicator of food security and resource-use efficiency. Several studies have shown that organic farming tends to have lower yields than conventional farming, mainly due to the lower availability of nutrients, the higher incidence of weeds, pests and diseases, and the lower adoption of improved varieties and breeds (Seufert *et al.*, 2012). According to a meta-analysis by Seufert *et al.* (2012), organic farming can produce 19–25% less yield than conventional farming, depending on the crop type, the management system and the environmental conditions. This yield gap implies that more land is needed to produce the same amount of food with organic farming, which can have negative consequences for land-use change, deforestation, biodiversity loss and greenhouse gas emissions (Debuschewitz and Sanders, 2022).

Another challenge is the variability and uncertainty of the environmental performance of organic ecosystem, depending on the context and the indicators used. Organic farming is not a uniform or static system, but rather a dynamic and diverse one that can vary widely in terms of practices, inputs, outputs and outcomes, depending on the local conditions, the farmer's decisions and the market opportunities. Moreover, the environmental impacts of organic farming are not always straightforward or consistent, but rather complex and context-dependent, depending on the scale, the perspective and the indicators used to measure them. For example, organic farming can have lower greenhouse gas emissions per unit of land area but higher emissions per unit of product, due to the lower yield and the higher methane emissions from ruminants (de Ponti *et al.*, 2012). Similarly, organic farming can have higher biodiversity per unit of land area, but lower biodiversity per unit of product, due to the lower land-use efficiency and the higher land demand. Therefore, the environmental assessment of organic ecosystem requires a comprehensive and holistic approach, that considers multiple dimensions, scales and trade-offs, and that accounts for the variability and uncertainty of the system (Tuomisto *et al.*, 2012).

Climate is another natural resource that significantly influences organic ecosystems. Climate patterns, such as temperature and precipitation, shape the distribution of species and determine the types of ecosystems that can thrive in a particular region (Comer *et al.*, 2022). Changes in climate, such as global warming, can have profound impacts on organic ecosystems, altering species composition, disrupting ecological processes and leading to habitat loss (Kershner *et al.*, 2020).

Natural Resources and Organic Practices in Organic Ecosystem

Agricultural production systems currently face multiple challenges, especially linked with environmental protection, climate change and degradation of natural resources. In this context, the promotion and development of an organic ecosystem in which environmentally friendly practices are applied represents a viable solution for the rational use of natural resources and for a greater sustainability of ecosystems. Main natural resources and components involved in the organic ecosystems are represented by healthy soils, water quality resources, organic substrates, organic fertilizers, biostimulators, organic plant protection, etc. (Fig. 16.2).

Organic farming has multiple benefits for organic ecosystems, maintaining and improving biodiversity and reducing the anthropogenic footprint on soil, air, water, wildlife and especially on the farming communities (Parizad and Bera, 2023).

Water, as a vital natural resource, is essential for the survival of all living organisms. It serves as a medium for various physiological processes, including nutrient transport, waste removal and temperature regulation (De Swaef *et al.*, 2022). In organic ecosystems, water availability and quality play a crucial role in determining the distribution and abundance of species, as well as the overall functioning of the ecosystem (Meran *et al.*, 2021). Water is a fundamental natural resource for agricultural systems and organic ecosystem. Water quality is significantly affected by the use of agrochemicals in agricultural production. Water resources and crop pollution is almost non-existent in organic ecosystems (Parizad and Bera, 2023).

Soil, often overlooked as a natural resource, also holds great importance in organic ecosystems. It serves as a medium for plant growth, providing essential nutrients, water and anchorage for roots. Moreover, soil acts as a habitat for a wide range of organisms, including bacteria, fungi and insects, which play crucial roles in nutrient cycling and decomposition processes. Additionally, soil acts as a carbon sink, storing significant amounts of carbon and helping to mitigate climate change (Tahat *et al.*, 2020). The resilience and functionality of agricultural ecosystems is affected by agricultural practices applied to the soil. It is appreciated that organic practices enhance soil quality and develop a better environment in organic ecosystems (Aulakh *et al.*, 2022). New technologies based on sensors, automatic machines, robots and digital applications will change the concept of farming, increase the resilience of agricultural ecosystems and have a great potential for organic ecosystem (Popescu *et al.*, 2022).

Biodiversity, another key aspect of organic ecosystems, is heavily reliant on natural resources. The availability of diverse habitats, such as forests, wetlands and coral reefs, provides niches for a wide range of species. These habitats offer various resources, including food, shelter and breeding grounds, which support the survival and reproduction of different organisms. Biodiversity, in turn, contributes to the stability and resilience of organic ecosystems, enhancing their ability to withstand disturbances and adapt to changing environmental conditions (Feledyn-Szewczyk *et al.*, 2016). Organic farming systems provide a higher biodiversity than conventional farming (Gong *et al.*, 2022).

Minerals, such as phosphorus and nitrogen, are vital natural resources for organic ecosystems. These elements are essential for plant growth and are often limiting factors in nutrient availability. Phosphorus, for example, is a key component of DNA, RNA, and ATP, while nitrogen is a crucial building block for proteins and other cellular components. The cycling and availability of these minerals in organic ecosystems are influenced by various processes, including weathering, nutrient uptake by plants and microbial transformations (Rawat *et al.*, 2021; Feng *et al.*, 2023).

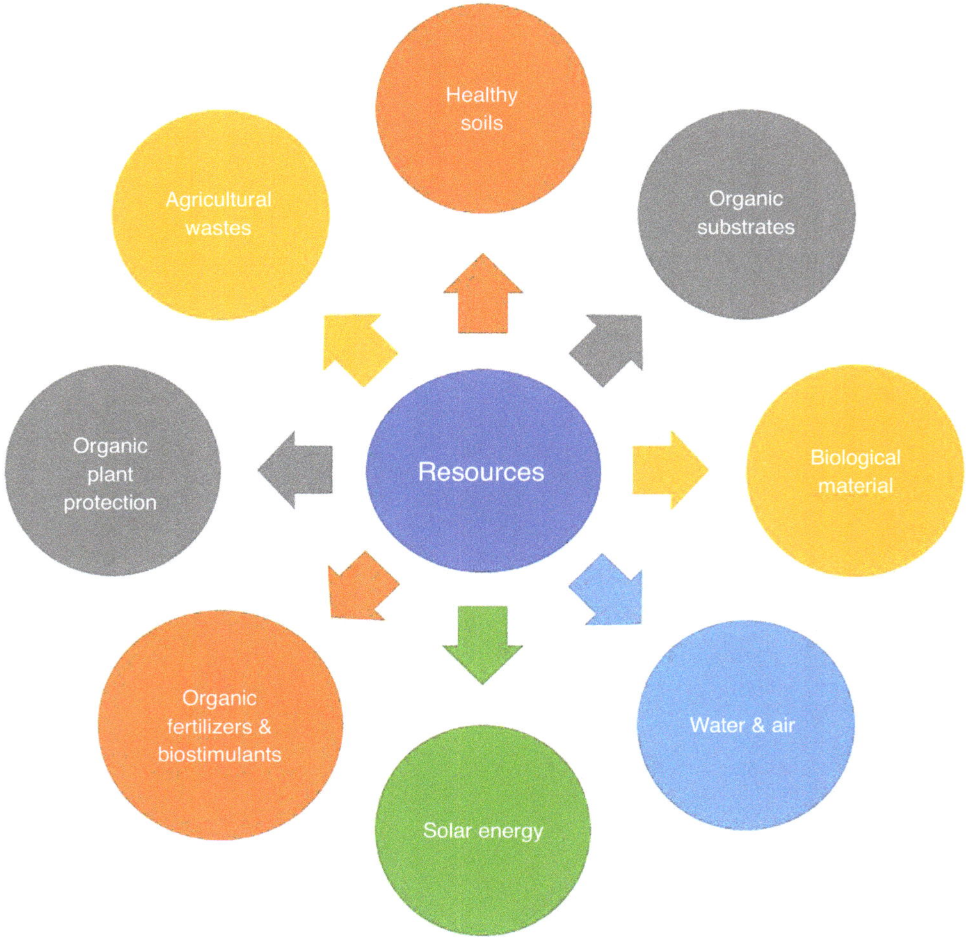

Fig. 16.2. Natural resources and components involved in organic ecosystem.

Air, although not directly consumed by organisms, is a crucial natural resource for organic ecosystems. Oxygen, a component of air, is essential for the respiration of organisms, while carbon dioxide is necessary for photosynthesis in plants (Tahat *et al.*, 2020). Additionally, air quality plays a significant role in the health and well-being of organisms within organic ecosystems (Manisalidis *et al.*, 2020). Pollution and contaminants in the air can have detrimental effects on ecosystem health, leading to reduced biodiversity and impaired ecosystem functioning (Ghorani-Azam *et al.*, 2016).

Ecosystem services are affected by agricultural practices, especially the use of chemical fertilizers and pesticides (Ioannidou *et al.*, 2022). Application of arbuscular aycorrhizal fungi as a sustainable practice in the management of orchards emphasized that organic orchards have a higher potential to provide ecosystem services than conventional management of orchards (Ioannidou *et al.*, 2022).

Agriculture puts pressure on mature ecosystems and the use of natural resources. In this context, agricultural production systems must permanently identify viable solutions to achieve sustainability standards in agricultural ecosystems. Nanotechnologies could represent an efficient solution for increasing the resilience of organic ecosystems. New sustainable formulas for the production of fertilizers and plant protection products using green nanoparticles from natural resources can improve sustainability in agricultural production systems (Kumar *et al.*, 2022; Popescu and Popescu, 2023).

A series of agroecological practices can play an important role to ensure the functionality of organic ecosystems including biological nitrogen fixation, intercropping, insectary plantings, border plantings, reduced tillage, riparian buffers, hedgerows, etc. For instance, hedgerow planting is an appropriate tool to optimize the ecological performance of ecosystems (Couthouis *et al.*, 2023). Also, biological nitrogen fixation of legume crops represents a major source and technological strategy to improve nitrogen level in organic ecosystems (Babu *et al.*, 2023; Barbieri *et al.*, 2023).

Modern agricultural practices widely use chemical fertilization and pesticides that negatively affect the nutrient cycle, soil erosion, carbon sequestration and many other ecosystem functions. Organic farming and the creation of organic ecosystems can minimize the environmental impact and exclude these practices from agricultural production (Gamage *et al.*, 2023).

The legislative, institutional and political framework has an essential role for the conservation and sustainable use of resources. In this context, rural development and food security policies must promote and support the organic ecosystem and rational use of natural resources (Popescu *et al.*, 2023). Agricultural policy should support farmers to apply agroecological practices (Liebert *et al.*, 2022). Relevant stakeholders that could be involved in the management of organic ecosystems are presented in Fig. 16.3. Co-operation and knowledge transfer among stakeholders play a significant role to promote sustainable use of natural resources in organic ecosystems.

Conclusions

Organic ecosystem is a term that captures the complex interactions between living organisms and their environment in an organic farming system. Organic farming is a form of agriculture that aims to enhance soil fertility, biodiversity, sustainability and food quality, while avoiding the use of synthetic inputs that may harm the environment or human health. This type of semi-natural ecosystem can offer several environmental benefits, such as improved soil health and carbon sequestration, and conserving and enhancing biodiversity and wildlife. However, organic ecosystem also faces some challenges and trade-offs, such as lower yield and land-use efficiency, and variability and uncertainty of the environmental performance. Therefore, the environmental assessment of organic ecosystem requires a comprehensive

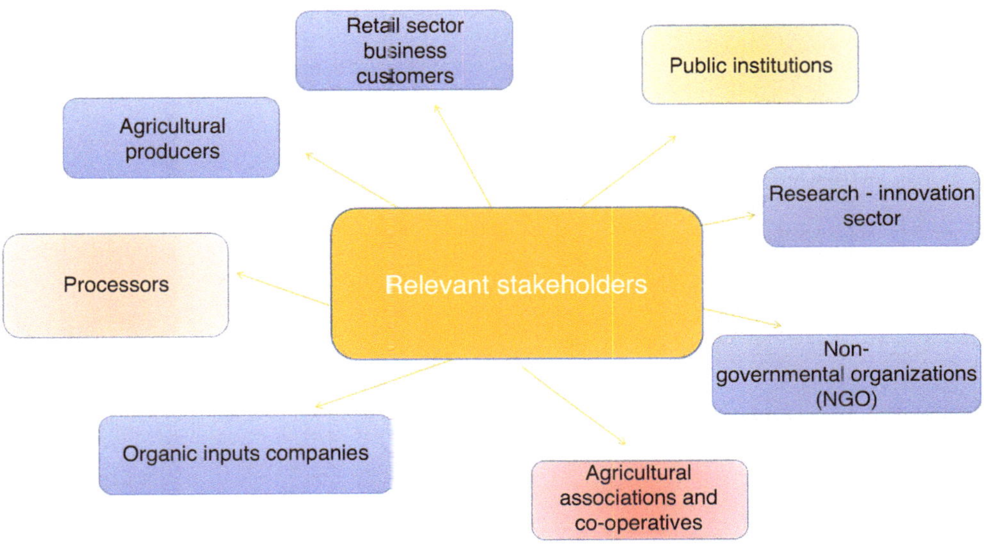

Fig. 16.3. Relevant stakeholders for efficient management of organic ecosystem and natural resources.

and holistic approach that considers multiple dimensions, scales and trade-offs, and that accounts for the variability and uncertainty of the system. Natural resources play a vital role in organic ecosystems, providing energy, water, nutrients and habitats for the survival and functioning of living organisms. These resources contribute to the maintenance of ecological balance, support biodiversity and provide various ecosystem services. Understanding the place and importance of natural resources in organic ecosystems is crucial for promoting their sustainable management and ensuring the well-being of both ecosystems and human societies. Therefore, it is essential to recognize the value of natural resources and implement measures to conserve and protect them for future generations.

References

Arora, N.K. (2018) Environmental sustainability: Necessary for survival. *Environmental Sustainability* 1(1), 1–2.

Aulakh, C.S., Sharma, S., Thakur, M. and Kaur, P. (2022) A review of the influences of organic farming on soil quality, crop productivity and produce quality. *Journal of Plant Nutrition* 45(12), 1884–1905.

Babu, S., Singh, R., Avasthe, R., Kumar, S., Rathore, S.S. *et al*. (2023) Soil carbon dynamics under organic farming: Impact of tillage and cropping diversity. *Ecological Indicators* 147, 109940.

Barbieri, P., Starck, T., Voisin, A.S. and Nesme, T. (2023) Biological nitrogen fixation of legumes crops under organic farming as driven by cropping management: A review. *Agricultural Systems* 205, 103579.

Bengtsson, J., Ahnström, J. and Weibull, A. (2005) The effects of organic agriculture on biodiversity and abundance: A meta-analysis. *Journal of Applied Ecology* 42(2), 261–269.

Comer, P.J., Hak, J.C. and McIntyre, P. (2022) Addressing climate change vulnerability in the IUCN red list of ecosystems: Results demonstrated for a cross-section of major vegetation-based ecosystem types in the united states. *Land* 11(2), 302.

Costanza, R., d'Arge, R., Groot, R., Farber, S., Grasso, M. *et al*. (1997) The value of the world's ecosystem services and natural capital. *Nature* 387(6630), 253–260.

Couthouis, E., Aviron, S., Pétillon, J. and Alignier, A. (2023) Ecological performance underlying ecosystem multifunctionality is promoted by organic farming and hedgerows at the local scale but not at the landscape scale. *Journal of Applied Ecology* 60(1), 17–28.

de Ponti, T., Rijk, B. and van Ittersum, M.K. (2012) The crop yield gap between organic and conventional agriculture. *Agricultural Systems* 108, 1–9.

De Swaef, T., Pieters, O., Appeltans, S., Borra-Serrano, I., Coudron, W. *et al*. (2022) On the pivotal role of water potential to model plant physiological processes. *In Silico Plants* 4(1), diab038.

Debuschewitz, E. and Sanders, J. (2022) Environmental impacts of organic agriculture and the controversial scientific debates. *Organic Agriculture* 12(1), 1–15.

FAO (2003) The state of food insecurity in the world 2003. Available at: http://www.fao.org/3/y1500e/y1500e00.htm (accessed 28 January 2025).

Feledyn-Szewczyk, B., Kus, J., Stalenga, J., Berbeć, A.K. and Radzikowski, P. (2016) *The role of biological diversity in agroecosystems and organic farming*. InTech Open.

Feng, W., Wang, T., Zhu, Y., Sun, F., Giesy, J.P. *et al*. (2023) Chemical composition, sources, and ecological effect of organic phosphorus in water ecosystems: A review. *Carbon Research* 2(1), 12.

Gamage, A., Gangahagedara, R., Gamage, J., Jayasinghe, N., Kodikara, N. *et al*. (2023) Role of organic farming for achieving sustainability in agriculture. *Farming System* 1(1), 100005.

Gattinger, A., Muller, A., Haeni, M., Skinner, C., Fliessbach, A. *et al*. (2012) Enhanced top soil carbon stocks under organic farming. *Proceedings of the National Academy of Sciences* 109(44), 18226–18231.

Ghorani-Azam, A., Riahi-Zanjani, B. and Balali-Mood, M. (2016) Effects of air pollution on human health and practical measures for prevention in Iran. *Journal of Research in Medical Sciences* 21, 65.

Gomiero, T., Pimentel, D. and Paoletti, M.G. (2011) Environmental impact of different agricultural management practices: Conventional vs. organic agriculture. *Critical Reviews in Plant Sciences* 30(1–2), 95–124.

Gong, S., Hodgson, J.A., Tscharntke, T., Liu, Y., van der Werf, W. *et al*. (2022) Biodiversity and yield trade-offs for organic farming. *Ecology Letters* 25(7), 1699–1710.

Groth, C. (2007) A new-growth perspective on non-renewable resources. In: Bretschger, L.A. and Smulders, S. (eds) *Sustainable Resource Use and Economic Dynamics*. Springer, Dordrecht, The Netherlands, pp. 127–163. DOI: 10.1007/978-1-4020-6293-3_7.

Hole, D.G., Perkins, A.J., Wilson, J.D., Alexander, I.H., Grice, P.V. *et al*. (2005) Does organic farming benefit biodiversity? *Biological Conservation* 122(1), 113–130.

Horwitz, P. and Finlayson, C.M. (2011) Wetlands as settings for human health: Incorporating ecosystem services and health impact assessment into water resource management. *BioScience* 61(9), 678–688.

IFOAM (2021) Principles of organic agriculture. Available at: https://www.ifoam.bio/en/organic-landmarks/principles-organic-agriculture (accessed 28 January 2025).

Ioannidou, S.C., Litskas, V.D., Stavrinides, M.C. and Vogiatzakis, I.N. (2022) Linking management practices and soil properties to ecosystem services in Mediterranean mixed orchards. *Ecosystem Services* 53, 101378.

Kershner, J., Woodward, A. and Torregrosa, A. (2020) *Integrating Climate Change Considerations into Natural Resource Planning: An Implementation Guide (no.6-C2)*. US Geological Survey.

Kumar, A., Singh, K., Verma, P., Singh, O., Panwar, A. *et al*. (2022) Effect of nitrogen and zinc nanofertilizer with the organic farming practices on cereal and oil seed crops. *Scientific Reports* 12(1), 6938.

Lal, R. (2004) Soil carbon sequestration impacts on global climate change and food security. *Science* 304(5677), 1623–1627.

Lee, R. (2019) Renewable resources and sustainable development. In: Filho, W.L. (ed.) *Encyclopedia of Sustainability in Higher Education*. Springer, pp. 1385–1393.

Liebert, J., Benner, R., Bezner Kerr, R., Björkman, T., De Master, K.T. *et al*. (2022) Farm size affects the use of agroecological practices on organic farms in the United States. *Nature Plants* 8(8), 897–905.

Mair, C. and Asada, R. (2019) Efficient use of natural resources. In: Filho, W.L., Azul, A., Brandli, L., Özuyar, P. and Wall, T. (eds) *Responsible Consumption and Production. Encyclopedia of the UN Sustainable Development Goals*. Springer, Cham, Switzerland. DOI: 10.1007/978-3-319-71062-4_13-1.

Manisalidis, I., Stavropoulou, E., Stavropoulos, A. and Bezirtzoglou, E. (2020) Environmental and health impacts of air pollution: A review. *Frontiers in Public Health* 8, 14.

Meran, G., Siehlow, M. and Hirschhausen, C. (2021) *The Economics of Water: Rules and Institutions*. Springer Nature.

Muller, A., Schader, C., El-Hage Scialabba, N., Brüggemann, J., Isensee, A. *et al*. (2017) Strategies for feeding the world more sustainably with organic agriculture. *Nature Communications* 8(1), 1290.

Odum, E.P., Libby, R., Sverker, S. and Paul, W. (eds) (2013) *Fundamentals of Ecology*. Yale University Press, New Haven, Connecticut, pp. 233–244.

Parizad, S. and Bera, S. (2023) The effect of organic farming on water reusability, sustainable ecosystem, and food toxicity. *Environmental Science and Pollution Research* 30(28), 71665–71676.

Pimentel, D., Hepperly, P., Hanson, J., Douds, D. and Seidel, R. (2005) Environmental, energetic, and economic comparisons of organic and conventional farming systems. *BioScience* 55(7), 573.

Popescu, G.C., Popescu, M., Khondker, M., Clay, D.E., Pampana, S. *et al*. (2022) Agricultural sciences and the environment: Reviewing recent technologies and innovations to combat the challenges of climate change, environmental protection, and food security. *Agronomy Journal* 114(4), 1895–1901.

Popescu, G.C. and Popescu, M. (2023) Agricultural nanotechnologies: Current scenario and future implications for global food security. In: Khan, Z. and Sutan, N.A. (eds) *Nanoformulations for Sustainable Agriculture and Environmental Risk Mitigation*. CAB International, Wallingford, UK, pp. 1–17.

Popescu, G.C., Popescu, M., Pampana, S., Khondker, M., Umehara, M. *et al*. (2023) Introduction: Sustainability as an agroecological strategy toward resilience in agricultural systems. *Agronomy Journal* 115(6), 2657–2664.

Rawat, P., Das, S., Shankhdhar, D. and Shankhdhar, S.C. (2021) Phosphate-solubilizing microorganisms: Mechanism and their role in phosphate solubilization and uptake. *Journal of Soil Science and Plant Nutrition* 21(1), 49–68.

Seufert, V., Ramankutty, N. and Foley, J.A. (2012) Comparing the yields of organic and conventional agriculture. *Nature* 485(7397), 229–232.

Skinner, E. and Castleden, H. (2014) Natural resource management (NRM). In: Michalos, A.C. (ed.) *Encyclopedia of Quality of Life and Well-Being Research*. Springer, Dordrecht, The Netherlands.

Suman, J., Rakshit, A., Ogireddy, S.D., Singh, S., Gupta, C. *et al*. (2022) Microbiome as a key player in sustainable agriculture and human health. *Frontiers in Soil Science* 2, 821589.

Tahat, M.M., Alananbeh, K.A., Othman, Y.A. and Leskovar, D.I. (2020) Soil health and sustainable agriculture. *Sustainability* 12(12), 4859.

Tuck, S.L., Winqvist, C., Mota, F., Ahnström, J., Turnbull, L.A. *et al*. (2014) Land-use intensity and the effects of organic farming on biodiversity: A hierarchical meta-analysis. *The Journal of Applied Ecology* 51(3), 746–755.

Tuomisto, H.L., Hodge, I.D., Riordan, P. and Macdonald, D.W. (2012) Does organic farming reduce environmental impacts? A meta-analysis of European research. *Journal of Environmental Management* 112, 309–320.

17 Land-use Carbon Changes and Monitoring

Wing K. P. Ng[1]*, Adrian P. Crew[2], Dayane L. Teixeira[1], Pete J. Maxfield[2] and Matt J. Bell[1]

[1]*Department of Animal and Agriculture, Hartpury University, Gloucester, UK;*
[2]*Department of Health and Applied Sciences, University of the West of England, Bristol, UK*

Abstract

About 40% of global habitable land area is used for agricultural-related activities. As farming practices can influence soil carbon stocks and the ability of the land to sequester carbon, there has been a growing interest in monitoring carbon fluxes and sequestration among land uses to mitigate atmospheric carbon emission. However, the lack of valuable benchmarking data on soil organic carbon (SOC) changes and carbon hotspots, due partly to a lack of regular monitoring, has been slow to develop and provide farmers and landowners with necessary data to improve their agricultural land management. To enhance and maintain soil carbon stocks, regular on-farm precision monitoring solutions are developing, such as modern laboratory, near-infrared spectroscopy (NIRS) and remote-sensing analysis. This chapter provides a brief insight on the effects of land-use change and the importance of on-farm monitoring of SOC to support land management practices with the current precision monitoring techniques.

Keywords: soil carbon, storage, sequestration, land-use change, precision monitoring

Introduction

More than one-third (38%) of the world's habitable land belongs to agriculture (FAO, 2020), which is about 5 billion ha. Agricultural practices have direct influence on field soil carbon stocks and soil organic matter (SOM) content (Batjes, 1996; Pulleman *et al.*, 2000; MacLaren *et al.*, 2022). Globally, a 1–2m depth topsoil can contain about 1500–2400 gigatonnes of SOC (Paustian *et al.*, 2019) through carbon cycle (Smith and Smith, 2012; Fig. 17.1). Adopting better land management practice can improve soil carbon cycle and encourage more SOC sequestration (Paustian *et al.*, 1997, 2016, 2019).

There are many innovative ecological approaches to improve land management practice, such as improved crop rotation, cover crops, manure and compost addition (Smith and Smith, 2012; Paustian *et al.*, 2019). All these approaches have strong and direct impact on the ability of agricultural land to retain soil nutrients and SOC content (Smith and Smith, 2012; NASEM, 2019).

This chapter therefore focuses on the basic concepts and effects of agricultural land-use change, the importance of on-farm monitoring of SOC to support land management practices, and the current and developing field precision monitoring approaches in laboratory, near-infrared spectroscopy (NIRS) and remote-sensing analysis.

Land Use Change

Many past studies have shown land SOC and OM contents to be either in a static (Lal and Kimble,

*Corresponding author: wing.ng@hartpury.ac.uk

 Advances in Organic Farming (ed. L. P. Awasthi)
DOI: 10.1079/9781800626850.0017

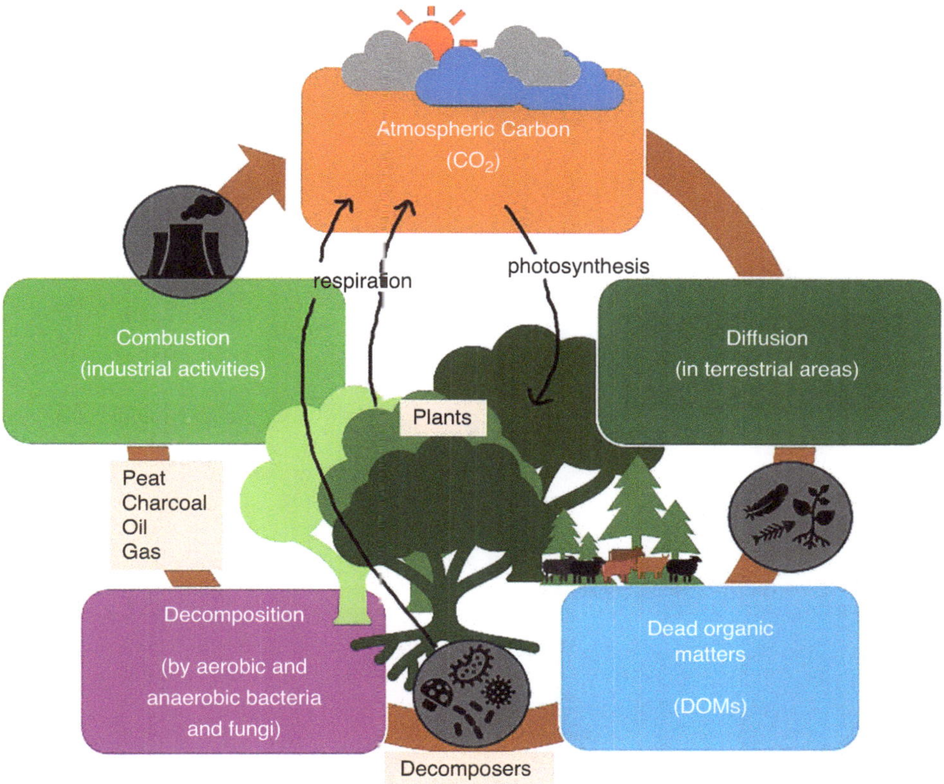

Fig. 17.1. Carbon cycle in terrestrial areas sharing three basic components: (i) atmospheric inputs of carbon; (ii) internal cycling of dead organic matter (DOM) such as litterfall, debris and decaying material; and (iii) outputs of soil nutrients for plant uptake and incorporation into plant tissues or loss due to burning.

1997) or dynamic form (Pulleman *et al.*, 2000; Baessler and Klotz, 2006). The fundamental nature of soil properties can vary spatially and temporally (Pulleman *et al.*, 2000; Popovici *et al.*, 2013; Bell *et al.*, 2020) through direct or indirect land-use change mainly from human activities (Prout *et al.*, 2020; Macedo *et al.*, 2023). Therefore, land practices play a critical role in influencing farmland soil and OM variability, particularly depending on the level of soil disturbance that occurs such as changing a grassland area to arable crops, or *vice versa* (Pulleman *et al.*, 2000; Bell *et al.*, 2018).

Typically, agricultural land-use categories consist of arable crops (AC), temporary leys (TL), permanent grassland (PG) and woodland (WL) fields (Popovici *et al.*, 2013; Prout *et al.*, 2020) (Fig. 17.2). Arable crops are regarded as fields under regular disturbance through differing

Fig. 17.2. *left:* field sampling in an arable field; *right:* field sampling in woodland area.

amounts of land cultivation (Brentrup *et al.*, 2002; Storkey *et al.*, 2012), from ploughing by machinery (e.g. tractor), to a more minimal or even no-tillage system to minimize soil disturbance nowadays (Bell *et al.*, 2018). They have relatively higher field traffic and cultivation (Popovici *et al.*, 2013) than other land-use types due to agricultural intensification (Baessler and Klotz, 2006). Both arable crops and temporary ley grasslands are considered to have more intensively managed land cover and often within a rotation of arable crops and grassland species (Angus *et al.*, 2009). Unlike AC and TL, permanent grassland and woodland sites are relatively well-established (Rackham, 2008) containing more woodland-grassland plants for animal grazing or foraging activities (Popovici *et al.*, 2013) under minimal human disturbance (Woodland Trust, 2012). Some past studies showed that WL-PG or moderately grazed PG had relatively higher SOC stocks and SOM content compared to AC and TL sites (Rackham, 2008; MacLaren *et al.*, 2022). This implies that WL and PG have high potential to be big land carbon sinks.

Types of Precision Monitoring Techniques

Innovative approaches using measurement tools with established remote-sensing and sampling technologies are developing and will provide large-scale and more frequent on-farm SOC monitoring (Paustian *et al.*, 2019) and soil nutrient fluxes. Further exploration on areas of potential carbon sinks and taking spectrum imaging on arable fields are highly beneficial when identifying strategies to refine carbon assessment. There are several critical precision monitoring techniques to quantify and determine both SOM and SOC changes, as discussed below.

Laboratory analysis

The Walkley-Black (WB) method, also known as principal wet combustion or oxidation analysis (Walkley and Black, 1934; Bell *et al.*, 2018), is a classical, affordable and considerably accurate method to measure SOC using a mixture of potassium dichromate and sulfuric acid, which can generate chemical hazardous waste (Salehi *et al.*, 2011). In general, quantified SOC measurements can be obtained through a traditional conversion of SOM with a correction factor of 1.724, as SOM counts about 58% organic carbon (Grewal *et al.*, 1991; Salehi *et al.*, 2011; El Mouridi *et al.*, 2023). Many recent studies considered the WB method to be relatively dangerous and erroneous because of toxic waste generation and incomplete oxidation of SOC (Salehi *et al.*, 2011; Wang *et al.*, 2012; El Mouridi *et al.*, 2023). Therefore, this robust method has been replaced gradually by more secure and time-saving laboratory methods like loss on ignition (LOI) (Salehi *et al.*, 2011) and electronic measurement (Wang *et al.*, 2012).

Loss on ignition is a common method for inexpensive estimation of SOMs in a good and safe testing environment under the absence of chromic acid waste (Schulte *et al.*, 1991; Salehi *et al.*, 2011). It is a laboratory-based testing method through combustion that generally involves oven-dried soil samples (overnight at around 105°C and then cool down in a desiccator) to be combusted under a designated high temperature (i.e. 350–400°C) for around two hours in a muffle furnace. The weight loss of the samples is then measured soon after combustion (Schulte *et al.*, 1991; Schulte and Hopkins, 1996; Salehi *et al.*, 2011). After the measurement, the estimates of SOM percentage are further calculated based on equation (1) (Schulte and Hopkins, 1996; Salehi *et al.*, 2011):

$$SOM(LOI) = \left(\frac{\text{soil weight after combustion} - \text{oven dried soil weight}}{\text{oven dried soil weight}}\right) \times 100 \tag{1}$$

Although LOI has become a widely adopted method to estimate SOM contents in a much safer and time-saving environment, sample combustion under such high or even higher temperatures can evaporate and destroy overall sample structural moisture, as well as other inorganic constituents (ICs), which is likely to influence SOM estimates (Yerokun *et al.*, 2007; Salehi *et al.*, 2011). To fulfil better accuracy on SOMs and SOC estimates for precision agricultural monitoring (Ben-Dor and Banin, 1989; Schulte *et al.*, 1991; Salehi *et al.*, 2011), other minimal disturbance methods on ICs have also been developed in recent years. One of the modern and low-disturbance methods for SOC estimates is through electronical measurement using an elemental carbon analyzer. This method is of high result accuracy once the system software is properly set up and calibrated (Salehi *et al.*, 2011). Only small amounts of liquid or solid soil samples (i.e. 10 mg) are required for the measurement (Wang *et al.*, 2012). This method can act as a 'golden' benchmarking method for rapid SOC stock estimates despite its relatively higher testing cost (Wang *et al.*, 2012).

Near infrared spectroscopy (NIRS)

Laboratory or mobile near infrared spectroscopy (NIRS) analysis is a recent non-invasive and non-destructive measuring technique to estimate both SOM and SOC stocks using a portable NIRS scanner. The NIRS device (Fig. 17.3) is calibrated with the use of laboratory wet chemistry data (Bell *et al.*, 2018, 2020). It measures a spectrum of infrared energy reflected from the sample illuminated by white light during multiple scanning (Bell *et al.*, 2018). Previous studies have adopted NIRS analysis to identify and estimate seasonal and spatial differences in forage nutrient levels (Bell *et al.*, 2020; Balehegn *et al.*, 2022). The analysis can aid rapid and effective real-time evaluation on critical on-farm variabilities (e.g. SOC, total nitrogen, total carbon, clay, pH, moisture, crude protein, neutral and acid detergent fibre, lignin and ether extract) and other key parameters altogether (e.g. biomass height, plant fresh and dry cover, nitrate, oil, protein) on the collected fresh forage, soil and crop samples (Bell *et al.*, 2014, 2018, 2020) with much less time investment (from 16 hours to less than 2 min per sample) and associated cost (Bell *et al.*, 2018). Moreover, NIRS analysis will bring the benefits of reducing involved cost greatly, not damaging the collected samples' structure, not using any reagents to provide prompt analytical results (Paz *et al.*, 2019), as well as providing comparably accurate results for soil and plant nutritive and OM value estimation (Balehegn *et al.*, 2022).

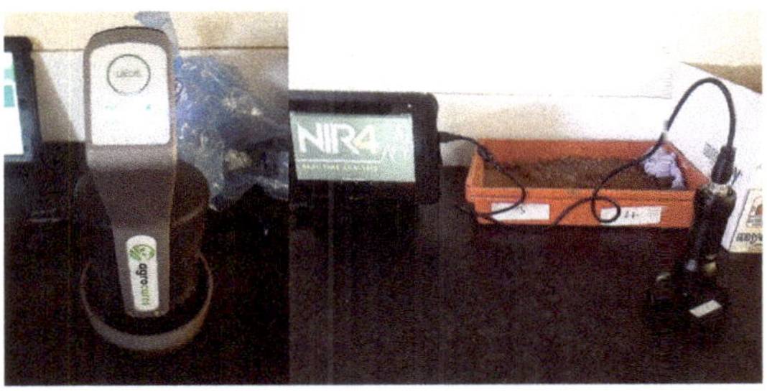

Fig. 17.3. Portable near infrared spectroscopy devices used for (*left*) soil nutrient analysis (Series E, AgroCares, Wageningen, The Netherlands); and (*right*) forage analysis (NIR4; Aunir, Towcester, UK).

Remote sensing (RS)

Since the 1800s, remote sensing (RS) was established in the form of photographic observations which are often being used for building a more time-saving and systematic process and framework when analyzing and recording collected terrestrial and field data. RS is a well-known remotely sensed data collection technique being classified into two major forms when observing and measuring land objects (Carter and Young, 2013) and changes, as well as different kinds of potential gas hotspots (i.e. carbon sinks, greenhouse gas fluxes on a region of interest) on earth surface (Näsi *et al.*, 2017; Gålfalk *et al.*, 2021). It can be in either passive sensor or active sensor system, depending on the type of energy sources. For a passive sensor system, it mainly includes typical energy source with low-level microwave radiation such as thermal energy (e.g. thermal infrared) and spectroradiometer (e.g. multispectral and hyperspectral wavelength), where such energy source cannot emit light to a designated survey target for energy reflectance unless under daytime period with sunlight and less cloud cover conditions (i.e. no more than 40% cloud coverage during daytime). For an active sensor system, the most typical example of energy source is Radarsat-1 synthetic aperture radar (SAR), which can emit short electromagnetic energy bursts or pulses towards the direction of the target during object detection (Carter and Young, 2013). Through decades of technological development and advancement, the common types of field survey methods to be adopted for RS are light detection and ranging (LiDAR), unmanned aerial vehicle (UAV) and artificial satellite imagery.

LiDAR is an airborne remote-sensing method using light in the form of a pulsed laser to measure ranges (variable distances) to the earth surface, combined with other data collected from a set airborne system during field mapping missions (NOAA, 2023). A LiDAR scanner with a global positioning system (GPS) can be mounted on an aircraft (usually fixed-wing) or satellite. The spatial resolution of LiDAR imagery data can usually range between 15m and 1000m above ground level, depending on the field-sampling requirements and purposes. It can be used to perform quick and initial field scans for further examination on both natural and man-made environments through multiple flights or satellite operations. Critical land and crop information such as land terrain, tree height and canopy, as well as digital elevation models (DEMs) and shoreline maps can also be derived and produced with flexibility, accuracy and precision after data collection (NOAA, 2023). Due to its readiness, reliability and convenience, LiDAR is being further adopted to precision monitoring activities among different land uses, particularly in soil and plant variability on small to larger plot-farm level in recent years (Florent *et al.*, 2019).

Despite the active system element of LiDAR, UAV is an airborne remote-sensing method to capture field aerial images with a passive camera system (Carter and Young, 2013; Fig. 17.4). Even though there is specific weather constraint (i.e. daytime) on the use of UAV, it has still been adopted in a wide range of practical applications to satisfy the growing need on geospatial data collection (Carter and Young, 2013; Christiansen *et al.*, 2017; Paustian *et al.*, 2019). Further, the use of UAV can be equipped with different types of colour sensors and scanners in order to fulfil specific aerial mapping tasks related to topographic or shoreline mapping in high-imagery resolution (Christiansen *et al.*, 2017); for instance, digital colour sensors like visible RGB colour sensor, multi-spectral sensor, hyper-spectral sensor, or a LiDAR scanner (Carter and Young, 2013). The adoption of UAV can be powerful in many agricultural missions for ground-truth measurements associated with plant health and growth (Christiansen *et al.*, 2017), such as sward height (Näsi *et al.*, 2017), biomass (Ghajar and Tracy, 2021), weed status and soil moisture (Paustian *et al.*, 2019). To a more unique extent, UAV has recently engaged in proximal sensing in ground-level carbon sinks from local to a wider landscape scale (Gålfalk *et al.*, 2021). As reducing carbon emissions also strongly depends on accuracy improvement to quantify carbon stock estimates (Paustian *et al.*, 2019), using UAV with geographical information system data can serve as another cost-effective, environmentally friendly and efficient alternative to monitor carbon fluxes or stores among land uses (Ghajar and Tracy, 2021).

Fig. 17.4. An example of an unmanned aerial vehicle with passive camera system including four spectral cameras in NIR, red, green and blue bands. (DJI Mavic M3M drone, DJI Agriculture, Shenzhen, China, 2024)

Due to the advantages of non-physical interaction with targeted survey objects or more remote areas (Carter and Young, 2013), satellite imagery and its related products have been widely employed in monitoring earth ecosystems and biodiversity conservation (Turner *et al.*, 2003), quantifying land cover change (Asner *et al.*, 2002; Boyle *et al.*, 2014) and detecting land degradation and deforestation (Carter and Young, 2013). In recent decades, the use of satellite imagery on measuring land carbon fluxes (Boyle *et al.*, 2014; Macedo *et al.*, 2023) has become frequent to investigate and monitor land changes. Through downloading satellite imagery data from currently existing and popular external sources (e.g. LandSat, Sentinel-2, MODIS), more field variability based on soil and plant spectral and spatial properties (Carter and Young, 2013) can be remotely tracked, verified and even assessed (NASEM, 2019) for making better land management decisions (Paustian *et al.*, 2019). Hence, this leads to a derivation of various common spectral vegetation indices (VIs), which are developed to classify and determine satellite images into different land management zones for on-farm vegetation response monitoring. Some representative examples of the indices are enhanced vegetation index (EVI), normalized difference vegetation index (NDVI), green normalized difference vegetation index (GNDVI) and green leaf index (GLI). The indices can act as quick and easy performance and productivity assessors for different preliminary on-farm conditions (Carter and Young, 2013; Lima-Cueto *et al.*, 2019), biomass cover and surface carbon hotspots (Gålfalk *et al.*, 2021) compared to traditional laboratory analysis on field soil and plants (NASEM, 2019; Macedo *et al.*, 2023), which will be further explored below.

Vegetation indices (VIs) are relative estimation indicators (Rasmussen *et al.*, 2016) developed to initially determine the presence or absence of field biomass and its distribution (Yeom *et al.*, 2019), monitor health status of agricultural crops (Raeva *et al.*, 2019; Yeom *et al.*, 2019), differentiate various soil and plant variabilities (Kareem *et al.*, 2023) and aid more in-depth studies on biotic and abiotic environmental changes (De Swaef *et al.*, 2021) among mixed farms (Bell *et al.*, 2018), forest and woodland (Chenari *et al.*, 2017), and key semi-arid regions (Kurbanov and Zakharova, 2020). As different VIs have their own sensitivity to specific spectrum and wavelengths (Kareem *et al.*, 2023), there are various types of improved VIs being developed by many researchers in order to cope with the growing demand for rapid detection and long-term precision monitoring on farms. There are many types of newly developed VIs, but the most common and frequently used VI is still normalized difference vegetation index (NDVI).

NDVI is one of the VIs which consists of different spectral bands (near-infrared (NIR) and red bands) of a multi-spectral sensor (Kurbanov and Zakharova, 2020; Gålfalk *et al.*, 2021). It is a useful vegetation response index to differentiate field greenness, and vegetation or biomass density and distribution (Kilwenge *et al.*, 2021). From the perspective of precision agricultural monitoring, some applications of NDVI are specifically extended to identify more reliable on-farm indicators (Näsi *et al.*, 2017) and plant characteristics (Swayze *et al.*, 2022), such as yield variation (Carter and Young, 2013), crop height (Lima *et al.*, 2021), leaf chlorophyll content (Kurbanov and Zakharova, 2020), nitrogen (N) concentrations (Caturegli *et al.*, 2016; Sozzi *et al.*, 2021), above-ground biomass volume and prediction (Bell *et al.*, 2018; Swayze *et al.*, 2022) and greenhouse gas measurements as field surface methane (Gålfalk *et al.*, 2021). Through the adoption of NDVI during precision in-farm monitoring, significant investment in time and manpower derived from conventional field sampling and analysis is very likely to be reduced (Ghajar and Tracy, 2021).

Although NDVI can demonstrate initial differentiation between soil and crops on fields, it has strong sensitivity to changes in soil, field reflectance spectra (red and green bands) and atmospheric influences (Eng *et al.*, 2019; Xiong *et al.*, 2023). Therefore, additional VIs have been further developed by many researchers to compensate the high sensitivity characteristic of NDVI during precision agricultural monitoring and field-mapping missions; for instance, enhanced vegetation index (EVI), green normalized difference vegetation index (GNDVI) and green leaf index (GLI). Table 17.1 summarizes the mentioned types of currently available and popular VIs being used in vegetation

Table 17.1. List of basic and common types of vegetation indices, associated spectra and wavelengths for the study of land and plant characteristics.

Indices/spectra	[a]Spectra / wavelengths	Equation	Key applications	Reference
Vegetation index (VI)	NIR and RED	$\frac{NIR}{RED}$	Healthy biomass cover, distribution and leaf chlorophyll content	Bell *et al.*, 2018
Enhanced vegetation index (EVI)	RED, BLUE and Near-infrared	$2.5\,x\left(\frac{(NIR-RED)}{NIR+(6RED-7BLUE+1)}\right)$	For higher distinctions among soil and crop with reduced atmospheric distortions	Jiang *et al.*, 2008
Normalized difference vegetation indices (NDVI)	NIR and RED	$\frac{\rho NIR-\rho RED}{\rho NIR+\rho RED}$	Soil and crop differentiation, object detection, on-farm precision monitoring	Carter and Young, 2013
Green normalized difference vegetation index (GNDVI)	NIR and GREEN	$\frac{NIR-GREEN}{NIR+GREEN}$	Forage biomass prediction and N concentrations	Carter and Young, 2013
Green leaf index (GLI)	RED, GREEN and BLUE	$\frac{2GREEN-RED-BLUE}{2GREEN+RED+BLUE}$	Vegetation monitoring with greater distinctions on fields	Eng *et al.*, 2019

[a]Wavelength of different spectral bands: RED (620–680 nm), GREEN (540–580 nm), BLUE (420–495 nm) and NIR (800–880 nm).
(Adapted from Kurbanov and Zakharova, 2020)

cover and precision agricultural monitoring across land-use change.

Combining aerial imaging and ground measurements

To further achieve continuous on-farm monitoring (Organic Valley, 2021), the use of combining aerial imaging and ground measurements can be extremely beneficial to farmers and land users (Castaldi *et al.*, 2019). Since NIRS technique can measure the *in situ* soil variabilities non-destructively (Muñoz and Kravchenko, 2011), collection of fresh soil samples from some remote fields is still inevitable (Bell *et al.*, 2018). By combining both NIRS and aerial-sensing techniques, more remote, site-specific and cross-sectional field measurements and imagery data, especially when assessing SOC variability (Macedo *et al.*, 2023), mapping carbon fluxes (Castaldi *et al.*, 2019) and observing associated pasture health and cover status (Organic Valley, 2021), can all be better streamlined and assessed at one shot. Not only the combination method can increase higher accuracy for on-farm monitoring, but it can also greatly ease labour, cost and time burdens during the monitoring by farmers and landowners on a regular basis (Bell *et al.*, 2018, 2020; Paustian *et al.*, 2019; Macedo *et al.*, 2023).

Conclusions

Through decades, many farmers and land users have been trying hard to make better agricultural land management in order to maintain the balance between on-farm productivity and sustainability land management. With the implementation of the precision monitoring techniques, no doubt they can aid farmers and land-interested parties to monitor farm soil among land uses, and make better agricultural management decisions with much lower cost, labour and time involvement. Land-use change therefore is very likely to act as one of the indispensable parameters to determine OM variabilities, SOC changes and land carbon hotspots for the farming industry. However, further field studies on farm data collection patterns and the practical combination of the different monitoring techniques are still desperately needed to improve terrestrial carbon cycle, to perform more accurate and regular on-farm monitoring, and to support higher field carbon storage for a more sustainable Earth.

References

Angus, A., Burgess, P.J., Morris, J. and Lingard, J. (2009) Agriculture and land use: Demand for and supply of agricultural commodities, characteristics of the farming and food industries, and implications for land use in the UK. *Land Use Policy* 26, S230–S242.

Asner, G.P., Keller, M., Pereira, R. and Zweede, J.C. (2002) Remote sensing of selective logging in Amazonia: assessing limitations based on detailed field observations, Landsat ETM+, and textural analysis. *Remote Sensing of Environment* 80(3), 483–496.

Baessler, C. and Klotz, S. (2006) Effects of changes in agricultural land-use on landscape structure and arable weed vegetation over the last 50 years. *Agriculture, Ecosystems & Environment* 115(1–4), 43–50.

Balehegn, M., Varijakshapanicker, P., Zampaligre, N., Blummel, M., Ayantunde, A. *et al.* (2022) Near-infrared reflectance spectroscopy for forage nutritive value analysis in Sub-Saharan African countries. *Agronomy Journal* 114(1), 100–114.

Batjes, N.H. (1996) Total carbon and nitrogen in the soils of the world. *European Journal of Soil Science* 47(2), 151–163.

Bell, M.J., Potterton, S.L., Craigon, J., Saunders, N., Wilcox, R.H. *et al.* (2014) Variation in enteric methane emissions among cows on commercial dairy farms. *Animal* 8(9), 1540–1546.

Bell, M.J., Mereu, L. and Davis, J. (2018) The use of mobile near-infrared spectroscopy for real-time pasture management. *Frontiers in Sustainable Food Systems* 2, 76.

Bell, M.J., Huggett, Z., Slinger, K.R. and Roos, F. (2020) Effect of pasture cover and height on nutrient concentrations in diverse swards in the UK. *Grassland Science* 67(3).

Ben-Dor, E. and Banin, A. (1989) Determination of organic matter content in arid-zone soils using a simple loss-on-ignition method. *Communications in Soil Science and Plant Analysis* 20, 1675–1695.

Boyle, S.A., Kennedy, C.M., Torres, J., Colman, K., Pérez-Estigarribia, P.E. *et al.* (2014) High-resolution satellite imagery is an important yet underutilized resource in conservation biology. *PloS One* 9(1), e86908.

Brentrup, F., Küsters, J., Lammel, J. and Kuhlmann, H. (2002) Life cycle impact assessment of land use based on the hemeroby concept. *The International Journal of Life Cycle Assessment* 7(6), 339–348.

Carter, P.G. and Young, S.L. (2013) Applications of remote sensing in precision agriculture for sustainable production. In: Oliver, M., Bishop, T. and Marchant, B. (eds) *Precision Agriculture for Sustainability and Environmental Protection*, 2nd edn. Routledge Taylor and Francis Group, New York, pp. 82–98.

Castaldi, F., Hueni, A., Chabrillat, S., Ward, G.B., Bomans, B. *et al.* (2019) Evaluating the capability of the sentinel 2 data for soil organic carbon prediction in croplands. *ISPRS Journal of Photogrammetry and Remote Sensing* 147, 267–282.

Caturegli, L., Corniglia, M., Gaetani, M., Grossi, N., Magni, S. *et al.* (2016) Unmanned aerial vehicle to estimate nitrogen status of turfgrasses. *PloS ONE* 11(6), e0158268.

Chenari, A., Erfanifard, Y., Dehghani, M. and Pourghasemi, H.R. (2017) Woodland mapping at single-tree levels using object-oriented classification of unmanned aerial vehicle (UAV) images. *The International Archives of the Photogrammetry, Remote Sensing and Spatial Information Sciences* XLII-4/W4, 43–49.

Christiansen, M.P., Laursen, M.S., Jørgensen, R.N., Skovsen, S. and Gislum, R. (2017) Designing and testing a UAV mapping system for agricultural field surveying. *Sensors* 17(12), 2703.

De Swaef, T., Aper, J., Baert, J., Cougnon, M., Reheul, D. *et al.* (2021) Applying RGB- and thermal-based vegetation indices from UAVs for high-throughput field phenotyping of drought tolerance in forage grasses. *Remote Sensing* 13, 147.

El Mouridi, Z., Ziri, R., Douaik, A., Bennani, S., Lembaid, I. *et al.* (2023) Comparison between Walkley-Black and loss on ignition methods for organic matter estimation in different Moroccan soils. *Ecological Engineering & Environmental Technology* 24(4), 253–259.

Eng, L.S., Ismail, R., Hashim, W. and Baharum, A (2019) The use of VARI, GLI, and vigreen formulas in detecting vegetation in aerial images. *International Journal of Technology* 10(7), 1385.

FAO (Food and Agriculture Organization of the United Nations) (2020) Land use in agriculture by the numbers. Sustainable Food and Agriculture. Available at: https://www.fao.org/sustainability/news/detail/en/c/1274219/ (accessed November 2022).

Florent, D., Galya, B., Tamas, J., Demelezi, I., Nagy, A. *et al.* (2019) Evaluation of soil water management properties based on lidar data and soil analyses, at farm level. *Natural Resources and Sustainable Development* 9(2), 160–173.

Gålfalk, M., Nilsson Påledal, S. and Bastviken, D. (2021) Sensitive drone mapping of methane emissions without the need for supplementary ground-based measurements. *ACS Earth & Space Chemistry* 5(10), 2668–2676.

Ghajar, S. and Tracy, B. (2021) Proximal sensing in grasslands and pastures. *Agriculture* 11(8), 740.

Grewal, K.S., Buchan, G.D. and Sherlock, R.R. (1991) A comparison of three methods of organic carbon determination in some New Zealand soils. *Journal of Soil Science* 42, 251–257.

Jiang, Z., Huete, A.R., Didan, K. and Miura, T. (2008) Development of a two-band enhanced vegetation index without a blue band. *Remote Sensing of Environment* 112(10), 3833–3845.

Kareem, H.H., Attaee, M.H. and Omran, Z.A. (2023) Evaluation the soil-adjusted vegetation indices SAVI and MSAVI for Bristol city, United Kingdom using landsat 8-OLI through geospatial technology. *Ecological Engineering & Environmental Technology* 24(7), 89–97.

Kilwenge, R., Adewopo, J., Sun, Z. and Schut, M. (2021) UAV-based mapping of banana land area for village-level decision-support in Rwanda. *Remote Sensing* 13(24), 4985.

Kurbanov, R.K. and Zakharova, N.I. (2020) Application of vegetation indexes to assess the condition of crops. *Agricultural Machinery and Technologies* 14(4), 4–11.

Lal, R. and Kimble, J.M. (1997) Conservation tillage for carbon sequestration. *Nutrient Cycling in Agroecosystems* 49, 243–253.

Lima, F., Blanco-Sepúlveda, R., Gómez-Moreno, M.L., Dorado, J. and Peña, J.M. (2021) Mapping tillage direction and contour farming by object-based analysis of UAV images. *Computers and Electronics in Agriculture* 187.

Lima-Cueto, F.J., Blanco-Sepúlveda, R., Gómez-Moreno, M.L. and Galacho-Jiménez, F.B. (2019) Using vegetation indices and a UAV imaging platform to quantify the density of vegetation ground cover in olive groves (*Olea europaea* L.) in southern Spain. *Remote Sensing* 11(21), 2564.

Macedo, F.L., Nóbrega, H., de Freitas, J.G.R., Ragonezi, C., Pinto, L. *et al.* (2023) Estimation of productivity and above-ground biomass for corn (*Zea mays*) via vegetation indices in Madeira Island. *Agriculture* 13(6), 1115.

MacLaren, C., Mead, A., Balen, D.V., Claessens, L., Etana, A. *et al.* (2022) Long-term evidence for ecological intensification as a pathway to sustainable agriculture. *Nature Sustainability* 5(9), 770–779.

Muñoz, J.D. and Kravchenko, A. (2011) Soil carbon mapping using on-the-go near infrared spectroscopy, topography and aerial photographs. *Geoderma* 166(1), 102–110.

NASEM (2019) Negative emissions technologies and reliable sequestration: A research agenda. National Academies of Sciences, Engineering, and Medicine, The National Academies Press, Washington, DC.

Näsi, R., Hakala, T., Viljanen, N., Honkavaara, E. and Peltonen-Sainio, P. (2017) How farmer can utilize drone mapping? Conference Paper, FIG Working Week, Helsinki, 29 May-2 June.

NOAA (2023) What is lidar? Available at: https://oceanservice.noaa.gov/facts/lidar.html (accessed March 2024).

Organic Valley (2021) Organic Valley to pioneer use of satellite technology to improve pasture grazing. Availableat:https://www.organicvalley.coop/newspress/organic-valley-pioneer-use-satellite-technology-improve-pasture-grazing/ (accessed March 2024).

Paustian, K., Andren, O., Janzen, H.H., Lal, R., Smith, P. *et al*. (1997) Agricultural soils as a sink to mitigate CO2 emissions. *Soil Use Management* 13, 230–244.

Paustian, K., Campbell, N., Dorich, C., Marx, E. and Swan, A. (2016) Assessment of potential greenhouse gas mitigation from changes to crop root mass and architecture. U.S. Department of Energy Office of Scientific and Technical Information. Available at: https://www.osti.gov/servlets/purl/1339423 (accessed November 2022).

Paustian, K., Larson, E., Kent, J., Marx, E. and Swan, A. (2019) Soil C sequestration as a biological negative emission strategy. *Frontiers in Climate* 1, 8.

Paz, C.C. da, e Silva, A.G.M. and Rêgo, A.C. do (2019) Use of near infrared spectroscopy for the evaluation of forage for ruminants. *Amazonian Journal of Agricultural and Environmental Sciences* 62.

Popovici, E.A., Balteanu, D. and Kucsicsa, G. (2013) Assessment of changes in land-use and land-cover pattern in Romania using CORINE and land cover database. *Carpathian Journal of Earth and Environmental Sciences* 8(4), 195–208.

Prout, J.M., Shepherd, K.D., McGrath, S.P., Kirk, G.J.D. and Haefele, S.M. (2020) What is a good level of soil organic matter? An index based on organic carbon to clay ratio. *European Journal of Soil Science* 1–11.

Pulleman, M.M., Bouma, J., Essen, E.A. and Meijles, E.W. (2000) Soil organic matter content as a function of different land use history. *Soil Science Society of America Journal* 64(2), 689–693.

Rackham, O. (2008) Ancient woodlands: Modern threats. *The New Phytologist* 180(3), 571–586.

Raeva, P., Šedina, J. and Dlesk, A. (2019) Monitoring of crop fields using multispectral and thermal imagery from UAV. *European Journal of Remote Sensing* 52(sup1), 192–201.

Rasmussen, J., Ntakos, G., Nielsen, J., Svensgaard, J., Poulsen, R.N. *et al*. (2016) Are vegetation indices derived from consumer-grade cameras mounted on UAVs sufficiently reliable for assessing experimental plots? *European Journal of Agronomy* 74, 75–92.

Salehi, M.H., Beni, O.H., Harchegani, H.B., Borujeni, I.E. and Motaghian, H.R. (2011) Refining soil organic matter determination by loss-on-ignition. *Pedosphere* 21(4), 473–482.

Schulte, E.E., Kaufmann, C. and Peters, J.B. (1991) The influence of sample size and heating time on soil weight loss-on-ignition. *Communication in Soil Science and Plant Analysis* 22, 159–168.

Schulte, E.E. and Hopkins, B.G. (1996) Estimation of organic matter by weight loss-on-ignition. In: Magdoff, F.R., Tabatabai, M.A. and Hanlon, E.A. (eds) *Estimation of Organic Matter by Weight Loss-on-Ignition*. Soil Science Society of America, Inc, Madison, Wisconsin, pp. 21–31.

Smith, T.M. and Smith, R.L. (2012) *Elements of Ecology*, 8th edn. Pearson Benjamin Cummings Publishing.

Sozzi, M., Kayad, A., Gobbo, S., Cogato, A., Sartori, L. *et al*. (2021) Economic comparison of satellite, plane and UAV-acquired NDVI images for site-specific nitrogen application: Observations from Italy. *Agronomy* 11(11), 2098.

Storkey, J., Meyer, S., Still, K.S. and Leuschner, C. (2012) The impact of agricultural intensification and land-use change on the European arable flora. *Proceedings. Biological Sciences* 279(1732), 1421–1429.

Swayze, N.C., Tinkham, W.T., Creasy, M.B., Vogeler, J.C., Hoffman, C.M. *et al*. (2022) Influence of UAS flight altitude and speed on aboveground biomass prediction. *Remote Sensing* 14(9), 1989.

Turner, W., Spector, S., Gardiner, N., Fladeland, M., Sterling, E. *et al*. (2003) Remote sensing for biodiversity science and conservation. *Trends in Ecology & Evolution* 18(6), 306–314.

Walkley, A. and Black, I.A. (1934) An examination of Degtjareff method for determining soil organic matter and proposed modification of the chromic acid titration method. *Journal of Soil Science* 37, 29–37.

Wang, X., Wang, J. and Zhang, J. (2012) Comparisons of three methods for organic and inorganic carbon in calcareous soils of northwestern China. *PLoS ONE* 7(8), e44334.

Woodland Trust (2012) Impacts of nearby development on ancient woodland - addendum. Available at: https://www.woodlandtrust.org.uk/media/43619/impacts-of-nearby-development-on-the-ecology-of-ancient-woodland-addendum.pdf (accessed 1 February 2024).

Xiong, C., Ma, H., Liang, S., He, T., Zhang, Y. *et al*. (2023) Improved global 250 m 8-day NDVI and EVI products from 2000-2021 using the LSTM model. *Scientific Data* 10(1), 800.

Yeom, J., Jung, J., Chang, A., Ashapure, A., Maeda, M. *et al*. (2019) Comparison of vegetation indices derived from UAV data for differentiation of tillage effects in agriculture. *Remote Sensing* 11(13), 1548.

Yerokun, O.A., Chikuta, S. and Mambwe, D. (2007) An evaluation of spectroscopic and loss on ignition methods for estimating soil organic carbon in Zambian soils. *International Journal of Agricultural Research* 2(11), 965–970.

18 Importance of Vegetation Cover in Organic Systems

Sito-Obong U. Udofia*, Lisa K. Williams, Alison P. Wills and Matt J. Bell

Animal and Agriculture Department, Hartpury University, Gloucester, UK

Abstract

The condition of the microclimate within agricultural fields, crops, livestock and soils is influenced by the landscape of the ecosystem. A farming landscape is characterized by vegetation covers such as crops, grasslands, hedgerows and woodlands, which are important for food production and wildlife habitat. The aim of this review was to explore existing research on quantifying the effects of vegetation cover in landscapes, explore the relationship between vegetation cover and soil factors, highlight areas for future research, and identify novel strategies for deployment in a changing climate. Research on a wide range of topics related to vegetation cover, including its impact on crop and livestock productivity, water-use efficiency, soil health and overall agricultural sustainability, was reviewed. Findings showed that seasonal changes, vegetation type and climate conditions influence agricultural systems. A comprehensive understanding of biomass covers and soil factor interactions can support sustainable farming practices and environmental management, but these relationships will require adequate systematic measures of quantification to guide informed land management decision making. Developments in farm measurement techniques and application of rapid sensors and remote-sensing technology to understand the effects of different vegetation covers on agricultural farmland and their interactions with soil is therefore important to guide land and habitat management practices for sustainable food production.

Keywords: vegetation cover, soil and plant measurements, sensor, remote-sensing technology

Introduction

There is a variety of different biomass types found in agricultural farmland such as grasses, arable crops, trees, shrubs and hedgerows, which play an important role in enhancing the resilience of the farm landscape to climate effects, its food productivity and functionality for wildlife habitation (Clark and Nicholas, 2013; Wilson and Lovell, 2016). In England, 68% of total land area is farmed, with 20% used for arable crop production and 50% as grassland (Defra, 2023). The state of vegetation cover can be influenced by climate, crop species, soil conditions, management practices and health problems such as disease and nutrient deficiency, and can be measured by its greenness level, photosynthetic status, growth, and yield performance (Lang *et al.*, 2019; Amiri and Pourghasemi, 2022). The association between these factors can be quantified to provide information on the dynamics of their relationships and interactions over seasons and time, to guide informed decision making in agricultural and environmental management.

Management of vegetation cover can offer potential solutions to mitigate the adverse effects on crops, soil and livestock due to diverse weather, and support resilience in agricultural farmlands, along with wildlife protection for sustainable land use (Gagliardi *et al.*, 2021). Understanding the factors that impact on vegetation in the farming landscape can guide their quantification (Mehmood *et al.*, 2024). Recently, there has been growing interest in sustainable land management,

*Corresponding author: Sito-Obong.Udofia@hartpury.ac.uk

DOI: 10.1079/9781800626850.0018

Fig. 18.1. Examples of vegetation cover found in the agricultural landscape including grass, trees, shrubs and arable crops.

and the need to adapt to changing environmental conditions has led to increased research assessing the importance of vegetation types and their effects in agricultural systems (Mehmood *et al.*, 2024). This chapter will review the literature on key aspects associated with selected vegetation types found in the farming landscape, and their potential influence on soil, plant, livestock and environmental properties.

Types of Vegetation Cover and Importance

Different types of vegetation, such as grass, trees and shrub plants (Fig. 18.1), can support crop and livestock health, soil quality and overall farm productivity (Ghafarian *et al.*, 2024).

Understanding the effects of different vegetation on plant growth and yield is essential for selection of options that can enhance agricultural productivity and farm landscape management and biomass cover in farmland (Fig. 18.2).

Agri-environmental effects of vegetation cover

Vegetation can act as shade and shelter in farmland to reduce soil erosion by acting as a barrier against strong winds, which might otherwise remove topsoil and decrease soil fertility, and can protect crops and livestock from extreme weather conditions such as high temperatures, strong winds, heavy rain and hail (Suryanto *et al.*, 2014). Studies have investigated the effects of shade and shelter on plant, soil and livestock in an agricultural landscape (Table 18.1).

Shading effects of vegetation on plants

Vegetation such as trees as well as artificial shade structures can offer cooling effects to crops during hot weather, improving their resilience and productivity (Ludwig *et al.*, 2005). This is particularly beneficial in regions with high temperatures, where heat stress can cause crop wilting and reduced yields due to insufficient soil moisture or excessive temperature (Walter *et al.*, 2013). In regions with high temperature, plants referred to as C4 plants have higher photosynthesis rates than those referred to as C3 plants and *vice versa* due to the variations in their pathways for the dark reaction of photosynthesis (Bhattacharya, 2022).The shade effect of vegetation can therefore be used to manage C4 versus C3 plants phenomenon for higher productivity (Bhattacharya, 2022). Shade provided by trees can help produce higher-quality fruits and vegetables by preventing sunburn and reducing plant stress, leading to better market prices (Molina-Flores *et al.*, 2020). Windbreaks can minimize evaporation and transpiration in crops (Cornelis and Gabriels, 2005). Brandle *et al.* (2021) found that trees can increase the longevity and health of crops, leading to better yields and potentially higher profits.

By moderating temperatures, healthy vegetation enhances photosynthesis and prevents damage to crops caused by excessive heat (Walter *et al.*, 2013). Comparing climate variables across different zones, researchers can assess how shading effects of vegetation can modify local environmental conditions and influence plant growth (Liu *et al.*, 2022). Evaluating the difference in yields, researchers can assess the direct impact of shelter on crop

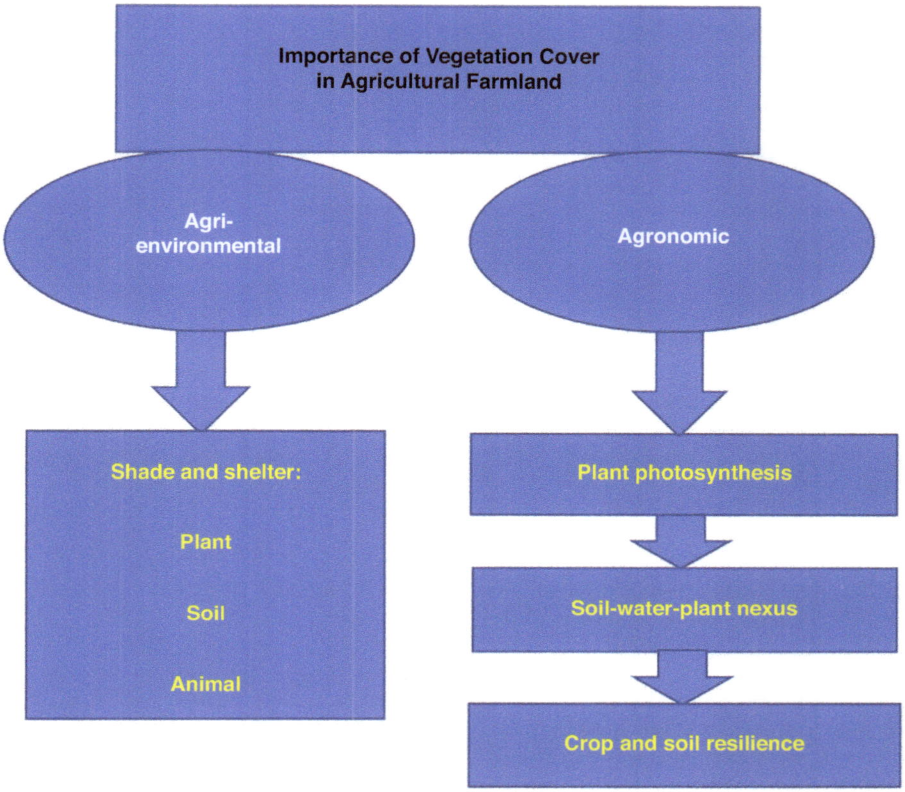

Fig. 18.2. Schematic diagram outlining the areas of importance in assessing the effect of vegetation cover in agricultural landscape.

productivity (Bennell and Verbyla, 2008). Overall, most research, including the study of Mira-García *et al.* (2022) and Ghafarian *et al.* (2024), has looked at trees and shrubs as a source of shade and shelter and windbreaks, but research on other important vegetation such as grassland ecosystem is lacking. Investigation of vegetation effects and their interactions with soil properties in grassland may provide useful information for sustainable land and livestock management (Bell *et al.*, 2020).

Shading effects of vegetation on soil

Vegetation providing shade protects the soil from erosion caused by heavy rain and strong winds thereby promoting soil health and fertility over time (Tongway and Ludwig, 2011). Vegetation can intercept and slow rainfall thus increasing water infiltration into the soil and preventing erosion of the soil surface, providing cover to keep soil from becoming too dry (Sharpley *et al.*, 2001; McMillan *et al.*, 2016). An example is the planting of vegetation as windbreaks to protect crops and soil from wind damage and associated effects (Sudmeyer *et al.*, 2002). By reducing ground temperature and sunlight exposure to evaporation, vegetation improves soil water-use efficiency by plants (McMillan *et al.*, 2016). Comparing soil moisture levels in shaded and unshaded areas can reveal the effectiveness of shade in conserving soil moisture and reducing water evaporation (Sample *et al.*, 2022). Field measurements of erosion can quantify the effectiveness of vegetation cover in protecting soil and maintaining agricultural productivity (Onyewotu *et al.*, 2003). Experiments to track changes in vegetation health, soil moisture and land

Table 18.1. Studies investigating the effect of shading by vegetation.

Source	Biomass cover as shade					
	Location	Vegetation/other shade	Animal/crop	Factor impacted	Impact	Measure of improvement
Suryanto *et al.* (2014)	South-east Asia	Trees	Maize, cassava	Productivity	Increased	Leaf area, biomass, growth rate
Ghafarian *et al.* (2024)	North Africa	Trees	Dairy cattle	Heat stress	Reduced	Feeding/milk production
(Gerlach *et al.*, 2023)	Guatemala	Shade plots	Crops	Dryness	Reduced	Water access/-use efficiency (Touil *et al.*, 2021)
Nesper *et al.* (2017)	India	Tree	Coffee	Sunlight Microclimate	Reduced intensity	Yield/quality
Qi *et al.* (2001)	USA	Windbreaks	Corn, soybean	Soil erosion Wind break	Reduced	Yield increase
Mira-García *et al.* (2022)	Spain	Wind break	Grape vine	Wind speed and damage	Reduced	Yield/quality

cover over time have been conducted to provide insights into the broader impact of the shading effects of vegetation on agricultural landscapes (Ewane *et al.*, 2023). Research shows tracking changes in vegetation cover, soil moisture and crop health over time can enhance the understanding of how different vegetation types can influence landscape productivity (Bell *et al.*, 2020; Sample *et al.*, 2022). However, there are no data on the seasonal variations in the effects of vegetation type on various soil and plant parameters of an agricultural ecosystem such as soil clay, soil organic matter, soil nitrogen ratio, plant ash, plant protein, plant oil and plant nitrogen. Information on this will be useful in guiding seasonal farming decision making for improved farmland productivity.

Shading effects of vegetation on pests and animals

Vegetation can enhance biodiversity on agricultural farmland by providing habitats for birds, insects and other wildlife, contributing to a balanced ecosystem, and promoting natural pest control and pollination (Jose, 2009). Windbreaks and other vegetation cover such as hedgerows can provide protection from heat stress to livestock, particularly during harsh weather events (Nuberg and Mylius, 2002; Bentrup, 2008). Reduced heat stress in livestock can subsequently lead to better milk production, increased longevity, overall health increase, lower mortality rates and potentially higher profits (Brandle *et al.*, 2021).

Usually, controlled environment experiments allow for precise control over the experimental conditions and enables researchers to study the direct effects of shade and shelter on both plant and animal physiology and growth (Bellot *et al.*, 2002). Behavioural observations in livestock experiments involve comparing animals kept in shaded areas with those in open areas to determine the effects of shade on feed intake, weight gain and reproductive performance (Sullivan *et al.*, 2011). Blaine and Nsahlai (2011) found that access to shade significantly improved milk production of dairy cattle and reduced signs of heat stress. Modelled scientific investigations can be conducted to provide useful explanations for the prediction and simulation of shade and shelter effect on livestock (Baker *et al.*, 2018).

Agronomic effects of vegetation cover

Plant biomass cover or density in agricultural farmland is associated with different agronomic benefits to crop and soil (González Giro *et al.*, 2023). Specifically, biomass cover can affect the photosynthesis of vulnerable crops, soil and plant water, and nutrient use, availability and efficiency, as well as crop and soil resilience to climate and stress (Zivcak *et al.*, 2014).

Photosynthesis of vulnerable crops

Shading by plants may affect the rate of photosynthesis of other plants by altering the availability of light and influencing other environmental conditions such as water interception and soil nutrients which are essential for proper growth and yield (Kabir, 2019). Photosynthesis relies on light energy to convert carbon dioxide and water into glucose and oxygen and lack of light hinders these factors thereby hindering photosynthesis (Al Murad *et al.*, 2021). Kabir (2019) found that moderate shading increased photosynthetic efficiency and improved water-use efficiency in tomato plants. In shaded areas, the light spectrum is often shifted towards the red and far-red wavelengths, which can affect the activation of specific photoreceptors involved in regulating photosynthesis (Zivcak *et al.*, 2014).

Crop modelling can predict how different levels of shade will impact crop productivity under various environmental conditions. For example, Cheng *et al.* (2022) used modelling to identify the optimal shade levels for maximizing soybean yield in different regions. Greenness or normalized difference vegetation index (NDVI) records, health and productivity have been noted to differ at different levels of biomass accumulation (Macagga *et al.*, 2025). Some plants have adapted to thrive in shaded environments by developing mechanisms to maximize light absorption; for example, shade-tolerant plants may have larger, thinner leaves to increase light capture and improve photosynthetic efficiency in low-light conditions (Jackson and Caldwell, 1992). These adaptations enable certain plant species to photosynthesize effectively even in shaded environments. Understanding how certain crop types respond to shade from plant biomass and interception of temperature, humidity and photosynthesis is crucial for optimizing plant growth and agricultural productivity.

Soil-water-plant nexus

Plant cover can reduce evaporation rates from the soil surface by blocking direct sunlight, retaining soil moisture, and making water more available for plant uptake (Lin, 2010). This helps maintain soil moisture levels, allowing plants to access more water in the soil and reducing the risk of water stress due to shortages (González Giro *et al.*, 2023). The reduction in evaporation can be particularly beneficial in arid and semi-arid regions where water conservation is crucial. In addition to reducing soil evaporation, biomass cover can also decrease transpiration rates, minimize water loss and improve water-use efficiency by plants for growth and photosynthesis (Kitta *et al.*, 2021). Plants shaded by vegetation may develop larger, thinner leaves to increase light capture while also reducing water loss through transpiration (Hatfield and Dold, 2019). This can promote cooler microclimates in agricultural ecosystems and cooler temperatures have been known to reduce the rate of water loss from the soil and plants, providing a more favourable environment for crop growth (Mira-García *et al.*, 2022).

The provision of shading can also influence plant root development, leading to deeper root systems that can access water deeper into the soil profile, which can aid resilience to drought conditions (Shrestha and Cole, 2013). Shade from vegetation can cause chlorosis or other signs of nutrient deficiency in crops, impact overall plant health and productivity, and prompt crops to invest more resources in root growth to compensate for reduced photosynthesis and energy production (Dunbabin *et al.*, 2002). This can result in a larger root system that is better able to explore the soil and absorb nutrients more efficiently (Jackson and Caldwell, 1992). However, in some cases, vegetation cover can also lead to less-extensive root systems due to limited energy availability, potentially affecting nutrient uptake.

Crop and soil resilience

Crop growth parameters, such as height, yield and leaf area, are key plant traits associated with production and resilience to climate impacts, and adverse conditions, and will help manage these important qualities (Brandle *et al.*, 2021). By altering wind patterns, shelter vegetation may influence where rainwater is deposited across a field and can filter sunlight, reducing the intensity of ultraviolet radiation and other harmful wavelengths (Dodd *et al.*, 2005). In wind tunnels or controlled growth chambers, researchers can simulate wind conditions and observe how crops respond with or without wind protection (Mira-García *et al.*, 2022). These methods help identify optimal locations and configurations for windbreaks and shelterbelts as researchers can assess changes in vegetation cover and crop health in areas with and without wind protection (Mira-García *et al.*, 2023). Assessing wind erosion rates using soil-loss measurements can provide further insights into the protective effects of windbreaks (Sudmeyer *et al.*, 2002).

Soil and crop plants can be shaded by vegetation cover to reduce transpiration rates by lowering leaf temperatures and minimizing the intensity of solar radiation on crops and soil (Asbjornsen *et al.*, 2014). By reducing direct sunlight exposure, plant biomass cover lowers soil and air temperatures, creating a cooler environment for crops (González-Salvatierra *et al.*, 2013). The use of mulching and planting of shade vegetation can lower wind speed, reduce the rate of evaporation from the soil surface and decrease transpiration rates (Edwards *et al.*, 2019). Biomass cover can help maximize irrigation water use by reducing evapotranspiration of water supplied to soil and crop by irrigation, thus ensuring that plants receive the necessary moisture (Wang *et al.*, 2020). This targeted approach can improve crop health and productivity while conserving water resources.

Methods, Indices, Tools and Technologies for Quantifying the Effects of Vegetation in Agricultural Ecosystem

The approaches and technology for vegetation and land assessment have advanced over the years from traditional and conventional field observation methods to now more sensitive, simple, easy and fast methods of analysis for identification of effective management options (Mollick *et al.*, 2023). The technologies and measurements that have been applied to quantity vegetation cover and its relationship with relevant biotic and abiotic factors of an agricultural ecosystem are summarized in Table 18.2.

In addition to field measurements, modelling approaches can be used to simulate the effects of plants (Cleugh, 2002). For example, advanced technologies such as remote sensing and rapid sensor technologies are being used with modelling techniques to assess changes in vegetation health, soil moisture and land cover in response to shelter structures (Arcidiacono and Porto, 2010; Cleugh, 2002).

Remote-sensing technology for quantifying vegetation cover

Remote-sensing technology applications enable land mapping with imagery and provide vital information for monitoring and predicting weather, climate change and land-use change over time (Yang, 2020; Eni *et al.*, 2021). Google Earth is used to explore changes in vegetation over space and time that may reflect biotic and abiotic changes leading to poor plant performance and disease incidence (Eni *et al.*, 2021; Amiri and Pourghasemi,

Table 18.2. Different technologies and measurements for quantifying vegetation cover and associated studies.

Technology used	Measurement	Source
Rising plate meter	Plant height	Bell *et al.* (2018)
Sensor (NIRS) scanners; remote-sensing technology	Leaf area, plant moisture, nutrients and biomass accumulation	Bell *et al.* (2018); Imran *et al.* (2021)
Biosensors; satellite remote sensing	Plant health and growth stage	Lang *et al.* (2019); Amiri and Pourghasemi (2022)
Biosensors; satellite remote sensing	Plant pathogen and disease incidence and severity	Yang *et al.* (2022)
Plant sensor such as SPAD and Greenseeker; remote sensing with satellite and UAV	Leaf greenness (NDVI), coloration, chlorosis or yellowing	Lang *et al.* (2019); Zhao *et al.* (2020)
UAV and satellite remote sensing; rapid sensors	Temperature, humidity, wind speed, and solar radiation	Oo and Phyu (2019)
Sensor probes; NIRS soil scanners	Soil moisture and erosion	Su *et al.* (2021)
Satellite remote sensing, NIRS scanners; portable sensors	Soil type, irrigation water, and fertilization	Amiri and Pourghasemi (2022).
Sensors (NIRS) scanners	Soil nutrients	Bell *et al.* (2018)
Conventional measures	Livestock feed intake, weight gain, and reproductive performance	Ainsworth *et al.* (2012)

Table 18.3. Challenges in assessing the benefits of vegetation cover identified in the literature.

Key aspects	Source
Physiological and environmental requirements of some crops and livestock differ	Rivest and Vézina (2015)
Varying effect per topography/soil type	Zulka *et al.* (2014)
Short-term research is unable to provide understanding of seasonal variations	Azzari *et al.* (2013)
Uneven distribution of differed biomass covers due to spatial configuration and shade intensity variation	Eggers *et al.* (2023)
Pest and disease attack on biomass	Yang *et al.* (2022)
Sensor variation	Bell *et al.* (2018); Su *et al.* (2021)
Accuracy and consistency of measurements	Azzari *et al.* (2013)

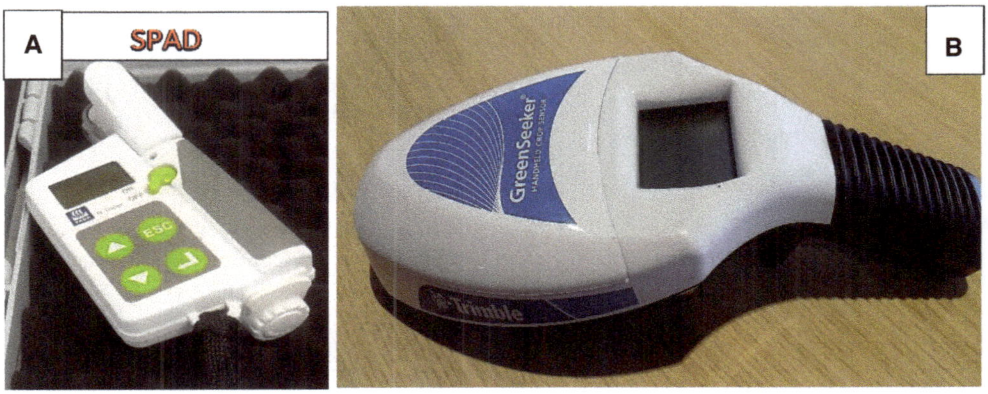

Fig. 18.3. Images showing (A) SPAD (Yara International, Oslo, Norway) and (B) Trimble GreenSeeker Hand-held Crop Sensor (Trimble, Westminster, USA) for plant greenness (chlorophyll) measurements.

2022). Imagery assessment of farmlands is carried out based on different red, blue and green colour variations and texture differentiations that promote the distinction of vegetation types from soil and other crop types, while determining varying plant canopies chlorophyl levels using remote sensing (Lang *et al.*, 2019).

The use of remote sensing tools and technology with satellite and unmanned aerial vehicles (UAV) in quantifying the effects of vegetative cover by greenness has been extensively applied in recent years by farmers, environmental managers and researchers worldwide (Kumar and Mutanga, 2017). Remote-sensing techniques using satellite imagery allow for large-scale monitoring of vegetative cover over time and can provide plant health measurements across different seasons and vegetation types such as NDVI measurements (García-Berná *et al.*, 2020). A UAV equipped with multi-spectral or thermal cameras can offer high-resolution images for more localized assessments of plant health, soil conditions and moisture levels (Fanigliulo *et al.*, 2020). Remote-sensing techniques have revolutionized the way researchers assess the impact of shade and shelter in agricultural farmland. By using advanced imaging technology from satellite imagery and drone-based surveys, researchers can collect comprehensive data on agricultural landscapes. Research applications of remote-sensing technology in quantifying vegetation cover and the associated relevance in agriculture is outlined in Table 18.3.

Rapid sensor technology for quantifying vegetation cover

Near-infrared spectroscopy (NIR) scanners (Figs 18.3 to 18.5), also referred to as rapid plant and soil sensors, can provide immediate and precise measurements of vegetation cover and their effects on plant nutrients and soil properties.

These tools enable real-time monitoring of crop and soil health and can support targeted

Fig. 18.4. Mobile near-infrared spectroscopy (NIRS) scanner (NIR4; Aunir, Towcester, UK) and associated tablet for sensing and analyzing Figs 18.3–18.5 plant nutrients in real time.

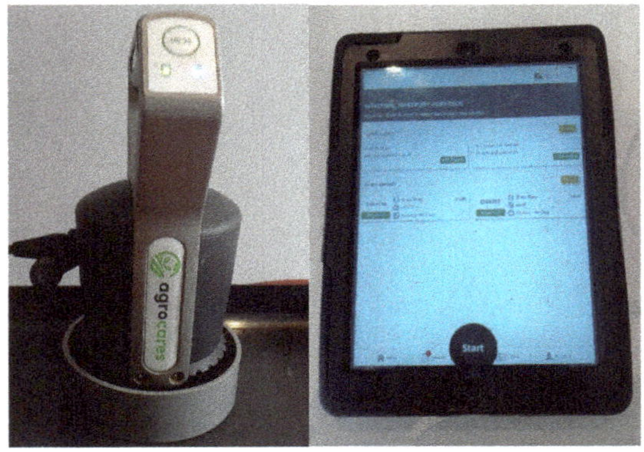

Fig. 18.5. An Agrocares near-infrared spectroscopy (NIRS) scanner (E Series; agrocares, Wageningen, The Netherlands) and associated tablet for sensing and analyzing soil properties in real time.

management practices (Bell *et al.*, 2018). The shape of plants, canopy and biomass measures can be estimated with the use of Kinect sensors (Azzari *et al.*, 2013). Ground cover estimation and biomass measurement are key parameters to assessing vegetation effect in farmland (Azzari *et al.*, 2013). Light detection and ranging (LiDAR) multi-temporal technique set-up by mounting to a ground-based mobile platform has been adopted for non-destructive ground and near-ground measurements of plant height, ground cover and biomass (Jimenez-Berni *et al.*, 2018). Tao *et al.* (2020) reported the use of smart phones to study nitrogen and chlorophyll levels in plants. Fig. 18.3 shows GreenSeeker and SPAD, portable smart plant sensors for measuring plant greenness (NDVI and chlorophyl levels, respectively), to inform nitrogen fertilizer application to crops and plant health monitoring.

Soil-sensing technology provides the opportunity for rapid measurements of soil conditions such as the use of real-time visible near-infrared soil sensors to detect soil texture, properties (e.g. organic matter content, nitrogen content, pH, moisture) and temperature (Rossel and Bouma, 2016). Increased use of smart sensors for relevant near-ground and ground quantification measures of shade effect on land can rapidly detect and measure growth and health indices such as moisture, chlorophyl and nitrogen levels and could benefit the agri-environmental industry by enabling more sensitive and accurate near-ground and ground measurements of abiotic and biotic effects (Lang *et al.*, 2019; Liu *et al.*, 2023).

Identifying and Addressing Challenges

Future direction on assessing the effects and quantification of vegetation cover in agricultural farmland should address several key areas highlighted in Table 18.3.

Several studies have identified further research needed regarding land management options (Table 18.4) against these challenges (Table 18.3). The opportunities highlighted in Table 18.4 could be practised and researched over time and season for improvement in agricultural ecosystem management.

Conclusions

Vegetation cover can enhance crop growth, productivity and animal welfare by reducing the potential risk of heat and water stress by providing a more stable growing environment. In addition, vegetation cover can increase water-use efficiency by improving soil and plant water retention, and reducing evaporation and transpiration rates. The quantification of vegetation cover effects in agricultural farmland by remote-sensing and sensor technology is practicable and can allow monitoring at a large scale. Focused research with the overall aim of investigating the different near-ground, ground and above-ground means of measuring the effects of changing vegetation cover in farmland is required to inform management options. Ongoing developments will inform better landscape management and farming practices for sustainable productivity

Table 18.4. Opportunities for further research regarding vegetation cover in farmland.

Practice	Source
Implications for agricultural management and planning	Lin (2007)
Optimization of biomass cover strategies for different crops	Ilić *et al.* (2018); Jungers *et al.* (2023)
Use of legume biomass cover for nitrogen-fixing bacteria	Whitehead and Isaac (2012)
Integration of shading by vegetation in farm design	Fekadu and Yunusa (2020)
Agroforestry systems, wind breaks and shelterbelts for protection of crop and soil health and carbon sequestration	Lin (2007)
Adaptable farming systems	Koutouleas *et al.* (2022)
Government incentives/financial support/evaluate the potential return on investment	García-Barrios and Ong (2004)

and improved plant resilience to biotic and abiotic changes.

References

Ainsworth, J.A.W., Moe, S.R. and Skarpe, C. (2012) Pasture shade and farm management effects on cow productivity in the tropics. *Agriculture, Ecosystems & Environment* 155, 105–110. DOI: 10.1016/j.agee.2012.04.005.

Al Murad, M., Razi, K., Jeong, B.R., Samy, P.M.A. and Muneer, S. (2021) Light emitting diodes (leds) as agricultural lighting: Impact and its potential on improving physiology, flowering, and secondary metabolites of crops. *Sustainability* 13(4). DOI: 10.3390/su13041985.

Amiri, M. and Pourghasemi, H.R. (2022) Mapping the NDVI and monitoring of its changes using google earth engine and sentinel-2 images. In: *Computers in Earth and Environmental Sciences: Artificial Intelligence and Advanced Technologies in Hazards and Risk Management*. pp. 127–136. DOI: 10.1016/B978-0-323-89861-4.00044-0.

Arcidiacono, C. and Porto, S.M.C. (2010) A model to manage crop-shelter spatial development by multi-temporal coverage analysis and spatial indicators. *Biosystems Engineering* 107(2), 107–122. DOI: 10.1016/j.biosystemseng.2010.07.007.

Asbjornsen, H., Hernandez-Santana, V., Liebman, M., Bayala, J., Chen, J. *et al.* (2014) Targeting perennial vegetation in agricultural landscapes for enhancing ecosystem services. *Renewable Agriculture and Food Systems* 29(2), 101–125. DOI: 10.1017/S1742170512000385.

Azzari, G., Goulden, M.L. and Rusu, R.B. (2013) Rapid characterization of vegetation structure with a microsoft kinect sensor. *Sensors* 13(2), 2384–2398. DOI: 10.3390/s130202384.

Baker, T.P., Moroni, M.T., Mendham, D.S., Smith, R. and Hunt, M.A. (2018) Impacts of windbreak shelter on crop and livestock production. *Crop and Pasture Science* 69(8), 785. DOI: 10.1071/CP17242.

Bell, M.J., Mereu, L. and Davis, J. (2018) The use of mobile near-infrared spectroscopy for real-time pasture management. *Frontiers in Sustainable Food Systems* 2, 76. DOI: 10.3389/fsufs.2018.00076.

Bell, M.J., Huggett, Z., Slinger, K.R. and Roos, F. (2020) Effect of pasture cover and height on nutrient concentrations in diverse swards in the UK. *Grassland Science* 67(3), 267–272. DOI: 10.1111/grs.1230.

Bellot, J., Urbina, J.O., Bonet, A. and Sánchez, J.R. (2002) The effects of treeshelters on the growth of *Quercus coccifera* L. seedlings in a semiarid environment. *Forestry* 75(1), 89–106. DOI: 10.1093/forestry/75.1.89.

Bennell, M.R. and Verbyla, A.P. (2008) Quantifying the response of crops to shelter in the agricultural regions of South Australia. *Australian Journal of Agricultural Research* 59(10), 950. DOI: 10.1071/AR08188.

Bentrup, G. (2008) Conservation buffers: Design guidelines for buffers, corridors, and greenways. Available at: https://www.fs.usda.gov/nac/buffers/docs/references.pdf (accessed 30 January 2025).

Bhattacharya, A. (2022) Effect of low temperature stress on photosynthesis and allied traits: A review. In: Bhattacharya, A. (ed.) *Physiological Processes in Plants Under Low Temperature Stress*. Springer, Singapore. DOI: 10.1007/978-981-16-9037-2_3.

Blaine, K.L. and Nsahlai, I.V. (2011) The effects of shade on performance, carcass classes and behaviour of heat-stressed feedlot cattle at the finisher phase. *Tropical Animal Health and Production* 43(3), 609–615. DOI: 10.1007/s11250-010-9740-x.

Brandle, J.R., Takle, E. and Zhou, X. (2021) Windbreak practices. In: Gene Garrett, H.E., Jose, S. and Gold, M.A. (eds) *North American Agroforestry*. Wiley, Chichester, UK, pp. 89–126. DOI: 10.1002/9780891183785.ch5.

Cheng, B., Wang, L., Liu, R., Wang, W., Yu, R. *et al.* (2022) Shade-tolerant soybean reduces yield loss by regulating its canopy structure and stem characteristics in the maize–soybean strip intercropping system. *Frontiers in Plant Science* 13, 848893. DOI: 10.3389/fpls.2022.848893.

Clark, K.H. and Nicholas, K.A. (2013) Introducing urban food forestry: A multifunctional approach to increase food security and provide ecosystem services. *Landscape Ecology* 28(9), 1649–1669. DOI: 10.1007/s10980-013-9903-z.

Cleugh, H.A. (2002) Parameterising the impact of shelter on crop microclimates and evaporation fluxes. *Australian Journal of Experimental Agriculture* 42(6), 859. DOI: 10.1071/EA02006.

Cornelis, W.M. and Gabriels, D. (2005) Optimal windbreak design for wind-erosion control. *Journal of Arid Environments* 61(2), 315–332. DOI: 10.1016/j.jaridenv.2004.10.005.

Defra (Department for Environment Food and Rural Affairs) (2023) Agricultural land use in England: National statistics. Available at: https://www.gov.uk/government/statistics/agricultural-land-use-in-england/agricultural-land-use-in-england-at-1-june-2023 (accessed 30 January 2025).

Dodd, M.B., McGowan, A.W., Power, I.L. and Thorrold, B.S. (2005) Effects of variation in shade level, shade duration and light quality on perennial pastures. *New Zealand Journal of Agricultural Research* 48(4), 531–543. DOI: 10.1080/00288233.2005.9513686.

Dunbabin, V.M., Diggle, A.J., Rengel, Z. and van Hugten, R. (2002) Modelling the interactions between water and nutrient uptake and root growth. *Plant and Soil* 239(1), 19–38. DOI: 10.1023/A:1014939512104.

Edwards, B.L., Webb, N.P., Brown, D.P., Elias, E., Peck, D.E. *et al.* (2019) Climate change impacts on wind and water erosion on US rangelands. *Journal of Soil and Water Conservation* 74(4), 405–418. DOI: 10.2489/jswc.74.4.405.

Eggers, J., Davis, S., Doscher, C. and Gregorini, P. (2023) Enhancing multifunctionality in agricultural landscapes with native woody vegetation. *Sustainability* 15(14), 11295. DOI: 10.3390/su151411295.

Eni, A.O., Efekemo, O.P., Onile-ere, O.A. and Pita, J.S. (2021) South West and North Central Nigeria: Assessment of cassava mosaic disease and field status of African cassava mosaic virus and East African cassava mosaic virus. *Annals of Applied Biology* 178(3), 466–479.

Ewane, E.B., Bajaj, S., Velasquez-Camacho, L., Srinivasan, S., Maeng, J. *et al.* (2023) Influence of urban forests on residential property values: A systematic review of remote sensing-based studies. *Heliyon* 9(10), e20408. DOI: 10.1016/j.heliyon.2023.e20408.

Fanigliulo, R., Antonucci, F., Figorilli, S., Pochi, D., Pallottino, F. *et al.* (2020) Light drone-based application to assess soil tillage quality parameters. *Sensors* 20(3), 728. DOI: 10.3390/s20030728.

Fekadu, M. and Yunusa, I. (2020) The impacts of shade trees on fodder and food crops integrated with or without coffee farming system in times of climate change. *Climate Change* 22, 177–185. Available at: https://www.discoveryjournals.org/climate_change/current_issue/v6/n22/A2.pdf

Gagliardi, S., Avelino, J., Virginio Filho, E. de M. and Isaac, M.E. (2021) Shade tree traits and microclimate modifications: Implications for pathogen management in biodiverse coffee agroforests. *Biotropica* 53(5), 1356–1367. DOI: 10.1111/btp.12984.

García-Barrios, L. and Ong, C.K. (2004) Ecological interactions, management lessons and design tools in tropical agroforestry systems. In: Nair, P.K.R., Rao, M.R. and Buck, L.E. (eds) *New Vistas in Agroforestry: A Compendium for 1st World Congress of Agroforestry, 2004*. Springer, The Netherlands, pp. 221–236. DOI: 10.1007/978-94-017-2424-1_16.

García-Berná, J.A., Ouhbi, S., Benmouna, B., Garcia-Mateos, G., Fernández-Alemán, J.L. *et al.* (2020) Systematic mapping study on remote sensing in agriculture. *Applied Sciences* 10(10), 3456. DOI: 10.3390/app10103456.

Gerlach, M.D., Lozano-Baez, S.E., Castellini, M., Guzman, N., Gomez, W.A. *et al.* (2023) Low cost and easy to implement physical and hydrological soil assessment of shade-grown coffee in santa rosa, guatemala. *Land* 12, 390.

Ghafarian, F., Ghazaryan, G., Wieland, R. and Nendel, C. (2024) The impact of small woody features on the land surface temperature in an agricultural landscape. *Agricultural and Forest Meteorology* 349, 109949. DOI: 10.1016/j.agrformet.2024.109949.

González Giro, Z., Pedraza Olivera, R., Lamadrid Mandado, R., Hu, J., Font Vila, L. *et al.* (2023) Invasive woody plants in the tropics: A delicate balance between control and harnessing potential benefits. *Agronomy for Sustainable Development* 43(3), 41. DOI: 10.1007/s13593-023-00892-x.

González-Salvatierra, C., Badano, E.I., Flores, J. and Rodas, J.P. (2013) Shade shelters increase survival and photosynthetic performance of oak transplants at abandoned fields in semi-arid climates. *Journal of Forestry Research* 24(1), 23–28. DOI: 10.1007/s11676-013-0321-5.

Hatfield, J.L. and Dold, C. (2019) Water-use efficiency: Advances and challenges in a changing climate. *Frontiers in Plant Science* 10, 103. DOI: 10.3389/fpls.2019.00103.

Ilić, Z.S., Milenković, L., Šunić, L. and Manojlović, M. (2018) Color shade nets improve vegetables quality at harvest and maintain quality during storage. *Contemporary Agriculture* 67(1), 9–19. DOI: 10.2478/contagri-2018-0002.

Imran, Q.M., Falak, N., Hussain, A., Mun, B.G. and Yun, B.W. (2021) Abiotic stress in plants: Stress perception to molecular response and role of biotechnological tools in stress resistance. *Agronomy* 11(8), 1579. DOI: 10.3390/agronomy11081579.

Jackson, R.B. and Caldwell, M.M. (1992) Shading and the capture of localized soil nutrients: Nutrient contents, carbohydrates, and root uptake kinetics of a perennial tussock grass. *Oecologia* 91(4), 457–462. DOI: 10.1007/BF00650316.

Jimenez-Berni, J.A., Deery, D.M., Rozas-Larraondo, P., Condon, A.T.G., Rebetzke, G.J. *et al.* (2018) High throughput determination of plant height, ground cover, and above-ground biomass in wheat with LiDAR. *Frontiers in Plant Science* 9, 237. DOI: 10.3389/fpls.2018.00237.

Jose, S. (2009) Agroforestry for ecosystem services and environmental benefits: An overview. *Agroforestry Systems* 76(1), 1–10. DOI: 10.1007/s10457-009-9229-7.

Jungers, J., Runck, B., Ewing, P.M., Maaz, T., Carlson, C. *et al.* (2023) Adapting perennial grain and oilseed crops for climate resiliency. *Crop Science* 63(4), 1701–1721. DOI: 10.1002/csc2.20972.

Kabir, M.Y. (2019) Abiotic factors affecting plant physiology and fruit yield and physiological disorders in bell pepper and tomato. Doctoral dissertation, ProQuest Dissertation & Theses Global. Available at: https://www.proquest.com/openview/426a9c20bbe9d249e33fb8a54cb8dafc/1?pq-origsite=gscholar&cbl=18750&diss=y

Kitta, E., Katsoulas, N. and Kittas, C. (2021) Effect of shading on photosynthesis in greenhouse hydroponic cucumber crops. In: *VIII South-Eastern Europe Symposium on Vegetables and Potatoes 1320*, pp. 167–172. DOI: 10.17660/ActaHortic.2021.1320.21.

Koutouleas, A., Sarzynski, T., Bordeaux, M., Bosselmann, A.S., Campa, C. *et al.* (2022) Shaded-coffee: a nature-based strategy for coffee production under climate change? *Frontiers in Sustainable Food Systems* 6. DOI: 10.3389/fsufs.2022.877476.

Kumar, L. and Mutanga, O. (2017) Remote sensing of above-ground biomass. *Remote Sensing* 9(9), 935. DOI: 10.3390/rs9090935.

Lang, Q., Zhiyong, Z., Longsheng, C., Hong, S., Minzan, L. *et al.* (2019) Detection of chlorophyll content in maize canopy from UAV imagery. *IFAC-PapersOnLine* 52(30), 330–335. DOI: 10.1016/J.IFACOL.2019.12.561.

Lin, B.B. (2007) Agroforestry management as an adaptive strategy against potential microclimate extremes in coffee agriculture. *Agricultural and Forest Meteorology* 144(1–2), 85–94. DOI: 10.1016/j.agrformet.2006.12.009.

Lin, B.B. (2010) The role of agroforestry in reducing water loss through soil evaporation and crop transpiration in coffee agroecosystems. *Agricultural and Forest Meteorology* 150(4), 510–518. DOI: 10.1016/j.agrformet.2009.11.010.

Liu, Q., Yao, F., Garcia-Garcia, A., Zhang, J., Li, J. *et al.* (2023) The response and sensitivity of global vegetation to water stress: A comparison of different satellite-based NDVI products. *International Journal of Applied Earth Observation and Geoinformation* 120, 103341. DOI: 10.1016/J.JAG.2023.103341.

Liu, Y., Li, H., Yuan, F., Shen, L., Wu, M. *et al.* (2022) Estimating the impact of shelterbelt structure on corn yield at a large scale using google earth and sentinel 2 data. *Environmental Research Letters* 17(4), 044060. DOI: 10.1088/1748-9326/ac58ab.

Ludwig, J.A., Wilcox, B.P., Breshears, D.D., Tongway, D.J. and Imeson, A.C. (2005) Vegetation patches and runoff-erosion as interacting ecohydrological processes in semiarid landscapes. *Ecology* 86(2), 288–297. DOI: 10.1890/03-0569.

Macagga, R., Sossa, G., Ayaribil, Y., Sanchez, P., Bayot, R. *et al.* (2025) A new, low-cost ground-based NDVI sensor for manual and automated crop monitoring. DOI: 10.2139/ssrn.5085232.

McMillan, H., Montanari, A., Cudennec, C., Savenije, H., Kreibich, H. *et al.* (2016) Panta Rhei 2013–2015: Global perspectives on hydrology, society and change. *Hydrological Sciences Journal* 61(7), 1174–1191. DOI: 10.1080/02626667.2016.1159308.

Mehmood, K., Anees, S.A., Muhammad, S., Hussain, K., Shahzad, F. *et al.* (2024) Analyzing vegetation health dynamics across seasons and regions through NDVI and climatic variables. *Scientific Reports* 14(1), 11775. DOI: 10.1038/s41598-024-62464-7.

Mira-García, A.B., Conejero, W., Vera, J. and Ruiz-Sánchez, M.C. (2022) Water status and thermal response of lime trees to irrigation and shade screen. *Agricultural Water Management* 272, 107843. DOI: 10.1016/j.agwat.2022.107843.

Mira-García, A.B., Conejero, W., Vera, J. and Ruiz-Sánchez, M.C. (2023) Effect of water stress and shading on lime yield and quality. *Plants* 12(3), 503. DOI: 10.3390/plants12030503.

Molina-Flores, B., Manzano-Baena, P. and Coulibaly, M.D. (2020) *The Role of Livestock in Food Security, Poverty Reduction and Wealth Creation in West Africa*. FAO, Accra. DOI: 10.4060/ca8385en.

Mollick, T., Azam, M.G. and Karim, S. (2023) Geospatial-based machine learning techniques for land use and land cover mapping using a high-resolution unmanned aerial vehicle image. *Remote Sensing Applications: Society and Environment* Available at, 100859. DOI: 10.1016/j.rsase.2022.100859.

Nesper, M., Kueffer, C., Krishnan, S., Kushalappa, C.G., Jaboury, C. *et al.* (2017) Shade tree diversity enhances coffee production and quality in agroforestry systems in the western ghats. *Agriculture, Ecosystems & Environment* 247 .

Nuberg, I.K. and Mylius, S.J. (2002) Effect of shelter on the yield and water use of wheat. *Australian Journal of Experimental Agriculture* 42(6), 773–780. DOI: 10.1071/EA02013.

Onyewotu, L.O.Z., Stigter, C.J., Abdullahi, A.M., Ariyo, J.A., Oladipo, E.O. *et al.* (2003) Reclamation of desertified farmlands and consequences for its farmers in semiarid Northern Nigeria: A case study of Yambawa rehabilitation scheme. *Arid Land Research and Management* 17(1), 85–101. DOI: 10.1080/15324980301590.

Oo, Z.Z. and Phyu, S. (2019) Microclimate prediction using cloud centric model based on IoT technology for sustainable agriculture. In: *2019 IEEE 4th International Conference on Computer and Communication Systems (ICCCS)*, pp. 660–663. DOI: 10.1109/CCOMS.2019.8821705.

Qi, X., Mize, C.W., Batchelor, W., Takle, E.S. and Litvina, V. (2001) SBELTS: a model of soybean production under tree shelter. *Agroforestry Systems* 52(1), 53–61. DOI: 10.1023/A:1010647118256.

Rivest, D. and Vézina, A. (2015) Maize yield patterns on the leeward side of tree windbreaks are site-specific and depend on rainfall conditions in Eastern Canada. *Agroforestry Systems* 89(2), 237–246.

Rossel, R.A.V. and Bouma, J. (2016) Soil sensing: A new paradigm for agriculture. *Agricultural Systems* 148, 71–74. DOI: 10.1016/j.agsy.2016.07.001.

Sample, M., Thode, A.E., Peterson, C., Gallagher, M.R., Flatley, W. *et al.* (2022) Adaptation strategies and approaches for managing fire in a changing climate. *Climate* 10(4), 58. DOI: 10.3390/cli10040058.

Sharpley, A.N., McDowell, R.W. and Kleinman, P.J. (2001) Phosphorus loss from land to water: Integrating agricultural and environmental management. *Plant and Soil* 237(2), 287–307. DOI: 10.1023/A:1013335814593.

Shrestha, A. and Cole, J.C. (2013) Shade intensity influences water use and growth of three viburnum species. *Journal of Environmental Horticulture* 31(4), 259–266. DOI: 10.24266/0738-2898.31.4.259.

Su, B., Su, Z. and Shangguan, Z. (2021) Trade-off analyses of plant biomass and soil moisture relations on the Loess Plateau. *CATENA* 197, 104946. DOI: 10.1016/j.catena.2020.104946.

Sudmeyer, R.A., Crawford, M.C., Meinke, H., Poulton, P.L. and Robertson, M.J. (2002) Effect of artificial wind shelters on the growth and yield of rainfed crops. *Australian Journal of Experimental Agriculture* 42(6), 841. DOI: 10.1071/EA02018.

Sullivan, M.L., Cawdell-Smith, A.J., Mader, T.L. and Gaughan, J.B. (2011) Effect of shade area on performance and welfare of short-fed feedlot cattle. *Journal of Animal Science* 89(9), 2911–2925. DOI: 10.2527/jas.2010-3152.

Suryanto, P., Putra, E.T.S., Kurniawan, S., Suwignyo, B. and Sukirno, D.A.P. (2014) Maize response at three levels of shade and its improvement with intensive agro forestry regimes in Gunung Kidul, Java, Indonesia. *Procedia Environmental Sciences* DOI, 370–376.

Tao, M., Ma, X., Huang, X., Liu, C., Deng, R. *et al.* (2020) Smartphone-based detection of leaf color levels in rice plants. *Computers and Electronics in Agriculture* 173, 105431. DOI: 10.1016/J.COMPAG.2020.105431.

Tongway, D.J. and Ludwig, J.A. (2011) *Restoring Disturbed Landscapes: Putting Principes into Practice*. Island Press, Washington, DC. Available at: https://tinyurl.com/5bf2438m (accessed 30 January 2025).

Touil, S., Richa, A., Fizir, M. and Bingwa, B. (2021) Shading effect of photovoltaic panels on horticulture crops production: A mini review. *Reviews in Environmental Science and Bio/Technology* 20(2), 281–296. DOI: 10.1007/s11157-021-09572-2.

Walter, J., Jentsch, A., Beierkuhnlein, C. and Kreyling, J. (2013) Ecological stress memory and cross stress tolerance in plants in the face of climate extremes. *Environmental and Experimental Botany* 94, 3–8. DOI: 10.1016/j.envexpbot.2012.02.009.

Wang, Y., Li, S., Qin, S., Guo, H., Yang, D. *et al.* (2020) How can drip irrigation save water and reduce evapotranspiration compared to border irrigation in arid regions in Northwest China. *Agricultural Water Management* 239, 106256. DOI: 10.1016/j.agwat.2020.106256.

Whitehead, M. and Isaac, M.E. (2012) Effects of shade on nitrogen and phosphorus acquisition in cereal-legume intercropping systems. *Agriculture* 2(1), 12–24. DOI: 10.3390/agriculture2010012.

Wilson, M.H. and Lovell, S.T. (2016) Agroforestry: The next step in sustainable and resilient agriculture. *Sustainability* 8(6), 574. DOI: 10.3390/su8060574.

Yang, C. (2020) Remote sensing and precision agriculture technologies for crop disease detection and management with a practical application example. *Engineering* 6(5), 528–532. DOI: 10.1016/j.tplants.2022.12.011.

Yang, L.-N., Ren, M. and Zhan, J. (2022) Modeling plant diseases under climate change: evolutionary perspectives. *Trends in Plant Science* 28(5), 519–526.

Zhao, H., Yang, C., Guo, W., Zhang, L. and Zhang, D. (2020) Automatic estimation of crop disease severity levels based on vegetation index normalization. *Remote Sensing* 12, 12. DOI: 10.3390/rs12121930.

Zivcak, M., Brestic, M., Kalaji, H.M. and Govindjee, A. (2014) Photosynthetic responses of sun- and shade-grown barley leaves to high light: Is the lower PSII connectivity in shade leaves associated with protection against excess of light? *Photosynthesis Research* 119(3), 339–354. DOI: 10.1007/s11120-014-9969-8.

Zulka, K.P., Abensperg-Traun, M., Milasowszky, N., Bieringer, G., Gereben-Krenn, B.-A. *et al.* (2014) Species richness in dry grassland patches of Eastern Austria: A multi-taxon study on the role of local, landscape and habitat quality variables. *Agriculture, Ecosystems & Environment* 182, 25–36. DOI: 10.1016/j.agee.2013.11.016.

19 Organic Nutrient Resources and Fortification: Restricted Nutrient Use in Organic Farming

Hector Valenzuela*

Department of Plant and Environmental Protection Sciences, University of Hawaii at Manoa, Honolulu, Hawaii

Abstract

Organic nutrient management practices are guided by standards established by national and international organic certification agencies. By and large, organic farmers are precluded from the use of synthetic chemical fertilizers. A list of natural products that may be used in organic farming is often provided by certification agencies. Guidelines for organic nutrient management call for an ecological management approach to build the soil health and quality by adopting cultural practices such as crop rotations, cover crops, organic amendments, germplasm selection and vegetation diversification. A central philosophy of organic farming, with respect to nutrient management, is thus to build soil health and biodiversity, rather than following a 'recipe' approach to 'feed' particular crops. Meeting a crop's nutritional demands, while minimizing environmental losses and without building soil nutrient imbalances, represents a major challenge for organic farmers. Important considerations and goals to better meet the crop's nutritional demands include: to determine nutrient budgets (inputs vs. outputs); to better understand internal nutrient cycles; and to better understand the timing of nutrient release or mineralization rates with the goal of better synchronizing the release of nutrients from organic amendments with the crop uptake demands throughout the growing season. An overview is here provided of some of the ecological management practices used in organic farming to meet the crop's nutritional demands for the primary, secondary and micronutrients. Indigenous millennial knowledge is highlighted as a foundation of organic farming, and this knowledge should thus be included, along with our modern understanding of agroecological principles, for the design of organic systems adapted to the local environmental, socio-economic and growing conditions.

Keywords: agroecology, indigenous knowledge, organic farming, nitrogen, potassium, phosphorus, secondary nutrients, micronutrients, biostimulants, microbial activity

Introduction

The nutrient management practices followed in organic farming are guided by national and international certification standards. Organic farmers are precluded from using synthetic nutrients and instead rely on products that are naturally available or processed solely from natural products, such as composts and animal by-products. Commercially available organic fertilizers need to meet the approval of local or international certification agencies, prior to application on organic farms, to ensure the absence of any synthetic ingredients during the manufacturing process. The nutrient management philosophy for organic farmers is distinct from that followed by conventional farmers because the primary focus for organic farmers is to build the long-term quality and productivity of the soil. In contrast, the principal goal for conventional farmers is to provide for the immediate nutrient needs of the crop via the targeted application of readily available chemical soluble fertilizers.

*Corresponding author: hector@hawaii.edu

DOI: 10.1079/9781800626850.0019

Because the nutrients in organic amendments or fertilizers are not always readily available, but are mineralized or released over a period of time, it is challenging for organic farmers to estimate the total amount of nutrients available for plant uptake at a specific period of time. To optimize the soil nutrient availability for crop uptake, organic farmers thus adopt an ecologically based and holistic management approach that consists of a range of integrated management practices, which altogether help to synchronize the amount of nutrients available, with the crop uptake demand, during the growing season.

Key ecological strategies used by organic farmers to build up the nutrient content, quality and fertility of their soils include: germplasm selection; soil moisture management; organic amendment applications; planting of cover crops; integrated crop-animal systems; vegetation diversification in time and space such as crop rotations, intercropping, alley crops and agroforestry; improvement of soil microbial activity; and application of commercial or locally produced organic fertilizers. Because organic farming practices are intimately tied to the local social and environmental conditions, the recognition and adoption of indigenous millennial knowledge is essential for the design of production systems that are adapted and cognizant of the local growing and social conditions.

Cultural Management Practices

A fundamental philosophy of organic farming in terms of nutrient management is the concept of 'feeding the soil', in contrast to the practice followed in conventional farms of 'feeding the plant' (Baldwin, 2006; Möller, 2018). In conventional farms, a prescribed amount of particular elements, such as NPK, is supplied to the crop, during the growing season. In organic farming, the central goal is to build soil health and quality, which subsequently will become a source of slow-release nutrients, available for plant uptake, during the growing season. A central aspect of the goal to 'feed the soil' consists of promoting the soil biology and microbial activity to improve the cycling and mineralization or release of nutrients, in synchrony with crop uptake demands (Drinkwater and Snapp, 2007; Schonbeck *et al.*, 2019; Shaji *et al.*, 2021).

Key cultural management practices, based on ecological principles (Drinkwater and Snapp, 2007), are followed by organic farmers to enrich and build the soil fertility over time (Table 19.1). Along the same lines, organic farmers select particular crops and crop varieties that match or are more adapted to the fertility of the soils on their farms. To conserve the available resources on the farm, another central concept of organic farming is the goal of recycling nutrients within the farm, to improve nutrient-use efficiency and to minimize any potential environmental losses, via leaching, runoff or greenhouse gas emissions (Moeller *et al.*, 2018; Ekyaligonza *et al.*, 2022; Valenzuela, 2023). Cultural management practices are thus adopted by organic farmers to build the soil fertility, and to recycle nutrients within the farm.

Nutrient budgets

A nutrient balance or budget is useful to estimate the total amount of individual elements available in the system for crop uptake (Setyorini and Hartatik, 2021). A tentative nutrient budget (outputs vs inputs) for vegetable crops is presented in Table 19.2. In organic farming, nutrients are normally provided by multiple sources including from residual levels in the soil, cover crops, amendment and residue applications, and organic fertilizers. It is useful for organic farmers to be able to determine nutrient budgets, to optimize yields and to minimize the likelihood of environmental losses. However, it is important to understand that the total amount of nutrients available in the system may not be readily available for crop uptake due to the different rates of release or mineralization. For example, it may take weeks, months and even years for all the nitrogen to be released from an initial compost or manure application. Thus, organic farmers should take into account the potential mineralization rate of organic amendments when preparing a nutrient budget for the current cropping season (Baldwin, 2006; Valenzuela, 2023). The calculation of nutrient budgets is also important to prevent the undesirable build-up of particular elements, such as high P soil levels, from historical high rates of annual manure applications (Oelofse *et al.*, 2010; Köninger *et al.*, 2021)

Nutrient soil calibration

Soil nutrient calibrations, based on soil tests, consist of field studies that correlate the amount of particular elements available in the soluble soil solution,

Table 19.1. A description of some key cultural management practices to improve soil quality and fertility in organic farms.

Cultural practice	Benefits on soil health	References
Vegetation diversification, intercropping, hedgerows	Improve nutrient cycles, promote microbial activity, restore ecosystem functions	Drinkwater and Snapp, 2007; Ekyaligonza *et al.*, 2022; Mårtensson *et al.*, 2022; Valenzuela, 2023
Organic amendments, composts	Soil quality, organic matter content, microbial activity, nutrient reservoir, pH buffering capacity, improved soil-water holding capacity	Haynes and Mokolobate, 2001; Baldwin, 2006; Drinkwater and Snapp, 2007; Wortman *et al.*, 2012; Schonbeck *et al.*, 2019
pH management	Nutrient uptake, microbial activity	Delate *et al.*, 2003; Rosen and Bierman, 2005; Gaskin *et al.*, 2014
Rotations	Nutrient cycling, microbial activity	Drinkwater and Snapp, 2007; Wortman *et al.*, 2012; Yfantopoulos *et al.*, 2022; De Notaris *et al.*, 2023
Germplasm selection, landraces, agrobiodiversity	Adaptation to local growing conditions and fertility of the soil	Drinkwater and Snapp, 2007; Setyorini and Hartatik, 2021
Cover crops, crop residues	Nutrient cycling, N fixation, nutrient (P) uptake, improved organic matter content, erosion control	Drinkwater and Snapp, 2007; Valenzuela, 2020; Ekyaligonza *et al.*, 2022
Promote microbial activity	Nutrient mineralization, nutrient cycling, growth promotion, soil quality	Drinkwater and Snapp, 2007; Schonbeck *et al.*, 2019
Nitrogen fixation (intercropping, hedgerows, cover crops, rotations)	Nitrogen uptake (most limiting nutrient in most organic farms)	Jiang *et al.*, 2022; Ekyaligonza *et al.*, 2022; Yfantopoulos *et al.*, 2022; Valenzuela, 2023
Agroforestry	Biodiversity, nutrient cycling, erosion control	Rosati *et al.*, 2021; Piato *et al.*, 2022

with yield response curves (Rosen and Bierman, 2005). The soil calibration studies are applicable for a specific soil type and location, in relation to specific crops. Nutrient calibration studies have traditionally been conducted for the production of conventional crops. However, the results from calibration studies conducted with chemical fertilizers may not translate directly for use by organic farmers (Baldwin, 2006). Thus, organic farmers should be cautious when using data from calibration studies conducted for conventional crops. For example, the recommended soil levels for P or K provided by local diagnostic laboratories were established with the use of synthetic fertilizers and may differ to some degree if the calibration studies had been performed under organic farming conditions.

Restrictions on the use of organic amendments

Organic certification agencies provide guidelines and limitations in terms of the use of some organic amendments, especially for environmental or food safety considerations. Some certifying agencies require that manure be applied by three months prior to the harvest of an edible product that does not come in contact with the soil and by four months if the edible product comes into contact with the soil (Baldwin, 2006). For compost preparation, organic farmers are required to follow a proper preparation protocol, including an initial 25:1 to 40:1 C:N ratio, and a 55–75°C temperature range for a prescribed period of time, to minimize the risks of microbial contamination (Rosen and Bierman, 2005). In Europe, manure obtained from factory farms is not allowed in organic farming, and the total amount of N applied in the amendment should not exceed 170 kg N Ha^{-1} (Gatsios *et al.*, 2021). Individual countries and private certifiers may have additional restrictions on the use of organic amendments in terms of its sourcing or annual application rates (Løes *et al.*, 2017). In general, it is recommended that manure be composted, prior to application (Delate *et al.*, 2003). While tables exist in the literature that list the nutrient content of a variety of manures and

Table 19.2. Tentative nutrient budgets (outputs vs inputs) and physiological functions in vegetable crops (adapted from Fox and Valenzuela, 1992).

Element	System inputs (range, kg ha^{-1})	Crop uptake (output), harvestable product (range, kg ha^{-1})	Tissue nutrient content (range, most recently matured leaves, %)	Key physiological functions
N	150–300	15–180 (mostly 100–150)	2.4–4.1	Key structural and physiological functions, plant growth and development
P	50–260 (avg. 97)	7–31 (avg. 17)	0.2–0.5	Plant growth and development, various physiological and biochemical functions, energy transport
K	75–260 (avg. 130)	60–265 (avg. 127)	1.9–4.7	Water and nutrient transport, photosynthesis, respiration, fruit quality
CaO	Often supplied with liming applications	4–225 (mostly 25–100)	0.4–3.2 (Ca)	Plant growth, cell structure and defence against pests
MgO	25 (deficient soils)	8–43	0.15–0.62 (Mg)	Important physiological and biochemical functions, plant growth, assimilate transport
S	6–50	5–40	0.1–0.3 (S)	Ingredient in amino acids and secondary metabolites

Note: System inputs (column 1) indicate the available amount of soluble nutrients in the soil from a range of possible sources, such as organic amendments, crop residues and residual soil levels. Numbers based on extension recommendations for the production of vegetable crops. The crop uptake (outputs, column 2) represents an uptake efficiency of 30–60% of the available soil soluble nutrient levels. These values are presented as a general guide, based on the recommendations for the production of vegetable crops, and will vary depending on crop species, cultivar, location, growing season, and field history, among other variables. Considerable research is necessary to fine-tune these values to develop crop- and location-specific nutrient budgets for organic farming, based on yield response curves.

organic amendments, due to the many variables involved in their preparation, handling and storage, it is recommended that individual batches be tested in diagnostic laboratories, for their nutrient content, for the preparation of more realistic nutrient budgets (Rosen and Bierman, 2005).

Macronutrients

Nitrogen

Nitrogen (N) is often the most limiting nutrient in organic production. The dynamics of N in the soil and environment are very complex due to its many chemical forms, which makes it very difficult to correlate the N soil status at any given time with final crop yields. As such, tissue N levels or NO_3^{-1} sap analysis represents one of the few ways to monitor plant N uptake. Soil fauna and microbial activity are the main drivers of the N cycle in the soil by decomposing organic matter, fixing atmospheric N, and by transforming N into plant available forms (Valenzuela, 2023).

A variety of management strategies are followed by organic farmers to improve the N supply for crop growth (Table 19.1). The principal source of N in most organic farms is N fixation. Cover crop rotations, alley cropping, intercropping and other diversification strategies with leguminous species are standard practices to increase the rates of N fixation in organic farms (Table 19.4) (Mårtensson *et al.*, 2022). Typical N fixation rates from leguminous cover crops are in the range of 40–200 kg ha^{-1} with 10% to over 50% of the N released for the following crop, depending on several variables (Valenzuela, 2023). Another principal strategy to increase N levels on organic farms is the build-up of organic matter from the periodic application of manure, organic amendments and crop residues from non-leguminous cover crops (Tables 19.3 and 19.4). Integrated livestock- or aquaculture-crop systems are another approach to increase the N

Table 19.3. Example of some nutrient sources used in organic farming.

Nutrient source	Nutrient content (% N-P_2O_5-K_2O)	References
Composts	1.5-1-1.5 NPK (0.7–2.5 N), other elements, increased microbial activity, CEC	Baldwin, 2006; Gaskin *et al*., 2014; Valenzuela, 2023
Manures	0.5–0.2-0.4 (NPK, cattle) to 3-3-2.8 (chicken), NPK, other elements, Increased microbial activity, CEC, pH buffering capacity	Baldwin, 2006; Oelofse *et al*., 2010; Gaskin *et al*., 2014; Shaji *et al*., 2021; Valenzuela, 2023
Crop residues, hedgerow clippings, *Azolla*	Crop residues: 1.5–4.0 N; 0.2–0.36 P; 0.1–5 K, other elements, microbial activity	Haynes and Mokolobate, 2001; Oelofse *et al*., 2010; Valenzuela, 2020; Setyorini and Hartatik, 2021; Valenzuela, 2023
Fish emulsions, seaweed (sea kelp), kelp meal	5-2-3, NPK, micronutrients (emulsions); 3 P, 2 K (fish meal); 1-0.5-8 (kelp meal)	Monaghan *et al*., 1994; Delate *et al*., 2003; Gaskin *et al*., 2014; Valenzuela, 2023
Animal by-products, e.g. feather meal, bone meal	11-0.4-1.5 (feather meal, slow release); 10-4-0 (fish meal); 7.5-9-0.2 NPK (bone meal), other elements	Monaghan *et al*., 1994; Delate *et al*., 2003; Baldwin, 2006; Gaskin *et al*., 2014; Möller and Schultheiß, 2015)
Plant by-products, e.g. castor bean, alfalfa and vetch seed meal, wood ash	5-1.8-1 (castor bean), 3-1-2 (alfalfa meal); 6.5-1.5-2.3 (soybean meal), N, other elements; 0-1-4 (wood ash)	Baldwin, 2006; Oelofse *et al*., 2010; Gaskin *et al*., 2014; Valenzuela, 2023
Rock phosphate	Phosphorus source (25–40% P; 2% available P) 29% Calcium	Monaghan *et al*., 1994; Oelofse *et al*., 2010
Biofertilizers	Growth promotion	Oelofse *et al*., 2010
Commercial blended fertilizers	3-2-3, 4-3-3 (NPK)	Monaghan *et al*., 1994
Wood ashes	0-2-5 NPK (range of 5–7% for K)	Baldwin, 2006; Gaskin *et al*., 2014

supply in organic farms (Bonaudo *et al.*, 2014). However, food safety regulations in some countries make it challenging to adopt integrated animal-crop systems of production.

A challenge for organic farmers is the need to synchronize the release or availability of N from N fixation or organic amendments, with the crop N uptake demand, especially during periods of high N demand (Valenzuela, 2023). Excessive amounts of N during periods of low crop demand may result in leaching or volatilization, especially with warm soil temperatures, while deficient soil N levels during periods of high uptake demand will lead to reduced quality and yields. The build-up of soil health through periodic organic amendments applications, N fixation, vegetation diversification such as alley cropping and intercropping, foliar applications during periods of high N demand, and the adoption of crop varieties adapted to organic systems are among the strategies that may be adopted to better synchronize soil N levels with crop uptake demands.

Phosphorus

Along with nitrogen, phosphorus (P) is among the most limiting nutrients in organic farms, especially in the tropics. Most tropical soils are considered to be 'severely' deficient in P, and most of the P in those soils is unavailable for plant uptake, due to their very inherent high P fixation rates – as P bonds tightly to iron and aluminum oxides, under acid conditions (Nwaga *et al.*, 2010). In soils with high P fixation rates, P becomes 'less' available for plant uptake but it is not necessarily 'unavailable'. Thus, management strategies may be adopted in soils with high P fixation rates to improve the concentration of soluble P in the soil. Because P is relatively immobile in the soil, available soluble soil P should be in close proximity to the roots. Management strategies to increase soil P levels include adjusting the soil pH; promoting P solubilizing microbial and mycorrhizal activity (Basak *et al.*, 2020); the planting of cover crops, such as cowpea, buckwheat and pigeon pea, that release organic acids via root exudates which help to solubilize soil P (Table

Table 19.4. Examples of some cover crops and hedgerows or alley crops used in organic farming.

Species	Benefits	Biomass (tons ha^{-1})	Nutrient content (% NPK)	References
Jackbean, sword bean, *Canavalia ensiformis*	Nitrogen, biomass	10–25	4–8 N	Valenzuela, 2023
Cowpea, *Vigna unguiculata*	Nitrogen, P scavenger, soil quality	2–6	1.5 N	Valenzuela, 2020, 2023
Faba bean, *Vicia faba* L.	Nitrogen	4.7	3.8 N	De Notaris *et al.*, 2023; Yfantopoulos *et al.*, 2022; Valenzuela, 2023
Gliricidia sepium	Nitrogen, biomass	10.5	2–5 N	Valenzuela, 2023
Mucuna pruriens	Nitrogen, nematode management	2–8	2.2–2.5 N	Reddy *et al.*, 1986; Valenzuela, 2023
Hairy vetch (*Vicia villosa*)	Nitrogen, weed suppression, P scavenger	2.8–7.7	3–4 N	Mirsky *et al.*, 2017; Valenzuela, 2023
Grasses: rye (*Secale cereale*), annual rye grass (*Lolium multiflorum*), sudex (*Sorghum bicolor x S. bicolor* var. *sudanense*), black oats (*Avena strigosa*)	N scavengers (catch crops); soil quality, erosion control, weed suppression; biomass	1.2–25	1.2–1.5 N; 0.3–0.4 P; 1.8–4 K; 15–22 C:N ratio	Monaghan *et al.*, 1994; Valenzuela and Smith, 2002
Pigeon pea, *Cajanus cajan*	Nitrogen, biomass; low fertility, drought and high temperature- tolerant	9–40	2.5–3.5 N	Reddy *et al.*, 1986; Valenzuela, 2023
Sunn hemp, *Crotalaria juncea*	Nitrogen, biomass, nematode management	6–24	2–4 N	Reddy *et al.*, 1986; Valenzuela, 2023

19.4); as well as adopting integrated crop-livestock systems. The selection of crop species and varieties that can tolerate soils low in P, and with more effective mycorrhizal associations, is another strategy to grow crops under conditions of low soil P (Elbon and Whalen, 2015; Leiser *et al.*, 2015).

A survey of organic farms in Germany found a steady decline of available soil P over the years, indicating a possible 'mining' of the available soil P reserves. Because soil P deficits are common in arable organic farms, as well as unbalanced nutrient applications, resulting in a soil surplus of some elements and a deficit of others, it is recommended that a prescribed combination of amendments be applied to provide a more balanced nutrient application (Reimer *et al.*, 2020).

Common organic P sources include rock phosphate, bone meal, wood ash, fish and seed meal, crop residues, and organic amendments (Table 19.3). In some cases, such as with rock phosphate, P is released over months or years, so the goal is to build the long-term P soil fertility. The over-application of P, resulting in surplus P soil levels, is possible on farms that receive annual applications of manure or composts. Surplus soil P levels may result in nutrient imbalances, as well as environmental losses. On farms where surplus P soil levels may occur, it is important to calculate system nutrient budgets (system P outputs minus inputs) to optimize future rates of amendment applications (Table 19.2) (Baldwin, 2006; Möller, 2018).

Potassium

Potassium (K) is the second major micronutrient, after N, in terms of the amount required by plants (Table 19.2). Potassium provides several important physiological functions, activating over 80 enzymes that affect plant growth and development, maintain plant turgor, reduce water losses, and improve crop quality and shelf life. Adverse effects from K deficiency include yield losses, poor root growth, increased disease and insect pressure, less tolerance to abiotic stress, and lower N and water-use efficiency. Potassium also interacts with other nutrients in the plant. For instance, high K levels may reduce calcium and magnesium uptake, while K deficiency may increase iron deficiency.

Organic sources of K include mined sources of potassium sulfate (50% K_2O), manures, composts and crop residues, all of which release K^+, which is available for plant uptake (Table 19.3). Greensand, used by some organic farmers, contains up to 5% K but it has a very slow release rate (< 1% soluble K), so it serves best as a long-term slow-release K source. The K content in organic amendments is variable, so it is important to be aware of the original sourcing material. Seaweed is a readily available source of K, but its relatively low K content (2%) and transportation costs may make it too costly for some organic operations. Wood ash is another potential K source, and it also has a liming effect, with pH that ranges from 9 to 13 (Mikkelsen, 2007). Organic mulches can be a valuable source of K. Most of the K available on manures becomes available in the first year of application; however, the individual batches should be tested to determine the actual rates of application. Additional organic K sources include alfalfa meal and kelp meal (Table 19.3). Because K is relatively immobile in the soil, amendments should be incorporated 15–20 cm into the soil for optimal root uptake (Rosen and Bierman, 2005). Microbial activity in the soil can help to increase soil K availability. Potassium solubilizing microorganisms help to release K from insoluble soil minerals, and may have other beneficial side effects by improving plant growth and suppressing plant diseases (Bashir *et al.*, 2016; Basak *et al.*, 2020).

Secondary Nutrients and Micronutrients

The secondary nutrients, calcium (Ca), magnesium (Mg) and sulfur (S), play important physiological functions in plants. Calcium is required for plant growth and is involved in cell and membrane structure, N metabolism, as intracellular messenger, and in resistance to diseases and abiotic stress. Calcium is the dominant cation that determines the cation exchange capacity in soils. Low Ca levels may result in poor pollen formation, delayed flowering and fruit maturation, and in reduced lignification or cell wall formation. In the soil solution, calcium is readily available for plant uptake. Foliar applications may be effective during periods of high crop demand. Magnesium is also an integral component of plant tissues, is involved in physiological and biochemical processes that affect plant growth, including the transport of assimilates, the activity of >300 enzymes, and in plant defence against abiotic stress and diseases (Huber and Jones, 2013; Senbayram *et al.*, 2015). Because Mg is very mobile in the soil, it has a propensity for leaching, especially in sandy soils. High K levels in the soil may reduce both Ca and Mg uptake. Mg deficiencies may also occur in soils that have received high N inputs. Sulfur (S) is an essential component of amino acids and metabolites. Sulfur deficiency results in reduced photosynthetic activity and imbalance in the composition of amino acids, resulting in reduced yields and quality. As a reflection of its importance for plant growth, many crops contain similar amounts of S and phosphorus (Table 19.2).

Micronutrients provide important physiological functions to support plant growth and resistance against diseases. It is important to monitor potential micronutrient deficiency or imbalanced nutrient levels, which may result in physiological disorders, susceptibility to diseases, and reduced quality and yields. In many instances, increasing micronutrient availability is obtained by adjusting the soil pH by liming acid soils.

The crop demand of secondary and micronutrients in organic farms is met in many cases with the periodic application of crop residues, organic amendments and commercially available organic fertilizers (Table 19.3) (Kuepper and Everett, 2004; Fischer *et al.*, 2005; Rosen and Bierman, 2005; Möller and Schultheiß, 2015). For example, the S content in manure ranges from 0.7 to 1.5 kg ton^{-1}. Soils that have received crop residue amendments showed elevated levels of calcium, magnesium and micronutrients (Holtz *et al.*, 2004). Experienced organic farmers also believe that proper nutrition, such as with kelp extracts in combination with fish emulsion or fish powder, will improve the ability of crops to uptake micronutrients (Cantisano, 2012).

Products such as mined Epsom salt (magnesium sulfate) and dolomitic limestone are a source of Ca and Mg; gypsum may be applied when no soil pH adjustment is required, to provide Ca and S; and potassium magnesium sulfate (KMag) can provide Mg and S. Seaweed extracts are used as a source of micronutrients. Soil microbial activity also helps to solubilize nutrients that are closely tied to the soil, such as Mg. The use of purchased micronutrient sources is restricted by some certifiers, and their use is allowed only in cases when there is a demonstrated deficiency on the farm.

Biostimulants and Microbial Inoculants

To supplement the use of organic amendments to improve soil health, there is increased interest in the use of biostimulants and microbial inoculants to promote crop growth and quality and to overcome abiotic and biotic stress (Pylak *et al.*, 2019; Guimarães *et al.*, 2023). To reflect on their popularity, the global market for biostimulants has increased at an annual rate of over 10% over the past few years (Conselvan *et al.*, 2017). Biological products, which may consist of plant growth promoting microorganisms, plant extracts, or animal by-products, are used to control plant pests, to stimulate plant growth or to enhance nutrient-use efficiency. Biostimulants, which are used to promote crop growth by enhancing plant hormonal and signalling responses, consist of microorganisms, seaweed extracts and other products (Pylak *et al.*, 2019). Their use has also been explored as seed-priming treatments to overcome heat stress (Anjos Neto *et al.*, 2020).

Products that may be used as biostimulants include fish extracts, humic acids, seaweed, kelp and vermicompost teas. Microbial inoculants used include *Azospirillum*, *Azotobacter*, *Bacillus*, *Beauveria*, *Pseudomonas* and *Trichogramma*. Prospective users need to verify whether particular microbial products are allowed by local organic certifiers. However, considerable research is needed to determine the particular conditions on the farm when the use of biostimulants and microbial inoculants would be most beneficial, and the economic value and practicality of applying these products on commercial operations. For example, an evaluation of a number of commercial bioproducts on two experiment stations and on 19 organic farms in the northern USA found no significant effect on yields from their use (Schonbeck *et al.*, 2019). The use of biological products and extracts may be most beneficial to overcome temporary periods of biotic or abiotic stress.

Conclusions

Nutrient deficits during the different stages of crop growth are among the major limiting factors to reach optimum productivity in organic farming. In organic farming, management practices to supply nutrients and to improve soil tilth and quality are conducted concurrently, and these practices are outlined as part of the certification process, by national and international organic certification standards. Rather than following a prescriptive approach, as is often practised under conventional agriculture, organic farmers tend to follow a more holistic and agroecological approach to nutrient management which includes germplasm selection of adapted varieties; building the soil tilth and quality through the periodic application of crop residues and organic amendments; habitat management or vegetational diversification such as alley cropping, intercropping and rotations; the planting of cover crops as 'catch crops' to capture residual nutrient levels from the previous crop, and to improve soil tilth; as well as the inclusion of leguminous crops to supplement nitrogen pools in the soil. Because of the intimate interaction between production practices and the local environment, recommended production techniques in organic farming are often location-specific and may not be directly transferable to other locations. Thus, locally based applied research is often required to evaluate new production techniques, for adoption in particular locations. The consideration of local socio-economic conditions and of traditional or millennial indigenous knowledge is also important for the implementation of organic nutrient management programmes that are closely attuned with the local social and growing conditions.

References

Anjos Neto, A.P. dos, Oliveira, G.R.F., Mello, S. da C., Silva, M.S. da, Gomes-Junior, F.G. *et al.* (2020) Seed priming with seaweed extract mitigate heat stress in spinach: Effect on germination, seedling growth and antioxidant capacity. *Bragantia* 79(4), 502–511.

Baldwin, K.R. (2006) *Soil Fertility on Organic Farms*. North Carolina Cooperative Extension 32.

Basak, B.B., Maity, A. and Biswas, D.R. (2020) Cycling of natural sources of phosphorus and potassium for environmental sustainability. In: Dontsova, K., Balogh-Brunstad, Z. and Roux, G.L. (eds) *Biogeochemical Cycles: Ecological Drivers and*

Environmental Impact. John Wiley, Hoboken, New Jersey, pp. 285–299.

Bashir, Z., Zargar, M.Y. and Vishwakarma, D.K. (2016) Potassium solubilizing microorganisms for sustainable agriculture. In: Kumar, R., Singh, V.P., Jhajharia, D. and Mirabbasi, R. (eds) *Applied Agricultural Practices for Mitigating Climate Change*. CRC Press, Boca Raton, Florida, pp. 17–28.

Bonaudo, T., Bendahan, A.B., Sabatier, R., Ryschawy, J., Bellon, S. *et al.* (2014) Agroecological principles for the redesign of integrated crop–livestock systems. *European Journal of Agronomy* 57, 43–51.

Cantisano, A. (2012) *Know Your Soil: A Handbook for Understanding and Utilizing A Soil Analysis for Growing Organic*, 3rd edn. Organic Ag. Advisors, San Juan, California.

Conselvan, G.B., Pizzeghello, D., Francioso, O., Di Foggia, M., Nardi, S. *et al.* (2017) Biostimulant activity of humic substances extracted from Leonardites. *Plant and Soil* 420(1–2), 119–134.

De Notaris, C., Enggrob, E.E., Olesen, J.E., Sørensen, P. and Rasmussen, J. (2023) Faba bean productivity, yield stability and N2-fixation in long-term organic and conventional crop rotations. *Field Crops Research* 295, 108894.

Delate, K., Cambardella, C. and Karlen, D. (2003) *Soil quality in organic agricultural systems*. Iowa State University Cooperative Extension Service, p. 8.

Drinkwater, L.E. and Snapp, S. (2007) Nutrients in agroecosystems: Rethinking the management paradigm. *Advances in Agronomy* 92, 163–186.

Ekyaligonza, D.M., Kahigwa, T.T., Dietrich, P., Akoraebirungi, B., Kagorora, J.P. *et al.* (2022) Biomass contribution and nutrient recycling of organic matter management practices in tropical smallholder annual farming systems. *Acta Agriculturae Scandinavica, Section B — Soil & Plant Science* 72(1), 945–956.

Elbon, A. and Whalen, J.K. (2015) Phosphorus supply to vegetable crops from arbuscular mycorrhizal fungi: A review. *Biological Agriculture & Horticulture* 31(2), 73–90.

Fischer, M., Raupp, J., Mäder, P., Dubois, D. and Römheld, V. (2005) Miconutrient status in two long-term trials with fertilisation treatments and different cropping systems. In: *Proceedings of the Conference 'Researching Sustainable Systems* 15th IFOAM Organic World Congress, Adelaide, Australia, pp. 522–525.

Fox, R. and Valenzuela, H. (1992) Vegetables grown under tropical and subtropical conditions. In: Wichmann, W. (ed.) *IFA World Fertilizer Use Manual*. International Fertilizer Industry Association, Germany, pp. 293–338. Available at: http://www.fertilizer.org/PUBLISH/PUBMAN/manual.htm (accessed 3 February 2025).

Gaskin, J., Kissel, D., Harris, G. and Boyhan, G. (2014) How to convert an inorganic fertilizer recommendation to an organic one. The University of Georgia, Cooperative Extension Service Circular 853.

Gatsios, A., Ntatsi, G., Yfantopoulos, D., Baltzoi, P., Karapanos, I.C. *et al.* (2021) Effects of different organic soil amendments on nitrogen nutrition and yield of organic greenhouse tomato crop. *Nitrogen* 2(3), 347–358.

Guimarães, G.S., Rondina, A.B.L., de Oliveira Jr, A.G., Jank, L. and Nogueira, M.A. (2023) Inoculation with plant growth-promoting bacteria improves the sustainability of tropical pastures with *Megathyrsus maximus*. *Agronomy* 13(3), 734.

Haynes, R.J. and Mokolobate, M.S. (2001) Amelioration of Al toxicity and P deficiency in acid soils by additions of organic residues: a critical review of the phenomenon and the mechanisms involved. *Nutrient Cycling in Agroecosystems* 59, 47–63.

Holtz, B.A., McKenry, M.V. and Caesar-TonThat, T.C. (2004) Wood chipping almond brush and its effect on the almond rhizosphere, soil aggregation and soil nutrients. *Acta Horticulture* 638, 127–134.

Huber, D.M. and Jones, J.B. (2013) The role of magnesium in plant disease. *Plant Soil* 368, 73–85.

Jiang, Y., Khan, M.U., Lin, X., Lin, Z., Lin, S. *et al.* (2022) Evaluation of maize/peanut intercropping effects on microbial assembly, root exudates and peanut nitrogen uptake. *Plant Physiology and Biochemistry* 171, 75–83.

Köninger, J., Lugato, E., Panagos, P., Kochupillai, M., Orgiazzi, A. *et al.* (2021) Manure management and soil biodiversity: Towards more sustainable food systems in the EU. *Agricultural Systems* 194, 103251.

Kuepper, G. and Everett, K. (2004) Potting mixes for certified organic production. In: *Horticulture Technical Note IP112. National Sustainable Agriculture Information Service*. ATTRA, Washington, DC.

Leiser, W.L., Rattunde, H.F.W., Piepho, H., Weltzien, E., Diallo, A. *et al.* (2015) Phosphorous efficiency and tolerance traits for selection of sorghum for performance in phosphorous-limited environments. *Crop Science* 55(3), 1152–1162.

Løes, A.K., Bünemann, E.K., Cooper, J., Hörtenhuber, S., Magid, J. *et al.* (2017) Nutrient supply to organic agriculture as governed by EU regulations and standards in six European countries. *Organic Agriculture* 7(4), 395–418.

Mårtensson, L.M.D., Barreiro, A., Li, S. and Jensen, E.S. (2022) Agronomic performance, nitrogen acquisition and water-use efficiency of the perennial grain crop *Thinopyrum intermedium* in a monoculture and intercropped with alfalfa in Scandinavia. *Agronomy for Sustainable Development* 42(2), 21.

Mikkelsen, R.L. (2007) Managing potassium for organic crop production. *HortTechnology* 17(4), 455–460.

Mirsky, S.B., Ackroyd, V.J., Cordeau, S., Curran, W.S., Hashemi, M. *et al.* (2017) Hairy vetch biomass across the eastern United States: Effects of latitude, seeding rate and date, and termination timing. *Agronomy Journal* 109(4), 1510–1519.

Moeller, K., Oberson, A., Bünemann, E.K., Cooper, J., Friedel, J.K. *et al*. (2018) Improved phosphorus recycling in organic farming: Navigating between constraints. *Advances in Agronomy* 147, 159–237.

Möller, K. (2018) Soil fertility status and nutrient input–output flows of specialised organic cropping systems: A review. *Nutrient Cycling in Agroecosystems* 112(2), 147–164.

Möller, K. and Schultheiß, U. (2015) Chemical characterization of commercial organic fertilizers. *Archives of Agronomy and Soil Science* 61(7), 989–1012.

Monaghan, P.F., Brinen, G.H. and Kostewicz, S.R. (1994) Organic vegetable production in Florida. *Proceeding of the Florida State Horticulture Society* 107, 377–379.

Nwaga, D., Jansa, J., Angue, M.A. and Frossard, E. (2010) The potential of soil beneficial micro-organisms for slash-and-burn agriculture in the humid forest zone of Sub-Saharan Africa. In: Dion, P. (ed.) *Soil Biology and Agriculture in the Tropics*. Springer-Verlag, Berlin Heidelberg, pp. 81–107.

Oelofse, M., Høgh-Jensen, H., Abreu, L.S., Almeida, G.F., El-Araby, A. *et al*. (2010) A comparative study of farm nutrient budgets and nutrient flows of certified organic and non-organic farms in China, Brazil and Egypt. *Nutrient Cycling in Agroecosystems* 87(3), 455–470.

Piato, K., Subía, C., Lefort, F., Pico, J., Calderón, D. *et al*. (2022) No reduction in yield of young robusta coffee when grown under shade trees in Ecuadorian Amazonia. *Life* 12(6), 807.

Pylak, M., Oszust, K. and Frąc, M. (2019) Review report on the role of bioproducts, biopreparations, biostimulants and microbial inoculants in organic production of fruit. *Reviews in Environmental Science and Bio/Technology* 18(3), 597–616.

Reddy, K.C., Soffes, A.R. and Prine, G.M. (1986) Tropical legumes for green manure. I. Nitrogen production and the effects on succeeding crop yields. *Agronomy Journal* 78(1), 1–4.

Reimer, M., Hartmann, T.E., Oelofse, M., Magid, J., Bünemann, E.K. *et al*. (2020) Reliance on biological nitrogen fixation depletes soil phosphorus and potassium reserves. *Nutrient Cycling in Agroecosystems* 118(3), 273–291.

Rosati, A., Borek, R. and Canali, S. (2021) Agroforestry and organic agriculture. *Agroforestry Systems* 95(5), 805–821.

Rosen, C.J. and Bierman, P.M. (2005) *Maintaining Soil Fertility in an Organic System*, Vol. 10. Minnesota Extension, University of Minnesota, College of Agricultural, Minneapolis, Minnesota.

Schonbeck, M., Jerkins, D. and Lowell, V. (2019) *Soil Health and Organic Farming*. Organic Farming Research Foundation, Santa Cruz, California.

Senbayram, M., Gransee, A., Wahle, V. and Thiel, H. (2015) Role of magnesium fertilisers in agriculture: Plant–soil continuum. *Crop and Pasture Science* 66(12), 1219.

Setyorini, D. and Hartatik, W. (2021) Nutrients balance under organic rice farming system in Central Java, Indonesia. In: *IOP Conference Series*, IOP Publishing (Vol. 648).

Shaji, H., Chandran, V., Mathew, L., Thomas, S. and Rakhimol, K.R. (2021) Organic fertilizers as a route to controlled release of nutrients. In: Lewu, F.B. and Volova, T. (eds) *Controlled Release Fertilizers for Sustainable Agriculture*. Academic Press, Amsterdam, The Netherlands, pp. 231–245.

Valenzuela, H. (2020) The use of crop residues on the farm. Hanai Ai Sustainable Ag Newsletter, CTAHR, Spring 2020. University of Hawaii Cooperative Extension Service, Honolulu, Hawaii. Available at: https://www.researchgate.net/publication/338980740_The_use_of_Crop_Residues_on_the_Farm#fullTextFileContent (accessed 3 February 2025).

Valenzuela, H. (2023) Ecological management of the nitrogen cycle in organic farms. *Nitrogen* 4(1), 58–84.

Valenzuela, H. and Smith, J. (2002) Rye. *Sustainable Agriculture Green Manure Crops*. SA-GM 9. Univerity of Hawaii Cooperative Extensive Service 3, Honolulu, Hawaii.

Wortman, S.E., Galusha, T.D., Mason, S.C. and Francis, C.A. (2012) Soil fertility and crop yields in long-term organic and conventional cropping systems in Eastern Nebraska. *Renewable Agriculture and Food Systems* 27(3), 200–216.

Yfantopoulos, D., Ntatsi, G., Gruda, N., Bilalis, D. and Savvas, D. (2022) Effects of the preceding crop on soil N availability, biological nitrogen fixation, and fresh pod yield of organically grown faba bean (*Vicia faba* L.). *Horticulturae* 8(6), 496.

20 Nutrient Management through Organic Practices

Amjad A. Ahmad[1]*, Maria E.O. Escobar[2] and Nguyen V. Hue[1]

[1]Tropical Plant and Soil Sciences Department, University of Hawaii at Manoa, USA; [2]Soil Science Department, Federal University of Ceara, Brazil

Abstract

Soils rarely have sufficient nutrients for optimal crop growth, and most organic fertilizers contain relatively low nutrients and release them slowly. Thus, successful management of crop nutrients in organic farming requires skill and relevant knowledge. Knowledge on availability and nutrient content of different organic fertilizers, for various nutrients, is crucial in order to ensure optimum crop yields under organic farming system. This chapter presents information on organic inputs for nitrogen (N), phosphorus (P), potassium (K) and, to a lesser extent, calcium (Ca) and magnesium (Mg), as crop nutrients for organic farming. First, we discuss N content and its release patterns in ground materials (i.e. meals) of meat, bone, blood and feather, as well as N in animal and green manures and that provided by Rhizobium bacteria. Next, P enhancements from phosphate rock, bone meal, manures, mycorrhizal fungi and P-solubilizing microbes are covered. Third, K sources from greensand, seaweeds, manures and biochar are briefly evaluated. Finally, practices are recommended to increase, or at least sustain, soil organic matter, which is essential to soil fertility and the success of organic farming.

Keywords: nitrogen, potassium, phosphorus, calcium, micronutrients

Introduction

Organic agriculture has a vision for energy-efficient and ecologically sound agricultural systems (Reganold and Wachter, 2016). Restoring soil fertility and improving the physical, chemical and biological functions of soils are essential to optimal plant growth, yield and quality under organic system (van der Werf *et al.*, 2020). Thus, organic inputs and proper managements are very important in ensuring successful biological decomposition processes and sufficient nutrient release for plant uptake (Diacono and Montemurro, 2011).

Organic farming follows a holistic approach to agriculture and avoids the use of synthetic fertilizers and pesticides (Radovich *et al.*, 2014). More specifically, it relies on mined substances of low solubility (e.g. rock phosphate and greensand) and organic local inputs (e.g. manures, compost) to maintain soil health and crop productivity (Soni and Yadav, 2019). With no chemical inputs, organic farming must be skillfully managed, since soil nutrient management poses a serious challenge to organic farming, especially if the farmed soils are nutrient-deficient as is often the case with 'old', highly weathered soils (Sanchez, 2019).

Soils rarely are able to support and have sufficient nutrients to provide for most crops to reach their optimum growth and yield. Therefore, the application of soil amendments containing macro- and micronutrients are needed (Ahmad *et al.*, 2016). Nutrient release from organic fertilizers is different to that of synthetic fertilizers. Knowledge of availability, quality, nutrient content and mineralization pattern are crucial

*Corresponding author: alobady@hawaii.edu

DOI: 10.1079/9781800626850.0020

in order to improve nutrient management, soil fertility and crop production, and reduce nutrient losses under organic farming system, while protecting the ecosystem from pollution. This chapter covers sources, quality and nutrient content of different organic fertilizers for nitrogen (N), phosphorus (P), potassium (K), calcium (Ca) and magnesium (Mg).

Nitrogen and Organic Agriculture

Nitrogen (N) is the most essential nutrient in crop production (Carr *et al.*, 2019) and is rather difficult to manage (Anas *et al.*, 2020; Cattell-Noll *et al.*, 2020). While many parts of the world do not have enough N to achieve food and nutrition security, others are facing N pollution and damaging consequences (Squire *et al.*, 2022). Nitrogen is an essential component of building the plant DNA (Doane, 2017) and plant chlorophyll, which are essential to the photosynthesis process (Schulte-Uebbing *et al.*, 2022). Around 78% of the Earth's atmosphere is made up of N gas or N_2, but it is mostly not accessible in this form. Atmospheric N must be 'fixed' before it can be used by plants (Grzyb *et al.*, 2021). For thousands of years, humans have relied on various organic sources, e.g. animal manures, as N fertilizer for food production (Ahmad *et al.*, 2016). Synthetic N fertilizer, in today's intensive agricultural systems, is crucial in order to meet the demand for more agricultural products (Randive *et al.*, 2021). However, since the early 1990s, there has been a rapid increase in demand for organically produced food driven by the belief that they are healthier, tastier and more environmentally friendly in comparison to conventionally produced food (Oduor *et al.*, 2023). Organic agriculture combines modern science and indigenous knowledge to achieve an efficient and more productive system by closing nutrient cycles and utilizing resources better (van der Werf *et al.*, 2020).

Practical nitrogen sources for organic farming

Meat and bone meal (MBM)

MBM is the solid by-product of animal waste rendering, after cooking meat, bone and all other slaughterhouse waste, and finally extracting fat, then drying and crushing the final product (Cascarosa *et al.*, 2012). Nutrient content of MBM varies with feedstock and storage time, averaging 9.8% N, 2.5% P, 0.75% K_2O and 5.0% Ca. MBM is National Organic Program (NOP)-compliant (Gurr, 2011). Table 20.1 shows nutrient content of ten samples collected over 30 months (Ahmad *et al.*, 2016). Approximately 2–4 tons/ha of MBM are recommended for annual vegetable growth, and 675–1750 kg/ha per crop cycle for most short-cycle crops (Radovich *et al.*, 2014).

Blood meal and feather meal

Blood meal is the by-product of animal blood collected from slaughterhouses and meat-packing facilities, then dried and made into a fertilizer (Mondini *et al.*, 2008). Nitrogen mineralization in blood meal is usually fast with highest N release occurring within 1–2 weeks of incorporation. The use of blood meal has largely been as a feed supplement for livestock or aquaculture because of its high (12%) N concentration in the form of protein (Hartz and Johnstone, 2006). Blood meal is an effective N fertilizer for crop growth and yield. However, it lacks other essential macro- and micronutrients for healthy plant growth and yield (Mondini *et al.*, 2008).

Table 20.1. Average and standard deviation for macronutrient (%) and micronutrient µg/g) content in ten meat and bone meal (MBM) samples collected over 30 months (2015–2018), locally, in Hawaii.

	Percentage (%)							µg/g					
MBM	N	C	P	K	Ca	Mg	Na	Fe	Mn	Zn	Cu	B
Average	10.0	45.0	3.0	0.80	5.90	0.17	0.75	761	9.7	90	3.1	3.3
St.dev.	1.1	1.4	0.2	0.1	0.3	0.01	0.08	120	1.5	8.7	1.5	0.75

Each value is a mean of 30 replicates (10 initial samples and three subsamples from each initial sampling).

Table 20.2. Macro- and micronutrients and C/N ratio in chicken and dairy manure collected in Hawaii.

	Percentage (%)							μg/g					
Manure type	N	C	P	K	Ca	Mg	Na	Fe	Mn	Zn	Cu	B	C:N ratio
Chicken	3.0	21.5	1.5	1.9	14.0	0.7	0.4	210	960	400	45	30	7.1
Dairy	1.8	15.0	0.5	1.8	2.0	1.0	0.5	1300	325	125	190	45	8.2

Feather meal is the by-product derived from processing the feather (contains ~87% protein) waste from poultry slaughterhouses. The feather protein is a complex structure and requires to be cooked under high pressure, which is necessary to break the keratin helix in the feather (Ahmad *et al.*, 2016). Feathers are 3–7% of a poultry weight and produced in large quantities. Feather meal has been used as an organic fertilizer or a source of high protein (N content ~15%) in animal and fish feed, and for better N mineralization it is important to incorporate (1–2 weeks for the nutrient to become available) it with the soil prior to planting (Ahmad *et al.*, 2016).

Animal manure and green manure

Livestock are a good source of organic fertilizer and animal manures have been used as fertilizer for thousands of years (Seefeldt, 2013). For example, on average, a 1000-pound cow may produce about 15 tons of manure annually, which may contain 100 kg of N, 95 kg of P_2O_5 and 125 kg of K_2O. Animal manures usually contain essentialmacro- and micronutrients needed for healthy crop growth and yield, see Table 20.2 two animal manures collected in Hawaii (Radovich *et al.*, 2014). Animal manures are commonly applied under irrigated agricultural operations to improve crop yields (Loss *et al.*, 2019).

Green manures can often be used as an organic N source. The N content in green manures is highly variable (20–132 kg N/ha) due to the variability in crops used as green manure and the crops' ability to accumulate N during the growth period. Green manures are usually grown till the flowering stage and then flowering plants mowed and tilled into the soil (Torma *et al.*, 2018; Lazcano *et al.*, 2023). Cover crop calculator is an important tool to utilize for both small- and large-scale farms (https://western.sare.org/resources/organic-fertilizer-and-cover-crop-calculator/ [accessed 3/2/25]).

Nitrogen fixing rhizobium bacteria (biofertilizer)

Rhizobium bacteria is a generic name of the Gram-negative group of *Alpha- and Beta-proteobacteria*, which forms nodules on the root or stem of legume crops and fixes atmospheric N in a symbiotic relationship (Lindström and Mousavi, 2020). *Frankia* and *Cyanobacteria* bacteria group is also known to nodulate a wide range of plant and fix atmospheric N (Op den Camp *et al.*, 2011). Total N fixed ranged from 50 kg to 300 kg N_2 fixed/ha/year, based on different legume crops, cropping system and environmental condition (Lindström and Mousavi, 2020). Inoculating legume seeds with the appropriate *Rhizobium* is an essential practice to ensure a successful nodulation (Dasgupta *et al.*, 2021).

Additionally, there are other microorganisms (*Azotobacter*, *Azospyrillum* inoculant, *Cyanobacteria* and *Azolla*) with the capacity to enhance plant nutrient availability and uptake through colonizing the plant rhizosphere, which is known collectively as 'biofertilizer' and are applied as seed or soil treatments (Glick and Gamalero, 2021; Zhang *et al.*, 2021).

Phosphorus and Organic Agriculture

Phosphorus (P) is an essential nutrient to all crops (Havlin *et al.*, 2017). Biologically, P is a component of genetic molecules (DNA, RNA) as well as adenosine triphosphate (ATP), which transfers energy during photosynthesis and respiration (Barker, 2010; Sanchez, 2019). In P-deficient plants, reduction in leaf expansion and number of leaves occurs, resulting in stunted growth (Fageria *et al.*, 2017; Hawkesford *et al.*, 2023). In advanced stage of P deficiency, leaves turn purple (Sharanappa, 2023). Unlike N, P is not renewable since there is no atmospheric P source. Total P in soil can be divided into five functional pools: (i) soil solution P; (ii) labile or weakly adsorbed P; (iii) strongly adsorbed and/or precipitated P;

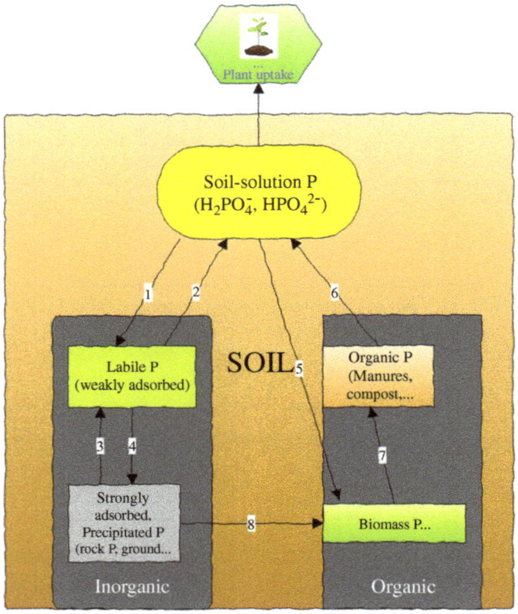

Fig. 20.1. Representation of the different P pools in soils. The different numbers refer to the processes affecting the P pools: (1) adsorption; (2) desorption; (3) dissolution; (4) precipitation; (5) assimilation; (6) mineralization; (7) decomposition; and (8) phosphorus solubilization by microbes.

(iv) manures and compost P; and (v) microbial/biomass P (Fig. 20.1). Pools ii and iii are inorganic solids; pools iv and v are obviously organic.

Soil-solution P pool is rather small; its P concentration ranges from 0.1 µM (or 0.003 mg/L) to 10 µM (Hinsinger, 2001). It averages 0.05 mg/L and exists predominantly as $H_2PO_4^-$ anion at pH 3.0–7.0 and as HPO_4^{2-} anion at pH > 7.2 (Sanchez, 2019). Both anions move by the slow diffusion process and are readily taken up by plants. However, the dynamics of soil-solution P are fast and are interconnected with the other P pools.

The weakly adsorbed or labile P pool and the strongly adsorbed or precipitated P pool are inorganic P attached to the soil solid phase (Nesme *et al.*, 2014). Although the inorganic pools are commonly dominant (> 50%) in cultivated soils, under organic farming practices, organic P can be high, ranging from 30 to 80%. Most soil organic P compounds have P bonded to oxygen in ester bonds (C-O-P). Over 50% organic P is monoesters, mainly inositol phosphate (phytate); about 5% is phospholipids, and 2% is nucleic acids (Condron and Tiessen, 2005).

Practical P sources for organic farming

Phosphate rock

Chemically, P rock is a combination of hydroxyapatite [$Ca_{10}(OH)_2(PO_4)_6$] and fluoroapatite [$Ca_{10}F_2(PO_4)_6$] with some minor substitution of OH^- or F^- by carbonate (Sanchez, 2019). The solubility of rock P is quite low, so its use requires some soil amending actions such as by keeping relatively low pH and/or low Ca^{2+} (Ca^{2+} can be lowered by chelation with organic-acid anions) as illustrated by the following chemical reaction:

$$\rightarrow\rightarrow Ca_{10}(OH)_2(PO_4)_6 + 14\ H^+\ 10\ Ca^{2+} + 6\ H_2PO_4^- + 2\ H_2O$$

Fine sizes (< 100 mesh) of P rock and its use in combination with other organic inputs such as manures or compost also help increase its P-releasing capacity because of small sizes and thus large surface areas. Bone meal (about 30% total P) and meat from cattle or fish (approximately 3–5% P) can be used as a source of P in organic farming. These materials consist mainly of apatite-like calcium phosphate minerals, so they should react in a manner similar to rock P.

Animal manures

Unlike P rocks, which may have 10–18% total P with low solubility and slow-release rate, P in manures is rather low (1–5% P) and the release rate is relatively fast. About 70% of P in most animal manures is available to the crop during the year of application (Shapiro *et al.*, 2021). Poultry manure generally has lower N/P ratios than cattle manure due to its higher P content. This may result in over-P fertilization when meeting crop N requirements.

Green manures

Most plant materials contain 0.2–0.4% P (Lazcano *et al.*, 2023), whereas their N can be as high as 4–6%. Green manures are more often used as an organic source for N than for P (Cherr *et al.*, 2006). According to Sanchez (2019), *Tithonia diversifolia*

(Mexican sunflower) can have up to 4% N and K, and 0.6% P in leaves, and releases its nutrients fast. It is one of the best green manures for resource-poor farmers in the tropics.

Compost

Compost has been a popular organic nutrient source (Lazcano *et al.*, 2023). Since compost is often made from plant materials, either alone or in combination with animal manures, P is also low (0.1–2.3%) in compost (Fageria, 2008). Plant-based compost usually contains less nutrients (N, P, K) than animal manure-based compost (Lazcano *et al.*, 2023).

Bioenhancers of soil P availability

Given the low P concentration in the soil solution, root architecture plays an important role in P acquisition (Lynch, 2007). Producing proteiod roots is a response to low P in some plant species such as macadamia (Hue, 2009). Other plant species can exude P-mobilizing compounds such as protons, carboxylates and phosphatases as triggered by P deficiency (Hinsinger, 2001). The rhizodeposition of carbon-rich compounds stimulates both naturally occurring and inoculated microorganisms in the rhizosphere, which can alter soil P availability (Richardson *et al.*, 2009).

Mycorrhizal fungi

While root hairs can extend up to 1–2 mm away from the root surface, mycorrhizal hyphae can extend beyond many cm (Thonar *et al.*, 2011). Mycorrhizae serve as root extenders. They infect the roots of most vascular plants, effectively increasing the soil volume from which plants can take up nutrients and water. Through mycorrhizal hyphae, the inflow of phosphate ions can be many orders faster than their diffusion through soil (Giri *et al.*, 2019).

Phosphorus solubilizing microbes

Although total P content may be large, most phosphates are in solid forms and unavailable for plant uptake. The P-solubilizing bacteria such as *Achromobacter*, *Micrococcus*, *Aerobacter* and *Pseudomonas* can increase P availability to plants by secreting organic acids such as citric, malic, oxalic, tartaric, gluconic, etc. The secreted acids can either directly replace solidly held P anions by their conjugate anions or by dissolving P minerals via chelating Ca in rock P or Al/Fe in minerals (Tarafdar, 2019). To use phosphate from its organic forms, the P compounds must first be hydrolyzed by phosphatases. Some fungi of the genera *Aspergillus* and *Penicillium* can efficiently cleave the C-O-P ester bond and release P for plant uptake (Tarafdar, 2019).

Biofertilizers

Biofertilizers consisting of bacteria and fungi that can solubilize solid P and/or hydrolyse organic P have been formulated and commercialized (Giri *et al.*, 2019). They can have solid or liquid formulations to prolong microbial viability and efficiency under variable conditions; and their nutrient-enhancing functions can go beyond P. For P nutrient, biofertilizers contain one or more of the following microbes: (i) phosphate solubilizer organisms; and (ii) phosphate mobilizer organisms (Wong and Teh, 2021).

Potassium (K), Calcium (CA) and Magnesium (MG) Nutrients for Organic Farming

Organic agriculture can be considered part of the solution to some global issues, such as climate change and food security, and this entails improving crop productivity and good soil health. Potassium accounts for approximately 2.6% and calcium 3.6% of the Earth's crust, and magnesium is slightly lower (Foth and Ellis, 1997).

Potassium (K), calcium (Ca) and magnesium (Mg) cycle in soils

The transfers and transformations of K in crop production are shown in Fig. 20.2 (Brouder and Volenec, 2021). This cycle helps clarify potential contributors to soil solution K during a growing season and potential sinks for K added as fertilizers, manures and returned residues. It is essential to fully understand the K cycle to properly manage K-fertilization, since the processes are controlled more by physical and chemical than biological factors, as with N and P. As with any other cation, exchangeable and solution Ca^{+2} are in equilibrium and provide the majority of plant available Ca^{+2}.

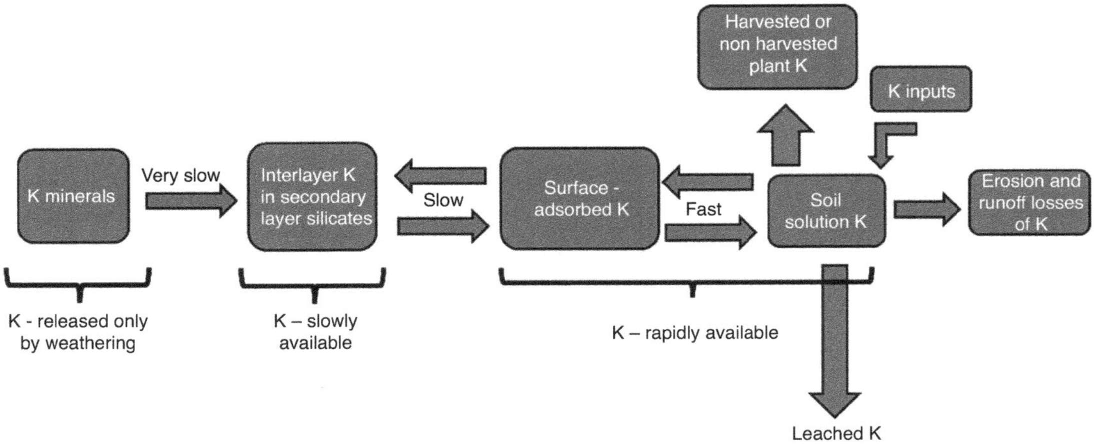

Fig. 20.2. The K cycle.

On the other hand, Mg absorption by plants depends on the amount of solution Mg^{+2}, soil pH, %Mg saturation on the mineral surface, quantity of other exchangeable ions, and type of clay (Havlin *et al.*, 2017).

Organic sources for K, Ca and Mg

Plant roots do not distinguish solutes due to the origin, whether from artificial fertilizers or from a natural source (Kirchmann *et al.*, 2008). Organic farming relies on organic matter to meet crop nutrient requirement in addition to improving soil properties, and enhancing crop productivity and quality (EgrinyaEneji *et al.*, 2003).

Animal manure

Most studies on decomposition and nutrient release of soil-applied manure focus on C, N, and P, while the mineralization of Ca, Mg and K are poorly documented (EgrinyaEneji *et al.*, 2003). Poultry manure contains a wide range of nutrients and often contains higher N-K-Ca than other animal manures (Dróżdż *et al.*, 2020). Cui *et al.* (2023) stated that the long-term addition of swine manure (over 30 years) alone or combined with synthetic fertilizers increased soil-exchangeable Ca^{2+} and decreased exchangeable Na^{+}.

Biochar

Biochar amendment is more beneficial than other organic materials in two ways: first, it has a strong resistance to decay, thus remaining longer in the soil; secondly, it has a greater ability to enhance carbon, water, nutrient retention and overall soil health (Teodoro *et al.*, 2020). Despite those benefits, biochar is nutrient-impoverished compared to manures and composts (Lehmann and Joseph, 2015). When biochar is produced at low temperature it tends to have more nitrogen, CEC and surface functional groups, while P, K and surface area increase with temperature (Suliman *et al.*, 2016). Large surface areas of biochar and negatively charged surfaces generally favour the retention of K in soils, with an average K in biochar of 2% with a range of 0.1–6% (Jindo *et al.*, 2014). Nearly 50% of total K in biochar is bioavailable (Hue, 2020).

Green manure and crop residues

Crop residues are good organic sources of N and K. The amount of nutrients in postharvest crop residues is highly variable among plant species (Torma *et al.*, 2018). Since most nutrients from organic sources are not readily available, raw manure should be applied more than 120 days before harvest if the edible portion of the crop will be in direct contact with soil, and 90 days before harvest if the edible portion of the crop will not be in direct contact with soil (Wortman *et al.*, 2017).

Seaweed Products

Seaweeds are a sea-growing biomass resource and primary producers in the marine food chain. Macro-algae comprise different seaweeds based on pigmentation and structural composition (Samarathunga *et al.*, 2022). They can be cultivated for fertilizer purposes, and thus constitute a method for transformation of diluted organic wastes, like wastewater from aquaculture systems, into concentrated plant fertilizers (Wuang *et al.*, 2016). Gangaiah *et al.* (2017) reported three seaweed species, with K content ranging from 14 to 20%. Based on a greenhouse experiment with Chinese cabbage, the authors suggested that the tested algae could replace synthetic K fertilizer in vegetable crop production. Agrawal *et al.* (2022) studied the biochemical constituents of selected seaweeds and reported macronutrient and micronutrient concentrations and highlighted the importance of the use of seaweeds as a novel application in soil.

Greensands

Greensands are composed primarily of glauconite, a potassium-iron-aluminum silicate (DGS (Delaware Geological Survey, 2020). Addition of greensand to soils enhances the ability of the soil to store exchangeable nutrients such as Ca, Mg, K, and micronutrients, being recommended as a natural K source (average 5% K) for organic agriculture (Heckman and Tedrow, 2004). Despite its benefits, the solubility of K-bearing minerals is low (Martins *et al.*, 2015). Nevertheless, the addition of greensand to soil has permanent benefits, contrary to the addition of organic matter (Heckman and Tedrow, 2004).

Building Up Soil Organic Matter in Organic Farming

Organic farming encourages the build-up of soil organic matter (SOM), which includes soil organic carbon (SOC), organic N, P, S, and micronutrients as well (although SOM is often used synonymously with SOC). Common practices in organic farming that can increase SOM include:

1. **Crop rotation and reduced tillage.** Changing crops from year to year would minimize nutrient and pest stresses on crops. Minimum- or no-till would prevent disruption of soil structure and oxidation (conversion to CO_2) of SOM.
2. **Cover crop and green manure.** Besides adding organic C and nutrients, cover crops, particularly legumes, can help reduce soil erosion and improve soil structure and water retention.
3. **Compost and biochar.** Well-matured compost has ~ 30% organic C and can serve as a good moisture-holding medium, plant nutrient, and food for soil microbiota. On the other extreme, biochar has more C (up to 80%; most of it is inert) and less nutrients (Hue, 2020).
4. **Animal manure.** In addition to organic C, animal manures are a good source of nutrients, particularly N, and, to a lesser extent, P. The rate of nutrient release is usually slow; thus, management skills and farming knowledge are often required for better nutrient synchronization (Valenzuela, 2023).

In summary, building up and maintaining the optimum level of SOM (the optimum level may vary from soil to soil and with environmental factors, such as temperature and rainfall) through various organic inputs would be essential to a successful management strategy for soil and the environment in organic farming.

Conclusions

Organic amendments are known for their low nutrient content and slow release rate in comparison to synthetic/chemical soil amendments. Knowledge of sources, content and nutrient release pattern are much needed for successful soil and crop nutrient management. Additionally, organic soil amendments may contain a wide range of macro- and micronutrients at various levels. This chapter provides useful information on organic sources for nitrogen (meat, bone, blood and feather, as well as N in animal and green manures, and that provided by *Rhizobium* bacteria), phosphorus (phosphate rock, bone meal, manures, mycorrhizal fungi and P-solubilizing microbes), potassium (greensand, seaweeds, manures and biochar), and briefly covers calcium and magnesium sources as crop nutrients for the organic farming system.

References

Agrawal, S., Siddiqui, S.A., Chaudhary, D.R. and Rathore, M.S. (2022) Bio-prospection and compositional multivariate analysis revealed the industrial potential of selected seaweeds along Gujarat seacoast. *Bioresource Technology Reports* 19, 101130.

Ahmad, A.A., Radovich, T.J., Nguyen, H.V., Uyeda, J., Arakaki, A. *et al.* (2016) Use of organic fertilizers to enhance soil fertility, plant growth, and yield in a tropical environment. In: Larramendy, M.L. and Soloneski, S. (eds) *Organic Fertilizers: From Basic Concepts to Applied Outcomes*. pp. 85–108. Available at: http://www.intechopen.com/books/organic-fertilizers-from-basic-concepts-to-applied-outcomes (accessed 4 February 2025).

Anas, M., Liao, F., Verma, K.K., Sarwar, M.A., Mahmood, A. *et al.* (2020) Fate of nitrogen in agriculture and environment: agronomic, eco-physiological and molecular approaches to improve nitrogen use efficiency. *Biological Research* 53(1), 47.

Barker, A.V. (2010) *Science and Technology of Organic Farming*. CRC Press, Boca Raton, Florida, pp. 17–80.

Brouder, S.M. and Volenec, J.J. (2021) The potassium cycle and its relationship to recommendation development. In: Murrel, T.S. (ed.) *Improving Potassium Recommendations for Agricultural Crops*. Springer.

Carr, P.M., Cavigelli, M.A., Darby, H., Delate, K., Eberly, J.O. *et al.* (2019) Nutrient cycling in organic field crops in Canada and the United States. *Agronomy Journal* 111(6), 2769–2785.

Cascarosa, E., Gea, G. and Arauzo, J. (2012) Thermochemical processing of meat and bone meal: A review. *Renewable and Sustainable Energy Reviews* 16(1), 942–957.

Cattell-Noll, C., Laura, L., Allison, M., Seufert, V., Galloway, J.N. *et al.* (2020) The nitrogen footprint of organic food in the United States. *Environmental Research Letter* 15.

Cherr, C.M., Scholberg, J.M.S. and McSorley, R. (2006) Green manure approaches to crop production: a synthesis. *Agronomy Journal* 98(2), 302–319.

Condron, L.M. and Tiessen, H. (2005) Interactions of organic P in terrestrial ecosystems. In: Turner, B.L., Frossard, E. and Baldwin, D.S. (eds) *Organic Phosphorus in the Environment*. CAB International, Wallingford, UK.

Cui, H., Liu, Q., Zhang, H., Zhang, Y., Wei, W. *et al.* (2023) Long-term manure fertilization increases rill erosion resistance by improving soil aggregation and polyvalent cations. *CATENA* 223, 106909.

Dasgupta, D., Kumar, K., Miglani, R., Mishra, R., Panda, A.K. *et al.* (2021) Microbial biofertilizers: Recent trends and future outlook. In: Mandal, S. and Passari, A.K. (eds) *Recent Advancement in Microbial Biotechnology*. Academic Press, Amsterdam, pp. 1–26.

DGS (Delaware Geological Survey (2020) Glauconite (greensand). Available at: https://www.dgs.udel.edu/delaware-geology/glauconite-greensand (accessed 1 February 2023).

Diacono, M. and Montemurro, F. (2011) Long-term effects of organic amendments on soil fertility. In: Lichtfouse, E., Hamelin, M., Navarrete, M. and Debaeke, P. (eds) *Sustainable Agriculture*, Vol. 2. Springer, Dordrecht.

Doane, T.A. (2017) The abiotic nitrogen cycle. *ACS Earth and Space Chemistry* 1(7), 411–421.

Dróżdż, D., Wystalska, K., Malińska, K., Grosser, A., Grobelak, A. *et al.* (2020) Management of poultry manure in Poland: current state and future perspectives. *Journal of Environmental Management* 264, 110327.

EgrinyaEneji, A., Irshad, M., Honna, T., Yamamoto, S., Endo, T. *et al.* (2003) Potassium, calcium, and magnesium mineralization in manure-treated soils. *Communications in Soil Science and Plant Analysis* 34(11–12), 1669–1679.

Fageria, N.K. (2008) Phosphorus. In: Fageria, N.K. (ed.) *The Use of Nutrients in Crop Plants*. CRC Press, Boca Raton, Florida, pp. 91–130.

Fageria, N.K., He, Z. and Baligar, V.C. (2017) Functions of phosphorus in crop plants. In: Fageria, N.K., He, Z. and Baligar, V.C. (eds) *Phosphorus Management in Crop Production*. CRC Press, Boca Raton, Florida, pp. 47–90.

Foth, H.D. and Ellis, B.G. (1997) *Potassium*. In: Foth, H.D. and Ellis, B.G. (eds) *Soil Fertility*, 2nd edn. CRC Press, Boca Raton, Florida, pp. 163–184.

Gangaiah, C., Ahmad, A., Nguyen, H.V., Wang, K.-H. and Radovich, T.J.K. (2017) Evaluating three invasive algal species as local organic sources of potassium for pak choi (*Brassica rapa*, *chinensis* group) growth. *HortScience* 52(3), 436–440.

Giri, B., Prasad, R., Wu, Q.-S. and Varma, A. (2019) *Biofertilizers for Sustainable Agriculture and Environment*. Springer Nature, Switzerland.

Glick, B.R. and Gamalero, E. (2021) Recent developments in the study of plant microbiomes. *Microorganisms* 9(7), 1533.

Grzyb, A., Wolna-Maruwka, A. and Niewiadomska, A. (2021) The significance of microbial transformation of nitrogen compounds in the light of integrated crop management. *Agronomy* 11(7). DOI: 10.3390/agronomy11071415.

Gurr, I. (2011) Evaluating vermicompost and rendered meat products as local media components in vegetable seedling production. MSc thesis, University of Hawaii at Manoa. Available at: https://scholarspace.manoa.hawaii.edu/server/api/core/bitstreams/9a07950e-ca75-40f4-8393-7d0aef5ede5b/content (accessed 4 February 2025).

Hartz, T.K. and Johnstone, P.R. (2006) Nitrogen availability from high-nitrogen-containing organic fertilizers. *HortTechnology* 16(1), 39–42.

Havlin, J.L., Tisdale, S.L., Nelson, W.L. and Beaton, J.D. (2017) Soil fertility and fertilizers: an introduction to nutrient management, 8th edn. Pearson India Education Services, Uttar Pradesh, India.

Hawkesford, M.J., Cakmak, I., Coskum, D., De-Kok, L.J., Lambers, H. *et al.* (2023) Functions of macronutrients. In: Rengel, Z. and Cakmak, I. (eds) *Marschner's Mineral Nutrtion of Plants*. Academic Press, London, UK, pp. 201–281.

Heckman, J.R. and Tedrow, J.C.F. (2004) Greensand as a soil amendment. *Better Crops* 88(2), 16–17. Available at: https://ucanr.edu/sites/nm/files/76652.pdf (accessed 4 February 2024).

Hinsinger, P. (2001) Bioavailability of soil inorganic P in the rhizosphere as affected by root-induced chemical changes: A review. *Plant and Soil* 237(2), 173–195.

Hue, N. (2009) Iron and phosphorus fertilizations and the development of proteoid roots in macadamia. *Plant and Soil* 318, 93–100. DOI: 10.1007/s11104-008-9820-0.

Hue, N. (2020) Soil health. In: Giri, B. and Varma, A. (eds) *Biochar for Maintaining Soil Health*. Springer Nature, Switzerland AG, pp. 21–46.

Jindo, K., Mizumoto, H., Sawada, Y., Sanchez-Monedero, M.A. and Sonoki, T. (2014) Physical and chemical characterization of biochars derived from different agricultural residues. *Biogeosciences* 11(23), 6613–6621.

Kirchmann, H., Kätterer, T. and Bergström, L. (2008) Nutrient supply in organic agriculture – plant availability, sources and recycling. In: Kirchmann, H. and Bergström, L. (eds) *Organic Crop Production – Ambitions and Limitations*. Springer, Dordrecht, The Netherlands, pp. 89–116.

Lazcano, C., Decock, C., Wong, C. and Garcia, B. (2023) Assessing the effects of compost on soil health. In: Horwath, W. (ed.) *Improving Soil Health*. Burleigh Dodds Science Publications, Cambridge, UK.

Lehmann, J. and Joseph, S. (2015) Biochar for environmental management: an introduction. In: *Biochar for Environmental Management*, 2nd edn. Taylor & Francis Group, pp. 1–12.

Lindström, K. and Mousavi, S.A. (2020) Effectiveness of nitrogen fixation in rhizobia. *Microbial Biotechnology* 13(5), 1314–1335.

Loss, A., Rosa Couto, R., Brunetto, G., Veiga, M., Toselli, M. *et al.* (2019) Animal manure as fertilizer: changes in soil attributes, productivity and food composition. *International Journal of Research – Granthaalayah* 7(9), 307–331.

Lynch, J.P. (2007) Roots of the second green revolution. *Australian Journal of Botany* 55(5), 493.

Martins, V., Gonçalves, A.S.F., Marchi, G., Guilherme, L.R.G. and Martins, É.D.S. (2015) Solubilização de potássio em misturas de verdete e calcário tratadas termoquimicamente. *Pesquisa Agropecuária Tropical* 45(1), 66–72.

Mondini, C., Cayuela, M.L., Sinicco, T., Sánchez-Monedero, M.A., Bertolone, E. *et al.* (2008) Soil application of meat and bone meal: Short-term effects on mineralization dynamics and soil biochemical and microbiological properties. *Soil Biology and Biochemistry* 40, 462–474.

Nesme, T., Colomb, B., Hinsinger, P. and Watson, C.A. (2014) *Soil phosphorous management in organic cropping systems*. In: Bellon, S. and Penvern, S. (eds). Springer Science, Dordrecht, The Netherlands, pp. 23–46.

Oduor, N.A., Munga, C.N., Ong'anda, H.O., Botwe, P.K. and Moosdorf, N. (2023) Nutrients and harmful algal blooms in Kenya's coastal and marine waters: a review. *Ocean & Coastal Management* 233, 106454.

Op den Camp, R., Streng, A., De Mita, S., Cao, Q., Polone, E. *et al.* (2011) Lysm-type mycorrhizal receptor recruited for rhizobium symbiosis in nonlegume parasponia. *Science* 331(6019), 909–912.

Radovich, T., Pant, A., Ahmad, A., Elevitch, C. and Hue, N. (2014) Enhancing soil function and plant health with locally available resources. In: Elevitch, C.R. (ed.) *Agroforestry Landscapes for Pacific Islands. Permanent Agriculture Resources*. Hawaii.

Randive, K., Raut, T. and Jawadand, S. (2021) An overview of the global fertilizer trends and India's position in 2020. *Mineral Economics* 34(3), 371–384.

Reganold, J.P. and Wachter, J.M. (2016) Organic agriculture in the twenty-first century. *Nature Plants* 2, 15221.

Richardson, A.E., Barea, J.-M., McNeill, A.M. and Prigent-Combaret, C. (2009) Acquisition of phosphorus and nitrogen in the *Rhizosphere* and plant growth promotion by microorganisms. *Plant and Soil* 321(1–2), 305–339.

Samarathunga, J., Wijesekara, I. and Jayasinghe, M. (2022) Seaweed proteins as a novel protein alternative: types, extractions, and functional food applications. *Food Reviews International* 1–26.

Sanchez, P.A. (2019) Properties and Management of Soils in the Tropics. 2nd edn. Cambridge University Press, Cambridge, UK, pp. 370–414.

Schulte-Uebbing, L.F., Beusen, A.H.W., Bouwman, A.F. and de Vries, W. (2022) From planetary to regional boundaries for agricultural nitrogen pollution. *Nature* 610(7932), 507–512.

Seefeldt, S. (2013) Animal manure as fertilizer. University of Alaska Fairbanks. Cooperative extension service. Available at: www.uaf.edu/ces (accessed February 2025).

Shapiro, C.A., Johnson, L., Schmidt, A.M. and Koelsch, R.K. (2021). Nebraska Extension, University of Nebraska, Lincoln, Nebraska. Available at: https://extensionpublications.unl.edu/assets/html/g1335/build/g1335.htm

Sharanappa. (2023) *Soil Fertility and Nutrient Management: Principles and Practices*. CRC Press/New India Publishing Agency, New Delhi.

Soni, R. and Yadav, S.K. (2019) Prospects of organic farming as financial sustainable strategy in modern agriculture. In: Panpatte, D.G. and Jhali, Y.K. (eds) *Soil Fertility Management for Sustainable Development*. Springer, Dordrecht, The Netherlands, pp. 251–265.

Squire, G.R., Young, M.W. and Hawes, C. (2022) Agroecological management and increased grain legume area needed to meet nitrogen reduction targets for greenhouse gas emissions. *Nitrogen* 3, 539–554.

Suliman, W., Harsh, J.B., Abu-Lail, N.I., Fortuna, A.M., Dallmeyer, I. *et al*. (2016) Influence of feedstock source and pyrolysis temperature on biochar bulk and surface properties. *Biomass Bioenergy* 84, 37–48.

Tarafdar, J.C. (2019) Fungal inoculants for native phosphorus mobilization. In: Giri, B., Prasad, R., Wu, Q.S. and Varma, A. (eds) *Biofertilizers for Sustainable Agriculture and Environment*. Springer Nature, Switzerland.

Teodoro, M., Trakal, L., Gallagher, B.N., Šimek, P., Soudek, P. *et al*. (2020) Application of co-composted biochar significantly improved plant-growth relevant physical/chemical properties of a metal contaminated soil. *Chemosphere* 242, 125255.

Thonar, C., Schnepf, A., Frossard, E., Roose, T. and Jansa, J. (2011) Traits related to differences in function among three arbuscular mycorrhizal fungi. *Plant and Soil* 339, 231–245.

Torma, S., Vilček, J., Lošák, T., Kužel, S. and Martensson, A. (2018) Residual plant nutrients in crop residues – an important resource. *Acta Agriculturae Scandinavica, Section B – Soil & Plant Science* 68(4), 358–366.

Valenzuela, H. (2023) Ecological management of the nitrogen cycle in organic farms. *Nitrogen* 4(1), 58–84.

van der Werf, H.M.G., Knudsen, M.T. and Cederberg, C. (2020) Towards better representation of organic agriculture in life cycle assessment. *Nature Sustainability* 3(6), 419–425.

Wong, C.K.F. and Teh, C.-Y. (2021) Impact of biofertilizers on horticultural crops. In: Inamuddin, A.M., Boddula, R. and Rezakazemi, M. (eds) *Biofertilizers: Study and Impact*. Scrivener Publication, Malaysia, pp. 39–104.

Wortman, S.E., Wortmann, C.S., Pine, A.L., Shapiro, C.A., Thompson, A.A. *et al*. (2017) Nutrient Management in Organic Farming. University of Nebraska.

Wuang, S.C., Khin, M.C., Chua, P.Q.D. and Luo, Y.D. (2016) Use of spirulina biomass produced from treatment of aquaculture wastewater as agricultural fertilizers. *Algal Research* 15, 59–64.

Zhang, J., Cook, J., Nearing, J.T., Zhang, J., Raudonis, R. *et al*. (2021) Harnessing the plant microbiome to promote the growth of agricultural crops. *Microbiological Research* 245, 126690.

21 The Role of Arbuscular Mycorrhizas in Organic Farming

Mary Cole[1]* and Rachel Cavallo[2]

[1]Director, Faculty of Veterinary & Agricultural Science, University of Melbourne, Australia; [2]Agpath Pty Ltd, Garfield, Victoria, Australia

Abstract

Mycorrhizal associations may be older than land plants. Organic or regenerative farming is more environmentally conservative than are conventional industrial methods of agriculture. Mycorrhizae have various strategies to benefit their plant hosts. In an organic or regenerative soil, the root exudates modify soil pH within the rhizosphere, attract beneficial microorganisms, and release toxic substances to manage pathogens. Mycorrhizal fungal relationships change the chemical composition of plant exudates. The activity of the fungal and bacterial communities benefits the mycorrhizosphere, along with the plant. This area around the plant root is highly diverse and chemically rich. Fungal:bacterial ratios are found to be a driving force in determining plant health.

Keywords: mycorrhizae, organic, regenerative, arbuscular fungi

Introduction

Mycorrhizal fungi have supported land plants by helping to modify the microbial activity around plant roots for the benefit of the microorganisms present and the plant. As the cost of inputs such as fertilizers, pesticides and fungicides increases and the environmental and human health issues become more apparent, AM fungi may provide a more sustainable and environmentally acceptable alternative in the present agriculture system worldwide.

Ploughing soils is unsustainable because soil aggregates are broken down. Compaction occurs, plant nutrients are lost, carbon is lost to the atmosphere, and the need for pesticides and herbicides increases as the system collapses biologically (Brown *et al.*, 2000; Hasan *et al.*, 2016; Meena *et al.*, 2020; Devi *et al.*, 2021).

The presence of strong rhizosphere microbial populations, including AM, directly affects plant growth and health through enhanced nutrient cycling and improved soil structure. Organic or regenerative farming is more environmentally conservative than conventional industrial methods of agriculture, as is shown in this chapter.

Role of Mycorrhizal Fungi in Agriculture

Mycorrhizal associations have various strategies such as ectomycorrhizae (ECM), endomycorrhizae/arbuscular (AM), ectendo, orchidaceous, ericoid mycorrhizae and non-mycorrhizal (NM) roots to benefit their host species. AM fungi are obligate mutualists with the widest host distribution and NM plants are found generally in extreme habitats not considered conducive for mycorrhizal associations. ECM fungi are widespread but are found to be associated with vascular plant species. Ectendomycorrhizae show characteristics of both ECM and AM fungi with a Hartig net and intracellular penetration of plant cells defined by the AM fungi. Orchid mycorrhizae appear to be *Basidiomycete* species. Ericoid mycorrhizal fungi

*Corresponding author: mary@agpath.com.au

DOI: 10.1079/9781800626850.0021

inhabit roots of ericaceous plants where the fungus grows on the surface of the root hairs forming a loose network of fine hyphal strands.

Mycorrhizal fungal hyphae are smaller than plant roots but can account for 300 km per 30 gm of soil (Mosaic AgriSight, 2023).

Fungi may have been the original roots for plants in the transition from the aquatic environment to the terrestrial environment (Taylor *et al.*, 1995; Taylor and Osborn, 1996; Öpik *et al.*, 2013; Edwards *et al.*, 2015; Delaux *et al.*, 2015; Berbee *et al.*, 2017; Brundrett and Tedersoo, 2018; Sportes *et al.*, 2021).

Work by Wang and Qiu (2006) and Wang *et al.* (2010) demonstrated that the association of arbuscular mycorrhizal (AM) fungal symbiosis with plants may be a trait that was established in more ancient liverworts (Bonfante and Selosse, 2010). Trees appeared on the terrestrial landscape in the Middle Devonian period (~395 million years ago) with fungal symbionts already established.

This burst of photosynthetic activity must have influenced the climate through carbon dioxide regulation and increased weathering of the land rocks into soil. Quirk *et al.* (2012) found in their experiments that increased fungal hyphal colonization across tree species translated into more efficient silica rock weathering showing that AM fungi were active participants in the biological accumulation of soil over eons (Taylor *et al.*, 1995; Hoorman, 2009; Quirk *et al.*, 2012; Edwards *et al.*, 2015).

Mycorrhizal fungi are different from fungi that cause disease (pathogens) or live on dead and dying plant material (saprophytes).

Over 80% of all land plants today have some type of mycorrhizal association including herbaceous legumes and Gramineae. Those that do not or rarely have mycorrhizae are common vegetable families such as the Brassicaceae, which include kale, cauliflower and broccoli, Chenopodiaceae, which include beetroot, quinoa and spinach, Cyperaceae, Junaceae and Caryophyllaceae. Plant species that are most mycorrhizal are those with thick, fleshy roots with limited root hairs such as cassava (*Manihot esculenta*, Crantz), citrus (*Citrus* species), grape (*Vitus vinifera*) and many legume species (Howeler *et al.*, 1987).

The region around the roots of plants is complex and interactive (Figs 21.1–21.3). The rhizosphere is a general term used to describe the area around the root that has high microbial populations influenced by root exudates. Non-mycorrhizal plants have rhizospheres. The mycorrhizosphere refers to the plant roots colonized by mycorrhizal fungi (Azcón-Aguilar and Barea, 1996; Muthukumar *et al.*, 2004; Linderman, 2008; Quirk *et al.*, 2012; Priyadharsini *et al.*, 2016). The hyposphere is the region around the plant roots and mycorrhizal fungal biomass where the fungi release exudates containing carbon into the surrounding soil. It is in this region that glomalin, a glycoprotein, has an influence on carbon sequestration by plants and greater transfer of carbon to roots, soil (Wright *et al.*, 2000) and aggregate water stability (Rillig *et al.*, 2001). These relationships are important to healthy soil and are destroyed or damaged in conventional/industrial synthetic agriculture due to soil disturbance. In an organic or regenerative soil, the root exudates modify soil pH within the rhizosphere, attract beneficial microorganisms and release toxic substances to manage pathogens. These exudates also attract the mycorrhizal fungi and plant growth-promoting rhizobacteria (PGRB).

Impact of Conventional Industrial Agriculture on Mycorrhizal Fungal Populations

Baon *et al.* (1992) found that cereal crops such as wheat, barley and rye responded differently to the presence of high or low populations of AM spores in agricultural soils. However, they did find that the colonization of cereal roots by indigenous AM fungi decreased with the addition of the synthetic phosphorus to the soil.

Since World War II and the increased use of synthetic fertilizers, the symbiotic relationship between plant roots and AM fungi has been broken so the natural flow of solubilized nutrients leading to increased plant growth, plant diversity, and pest and disease protection has been lost. Ploughing soils is unsustainable because soil aggregates are broken down. Compaction occurs, plant nutrients are lost, carbon is lost to the atmosphere, and the need for pesticides and herbicides increases as the system collapses biologically (Brown *et al.*, 2000; Hasan *et al.*, 2016; Meena *et al.*, 2020; Devi *et al.*, 2021).

In farming systems where nutrients are low and when nutrients are locked into organic matter, use of AM fungi can improve quality of crop, profitability and sustainability of agricultural systems (Devi *et al.*, 2021). Industrial farming systems are failing because of excess fixed nitrogen fertilizer and phosphorus use (Socolow, 1999; Hasan *et al.*, 2016).

Bacterial population density and diversity decreases with external inorganic nitrogen inputs resulting in a less-stable bacterial community structure over time. By contrast, more-stable bacterial and microbial communities are provided by organic treatments (Murphy *et al.*, 2015; Qin *et al.*, 2015; Kavamura *et al.*, 2018).

Unintended side effects of the use of nitrogen include greenhouse gas impacts, contaminated ground water and drinking water, acidified rain, eutrophied estuarine environments and stressed ecosystems (Socolow, 1999; Kavamura *et al.*, 2018). Pesticides and fungicides used in conventional industrial agriculture also affect microbial biodiversity on leaves of perennial and annual plants as well as affecting the ecto- and endomycorrhizae on roots in the soil. This results in a loss of diversity of beneficial microbes on the plant surfaces, exposing the plant to attack by pests and disease organisms (Dawoud *et al.*, 2017).

Intensive industrial agriculture requires extensive ploughing of the soil to provide a weed-free tilth in which to plant seeds or seedlings. This practice is accompanied by use of pre-emergence and post-emergence fungicides, pesticides and herbicides. The AM fungal biomass of mycelium is destroyed through these methods so that benefit derived from the AM symbiosis is lost.

Long-term conventional industrial agriculture reduces the diversity of AM species or reduces them altogether (Gosling *et al.*, 2006; Davis and Abbott, 2006; Zhong and Cai, 2007; Rafique *et al.*, 2019). Johnson (1993) showed that AM species change over time where fertilizer is added to species that may be inferior symbionts. This may account for observations that non-fertilized crops are sometimes less productive than crops grown with synthetic fertilizers because the less-fit AM fungi may exert a greater carbon cost on their host (Johnson, 1993; Gosling *et al.*, 2006).

Impact of Organic/Regenerative Agriculture on Mycorrhizal Fungal Populations

Organic or regenerative farming does not mean that farms do not use imputes. Soil does not become less fertile as suggested in a review (Conacher and Conacher, 1998). Australia is a world leader in organic/regenerative agriculture, based on certified organic hectares, accounting for 54% of the world's certified area (Paull and Hennig, 2018; Paull, 2019). Organic or regenerative systems around the world rely on a healthy, active soil microbial population. Within this paradigm, AM fungi have an important role and it is shown that they compensate for the reduction in synthetic phosphorus (Arias *et al.*, 1991).

Increase in organic matter is necessary for active soil biota as well as that of earthworms. These soil faunae are essential to driving nutrients into the soil allowing air and water to move down the earthworm channels. AM fungi and protozoa produce plant-available nitrogen and phosphorus and facilitate plant roots moving deeper into the soil profile giving the above-ground photosynthetic leaves more resilience in the face of stress.

Biological control is considered a directed, accurate management of the many components of the soil ecosystem that will protect plants against pathogens (Azcón-Aguilar and Barea, 1996). The relationship is mostly mutualistic where the beneficial fungi exchange sugars/carbohydrates from plants for moisture and nutrients such as phosphorus from the soil to the plant via fungal hyphae. Mycorrhizae, mostly AM and ECM, have a positive effect on plant growth leading to their use in sustainable agriculture. Leifheit *et al.* (2015) found AM fungi exerted a level of control over decomposers in litter decomposition under experimental conditions. This could have implications in nutrient cycling for woody plants and perennials that develop lignified roots in grasslands or pastures.

The fungal hyphal network is substantial in healthy, largely undisturbed, soil and the mycorrhizal association increases the radial volume of soil from which the plant can access nutrients.

The mycorrhizal fungi promote plant growth by inducing resistance against disease-causing pathogens, solubilize nutrients making them plant available, and regulate many of the nutritional balances found in healthy soil.

Mycorrhizal fungal relationships change the chemical composition of plant exudates. The activity of the fungal and bacterial communities benefits the mycorrhizosphere, along with the plant. This area around the plant root is highly diverse and chemically rich (Azcón and Ocampo, 1981; Alloush and Clark, 2007; Devi *et al.*, 2021; Sangwan and Prasanna, 2022). Fungal to bacterial ratios are found to be a driving force in determining plant health. Fungal to bacterial ratios of 2–5:1 for most horticultural crops and 5–10:1 for trees and long-lived perennials are aimed for in biological regenerative systems (Ingham *et al.*, 2000; Artursson *et al.*, 2006; Liu *et al.*, 2021. Spinach (*Spinacia oleracea* L.) grown in organic systems shows lower levels of nitrates and higher levels of flavonoids and ascorbic acid, making it better from a human health and gut biota perspective (Mitchell *et al.*, 2007; Koh *et al.*, 2012).

AM fungi are beneficial in low-fertility agricultural soils resulting in increased crop yields. In legume crops, AM optimize growth and effective nodule formation and nitrogen fixation (Howeler *et al.*, 1987). Response to AM fungal colonization varies with soil types but in all soils, presence of AM fungi improved the health and productivity of the crop plants (Atkinson *et al.*, 2002; Sanmartín *et al.*, 2014; Hu *et al.*, 2015; Kim *et al.*, 2017). Low-input production systems increase sporulation and diversity of the AM species, more so with inter-row plantings of cover crops aiming for plant cover (Douds *et al.*, 1993, 1995; Hamel, 1996; Ryan and Graham, 2002; Jeffries and Barea, 2012; Schneider *et al.*, 2015).

There is evidence of differences in the response of AM fungi in their sensitivity of input phosphorus but it is clear that the ability of AM fungi to colonize plant roots is negatively correlated with increased application of nitrogen and phosphorus fertilizer and pesticides.

Fertilizers commonly used in industrial agriculture, such as mono-ammonium phosphate (MAP), di-ammonium phosphate (DAP) and superphosphate, suppress indigenous AM fungal populations more than introduced inoculant AM species. Long-term application of phosphate can select a phosphate-tolerant endophyte population and species native to high-phosphate soils may show higher phosphate tolerance than those in lower-phosphate soils (Arias *et al.*, 1991; Azcón-Aguilar *et al.*, 1997; Augé, 2001; Azcón *et al.*, 2002).

Substantial mycorrhizal biomass in the root zones of a plant results in good soil aggregation with increased water-holding capacity. This allows soil to provide soil moisture to plants longer than those with no or little mycorrhizal population. Plants will withstand periods of drought for longer and are able to tolerate temperatures that would lead to physiological shutdown in a conventional farming system (Augé *et al.*, 2001; Rillig *et al.*, 2001; Shahzad *et al.*, 2015).

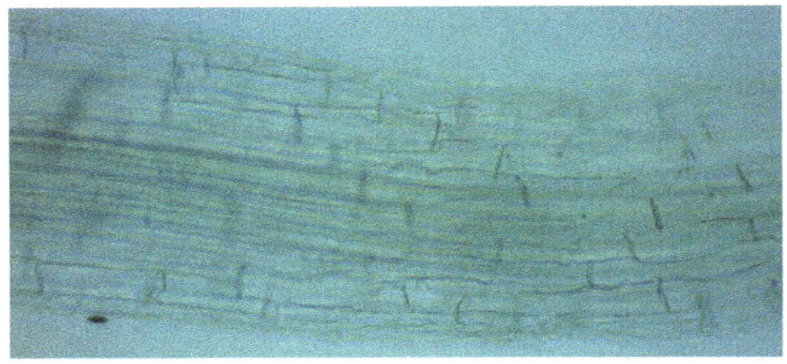

Fig. 21.1. Root of a non-mycorrhizal plant species. No mycorrhizal fungal hyphae or vesicles. X100 mag. (Photo: Agpath Pty Ltd).

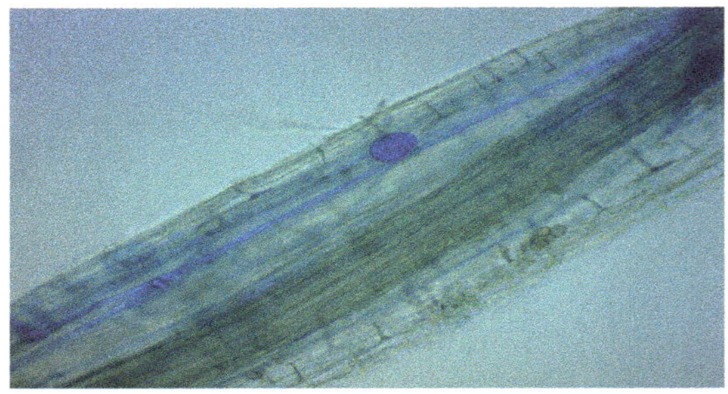

Fig. 21.2. Root of a mycorrhizal plant from a conventional industrial agricultural farming system. Few mycorrhizal vesicles and hyphae. X200 mag. (Photo: Agpath Pty Ltd).

AM plants display higher rates of photosynthesis than non-mycorrhizal plants (Augé *et al.*, 2001) and can hold their photosynthetic leaves longer than a non-mycorrhizal plant. Nodules on legume roots have higher water content than non-mycorrhizal roots allowing them to survive longer into drought conditions (Goicoechea *et al.*, 1997; Augé *et al.*, 2001; Al-Karaki and Williams, 2021).

All mycorrhizal types enhance the growth characteristics of their host plants by facilitating an increase in the uptake of nutrients – phosphorus in particular – by the use of solubilizing enzymes. There are excellent reviews that show the importance of mycorrhizal fungi to the health of agricultural plants (Smith *et al.*, 1980; Bolan, 1991; Brundrett, 2009). Stringolactones are a new class of bioactive molecules, phytohormones, released as root exudates that appear to stimulate AM fungal metabolism and branching to increase the opportunity for the AM fungus to find its host and increase

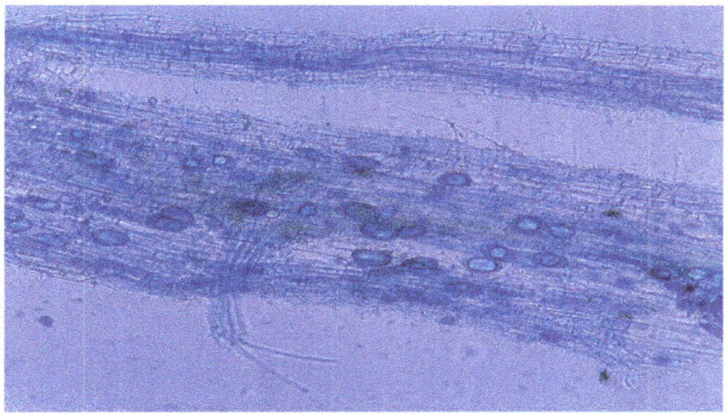

Fig. 21.3. Root of a mycorrhizal plant from an organic, regenerative farming system. X100 mag. (Photo: Agpath Pty Ltd).

nutrient transfer between the two (Bonfante and Genre, 2010; Yoneyama *et al.*, 2012; Carvalhais *et al.*, 2019; Wang *et al.*, 2021).

Organic regenerative farming has a basic tenet, proposed in 1943 by Sir Albert Howard (Berry, 1964), that is as true today and equally important:

> The main characteristic of Nature's farming can.... be summed up in a few words. Mother Earth never attempts to farm without livestock; she always raises mixed crops; great pains are taken to preserve the soil and to prevent erosion; the mixed vegetable and animal wastes are converted into humus; there is no waste; the processes of growth and the processes of decay balance one another; ample provision is made to maintain large reserves of fertility; the greatest care is taken to store the rainfall; both plants and animals are left to protect themselves against disease.

Nitrogen and phosphorus availability is enhanced with the presence of mycorrhizal fungi along with nitrogen-fixing bacteria and phosphate-solubilizing bacteria. Cruz *et al.* (2007) found evidence that enzyme action of arginine of the urea cycle was a major driver in nitrogen translocation in AM mycelium.

Plants have two major strategies for disease and pest protection: the first results from direct attack stimulating a systemic acquired resistance (SAR) throughout the infected plant; the second is activated by nutritional or biological elicitors resulting in an array of defence compounds providing induced systemic resistance (ISR).

Mycorrhizal fungi and bacteria entering into the cellular compartments of plant cells indirectly stimulate pathways that increase the levels of these resistance metabolites throughout the host plant. Plants with good mycorrhizopheres are more able to withstand attack by fungal pathogens and pests. A complex set of reactions develops to protect the plant against attack. The jasmonic acid and methyl jasmonate and ethylene pathways are integral in warning neighbouring plants of impending attack. The mycorhizosphere effects can modify microbial populations, adjust competition for colonization sites on roots and leaves, and activate defence mechanisms such as SAR and ISR. Plants growing in soils with actively diverse microbial ecosystems in partnership with AM fungi have less damage caused by soil-borne pathogens and less leaf damage caused by pests and foliar pathogens.

AM fungi provide the highway linkage between plant roots, providing an early-warning system for insect attack. Insect attack causes systemic changes in production of plant volatiles – for example methyl salicylate – making bean plants repellent to aphids but attractive to parasitoids that attack the aphids. This underground messaging mycorrhizal network prewarns neighbouring plants of an impending attack allowing them time to invoke herbivore defenses (Barea and Jeffries, 1995; Azcón *et al.*, 2002; Jeffries and Barea, 2012; Babikova *et al.*, 2013; Johnson and Gilbert, 2015; Ouledali *et al.*, 2019).

As cost of inputs such as fertilizers, pesticides and fungicides increases and the environmental and human health issues become more apparent, AM fungi may provide a more sustainable and environmentally acceptable alternative in the present agriculture system worldwide. AM fungi are shown to be effective biocontrol agents against phytopathogenic fungi (Azcón *et al.*, 2002; Harrier and Watson, 2004; Jeffries and Barea, 2012), phytopathogenic bacteria (Filion *et al.*, 1999) and plant-parasitic nematodes (Jones *et al.*, 2013). These effects have many sources such as competition for nutrients and space, altered root morphology, altered rhizosphere interactions, host response phytoalexins and phytoanticipins, and antibiosis (Jeffries and Barea, 2012; Jung *et al.*, 2012; Jain and Pundir, 2019).

Root exudates are rich in carbohydrates and proteins and provide nutrition for the rhizosphere biota. Plants direct around two-thirds of the photosynthetic sugars to root-inhabiting mycorrhizal fungi and into the rhizosphere to stimulate microbial interaction. AM fungi derive most, and sometimes all, of their carbon from the host plant and in return assist in increasing plant biomass and photosynthetic rate through the symbiotic relationship (Bago *et al.*, 2000; Ortas and Ustuner, 2014; Vives-Peris *et al.*, 2020).

Biodiversity loss is often seen with changes in the functional composition of the vegetation in an area. Intensive industrial pasture management favours fast-growing grass species over slower-growing conservative species. This change in land use with increased biomass production requires faster nutrient-cycling rates that can alter some of the natural ecosystem functions. Plant communities are altered and the loss in biodiversity leads to less-productive pasture and grassland.

AM fungi are seen as the major component of most agroecosystems where they absorb mineral nutrients and feed plants in exchange for carbohydrates. Re-establishment of a stable, diverse and viable soil microbial population may take up to 15 years in a cropping system but much less in a perennial system where there is little or no soil disturbance. Carbon:nitrogen ratio is higher in non-tilled soil than in tilled soil. It is found that organic carbon is higher in non-tilled than in tilled soils, in part due to the presence of the AM fungal network in the plant root profile.

Mycorrhizal fungal biomass with its glomalin exoprotein component sequesters carbon better than does humus.

Ectomycorrhizae allocate proportionally more carbon to roots in low-fertility soils. However, as high-fertility soils result in more photosynthetic above-ground biomass, the total amount of carbon allocated below ground may sometimes be higher than in a less-fertile soil (Agricultural Research Magazine, 2002; Ekblad *et al.*, 2013).

A conventional high-input agricultural system tends to be a simple, linear management model driven mostly by external inputs that must be purchased in increasing amounts and with rising prices. Nitrogeneous fertilizers used in conventional industrial farming systems, particularly in regions of high rainfall, show up to 85% loss when applied at planting. This leads to increased emissions into the atmosphere with a significant cost to farmers and the environment (Muhammad *et al.*, 2022).

Organic and regenerative systems tend to use what is available on the farm, such as compost, with inputs that are beneficial to soil biota as well as plants. Plant-available nitrogen can be a limiting element during the growth of a plant. In an organic-regenerative soil, protozoa feed on bacteria to produce plant-available nitrogen (Clarholm, 1981; Hoorman, 2009; Rønn *et al.*, 2012). Populations of N-transforming bacteria vary during plant growth and increase in response to soluble organic exudates from the plant. The presence of a good population of AM enhances the bacterial populations across the whole plant-growth period. AM fungi may be responsible for up to 50% of nutrient uptake in crops, e.g. zinc. Protozoa account for 8–10% of nitrogen mineralization annually in an organic-regenerative farming enterprise (Mosaic AgriSight, 2023).

In an organic-regenerative farming system, which would be fungal-dominated, presence of AM fungi leads to an increase in ammonium-oxidizing bacterial species and a decrease in numbers of ammonifying and denitrifying bacteria. Although nitrification in the soil is largely bacterially driven, there is evidence that a healthy mycorrhizal population may change the number and activity of particular N-transforming species because changes in the soil environment will affect all the microbial groups. This relationship between AM and bacteria is thought to affect plant responses such as resistance to disease and to stress, including increased salinity stress (Artursson *et al.*, 2006; Mosaic AgriSight, 2023).

Agricultural soil is a complex of physical, chemical and biological factors that influence soil quality. The presence of strong rhizosphere microbial populations, including AM, directly affects plant growth and health through enhanced nutrient cycling and improved soil structure. The most obvious impact of microbial nutrition is only evident when there are limits on the availability of nitrogen and phosphorus, so it is obvious that synthetic fertilizer application reduces, and may even eliminate, the benefits of mycorrhizal fungi in the soil (Azcón-Aguilar *et al.*, 1997; Azcón *et al.*, 2002). Azcón *et al.* (2002) demonstrated that high nitrogen and phosphorus applications to agricultural soils had a negative effect on the acquisition of nitrogen, phosphorus, potassium, iron, manganese and zinc for mycorrhizal plants. These fertilizer levels reduced and inhibited AM-intra- and extra-radical hyphae and colonization. It is the extra-radical hyphal mass that spreads into a wider soil volume connecting plant roots allowing those plants to access nutrient- and water-deficient areas. Reduced accumulation of minerals in AM plants has been observed in high nutrient environments (Azcón *et al.*, 2002) so increased application of synthetic chemistry reduces AM activity on and in roots.

The ability of AM fungi to remediate contaminated soil is important in agriculture where farming systems are transitioning from industrial synthetic chemical input to more sustainable systems such as organic, regenerative agriculture or biodynamics.

Ambrosini *et al.* (2015) demonstrated the ability of AM fungal species to contribute to the establishment of young grapevines exposed to high Cu levels in an old vineyard. One of the strategies AM fungi have is to release glomalin into the surrounding soil organic matter.

Averill *et al.* (2014) demonstrated that ectomycorrhizal and ericoid mycorrhizal fungi produce nitrogen-degrading enzymes, giving them greater access to sources of organic nitrogen than AM fungi. Theoretically, this would result in greater soil carbon storage in ecosystems dominated by ectomycorrhizal fungi than by AM fungi. In an agricultural system, this would suggest that an undisturbed/underdisturbed system of perennial plant species carrying ectomycorrhizal fungi can contribute significantly to carbon capture on a global level.

AM fungi associated with important food crops can alleviate public health deficiencies from elements in areas of the world (Miransari and Smith, 2007; Pagano *et al.*, 2020; Chen *et al.*, 2020). Soil organic matter is important in enhancing crop production and improving soil structural stability. Intensive industrial agriculture causes rapid oxidation of soil organic matter with a deterioration in aggregate stability and overall physical properties. As soil structure deteriorates with constant cultivation, compaction increases, seedling emergence and root elongation are affected, and soil carbon levels decrease.

Soil health in regenerative organic systems is improved with 100% plant cover and minimal disturbance together with multi-species cropping.

It is soil porosity and good aggregate size that allow airflow into the soil with increased fungal and aerobic protozoan activity. Work in Western Australia showed in a GRDC (Grains Research and Development Corporation) project (Murphy *et al.*, 2015), that levels of soil carbon were a balance between inputs such as plant residues, manures and composts, and losses due to microbial decomposition and erosion, and that soil carbon changes on a decadal timeframe.

Nitrogen-fixing bacteria, plant growth-promoting rhizobacteria (PGPR) and mycorrhizal helper bacteria (MHB) all assist in stimulating development of mycorrhizal symbiosis in good soil in organic-regenerative systems. It is this mycorrhizosphere complex that is damaged or destroyed in repeatedly cultivated agricultural soil by a change in the fungal:bacterial ratio from fungal-dominated to bacteria-dominated soil.

Adaption Practices Using Mycorrhizal Fungi to Mitigate Climate Change

Soil holds more carbon than do the atmosphere and plants combined (Schmidt *et al.*, 2011; Averill *et al.*, 2014).

Ecosystems that have ectomycorrhizal symbionts, such as old growth forests and stable perennial plant communities, contain 70% more carbon per unit of nitrogen than soils, particularly agricultural soils, dominated by plants with AM associations (Averill *et al.*, 2014; Lone *et al.*, 2017). Soil erosion, increased use of synthetic fertilizers, increased unpredictability of rainfall, availability of clean water and clean air, are just some of the difficulties facing the world's populations as healthy soil becomes more damaged. Deforestation, overgrazing, ploughing and inappropriate farming practices are well known and widely documented as being destructive of soil quality and loss of soil AM fungal populations.

Global food security requires global farming methods to consider soil health to preserve what agricultural soil is available, to use it more sensitively and efficiently so that farmers can see a future (Borrelli *et al.*, 2017). One of the latest documents from the United Nations defined global soil resources as fair, poor or very poor, and one-third of arable land has been lost due to pollution and erosion in the past 40 years (Amundson *et al.*, 2015; Koch *et al.*, 2015; Milman, 2015). Damaged soil does not have the soil biota (AM fungi biomass) that can hold together the fertility that will be necessary for future generations. It takes around 500 years to create 2.5 cm of topsoil in the current ecological changes (Milman, 2015). Lack of available water contributes more than 70% to crop yield losses in many agriculture regions of the world (Ruiz-Lozano *et al.*, 2012).

AM fungi in the plant biomass hold water and could help to reduce unreliable rainfall effects on cropping and animal enterprises, thus alleviating some of the potential drought stress in the future. Plants also produce phytohormones in response to drought stress. AM fungi are known to influence stomate functioning by regulation of abscisic acid (Diagne *et al.*, 2020).

Hashem *et al.* (2019) and Al-Karaki *et al.* (2004) found inoculation of chickpeas and wheat, respectively, with AM fungi, together with biochar under drought conditions responded positively with increased shoot and root length. Keith *et al.* (2015) found in their studies on biochar additions to different soil types that the rhizosphere priming effect was influenced by soil type but ultimately contributed to long-term carbon storage in soils (Mu *et al.*, 2020).

Adoption of organic-regenerative farming practices where soil health becomes the driver (Wheeler, 2008) rather than just economic results will require education, stewardship and the creation of socio-economic and institutional conditions that appreciate and favour sustainable soil management by helping to remove obstacles in some regions (FAO, 2015; Pennock, 2019). The report card on sustainable nature resource use in agriculture shines a clear light on the status of agricultural areas in south-west Western Australia (Department of Agriculture and Food, 2013). This report shows that soil acidity and water repellence are increasing, and higher inputs of synthetic chemistry are needed in the future to optimize production under the current industrial agriculture paradigm. Organic-regenerative agriculture practices need to be implemented to sustain farmers on the land and improve productivity with an economic return to growers.

AM fungi can be the next-generation biofertilizer with their ability to enable plants to use mineral elements such as nitrogen and phosphorus more effectively. They can be used as a biofertilizer and can open new strategies to support organic-regenerative and industrial agriculture and increase global food safety and security while, at the same time, decreasing synthetic chemical inputs (Kuila and Ghosh, 2022; Sharma *et al.*, 2022). Arneth *et al.* (2021) describe succinctly the relationship between land degradation and climate change and the need for stewardship of land resources to help communities adapt their agricultural systems to climate change.

In an organic-regenerative farming system, root biomass with its extensive attached mycorrhizal biomass generated by a consistent 100% ground cover will hold more moisture in the plant root system, reduce the general soil temperature, hold nutrients in the radial region of the plant roots, assist plants to access more mineralized plant-available nutrients and increase the carbon capture capacity of the soil. All these aspects together will have a positive impact on climate change mitigation while still feeding the world's increasing population.

Conclusions

Mycorrhizal associations may be older than land plants. Mycorrhizal fungi infest more than 80% of all land plants but are different from fungi that cause disease (pathogens) or live on dead and dying plant material (saprophytes).

Mycorrhizae have various strategies to benefit their plant hosts and change the chemical composition of plant exudates. In an organic or regenerative soil, the root exudates modify soil pH within the rhizosphere, attract beneficial microorganisms, and release toxic substances to manage pathogens. The activity of the fungal and bacterial communities benefits the mycorrhizosphere, along with the plant. This area around the plant root is highly diverse and chemically rich. Fungal:bacterial ratios are found to be a driving force in determining plant health.

Mycorrhizal fungal biomass with its glomalin exoprotein component sequesters carbon better than does humus. Ecosystems that have ectomycorrhizal symbionts such as old growth forests and stable perennial plant communities, contain 70% more carbon per unit nitrogen than soils, particularly agricultural soils, dominated by plants with AM associations (Averill *et al.*, 2014; Lone *et al.*, 2017). Soil erosion, increased use of synthetic fertilizers, increased unpredictability of rainfall, availability of clean water and clean air, are just some of the difficulties facing the world's populations as healthy soil becomes more damaged. Organic or regenerative farming is more environmentally conservative than conventional industrial methods of agriculture.

Since World War II and the increased use of synthetic fertilizers, the symbiotic relationship between plant roots and AM fungi has been broken so the natural flow of solubilized nutrients leading to increased plant growth, plant diversity, and pest and disease protection has been lost. Ploughing soils is unsustainable because soil aggregates are broken down. Compaction occurs, plant nutrients are lost, carbon is lost to the atmosphere, and the need for pesticides and herbicides increases as the system collapses biologically (Brown *et al.*, 2000; Hasan *et al.*, 2016; Meena *et al.*, 2020; Devi *et al.*, 2021).

In farming systems where nutrients are low and when nutrients are locked into organic matter, use of AM fungi can improve quality of crop, profitability and sustainability of agricultural systems (Devi *et al.*, 2021).

References

Agricultural Research Magazine (2002) Glomalin: Hiding place for a third of the world's stored soil carbon. September.

Al-Karaki, G.N. and Williams, M. (2021) Mycorrhizal mixtures affect the growth, nutrition, and physiological responses of soybean to water deficit. *Acta Physiologiae Plantarum* 43(5), 5.

Al-Karaki, G.N., McMichael, B. and Zak, J. (2004) Field response of wheat to arbuscular mycorrhizal fungi and drought stress. *Mycorrhiza* 14(4), 263–269.

Alloush, G.A. and Clark, R.B. (2007) Maize response to phosphate rock and arbuscular mycorrhizal fungi in acidic soil. *Communications in Soil Science and Plant Analysis* 32(1–2), 231–254.

Ambrosini, V.G., Voges, J.G., Canton, L., Couto, R.D.R., Ferreira, P.A.A. *et al.* (2015) Effect of arbuscular mycorrhizal fungi on young vines in copper-contaminated soil. *Brazilian Journal of Microbiology* 46(4), 1045–1052.

Amundson, R., Berhe, A.A., Hopmans, J.W., Olson, C., Sztein, A.E. *et al.* (2015) Soil science, soil and human security in the 21st century. *Science* 348(6235), 1261071.

Arias, I., Koomen, I., Dodd, J.C., White, R.P. and Hayman, D.S. (1991) Growth responses of mycorrhizal and non-mycorrhizal tropical forage species to different levels of soil phosphate. *Plant and Soil* 132(2), 253–260.

Arneth, A., Olsson, L., Cowie, A., Erb, K.-H., Hurlbert, M. *et al.* (2021) Restoring degraded lands. *Annual Review of Environment and Resources* 46(1), 569–599.

Artursson, V., Finlay, R.D. and Jansson, J.K. (2006) Interactions between arbuscular mycorrhizal fungi and bacteria and their potential for stimulating plant growth. *Environmental Microbiology* 8(1), 1–10.

Atkinson, D., Baddeley, J.A., Goicoechea, N., Green, J., Sanchez-Diaz, M. *et al.* (2002) Arbuscular mycorrhizal fungi in low input agriculture. In: Gianinazzi, S., Schüepp, H., Barea, J.M. and Haselwandter, K. (eds) *Mycorrhizal Technology in Agriculture: From Genes to Bioproducts*. Springer, pp. 211–222.

Augé, R.M. (2001) Water relations, drought and vesicular-arbuscular mycorrhizal symbiosis. *Mycorrhiza* 11, 3–42.

Augé, R.M., Stodola, A.J.W., Tims, J.E. and Saxton, A.M. (2001) Moisture retention properties of a mycorrhizal soil. *Plant and Soil* 230, 87–97.

Averill, C., Turner, B.L. and Finzi, A.C. (2014) Mycorrhiza-mediated competition between plants and decomposers drives soil carbon storage. *Nature* 505(7484), 543–545.

Azcón, R. and Ocampo, J.A. (1981) Factors affecting the vesicular-arbuscular infection and mycorrhizal dependency of thirteen wheat cultivars. *New Phytologist* 87(4), 677–685.

Azcón, R., Ambrosano, E. and Charest, C. (2002) Nutrient acquisition in mycorrhizal lettuce plants under different phosphorus and nitrogen concentration. *Plant Science* 165(5), 1137–1145.

Azcón-Aguilar, C. and Barea, J.M. (1996) Arbuscular mycorrhizas and biological control of soil-borne plant pathogens: An overview of the mechanisms involved. *Mycorrhiza* 6, 457–464.

Azcón-Aguilar, C., Cantos, M., Troncoso, A. and Barea, J.M. (1997) Beneficial effect of arbuscular mycorrhizas on acclimatization of micropropagated cassava plantlets. *Scientia Horticulturae* 72(1), 63–71.

Babikova, Z., Gilbert, L., Bruce, T.J.A., Birkett, M., Caulfield, J.C. *et al.* (2013) Underground signals carried through common mycelial networks warn neighbouring plants of aphid attack. *Ecology Letters* 16(7), 835–843.

Bago, B., Pfeffer, P.E. and Shachar-Hill, Y. (2000) Carbon metabolism and transport in arbuscular mycorrhizas. *Plant Physiology* 124(3), 949–958.

Baon, J.B., Smith, S.E., Alston, A.M. and Wheeler, R.D. (1992) Phosphorus efficiency of three cereals as related to indigenous mycorrhizal infection. *Australian Journal of Agricultural Research* 43(3), 479.

Barea, J.M. and Jeffries, P. (1995) Arbuscular mycorrhizas in sustainable soil-plant systems. In: Varma, A. and Hock, B. (eds) *Mycorrhiza: Structure, Function, Molecular Biology and Biotechnology*. Springer, pp. 521–560.

Berbee, M.L., James, T.Y. and Strullu-Derrien, C. (2017) Early diverging fungi: Diversity and impact at the dawn of terrestrial life. *Annual Review of Microbiology* 71, 41–60.

Berry, W. (1964) Introduction. In: Howard, A. (ed.) *The Soil and Health: A Study of Organic Agriculture*.

Bolan, N.S. (1991) A critical review on the role of mycorrhizal fungi in the uptake of phosphorus by plants. *Plant and Soil* 134(2), 189–207.

Bonfante, P. and Genre, A. (2010) Mechanisms underlying beneficial plant-fungus interactions in mycorrhizal symbiosis. *Nature Communications* 1, 48.

Bonfante, P. and Selosse, M.A. (2010) A glimpse into the past of land plants and of their mycorrhizal affairs: From fossils to evo-devo. *The New Phytologist* 186(2), 267–270.

Borrelli, P., Robinson, D.A., Fleischer, L.R., Lugato, E., Ballabio, C. *et al.* (2017) An assessment of the global impact of 21st century land use change on soil erosion. *Nature Communications* 8(1).

Brown, G.G., Barois, I. and Lavelle, P. (2000) Regulation of soil organic matter dynamics and microbial activity in the drilosphere and the role of interactions with other edaphic functional domains. *European Journal of Soil Biology* 36(3–4), 177–198.

Brundrett, M.C. (2009) Mycorrhizal associations and other means of nutrition of vascular plants: Understanding the global diversity of host plants by resolving conflicting information and developing reliable means of diagnosis. *Plant and Soil* 320(1–2), 37–77.

Brundrett, M.C. and Tedersoo, L. (2018) Evolutionary history of *mycorrhizal symbioses* and global host plant diversity. *New Phytologist* 220(4), 1108–1115.

Carvalhais, L.C., Rincon-Florez, V.A., Brewe, P.B., Beveridge, C.A., Dennis, P.G. *et al*. (2019) The ability of plants to produce strigolactones affects rhizosphere community composition of fungi but not bacteria. *Rhizosphere* 9, 18–26.

Chen, X., Zhang, Z., Gu, M., Li, H., Shohag, M.J.I. *et al*. (2020) Combined use of arbuscular mycorrhizal fungus and selenium fertilizer shapes microbial community structure and enhances organic selenium accumulation in rice grain. *Science of the Total Environment* 748, 141166.

Clarholm, M. (1981) Protozoan grazing of bacteria in soil: Impact and importance. *Microbial Ecology* 7(4), 343–350.

Conacher, J. and Conacher, A. (1998) Organic farming and the environment, with particular reference to Australia: A review. *Biological Agriculture & Horticulture* 16(2), 145–171.

Cruz, C., Egsgaard, H., Trujillo, C., Ambus, P., Requena, N. *et al*. (2007) Enzymatic evidence for the key role of arginine in nitrogen translocation by arbuscular mycorrhizal fungi. *Plant Physiology* 144(2), 782–792.

Davis, J. and Abbott, L. (2006) Soil fertility in organic farming systems. In: Kristiansen, P., Taji, A. and Reganold, J. (eds) *Organic Agriculture: A Global Perspective*. CSIRO Publishing, Melbourne, Australia, pp. 25–51.

Dawoud, M., Bundschuh, M., Goedkoop, W. and McKie, B.G. (2017) Interactive effects of an insecticide and a fungicide on different organism groups and ecosystem functioning in a stream detrital food web. *Aquatic Toxicology* 186, 215–221.

Delaux, P.-M., Radhakrishnan, G.V., Jayaraman, D., Cheema, J., Malbreil, M. *et al*. (2015) Algal ancestor of land plants was preadapted for symbiosis. *Proceedings of the National Academy of Sciences of the United States of America* 112(43), 13390–13395.

Department of Agriculture and Food (2013) Report card on sustainable natural resource use in australia.

Devi, H., Soibam, I.B., Soibam Sinyorita, S.K.C. and Devi, E.L. (2021) Mycorrhizal fungi and sustainable agriculture. In: Ohyama, T. and Inubushi, K. (eds) *Nitrogen in Agriculture: Physiological, Agricultural and Ecological Aspects*. InTech Open.

Diagne, N., Ngom, M., Fall, D., Hocher, V. and Svistoonoff, S. (2020) Roles of arbuscular mycorrhizal fungi on plant growth and performance: Importance in biotic and abiotic stressed regulation. *Diversity* 12(10), 370.

Douds, D.D., Jr, Janke, R.R. and Peters, S.E. (1993) VAM fungus spore populations and colonization of roots of maize and soybean under conventional and low-input sustainable agriculture. *Agriculture, Ecosystems & Environment* 43(3–4), 325–335.

Douds, D.D., Jr, Galvez, L., Janke, R.R. and Wagoner, P. (1995) Effect of tillage and farming system upon populations and distribution of vesicular-arbuscular mycorrhizal fungi. *Agriculture, Ecosystems & Environment* 52(2–3), 111–118.

Edwards, D., Cherns, L., Raven, J.A. and Smith, A. (2015) Could land-based early photosynthesizing ecosystems have bioengineered the planet in mid-palaeozoic times. *Palaeontology* 58(5), 803–837.

Ekblad, A., Wallander, H., Godbold, D.L., Cruz, C., Johnson, D. *et al*. (2013) The production and turnover of extrametrical mycelium of ectomycorrhizal fungi in forest soils: Role in carbon cycling. *Plant and Soil* 366(1–2), 1–27.

FAO (Food and Agriculture Organization of the United Nations) (2015) Revised World Soil Charter.

Filion, M., Fortin, J.A. and St-Arnaud, M. (1999) Direct interaction between the arbuscular mycorrhizal fungus Glomus intraradices and different rhizosphere microorganisms. *The New Phytologist* 141(3), 525–533.

Goicoechea, N., Antolín, M.C. and Sánchez-Díaz, M. (1997) Influence of arbuscular mycorrhizae and rhizobium on nutrient content and water relations in drought stressed alfalfa. *Plant and Soil* 192(2), 261–268.

Gosling, P., Hodge, A., Goodlass, G. and Bending, G.D. (2006) Arbuscular mycorrhizal fungi and organic farming. *Agriculture, Ecosystems & Environment* 113(1–4), 17–35.

Hamel, C. (1996) Prospects and problems pertaining to the management of arbuscular mycorrhizae in agriculture. *Agriculture, Ecosystems & Environment* 60(2–3), 197–210.

Harrier, L.A. and Watson, C.A. (2004) The potential role of arbuscular mycorrhizal (AM) fungi in the bioprotection of plants against soil-borne pathogens in organic and/or other sustainable farming systems. *Pest Management Science* 60(2), 149–157.

Hasan, M.M., Hasan, M.M., Silva, J.A. and Li, X. (2016) Regulation of phosphorus uptake and utilization: Transitioning from current knowledge to practical strategies. *Cellular & Molecular Biology Letters* 21, 1–19.

Hashem, A., Kumar, A., Al-Arjani, A.-B.F. and Singh, G. (2019) Arbuscular mycorrhizal fungi and biochar improves drought tolerance in chickpea. *Saudi Journal of Biological Sciences* 26(3), 614–624.

Hoorman, J.J. (2009) *Role of Soil Fungus: Update and Revision*. Ohio State University Extension.
Howeler, R.H., Sieverding, E. and Saif, S. (1987) Practical aspects of mycorrhizal technology in some tropical crops and pastures. *Plant and Soil* 100(1–3), 249–283.
Hu, J., Yang, A., Wang, J., Zhu, A., Dai, J. *et al*. (2015) Arbuscular mycorrhizal fungal species composition, propagule density, and soil alkaline phosphatase activity in response to continuous and alternate no-tillage in Northern China. *Catena* 133, 215–220.
Ingham, E.R., Moldenke, A.R. and Edwards, C.A. (2000) Soil biology primer. Soil and water conservation society and USDA national resources conservation service.
Jain, P. and Pundir, R.K. (2019) Biocontrol of soil phytopathogens by arbuscular mycorrhiza – a review. In: Choudhary, D.K. and Varma, A. (eds) *Mycorrhizosphere and Pedogenesis*. Springer, pp. 221–237.
Jeffries, P. and Barea, J.M. (2012) Arbuscular mycorrhiza: A key component of sustainable plant–soil ecosystems. In: Hsueh, Y.-P. and Blackwell, M. (eds) *Fungal Associations*. Springer.
Johnson, N.C. (1993) Can fertilization of soil select less mutualistic mycorrhizae? *Ecological Applications* 3(4), 749–757.
Johnson, D. and Gilbert, L. (2015) Interplant signaling through hyphal networks. *New Phytologist* 205(4), 1448–1453.
Jones, J.T., Haegeman, A., Danchin, E.G.J., Gaur, H.S., Helder, J. *et al*. (2013) Top 10 plant-parasitic nematodes in molecular plant pathology. *Molecular Plant Pathology* 14(9), 946–961.
Jung, S.C., Martinez-Medina, A., Lopez-Raez, J.A. and Pozo, M.J. (2012) Mycorrhiza-induced resistance and priming of plant defenses. *Journal of Chemical Ecology* 38, 651–664.
Kavamura, V.N., Hayat, R., Clark, I.M., Rossmann, M., Mendes, R. *et al*. (2018) Inorganic nitrogen application affects both taxonomical and predicted functional structure of wheat rhizosphere bacterial communities. *Front Microbiology* 9, 1074.
Keith, A., Singh, B. and Dijkstra, F.A. (2015) Biochar reduces the rhizosphere priming effect on soil organic carbon. *Soil Biology and Biochemistry* 88, 372–379.
Kim, S.J., Eo, J.K., Lee, E.H., Park, H. and Eom, A.H. (2017) Effects of arbuscular mycorrhizal fungi and soil conditions on crop plant growth. *Mycobiology* 45(1), 20–24.
Koch, A., Chappell, A., Eyres, M. and Scott, E. (2015) Monitor soil degradation or triage for soil security? An Australian challenge. *Sustainability* 7(5), 4870–4892.
Koh, E., Charoenprasert, S. and Mitchell, A.E. (2012) Effect of organic and conventional cropping systems on ascorbic acid, vitamin C, flavonoids, nitrate, and oxalate in 27 varieties of spinach (*Spinacia oleracea* L.). *Journal of Agricultural and Food Chemistry* 60(12), 3144–3150.
Kuila, D. and Ghosh, S. (2022) Aspects, problems and utilization of arbuscular mycorrhizal (AM) application as bio-fertilizer in sustainable agriculture. *Current Research in Microbial Sciences* 3, 100107.
Leifheit, E.F., Verbruggen, E. and Rillig, M.C. (2015) Arbuscular mycorrhizal fungi reduce decomposition of woody plant litter while increasing soil aggregation. *Soil Biology and Biochemistry* 81, 323–328.
Linderman, R.G. (2008) *The Mycorrhizosphere Phenomenon*. Mycorrhiza Works, Deutsche Phytomedizinische Gesellschaft, Braunschweig, pp. 341–355.
Liu, L., Yan, Y., Ding, H., Zhao, J., Cai, Z. *et al*. (2021) The fungal community outperforms the bacterial community in predicting plant health status. *Applied Microbiology and Biotechnology* 105, 6499–6513.
Lone, R., Shuab, R., Khan, S., Ahmad, J. and Koul, K.K. (2017) Arbuscular mycorrhizal fungi for sustainable agriculture. In: Kumar, V., Kumar, M., Sharma, S. and Prasad, R. (eds) *Probiotics and Plant Health*. Springer, pp. 553–577.
Meena, R.S., Kumar, S., Datta, R., Lal, R., Vijayakumar, V. *et al*. (2020) Impact of agrochemicals on soil microbiota and management: A review. *Land* 9(2), 34.
Milman, O. (2015) Earth has lost a third of arable land in the past 40 years, scientists say. Available at: https://www.theguardian.com/environment/2015/dec/02/arable-land-soil-food-security-shortage (accessed 5 February 2025).
Miransari, M. and Smith, D.L. (2007) Overcoming the stressful effects of salinity and acidity on soybean nodulation and yields using signal molecule genistein under field conditions. *Journal of Plant Nutrition* 30(12), 1967–1992.
Mitchell, A.E., Hong, Y.J., Koh, E., Barrett, D.M., Bryant, D.E. *et al*. (2007) Ten-year comparison of the influence of organic and conventional crop management practices on the content of flavonoids in tomatoes. *Journal of Agricultural and Food Chemistry* 55(15), 6154–6159.
Mosaic AgriSight (2023) Beyond bacteria: The importance of fungi, protozoa, and nematodes in soil microbiology. Available at: https://www.cropnutrition.com/resource-library/beyond-bacteria-the-importance-of-fungi-protozoa-and-nematodes-in-soil (accessed 5 February 2025).
Mu, D., Du, N. and Zwiazek, J.J. (2020) Inoculation with ericoid mycorrhizal associations alleviates drought stress in lowland and upland velvetleaf blueberry (*Vaccinium myrtilloides*) seedlings. *Plants* 10(12), 2786.
Muhammad, I., Lv, J.Z., Yang, L., Ahmad, S., Farooq, S. *et al*. (2022) Low irrigation water minimizes the nitrate nitrogen losses without compromising the soil fertility,

enzymatic activities and maize growth. *BMC Plant Biology* 22(1), 159.

Murphy, C.J., Baggs, E.M., Morley, N., Wall, D.P. and Paterson, E. (2015) Rhizosphere priming can promote mobilisation of N-rich compounds from soil organic matter. *Soil Biology and Biochemistry* 81, 236–243.

Muthukumar, T., Udaiyan, K. and Shanmughavel, P. (2004) Mycorrhiza in sedges: An overview. *Mycorrhiza* 14(2), 65–77.

Öpik, M., Zobel, M., Cantero, J.J., Davison, J., Facelli, J.M. *et al.* (2013) Global sampling of plant roots expands the described molecular diversity of arbuscular mycorrhizal fungi. *Mycorrhiza* 23(5), 411–430.

Ortas, I. and Ustuner, O. (2014) The effects of single species, dual species and indigenous mycorrhiza inoculation on citrus growth and nutrient uptake. *European Journal of Soil Biology* 63, 64–69.

Ouledali, S., Ennajeh, M., Ferrandino, A., Khemira, H., Schubert, A. *et al.* (2019) Influence of arbuscular mycorrhizal fungi inoculation on the control of stomata functioning by abscisic acid (ABA) in drought-stressed olive plants. *South African Journal of Botany* 121, 152–158.

Pagano, M.C., Freitas Duarte, N. and Corrêa, E.J.A. (2020) Effect of crop and grassland management on mycorrhizal fungi and soil aggregation. *Applied Soil Ecology* 147, 103385.

Paull, J. (2019) Organic agriculture in australia: Attaining the global majority. *Journal of Environment Protection and Sustainable Development* 5, 70–74.

Paull, J. and Hennig, B. (2018) Maps of organic agriculture in Australia. *Journal of Organics* 5, 29–39.

Pennock, D. (2019) *Soil Erosion: The Greatest Challenge for Sustainable Soil Management*. FAO, Rome.

Priyadharsini, P., Rojamala, K., Ravi, R.K., Muthuraja, R., Nagaraj, K. *et al.* (2016) Mycorrhizosphere: The extended rhizosphere and its significance. In: Choudary, D.K., Varma, A. and Tuteja, N. (eds) *Plant-Microbe Interaction: An Approach to Sustainable Agriculture*. Springer, pp. 97–124.

Qin, H., Lu, K., Strong, P.J., Xu, Q., Wu, Q. *et al.* (2015) Long-term fertilizer application effects on the soil, root arbuscular mycorrhizal fungi and community composition in rotation agriculture. *Applied Soil Ecology* 89, 35–43.

Quirk, J., Beerling, D.J., Banwart, S.A., Kakonyi, G., Romero-Gonzalez, M.E. *et al.* (2012) Evolution of trees and mycorrhizal fungi intensifies silicate mineral weathering. *Biology Letters* 8(6), 1006–1011.

Rafique, M., Ortas, I., Ahmed, I.A.M., Rizwan, M., Afridi, M.S. *et al.* (2019) Potential impact of biochar types and microbial inoculants on growth of onion plant in differently textured and phosphorus limited soils. *Journal of Environmental Management* 247, 672–680.

Rillig, M.C., Wright, S.F., Kimball, B.A., Pinter, P.J., Wall, G.W. *et al.* (2001) Elevated carbon dioxide and irrigation effects on water stable aggregates in a sorghum field: A possible role for arbuscular mycorrhizal fungi. *Global Change Biology* 7(3), 333–337.

Rønn, R., Vestergård, M. and Ekelund, F. (2012) Interactions between bacteria, protozoa and nematodes in soil. *Acta Protozoologica* 5(3).

Ruiz-Lozano, J.M., Porcel, R., Bárzana, G., Azcón, R. and Aroca, R. (2012) Contribution of arbuscular mycorrhizal symbiosis to plant drought tolerance: State of the art. In: Aroca, R. (ed.) *Plant Responses to Drought Stress*. Springer.

Ryan, M.H. and Graham, J.H. (2002) Is there a role for arbuscular mycorrhizal fungi in production agriculture? *Plant and Soil* 244, 263–271.

Sangwan, S. and Prasanna, R. (2022) Mycorrhizae helper bacteria: Unlocking their potential as bioenhancers of plant-arbuscular mycorrhizal fungal associations. *Microbial Ecology* 84(1), 1–10.

Sanmartín, C., Garmendia, I., Romano, B., Díaz, M., Palop, J.A. *et al.* (2014) Mycorrhizal inoculation affected growth, mineral composition, proteins and sugars in lettuces biofortified with organic or inorganic selenocompounds. *Scientia Horticulturae* 180, 40–51.

Schmidt, M.W.I., Torn, M.S., Abiven, S., Dittmar, T., Guggenberger, G. *et al.* (2011) Persistence of soil organic matter as an ecosystem property. *Nature* 478, 49–56.

Schneider, K.D., Lynch, D.H., Dunfield, K., Khosla, K., Jansa, J. *et al.* (2015) Farm system management affects community structure of arbuscular mycorrhizal fungi. *Applied Soil Ecology* 96, 192–200.

Shahzad, T., Chenu, C., Genet, P., Barot, S., Perveen, N. *et al.* (2015) Contribution of exudates, arbuscular mycorrhizal fungi and litter depositions to the rhizosphere priming effect induced by grassland species. *Soil Biology and Biochemistry* 80, 146–155.

Sharma, A., Negi, N.P., Narwal, P., Kumari, P. and Kumar, D. (2022) Arbuscular mycorrhizal fungi: A next-generation biofertilizer for sustainable agriculture. In: Prasad, R. and Zhang, S.-H. (eds) *Beneficial Microorganisms in Agriculture*. Springer, pp. 161–176.

Smith, J.L., Schnabel, R.R., McNeal, B.L. and Campbell, G.S. (1980) Potential errors in the first-order model for estimating soil nitrogen mineralization potentials. *Soil Science Society of America Journal* 44(5), 996–1000.

Socolow, R.H. (1999) Nitrogen management and the future of food: Lessons from the management of energy and carbon. *Proceedings of the National Academy of Sciences* 96(11), 6001–6008.

Sportes, A., Hériché, M., Boussageon, R., Noceto, P.-A., Tuinen, D. *et al.* (2021) A historical perspective on mycorrhizal mutualism emphasizing arbuscular mycorrhizas and their emerging challenges. *Mycorrhiza* 31(6), 637–653.

Taylor, T.N. and Osborn, J.M. (1996) The importance of fungi in shaping the paleoecosystem. *Review of Palaeobotany and Palynology* 90(3–4), 249–262.

Taylor, T.N., Remy, W., Hass, H. and Kerp, H. (1995) Fossil arbuscular mycorrhizae from the early Devonian. *Mycologia* 87(4), 560–573.

Vives-Peris, V., Ollas, C., Gómez-Cadenas, A. and Pérez-Clemente, R.M. (2020) Root exudates: From plant to rhizosphere and beyond. *Plant Cell Reports* 39(1), 3–17.

Wang, B. and Qiu, Y.L. (2006) Phylogenetic distribution and evolution of mycorrhizas in land plants. *Mycorrhiza* 16(5), 299–363.

Wang, B., Yeun, L.H., Xue, J.-Y., Liu, Y., Ané, J.-M. *et al.* (2010) Presence of three mycorrhizal genes in the common ancestor of land plants suggests a key role of mycorrhizas in the colonization of land by plants. *The New Phytologist* 186(2), 514–525.

Wang, Y., Duran, H.G.S., Haarst, J.C., Schijlen, E.G.W.M., Ruyter-Spira, C. *et al.* (2021) The role of strigolactones in P deficiency induced transcriptional changes in tomato roots. *BMC Plant Biology* 21(1), 349.

Wheeler, S.A. (2008) The barriers to further adoption of organic farming and genetic engineering in Australia: Views of agricultural professionals and their information sources. *Renewable Agriculture and Food Systems* 23(2), 161–170.

Wright, S.F., Rillig, M.C. and Nichols, K.A. (2000) Glomalin: A soil protein important in carbon sequestration. In: *Proceedings of American Chemical Society Annual Meeting Symposium*, pp. 721–725.

Yoneyama, K., Xie, X., Kim, H.I., Kisugi, T., Nomura, T. *et al.* (2012) How do nitrogen and phosphorus deficiencies affect strigolactone production and exudation? *Planta* 235, 1197–1207.

Zhong, W.H. and Cai, Z.C. (2007) Long-term effects of inorganic fertilization microbial biomass and community functional diversity in a paddy soil derived from quaternary red clay. *Applied Soil Ecology* 36, 84–91.

22 Nitrogen Fixation and Dynamics in Organic Agriculture in Temperate Climates

JOCHEN MAYER*

Agroscope, Department Agroecology and Environment, Zurich, Switzerland

Abstract

Sufficient and stable crop yields build the basis for feeding a growing world population. The future challenge in organic agriculture will be to improve crop yields without trade-offs between productivity and sustainability of agricultural management. In organic agriculture, nitrogen (N) supply is based on symbiotic N_2-fixation (N_{fix}), mainly by cropping of legumes, which limits the yield potential of non-legumes. Including below-ground N, grain legumes fix 200 kg N ha^{-1} (60–420 kg ha^{-1} year^{-1}). A grass-clover ley fixes 230 kg N ha^{-1} (90–470 kg N ha^{-1} year^{-1}) if N transfer from clover to associated grass is taken into account and cover crops fix 80 kg N ha^{-1} (20–150 kg N ha^{-1} year^{-1}). The direct contribution of a preceding one-year grass-clover ley to feed a succeeding non-legume crop is about 25 kg N ha^{-1} and the potential N contribution to the crop rotation is about 75 kg N ha^{-1}. These values are largely variable depending on the realized N_{fix} on a specific site. A two-year grass-clover ley could contribute up to 50 kg N ha^{-1} or more if conditions are favourable. With a proportion of fodder legumes in the rotation of 33% and an optimistic N_{fix} of about 280 kg N ha^{-1} per year, the yield potential of organic cereals is 60% of conventional in high-yielding regions like central Europe.

Keywords: symbiotic nitrogen fixation, clover, grain-legumes, nitrogen transfer, preceding crop value

Introduction

Sufficient and stable crop yields build the basis for feeding a growing world population. Limited cropland, climate change, loss of soil quality and biodiversity, coupled with excessive use of non-renewable resources, require new solutions for future cropping systems beyond existing management practices. World population is expected to peak at about 2065 with large regional variations (Vollset *et al.*, 2020). The available cropland per capita decreased in the last 50 years by 50% from 0.45 ha/person (1961) to 0.21 ha/person (2016). The future challenge will be to reduce the yield gap between organic and conventional systems by a substantial yield increase in organic systems without trade-offs between productivity and sustainability of agricultural management. The main drivers are an improvement of nitrogen (N) availability and a synchronization of N supply and crop demand. In organic agriculture, N supply for non-legumes is based on symbiotic N_2-fixation (N_{fix}), mainly by cropping of legumes. Mineral N fertilizers from the Haber-Bosch process are banned and other external inputs are restricted, which limits the yield potential of non-legumes. This chapter focuses on the potential of N derived from symbiotic N_2-fixation as main N source of organic cropping systems in temperate climate zones.

The Potential of Symbiotic N_2-Fixation

In temperate-climate zones, legumes integrated in organic crop rotations are mainly used as animal feed. Two types are most important: (i) temporary grass-clover or grass-lucerne leys; and (ii) grain

*Corresponding author: jochen.mayer@agroscope.admin.ch

DOI: 10.1079/9781800626850.0022

Table 22.1. N_2-fixation of red (*Trifolium pratense*) and white clover (*Trifolium repens*) in pure and mixed stands with grass. (From Schmidtke,1997, extended)

Reference	Legume proportion	Fixed N and % Ndfa		
		1. year (sowing year)	2. year	3. year
	% dry matter	kg N ha^{-1} (% Ndfa)		
Pure stand				
Heichel et al. 1985[1)]		133 (65%)	69 (35%)	87 (48%)
Boller 1988[2)]		156 (70%)		
Schnotz 1995[2)]		274 (71%)	196 (66%)	
Grass-clover ley				
Schnotz 1995[2)]		144 (93%)	77 (70%)	
Boller & Nösberger 1987[2)]	61% / 92%	165 (93%)	373 (85%)	
	21% / 70%	49 (87 %)	307 (85%)	
Boller & Nösberger 1994[2)]	46%	145 (80%)		
	47%	215 (85%)		
Farnham & George 1993[2)]	73% / 40%	153 (96%)	93 (98%)	
Heichel & Henjum 1991[1)]	73% / 71% / 34%	10 (36%)	140 (91%)	120 (88%)
Nesheim & Øyen 1994[2)]		56 (98%)	258 (91%)	120 (88%)
		7 (89%)	174 (98%)	83 (96%)
Oberson et al. 2013[2)]	46% / 41%	128 (87%)	132 (n.a.)	
	53% / 38%	161 (85%)	137 (n.a.)	
	28% / 25%	104 (87%)	92 (n.a.)	
	29% / 30%	99 (84%)	114 (n.a.)	

legumes. There is also a small percentage of grain legumes used as human food or as base for food products like beans, peas, lentils or soybeans. Relevant amounts of symbiotically fixed N can also be provided by legume-based cover crops or legume/non-legume cover crop mixtures. The potential of fixed N per area differs largely depending on the type, management, site conditions and the farm internal use of the legume.

Grass-clover Leys

The potential of N_{fix} in grass-clover leys based on above-ground yield surveys is huge, up to 280 kg N ha^{-1} y^{-1} (90% percentile of studies, Table 22.1), with a large variation and an average of 140 kg N ha^{-1} (median of studies, Table 22.1). The main driver of N_{fix} in leys is the productivity (dry-matter production) and, thus, the total N uptake of the legume partner (clover or lucerne species), which is mainly determined by the legume proportion in the ley. The proportion of N derived from air (Ndfa) has a comparatively small variation and is in most cases between 85% and 95% NdfaTable 22.1. Thus, the impact on N_{fix} is smaller than the legume proportion of the ley. The management and internal use of the above-ground biomass affects N_{fix} largely. One example is the common use as animal feed in mixed organic farms compared to the use as soil surface mulch in stockless organic farms (Table 22.2). In the mixed farm, the above-ground biomass is cut and removed from the field; in the mulched treatment

Table 22.2. Effect of grass-clover management on symbiotic N_2-fixation (From Helmert *et al.*, 2003)

	Grass-clover ley		
	Cut	Mulching	Mulching/ cut
Dry matter (g m)	1353	1233	91%
Legume proportion (% DM)	75	60	80%
% Ndfa	97	91	94%
N_2-fixation (g m)	32	22	69%

Table 22.3. Total crop N uptake, proportion of N derived from air (Ndfa), symbiotic N_2-fixation, grain N and field N-balance of grain legumes.

Legume	Total crop N	N_2-fixation		N in grain	Field N balance	Reference
		Ndfa	N_{fixed}			
	(kg N ha^{-1})	(%)	(kg N ha^{-1})	(kg N ha^{-1})	(kg N ha^{-1})	
Field pea	220 - 227	60 - 81	133 - 183	135 - 162	-2 to +21	Peoples et al. (1995)
(Pisum sativum)	146 - 166	57 - 65	96 - 107	105 - 125	-29 to -4	Schmidtke (1996)
	314	74	232	265	-33	Jensen (1987)
Lupin *(L. angustifolius)*	241 - 372	78 - 88	188 - 327	126 - 170	+62 to +157	Unkovich et al. (1994)
(Lupinus albus)	160 - 410	75 - 85	120 - 350	80 - 280	+ 40 to +70	Hartmann u. Aldag (1989)
Faba bean	61 - 171	62 - 76	38 - 130	42 - 74	-4 to +56	Peoples et al. (1995)
(Vicia faba)	307 - 505	75 - 81	232 - 408	200 - 292	+29 to +115	Schmidtke (1996)
	260 - 380	77 - 87	200 - 330	160 - 250	+40 to +80	Hartmann u. Aldag (1989)
Soybean	70 - 210	57 - 52	40 - 110	50 - 170	-10 to -60	Hartmann u. Aldag (1989)
(Glycine max)	363 - 417	34 - 67	143 - 244	187 - 205	-44 to +39	Bergersen et al. (1985)
	330 - 348	13 - 72	44 - 250	178 - 181	-134 to +69	Peoples et al. (1995)

it is left as surface mulch on the field. Overall, the dry-matter production of the mulched treatment was lower, with 91% of the cut treatment. Also, the proportion of the clover was only 80% of that in the cut treatment. The % Ndfa was only slightly affected by mulching and constituted 94% of the cut treatment. Combining these factors, mulching reduced N_{fix} by 30% compared to the cutting management. The main reason is a mineralization of N from the mulch which leads to a displacement of clover by grass and a suppression of Ndfa by suppressing nitrogenase activity in nodules by soil nitrate. The way to improve N_{fix} and N-use efficiency in stockless organic cropping systems is to cut and remove above-ground biomass from the ley and apply it as organic fertilizer to non-legume crops.

Grain Legumes

The N_{fix} potential of grain legumes is also high with 330 kg N ha^{-1} y^{-1} (90% percentile of studies, Table 22.3) and a large variation with an average of 160 kg N ha^{-1} (median of studies, Table 22.3). The potential depends largely on site conditions, nutrient availability, species and cultivars Table 22.3. The net N_{fix} contribution to an arable crop rotation differs, largely compared to grass-clover leys. Grass-clover harvests are mainly used as feed. The farm internal use as feed keeps a relevant part of the N within the farm nutrient cycle and recycles about 35–50% of the excreted N as manure, which can be used as organic fertilizer. This is also the case if grains from grain legumes are used as feed. However, if the grains are not used farm-internally and sold, the net contribution to the crop rotation is much lower and can become negative. This is illustrated in Table 22.3 . The N uptake of grain legumes is in most cases very high, but the %Ndfa is lower than in clover or lucerne and underlies a large variation. The N_{fix} reflects that pattern, but can reach very high values as shown for faba bean, with 408 kg N_{fix} ha^{-1}. Grain legumes are crops with a very long breeding history and have very high N harvest indices (N in grains/total crop N). In many cases, more than 70% of total of N are harvested with grains and exported from the field. Simple field balances reveal the effect on net inputs within the crop rotation, which can be very low or, in the worst case, become negative. Large differences in the grain-legume field balance were found between species, but also within species between cultivars. The knowledge on the N_{fix} potential and the harvest index is crucial for cultivar selection and a positive net N contribution to organic crop rotations. If the N harvest index is less than the %Ndfa, the field balance turns to negative values.

Cover Crops

In temperate-climate regions, cover crops bridge the vegetation periods, protect the soil from erosion, improve soil physical structure, enhance soil organic matter and, in the case of legume-based cover crops, provide an additional N_{fix} source which supports the N nutrition of crop rotations. The N_{fix} of cover crops is largely variable due to a very large number of cover crop species, type of mixtures, their legume proportion and the duration of the cropping period, which can last a few weeks or more than half a year. This large variability determines dry-matter yield as well as the N_{fix} potential. A comprehensive study in Switzerland investigated 19 leguminous cover crop species and determined dry-matter yields, total N uptake and %Ndfa on two Swiss sites with a growing period of three months (Büchi *et al.*, 2015). Some species produced an important amount of biomass, up to 6.9t ha^{-1} for *Vicia faba*. N_{fix} was between 0 and 143 kg N ha^{-1}, whereas five species, *Lathyrus sativus*, *Pisum sativum*, *Vicia sativa*, *Vicia villosa and V. faba*, fixed more than 100 kg N ha^{-1}. However, the average was 42 kg N_{fix} ha^{-1} (median of 19 legumes, Fig. 22.1A). The %Ndfa was also variable, but overall high, with an average of 72% (median of 19 legumes). Thus, the N_{fix} is strongly related to the N uptake of the legume and can roughly be estimated from that parameter (Fig. 22.1B). Including legume-based cover crops in organic crop rotations improves the N balance, enhances the yield potential for non-legumes and reduces N losses from the system by nitrate leaching (Adetunji *et al.*, 2020).

Below-ground N and N Transfer

Below-ground N (BGN) constitutes of root N and N derived from rhizodeposition (NdfR). Rhizodeposition is defined as 'the release of all kinds of compounds lost from living plant roots, including ions and volatile compounds' (Uren, 2000; Wichern *et al.*, 2008). Rhizodeposition contributes a significant amount to below-ground

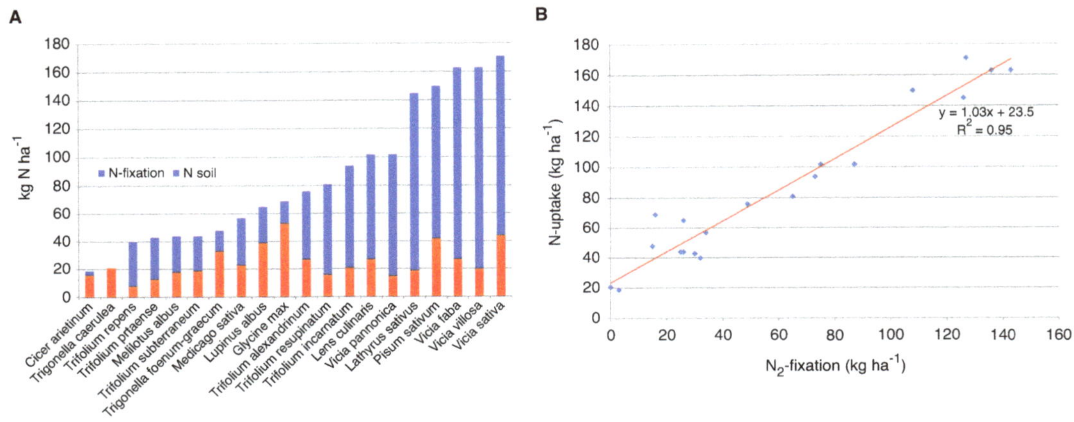

Fig. 22.1. (A) N derived from soil and symbiotic N_2-fixation of different cover crops in Switzerland (mean of two sites, Changins, Zollikofen); and (B) relation between N-uptake and symbiotic N_2-fixation. (Data from Büchi *et al.*, 2015)

N and total plant N. The review of Wichern *et al.* (2008) revealed that NdfR ranged from 4% to 71% of total assimilated plant N for legumes. The median was 16%. The proportion of NdfR of BGN was high with 15–96% of BGN, with a median of 73%. Below-ground N contributed 14–74% to total N. The median was 34%.

The knowledge of BGN is quite important for estimating N_{fix} because it comprises approximately one-third of total plant N assimilation. Hammelehle *et al.* (2018) showed that BGN of red clover in a temporal grass-clover ley can be estimated by multiplying above-ground N with a factor of 0.4. Rhizodeposition comprises soluble N forms as well as easily decomposable forms (Mayer *et al.*, 2003a; Wichern *et al.*, 2008) and is therefore an N source for N transfer to associated grass in grass-clover leys. Oberson *et al.* (2013) and Hammelehle *et al.* (2024) showed in ^{15}N tracer studies that 40% of the N taken up by the associated grass derives from the clover in the ley. Both BGN and N transfer changes markedly the picture of the amount of N_{fix} in grass-clover leys. The model calculation in Fig. 22.2 illustrates the importance of taking these plant pools into account. Based on clover N_{fix} of 140 kg N ha^{-1} above-ground, N_{fix} below-ground N amounts to 56 kg ha^{-1}. Forty per cent of N uptake of associated grass derives from clover and amounts to 54 kg N ha^{-1} (39 kg above-ground, 15 kg below-ground). Summing up fixed clover above-ground and below-ground N and N transfer, total N_{fix} of the ley increases to 249 kg N_{fix} ha^{-1}. This is 80% higher compared to N_{fix} estimates based on solely above-ground clover N.

N Contribution to Succeeding Crops

Legumes contribute significantly to N supply of organic crop rotations. A smart integration in the rotation allows a direct N supply of a succeeding non-legume like cereals, maize or potatoes by the preceding legume. But what is the direct contribution to a succeeding crop and what is the overall potential by legume crop residues? Based on the example in Fig. 22.2, the model illustrated in Fig. 22.3 gives an estimate of the potential N contribution. A relevant part of NdfR is directly transferred to associated grass, another part is immobilized by microbial biomass (Mayer *et al.*, 2003a; Wichern *et al.*, 2007) and further stabilized within the soil organic N. A part is bound in above-ground stubble of clover and grass and in non-decomposed root residues. These pools account for about 75 kg N ha^{-1} which is the potential direct N contribution to a subsequent crop. However, several studies show that only a part of this N can directly be used by the subsequent crop.

Mayer *et al.* (2003a) and Wichern *et al.* (2007) showed that 14–18% of NfdR from grain legumes were immobilized by soil microorganisms at maturity of the crops. Only 9–12% of grain legume BGN and above-ground crop residues incorporated

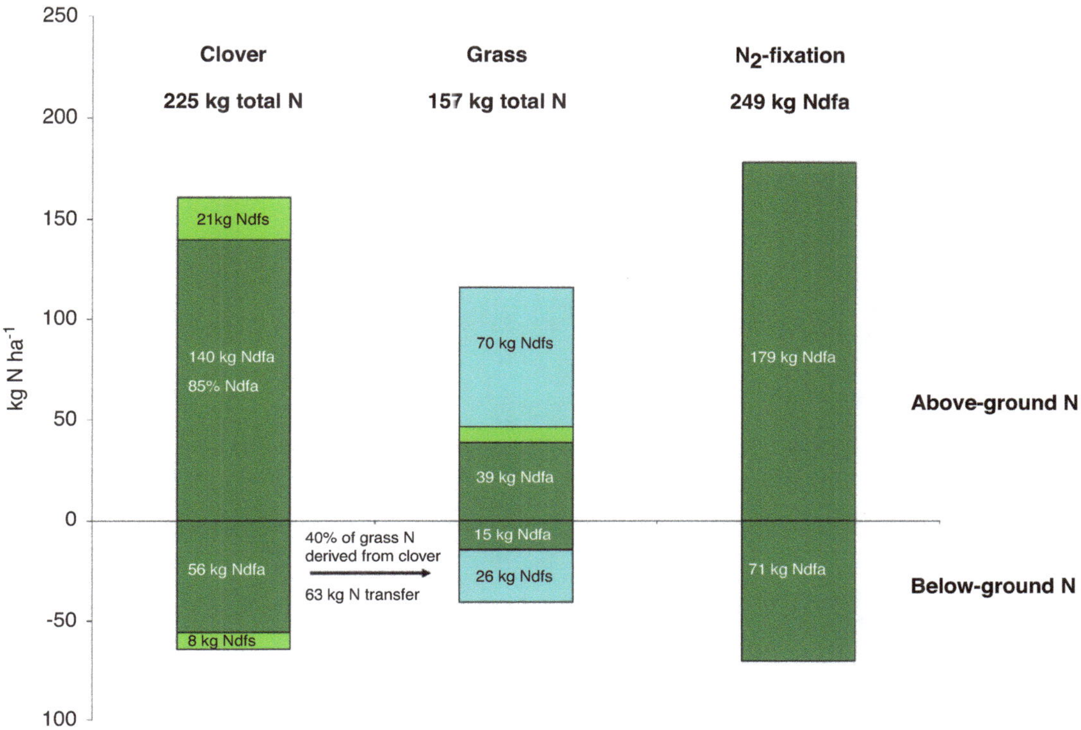

Fig. 22.2. Model calculation of symbiotic N_2-fixation of a grass-clover ley taking into account below-ground N and N transfer from clover to associated grass. Calculation based on above-ground average N_{fix} of 140 kg ha^{-1} in the first cultivation year. (Assumptions based on Hammelehle *et al.*, 2018, 2024; Oberson *et al.*, 2013)

to the soil in a pot experiment were recovered by succeeding wheat and oilseed rape. The rest was either undecomposed or stabilized in the soil organic matter (Mayer *et al.*, 2003b). Field studies with incorporated ^{15}N-labelled pea or lucerne crop residues reveal a recovery of pea N of 7–15% in barley in the first subsequent year; in the second year the recovery in oilseed rape was only 2–4% and in winter-wheat in the third year the N contributions from pea were <2% (Jensen, 1994). Maize recovered only 17–25% of incorporated lucerne residue N in the first year and barley only 1.2% in the second subsequent year (Harris and Hesterman, 1990). This supports the observation of a fast incorporation of residue N in soil organic matter pools and, afterwards, a comparatively slow remineralization from theses pools. Overall, the potential N uptake from legume N in a subsequent crop is <30% (Peoples *et al.*, 2009). However, a field study in Switzerland revealed a strong preceding crop effect of legumes compared to non-legumes (Fig. 22.4). Dry-matter yield and harvested N were 31% and 35% higher in a pure ryegrass ley if the preceding ley was a two-year clover-grass or a pure clover stand compared to a pure grass. The N effect was +53 kg N ha^{-1} if the preceding crop contained a legume. Surprisingly, the effect was the same for a pure clover stand or a grass-clover mixed stand with 50% clover (Fox *et al.*, 2020).

Synthesis

Overall, taking into account below-ground N and N transfer, the potential of N_{fix} is high in temperate climate regions. Grain legumes fix, including BGN, 200 kg N ha^{-1} (60–420 kg ha^{-1} $year^{-1}$). Clover-grass mixtures, including BGN and N transfer to associated grass, fix 230 kg N ha^{-1} (90–470 kg N ha^{-1} $year^{-1}$) and cover crops without BGN and N transfer fix 80 kg N ha^{-1} (20–150 kg N ha^{-1} $year^{-1}$). However,

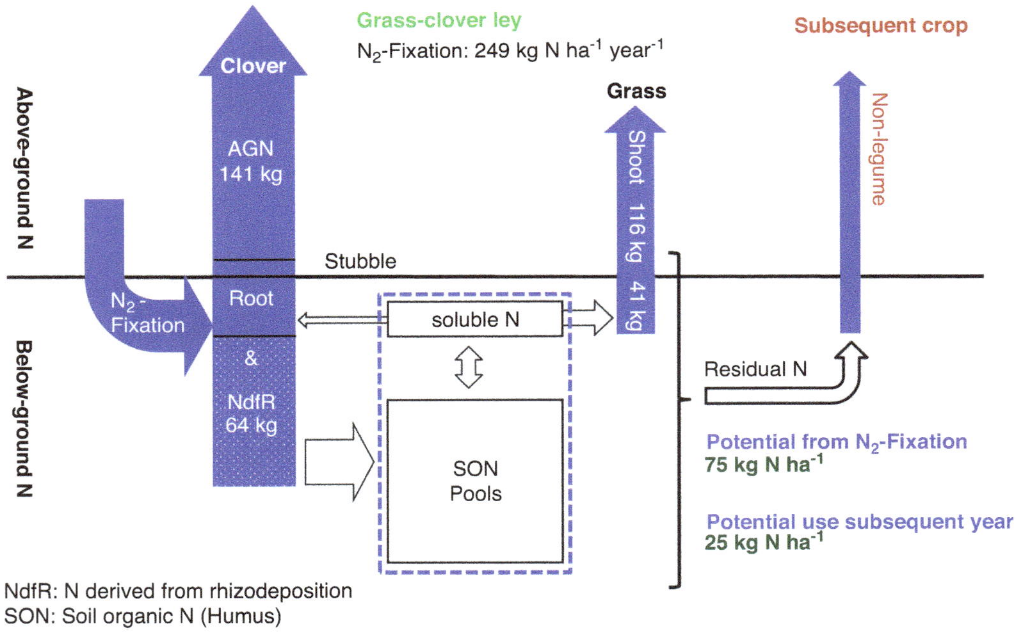

Fig. 22.3. N_2-fixation in clover-grass leys and potential residual N contribution to subsequent crops. (Assumptions based on Hammelehle *et al*., 2018, 2024; Mayer *et al*., 2003a, b; Oberson *et al*., 2013; Wichern *et al*., 2007)

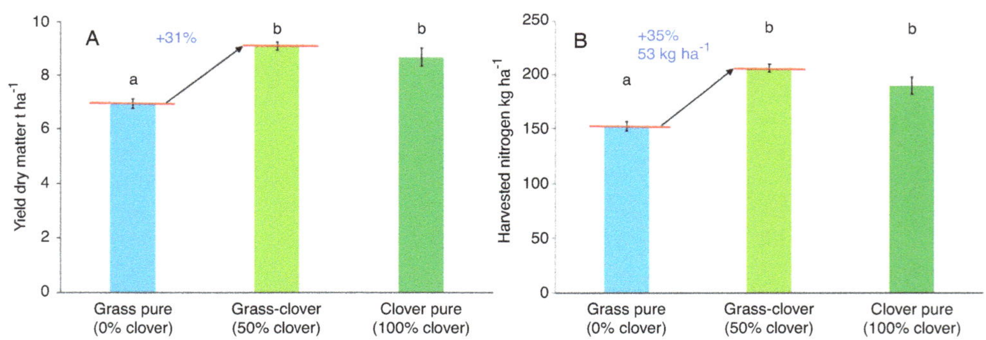

Fig. 22.4. Effect of a preceding non-legume grass ley, grass-clover ley and pure clover stand on (A) dry-matter yield of succeeding *Lolium multiflorum*; and (B) harvested N. (Data from Fox *et al*., 2020)

data especially on below-ground inputs are scarce for cover crops.

The direct contribution of a preceding one-year grass-clover ley to feed a succeeding non-legume crop is about 25 kg N ha^{-1} and the potential N contribution to the crop rotation is about 75 kg N ha^{-1}. These values are largely variable depending on the realized N_{fix} on a specific site. A two-year grass-clover ley could contribute up to 50 kg N ha^{-1} or more if conditions are favourable.

Conclusions

Legumes build the major source for the N nutrition in organic agriculture and limit the yield potential. In arable mixed organic crop rotations, the direct input of legume-N contributes 50–60% of all inputs like organic fertilizers or atmospheric deposition if about 45% of crops in the rotation are legumes (Oberson *et al.*, 2024). With a proportion of fodder legumes in the rotation of 33% and an optimistic N_{fix} of about 280 kg N ha^{-1} per year, the yield potential of organic cereals is 60% of conventional in high-yielding regions like central Europe.

References

Adetunji, A.T., Ncube, B., Mulidzi, R. and Lewu, F.B. (2020) Management impact and benefit of cover crops on soil quality: A review. *Soil and Tillage Research* 204, 104717.

Bergersen, F., Turner, G., Gault, R., Chase, D. and Brockwell, J. (1985) The natural abundance of 15n in an irrigated soybean crop and its use for the calculation of nitrogen fixation. *Australian Journal of Agricultural Research* 36(3), 411.

Boller, B.C. (1988) Biologische stickstofffixierung von weiss- und rotklee unter feldbedingungen. *Landwirtschaft Schweiz* 1, 251.

Boller, B.C. and Nösberger, J. (1987) Symbiotically fixed nitrogen from field-grown white and red clover mixed with ryegrasses at low levels of 15n-fertilization *Plant and Soil* 104(2), 219–226.

Boller, B.C. and Nösberger, J. (1994) Differences in nitrogen fixation among field-grown red clover strains at different levels of 15n fertilization. *Euphytica* 78(3), 67–174.

Büchi, L., Gebhard, C.-A., Liebisch, F., Sinaj, S., Ramseier, H. *et al.* (2015) Accumulation of biologically fixed nitrogen by legumes cultivated as cover crops in Switzerland. *Plant and Soil* 393(1–2), 163–175.

Döring, T.F. and Neuhoff, D. (2021) Upper limits to sustainable organic wheat yields. *Scientific Reports* 11(1), 12729.

Farnham, D.E. and George, J.R. (1993) Dinitrogen fixation and nitrogen transfer among red clover cultivars. *Canadian Journal of Plant Science* 73(4), 1047–1054.

Fox, A., Suter, M., Widmer, F. and Lüscher, A. (2020) Positive legacy effect of previous legume proportion in a ley on the performance of a following crop of *Lolium multiflorum*. *Plant and Soil* 447(1–2), 497–506.

Hammelehle, A., Oberson, A., Lüscher, A., Mäder, P. and Mayer, J. (2018) Above- and belowground nitrogen distribution of a red clover-perennial ryegrass sward along a soil nutrient availability gradient established by organic and conventional cropping systems *Plant and Soil* 425(1–2), 507–525.

Hammelehle, A., Mayer, J., Lüscher, A. and Oberson, A. (2024) Nitrogen-15 natural abundance is robust to quantify nitrogen transfer from clover to grass in temporary grassland. *Soil Biology and Biochemistry* 192, 109290.

Harris, G.H. and Hesterman, O.B. (1990) Quantifying the nitrogen contribution from alfalfa to soil and two succeeding crops using nitrogen-15. *Agronomy Journal* 82(1), 129–134.

Hartmann, C. and Aldag, R. (1989) N_2-Fixierung und Ertragsstruktur der Weißen Lupine (*Lupinus albus* L.) im Vergleich zu *Vicia faba* L. und *Glycine max* (L.) merr. auf verschiedenen Standorten. *Journal of Agronomy and Crop Science* 163, 201–211.

Heichel, G.H., Vance, C.P., Barnes, D.K. and Henjum, K.I. (1985) Dinitrogen fixation, and N and dry matter distribution during 4 year stands of birdsfoot trefoil and red clover. *Crop Science* 25(1), 101–105.

Heichel, G.H. and Henjum, K.I. (1991) Dinitrogen fixation, nitrogen transfer, and productivity of forage legume-grass communities. *Crop Science* 31(1), 202–208.

Helmert, M., Pommer, G., Gutser, R. and Schmidhalter, U. (2003) N-flüsse in gemulchtem und geschnittenem kleegras: Warum kleegras-brache im ökologischen landbau die erträge der folgefrucht nicht erhöht. *Mittteilungen Der Deutschen Bodenkundlichen Gesellschaft* 102, 347–348.

Jensen, E.S. (1987) Seasonal patterns of growth and nitrogen fixation in field-grown pea. *Plant and Soil* 101(1), 29–37.

Jensen, E.S. (1994) Availability of nitrogen in 15n-labelled mature pea residues to subsequent crops in the field. *Soil Biology and Biochemistry* 26(4), 465–472.

Mayer, J., Buegger, F., Jensen, E.S., Schloter, M. and Heß, J. (2003a) Estimating N rhizodeposition of grain legumes using a ^{15}N in situ stem labelling method. *Soil Biology and Biochemistry* 35(1), 21–28.

Mayer, J., Buegger, F., Jensen, E.S., Schloter, M. and Heß, J. (2003b) Residual nitrogen contribution from grain legumes to succeeding wheat and rape and related microbial process. *Plant and Soil* 255(2), 541–554.

Nesheim, L. and Øyen, J. (1994) Nitrogen fixation by red clover (*trifolium pratense* l.) grown in mixtures with timothy (*phleum pratense* l.) at different levels of nitrogen fertilization. *Acta Agriculturae Scandinavica, Section B — Soil & Plant Science* 44(1), 28–34. DOI: 10.1080/09064719409411254.

Oberson, A., Frossard, E., Bühlmann, C., Mayer, J., Mäder, P. *et al.* (2013) Nitrogen fixation and transfer in grass-clover leys under organic and conventional cropping systems. *Plant and Soil* 371(1–2), 237–255.

Oberson, A., Jarosch, K.A., Frossard, E., Hammelehle, A., Fliessbach, A. *et al.* (2024) Higher than expected: Nitrogen flows, budgets, and use efficiencies over 35 years of organic and conventional cropping. *Agriculture, Ecosystems & Environment* 362, 108802.

Peoples, M.B., Herridge, D.F. and Ladha, J.K. (1995) Biological nitrogen fixation: An efficient source of nitrogen for sustainable agricultural production. In: *Management of biological nitrogen fixation for the development of more productive and sustainable agricultural systems*, Extended versions of papers presented at the Symposium on Biological Nitrogen Fixation for Sustainable Agriculture at the 15th Congress of Soil Science, 1994, pp. 3–28.

Peoples, M.B., Brockwell, J., Herridge, D.F., Rochester, I.J., Alves, B.J.R. *et al*. (2009) The contributions of nitrogen-fixing crop legumes to the productivity of agricultural systems. *Symbiosis* 48(1–3), 1–17.

Schmidtke, K. (1997) *Dissertation. Einfluß von rotklee (Trifolium pratense L.) in reinsaat und gemenge mit poaceen auf symbiontische N2-fixierung, bodenbürtige N-aufnahme und CaCl2-extrahierbare N-fraktionen im boden*. Göttingen, Germany.

Schnotz, G. (1995) *Stickstoff-Fixierungsleistungvermögen mehrjähriger Leguminosen des Dauergrünlandes*. Verlag Ulrich E, Grauer, Stuttgart, Germany.

Unkovich, M., Pate, J. and Hamblin, J. (1994) The nitrogen economy of broadacre lupin in southwest Australia. *Australian Journal of Agricultural Research* 45(1), 149–164.

Uren, N.C. (2000) Types, amounts, and possible functions of compounds released into the rhizosphere by soil-grown plants. In: Willig, S., Varanini, Z. and Nannipieri, P. (eds) *The Rhizosphere*. CRC Press, Boca Raton, Florida, pp. 35–56.

Vollset, S.E., Goren, E., Yuan, C.-W., Cao, J., Smith, A.E. *et al*. (2020) Fertility, mortality, migration, and population scenarios for 195 countries and territories from 2017 to 2100: A forecasting analysis for the global burden of disease study. *The Lancet* 396(10258), 1285–1306.

Wichern, F., Eberhardt, E., Mayer, J., Joergensen, R.G. and Müller, T. (2008) Nitrogen rhizodeposition in agricultural crops: Methods, estimates and future prospects. *Soil Biology and Biochemistry* 40(1), 30–48.

Wichern, F., Mayer, J., Joergensen, R.G. and Müller, T. (2007) Rhizodeposition of C and N in peas and oats after ^{13}C–^{15}N double labelling under field conditions. *Soil Biology and Biochemistry* 39(10), 2527–2537.

23 Soil Management: The Basic Requirement of Organic Farming

Shehu F. G. Yusuf[1]*, Eyinade G. Adewunmi[2] and Abbyssina Mushunje[2]

[1]Sustainable Livelihood Supports and Development Network for Africa (SLIDEN AFRICA), Abeokuta, Nigeria; [2]Department of Agriculture, Economics and Extension, University of Fort Hare, South Africa

Abstract

Organic farming is characterized by the use of natural processes and inputs to promote soil health and ecological unit services. A growing body of scientific literature proposes that the soil management approaches used in organic farming enhance general soil health, but little has been studied on the exact techniques used to build healthy soils over time. In an attempt to address the water scarcity affecting the resource-poor farmers and home-gardeners in the Eastern Cape of South Africa, an agricultural extension pilot project was designed as an intervention to address the challenge of both drought and poor soil quality. The approach was two-pronged, in that the project targeted both resource-poor households and students of the Agricultural Extension Honours programme, with students participating as part of their extension outreach experiential learning. The programme was infused with pragmatic pedagogical knowledge and a strong skills development component, designed to equip the Agricultural Extension graduates with the knowledge and skills that would most benefit local resource-poor household gardeners. Results showed that cabbage fertilized with animal-based liquid fertilizers flourished, developing a profuse growth of new, shiny leaves, and that with an enhanced, more efficient use of scarce water, fewer weeds and pests were observed. Therefore, liquid organic fertilizers could be used as an alternative to the direct application of cow manure on the vegetable garden, since it has multiple advantages, forming a means of irrigation and soil enrichment at the same time.

Keywords: organic manure, organic practices, soil health, soil productivity

Introduction

The continuity of crop production in the agrarian sector relies on the health of the soil. Soil is regarded as an ultimate resource support for agricultural production systems (Mitchell *and Wali, 1995*). Soil also serves as a medium for crop development, supports crop productivity, conserves ecological quality, and caters for plant, animal and human well-being. The terms 'soil characteristics' and 'soil health' refer to soil's ability to achieve these crucial tasks. Soil quality is broadly regarded as the basis of effective organic production methods. Conserving and enhancing soil quality over a long period is one of the primary management objectives of organic farmers. Most imperative ecological unit services are provided by healthy soils; these include carbon sequestration, nutrient supply and water regulation (Norris and Congreves, 2018). Ecosystem unit services are easily disturbed by modification of soil's physical, chemical and biological conditions, especially with cropping intensification (Power, 2010; Tiemann *et al.*, 2015). However, decades of research have led to the use of enhanced crop production systems which limit deleterious effects on soil health. These include the practices of reduced tillage, reduced frequency of fallowing in communities where

*Corresponding author: 4layusuf@gmail.com

DOI: 10.1079/9781800626850.0023

agricultural land is scarce, reduced use of fertilizer, and planned rotational cropping (Hobbs, 2007; Knowler and Bradshaw, 2007). All of these have helped maintain soil integrity, albeit to a limited extent in the face of widespread fertilizer use.

Organic farming is practised from a point of view that values conservation and ecological balance (Scialabba and Müller-Lindenlauf, 2010). While the approach has gained popularity and is often touted as a more sustainable and environmentally friendly approach to agriculture than conventional means, it is not without its critics. Organic yields are typically lower than yields produced conventionally (Seufert *et al.*, 2012). Organic farming relies on natural inputs and prohibits the use of synthetic fertilizers and pesticides (Pasupulla *et al.*, 2021). It therefore requires more land than conventional farming to achieve the same level of productivity (Tscharntke *et al.*, 2021). It is also highly time- and labour-intensive, which makes the final produce more expensive than conventionally grown produce (Thiemann and Roman-Alcalá, 2019). The certification process, which requires compliance with particular organic standards, is an additional expense for farmers, making it difficult for small-scale farmers to enter the organic market (Meemken *et al.*, 2021). The use of organic fertilizers, such as manure and compost, can lead to nutrient imbalances in the soil (Zhang *et al.*, 2022). These can cause organic produce to spoil more quickly than conventionally grown produce, potentially resulting in higher levels of food waste during transport and storage (Heng and House, 2022; Liu *et al.*, 2022). However, Scialabba and Müller-Lindenlauf (2010) argue that even with its substantial yield concerns, organic techniques have a potent ability to build soil health and enhance soil richness through the introduction of organic material. Tully and McAskill (2020) state that organic farming techniques can prevent soil health dilapidation and rebuild the health of degraded soils, since organic techniques restock the soil with organic carbon and safeguard underground biodiversity.

Water availability is a key consideration in farming, especially for resource-poor farmers who depend mainly on rainfall. The relationship between organic farming and water availability is complex, since water availability depends on various factors such as agricultural practices, local climate, water management techniques, and the specific context in which farming takes place (Zarei *et al.*, 2020; Zobeidi *et al.*, 2022). Organic farming generally emphasizes practices that promote soil health and water conservation. By focusing on soil health and organic matter content, organic farming enhances the soil's water-holding capacity and reduces water loss through evaporation (Ghorbani *et al.*, 2023). In addition, by eschewing the use of synthetic fertilizers and pesticides, organic farming preserves moisture in the soil, since chemical fertilizers and pesticides have adverse effects on water quality (Baweja *et al.*, 2020; Benbrook *et al.*, 2021). Organic farming's reliance on natural fertilizers and pest management practices also reduces the risk of water pollution from agricultural runoff (Sivaranjani and Rakshit, 2019).

South Africa is a drought-prone and water-stressed region (Department of Agriculture Forestry and Fisheries [DAFF], 2015). The average rainfall in the country is 450 mm per annum, well below the world average of 860 mm per annum (Department of Water and Sanitation [DWS], 2015). The Eastern Cape (EC) is regarded as both an arid and a semi-arid climatic zone, receiving 550–700 mm of rainfall per year (Mtyelwa *et al.*, 2022). As a result, water scarcity is profound in the province, directly affecting the agricultural sector (Popoola *et al.*, 2020a). The majority of smallholder farmers in the EC province of the country are resource-poor and largely dependent on rainfed agriculture (Popoola *et al.*, 2020b), and rainfed agriculture is one of the sectors most vulnerable to climate change (McCosh *et al.*, 2017). As a result, smallholder production from resource-poor farmers in the province is characterized by high on-farm losses and low productivity (Innazent *et al.*, 2022). Increasing inconsistency in rainfall patterns induced by climate change may well be exacerbating the challenges already faced by this farming cohort, thereby threatening household food security. In an attempt to address the water scarcity affecting resource-poor farmers and home-gardeners in the province, an agricultural extension pilot project making use of organic methods was designed as an intervention to assist farmers.

The Project Design

The project was designed around the use of liquid fertilizer produced from cow manure, applied via two- and five-litre plastic bottles for drip irrigation.

The Project Objectives

The project had the following objectives:

- to impart simple technologies for efficient irrigation making use of locally available resources;
- to use water efficiently and productively;
- to produce vegetables using non-chemical inputs; and
- to increase yields from small plots.

Literature Review

Soil management is a crucial component of organic farming. Organic farming emphasizes the use of natural processes and inputs to promote soil health and fertility. A great deal of scientific literature shows evidence that organic soil management techniques enhance soil health and preserve soil moisture content (Tully and McAskill, 2020). However, the literature does not offer much detail on the precise procedures used to build healthy soils. An exception is the work of Tully and McAskill (2020), who give a comprehensive account of techniques used in organic farming to improve soil health. These techniques include: soil health management; rotational cropping; use of organic fertilizers; and weed, pest and disease management.

Soil health/management

The basis of organic crop production is good soil quality. Soil testing is the simplest way to ascertain the state of a soil, and the adoption of methods that enhance soil health is the way to improve and maintain a healthy soil over time. Thus, good land-use practices are crucial for soil management. A healthy soil has several characteristics: it contains a relatively high proportion of organic matter (Cotrufo and Lavallee, 2022); it cycles nutrients to make them available for crops; it has high infiltration to optimize the use of rainfall and irrigation water; it provides a good environment for root growth; and it maintains a diverse below-ground ecosystem of microbes, fungi and invertebrates. Building organic matter in soil entails reducing soil disturbance, using cover crops, using a complex system of crop rotation, and using composts and animal manures.

Crop rotation

Introducing biological diversity is one of several ways to boost crops' resilience to pests and possibly to severe climatic weather conditions (Yu *et al.*, 2022). A number of studies have shown that crop rotation boosts biological diversity, in that it entails the planting of crops from different families of crop over a certain period of time (Van Bruggen *et al.*, 2016; Shah *et al.*, 2021). Crop rotation has shown the ability to build organic material in the soil, improve yields and conserve soil quality through the planting of diverse residual crops with their different root formations (Norris and Congreves, 2018). The practice is evident among most small organic vegetable farmers who rely on crop rotation, ensuring that they plant crops from different families on a plot over a period of time (Baldwin, 2006). Consistent crop rotation facilitates the breaking of pest and disease progressions and aids in managing nutrient availability in the soil.

Organic fertilizers

Organic fertilizers differ from chemical fertilizers, being animal-, plant- or mined mineral-based. Decaying material from these sources breaks down naturally, providing nutrients to the soil. Assefa and Tadesse (2019) state that one of the benefits of organic fertilizers is that they release nutrients into the soil more slowly than chemical fertilizers. Sarkar *et al.* (2003) view this slowness as a strength, as it allows the plant to process the fertilizer in a more natural way and does not result in over-fertilizing, which can damage plants. Organic fertilizer has also been found to improve soil structure, improving the circulation of air, which sustains beneficial micro-organisms that in turn release nutrients into the soil (Assefa and Tadesse, 2019). Well-managed organic systems with improved nutrient-retaining capabilities greatly reduce the menace of groundwater pollution. Organic fertilizers thus enhance soil nutrient density, soil structure and water infiltration.

Weed, pest and disease management

Weed, pest and disease management in organic production demands a systems approach, which entails the use of a variety of tools to solve a problem rather than reliance on one method alone, such as a spray. The systems approach is known as integrated pest management (IPM). IPM employs crop rotation, the use of resistant plants, and a deliberate effort to safeguard helpful predator insects and parasite species. Other key points of IPM are: (i) it does not use artificial pesticides; (ii) it uses naturally derived substances, such as vinegar, neem oil or other oil-based herbicides, kaolin and soaps; (iii) it uses biological controls such as predatory nematodes, parasitic wasps, ladybirds, *Bacillus thuringiensis* and *Trichoderma* spp.; (iv) it may involve the use of physical approaches such as sticky traps, hand removal, vacuuming, trap crops, coloured mulches, exclusion devices, flame-weeding and solarization; and (v) it may also involve cultural approaches; these include the installation of drip irrigation to reduce leaf wetness and weed germination, managing crop spacing in order to provide good air circulation, maintaining optimal soil nutrient levels, implementing timely planting, and sanitation of tools.

Methodology

The project approach

The approach was two-pronged, targeting both resource-poor households and students of the Agricultural Extension Honours' programme as part of the extension outreach component for experiential learning. The programme was infused with pragmatic pedagogical knowledge and a strong focus on skills development, aiming to equip the agricultural extension graduates with the necessary knowledge and skills to benefit local resource-poor household gardeners. The programme was based on the Participatory Action Research Approach (PARA), as recognized in the literature (Amaya and Yeates 2015). It is an appropriate approach for collective problem identification and solving, empowerment and shared learning, and has been strongly linked with knowledge for action that directly benefits people, particularly the socially vulnerable and disadvantaged.

The agricultural extension outreach programme took place in three locations: two resource-poor home gardens and an old people's home with a garden, all in Tselamanzi, in Alice village, Raymond Mhlaba Local Municipality (RMLM). The municipality is situated in the Eastern Cape Province of South Africa, in the Amathole district, and covers about 6474 km^2. It is a mostly rural district, with agriculture being the main economic activity of the region.

Each of the organic farming projects was implemented in its respective setting for one week. The projects were birthed as a response to the critical issue of water scarcity, and focused on irrigation, making use of two- and five-litre water bottles to create a drip irrigation system across the vegetable beds. The first author (SFGY), a senior lecturer in the Department of Agricultural Economics and Extension, University of Fort Hare, designed the project, along with the Agricultural Extension Honours students of 2022.

Input material

The input material was several two- and five-litre plastic bottles, cow manure, large water containers, sacks, diggers, garden spades, garden forks, hand trowels, hand forks, measuring tape, pegs, twine and seedlings.

Methods

The method involved land preparation and the creation of liquid fertilizer using the extraction method.

The extraction method

The sack method was used. A porous sack was filled with manure, tied securely with twine, suspended in a container of water and allowed to seep for three days. The sack was gently shaken in the water container daily.

Straining and storage

The liquid fertilizer obtained from the extraction method was transferred to the plastic bottles to be used for drip irrigation. The bottles were punctured at the side, close to their base, in two places, and embedded at regular intervals in the soil between rows of seedlings, as shown in Fig 23.5. The bottles contained a litre of the liquid fertilizer, topped up daily with plain water after a week. The lids on the bottles were loosely closed to allow water to exit at a rate of a few drops a minute. The vegetable beds were planted with the drumhead variety of cabbage seedling (a Starke Ayres product), spaced 60 cm × 60 cm. The variety produces a grey-green cabbage with a large flat head, reaching maturity 121 days after transplanting.

The householders were encouraged to store the strained liquid fertilizer in containers such as bottles, buckets or barrels with airtight lids in a cool, dark place to prevent the growth of algae or bacteria. They were reminded to use the solution within a few weeks to maintain its effectiveness.

Pictorial Illustrations of the Activities

Fig. 23.1. Land preparation

Fig. 23.2. Land preparation and layout

Fig. 23.3. Farm layout and planting.

Fig. 23.4. Farm layout after planting.

Conclusion

Improving soil health in an organic farm relies on several soil management practices such as rotational cropping; use of organic fertilizers; and weed, pest and disease management. The study has shown the effective use of organic liquid fertilizer in enhancing crop yield and boosting soil fertility. Moreover, the organic farming initiative for the resource-poor farmers and farming households employed the principles of using animal-based liquid fertilizer farming as one of several key, low-technology ways to improve the use of the area's limited water resources and increase soil fertility. It had the added benefit of making optimum use of the plastic bottles that contribute to plastic pollution in the environment.

One of the immediate advantages of the irrigation system used was that it allowed farmers with limited resources to irrigate their land less frequently, which saved labour and conserved water. On the farm, weeds and pests were successfully controlled, since irrigation was confined to the seedlings. Students learned a practical technique for agriculture extension services which provides a real solution to farmers' challenges, using a participatory approach. The project therefore served the dual purpose of providing a service to homes with few resources and providing the

Fig. 23.5. Beds with plastic bottles for drip irrigation.

Fig. 23.7. A healthy vegetable farm.

Agricultural Extension Honours students with experiential learning. The project was designed to equip the graduates of the Agricultural Extension programme with the practical pedagogical knowledge and skills necessary to advance their abilities in the field while benefiting resource-poor households.

The results were that crop yields for these households were higher than had been the case when the farmers used watering-cans to irrigate their plots. Liquid fertilizer made from cow manure stimulated the growth of fresh, glossy leaves in cabbages and the entire head of cabbage was dense, robust and healthy. On the farm, vegetation regrowth was effectively controlled, pest infestations were kept to a minimum, and scarce water was used effectively and productively.

Fig. 23.6. Liquid fertilizer preparation.

Fig. 23.8. The Agricultural Extension students.

As a result, we conclude that the use of liquid organic fertilizers has several benefits, and may be considered preferable to the use of manure applied directly to the soil as a fertilizer for vegetable crops.

References

Amaya, A.B. and Yeates, N. (2015) Participatory Action Research: New Uses, New Contexts, New Challenges. *PRARI Working Paper* 15-6. Available at:

https://oro.open.ac.uk/98476/1/98476.pdf (accessed 16 June 2025)

Assefa, S. and Tadesse, S. (2019) The principal role of organic fertilizer on soil properties and agricultural productivity: A review. *Agricultural Research & Technology* 22(2).

Baldwin, K.R. (2006) Crop rotations on organic farms. Center for Environmental Farming Systems 16. Available at: http://carolinafarmstewards.org/wp-content/uploads/2012/12/7-CEFS-Crop-Rotation-on-Organic-Farms.pdf (accessed 16 February 2025).

Baweja, P., Kumar, S. and Kumar, G. (2020) Fertilizers and pesticides: Their impact on soil health and environment. In: Giri, B. and Varma, A. (eds) *Soil Health*. Springer, pp. 265–285.

Benbrook, C., Kegley, S. and Baker, B. (2021) Organic farming lessens reliance on pesticides and promotes public health by lowering dietary risks. *Agronomy* 11(7), 1266.

Cotrufo, M.F. and Lavallee, J.M. (2022) Soil organic matter formation, persistence, and functioning: A synthesis of current understanding to inform its conservation and regeneration. *Advances in Agronomy* 172, 1–66.

Ghorbani, M., Neugschwandtner, R.W., Konvalina, P., Asadi, H., Kopecký, M. *et al.* (2023) Comparative effects of biochar and compost applications on water holding capacity and crop yield of rice under evaporation stress: A two-years field study. *Paddy and Water Environment* 21(1), 47–58.

Heng, Y. and House, L. (2022) Consumers' perceptions and behavior toward food waste across countries. *International Food and Agribusiness Management Review* 25(2), 197–209.

Hobbs, P.R. (2007) Conservation agriculture: What is it and why is it important for future sustainable food production. *The Journal of Agricultural Science* 145(2), 127.

Innazent, A., Jacob, D., Bindhu, J.S., Joseph, B., Anith, K.N. *et al.* (2022) Farm typology of smallholders integrated farming systems in Southern Coastal Plains of Kerala, India. *Scientific Reports* 12(1), 1–14.

Knowler, D. and Bradshaw, B. (2007) Farmers' adoption of conservation agriculture: A review and synthesis of recent research. *Food Policy* 32(1), 25–48.

Kopittke, P.M., Menzies, N.W., Wang, P., McKenna, B.A. and Lombi, E. (2019) Soil and the intensification of agriculture for global food security. *Environment International* 132, 105078.

Liu, C., Jiang, H., Badulescu, D. and Bac, D.P. (2022) Achieving zero hunger goal through minimizing waste in food supply chain: Evidence from Asian emerging region. *Sustainability* 14(10), 5930.

McCosh, L., Dedekind, L., Ntombela, Z., Khuzwayo, M., Letty, B. *et al.* (2017) *Upscaling of Rainwater Harvesting and Conservation on Communal Crop and Rangeland Through Intergrated Crop And Livestock Production for Increased Water Use Productivity*. Available at: https://www.wrc.org.za/wp-content/uploads/mdocs/TT%20712-17%20web.pdf (accessed 16 June 2025).

Meemken, E.-M., Barrett, C.B., Michelson, H.C., Qaim, M., Reardon, T. *et al.* (2021) Sustainability standards in global agrifood supply chains. *Nature Food* 2(10), 758–765.

Miller, F.P. and Wali, M.K. (1995) Soils, land use and sustainable agriculture: A review. *Canadian Journal of Soil Science* 75(4), 413-422.

Mtyelwa, C., Yusuf, S.F.G. and Popoola, O.O. (2022) Adoption of in-field rainwater harvesting: Insights from smallholder farmers in Raymond Mhlaba local municipality, Eastern Cape Province, South Africa. *South African Journal of Agricultural Extension* 50(2), 81–100.

Norris. C.E. and Congreves. K.A. (2018) Alternative Management Practices Improve Soil Health Indices in Intensive Vegetable Cropping Systems: A Review. *Frontiers in Environmental Science* 6. https://doi.org/10.3389/fenvs.2018.00050

Pasupulla, A.P., Pallathadka, H., Nomani, M.Z.M., Salahuddin, G. and Rauf, M. (2021) A Survey on Challenges in Organic Agricultural Practices for Sustainable Crop Production. *Annals of the Romanian Society for Cell Biology*, 338–347.

Popoola, O.O., Yusuf, S.F.G. and Monde, N. (2020a) Information sources and constraints to climate change adaptation amongst smallholder farmers in Amathole District Municipality, Eastern Cape Province, South Africa. *Sustainability* 12(14), 5846.

Popoola, O.O., Yusuf, S.F.G. and Monde, N. (2020b) South African national climate change response policy sensitization: An assessment of smallholder farmers in Amathole District Municipality, Eastern Cape Province. *Sustainability* 12(7), 2616.

Power, A.G. (2010) Ecosystem services and agriculture: Tradeoffs and synergies. *Philosophical Transactions of the Royal Society of London. Series B, Biological Sciences* 365(1554), 2959–2971.

Sarkar, S., Singh, S.R. and Singh, R.P. (2003) The effect of organic and inorganic fertilizers on soil physical condition and the productivity of a rice–lentil cropping sequence in India. *The Journal of Agricultural Science* 140(4), 419–425.

Scialabba, N.E.-H. and Müller-Lindenlauf, M. (2010) Organic agriculture and climate change. *Renewable Agriculture and Food Systems* 25(2), 158–169.

Seufert, V., Ramankutty, N. and Foley, J.A. (2012) Comparing the yields of organic and conventional agriculture. *Nature* 485(7397), 229–232.

Shah, K.K., Modi, B., Pandey, H.P., Subedi, A., Aryal, G. *et al.* (2021) Diversified crop rotation: An approach for sustainable agriculture production. *Advances in Agriculture* 1–9.

Sivaranjani, S. and Rakshit, A. (2019) Organic farming in protecting water quality. In: Chandran, C.S., Thomas, S. and Unni, M.R. (eds) *Organic Farming: New Advances Towards Sustainable Agricultural Systems*. Springer, pp. 1–9.

Thiemann, L. and Roman-Alcalá, A. (2019) Fast food sovereignty: Contradiction in terms or logical next step? *Journal of Agricultural and Environmental Ethics* 32(5, 6), 813–834.

Tiemann, L.K., Grandy, A.S., Atkinson, E.E., Marin-Spiotta, E. and McDaniel, M.D. (2015) Crop rotational diversity enhances belowground communities and functions in an agroecosystem. *Ecology Letters* 18(8), 761–771.

Tscharntke, T., Grass, I., Wanger, T.C., Westphal, C. and Batáry, P. (2021) Beyond organic farming – harnessing biodiversity-friendly landscapes. *Trends in Ecology & Evolution* 36(10), 919–930.

Tully, K.L. and McAskill, C. (2020). Promoting Soil Health in Organically Managed Systems: A Review. *Organic Agriculture* 10, 339-358. https://doi.org/10.1007/s13165-019-00275-1

Van Bruggen, A.H.C., Gamliel, A. and Finckh, M.R. (2016) Plant disease management in organic farming systems. *Pest Management Science* 72(1), 30–44.

Yu, T., Mahe, L., Li, Y., Wei, X., Deng, X. *et al*. (2022) Benefits of crop rotation on climate resilience and its prospects in China. *Agronomy* 12(2), 436.

Zarei, Z., Karami, E. and Keshavarz, M. (2020) Co-production of knowledge and adaptation to water scarcity in developing countries. *Journal of Environmental Management* 262, 110283.

Zhang, Z., Liu, H., Liu, X., Chen, Y., Lu, Y. *et al*. (2022) Organic fertilizer enhances rice growth in severe saline–alkali soil by increasing soil bacterial diversity. *Soil Use and Management* 38(1), 964–977.

Zobeidi, T., Yaghoubi, J. and Yazdanpanah, M. (2022) Farmers' incremental adaptation to water scarcity: An application of the model of private proactive adaptation to climate change (MPPACC). *Agricultural Water Management* 264, 107528.

24 Soil Tillage and Crop Rotation: Pivotal Practices of the Organic Cropping System Management

Roberto Mancinelli[1*], Mariam Atait[1], Mohamed Allam[1,2] and Emanuele Radicetti[3]

[1]Department DAFNE, University of Tuscia, Viterbo, Italy; [2]Department of Agronomy, Faculty of Agriculture, Assuit University, Egypt; [3]Department of Chemical, Pharmaceutical and Agricultural Sciences, University of Ferrara, Italy

Abstract
Organic farming is the backbone of sustainable agriculture to ensure environmental, social and economic stability for future generations. In order to achieve success in the domain of organic crop management system, the adoption of efficient and correct agronomic practices is of great importance. Different soil tillage methods and crop rotation techniques are widely adopted worldwide to grow organic food. Studies have shown that the use of reduced tillage practices not only improves soil health but also helps in producing high-quality food with increased yield. Moreover, use of crop rotation methods in combination with conservation tillage can serve the purpose of improved crop production without sacrificing the sanctity of environmental and human health.

Keywords: alternative cropping systems, conservation tillage, crop rotation

Introduction

Organic agriculture is a food production method based on the adoption of integrated sustainable agricultural practices, by using regenerative native resources in the system for agroecological components benefits (Patle *et al.*, 2020). In organic management systems, synthetic chemical products such as insecticides, fungicides, herbicides and fertilizers are not used. The chemically synthesized products not allowed in organic farming systems are environmentally, socially and economically viable. According to the FAO definition, 'Organic agriculture is a holistic production management system which promotes and enhances agroecosystem health, including biodiversity, biological cycles, and soil biological activity' (Joint FAO/WHO Codex Alimentarius Commission, 2007; Kumar *et al.*, 2023). This farming management method determines high-quality food crop production to meet consumer demand without, or with highly reduced, environmental impacts to the Earth (Ferreira *et al.*, 2020). Moreover, the Green Paper on agricultural product quality (Brussels, 15.10.2008, COM(2008) 641 final) of the European Commission, highlighted that high levels of food quality are obtained from organic agriculture. This certified production system identifies consumers' products having specific qualities resulting from a particular guaranteed farming method. Consequently, organic certified food products are highly in demand throughout the world because of many health and environmental benefits.

The increased awareness and higher interest in the consumption of safe and healthy food by consumers are the force which is determining the growing of organic food demand (Patle *et al.*,

*Corresponding author: mancinel@unitus.it

DOI: 10.1079/9781800626850.0024

2020). In the EU, the data show continued retail sales growth in most countries. In 2021, retail sales grew to €46.7 billion, increasing to 3.6%, whereas organic farmland grew faster, reaching 15.6 million ha, increasing to 9.6%. For EU organic agriculture to reach the 25% goal by 2030, as set out by the European Commission, stronger annual growth will be needed than in 2021 (Willer *et al.*, 2023). In the USA, 2% of the food supply is grown using organic methods. The sale of organic products over the past decade has shown an annual increase of at least 20%. Organic agriculture has significant potential to overcome the challenges related to food security and safety, climate change and conservation of natural resources. Hence, the organic farming system is emerging as a fast-growing sector of world agriculture (Singh and Yadav, 2022).

It is now generally known that the agronomic practices adopted in organic farming systems help in maintaining soil fertility, soil biodiversity, soil structure, GHG sequestration, and food production free of synthetic chemical compounds (Seitz *et al.*, 2019). The organic farming effectiveness is based on its agronomic management system. The adoption of organic matter, biofertilizers, integrated pest management techniques, reduced tillage practices and cropping rotation are the highlights of organic management system (Ferreira *et al.*, 2020). Among these, crop rotation and soil tillage carry out important roles in the organic management system for reaching the organic farming objectives for food production. Crop rotation and reduced soil tillage practices are used as an alternative to synthetic inputs for maintaining the crop production cycle to provide sufficient and nutritious food for all, while minimizing environmental impacts and helping producers earn a decent living. In other words, organic farming is a way forward to achieve many of the Sustainable Development Goals (Eyhorn *et al.*, 2019).

Soil Tillage and Crop Rotation

An organic farming objective, ecologically, is based on developing the sustainable agriculture production system to reduce climate change negative effects on the environment, food production and human health. Each chosen agronomic practice for the cropping system is a critical step while planning an organic farm system. Agricultural management techniques such as soil fertilization and soil tillage have direct impact on the soil's biological, physical and chemical properties, and on potential crops' productivity in the long term. Therefore, these agronomic practices should be rationally selected to minimize the negative effects of agricultural activities on the agroecosystems (Sokolowski *et al.*, 2020).

Soil tillage is referred to as the practice of mechanically turning over and breaking up the soil to prepare land for planting (Ginting, 2019). In conventional agriculture, aggressive soil tillage methods are often those normally adopted, which mainly involve inversion or flipping soil layers by mould-board plow. Modern, sustainable and organic agriculture increasingly uses no-tillage and/or conservation tillage systems to minimize negative effects on soil erosion, soil quality, greenhouse gas emissions, water pollution, etc. In sustainable and organic farming, the adoption of conservation tillage practices is preferred for maintaining good soil quality and health. Conservation tillage practices are based on minimum or no tillage followed by crop rotation and permanent soil cover. These agronomic actions help in maintaining soil moisture, promoting biological processes, increasing soil organic matter accumulation, reducing soil erosion, mitigating CO_2 emissions, etc. Soil tillage and crop rotation are fundamental factors affecting soil quality and thus cropping system sustainability. Then, these innovative soil management methods are imperative to the successful adaptation of sustainable and organic agriculture trends for conservation of agroecosystems (Pagnani *et al.*, 2019).

Sustainability is promoted when conservation tillage practices are well combined with crop rotation, cover crops and good crop residue management. So, diverse crop rotations are needed for reaching optimal performance of no-tillage and/or reduced soil tillage (Munkholm *et al.*, 2013) and at different timings to maintain the soil fertility, to control weeds, pests, diseases, etc. In crop rotation, the selection of a succeeding crop is a very important step, because it must be from a different genus, species, subspecies or variety than the previous crop for the optimum native resource's utilization. Crop rotation is the solution to multiple crop management issues related to organic farming. Therefore, it is impracticable to manage and fulfil crop growing requirements in an organic field without crop rotation practice (Bowles *et al.*, 2020).

The adoption of sustainable agricultural practices holds a key position in mitigating the effects of climate change and global warming. The use of sustainable agricultural practices is of great interest and is given high consideration by both developed and developing countries. Sustainable agriculture and organic agriculture are dependent upon crop production methods. The most important factors determining agriculture sustainability are related to the study, design and management of cropping systems. Cropping pattern and management schemes used in crop production systems have direct impact on the sustainability of agriculture management (Singha *et al.*, 2020). Therefore, sustainable crop management practices adoption should be designed to promote plant growth and maintain crop productivity without causing harm to the agroecosystem. Some studies found that the Mediterranean environment relies on the adoption of conservation tillage practices and organic fertilization. The subsoiling and/or spading tillage adoption along with organic fertilization could effectively improve crop production and the total carbon in the soil (Mancinelli *et al.*, 2023; Fig. 24.1).

Organic Matter Soil Management through Soil Tillage and Crop Rotation

Adoption of the conservation agricultural system has proved very effective in improving physical, chemical and biological soil properties without disturbing the agroecosystem characteristics and functioning, but improving them. In organic farming the best way to build soil health is through cover crops, crop rotation and conservation tillage. All these organic management practices increase soil organic matter, water holding capacity, microbial population, overall soil structure etc. Then, as a result of improved soil structure, the problem of soil erosion, nutrient leaching and release of GHGs is reduced (Singh *et al.*, 2021) (Fig. 24.2).

Soil fertility

Several studies showed that the combination of conservation tillage, cover crops and crop rotation has a positive effect on soil available nutrients.

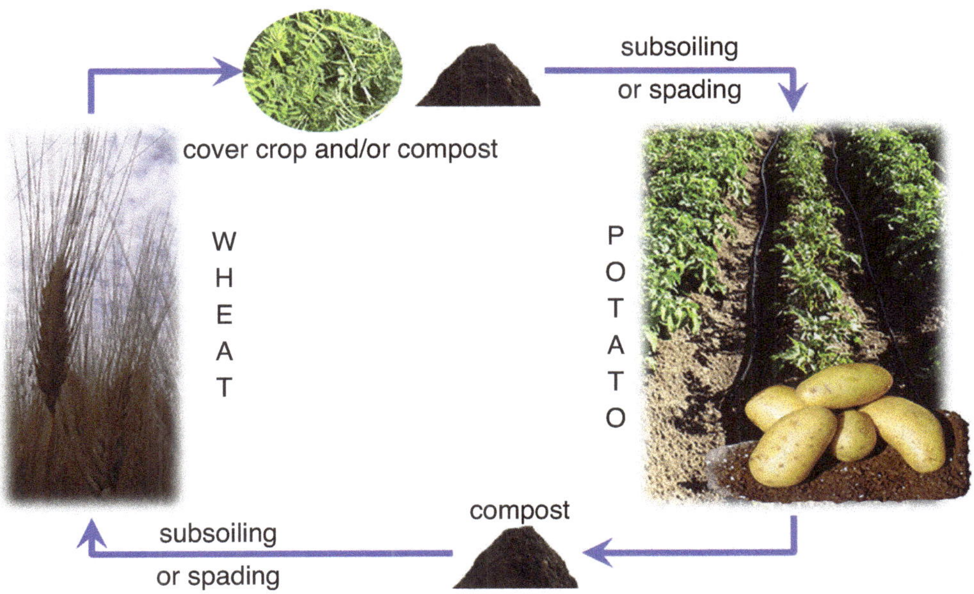

Fig. 24.1. A simple hypothesis of crop rotation combined with alternative soil tillage.

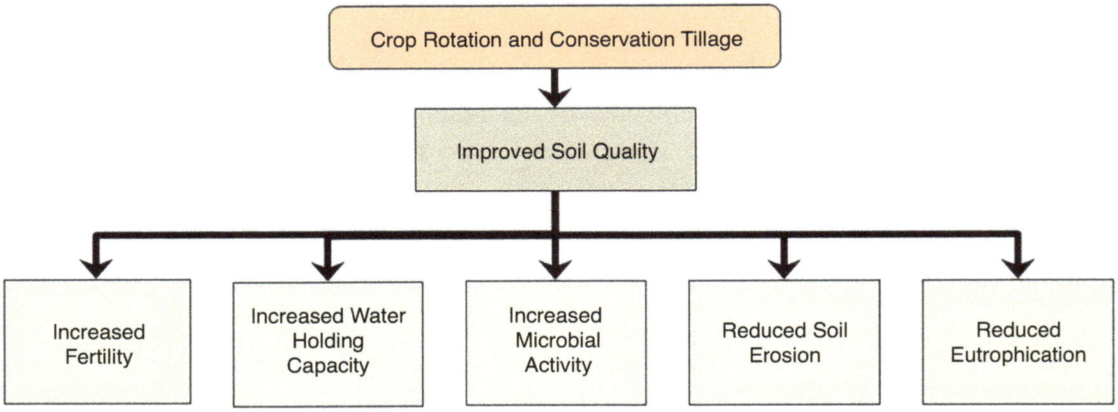

Fig. 24.2. Crop rotation and conservation tillage effects on soil characteristics.

This means that the need to add mineral fertilizer can be greatly reduced or eliminated, which in return promotes a sustainable agriculture management system (Haruna and Nkongolo, 2020). Conservation tillage techniques can effectively improve the general soil fertility by increasing soil organic carbon and other important parameters such as available mineral elements. The practice of reduced tillage and/or no tillage over a long period can significantly increase the soil organic matter content and then macro- and micronutrients which are the backbone of an organic management agroecosystem for its potential soil productivity. The reduced tillage concept includes non-inversion tillage, shallow soil tillage, superficial tillage with pulled or power-driven machinery, as it is less intensive than traditional mouldboard ploughing (Cooper *et al.*, 2016; Basavanneppa *et al.*, 2017; Yadav *et al.*, 2017; Zikeli and Gruber, 2017; Krauss *et al.*, 2022). Reduced tillage is often practised along with crop residue management to increase global soil fertility and consequently to have increased potential production and better crop agrifood yields (CTIC, 2020).

In the crop rotation technique, cash crops are sometimes intercropped with fertility building crops (leguminous species) to fix atmospheric nitrogen and maintain soil fertility. Atmospheric nitrogen fixed by leguminous species through legume rhizobium symbiosis is made available to the intercropped cash crop and/or to subsequent cash crop after nitrogen mineralization by soil microorganisms. This is one of the most effective ways for maintaining soil fertility in an organic management system and consequent optimum crop growth without causing environmental impact.

Water holding capacity

Water scarcity is one of the biggest challenges faced by many European countries along with rest of the world due to climate change (EEA, 2018). The rising temperature, low precipitation and intensive conventional farming are the driving forces behind water unavailability. In future, the demand for water for agriculture activities will increase because of increasing population and food demand (European Commission, 2018; Hristov *et al.*, 2021). Soil hydraulic properties are directly or indirectly affected by tillage methods adopted. Then, water infiltration, water retention, hydraulic conductivity determines the soil's ability in capturing and storing rainwater and/or irrigation water. Soil tillage methods affect water flow rate and path by modifying aggregate size distribution. Some specific soil tillage methods conserve water in soil and for this reason they are important for crop growth under arid and semi-arid conditions (Blanco-Canqui *et al.*, 2017).

To overcome the water crisis issue there is a need to adopt a farm management system that can increase water-using efficiency of the crops. In this regard, organic farming can play a crucial role by maintaining soil water holding capacity through sustainable agronomic practices. The conservation tillage can surely help in maintaining and/or increasing soil water holding capacity by maintaining soil structure and improving soil organic matter.

Conservation tillage helps in increasing rainfall infiltration, reducing evaporation, and enhancing water-use efficiency of the crop through moisture conservation in the soil (Khan *et al.*, 2017; Jakab *et al.*, 2017; Peng *et al.*, 2020). Similarly, the increase of soil organic matter and provision of ground cover throughout the year with the help of crop rotation mean the rate of moisture evaporation from the soil can be reduced. The rational selection of a suitable cover crop for a particular area can improve soil structure and fertility; in turn, this can reduce erosion and runoff, provide organic mulch, and increase infiltration capacity. It can also help in reducing downstream flood risk. The increase in soil water storage through the tillage and cropping system can effectively reduce the water needs for growing a good-quality crop in an organic field (European Commission, 2018; Sun *et al.*, 2018; Selim, 2019).

Microbial and enzymatic activity

In organic management systems, the conservation and maintenance of the microbial population is very important to facilitate nutrient availability to plants. The use of crop rotation and conservation tillage is an effective method of promoting soil microbial life. Conservation agricultural practices also affect the diversity of soil microbes that are essential for improving the quality of soil and crop production (Shrestha *et al.*, 2020). Previous studies show that long-term use of pulse crops such as pigeon pea, chickpea and mung bean in crop rotation with maize and wheat can effectively increase the soil microbial population, enzyme activity (acid/alkaline phosphatase, β-glucosidase, and arylsulfatase), and soil microbial biomass carbon (Borase *et al.*, 2020). Crop rotation could enhance soil microbial biomass and bacterial diversity. With fertility-building crops, such as legume species, included in the rotation, and practices of reduced tillage and agronomic techniques addressed to improve soil moisture conditions, the advantage of crop rotation on soil microorganisms can be increased (Liu *et al.*, 2023).

Conservation tillage practices such as zero tillage can affect the soil microbial and enzyme activity to a great extent. The comparison between zero tillage, no tillage with residue retention and conventional tillage showed that the availability of soil organic carbon and invertase activity in the soil was increased in zero tillage and no tillage, with residue retention, as compared to conventional tillage. The disturbance caused by conventional tillage changed the soil microbial phylogenetic composition which is not desirable in organic farming. Therefore, use of conservation tillage in organic farming is the solution to maintain healthy soil for sustainable agriculture (Li *et al.*, 2020; Wang *et al.*, 2020).

Soil erosion

Soil erosion has very strong effects on terrestrial and fluvial ecosystems, and it is one of the major environmental problems facilitated by conventional farming methods, particularly in Mediterranean environments (Smith *et al.*, 2016). Some studies reported that around 3–40t ha^{-1} of soil material is eroded in Europe each year. As a consequence of increased soil erosion, loss of organic matter, nutrient depletion, removal of topsoil and soil compaction can happen. To control and reduce soil erosion, the adoption of whole-agronomic set techniques applied in the organic management system is the best solution (Bünemann *et al.*, 2018).

The adoption of conservation tillage techniques applied in the agroecosystem, organically managed, can effectively improve the soil structure by stabilizing soil aggregation and enhancing soil surface cover. This can control soil erosion, surface runoff and nutrient loss to a greater extent (Seitz *et al.*, 2019). Also, the use of mulch-tillage and no-tillage techniques in organic farming can lower the rate of soil erosion and runoff of nitrogen and phosphorus very effectively (Klik and Rosner, 2020). Moreover, the crop rotation method is equally efficient in controlling soil erosion by providing ground cover and improving soil structure. The combination of no tillage with crop rotation has also shown excellent results to conserve soil structure through reduced erosion, higher bulk density, gravimetric moisture, and aggregate stability, even in heavy rain (Sokolowski *et al.*, 2020). The crop rotation of corn-soybean has the potential to reduce soil erosion by 60% as compared to conventional farming (Hunt *et al.*, 2019).

The application of reduced tillage in organic farming can further decrease sediment delivery and improve soil erosion control in agroecosystems. Reduced soil erosion in organic agriculture is generally mainly driven by soil surface cover and soil organic matter (Seitz *et al.*, 2019).

Eutrophication

The excessive available and water-soluble nutrients and chemical residues in agricultural soil can contaminate the environment through leaching in conventional agriculture systems. The nutrients leaching and runoff pollution can lead to severe environmental issues like eutrophication. The eutrophication process can contaminate the water resources and reduce the availability of drinking water along with water required for agricultural and human use. In organic farming management systems, there is no use of chemical inputs for crop production, resulting in less chance of water contamination through eutrophication (Ansari and Gill, 2014; Cooper *et al.*, 2020).

Sustainable agriculture systems and then agroecosystems organically managed are more environmentally friendly and can effectively overcome the challenge of water pollution through agriculture activities. The agronomic practice of over-fertilization in conventional agriculture leads to nitrate leaching. Crop rotation can have the ability to mitigate this process much more if cover crops are included by catch-up nutrients before they are leached out. Specifically, the cover crops can retain N in their biomass, either above- or below-ground (García-González *et al.*, 2018; Abdalla *et al.*, 2019).

Pest and disease control through soil tillage and crop rotation

Pest management in organic farming systems is a very important step. The use of an integrated management system is what we commonly observe in the agroecosystems organically managed. The strategic adoption of sustainable soil tillage practices along with other agronomic management techniques can effectively help in managing pests in organic farming systems (Conyers *et al.*, 2019). The evidence from previous studies shows that conservation tillage can overcome the problem of pest management in agricultural fields by increasing the activity of natural-enemy populations. The undisturbed soil structure and low anthropogenic inputs help in maintaining the habitat of natural predators. The conservation practices in vegetable crop production affect positively pest control and, according to some recent studies, in potato and onion they have shown excellent results in improving natural pest predators (Rowen *et al.*, 2020; Alyokhin *et al.*, 2020).

Similarly, crop rotations also help in pest phytophagous control associated with crop rotation by creating unfavourable conditions for the pests and favourable conditions for the crop. Previous studies show that pest control and diseases through biological means has effectively been achieved through crop rotation since early times (Tariq *et al.*, 2019). In crop rotation, pest control could be carried out, for instance, through the management of planting patterns. These methods include the planting of susceptible and non-susceptible crops within the same field in alternate ways. The greater the differences among the crops in rotation sequence, the better the control of pests could be expected (Francis and Clegg, 2020). In cornfields, damage through rootworms is commonly observed. Several studies show that the practice of crop rotation can effectively control rootworm damage in cornfields (Carrière *et al.*, 2020).

Conservation tillage, along with crop rotation, can be a way to help maintain crop cycle management in organic farming. Along with improving soil health and managing pest control, it can be also a way of controlling diseases in agroecosystems managed organically. Conservation tillage, adopted in a systematic way, can control crown rot, bare patch and yellow spot in wheat (Conyers *et al.*, 2019). There is also a common perception that fusarium disease can be controlled by conventional tillage in cereals, but several studies show that the tillage practice is not affecting the development of fusarium. Some studies also report that fusarium root rot disease reduced in fields with zero and reduced tillage. Therefore, conservation tillage could be better than conventional tillage in controlling plant diseases (Yin *et al.*, 2020).

Moreover, for these aspects, when conservation soil tillage practice is applied in combination with crop rotation, the results obtained are more effective and positive (Fan *et al.*, 2022). Crop rotation can help in reducing wheat stem-based diseases caused by fungus more effectively than conservation tillage. So, conservation tillage alone cannot help in controlling all diseases, but combined with crop rotation it could reduce the disease rate more effectively (Bankina *et al.*, 2019; Larkin and Brewer, 2020). Crop rotation methods increase the soil fertility and also promote microbiomes. Microbiomes are the disease-suppressing microorganisms that help the host plant fight against infections and disease (Peralta *et al.*, 2018).

Weed control through soil tillage and crop rotation

The management of weed control in organic farming is one of the most challenging steps. There is not enough information available about the conservation tillage effects on weed control in Mediterranean environments or in other climatic conditions. Different studies have been carried out to examine the effects of conservation tillage practices on weed control management, but there is no significant definitive evidence of weed control through zero tillage, minimum tillage and subsoil tillage. However, it has been reported that the integration of conservation tillage practices with other organic management agronomic techniques can effectively control weeds in an agroecosystem organically managed (Sims *et al.*, 2018; Alarcón *et al.*, 2018; Hofmeijer *et al.*, 2019; Fonteyne *et al.*, 2020). The integration of soil tillage with the crop rotation method has proved to be an effective weed control method in the adoption of conservation agriculture. The diversified crop production on the same land along with other associated organic management agronomic practices can control weeds and effectively change the weed dynamics (Weisberger *et al.*, 2019; Ghosh *et al.*, 2021). It is recommended that the use of the crop rotation technique with no tillage, minimum tillage and conventional tillage has more positive effects on weed control in wheat (MacLaren *et al.*, 2021). Previous research studies also showed that the production of wheat in a crop rotation with pea effectively reduced weeds (Woźniak, 2019). Strategic soil tillage could provide a mitigation option for weeds in reduced tillage crop production with insignificant effects on crop yields and improvement of soil properties. The future of agriculture should be addressed to shift from conventional farming to a more sustainable management way of production to increase food, fibre and livestock with reduced environmental impact. In these agroecosystems, weed management will likely involve integrated management by combining reduced soil tillage, less or no herbicide use and large crop rotations.

Yield and food quality of organic produce through soil tillage and crop rotation

In 2020, the EU started the 'Farm to Fork' strategy to assure a high level of food security, safety and quality by paying attention to the protection of ecosystems. In this context, the sustainable management of agroecosystems is the main pillar to reach the objective of high-quality food crops in larger quantities. The conservation tillage adoption in sustainable agriculture can increase the crop productivity by increasing the soil fertility and reducing nitrogen leaching (Ceglar *et al.*, 2018; Jug *et al.*, 2019). The use of conservation tillage in rice fields can be a helpful tool to promote sustainable crop production as compared to conventional tillage (Nandan *et al.*, 2019). Some studies show that the conservation tillage adoption in soybean and wheat increased grain yield, water productivity and biomass water productivity, along with increased yield (Yang *et al.*, 2018). The other important agronomic technique to be applied in sustainable agriculture along with conservation tillage for optimum crop production is crop rotation. The results from some studies showed that the use of crop rotation increased corn yield by 20% while reducing the emissions of N_2O by 35% (Behnke *et al.*, 2018). In addition, the use of crop rotation between spring maize and winter wheat increased the yield of both crops (Sun *et al.*, 2018).

Sustainable agriculture systems such as organic farming can be the path to safe and healthy food production. Organic farming has a significant impact on human health, animals, food security and preserving the environment. The organic management adoption eliminates chemical-based inputs in agricultural farms. The reduction and/or elimination of synthetic fertilizers, insecticides, fungicides and herbicides reduces food contamination by unwanted chemicals. Pesticide exposure from conventional production may lead to important diseases, while in organic agriculture their use is severely restricted. Independent to organic or conventional farm management, crop rotation techniques can help to reduce fertilizers and herbicide use. As a result, food contamination with agrochemical residue is decreased as soil microbial communities are increased. The nutritional value of the crops produced through organic farming is known to be higher (Selim, 2019).

Conclusion

Soil tillage and crop rotation are the sustainable agriculture building blocks to assure saving of unrenewable resources and saving of renewable resources. Reduced soil disturbance, large crop rotation, including cover crops, are pivotal agronomic strategies. Farmers are strongly encouraged to apply sustainable agronomic practices to ensure increased soil quality and environmental sustainability for food security and quality.

Sustainable agronomic techniques for the organic management of farms require strong modifications of conventional farming in the approach to organizing all agroecosystem components. Obviously, all agronomically adopted practices must be well optimized with flexibility, depending on the pedoclimatic characteristics and taking into account the farmer's objectives.

References

Abdalla, M., Hastings, A., Cheng, K., Yue, Q., Chadwick, D. *et al.* (2019) A critical review of the impacts of cover crops on nitrogen leaching, net greenhouse gas balance and crop productivity. *Global Change Biology* 25(8), 2530–2543.

Alarcón, R., Hernández-Plaza, E., Navarrete, L., Sánchez, M.J., Escudero, A. *et al.* (2018) Effects of no-tillage and non-inversion tillage on weed community diversity and crop yield over nine years in a Mediterranean cereal-legume cropland. *Soil and Tillage Research* 179, 54–62.

Alyokhin, A., Nault, B. and Brown, B. (2020) Soil conservation practices for insect pest management in highly disturbed agroecosystems – a review. *Entomologia Experimentalis et Applicata* 168(1), 7–27.

Ansari, A.A. and Gill, S.S. (2014) *Eutrophication: Causes Consequences and Control*, Vol. 2. Springer.

Bankina, B., Bimšteine, G., Paulovska, L., Paura, L., Pavlovič̌a, O. *et al.* (2019) Effects of soil tillage and crop rotation on the development of wheat stem base diseases. *Canadian Journal of Plant Pathology* 41(3), 435–442.

Basavanneppa, M.A., Gaddi, A.K., Chittapur, B.M., Biradar, D.P. and Basavarajappa, R. (2017) Yield maximization through resource conservation technologies under maize-chickpea cropping system in *vertisols* of Tunga Bhadra command project area of Karnataka. *Research on Crops* 18(2), 225.

Behnke, G.D., Zuber, S.M., Pittelkow, C.M., Nafziger, E.D. and Villamil, M.B. (2018) Long-term crop rotation and tillage effects on soil greenhouse gas emissions and crop production in Illinois, USA. *Agriculture, Ecosystems & Environment* 261, 62–70.

Blanco-Canqui, H., Wienhold, B.J., Jin, V.L., Schmer, M.R. and Kibet, L.C. (2017) Long-term tillage impact on soil hydraulic properties. *Soil and Tillage Research* 170, 38–42.

Borase, D.N., Nath, C.P., Hazra, K.K., Senthilkumar, M., Singh, S.S. *et al.* (2020) Long-term impact of diversified crop rotations and nutrient management practices on soil microbial functions and soil enzymes activity. *Ecological Indicators* 114, 106322.

Bowles, T.M., Mooshammer, M., Socolar, Y., Calderón, F., Cavigelli, M.A. *et al.* (2020) Long-term evidence shows that crop-rotation diversification increases agricultural resilience to adverse growing conditions in North America. *One Earth* 2(3), 284–293.

Bünemann, E.K., Bongiorno, G., Bai, Z., Creamer, R.E., Deyn, G. *et al.* (2018) Soil quality – a critical review. *Soil Biology and Biochemistry* 120, 105–125.

Carrière, Y., Brown, Z., Aglasan, S., Dutilleul, P., Carroll, M. *et al.* (2020) Crop rotation mitigates impacts of corn rootworm resistance to transgenic Bt corn. *Proceedings of the National Academy of Sciences of the United States of America* 117(31), 18385–18392.

Ceglar, A., Croitoru, A.-E., Cuxart, J., Djurdjevic, V., Güttler, I. *et al.* (2018) PannEx: The pannonian basin experiment. *Climate Services* 11, 78–85.

Conyers, M., Rijt, V., Oates, A., Poile, G., Kirkegaard, J. *et al.* (2019) The strategic use of minimum tillage within conservation agriculture in southern New South Wales, Australia. *Soil and Tillage Research* 193, 17–26.

Cooper, J., Baranski, M., Stewart, G., Nobel-de Lange, M., Bàrberi, P. *et al.* (2016) Shallow non-inversion tillage in organic farming maintains crop yields and increases soil C stocks: A meta-analysis. *Agronomy for Sustainable Development* 36(1), 1–20.

Cooper, R.J., Hama-Aziz, Z.Q., Hiscock, K.M., Lovett, A.A., Vrain, E. *et al.* (2020) Conservation tillage and soil health: Lessons from a 5-year UK farm trial (2013–2018). *Soil and Tillage Research* 202, 104648.

CTIC (2020) Tillage type definitions. Conservation Technology Information Center. Purdue University, West Lafayette, Indiana.

EEA (2018) European waters: Assessment of status and pressures. In: *European Environment Agency Report No.7/2018. European Environment Agency*. Copenhagen, Denmark.

European Commission (2018) Proposal for a regulation of the European parliament and of the council on minimum requirements for water reuse. COM/2018/337 final – 2018/0169 (COD).

Eyhorn, F., Muller, A., Reganold, J.P., Frison, E., Herren, H.R. *et al.* (2019) Sustainability in global agriculture driven by organic farming. *Nature Sustainability* 2(4), 253–255.

Fan, P., Lai, C., Yang, J., Hong, S., Yang, Y. *et al.* (2022) Crop rotation suppresses soil-borne fusarium wilt of banana and alters microbial communities. *Archives of Agronomy and Soil Science* 68(4), 447–459.

Ferreira, S., Oliveira, F., Silva, F., Teixeira, M., Gonçalves, M. *et al.* (2020) Assessment of factors constraining organic farming expansion in Lis Valley, Portugal. *AgriEngineering* 2(1), 111–127.

Fonteyne, S., Singh, R.G., Govaerts, B. and Verhulst, N. (2020) Rotation, mulch and zero tillage reduce weeds in a long-term conservation agriculture trial. *Agronomy* 10(7), 962.

Francis, C.A. and Clegg, M.D. (2020) Crop rotations in sustainable production systems. In: Edwards, C.A. (ed.) *Sustainable Agricultural Systems*. CRC Press, Boca Raton, Florida, pp. 107–122.

García-González, I., Hontoria, C., Gabriel, J.L., Alonso-Ayuso, M. and Quemada, M. (2018) Cover crops to mitigate soil degradation and enhance soil functionality in irrigated land. *Geoderma* 322, 81–88.

Ghosh, D., Chethan, C.R., Chander, S., Kumar, B., Dubey, R.P. *et al.* (2021) Conservational tillage and weed management practices enhance farmers' income and system productivity of rice–wheat cropping system in central India. *Agricultural Research* 10(3), 398–406.

Ginting, J. (2019) Effect of soil tillage methods and cattle fertilizer of cow on increasing sweet corn (Zea mays saccharate sturt) production. *IOP Conference Series* 260(1), 012176.

Haruna, S.I. and Nkongolo, N.V. (2020) Influence of cover crop, tillage, and crop rotation management on soil nutrients. *Agriculture* 10(6), 225.

Hofmeijer, M.A.J., Krauss, M., Berner, A., Peigné, J., Mäder, P. *et al.* (2019) Effects of reduced tillage on weed pressure, nitrogen availability and winter wheat yields under organic management. *Agronomy* 9(4), 180.

Hristov, J., Barreiro-Hurle, J., Salputra, G., Blanco, M. and Witzke, P. (2021) Reuse of treated water in European agriculture: Potential to address water scarcity under climate change. *Agricultural Water Management* 251, 106872.

Hunt, N.D., Hill, J.D. and Liebman, M. (2019) Cropping system diversity effects on nutrient discharge, soil erosion, and agronomic performance. *Environmental Science & Technology* 53(3), 1344–1352.

Jakab, G., Madarász, B., Szabó, J.A., Tóth, A., Zacháry, D. *et al.* (2017) Infiltration and soil loss changes during the growing season under ploughing and conservation tillage. *Sustainability* 9(10), 1726.

Joint FAO/WHO Codex Alimentarius Commission (2007) Codex alimentarius: cereals, pulses, legumes and vegetable proteins. World Health Organization/Food and Agriculture Organization of the United Nations, Rome.

Jug, D., Đurđević, B., Birkás, M., Brozović, B., Lipiec, J. *et al.* (2019) Effect of conservation tillage on crop productivity and nitrogen use efficiency. *Soil and Tillage Research* 194, 104327.

Khan, H.Z., Shabir, M.A., Akbar, N., Iqbal, A., Shahid, M. *et al.* (2017) Effect of different tillage techniques on productivity of wheat (*Triticum aestivum* L). *Journal of Agriculture and Basic Sciences* 2(1), 44–49.

Klik, A. and Rosner, J. (2020) Long-term experience with conservation tillage practices in Austria: impacts on soil erosion processes. *Soil and Tillage Research* 203, 104669.

Krauss, M., Wiesmeier, M., Don, A., Cuperus, F., Gattinger, A. *et al.* (2022) Reduced tillage in organic farming affects soil organic carbon stocks in temperate Europe. *Soil and Tillage Research* 216, 105262.

Kumar, D., Willer, H., Ravisankar, N. and Kumar, A. (2023) Current scenario of organic farming worldwide. In: Kumar, D. and Willer, H. (eds) *Transforming Organic Agri-Produce into Processed Food Products*. Apple Academic Press, Burlington, Canada, pp. 1–24.

Larkin, R.P. and Brewer, M.T. (2020) Effects of crop rotation and biocontrol amendments on *Rhizoctonia* disease of potato and soil microbial communities. *Agriculture* 10(4), 128.

Li, Y., Zhang, Q., Cai, Y., Yang, Q. and Chang, S.X. (2020) Minimum tillage and residue retention increase soil microbial population size and diversity: Implications for conservation tillage. *Science of the Total Environment* 716, 137164.

Liu, Q., Zhao, Y., Li, T., Chen, L., Chen, Y. *et al.* (2023) Changes in soil microbial biomass, diversity, and activity with crop rotation in cropping systems: A global synthesis. *Applied Soil Ecology* 186, 104815.

MacLaren, C., Labuschagne, J. and Swanepoel, P.A. (2021) Tillage practices affect weeds differently in monoculture vs. crop rotation. *Soil and Tillage Research* 205, 104795.

Mancinelli, R., Allam, M., Petroselli, V., Atait, M., Jasarevic, M. *et al.* (2023) Durum wheat production as affected by soil tillage and fertilization management in a Mediterranean environment. *Agriculture* 13(2), 433.

Munkholm, L.J., Heck, R.J. and Deen, B. (2013) Long-term rotation and tillage effects on soil structure and crop yield. *Soil and Tillage Research* 127, 85–91.

Nandan, R., Singh, V., Singh, S.S., Kumar, V., Hazra, K.K. *et al.* (2019) Impact of conservation tillage in rice-based cropping systems on soil aggregation, carbon pools and nutrients. *Geoderma* 340, 104–114.

Pagnani, G., Galieni, A., D'Egidio, S., Visioli, G., Stagnari, F. *et al.* (2019) Effect of soil tillage and crop sequence on grain yield and quality of durum wheat in Mediterranean areas. *Agronomy* 9(9), 488.

Patle, G.T., Kharpude, S.N., Dabral, P.P. and Kumar, V. (2020) Impact of organic farming on sustainable agriculture system and marketing potential: A review. *International Journal of Environment and Climate Change* 10(11), 100–120.

Peng, Z., Wang, L., Xie, J., Li, L., Coulter, J.A. *et al.* (2020) Conservation tillage increases yield and precipitation use efficiency of wheat on the semi-arid Loess plateau of China. *Agricultural Water Management* 231, 106024.

Peralta, A.L., Sun, Y., McDaniel, M.D. and Lennon, J.T. (2018) Crop rotational diversity increases disease suppressive capacity of soil microbiomes. *Ecosphere* 9(5), e02235.

Rowen, E.K., Regan, K.H., Barbercheck, M.E. and Tooker, J.F. (2020) Is tillage beneficial or detrimental for insect and slug management? A meta-analysis?. *Agriculture, Ecosystems & Environment* 294, 106849.

Seitz, S., Goebes, P., Puerta, V.L., Pereira, E.I.P., Wittwer, R. *et al.* (2019) Conservation tillage and organic farming reduce soil erosion. *Agronomy for Sustainable Development* 39(1), 1–10.

Selim, M. (2019) A review of advantages, disadvantages and challenges of crop rotations. *Egyptian Journal of Agronomy* 41(1), 1–10.

Shrestha, J., Subedi, S., Timsina, K.P., Chaudhary, A., Kandel, M. *et al.* (2020) Conservation agriculture as an approach towards sustainable crop production: A review. *Farming & Management* 5(1), 7–15.

Sims, B., Corsi, S., Gbehounou, G., Kienzle, J., Taguchi, M. *et al.* (2018) Sustainable weed management for conservation agriculture: Options for smallholder farmers. *Agriculture* 8(8), 118.

Singh, B. and Yadav, J.P. (2022) Sustainable agriculture development through organic farming in india. *ECS Transactions* 107(1), 14919–14929.

Singh, P., Singh, G., Sodhi, G.P.S. and Benbi, D.K. (2021) Accounting carbon footprints and applying data envelopment analysis to optimize input-induced greenhouse gas emissions under rice–wheat cropping system in north-western India. *Journal of Soil Science and Plant Nutrition* 21(4), 3030–3050.

Singha, C., Swain, K.C. and Swain, S.K. (2020) Best crop rotation selection with GIS-AHP technique using soil nutrient variability. *Agriculture* 10(6), 213.

Smith, P., House, J.I., Bustamante, M., Sobocká, J., Harper, R. *et al.* (2016) Global change pressures on soils from land use and management. *Global Change Biology* 22(3), 1008–1028.

Sokolowski, A.C., Prack McCormick, B., Grazia, J., Wolski, J.E., Rodríguez, H.A. *et al.* (2020) Tillage and no-tillage effects on physical and chemical properties of an argiaquoll soil under long-term crop rotation in Buenos Aires, Argentina. *International Soil and Water Conservation Research* 8(2), 185–194.

Sun, L., Wang, S., Zhang, Y., Li, J., Wang, X. *et al.* (2018) Conservation agriculture based on crop rotation and tillage in the semi-arid Loess plateau, China: Effects on crop yield and soil water use. *Agriculture, Ecosystems & Environment* 251, 67–77.

Tariq, M., Ali, H., Hussain, N., Nasim, W., Mubeen, M. *et al.* (2019) Fundamentals of crop rotation in agronomic management. In: Hasanuzzaman, M. (ed.) *Agronomic Crops*, Vol. 1. Production Technologies. Springer, pp. 545–559.

Wang, Z., Li, T., Li, Y., Zhao, D., Han, J. *et al.* (2020) Relationship between the microbial community and catabolic diversity in response to conservation tillage. *Soil and Tillage Research* 196, 104431.

Weisberger, D., Nichols, V. and Liebman, M. (2019) Does diversifying crop rotations suppress weeds? A meta-analysis. *PLoS ONE* 14(7), e0219847.

Willer, H., Schlatter, B. and Trávníček, J. (eds) (2023) The world of organic agriculture: Statistics and emerging trends 2023. Research Institute of Organic Agriculture FiBL, Frick, and IFOAM – Organics International, Bonn.

Woźniak, A. (2019) Effect of crop rotation and cereal monoculture on the yield and quality of winter wheat grain and on crop infestation with weeds and soil properties. *International Journal of Plant Production* 13(3), 177–182.

Yadav, M.R., Parihar, C.M., Kumar, R., Yadav, R.K., Jat, S.L. *et al.* (2017) Conservation agriculture and soil quality – an overview. *International Journal of Current Microbiology and Applied Sciences* 6(2), 707–734.

Yang, X., Zheng, L., Yang, Q., Wang, Z., Cui, S. *et al.* (2018) Modelling the effects of conservation tillage on crop water productivity, soil water dynamics and evapotranspiration of a maize-winter wheat-soybean rotation system on the Loess plateau of China using APSIM. *Agricultural Systems* 166, 111–123.

Yin, C., McLaughlin, K., Paulitz, T.C., Kroese, D.R. and Hagerty, C.H. (2020) Population dynamics of wheat root pathogens under different tillage systems in Northeast Oregon. *Plant Disease* 104(10), 2649–2657.

Zikeli, S. and Gruber, S. (2017) Reduced tillage and no-till in organic farming systems, Germany: status quo, potentials and challenges. *Agriculture* 7(4), 35.

25 Soil Organic Matter Dynamics in Organic Farming Systems: Assessment Mechanisms and Implications for Sustainable Management

Yves Theoneste Murindangabo*, Marek Kopecký and Petr Konvalina

Department of Agroecosystems, Faculty of Agriculture and Technology, University of South Bohemia in České Budějovice, Czech Republic

Abstract

Soil organic matter (SOM) dynamics, encompassing the processes of accumulation, decomposition and transformation of organic materials in the soil, play a crucial role in soil health, nutrient cycling and overall ecosystem sustainability. The depletion of SOM, often resulting from intensive agricultural practices and land degradation, poses significant challenges to soil fertility, productivity and the environment. In contrast, the enhancement and maintenance of SOM stocks contribute to improved soil structure, water holding capacity, nutrient availability and carbon sequestration. Therefore, studying SOM dynamics is vital for developing strategies that mitigate soil degradation, optimize crop production and mitigate climate change impacts.

Farming approaches prioritizing the enhancement and maintenance of SOM are the cornerstone to sustainable agriculture. For example, organic farming principles, such as the avoidance of synthetic fertilizers and pesticides, the promotion of biodiversity and the use of organic amendments, contribute to the accumulation of SOM and its beneficial effects on soil and environment quality. As organic matter is continuously added to the soil through practices such as crop residues incorporation, cover cropping, crop rotation, and the use of organic fertilizers, organic farming systems can effectively increase SOM stocks. Moreover, understanding the implications of SOM dynamics in sustainable farming approaches extends beyond individual farm boundaries. Sustainable agriculture practices that promote SOM enrichment can contribute to mitigating climate change by sequestering atmospheric carbon dioxide and reducing greenhouse gas emissions associated with synthetic inputs. Additionally, sustainable farming systems can improve soil resilience, water infiltration and nutrient cycling, thereby reducing environmental pollution and enhancing overall ecosystem health.

This chapter sheds light on the mechanisms driving SOM dynamics and emphasizes the importance of sustainable soil management practices. It provides an overview of the approaches and methods used to monitor the changes of SOM, including identifying relevant processes describing and estimating its decomposition, lability and humification for sustainable management. By improving our understanding of these dynamics, we can develop strategies that promote soil health, enhance agricultural productivity, and contribute to long-term environmental sustainability.

Keywords: quantitative and qualitative assessment, soil organic matter dynamics, sustainable management

*Corresponding author: yatebo99@gmail.com

DOI: 10.1079/9781800626850.0025

Introduction

SOM is a heterogeneous mixture of macro- and microorganisms, living and non-living organic molecules from animals and plants, at various stages of decomposition and transformation (Bot and Benites, 2005). The living fraction of SOM is relatively small and consists of plant roots and soil organisms, while the non-living fraction is more significant and includes light fractions primarily derived from plants and humus. Humus is composed of humic substances (humic acid, fulvic acid and humin) and non-humic substances (carbohydrates, lipids, proteins, lignin, alkyls and more)(Gregorich *et al.*, 2006). In order to enhance understanding of SOM, researchers have divided it into functional pools based on the levels of decomposition, distinguishing between labile and stable fractions (Kopecký *et al.*, 2021). Since the physico-chemical properties and changing behaviour of these fractions (functional pools) are not easily distinguishable, SOM is described according to its pools' decomposition and resistance instead of their chemical characteristics. The labile fraction, also known as the active fraction, is light, a source of energy and a substrate for soil microorganisms. Importantly, it determines the soil's fertility regime. In contrast, the stable fraction is resistant to oxidation and decomposition and determines the long-term carbon stocks in the soil (Fiedler *et al.*, 2015). The stable fraction results from the degradation of biopolymers of organic materials in the soil and the progressive accumulation of more hydrophobic and recalcitrant molecules. The stable fraction comprises humic substances as well as non-humic organic compounds (Nieder and Benbi, 2008).

The significance of SOM in maintaining and improving soil fertility, productivity, agroecosystem functioning and climate stability has been recognized since the early days of agriculture. Recently, much attention has been given to the effects of land use/land cover on the change in quality, productivity and stability of resources around the world (Sainepo *et al.*, 2018; Murindangabo *et al.*, 2021, 2023; Kopecký *et al.*, 2021). One of the many reasons put forward to explain this dynamicity is the depletion in the quantity and/or quality of SOM (Bot and Benites, 2005).

To explore and estimate these effects, it is important to develop techniques, models and indicators that would accurately describe and evaluate the cause–effect relationship, the impact on the whole ecosystem, and the potential remedial actions. Early SOM quantitative and qualitative evaluation studies included analyzing substances extracted from the soil (Zaccone *et al.*, 2018) and measuring carbon dioxide fluxes at the soil and plant interaction level (McDowell and Likens, 1988). Later, studies on SOM dynamics were developed using improved physical, chemical, biological, spectroscopic and thermal approaches (Schnitzer, 2015; De Nobili, 2019). Assessing soil organic carbon stocks at the global level and mathematical modelling were then introduced. Several advanced analytical and simulation models are developed today to better understand SOM decomposition and transformation, including well-known models such as those developed by Hénin and Dupuis (1945) and Parnas (1975). These models were first based on simple exponential decomposition functions and later on complex functions with concentration-dependent rates of decomposition, and with simple or complex regression equations designs with the advancements in knowledge about computer modelling and simulation (Falloon and Smith, 2010). To model and study the decomposition, lability, stability and humification rate of SOM stocks, different kinetics' equations are used, including zero-order, first-order and Michaelis–Menten equations (Bekins *et al.*, 1998).

Measurement Techniques for SOM Monitoring

SOM fractionation

SOM can be physically, chemically and biologically fractionated to help in better understanding its composition, properties and functions in soil (Gregorich *et al.*, 2006). Physical fractionation involves separating SOM basing on its particle size, density and spatial distribution, where smaller particles usually have higher decomposition rates and contain more labile SOM fractions. One of the commonly used physical fractionation methods is density fractionation, which separates SOM into different fractions based on their densities using a heavy liquid (e.g. sodium polytungstate) (Golchin *et al.*, 1994). Chemical fractionation separates SOM based on its chemical properties, where different fractions are at different degrees of decomposition and stability, solubility, hydrolizability and chemical reactivity. The most common chemical fractionation method is the acid hydrolysis or oxidation method, which separates SOM into different fractions based on their oxidation or solubility levels in acid solutions of varying strengths (Blair *et al.*, 1995; Shirato and Yokozawa, 2006). Biological fractionation separates SOM based on its microbial accessibility and biomass, where different fractions have different microbial decomposability and utilization. One of the commonly used biological fractionation methods is the substrate-induced respiration (SIR) method, which measures the microbial respiration rate of SOM fractions incubated with a specific substrate (e.g. glucose). This method measures fungal, bacterial and total microbial contributions to glucose-induced respiration and the potentially active microbial biomass on decaying plant residues of differing composition (Bossuyt *et al.*, 2005).

Quantitative techniques for SOM evaluation

Quantitative SOM measurement techniques can be broadly classified into two categories, dry combustion methods and wet oxidation methods (Apesteguia *et al.*, 2018). Dry combustion methods include thermal exchange, loss on ignition (LOI) and Walkley-Black (WB) methods. These techniques rely on the complete combustion of soil samples to determine the organic carbon content. Thermal exchange involves heating a soil sample in a furnace under an inert atmosphere and measuring the evolved CO_2. The LOI method involves heating a soil sample to a high temperature to burn off organic matter, and measuring the weight loss. The WB method involves adding a dichromate-sulfuric acid reagent to a soil sample, which oxidizes the organic matter and releases CO_2, which is then measured (De Vos *et al.*, 2007; Shamrikova *et al.*, 2022). Wet oxidation methods involve the oxidation of dissolved organic material with dissolved oxygen at high temperatures. A strong oxidizing agent is used to release carbon dioxide or to change the absorbance properties of the soil sample. Wet oxidation methods include wet digestion and near-infrared reflectance (NIR) among others (Chang *et al.*, 2001). Digestion methods breakdown soil samples using acid (such as hydrochloric acid and hydrogen peroxide) digestion to release SOM, while the NIR is a non-destructive method that uses the interaction of infrared light with organic matter to estimate the quantity and quality of organic matter by measuring using a spectrophotometer the reflectance or absorbance of near-infrared light by the soil sample (Roberts *et al.*, 2004; Zornoza *et al.*, 2008). The choice between dry and wet methods depends on accuracy and precision needed, time, resources and all organic matter properties needing to be measured.

Qualitative techniques for SOM assessment

Qualitative techniques used to assess SOM are descriptive. They involve visual examination, which is a technique entailing assessing physical properties of the soil such as colour, texture and structure (Gregorich *et al.*, 1994). A positive correlation has been observed between soils with high organic matter content and darker colour, crumbly texture and granular structure. In addition to visual examination, the smell test can be employed to identify organic matter quality, as soils with high organic matter content often have a rich, earthy smell. Furthermore, the number of microorganisms, soil crumb structure test, infiltration rate, plant growth rate, organic matter colour, CEC, and crop residue decomposition rate can all serve as qualitative indicators of SOM quality (Váchalová *et al.*, 2014; Andriuzzi, 2015). Other important qualitative techniques include: **infrared spectroscopy**, where infrared radiation identifies and quantifies functional groups in SOM. It can provide information on the composition and structure of SOM. It measures the absorption and transmission of infrared light by SOM functional groups, providing information on SOM quality, quantity and composition. IR spectra can be collected from bulk soil samples, or from specific SOM fractions obtained by soil fractionation (Steffens *et al.*, 2021); **fluorescence spectroscopy**, which uses the fluorescence properties of SOM to characterize its composition and structure. It can provide information on the humification degree, aromaticity and molecular weight of SOM. It measures the emission of light from SOM after excitation with ultraviolet or visible light. Fluorescence spectra are sensitive to SOM quality and can be used to assess changes in SOM quantity and quality due to management practices or environmental factors (Freitas *et al.*, 2022); **pyrolysis mass spectroscopy** (PyMS), which uses high temperatures to decompose SOM into smaller fragments, which are then analyzed using mass spectrometry. It measures the mass and abundance of pyrolysis products generated from SOM upon heating to high temperatures in the absence of oxygen. PyMS provides information on SOM functional groups and the distribution of carbon and nitrogen within SOM molecules (Leinweber and Schulten, 1993); **nuclear magnetic resonance** (NMR), which uses magnetic fields to analyze the structure and composition of SOM. It can provide information on the molecular structure, functional groups and chemical bonding of SOM. It measures the relaxation times of nuclei within SOM molecules in response to a magnetic field (Simpson and Simpson, 2009); **high performance size exclusion chromatography**, which separates SOM into fractions based on their size and chemical composition. It can provide information on the molecular weight, size distribution and chemical composition of soil (Brezinski and Gorczyca, 2019); **electrospray ionization fourier transform ion cyclotron resonance mass spectrometry** (ESI-FTICR-MS), which measures the mass-to-charge ratios of SOM molecules. It combines electrospray ionization with fourier transform ion cyclotron resonance mass spectrometry. It informs about the molecular weight, elemental composition and functional group composition of individual organic molecules, which helps in processes that control the SOM dynamics (Bahureksa *et al.*, 2021). Likewise, it should be noted that the use of both quantitative and qualitative methods is often advised for a more comprehensive assessment of soil health and organic matter dynamics.

Lability and stability of organic matter in soils

Understanding the relationships between SOM fractions can express the relative degree of lability, stability and humification of overall SOM, hence carbon management index development. It should also be added that variations in these sub-pools (fractions) have been used in developing and understanding SOM dynamics models (Parton *et al.*, 2015). The constant supply of organic matter carbon depends on the available stock size and estimated turnover speed. Hence, quantitative assessment of SOM dynamics, its lability and stability could be early monitored using specific parameters such as carbon pool index (CPI), lability (L), lability index (LI), carbon management index (CMI) (Blair *et al.*, 1995); carbon stock (Sisti *et al.*, 2004); and humification index (HI), humification degree (HD) and humification rate (HR) (Kwiatkowska-Malina, 2018):

$$\text{Carbon stock} = \frac{\text{TOC}*\text{Ds}*\text{e}}{10} \quad (1)$$

$$\text{CPI} = \frac{\text{TC in sample}}{\text{TC in reference}} \quad (2)$$

$$\text{L} = \frac{\text{Labile carbon (oxidized)}}{\text{Non-labile carbon (non-oxidized)}} \quad (3)$$

$$\text{LI} = \frac{\text{Lability in sample}}{\text{Lability in reference}} \quad (4)$$

$$\text{CMI} = \text{CPI}*\text{LI}*100 \quad (5)$$

TOC = total organic carbon; TC = total carbon; Ds = soil bulk density; e = thickness of the layer (cm); CPI = carbon pool index; L = lability; LI = lability index; CMI = carbon management index.

$$\text{HI} = \frac{\text{NH}}{\text{H}} \quad (6)$$

$$\text{HD} = \frac{\text{H}}{\text{TC}} * 100 \quad (7)$$

$$\text{HR} = \frac{\text{H}}{\text{TOC}} * 100 \quad (8)$$

HI = humification index; HD = humification degree; HR = humification ratio; NH = non-humified (labile) fraction; H = humified (humic acid + fulvic acid = non-labile) fraction; TOC = total organic carbon; TC = total extracted carbon.

Modelling SOM Dynamics

SOM models are mathematical representations that describe the changes of organic matter in soils over time. These models can be used to analyze or simulate the effects of management practices, climate change and other environmental factors on available SOM, turnover rates and other related soil properties. SOM models are typically based on empirical or mechanistic approaches that help in predicting the future SOM trends (Woolf and Lehmann, 2012; Lehmann and Kleber, 2015). The development and advancement in SOM modelling from simple exponential decay functions; more complex functions with substrate concentration and time-dependent relative decomposition rate; simple regression equations; to more complex processes-based models, have been a success in understanding and designing sustainable soils, SOM, crops and climate management practices (Shibu *et al.*, 2006). The progress in mathematical, computer and simulation skills has allowed the development of different SOM analytical and simulation models, and as the measurement of the impact of a given management practice to SOM changes requires long-term or many short-term data, it is the same for modelling data for reliable results (Powlson *et al.*, 1996).

SOM models can be either mono- or multi-compartmental and the choice depends on the objectives of the study, the amount and quality of available data, and the complexity of the system being modelled. Mono-compartmental models assume that all organic matter in the soil is equivalent and represented by a single pool, with a single set of parameters describing its decomposition and turnover; whereas, multi-compartmental models recognize that organic matter in the soil is composed of different fractions (compartments), each with distinct properties and turnover rates. These models represent the different compartments, and their decomposition and turnover are modelled separately, with parameters that describe the properties of each fraction (Shibu *et al.*, 2006).

Analytical models

SOM analytical models are based on mathematical equations and formulas (such as calculus or linear algebra) that describe the changes in SOM over time. These models use assumptions and simplifications to predict SOM dynamics based on a few input parameters. They can provide rapid and efficient estimates of SOM dynamics, but their accuracy can be limited by the simplifying assumptions made during model development. Analytical models predict the dynamics of a system in certain conditions, and make accurate predictions if the underlying assumptions are met. Generally, they are also easier to use and interpret than simulation models (Parton *et al.*, 2015). They include Hénin and Dupuis (1945), Kortleven (1963), Godshalk (1977), ICBM, Andriulo *et al.* (1999), SOMM models, and others (Shibu *et al.*, 2006; Kwiatkowska-Malina, 2018).

Simulation models

Simulation models (comprehensive and summary) are based on computer programs that simulate the behaviour of the system being modelled. These models operate on realistic assumptions and are more complex than analytical models. They are able to deal with a wide range of conditions and handle large and complex data (Lawrence *et al.*, 2009). They can provide more realistic and accurate predictions of SOM dynamics in response to different environmental conditions and management practices (Campbell and Paustian, 2015). They are based on a more detailed approach and representation of soil ecosystems and use computer algorithms to simulate the biophysical and biochemical processes that control SOM dynamics, including microbial activity, plant residue decomposition and soil structure formation to explain the flux and distribution of carbon over a given time and scale (Shibu *et al.*, 2006). They include the Pernas model (1975), the Century model, the RothC model, the Van Veen and Paul model (1981), the DNDC model, the DayCent model, the Yasso model, the ANIMO model, the CANDY model, the Root Zone Water Quality Model, the PAPRAN model, the NCSOIL model, the DAISY model, the SUNDIAL model, the ECOSYS model, the APSIM model, NICCCE, the EPIC model, the Verberne model, and others (Shibu *et al.*, 2006).

Approaches for Sustainable Soil Organic Matter Management

For sustainable SOM management, two categories of factors affecting the balance of SOM, natural and anthropogenic, need to be considered. Natural factors, such as climate, soil parent material, biota and topography regulate the input, output and rates of decomposition of organic matter (Plante and Parton, 2007; Aranda *et al.*, 2011), while anthropogenic practices include those that reduce SOM, such as reduction in biomass production, organic matter input and increased rates of decomposition and mineralization. These practices include conventional tillage, overgrazing, and crop residue removal and burning practices. Practices that increase SOM are those that increase biomass production and organic matter input, and decrease decomposition rates, such as balanced fertilization, cover crops, agroforestry, afforestation, regenerative agriculture and pastoral practices (Bot and Benites, 2005). Contrary to certain practices like conventional farming, sustainable regenerative activities support and maintain soil biological processes, increase biomass production, and help build soil organic matter. It is important to focus on reducing the rate of organic matter decomposition and increasing the input of organic matter by managing crop residues, using animals and green manures, avoiding burning, and adopting minimal soil disturbance practices. By implementing practices that increase soil organic matter, we can help preserve soil fertility, prevent soil degradation and promote sustainability (Hijbeek *et al.*, 2018; Ramesh *et al.*, 2019; Lal, 2022).

Conclusion

SOM is widely recognized as a key indicator of soil quality and serves as both a source and sink of carbon and a substrate for various organisms. Understanding the quality, quantity and dynamics of SOM pools is crucial in comprehending the roles they play in nutrient cycling, soil health, environmental changes, and water storage capacity. However, the conflict between maintaining or increasing SOM and the effects of management practices on SOM require further investigation and resolution in different land-use systems. This is particularly crucial for sustainable soil management and for developing models that can accurately capture the quantitative and qualitative changes of SOM. Understanding and characterizing SOM quality, composition, and quantity requires, on one hand, examining the physical, biological and chemical properties of the soil and its organic compounds, which helps to explore SOM fractions and characterize their molecular structure and composition. This helps to identify the processes that contribute to SOM formation and decomposition and their effects on soil functions. On the other hand, it also involves measuring the total amount of soil carbon (quantitative methods) as well as evaluating its dynamics over time. However, as the quantitative methods do not inform about SOM quality or composition, a combination of both methods is necessary for a comprehensive analysis of SOM and its importance in maintaining soil health and sustainability. In the same vein, modelling has been proposed as a tool for analyzing and predicting SOM dynamics to develop sustainable soil management strategies. Developing accurate models can help researchers and land managers understand how different management practices and environmental factors affect available SOM, composition and turnover; hence, finding out sustainable corrective measures.

References

Andriulo, A., Mary, B. and Guerif, J. (1999) Modelling soil carbon dynamics with various cropping sequences on the rolling pampas. *Agronomie* 19(5), 365–377. DOI: ffhal-00885937f.

Andriuzzi, W.S. (2015) Ecological functions of earthworms in soil. PhD thesis, Wageningen University, Wageningen, The Netherlands.

Apesteguia, M., Plante, A.F. and Virto, I. (2018) Methods assessment for organic and inorganic carbon quantification in calcareous soils of the Mediterranean region. *Geoderma Regional* 12, 39–48.

Aranda, V., Ayora-Cañada, M.J., Domínguez-Vidal, A., Martín-García, J.M., Calero, J. *et al.* (2011) Effect of soil type and management (organic vs. conventional) on soil organic matter quality in olive groves in a semi-arid environment in Sierra Mágina Natural Park (S. Spain). *Geoderma* 164(1–2), 54–63.

Bahureksa, W., Tfaily, M.M., Boiteau, R.M., Young, R.B., Logan, M.N. *et al.* (2021) Soil organic matter characterization by Fourier transform ion cyclotron resonance mass spectrometry (FTICR MS): A critical review of sample preparation, analysis, and data interpretation. *Environmental Science & Technology* 55(14), 9637–9656.

Bekins, B.A., Warren, E. and Godsy, E.M. (1998) A comparison of zero-order, first-order, and monod biotransformation models. *Groundwater* 36(2), 261–268.

Blair, G., Lefroy, R. and Lisle, L. (1995) Soil carbon fractions based on their degree of oxidation, and the development of a carbon management index for agricultural systems. *Australian Journal of Agricultural Research* 46(7), 1459.

Bossuyt, H., Six, J. and Hendrix, P.F. (2005) Protection of soil carbon by microaggregates within earthworm casts. *Soil Biology and Biochemistry* 37(2), 251–258.

Bot, A. and Benites, J. (2005) The importance of soil organic matter: Key to drought-resistant soil and sustained food production. *FAO Soils Bulletin*.Food and Agriculture Organization of the United Nations,Rome.

Brezinski, K. and Gorczyca, B. (2019) An overview of the uses of high performance size exclusion chromatography (HPSEC) in the characterization of natural organic matter (NOM) in potable water, and ion-exchange applications. *Chemosphere* 217, 122–139.

Campbell, E.E. and Paustian, K. (2015) Current developments in soil organic matter modeling and the expansion of model applications: A review. *Environmental Research Letters* 10(12), 123004.

Chang, C.-W., Laird, D.A., Mausbach, M.J. and Hurburgh, C.R. (2001) Near-infrared reflectance spectroscopy – principal components regression analyses of soil properties. *Soil Science Society of America Journal* 65(2), 480–490.

De Nobili, M. (2019) Comment on 'Humic substances extracted by alkali are invalid proxies for the dynamics and functions of organic matter in terrestrial and aquatic ecosystems', by Kleber and Lehmann. *Journal of Environmental Quality* 48(4), 787–789.

De Vos, B., Lettens, S., Muys, B. and Deckers, J.A. (2007) Walkley-black analysis of forest soil organic carbon: recovery, limitations and uncertainty. Walkley-Black analysis of forest soil organic carbon: Recovery. *Soil Use and Management* 23, 221–229.

Falloon, P. and Smith, P. (2010) Modelling soil carbon dynamics. In: Kutsch, W.L., Bahn, M. and Heinemeyer, A. (eds) *Soil Carbon Dynamics*. Cambridge University Press, Cambridge, UK, pp. 221–244. DOI: 10.1017/CBO9780511711794.013.

Fiedler, S.R., Buczko, U., Jurasinski, G. and Glatzel, S. (2015) Soil respiration after tillage under different fertiliser treatments – implications for modelling and balancing. *Soil and Tillage Research* 150, 30–42.

Freitas, V. da S., de Babos, D.V., Guedes, W.N., Silva, F.P., de Lima Tozo, M.L. *et al.* (2022) Assessing soil organic matter quality with laser-induced fluorescence (LIFS) and its correlation to soil carbon stock. In: *Latin America Optics and Photonics Conference*, Optica Publishing Group, Recife, Brazil.

Godshalk, G.L. (1977) Decomposition of aquatic plants in lakes (No. COO-1599-121, 7218604). DOI: 10.2172/7218604.

Golchin, A., Oades, J., Skjemstad, J. and Clarke, P. (1994) Study of free and occluded particulate organic matter in soils by solid state 13C CP/MAS NMR spectroscopy and scanning electron microscopy. *Soil Research* 32(2), 285.

Gregorich, E.G., Carter, M.R., Angers, D.A., Monreal, C.M. and Ellert, B.H. (1994) Towards a minimum data set to assess soil organic matter quality in agricultural soils. *Canadian Journal of Soil Science* 74(4), 367–385.

Gregorich, E.G., Beare, M.H., McKim, U.F. and Skjemstad, J.O. (2006) Chemical and biological characteristics of physically uncomplexed organic matter. *Soil Science Society of America Journal* 70(3), 975–985.

Hénin, S. and Dupuis, M. (1945) Essai de bilan de la matière organique du sol. *Dudod*.

Hijbeek, R., Pronk, A.A., Ittersum, M.K., Berge, H.F.M., Bijttebier, J. *et al.* (2018) What drives farmers to increase soil organic matter? Insights from The Netherlands. *Soil Use and Management* 34(1), 85–100.

Kopecký, M., Kolář, L., Perná, K., Váchalová, R., Mráz, P. *et al.* (2021) Fractionation of soil organic matter into labile and stable fractions. *Agronomy* 12(1), 73.

Kortleven, J. (1963) Quantitative aspects of humus accumulation and decomposition. *Verslagen Landbouwkundige Onderzoekingen* 69.

Kwiatkowska-Malina, J. (2018) Qualitative and quantitative soil organic matter estimation for sustainable soil management. *Journal of Soils and Sediments* 18(8), 2801–2812.

Lal, R. (2022) Reducing carbon footprints of agriculture and food systems. *Carbon Footprints* 1(1), 3.

Lawrence, C.R., Neff, J.C. and Schimel, J.P. (2009) Does adding microbial mechanisms of decomposition improve soil organic matter models? A comparison of four models using data from a pulsed rewetting experiment. *Soil Biology and Biochemistry* 41(9), 1923–1934.

Lehmann, J. and Kleber, M. (2015) The contentious nature of soil organic matter. *Nature* 528(7580), 60–68.

Leinweber, P. and Schulten, H.-R. (1993) Dynamics of soil organic matter studied by pyrolysis: Field ionization mass spectrometry. *Journal of Analytical and Applied Pyrolysis* 25, 123–136.

McDowell, W.H. and Likens, G.E. (1988) Origin, composition, and flux of dissolved organic carbon in the Hubbard Brook Valley. *Ecological Monographs* 58(3), 177–195.

Murindangabo, Y.T., Kopecký, M. and Konvalina, P. (2021) Adoption of conservation agriculture in Rwanda: A case study of Gicumbi District Region. *Agronomy* 11(9), 1732.

Murindangabo, Y.T., Kopecký, M., Perná, K., Nguyen, T.G., Konvalina, P. *et al.* (2023) Prominent use of lactic acid bacteria in soil-plant systems. *Applied Soil Ecology* 189, 104955.

Nieder, R. and Benbi, D.K. (2008) Soil organic matter characterization. In: Benbi, D.K. and Nieder, R. (eds) *Carbon and Nitrogen in the Terrestrial Environment*. Springer, Dordrecht, The Netherlands.

Parnas, H. (1975) Model for decomposition of organic material by microorganisms. *Soil Biology and Biochemistry* 7(2), 161–169.

Parton, W.J., Del Grosso, S.J., Plante, A.F., Adair, E.C. and Lutz, S.M. (2015) Modeling the dynamics of soil organic matter and nutrient cycling. *Soil Microbiology, Ecology and Biochemistry*. Elsevier 505–537.

Plante, A.F. and Parton, W.J. (2007) The dynamics of soil organic matter and nutrient cycling. In: Paul, E.A. (ed.) *Soil Microbiology, Ecology and Biochemistry*. pp. 433–467.

Powlson, D.S., Smith, P. and Smith, J.U. (eds) (1996) *Evaluation of Soil Organic Matter Models: Using Existing Long-Term Datasets*. Springer, Berlin, Heidelberg.

Ramesh, T., Bolan, N.S., Kirkham, M.B., Wijesekara, H., Kanchikerimath, M. *et al.* (2019) Soil organic carbon dynamics: Impact of land use changes and management practices: A review. In: *Advances in Agronomy*. Elsevier, pp. 1–107.

Roberts, C.A., Workman, J. and Reeves, J.B. (2004) *Near-infrared Spectroscopy in Agriculture*. Agronomy Monographs. American Society of Agronomy, Crop Science Society of America, Soil Science Society of America, Madison, Wisconsin. DOI: 10.2134/agronmonogr44. (accessed 8 February 2025).

Sainepo, B.M., Gachene, C.K. and Karuma, A. (2018) Assessment of soil organic carbon fractions and carbon management index under different land use types in Olesharo Catchment, Narok County, Kenya.

Carbon Balance and Management 13(4). DOI: 10.1186/s13021-018-0091-7.

Schnitzer, M. (2015) Selected methods for the characterization of soil humic substances. In: MacCarthy, P., Clapp, C.E., Malcolm, R.L. and Bloom, P.R. (eds) *Humic Substances in Soil and Crop Sciences: Selected Readings*. Soil Science Society of America, Madison, WI, USA, pp. 65–89.

Shamrikova, E.V., Kondratenok, B.M., Tumanova, E.A., Vanchikova, E.V., Lapteva, E.M. *et al.* (2022) Transferability between soil organic matter measurement methods for database harmonization. *Geoderma* 412, 115547.

Shibu, M.E., Leffelaar, P.A., Keulen, H. and Aggarwal, P.K. (2006) Quantitative description of soil organic matter dynamics: A review of approaches with reference to rice-based cropping systems. *Geoderma* 137(1–2), 1–18.

Shirato, Y. and Yokozawa, M. (2006) Acid hydrolysis to partition plant material into decomposable and resistant fractions for use in the Rothamsted carbon model. *Soil Biology and Biochemistry* 38(4), 812–816.

Simpson, A.J. and Simpson, M.J. (2009) Nuclear magnetic resonance analysis of natural organic matter. In: Senesi, N., Xing, B. and Huang, P.M. (eds) *Biophysico-Chemical Processes Involving Natural Nonliving Organic Matter in Environmental Systems*. John Wiley & Sons, Inc, Hoboken, New Jersey, pp. 589–650.

Sisti, C.P.J., Santos, H.P., Kohhann, R., Alves, B.J.R., Urquiaga, S. *et al.* (2004) Change in carbon and nitrogen stocks in soil under 13 years of conventional or zero tillage in southern Brazil. *Soil and Tillage Research* 76(1), 39–58.

Steffens, M., Zeh, L., Rogge, D.M. and Buddenbaum, H. (2021) Quantitative mapping and spectroscopic characterization of particulate organic matter fractions in soil profiles with imaging VisNIR spectroscopy. *Scientific Reports* 11(1), 16725.

Váchalová, R., Borová-Batt, J., Kolář, L. and Váchal, J. (2014) Selectivity of ion exchange as a sign of soil quality. *Communications in Soil Science and Plant Analysis* 45(20), 2673–2679.

Woolf, D. and Lehmann, J. (2012) Modelling the long-term response to positive and negative priming of soil organic carbon by black carbon. *Biogeochemistry* 111(1–3), 83–95.

Zaccone, C., Plaza, C., Ciavatta, C., Miano, T.M. and Shotyk, W. (2018) Advances in the determination of humification degree in peat since: Applications in geochemical and paleoenvironmental studies. *Earth-Science Reviews* 185, 163–178.

Zornoza, R., Guerrero, C., Mataix-Solera, J., Scow, K.M., Arcenegui, V. *et al.* (2008) Near infrared spectroscopy for determination of various physical, chemical and biochemical properties in Mediterranean soils. *Soil Biology & Biochemistry* 40(7), 1923–1930.

26 Organic Farming Practices for Improving Soil Organic Matter and Soil Health

Mary Cole[1]* and Rachel Cavallo[2]

[1]*Faculty of Veterinary & Agricultural Science, University of Melbourne, Australia;*
[2]*Agpath Pty Ltd, Garfield, Australia*

Abstract

This chapter discusses the important role of soil organic matter (SOM) in creating soil health. It highlights the benefits of organic farming practices in enhancing SOM. SOM is the foundation of fertile productive soil and supports life on Earth. Globally, the top one metre of soil contains two to three times more organic matter than is found above-ground and is the final decomposition product of any living microbes, plants or animals and their excrement, plant litter, manure, stubble, leaves, mulch and vegetable residue broken down by microbes and earthworms. SOM stabilizes soil particles and reduces erosion, stores and purifies water, retains carbon from the atmosphere and other sources, and helps to ameliorate the negative impacts of industrial agriculture systems. Strategies need to be adopted that will strengthen the soil carbon sinks in the manner that will lead to commodification of soil carbon as a form of farm product by determining the value of soil carbon to the farmer or land manager and to society. The only long-term solution to soil health lies in increasing SOM that has been depleted from past land misuse. A more ecologically responsible farming system is essential for quality of life on Earth.

Keywords: SOM, SOC, organic, carbon, labile, recalcitrant, moisture, climate change, tillage

Introduction

Soil Organic Matter (SOM) is the foundation of fertile productive soil and supports life on Earth. SOM allows soil particles to be held together to stabilize and reduce erosion; provides the organic soil with carbon and energy for soil microbes; allows soil to store, purify and transfer air and water; stores and cycles nutrients like phosphorus, nitrogen and sulfur; retains carbon from the atmosphere and other sources; maintains soil in an uncompacted condition; restores resilience of ecosystems, and helps to reduce the negative impacts of pesticides and other pollutants (Schnitzer, 1991; Lal, 2000; Wood *et al.*, 2000; Syltie, 2002; Pretty, 2008).

Optimizing photosynthesis, stimulating root growth and support for the fungal energy channel, resulted in soil carbon building at 5–30 times faster than C from above-ground biomass or animal manure (Schmidt *et al.*, 2011; Macdonald *et al.*, 2015); and, possibly, 60–92% of contemporary agriculturally generated greenhouse gases could be offset if all land were to be converted to organic and regenerative practices (Leifeld, 2012).

Globally, the top one metre of soil contains two to three times more organic matter than is found above-ground (Anderson, 1995; Sanderman *et al.*, 2017). The efficiency of storage of SOM and soil carbon in organic and conventional farming systems is greatly influencing: climate; forms of tillage and soil disturbance; presence and composition of cover crops; presence and type of surface litter; widespread use of synthetic inputs such as fertilizers, herbicides and insecticides; and presence of earthworms, invertebrates and insects.

*Corresponding author: mary@agpath.com.au

DOI: 10.1079/9781800626850.0026

Soil Organic Matter and Soil Carbon (C)

SOM is the foundation of fertile productive soil and supports life on Earth. It is highly heterogeneous comprising fractions of plant and animal residues, live microorganism bodies and small animals that are decomposed and mineralized in the present or over many centuries (Spain *et al.*, 1983; Biederbeck *et al.*, 1994; Lal, 2000; Kleber, 2010). SOM consists of two fairly independent pools and is the total of all the elements that are components of organic compounds, of which most is carbon (C), but also includes water and other mineral nutrients. Further definitions include dissolved organic matter (DOM), the fraction of organic matter that will pass through a fine filter, leaving a residue of particulate organic matter (POM). Mineral associated organic matter (MAOM) is small molecular weight, microbially processed compounds that attach to mineral particle surfaces in the soil, and includes POM. The smaller labile pool, where carbon:nitrogen ratios control the speed of decomposition (Hoorman and Islam, 2010), depends more on land-use management practices and includes plant residues and their decomposition products, while the larger stable or recalcitrant pool is managed by soil-forming factors and is less influenced by soil management and land use (Strosser, 2010; Shahbaz *et al.*, 2017).

Beneficial attributes include:

- stabilizing and holding together soil particles, reducing erosion;
- providing the organic soil with carbon and energy for soil microbes;
- allowing soil to store, purify and transfer air and water;
- storing and cycling nutrients like phosphorus, nitrogen and sulfur;
- retaining carbon from the atmosphere and other sources;
- maintaining soil in an uncompacted condition and restoring resilience of ecosystems; and
- helping to reduce the negative impacts of pesticides and other pollutants.

(Schnizer,1991; Lal, 2000; Wood *et al.*, 2000; Pretty, 2008)

Fig. 26.1 illustrates the complexity of the natural ecological system and how human presence impacts these systems. All these processes depend on the soil C pool being present in two forms. Firstly, SOC is the measurable organic C fraction of SOM that makes up around 2–10% of most soil mass and comprises the decomposing products including: soluble organic compounds such as sugars, proteins and other metabolic compounds; amorphous organic compounds such as humus acids, fats, waxes, oils, lignin and polyuronides (Rimmer, 2006). Secondly, soil inorganic carbon (SIC) is the organo-mineral complexes that involve hybrid compounds of organic molecules attached to clay particles through cations such as aluminum Al^{3+}, calcium Ca^{2+} and magnesium Mg^{2+} and from reactions by soil minerals with atmospheric CO_2.

The SOC pool includes high levels of active humus and relatively inert charcoal C with large surface area, high charge density and bound to clay (Nelson and Baldock, 2000; Lal, 2000). Lehmann and Kleber (2015) argue that evidence does not support the formation of large-molecular-size and persistent humic substances in soils, but rather SOC is a continuum of progressively decomposing organic compounds. This pool (SOC) is generally close to the soil surface and is seriously negatively impacted by erosion (Smith *et al.*, 1993; Lal, 2000, 2003, 2004; Brookes *et al.*, 2008; De Nobili *et al.*, 2008; Peixoto *et al.*, 2020). SIC includes elemental C and carbonate minerals such as calcite, dolomite and gypsum, and is an important constituent of C sequestration in subsurface horizons in soils in arid and semi-arid regions (Rasmussen *et al.*, 1998; Lal, 2000). Implementation of no-till cropping systems in dryland systems increased residue biomass, which then increased effective porosity and water capture (Shaver *et al.*, 2003) and it primarily occurs in arid regions that may contain 2–5 times more SIC than SOC (Lal, 2000).

Significantly greater soil microbial biomass is seen where there has been continuous input of organic matter (OM). Soil bacteria start mineralization of OM into the inorganic form for plant growth. Mineralization continues with protozoa and nematodes feeding on bacteria and fungi; trophic-level interaction is essential for energy and nutrient flow. Although living bacteria dominate living fungi by virtue of size in a living soil, the fungal-derived residue is greater than the bacterial-derived residue (Young, 1999; Decaëns and Jiménez, 2002; Barrios, 2007; Khan *et al.*, 2016; Morriën *et al.*, 2017; Nunes *et al.*, 2020).

The decomposition rate of POM has been linked to its intrinsic chemical complexity, whereas the decomposition rate of stabilized OM relates to its extensive surface areas of clay and crystal particles

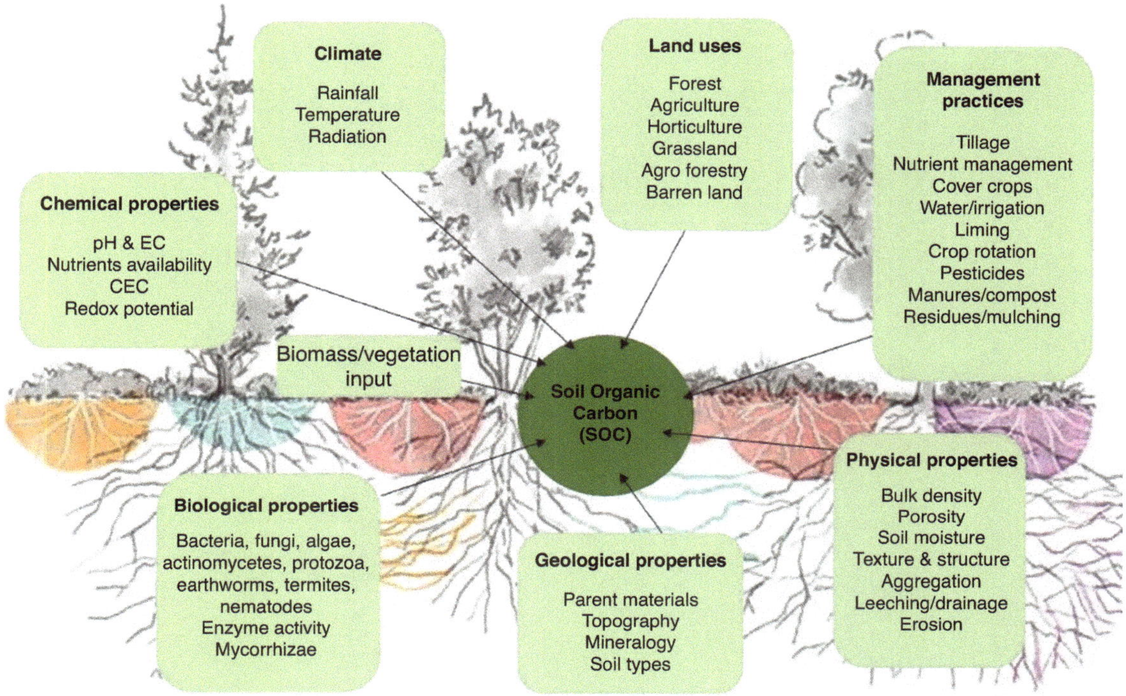

Fig. 26.1. Environmental and biological influences on SOM. (Agpath Pty Ltd, 2023, after Holistic Orchard Network, 2018)

(Oades, 1993; Skjemstad *et al.*, 1998; Shepherd *et al.*, 2002; Li *et al.*, 2015; Veloso *et al.*, 2020).

Globally, the top one metre of soil contains two to three times more organic matter than is found above-ground (Tisdall and Oades, 1979; Anderson, 1995; Sanderman *et al.*, 2017) and is the final decomposition product of any living microbe, plant or animal and their excrement (Wolters, 2000; Decaëns and Jiménez, 2002; Barrios, 2007; Hennig, 2016), plant litter (Tiessen *et al.*, 1994; Anderson, 1995; Almeida *et al.*, 2018; Córdova *et al.*, 2018), manure, stubble, leaves (Almeida *et al.*, 2018), and mulch and vegetable residue broken down by microbes and earthworms (Anderson, 1995; Bernard *et al.*, 2022). Fig. 26.2 shows the soil food web as a totally functional ecosystem where all components are in a cycle of harmony.

Conservation and restoration of SOM are important as a terrestrial sink and carbon (C) and nitrogen (N) storage, because of the many benefits of soil fertility to plant health and crop production (Tiessen *et al.*, 1994; Sparling *et al.*, 2006; Weigelt *et al.*, 2010; Sheng *et al.*, 2015) and because SOM is directly affected by agro-ecosystems such as crop varieties, rotation practices, availability of crop residue between crops and synthetic (organic or inorganic) inputs, all of which affect the health of the microbial populations (Wood *et al.*, 2000). Most agricultural soils have lost 50–70% of their original SOC pool and the depletion is exacerbated by further soil degradation and desertification (Oades, 1993). By optimizing photosynthesis, stimulating root growth, and supporting the fungal–root relationship, it was found that soil C could be built 5–30 times faster than C from above-ground biomass or animal manure with more microbial diversity (Schmidt *et al.*, 2011; Arnhold *et al.*, 2014; Macdonald *et al.*, 2015; Sun *et al.*, 2015; Alhameid *et al.*, 2017).

Biederbeck *et al.* (1994) demonstrated that continuous ploughing and tillage reduced SOM by 21–36% through spikes in microbial activity (Franzluebbers, 2002; Zhang *et al.*, 2014; Arriaga

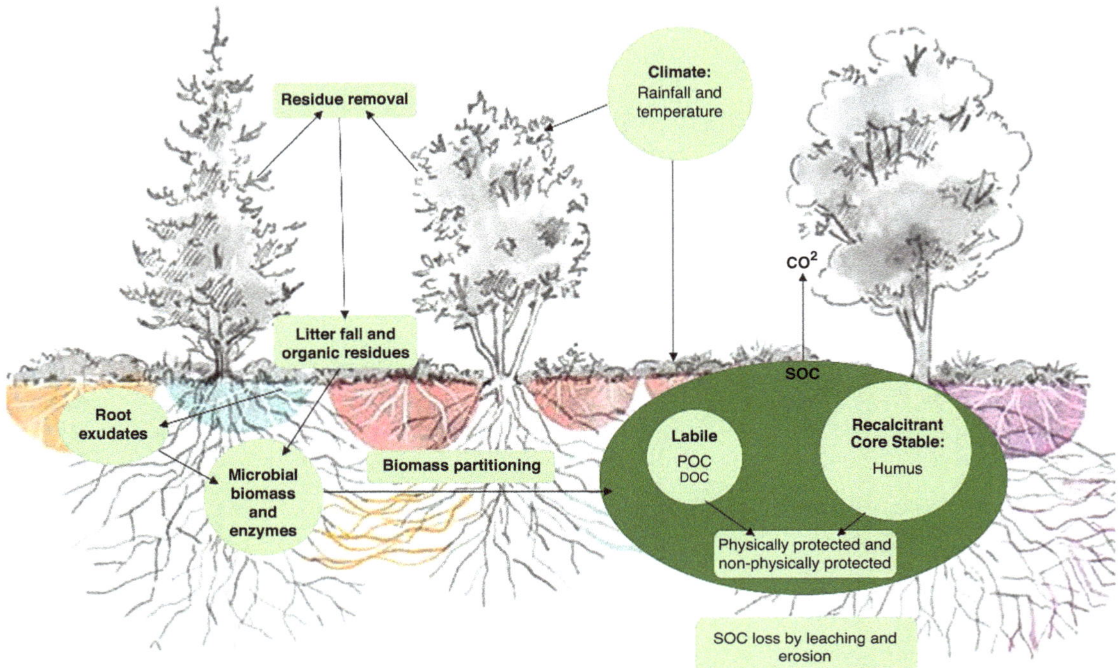

Fig. 26.2. The cycle of organic matter and carbon from terrestrial to soil (POC = particulate organic carbon; DOC = dissolved organic carbon; SOC = soil organic carbon). (Agpath Pty Ltd, 2023, after Holistic Orchard Network, 2018)

Table 26.1. Estimates of soil organic carbon pool. (From IPCC, 2000; Prentice *et al.*, 2001)

Ecosystem	Area (10^9/ha)	SOC pool (PgC) [PgC = petagrams of carbon or 10^{15}g of carbon. 1 PgC = 1 gigatonne of carbon]
Forests:		
• Tropical	1.76	213–216
• Temperate	1.04	100–153
• Boreal (cold temperate)	1.37	338–471
Tropical savannas & grasslands	2.25	247–264
Temperate grassland & scrubland	1.25	176–295
Tundra	0.95	115–121
Desert & semi-desert	4.55	159–191
Cropland	1.60	128–165
Wetlands	0.35	225

et al., 2017; Peltoniemi *et al.*, 2021) and a decrease in the total soil organic carbon (TOC) (Benbi *et al.*, 2015) compared with uncultivated fields (Lal, 2003). SOM responds to environmental conditions globally. Temperature; soil moisture; tillage; and external additions, such as herbicides, pesticides, fertilizers and use of cover crops, are examples (Scholes *et al.*, 1994). Table 26.1 demonstrates the SOC pool among the various climatic and vegetation types globally in 2000.

Soil C *four per mille* was launched at COP21 with an aspiration to increase global SOM stocks by 4 per 1000 (or 0.4%) per year to compensate for anthropogenically generated greenhouse gas emissions globally. An increase in SOM will impact on SOC globally (Ding *et al.*, 2017). The greatest potential for SOC increase is mostly in managed agricultural land. If *four per mille* is considered for the top 1m of global agricultural soils, SOC sequestration is between 2 and 3 GtC/year, which could offset 20–34% of global anthropogenic greenhouse gas emissions (Minasny *et al.*, 2017; VandenBygaart, A.J., 2018; Guillaume *et al.*, 2021).

Soil quality and function resulting from different land uses and soil management techniques have led to a reduction in the quality and quantity of ecosystem services such as a changed terrestrial C balance (Ågren and Wetterstedt, 2007; Brunori *et al.*, 2016; Sanderman *et al.*, 2017; Wang *et al.*, 2019). Adding new organic matter to soil stimulated microbial activity and altered SOC storage by modifying mineralization of native SOC (Oades, 1993; Derrien *et al.*, 2014). Organic farming increased richness, decreased evenness, reduced dispersion and shifted structure of soil microbiota when compared with conventionally managed soils under excessive mineral fertilization. Arbuscular mycorrhizal fungi (AMF) were more abundant in direct-drilled plots and pathogenic fungi were more abundant in ploughed plots (Detheridge *et al.*, 2016; Veloso *et al.*, 2020; Peltoniemi *et al.*, 2021). Long-term agricultural management appeared to select for system-specific community patterns (Hartmann *et al.*, 2015).

Effect of Climate/Temperature on SOM and SOC in Organic and Conventional Farming Systems

Climate is an important and active driver of soil formation, so soil properties such as the SOC pool and soil organic nitrogen (SON) pool are in dynamic equilibrium with climate, particularly with precipitation and air temperature (Lal, 2003; Osanai *et al.*, 2015; Martínez *et al.*, 2016; Li *et al.*, 2017). Many secondary factors including parent material, soil age, topography, potential biota losses, and relative losses of C and N during cultivation are contributing factors to a soil's organic matter response to a reversal of land management to natural flora (Rasmussen *et al.*, 1998; Kleber, 2010; Yemadje *et al.*, 2017; Ramesh *et al.*, 2019; Wulanningtyas *et al.*, 2021).

Ghee *et al.* (2013) question air and soil temperature sensitivity of SOC decomposition and suggest that the persistence of reduced organic C in the environment is co-determined by the interaction between substrates, microbes and abiotic conditions. King *et al.* (1997) suggest C storage in terrestrial systems results from balance or imbalance between processes of net primary production and heterotrophic respiration, particularly during decomposition of dead organic matter, and are influenced by environmental factors like air temperature and moisture. Smith *et al.* (1993) and King *et al.* (1997), in modelling global vegetation and biome distribution to response of terrestrial C storage to future climate change, used different approaches based on correlations between vegetation and climate between and within vegetation ecosystems, respectively, and found the CO_2-stimulated increase in organic matter input to the soil more than compensated for the losses due to climate change.

Root hair proliferation and soil warming strongly influence exudate release, enzyme activities and microbial substrate use in the rhizosphere, so it can be assumed that root hairs may play an important role in regulating enzyme systems and microbial growth to adapt to climate change (Zhang *et al.*, 2020). Decomposition of stable SOM is considered more sensitive to changes in temperature than that of labile organic matter (Jing *et al.*, 2019; Karhu *et al.*, 2019).

Temperature-dependence of litter and SOM mineralization is important because it determines how strongly the feedback from the expected warmer, wetter and more variable climate may be on atmospheric CO_2 concentration (King *et al.*, 1997; Ågren and Wetterstedt, 2007; Jing *et al.*, 2019) and it drives microbial composition, activity and C mineralization (Lal, 2003). However, moisture drives aggregation and C storage which is enhanced by dried conditions (Lal, 1997; Lal, 2003; Cates *et al.*, 2019).

Greatest SOM loss is expected in soils where the present mean annual temperature is less than 4°C (Ekeberg and Riley, 1997; Poeplau *et al.*, 2018; Jing *et al.*, 2019; Karhu *et al.*, 2019). Studies by Fang *et al.* (2005) do not support the opinion that resistant C pools are significantly less responsive to temperature variations than labile C pools, and Giardina and Ryan (2000) and Osanai *et al.* (2015)

suggest overall patterns of C decomposition were unchanged.

Influences of Tillage on SOM and SOC in Organic and Conventional Farming Systems

Continuous ploughing disturbs the soil and increases air introduced in the soil which increases mineralization, reduces OM input from roots and loss of amino acids (Dou *et al.*, 2008; Gosling *et al.*, 2013; Zhang *et al.*, 2014), increases CO_2 emissions by mixing mixed crop residue into the soil, and reduces SOC (Tisdall and Oades, 1982; Osanai *et al.*, 2015; Crews and Rumsey, 2017; Ramesh *et al.*, 2019; Wulanningtyas *et al.*, 2021).

Sustainable intensification often reduces mechanical and chemical inputs and increases biological biodiversity through diverse crop rotations, cover crops, no tillage and live plant residue integration onto cropping land (Reicosky, 2003; Tittonell, 2014; Alhameid *et al.*, 2017; Arriaga *et al.*, 2017; Novara *et al.*, 2018). In undisturbed ecosystems there is a near equilibrium between C inputs and C losses through microbial respiration (Brookes *et al.*, 2008).

Labile fractions of SOM can be reduced by 21–36% through the spike in microbial activity (Arriaga *et al.*, 2017) and decrease in TOC following cultivation (Benbi *et al.*, 2015), compared with uncultivated fields (Crews and Rumsey, 2017). Spring wheat grown with tilled base fallow increased labile carbon SOC while a break crop with 100% cover in place of fallow not only reduced labile carbon SOM loss but also improved moisture and temperature variabilities in the soil (Biederbeck *et al.*, 1994; Cates *et al.*, 2019).

The reduction in fungal:bacterial ratio across all systems emphasizes the need for reduced tillage options for high-quality vegetable systems (Brennan and Acosta-Martinez, 2017). Wu *et al.* (2008) showed that bacterial communities were similar in organic systems but varied in conventional agricultural systems when comparing a range of agricultural land management and crop production practices. The rate of depletion of SOC by tillage is especially high in soils of the tropics and results generally in a decrease in humus substances because of preferential degradation of large molecular-size humic substances and disrupting aggregate activity (Schnitzer, 1991; Woomer and Swift, 1994; Mhazo *et al.*, 2016).

Lee *et al.* (1993) predicted that if the current mix of tillage practices in the US corn belt were to continue (35% conservation tillage), soils used for corn-soybean production would lose 3.02 TgC/yr for the next 100 years. Conversion from plough to no-till (plus winter cover crop and eliminating summer fallow) could sequester 3.3 TgC/yr for the next 100 years. But conservation tillage may need additional herbicides and fertilizers, both of which are based on fossil fuels. Thus, a careful energy balance is needed to assess the net C gains by conversion from plough to no-till. While no-till reduced CO_2 loss through microbial respiration, tillage methods had no effect on N_2O (nitrous oxide) and CH_4 (methane) fluxes (Ferreira *et al.*, 2019, 2020).

Conservation tillage and integration with cover crop concentrations (Jones, 2011; Alhameid *et al.*, 2017; Crystal-Ornelas *et al.*, 2021) can: reduce the extent, frequency and magnitude of mechanical and chemical disturbance; significantly increase the size of SOC and all labile SOC pools, including the mineralized C as a fraction of SOC, the POC and DOM (Gaba *et al.*, 2014; Singh *et al.*, 2015; Stewart *et al.*, 2017; Novara *et al.*, 2018; Reinsch *et al.*, 2018; Crystal-Ornelas *et al.*, 2021; Ruis *et al.*, 2022); reduce the large air-filled pores to slow the rate of gas exchange and C oxidation and improve water efficiency (Hassink *et al.*, 1997; Dow and Carbone, 2007; Hatton *et al.*, 2014; Sun *et al.*, 2015; Jensen *et al.*, 2019; Ullah *et al.*, 2019), but they still fall short of accumulating SOM maintained in grasslands, forests and other natural ecosystems that agriculture replaced (Reicosky and Lindstrom, 1993; Reicosky, 2003; Shi *et al.*, 2013; Renton and Flower, 2015; Alhameid *et al.*, 2017; Novara *et al.*, 2018).

Zero-tillage leaves weed seeds on or near the soil surface resulting in a higher weed population through increased and faster emergence (Singh *et al.*, 2015; Stewart *et al.*, 2017). Loss from disturbance is greater in sandy soils (Hagedorn *et al.*, 2003; Larsen *et al.*, 2014; Sradnick *et al.*, 2014), though the labile pool size and management impacts decreased with depth (Lal, 1997; Dow and Carbone, 2007; Larsen *et al.*, 2014; Singh *et al.*, 2020). The reliance on herbicides for weed control leads to weed resistance.

In countries where agriculture is carried out in hilly terrain and where soils are prone to erosion, alternative management, such as reduced tillage, mulching and cover crops, developed to protect soils from erosion and increased SOM (Lal, 2003; Arnhold *et al.*, 2014; García-Díaz *et al.*, 2016; Vicente-Vicente *et al.*, 2016; Novara *et al.*, 2018). In dry conditions such as the Mediterranean region vineyards, where low OM content results from climate-induced high mineralization rates, tillage leads to high erosion rates (Novara *et al.*, 2018).

Traditional indigenous farmers, knowingly or not, were careful managers of SOM (Pereira *et al.*, 2018). Levels of 58% have been recorded, as fertility-generating cover crop rotations, pasture and shrub fallows, and farmyard manure applications were essential to maintaining fertility (Lal *et al.*, 2019). Intensive and conventional agricultural land use has doubled world grain harvest (1960–2005) exceeding approximately 2 billion tons per year attributed to higher yield cultivars, synthetic chemicals, mechanization and irrigation. However, this success has resulted in degraded water quality and loss of approximately 1.5 million ha of arable land/year. Forty per cent of global croplands show soil erosion, reduced fertility, loss of native habitat, loss of SOM, and increased crop disease (Foley *et al.*, 2005).

Influence of Cover Crops on SOM and SOC in Organic and Conventional Farming Systems

Cover crop or ground cover species should be multi-species and used in rotation where legumes appear to increase SOC better than grasses (Chahal and Van Eerd, 2019; Ramesh *et al.*, 2019), and be selected for their adaption to local conditions and the intended purpose of managing the soil. Such characters should include the ability to:

- easily establish, develop quickly over the soil;
- supply enough biomass in the short term in residue maintenance;
- resist disease and not act as a host for cash crop diseases (Adetunji *et al.*, 2020; Crystal-Ornelas *et al.*, 2021);
- have different rooting densities and branching at different soil depths;
- result in more efficient extraction of nutrients from all soil levels;
- be easy to terminate;
- build C and N fixation;
- have controlled height of standing residues; and
- be economically viable.

(Alhameid *et al.*, 2017; Adetunji *et al.*, 2020)

De Oliveira *et al.* (2022) found many years were needed to show a consistent increase in SOC which suggests that soil health benefits from best-practice management accumulate over time and highlight the need for adoption of long-term best practice to achieve agricultural sustainability so there exists the potential to mitigate the increase in CO_2 levels through C sequestration in soils (Larsen *et al.*, 2014; Brennan and Acosta-Martinez, 2017; Mathew *et al.*, 2020; Crystal-Ornelas *et al.*, 2021). Soils around the world provide an important active pool of organic C and play a major role in the global carbon cycle, and crop residue is a major renewable resource that impacts on the global carbon cycle. The positive impact on diverse functional groups like earthworms, pests and their natural predators, elemental transformers, decomposers and microsymbionts, is lost in intensively managed agricultural systems with parallel loss in yield and soil stability (Lal, 1997; Brookes *et al.*, 2008; De Nobili *et al.*, 2008; Bommarco *et al.*, 2013; Larsen *et al.*, 2014; Brennan and Acosta-Martinez, 2017).

Brennan and Acosta-Martinez (2017) found bacterial population composition was largely unaffected by management differences between organic systems but there is evidence that the species of cover crop and frequency did alter the relative abundance of some genera such as *Pseudomonas* that include important biological control agent species and growth-promoting rhizobacteria.

Cover crops, particularly winter legumes:

- reduced nitrate-N concentrations;
- built soil biodiversity and nutrient cycling;
- improved soil quality and crop yield, at the same time increasing stable OM and labile (active C);
- sequestered C and aided in climate change mitigation (Jones, 2011; Sequeira *et al.*, 2011; Verzeaux *et al.*, 2016; Alhameid *et al.*, 2017; Merante *et al.*, 2017; Poeplau *et al.*, 2018; Chahal and Van Eerd, 2019; Haghverdi and Kooch, 2020; Zhu *et al.*, 2020).

Ramesh *et al.* (2019) suggest that some of the weed seed germination suppression seen when cover crops are used may result from the release of biochemicals as a form of allelopathy (Sardana *et al.*, 2017; Korres *et al.*, 2019).

It is generally assumed that SOM levels are linearly proportional to the exogenous carbon input, but other studies demonstrate a non-linear relationship between C input quantity and SOM improvement (Sacco *et al.*, 2015; Colbach *et al.*,

2019; Adetunji *et al.*, 2020; Saleem *et al.*, 2020; Wulanningtyas *et al.*, 2021). Jian *et al.* (2020) estimated that cover cropping in agriculture resulted in a 15.5% increase in SOC (95% confidence interval of 13.8–17.3%) in near surface soils, indicating that cover crop inclusion in agricultural rotations has the potential to increase soil carbon sequestration. If 15% of the world's cropland incorporated cover crops, approximately 0.16 ± 0.06 Pg C could be sequestrated annually, representing 1–2% of yearly emissions from fossil fuel combustion (Jian *et al.*, 2020). Most agricultural soils have lost 50–70% of their original SOC pool and depletion is exacerbated by further soil degradation and sand desertification (Franzluebbers, 2002; Zhang *et al.*, 2014; Mathew *et al.*, 2020; Peltoniemi *et al.*, 2021).

Influences of Litter/Above-Ground Residue on SOM and SOC in Organic and Conventional Farming Systems

Early in decomposition, litter or surface organic plant material loses mostly non-structural compounds rapidly through the organic matter-microbial pathway. This results in incorporation into microbial biomass at high rates, efficient SOM formation and fast association with silt and clay fractions in the topsoil mineral layer (Lal, 1997; Gosling *et al.*, 2013; Benbi *et al.*, 2015; Qafoku, 2015; Gómez-Muñoz *et al.*, 2016; Martínez-García *et al.*, 2018). Residue litter fragments move into the soil and have low N and high lignin content generally described as recalcitrant, showing low decomposition rates and high relative amounts of remaining residue (Franzluebbers, 2002; Cotrufo *et al.*, 2015; Poeplau *et al.*, 2018).

The organic matter-microbial pathway is stabilized by organo-mineral interactions such as:

- having rhizosphere bacterial communities as ecological indicators for soil health in arable land;
- having healthier plants resulting under conservation agriculture that combines zero tillage;
- having permanent raised beds;
- having biofertilizers and crop residue retention with minimal soil disturbance;
- having good correlation with nitrogen fixation, nitrification and denitrification in the soil (Singh *et al.*, 2020);
- where cellulose and phosphate addition to soil hardly enhances microbial utilization of mineral-associated N but where mineral-associated N was highly bioavailable in mineral topsoils (Diaz *et al.*, 2006; Turner *et al.*, 2017).

Biodiversity loss from C in dead and living roots, fauna and microbes (Sanderman *et al.*, 2017; Edwards, 2022); eutrophication; burning; soil erosion (loss of SOM); flooding etc., have a direct relationship with human well-being (Lal, 1997, 2003; Ekschmitt *et al.*, 2005; Diaz *et al.*, 2006; Mhazo *et al.*, 2016; Udawatta *et al.*, 2017; Córdova *et al.*, 2018; Nunes *et al.*, 2020; Mason *et al.*, 2023).

Hatton *et al.* (2014) and Lam *et al.* (2022) found exogenous C sources increase SOM mineralization/positive priming (addition of fresh OM) and nutritional additions, and reduced the size of positive priming in non-fumigated soils but did not affect distribution of substrate-derived C in microbial communities (Zhou *et al.*, 2021). These findings support the view that microbial community composition is a determinant of SOM mineralization. There is evidence that utilization of a labile substrate by fungi and actinobacteria (*not* Gram-negative populations) promote positive SOM priming. This shows stable SOC is more vulnerable to priming than labile SOC once microbes are provided with exogenous C substrate. This high vulnerability of stable SOC to priming has implications on SOC cycling and climate change (Ågren and Wetterstedt, 2007; von Lützow and Kögel-Knabner, 2009; García-Pausas and Paterson, 2011; Frasier *et al.*, 2016; Vestergård *et al.*, 2016; Li *et al.*, 2017; Zhang *et al.*, 2023). There is no priming effect when ash or biochar was applied to soil, but priming was positive when soils were amended with unburnt prunings, and it was significantly correlated with water soluble C (Liang and Balser, 2011).

Mineral-associated organic matter (MAOM) typically accounts for more than 50% of total SOM and high accumulation of MAOM in soils without nutrient addition suggests plant biomolecules may be important contributors to MAOM in nutrient-poor environments where microbes depolymerize large amounts of plant compounds in search of nutrients (Ekschmitt *et al.*, 2005; Cotrufo *et al.*, 2015; Córdova *et al.*, 2018).

Influence of Organic and Synthetic Inputs on SOM and SOC in Organic and Conventional Farming Systems

Conti *et al.* (2014) found that strawberries grown in trials in Campania (southern Italy) generally yielded less produce when produced by organic agricultural methods than by conventional industrial-style agriculture. However, the organic produce gave increased berry weight, dry weight, higher sugars, vitamin C and β-carotine, and lower nitrates. By contrast, Suja *et al.* (2017) found higher yields (29%) of taro in organic than in conventional with more available phosphorus (P), higher organic C, and exchangeable calcium (Ca), magnesium (Mg), iron (Fe), manganese (Mn), zinc (Zn) and copper (Cu). Arnhold *et al.* (2014) found that where herbicides were used in intensive organic systems before planting, residue mulching was essential in mountainous farmland in South Korea. Long-term organic cropping and forest systems showed changes in SOC, related mainly to compost and microbial biomass C and N rather than cover crop frequency (Wander *et al.*, 1994; Lal, 2003; Marriott and Wander, 2006; Tittarelli *et al.*, 2007; Martínez *et al.*, 2016; Brennan and Acosta-Martinez, 2017).

While invertebrates and Gram-positive bacteria increased across the systems, the fungal:bacterial ratio decreased over time (Larsen *et al.*, 2014) and the bacteria were the most important in mineral weathering and nutrient cycling in low-nutrient and acidic soils (Anderson, 1995; Wardle *et al.*, 2004; Uroz *et al.*, 2015; Meena *et al.*, 2018; Jensen *et al.*, 2019; Mosier *et al.*, 2019; Kaushal *et al.*, 2021).

When Azeem *et al.* (2020) investigated legume-cereal crop rotation with biochar and nitrogen-phosphorus-potassium (NPK) additions, they found the crop type was more important in shaping the microbial communities during crop rotation than the biochar or fertilizer (Chadwick *et al.*, 1994). Arbuscular mycorrhizae appear to respond positively to biochar soil additions while diversity in actinobacteria, which is involved in OM decomposition, decreased. Carbon dioxide emissions spiked with the addition of biochar to soils because it is the Gram-positive bacterial pool that is usually involved in OM and litter decomposition and use of the bioavailable C fraction in biochar. Gram-negative bacteria prefer plant-derived labile C sources and were unaffected (Azeem *et al.*, 2020). They found that biochar led to plant growth stimulation and increased arbuscular mycorrhizal activity (Phillips, 2017) but a decrease in actinobacteria and OM decomposition.

The highly productive Amazonian dark earths, *terra preta*, are thought to result from human burning to clear forest for cultivation. The resulting black earth is biochar-like and has increased microbiological activity, and adds colloidal-sized organic decomposition products to the soil matrix. Together with by-products of incomplete combustion, these provide charged surfaces largely absent in local soils and increase in nutrient-retention capacity, setting up a synergistic cycle of continuous fertility in the same way as does biochar (Glaser and Woods, 2004; Schmidt *et al.*, 2011; Crews and Rumsey, 2017).

Long-term field research from Rothamstead, UK, has shown, over 180 years, that inorganic fertilizer reduced microbial diversity and microbial function. After 180 years of continuous wheat, P levels in plots receiving zero-fertilizer inputs remained the same and have not fallen. Applying large amounts of organic matter only resulted in small changes in soil C. Breakdown of organic matter to CO_2 resulted in no net gain. Optimizing photosynthesis, stimulating root growth and support for the fungal energy channel resulted in soil carbon building at 5–30 times faster than C from above-ground biomass or animal manure (Schmidt *et al.*, 2011; Macdonald *et al.*, 2015) and possibly 60–92% of contemporary agriculturally generated greenhouse gases could be offset if all land were to be converted to organic and regenerative practices (Leifeld, 2012).

Liu *et al.* (2020) found mineral P fertilization on the soil bacterial community in the wheat rhizosphere soils increased SOC and microbial biomass and available P, compared with the control (P0), but the total bacterial community in the wheat rhizosphere soil reduced bacterial operational taxonomic groups, diversity and the connections, which may affect soil biogeochemical cycles. Miller *et al.* (2003) showed that high N fertilizer in conventional agriculture reduced microbial contribution to SOC pools, particularly the fungal contribution at application rates of 150 kgN/ha/yr (Hatton *et al.*, 2014; Zhang *et al.*, 2016).

Instead of applying high annual doses of NPK only, alternation with mineral and organic fertilizers should be used to host a more efficient microbial network, e.g. decomposers, microsymbionts, elemental transformers, engineers such as earthworms, pests and their natural enemies (Lal, 1997; Bommarco *et al.*, 2013; Semenov *et al.*, 2020; Yao *et al.*, 2020). Fertilization is a more important factor in determining the abundance and diversity of total and active bacterial communities in both the rhizosphere and bulk soil than plant species grown. While the traditional concept of the key role of plant species in shaping bacterial community composition in the rhizosphere is valid for non-cultivated plant species, it is not confirmed for agricultural ecosystems (Yao *et al.*, 2020; Mason *et al.*, 2023). Manure applications annually greatly increased the contribution of microbial residues to SOC, reduced runoff by 2–62%, especially on sloping land, and reduced soil loss by 15–65% (Ding *et al.*, 2015). Long-term applications were found to hold both an active pool of stable SOM and a pool of relatively stable SOM in terms of microbial-derived organic matter on microaggregates (Oades, 1984; Gilley and Risse, 2000; Lal, 2003; Maillard *et al.*, 2015; García-Díaz *et al.*, 2016).

Influence of Soil Invertebrates, Earthworms and Insects on SOM and SOC in Organic and Conventional Farming Systems

The soil microfauna, which are abundant and species-rich in most terrestrial habitats, have an equally important role in the generation of SOC and SOM and soil health. They stabilize aggregates, burrow and bury OM – artifacts that regulate C and nutrient mineralization, and should be considered in ecosystem studies (Oades, 1993; Anderson, 1995; Jing *et al.*, 2019). Climate, soil condition, vegetation and human interaction determine the composition of the soil fauna community from local to global scale (Wolters, 2000; Swift *et al.*, 2004).

How crops are managed (organic vs conventional) impacts crop pests on both their presence and abundance. Puech *et al.* (2014) show that ladybirds and carabid beetles are negatively influenced by tillage frequency, also by crop varieties and pesticides, while parasitoids (e.g. aphids) were not affected by farming practices. Arthropod numbers varied within farming systems and invertebrate seed-predator activity density was higher in non-till cropping systems (Ramesh, 2015).

Influences of Herbicides and Insecticides on SOM and SOC in Organic and Conventional Farming Systems

In developed regions such as the USA, Canada, Australia, Europe and Japan, chemical weed control via herbicide application is favoured because of uniform application, high efficiency and cost reductions associated with mechanization and time. In the USA, herbicides are used routinely to eliminate weeds in more than 90% of crop acreage (Ziska, 2016; Ziska *et al.*, 2019).

Climate-change induced inconsistencies in rainfall, wind, relative humidity, or soil or air temperatures could reduce spray coverage or retention times of active ingredients following applications. It is highly likely that weed populations will change in parallel with cropping systems as temperature, climate and CO_2 change (Varanasi *et al.*, 2016).

Intensive use of herbicides for weed management in conventional and intensive agriculture and excessive use of N fertilizers have the greatest negative impact on biological diversity, soil SOM and acidification (Ziska, 2016; Arriaga *et al.*, 2017; Ukalska-Jaruga *et al.*, 2023).

Innovative non-chemical methods such as night soil cultivation prevents light-triggered germination of some weed seeds; and increased mulching with increase in SOM creates an environment on the soil surface unsuitable for weed emergence. Increased seeding rates provide a dense crop canopy to outcompete weedy species. Multi-species cropping, such as green manure and companion plants side-by-side with crop plants, where the green manure adds to SOM because it is not removed with crop plants, and encouraging weed seed predation by insects and rodents to deplete seed banks, have been introduced into fragile regions like sub-Saharan Africa (Crowder *et al.*, 2010; Wortman *et al.*, 2013; Bàrberi, 2019).

Conclusion

The use of organic farming practices offers many benefits for enhancing SOM, SOC and C sequestration. These include enhanced soil quality and biomass productivity; improved water and air quality by reducing risks of soil erosion, and decreasing the concentration of sediments and sediment-borne pollutants transported by water and wind. Additionally, it reduces the risks of accelerated greenhouse effects by decreasing the rate of enrichment of atmospheric CO_2. Strategies that consider the impact of a changing climate and tillage methods together with appropriate cover cropping and reduced synthetic inputs on SOM and SOC need to be adopted to strengthen the soil C sinks in the manner that convinces farmers to make changes to their operating methods. Soil C could be commodified as a form of farm product by determining the value of soil C to the farmer or land manager and to society. Farmers need to see a visible future before they make changes to their operating methods (Lal, 2000; Sparling *et al.*, 2006).

The restoration of degraded soils, conversion of agriculturally marginal lands to appropriate land use, and the adoption of recommended management practices on agricultural soils can reverse degradative trends and lead to SOC sequestration. All the conservation techniques will be needed to meet the food demands of the growing population regardless of the threat of global warming (Lal, 2000; Lal, 2003; Maggio *et al.*, 2013; Peltoniemi *et al.*, 2021). The only long-term solution lies in increasing SOM and SOC that has been depleted from past land misuse. Change is essential for quality of life on Earth.

References

Adetunji, A.T., Ncube, B., Mulidzi, R. and Lewu, F.B. (2020) Management impact and benefit of cover crops on soil quality: A review. *Soil and Tillage Research* 204, 104717.

Ågren, G.I. and Wetterstedt, J.Å.M. (2007) What determines the temperature response of soil organic matter decomposition? *Soil Biology and Biochemistry* 39(7), 1794–1798.

Alhameid, A., Tobin, C., Maiga, A., Kumar, S., Osborne, S. *et al.* (2017) Intensified agroecosystems and changes in soil carbon dynamics. In: Al-Kaisi, M.M. and Lowery, B. (eds) *Soil Health and Intensification of Agroecosystems*. Elsevier, pp. 195–214.

Almeida, L.F.J., Hurtarte, L.C.C., Souza, I.F., Soares, E.M.B., Vergütz, L. *et al.* (2018) Soil organic matter formation as affected by eucalypt litter biochemistry – evidence from an incubation study. *Geoderma* 312, 121–129.

Anderson, J.M. (1995) Soil organisms as engineers: Microsite modulation of macroscale processes. In: Jones, C.G. and Lawton, J. (eds) *Linking Species & Ecosystems*. Springer, pp. 94–106.

Arnhold, S., Lindner, S., Lee, B., Martin, E., Kettering, J. *et al.* (2014) Conventional and organic farming: Soil erosion and conservation potential for row crop cultivation. *Geoderma* 219–220, 89–105.

Arriaga, F.J., Guzman, J. and Lowery, B. (2017) Conventional agricultural production: Systems and soil functions. In: Al-Kaisi, M.M. and Lowery, B. (eds) *Soil Health and Intensification of Agroecosystems*. Elsevier, pp. 109–125.

Azeem, M., Sun, D., Crowley, D., Hayat, R., Hussain, Q. *et al.* (2020) Crop types have stronger effects on soil microbial communities and functionalities than biochar or fertilizer during two cycles of legume-cereal rotations of dry land. *The Science of the Total Environment* 715, 136958.

Bàrberi, P. (2019) Ecological weed management in Sub-Saharan Africa: Prospects and implications on other agroecosystem services. In: *Advances in Agronomy*, Vol. 156. pp. 219–264.

Barrios, E. (2007) Soil biota, ecosystem services and land productivity. *Ecological Economics* 64(2), 269–285.

Benbi, D.K., Brar, K., Toor, A.S. and Singh, P. (2015) Total and labile pools of soil organic carbon in cultivated and undisturbed soils in Northern India. *Geoderma* 237–238, 149–158.

Bernard, L., Basile-Doelsch, I., Derrien, D., Fanin, N., Fontaine, S. *et al.* (2022) Advancing the mechanistic understanding of the priming effect on soil organic matter mineralisation. *Functional Ecology* 36(6), 1355–1377.

Biederbeck, V.O., Janzen, H.H., Campbell, C.A. and Zentner, R.P. (1994) Labile soil organic matter as influenced by cropping practices in an arid environment. *Soil Biology & Biochemistry* 26(12), 1647–1656.

Bommarco, R., Kleijn, D. and Potts, S.G. (2013) Ecological intensification: Harnessing ecosystem services for food security. *Trends in Ecology & Evolution* 28(4), 230–238.

Brennan, E.B. and Acosta-Martinez, V. (2017) Cover cropping frequency is the main driver of soil microbial changes during six years of organic vegetable production. *Soil Biology and Biochemistry* 109, 188–204.

Brookes, P.C., Cayuela, M.L., Contin, M., De Nobili, M., Kemmitt, S.J. *et al.* (2008) The mineralisation of fresh and humified soil organic matter by the soil microbial biomass. *Waste Management* 28(4), 716–722.

Brunori, E., Farina, R. and Biasi, R. (2016) Sustainable viticulture: The carbon-sink function of the vineyard agro-ecosystem. *Agriculture, Ecosystems & Environment* 223, 10–21.

Cates, A.M., Braus, M.J., Whitman, T.L. and Jackson, R.D. (2019) Separate drivers for microbial carbon mineralization and physical protection of carbon. *Soil Biology and Biochemistry* 133, 72–82.

Chadwick, O.A., Kelly, E.F., Merritts, D.M. and Amundson, R.G. (1994) Carbon dioxide consumption during soil development. *Biogeochemistry* 24(3).

Chahal, I. and Van Eerd, L.L. (2019) Quantifying soil quality in a horticultural-cover cropping system. *Geoderma* 352, 38–48.

Colbach, N., Gardarin, A. and Moreau, D. (2019) The response of weed and crop species to shading: Which parameters explain weed impacts on crop production? *Field Crops Research* 238, 45–55.

Conti, S., Villari, G., Faugno, S., Melchionna, G., Somma, S. *et al.* (2014) Effects of organic vs. conventional farming system on yield and quality of strawberry grown as an annual or biennial crop in southern Italy. *Scientia Horticulturae* 180, 63–71.

Córdova, S.C., Olk, D.C., Dietzel, R.N., Mueller, K.E., Archontouilis, S.V. *et al.* (2018) Plant litter quality affects the accumulation rate, composition, and stability of mineral-associated soil organic matter. *Soil Biology and Biochemistry* 125, 115–124.

Cotrufo, M.F., Soong, J.L., Horton, A.J., Campbell, E.E., Haddix, M.L. *et al.* (2015) Formation of soil organic matter via biochemical and physical pathways of litter mass loss. *Nature Geoscience* 8(10), 776–779.

Crews, T. and Rumsey, B. (2017) What agriculture can learn from native ecosystems in building soil organic matter: A review. *Sustainability* 9(4), 578.

Crowder, D.W., Northfield, T.D., Strand, M.R. and Snyder, W.E. (2010) Organic agriculture promotes evenness and natural pest control. *Nature* 466(7302), 109–112.

Crystal-Ornelas, R., Thapa, R. and Tully, K.L. (2021) Soil organic carbon is affected by organic amendments, conservation tillage, and cover cropping in organic farming systems: A meta-analysis. *Agriculture, Ecosystems & Environment* 312, 107356.

De Nobili, M., Contin, M., Mahieu, N., Randall, E.W. and Brookes, P.C. (2008) Assessment of chemical and biochemical stabilization of organic C in soils from the long-term experiments at Rothamsted (UK). *Waste Management* 28(4), 723–733.

de Oliveira, D.C., Maia, S.M.F., Freitas, R. de C.A. and Cerri, C.E.P. (2022) Changes in soil carbon and soil carbon sequestration potential under different types of pasture management in Brazil. *Regional Environmental Change* 22(3), 3.

Decaëns, T. and Jiménez, J.J. (2002) Earthworm communities under an agricultural intensification gradient in Colombia. *Plant and Soil* 240(1), 133–143.

Derrien, D., Plain, C., Courty, P.-E., Gelhaye, L., Moerdijk-Poortvliet, T.C.W. *et al.* (2014) Does the addition of labile substrate destabilise old soil organic matter? *Soil Biology and Biochemistry* 76, 149–160.

Detheridge, A.P., Brand, G., Fychan, R., Crotty, F.V., Sanderson, R. *et al.* (2016) The legacy effect of cover crops on soil fungal populations in a cereal rotation. *Agriculture, Ecosystems & Environment* 228, 49–61.

Diaz, S., Fargione, J., Chapin, 3rd, F.S. and Tilman, D. (2006) Biodiversity loss threatens human well-being. *PLoS Biology* 4(8), 277.

Ding, X., Liang, C., Zhang, B., Yuan, Y. and Han, X. (2015) Higher rates of manure application lead to greater accumulation of both fungal and bacterial residues in macroaggregates of a clay soil. *Soil Biology and Biochemistry* 84, 137–146.

Ding, X., Zhang, B., Lü, X., Wang, J. and Horwath, W.R. (2017) Parent material and conifer biome influence microbial residue accumulation in forest soils. *Soil Biology and Biochemistry* 107, 1–9.

Dou, F., Wright, A.L. and Hons, F.M. (2008) Sensitivity of labile soil organic carbon to tillage in wheat-based cropping systems. *Soil Science Society of America Journal* 72(5), 1445–1453.

Dow, K. and Carbone, G. (2007) Climate science and decision making. *Geography Compass* 1(3), 302–324.

Edwards, T. (2022) *What Is Soil Organic Carbon? Department of Agriculture and Food*. Government of Western Australia.

Ekeberg, E. and Riley, H.C.F. (1997) Tillage intensity effects on soil properties and crop yields in a long-term trial on morainic loam soil in southeast Norway. *Soil and Tillage Research* 42(4), 277–293.

Ekschmitt, K., Liu, M., Vetter, S., Fox, O. and Wolters, V. (2005) Strategies used by soil biota to overcome soil organic matter stability – Why is dead organic matter left over in the soil? *Geoderma* 128(1–2), 167–176.

Fang, C., Smith, P., Moncrieff, J.B. and Smith, J.U. (2005) Similar response of labile and resistant soil organic matter pools to changes in temperature. *Nature* 433(7021), 57–59.

Ferreira, C. dos R., Nascimento Guedes, J. and Pereira, M.G. (2019) Diversity of the edaphic macrofauna in areas managed under no-tillage for different periods. *Semina: Ciências Agrárias* 40(2).

Ferreira, C. dos R., Pereira, M.G. and Nascimento Guedes, J. (2020) Dynamics of soil aggregation and organic carbon fractions over 23 years of no-till management. *Soil and Tillage Research* 198.

Foley, J.A., Defries, R., Asner, G.P., Barford, C., Bonan, G. *et al.* (2005) Global consequences of land use. *Science* 309(5734), 570–574.

Franzluebbers, A.J. (2002) Soil organic matter stratification ratio as an indicator of soil quality. *Soil & Tillage Research* 66, 95–106.

Frasier, I., Noellemeyer, E., Figuerola, E., Erijman, L., Permingeat, H. *et al.* (2016) High quality residues from cover crops favor changes in microbial community and enhance C and N sequestration. *Global Ecology and Conservation* 6, 242–256.

Gaba, S., Bretagnolle, F., Rigaud, T. and Philippot, L. (2014) Managing biotic interactions for ecological intensification of agroecosystems. *Frontiers in Ecology and Evolution* 2.

García-Díaz, A., Allas, R.B., Gristina, L., Cercà, A., Pereira, P. *et al*. (2016) Carbon input threshold for soil carbon budget optimization in eroding vineyards. *Geoderma* 271, 144–149.

García-Pausas, J. and Paterson, E. (2011) Microbial community abundance and structure are determinants of soil organic matter mineralisation in the presence of labile carbon. *Soil Biology and Biochemistry* 43(8), 1705–1713.

Ghee, C., Neilson, R., Hallett, P.D., Robinson, D. and Paterson, E. (2013) Priming of soil organic matter mineralisation is intrinsically insensitive to temperature. *Soil Biology and Biochemistry* 66, 20–28.

Giardina, C.P. and Ryan, M.G. (2000) Evidence that decomposition rates of organic carbon in mineral soil do not vary with temperature. *Nature* 404(6780), 858–861.

Gilley, J.E. and Risse, L.M. (2000) Runoff and soil loss as affected by the application of manure. *Transactions of the ASAE* 43(6), 1583–1588.

Glaser, B. and Woods, W.I. (2004) *Amazonian Dark Earths: Explorations in Space and Time*. Springer.

Gómez-Muñoz, B., Valero-Valenzuela, J.D., Hinojosa, M.B. and García-Ruiz, R. (2016) Management of tree pruning residues to improve soil organic carbon in olive groves. *European Journal of Soil Biology* 74, 104–113.

Gosling, P., Parsons, N. and Bending, G.D. (2013) What are the primary factors controlling the light fraction and particulate soil organic matter content of agricultural soils? *Biology and Fertility of Soils* 49(8), 1001–1014.

Grow organic apples with the holistic orchard network (2018). Available at: https://www.facebook.com/growo rganicapples/

Guillaume, T., Bragazza, L., Levasseur, C., Libohova, Z. and Sinaj, S. (2021) Long-term soil organic carbon dynamics in temperate cropland-grassland systems. *Agriculture, Ecosystems & Environment* 305.

Hagedorn, F., Spinnler, D. and Siegwolf, R. (2003) Increased N deposition retards mineralization of old soil organic matter. *Soil Biology and Biochemistry* 35(12), 1683–1692.

Haghverdi, K. and Kooch, Y. (2020) Soil carbon and nitrogen fractions in response to land use/cover changes. *Acta Oecologica* 109, 103659.

Hartmann, M., Frey, B., Mayer, J., Mäder, P. and Widmer, F. (2015) Distinct soil microbial diversity under long-term organic and conventional farming. *The ISME Journal* 9(5), 1177–1194.

Hassink, J., Whitmore, A.P. and Kubát, J. (1997) Size and density fractionation of soil organic matter and the physical capacity of soils to protect organic matter. *European Journal of Agronomy* 7(1–3), 189–199.

Hatton, P.-J., Bodé, S., Angeli, N., Boeckx, P., Zeller, B. *et al*. (2014) Assimilation and accumulation of C by fungi and bacteria attached to soil density fractions. *Soil Biology and Biochemistry* 79, 132–139.

Hennig, E. (2016) *Secrets of Fertile Soils Humus as the Guardian of the Fundaments of Natural Life*. Organischer Landbau.

Holistic Orchard Network (2018) *Grow organic apples with the holistic orchard network*.

Hoorman, J.J. and Islam, R. (2010) Understanding soil microbes and nutrient recycling. *Actinomycetes* 107(108), 40–500.

IPCC (2000) IPCC special report: *Emissions Scenarios*. Intergovernmental Panel on Climate Change. Available at: Emissions Scenarios - IPCC (accessed 10 February 2025).

Jensen, J.L., Schjønning, P., Watts, C.W., Christensen, B.T., Peltre, C. *et al*. (2019) Relating soil C and organic matter fractions to soil structural stability. *Geoderma* 337, 834–843.

Jian, J., Du, X., Reiter, M.S. and Stewart, R.D. (2020) A meta-analysis of global cropland soil carbon changes due to cover cropping. *Soil Biology and Biochemistry* 143, 107735.

Jing, Y., Wang, Y., Liu, S., Zhang, X., Wang, Q. *et al*. (2019) Interactive effects of soil warming, throughfall reduction, and root exclusion on soil microbial community and residues in warm-temperate oak forests. *Applied Soil Ecology* 142, 52–58.

Jones, C. (2011) Carbon that counts. Available at: https://soilcarboncoalition.org/files/JONES-Carbon-that-counts-20Mar11.pdf (accessed 10 February 2025).

Karhu, K., Hilasvuori, E., Järvenpää, M., Arppe, L., Christensen, B.T. *et al*. (2019) Similar temperature sensitivity of soil mineral-associated organic carbon regardless of age. *Soil Biology and Biochemistry* 136, 107527.

Kaushal, R., Tewari, S., Thapliyal, S.D., Kumar, A., Roy, T. *et al*. (2021) Build-up of labile, non-labile carbon fractions under fourteen-year-old bamboo plantations in the Himalayan foothills. *Heliyon* 7(8), e07850.

Khan, K.S., Mack, R., Castillo, X., Kaiser, M. and Joergensen, R.G. (2016) Microbial biomass, fungal and bacterial residues, and their relationships to the soil organic matter C/N/P/S ratios. *Geoderma* 271, 115–123.

King, A.W., Post, W.M. and Wullschleger, S.D. (1997) The potential response of terrestrial carbon storage to changes in climate and atmospheric CO2. *Climatic Change* 35(2), 199–227.

Kleber, M. (2010) What is recalcitrant soil organic matter?. *Environmental Chemistry* 7(4), 320.

Korres, N.E., Burgos, N.R., Travlos, I., Vurro, M., Gitsopoulos, T.K. *et al.* (2019) New directions for integrated weed management: Modern technologies, tools and knowledge discovery. In: Sparks, D.L. (ed.) *Advances in Agronomy* 155. pp. 245–319.

Lal, R. (1997) Residue management, conservation tillage and soil restoration for mitigating greenhouse effect by CO-enrichment. *Soil & Tillage Research* 43, 81–107.

Lal, R. (2003) Global potential of soil carbon sequestration to mitigate the greenhouse effect. *Critical Reviews in Plant Sciences* 22(2), 151–184.

Lal, R. (2004) Soil carbon sequestration impacts on global climate change and food security. *Science* 304(5677), 1623–1627.

Lal, R. (2000) World cropland soils as a source or sink for atmospheric carbon. In: Sparks, D.L. (ed.) *Advances in Agronomy* 71.

Lal, B., Gautam, P., Nayak, A.K., Panda, B.B., Bihari, P. *et al.* (2019) Energy and carbon budgeting of tillage for environmentally clean and resilient soil health of rice-maize cropping system. *Journal of Cleaner Production* 226, 815–830.

Lam, S.K., Xia, L. and Chen, D. (2022) Boosting the benefits of compost. *Nature Food* 3(9), 682–683.

Larsen, E., Grossman, J., Edgell, J., Hoyt, G., Osmond, D. *et al.* (2014) Soil biological properties, soil losses and corn yield in long-term organic and conventional farming systems. *Soil and Tillage Research* 139, 37–45.

Lee, J.J., Phillips, D.L. and Liu, R. (1993) The effect of trends in tillage practices on erosion and carbon content of soils in the US corn belt. *Water, Air, & Soil Pollution* 70(1–4), 389–401.

Lehmann, J. and Kleber, M. (2015) The contentious nature of soil organic matter. *Nature* 528(7580), 60–68.

Leifeld, J. (2012) How sustainable is organic farming? *Agriculture, Ecosystems & Environment* 150, 121–122.

Li, N., Xu, Y.-Z., Han, X.-Z., He, H.-B., Zhang, X. *et al.* (2015) Fungi contribute more than bacteria to soil organic matter through necromass accumulation under different agricultural practices during the early pedogenesis of a Mollisol. *European Journal of Soil Biology* 67, 51–58.

Li, Q., Tian, Y., Zhang, X., Xu, X., Wang, H. *et al.* (2017) Labile carbon and nitrogen additions affect soil organic matter decomposition more strongly than temperature. *Applied Soil Ecology* 114, 152–160.

Liang, C. and Balser, T.C. (2011) Microbial production of recalcitrant organic matter in global soils: Implications for productivity and climate policy. *Nature Reviews. Microbiology* 9(1), 75.

Liu, J., Ma, Q., Hui, X., Ran, J., Ma, Q. *et al.* (2020) Long-term high-P fertilizer input decreased the total bacterial diversity but not phoD-harboring bacteria in wheat *Rhizosphere* soil with available-P deficiency. *Soil Biology and Biochemistry* 149, 107918.

Macdonald, A., Poulton, P., Clark, I., Scott, T., Glendining, M. *et al.* (2015) Guide to the classical and other long-term experiments, datasets and sample archive. Rothamstead Research, Harpenden, UK.

Maggio, A., De Pascale, S., Paradiso, R. and Barbieri, G. (2013) Quality and nutritional value of vegetables from organic and conventional farming. *Scientia Horticulturae* 164, 532–539.

Maillard, É., Angers, D.A., Chantigny, M., Bittman, S., Rochette, P. *et al.* (2015) Carbon accumulates in organo-mineral complexes after long-term liquid dairy manure application. *Agriculture, Ecosystems & Environment* 202, 108–119.

Marriott, E.E. and Wander, M.M. (2006) Total and labile soil organic matter in organic and conventional farming systems. *Soil Science Society of America Journal* 70(3), 950–959.

Martínez, I., Chervet, A., Weisskopf, P., Sturny, W.G., Etana, A. *et al.* (2016) Two decades of no-till in the Oberacker long-term field experiment. Part I: Crop yield, soil organic carbon and nutrient distribution in the soil profile. *Soil and Tillage Research* 163, 141–151.

Martínez-García, L.B., Korthals, G., Brussaard, L., Jørgensen, H.B. and Gerlinde, B.D.D. (2018) Organic management and cover crop species steer soil microbial community structure and functionality along with soil organic matter properties. *Agriculture, Ecosystems & Environment* 263, 7–17.

Mason, A.R.G., Cavagnaro, T.R., Guerin, G.R. and Lowe, A.J. (2023) Soil bacterial assemblage across a production landscape: Agriculture increases diversity while revegetation recovers community composition. *Microbial Ecology* 85(3), 1098–1112.

Mathew, I., Shimelis, H., Mutema, M., Minasny, B. and Chaplot, V. (2020) Crops for increasing soil organic carbon stocks – a global meta analysis. *Geoderma* 367, 114230.

Meena, V.S., Mondal, T., Pandey, B.M., Mukherjee, A., Yadav, R.P. *et al.* (2018) Land use changes: Strategies to improve soil carbon and nitrogen storage pattern in the mid-Himalaya ecosystem, India. *Geoderma* 321, 69–78.

Merante, P., Dibari, C., Ferrise, R., Sánchez, B., Iglesias, A. *et al.* (2017) Adopting soil organic carbon management practices in soils of varying quality: Implications and perspectives in Europe. *Soil and Tillage Research* 165, 95–106.

Mhazo, N., Chivenge, P. and Chaplot, V. (2016) Tillage impact on soil erosion by water: Discrepancies due to climate and soil characteristics. *Agriculture, Ecosystems & Environment* 230, 231–241.

Miller, A.J., Amundson, R., Burke, I.C. and Yonker, C. (2003) The effect of climate and cultivation on soil organic C and N. *Biogeochemistry* 67(1), 57–72.

Minasny, B., Malone, B.P., McBratney, A.B., Angers, D.A., Arrouays, D. *et al.* (2017) Soil carbon 4 per mille. *Geoderma* 292, 59–86.

Morriën, E., Hannula, S.E., Snoek, L.B., Helmsing, N.R., Zweers, H. *et al.* (2017) Soil networks become more connected and take up more carbon as nature restoration progresses. *Nature Communications* 8, 14349.

Mosier, S., Paustian, K., Davies, C., Kane, M. and Cotrufo, M.F. (2019) Soil organic matter pools under management intensification of loblolly pine plantations. *Forest Ecology and Management* 447, 60–66.

Nelson, P.N. and Baldock, J.A. (2000) Soil organic matter. In: Malcolm, E. (ed.) *Handbook of Soil Science*. CRC press, Boca Raton, Florida, pp. 25–84.

Novara, A., Cerdà, A. and Gristina, L. (2018) Sustainable vineyard floor management: An equilibrium between water consumption and soil conservation. *Current Opinion in Environmental Science & Health* 5, 33–37.

Nunes, M.R., Karlen, D.L., Veum, K.S., Moorman, T.B. and Cambardella, C.A. (2020) Biological soil health indicators respond to tillage intensity: A US meta-analysis. *Geoderma* 369, 114335.

Oades, J.M. (1984) Soil organic matter and structural stability: Mechanisms and implications for management. *Plant and Soil* 76(1–3), 319–337.

Oades, J.M. (1993) The role of biology in the formation, stabilization and degradation of soil structure. *Geoderma* 56(1–4), 377–400.

Osanai, Y., Janes, J.K., Newton, P.C.D. and Hovenden, M.J. (2015) Warming and elevated CO_2 combine to increase microbial mineralisation of soil organic matter. *Soil Biology and Biochemistry* 85, 110–118.

Peixoto, L., Elsgaard, L., Rasmussen, J., Kuzyakov, Y., Banfield, C.C. *et al.* (2020) Decreased rhizodeposition, but increased microbial carbon stabilization with soil depth down to 3.6 m. *Soil Biology and Biochemistry* 150, 108008.

Peltoniemi, K., Velmala, S., Fritze, H., Lemola, R. and Pennanen, T. (2021) Long-term impacts of organic and conventional farming on the soil microbiome in boreal arable soil. *European Journal of Soil Biology* 104, 103314.

Pereira, P., Francos, M., Brevik, E.C., Ubeda, X. and Bogunovic, I. (2018) Post-fire soil management. *Current Opinion in Environmental Science & Health* 5, 26–32.

Phillips, M. (2017) *Mycorrhizal Planet: How Symbiotic Fungi Work with Roots to Support Plant Health and Build Soil Fertility*. Chelsea Green Publishing, White River Junction, Vermont.

Poeplau, C., Don, A., Six, J., Kaiser, M., Benbi, D. *et al.* (2018) Isolating organic carbon fractions with varying turnover rates in temperate agricultural soils – a comprehensive method comparison. *Soil Biology and Biochemistry* 125, 10–26.

Prentice, I.C., Farquhar, G.D., Fasham, M.J.R., Goulden, M.L., Martin Heimann, V.J.J. *et al.* (2001) The carbon cycle and atmospheric carbon dioxide. IPCC report. *TAR Climate Change 2001: The Scientific Basis*.

Pretty, J. (2008) Agricultural sustainability: Concepts, principles and evidence. *Philosophical Transactions of the Royal Society of London. Series B, Biological Sciences* 363(1491), 447–465.

Puech, C., Baudry, J., Joannon, A., Poggi, S. and Aviron, S. (2014) Organic vs. conventional farming dichotomy: Does it make sense for natural enemies. *Agriculture, Ecosystems & Environment* 194, 48–57.

Qafoku, N.P. (2015) Climate-change effects on soils: Accelerated weathering, soil carbon, and elemental cycling. In: Sparks, D.L. (ed.) *Advances in Agronomy* 131.

Ramesh, K. (2015) Weed problems, ecology, and management options in conservation agriculture: Issues and perspectives. In: Sparks, D.L. (ed.) *Advances in Agronomy* 131.

Ramesh, T., Bolan, N.S., Kirkham, M.B., Wijesekara, H., Kanchikerimath, M. *et al.* (2019) Soil organic carbon dynamics: Impact of land use changes and management practices: A review. *Advances in Agronomy* 156, 1–107.

Rasmussen, P.E., Albrecht, S.L. and Smiley, R.W. (1998) Soil C and N changes under tillage and cropping systems in semi-arid pacific northwest agriculture. *Soil and Tillage Research* 47(3–4), 197–205.

Reicosky, D.C. (2003) Intensive tillage as a mechanism for CO_2 emission from agricultural soils. In: *International Workshop on Conservation Agriculture for Sustainable Wheat Production in Rotation with Cotton in Limited Water Resource Areas*. 14-18 October 2022, Tashkent.

Reicosky, D.C. and Lindstrom, M.J. (1993) Fall tillage method: Effect on short-term carbon dioxide flux from soil. *Agronomy Journal* 85(6), 1237–1243.

Reinsch, T., Loges, R., Kluß, C. and Taube, F. (2018) Effect of grassland ploughing and reseeding on CO_2 emissions and soil carbon stocks. *Agriculture, Ecosystems & Environment* 265, 374–383.

Renton, M. and Flower, K.C. (2015) Occasional mouldboard ploughing slows evolution of resistance and reduces long-term weed populations in no-till systems. *Agricultural Systems* 139, 66–75.

Rimmer, D.L. (2006) Free radicals, antioxidants, and soil organic matter recalcitrance. *European Journal of Soil Science* 57(2), 91–94.

Ruis, S.J., Blanco-Canqui, H., Jasa, P.J. and Jin, V.L. (2022) No-till farming and greenhouse gas fluxes: Insights from literature and experimental data. *Soil and Tillage Research* 220, 105359.

Sacco, D., Moretti, B., Monaco, S. and Grignani, C. (2015) Six-year transition from conventional to organic farming: Effects on crop production and soil quality. *European Journal of Agronomy* 69, 10–20.

Saleem, M., Pervaiz, Z.H., Contreras, J., Lindenberger, J.H., Hupp, B.M. *et al*. (2020) Cover crop diversity improves multiple soil properties via altering root architectural traits. *Rhizosphere* 16, 100248.

Sanderman, J., Hengl, T. and Fiske, G.J. (2017) Soil carbon debt of 12,000 years of human land use. *Proceedings of the National Academy of Sciences of the United States of America* 114(36), 9575–9580.

Sardana, V., Mahajan, G., Jabran, K. and Chauhan, B.S. (2017) Role of competition in managing weeds: An introduction to the special issue. *Crop Protection* 95, 1–7.

Schmidt, M.W., Torn, M.S., Abiven, S., Dittmar, T., Guggenberger, G. *et al*. (2011) Persistence of soil organic matter as an ecosystem property. *Nature* 478(7367), 49–56.

Schnitzer, M. (1991) Soil organic matter: The next 75 years. *Soil Science* 151(1), 41–58.

Scholes, R.J., Dalal, R. and Singer, S. (1994) Soil physics and fertility: the effects of water, temperature and texture. In: Woomer, P.L. and Swift, M.J. (eds) *The Biological Management of Tropical Soil Fertility*. Wiley.

Semenov, M.V., Krasnov, G.S., Semenov, V.M. and van Bruggen, A.H.C. (2020) Long-term fertilization rather than plant species shapes rhizosphere and bulk soil prokaryotic communities in agroecosystems. *Applied Soil Ecology* 154, 103641.

Sequeira, C.H., Alley, M.M. and Jones, B.P. (2011) Evaluation of potentially labile soil organic carbon and nitrogen fractionation procedures. *Soil Biology and Biochemistry* 43(2), 438–444.

Shahbaz, M., Kuzyakov, Y. and Heitkamp, F. (2017) Decrease of soil organic matter stabilization with increasing inputs: mechanisms and controls. *Geoderma* 304, 76–82.

Shaver, T.M., Peterson, G.A. and Sherrod, L.A. (2003) Cropping intensification in dryland systems improves soil physical properties: Regression relations. *Geoderma* 116(1–2), 149–164.

Sheng, H., Zhou, P., Zhang, Y., Kuzyakov, Y., Zhou, Q. *et al*. (2015) Loss of labile organic carbon from subsoil due to land-use changes in subtropical China. *Soil Biology and Biochemistry* 88, 148–157.

Shepherd, M.A., Harrison, R. and Webb, J. (2002) Managing soil organic matter – implications for soil structure on organic farms. *Soil Use and Management* 18(3), 284–292.

Shi, S., Zhang, W., Zhang, P., Yu, Y. and Ding, F. (2013) A synthesis of change in deep soil organic carbon stores with afforestation of agricultural soils. *Forest Ecology and Management* 296, 53–63.

Singh, M., Bhullar, M.S. and Chauhan, B.S. (2015) Seed bank dynamics and emergence pattern of weeds as affected by tillage systems in dry direct-seeded rice. *Crop Protection* 67, 168–177.

Singh, S., Nouri, A., Singh, S., Anapalli, S., Lee, J. *et al*. (2020) Soil organic carbon and aggregation in response to thirty-nine years of tillage management in the southeastern US. *Soil and Tillage Research* 197, 104523.

Skjemstad, J.O., Janik, L.J. and Taylor, J.A. (1998) Non-living soil organic matter: What do we know about it? *Australian Journal of Experimental Agriculture* 38, 7.

Smith, T.M.W.C., Dixon, R., Rik Leemans, R.N. and Solomon, A.M. (1993) The global terrestrial carbon cycle. *Water, Air, and Soil Pollution* 70, 19–37.

Spain, A.V., Isbell, R.F. and Probert, M.E. (1983) Soil organic matter. In: *Soils: An Australian Viewpoint*. CSIRO/Academic Press, pp. 551–563.

Sparling, G.P., Wheeler, D., Vesely, E.-T. and Schipper, L.A. (2006) What is soil organic matter worth? *Journal of Environmental Quality* 35(2), 548–557.

Sradnick, A., Oltmanns, M., Raupp, J. and Joergensen, R.G. (2014) Microbial residue indices down the soil profile after long-term addition of farmyard manure and mineral fertilizer to a sandy soil. *Geoderma* 226–227, 79–84.

Stewart, C.E., Halvorson, A.D. and Delgado, J.A. (2017) Long-term N fertilization and conservation tillage practices conserve surface but not profile SOC stocks under semi-arid irrigated corn. *Soil and Tillage Research* 171, 9–18.

Strosser, E. (2010) Methods for determination of labile soil organic matter: An overview. *Journal of Agrobiology* 27(2), 49–60.

Suja, G., Byju, G., Jyothi, A.N., Veena, S.S. and Sreekumar, J. (2017) Yield, quality and soil health under organic vs conventional farming in taro. *Scientia Horticulturae* 218, 334–343.

Sun, Y., Zeng, Y., Shi, Q., Pan, X. and Huang, S. (2015) No-tillage controls on runoff: A meta-analysis. *Soil and Tillage Research* 153, 1–6.

Swift, M.J., Izac, A.-M.N. and van Noordwijk, M. (2004) Biodiversity and ecosystem services in agricultural landscapes: Are we asking the right questions? *Agriculture, Ecosystems & Environment* 104(1), 113–134.

Syltie, P.W. (2002) *How Soils Work*. Xulon Press, Maitland, Florida.

Tiessen, H., Cuevas, E. and Chacon, P. (1994) The role of soil organic matter in sustaining soil fertility. *Nature* 371(6500), 783–785.

Tisdall, J.M. and Oades, J.M. (1979) Stabilization of soil aggregates by the root systems of ryegrass. *Soil Research* 17(3), 429.

Tisdall, J.M. and Oades, J.M. (1982) Organic matter and water-stable aggregates in soils. *Journal of Soil Science* 33(2), 141–163.

Tittarelli, F., Pettruzzelli, G., Pezzarossa, B., Civilini, M., Beneditti, A. *et al*. (2007) Quality and agronomic use of compost. In: de Bertoldi, M., Bidlingmaier, W. and Stentiford, E. (eds) *Compost Science and Technology*. Waste Management Series 8, Elsevier, Amsterdam.

Tittonell, P. (2014) Ecological intensification of agriculture: Sustainable by nature. *Current Opinion in Environmental Sustainability* 8, 53–61.

Turner, S., Meyer-Stüve, S., Schippers, A., Guggenberger, G., Schaarschmidt, F. *et al*. (2017) Microbial utilization of mineral-associated nitrogen in soils. *Soil Biology and Biochemistry* 104, 185–196.

Udawatta, R.P., Clark, J.G. and Jose, S. (2017) Agroforestry practices and soil ecosystem services. In: Al-Kaisi, M.M. and Lowery, B. (eds) *Soil Health and Intensification of Agroecostystems*. Academic Press, Amsterdam, pp. 305–334.

Ukalska-Jaruga, A., Bejger, R., Smreczak, B. and Podlasiński, M. (2023) Sorption of organic contaminants by stable organic matter fraction in soil. *Molecules* 28(1), 1.

Ullah, H., Santiago-Arenas, R., Ferdous, Z., Attia, A., Datta, A. *et al*. (2019) Improving water use efficiency, nitrogen use efficiency, and radiation use efficiency in field crops under drought stress: A review. In: Sparks, D.L. (ed.) *Advances in Agronomy* 156. Elsevier.

Uroz, S., Kelly, L.C., Turpault, M.P., Lepleux, C. and Frey-Klett, P. (2015) The mineralosphere concept: Mineralogical control of the distribution and function of mineral-associated bacterial communities. *Trends in Microbiology* 23(12), 751–762.

VandenBygaart, A.J. (2018) Comments on soil carbon 4 per mille by Minasny *et al* 2017. *Geoderma* 309, 113–114.

Varanasi, A., Prasad, P.V. and Jugulam, M. (2016) Impact of climate change factors on weeds and herbicide efficacy. In: Sparks, D.L. (ed.) *Advances in Agronomy* 135. Elsevier.

Veloso, M.G., Angers, D.A., Chantigny, M.H. and Bayer, C. (2020) Carbon accumulation and aggregation are mediated by fungi in a subtropical soil under conservation agriculture. *Geoderma* 363, 114159.

Verzeaux, J., Alahmad, A., Habbib, H., Nivelle, E., Roger, D. *et al*. (2016) Cover crops prevent the deleterious effect of nitrogen fertilisation on bacterial diversity by maintaining the carbon content of ploughed soil. *Geoderma* 281, 49–57.

Vestergård, M., Reinsch, S., Bengtson, P., Ambus, P. and Christensen, S. (2016) Enhanced priming of old, not new soil carbon at elevated atmospheric CO_2. *Soil Biology and Biochemistry* 100, 140–148.

Vicente-Vicente, J.L., García-Ruiz, R., Francaviglia, R., Aguilera, E. and Smith, P. (2016) Soil carbon sequestration rates under Mediterranean woody crops using recommended management practices: A meta-analysis. *Agriculture, Ecosystems & Environment* 235, 204–214.

von Lützow, M. and Kögel-Knabner, I. (2009) Temperature sensitivity of soil organic matter decomposition: What do we know? *Biology and Fertility of Soils* 46(1), 1–15.

Wander, M.M., Traina, S.J., Stinner, B.R. and Peters, S.E. (1994) Organic and conventional management effects on biologically active soil organic matter pools. *Soil Science Society of America Journal* 58(4), 1130–1139.

Wang, X., Jelinski, N.A., Toner, B. and Yoo, K. (2019) Long-term agricultural management and erosion change soil organic matter chemistry and association with minerals. *The Science of the Total Environment* 648, 1500–1510.

Wardle, D.A., Bardgett, R.D., Klironomos, J.N., Setälä, H., van der Putten, W.H. *et al*. (2004) Ecological linkages between aboveground and belowground biota. *Science* 304(5677), 1629–1633.

Weigelt, A., Marquard, E., Temperton, V.M., Roscher, C., Scherber, C. *et al*. (2010) The Jena experiment: Six years of data from a grassland biodiversity experiment. *Ecology* 91(3), 930–931.

Wolters, V. (2000) Invertebrate control of soil organic matter stability. *Biology and Fertility of Soils* 31(1), 1–19.

Wood, S., Sebastian, K. and Scherr, S.J. (2000) *Pilot Analysis of Global Ecosystems: Agroecosystems*. A joint study by International Food Policy Research Institute and World Resources Institute.

Woomer, P.L. and Swift, M.J. (1994) *The Biological Management of Tropical Soil Fertility*. John Wiley & Sons.

Wortman, S.E., Drijber, R.A., Francis, C.A. and Lindquist, J.L. (2013) Arable weeds, cover crops, and tillage drive soil microbial community composition in organic cropping systems. *Applied Soil Ecology* 72, 232–241.

Wu, T., Chellemi, D.O., Graham, J.H., Martin, K.J. and Rosskopf, E.N. (2008) Comparison of soil bacterial communities under diverse agricultural land management and crop production practices. *Microbial Ecology* 55(2), 293–310.

Wulanningtyas, H.S., Gong, Y., Li, P., Sakagami, N., Nishiwaki, J. *et al*. (2021) A cover crop and no-tillage system for enhancing soil health by increasing soil organic matter in soybean cultivation. *Soil and Tillage Research* 205.

Yao, T., Chen, R., Zhang, J., Feng, Y., Huang, M. *et al.* (2020) Divergent patterns of microbial community composition shift under two fertilization regimes revealed by responding species. *Applied Soil Ecology* 154, 103590.

Yemadje, P.L., Chevallier, T., Guibert, H., Bertrand, I. and Bernoux, M. (2017) Wetting-drying cycles do not increase organic carbon and nitrogen mineralization in soils with straw amendment. *Geoderma* 304, 68–75.

Young, A. (1999) Environment. *Environment, Development and Sustainability* 1(1), 3–18.

Zhang, B., Drury, C.F., Yang, X., Reynolds, W.D. and Zhang, X. (2014) Effects of long-term and recently imposed tillage on the concentration and composition of amino sugars in a clay loam soil in Ontario, Canada. *Soil and Tillage Research* 135, 9–17.

Zhang, Q., Feng, J., Li, J., Huang, C.-Y., Shen, Y. *et al.* (2023) A distinct sensitivity to the priming effect between labile and stable soil organic carbon. *The New Phytologist* 237(1), 88–99.

Zhang, W., Cui, Y., Lu, X., Bai, E., He, H. *et al.* (2016) High nitrogen deposition decreases the contribution of fungal residues to soil carbon pools in a tropical forest ecosystem. *Soil Biology and Biochemistry* 97, 211–214.

Zhang, X., Kuzyakov, Y., Zang, H., Dippold, M.A., Shi, L. *et al.* (2020) Rhizosphere hotspots: Root hairs and warming control microbial efficiency, carbon utilization and energy production. *Soil Biology and Biochemistry* 148.

Zhou, J., Wen, Y., Shi, L., Marshall, M.R., Kuzyakov, Y. *et al.* (2021) Strong priming of soil organic matter induced by frequent input of labile carbon. *Soil Biology and Biochemistry* 152.

Zhu, X., Xie, H., Masters, M.D., Luo, Y., Zhang, X. *et al.* (2020) Microbial trade-off in soil organic carbon storage in a no-till continuous corn agroecosystem. *European Journal of Soil Biology* 96, 103146.

Ziska, L.H. (2016) The role of climate change and increasing atmospheric carbon dioxide on weed management: Herbicide efficacy. *Agriculture, Ecosystems & Environment* 231, 304–309.

Ziska, L.H., Blumenthal, D.M. and Franks, S.J. (2019) Understanding the nexus of rising CO2, climate change, and evolution in weed biology. *Invasive Plant Science and Management* 12(2), 79–88.

27 Composting, Vermicomposting, Green Manuring, Liquid Organic Manures and ITK for Nutrient Management

Thangapandian Dhanikodi[1]*, Uma Maheswari Rajadurai[2] and Rameshwari Ramaswamy[2]

[1]Adhavan College of Arts & Science, Bharathidasan University, Tiruchirappalli, India; [2]Department of Biotechnology, Cauvery College for Women, Tiruchirappalli, India

Abstract

Composting and vermicomposting offer effective ways to recycle organic waste, generating nutrient-rich compost that enhances soil health and fertility. Green manuring techniques contribute to soil improvement through the incorporation of organic matter and essential nutrients, promoting optimal soil structure and plant growth. The utilization of liquid organic manures proves valuable in supplying bioavailable nutrients to crops, optimizing nutrient uptake, and fostering healthy plant development. Additionally, this chapter highlights the significance of indigenous technical knowledge (ITK) as an invaluable resource, providing traditional practices that complement modern organic farming approaches, promoting cultural and ecological diversity in agriculture. By adopting these organic nutrient management strategies, farmers can reduce reliance on synthetic fertilizers and chemical inputs, mitigating environmental impacts, advancing regenerative agricultural practices, and contributing to a greener and more resilient future for agriculture and the environment.

Keywords: vermicompost, liquid manure, biodynamic

> '*Yaerinum nanraal yeruviduthal!*' ('To cast manure is better than to plough.')
> – Tamil poet Thiruvalluvar, born 31 BC

What Is Composting?

Composting is a natural and sustainable process that involves the decomposition of organic materials like kitchen scraps, yard waste, animal fodder rejection and biodegradable items, resulting in a nutrient-rich substance known as compost. This process relies on microorganisms such as bacteria, fungi and decomposers, which break down the organic materials through chemical reactions and biological processes, converting complex compounds into simpler forms.

Composting can be done on both a small scale, like backyard compost bins, and large-scale in commercial facilities, following the principles of providing the right mix of organic materials, moisture, oxygen and temperature for optimal decomposition in addition to mineral supplements such as rock phosphate etc (Fig. 27.1).

Suitable materials for composting include fruit and vegetable scraps, coffee grounds, tea bags, leaves, grass clippings, straw, shredded paper and yard trimmings, but it is crucial to avoid meat, dairy products, oily foods and pet waste, as these may attract pests or introduce pathogens.

*Corresponding author: farmindia@gmail.com

DOI: 10.1079/9781800626850.0027

Fig. 27.1. The School of Biodynamic Farming, Vinobhajipuram, India.

Composting offers numerous benefits, reducing organic waste in landfill sites, thus easing the strain on waste management systems. Additionally, it enriches soil by adding organic matter and nutrients, enhancing its structure, water retention and fertility, making it ideal for gardens, lawns, potted plants and agricultural fields to promote healthier plant growth.

Compost and Soil Health

Compost plays a crucial role in enhancing soil health through various beneficial effects. Firstly, its rich organic matter content increases the soil's carbon levels, promoting water and nutrient retention, while serving as a food source for vital soil microorganisms, thereby fostering their activity and diversity. Moreover, compost provides essential nutrients like nitrogen, phosphorus, potassium and micronutrients to plants gradually, acting as a natural slow-release fertilizer, and it aids in balancing nutrient levels and enhancing nutrient cycling within the soil. Additionally, compost contributes to improved soil structure by enhancing aggregation, leading to better root penetration, nutrient absorption and overall plant growth. Furthermore, its water-retaining capacity helps prevent water runoff and erosion while facilitating drainage in clay soils, promoting aeration.

Compost also plays a role in maintaining optimal pH levels within the soil, acting as a natural buffer against extreme fluctuations and creating a conducive environment for plant growth. The abundance of beneficial microorganisms in compost further contributes to soil's biological activity, breaking down organic matter, releasing nutrients and suppressing diseases, leading to improved soil fertility. Additionally, some compost contains microorganisms that help control plant diseases and pests by outcompeting harmful pathogens and increasing

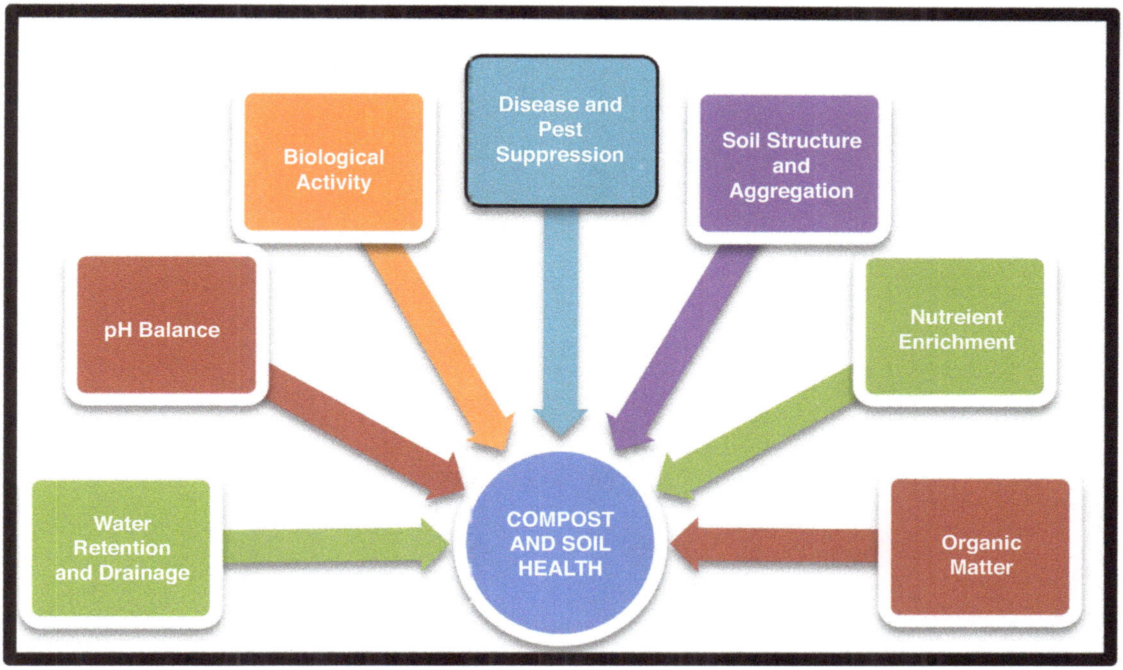

Fig. 27.2. Compost and soil health.

plant resistance, thereby reducing the need for chemical pesticides. In conclusion, composting and the application of compost contribute to healthier soil by increasing organic matter, enhancing nutrient availability, improving soil structure, supporting water retention and drainage, balancing pH, promoting beneficial microorganisms, and bolstering plant growth (Fig. 27.2).

Incorporating compost into soil management practices presents a sustainable approach for gardening, agriculture, and land restoration endeavor.

History of Composting

The history of composting can be traced back to the late 19th century when Sir Albert Howard, a British agronomist, studied Indian farmers' traditional composting practices. Recognizing its significance in maintaining soil fertility, Howard developed the Indore method of composting, emphasizing controlled decomposition to produce high-quality compost. However, in the mid-20th century, synthetic fertilizers and chemical pesticides gained dominance with the rise of industrial agriculture (Zheng *et al.*, 2020). Despite this shift, advocates like J.I. Rodale continued to promote composting and organic gardening through their publications in the USA.

The environmental movement of the 1970s brought increased awareness of sustainable practices, leading to the establishment of community composting programmes, co-operatives and backyard composting initiatives. Many cities and municipalities adopted composting as part of their waste management strategies to reduce waste and recycle organic materials. As landfill capacity and environmental concerns grew in the 1990s and 2000s, composting gained more attention. Governments, businesses and individuals recognized its benefits in diverting organic waste from landfill sites and reducing methane emissions, leading to the emergence of large-scale composting facilities and innovative composting technologies.

In the 21st century, composting continued to evolve and gain momentum as an essential component of sustainable waste management and soil health improvement. Governments and organizations worldwide implemented composting programmes, expanding the scope to include various organic materials like food scraps and agricultural residues. Innovative composting methods were developed to suit different scales and contexts, such as in-vessel composting, windrow composting, and bokashi composting. Today, composting is widely acknowledged for its role in reducing waste, mitigating climate change and promoting regenerative agriculture practices that prioritize healthy soils. It is practised by individuals, communities and commercial operations, solidifying its position as a vital tool in building a more sustainable future.

Science Inside the Composting Process

Science is at the core of understanding and optimizing the composting process, which involves the complex interactions of various microorganisms, fungi and other organisms in decomposing organic material. The microbial activity in composting is a central focus of scientific study, as different microorganisms play vital roles in breaking down organic matter through enzymatic reactions. Understanding their populations, metabolic activities, and the factors influencing their performance is essential in the composting process.

Temperature and oxygen availability are critical aspects of composting, and science helps in monitoring and controlling these factors to support optimal microbial activity and decomposition rates. The carbon:nitrogen (C:N) ratio is another crucial scientific consideration, as it affects the efficiency of organic matter decomposition. Scientists provide guidelines for achieving and maintaining the appropriate C:N ratio through proper ingredient selection and mixture. Moreover, science guides composters in maintaining the right moisture levels, as adequate moisture is essential for microbial activity and decomposition. Nutrient cycling is another key area of study, where scientific insights help in understanding the chemical transformations that occur during composting, optimizing nutrient availability, and reducing nutrient loss.

Process optimization techniques in composting are also rooted in scientific research and experimentation, where factors such as pile size, turning frequency, aeration methods, composting additives, and system design are studied for efficient composting outcomes. Furthermore, science plays a significant role in evaluating and assessing compost quality and stability through various analyses that measure parameters such as organic matter content, nutrient content, pH, maturity, and the presence of contaminants. By leveraging scientific principles, researchers, compost producers and practitioners can make informed decisions and adopt best practices to enhance the composting process, improve compost quality, and maximize the environmental and agronomic benefits of compost. Composting, driven by scientific understanding, remains a critical component of sustainable waste management and soil health improvement efforts.

Types of Composting

Composting offers several methods for decomposing organic materials and creating compost. The chosen method depends on factors like scale, space, time and the materials involved. Here are some common types of composting:

Backyard composting. Suitable for households, this basic method involves setting up a compost pile or bin in the backyard. Organic materials like kitchen scraps and yard waste decompose gradually, with periodic turning to speed up the process.

Windrow composting. Employed in commercial or municipal facilities, organic materials are arranged in long piles called windrows, and periodically turned to promote decomposition. This method requires ample space and equipment.

In-vessel composting. Large-scale operations use containers or enclosed systems to compost organic materials, providing better control over factors like temperature and moisture. It allows faster decomposition using rotating drums or enclosed tunnels.

Bokashi composting. This fermentation-based method uses specialized bacteria to break down organic matter. Organic waste is layered with effective microorganisms (EM) and fermented in an airtight container. It handles a broader range of materials, including meat and dairy.

***In situ* composting.** *In situ* composting, also known as on-site composting or composting in place, refers to the practice of composting organic materials directly at the location where they are generated. Instead of transporting the organic waste to an off-site composting facility, the composting process takes place on-site, typically in a designated area or composting bin.

Leaf-mould composting. Leaf-mould composting involves allowing the collected leaves to break down over time, mainly through microbial activity. The leaves are piled or stacked in a compost bin, compost pile or dedicated leaf composting area. The composting process is predominantly aerobic, meaning it relies on the presence of oxygen to facilitate decomposition (Flannery, 1991; RHS, 2025).

Various combinations and variations of these methods can be used based on specific needs and resources available to individuals, communities or organizations interested in composting organic materials.

Vermicomposting

Vermicomposting is a sustainable and eco-friendly process that harnesses the natural power of earthworms to convert organic waste into nutrient-rich manure. The materials required for vermicomposting are simple and readily available, making it accessible to farmers and gardeners alike. The primary inputs include organic waste such as kitchen scraps (vegetable peelings, fruit cores, coffee grounds), yard trimmings, and other decomposable materials. The key component, of course, is the earthworms themselves, with the most commonly used species being *Eisenia fetida* or Red Wigglers due to their voracious appetite and rapid composting abilities.

The process of vermicomposting involves setting up a controlled environment for the earthworms to thrive. A suitable container like a composting bin or a pit is filled with layers of organic waste materials. The worms are introduced into this bedding, and they start consuming the organic matter. As the earthworms digest the waste, their excretions, known as castings, are released. These castings are highly enriched with nutrients, such as nitrogen, phosphorus, potassium and micronutrients, making the resulting vermicompost a potent and balanced organic fertilizer. Throughout the composting process, it is essential to maintain optimal conditions for the worms, including proper moisture levels, and avoid extremes of temperature. Within a few weeks to months, depending on the scale of vermicomposting and environmental conditions, the organic waste is transformed into a dark, crumbly and earthy-smelling vermicompost, ready to be used as a valuable soil amendment.

The advantages of vermicomposting are manifold and it is a valuable technique in organic farming and gardening. It offers a sustainable solution to manage organic waste, reducing the burden on landfill and promoting a circular economy. Instead of wasting valuable nutrients, vermicomposting recycles them back into the soil, closing the nutrient loop. Additionally, vermicompost enhances soil structure and improves its water-retention capacity, reducing the risk of erosion and increasing overall soil fertility. Unlike some chemical fertilizers, vermicompost releases nutrients slowly over time, preventing nutrient leaching and ensuring a steady supply of nourishment to plants without causing harm to the environment. Moreover, vermicompost is teeming with beneficial microorganisms, enzymes and plant growth hormones, fostering a healthy soil ecosystem and supporting robust plant growth (Table 27.1).

In the context of regenerative agriculture, vermicompost plays a vital role in promoting soil health and biodiversity. By enriching the soil with organic matter and essential nutrients, it restores the soil's natural fertility and resilience, reducing reliance on synthetic inputs. Healthy

Table 27.1. Nutrient composition of vermicompost and farmyard manure (based on vermicompost analysis in Kadapa, Nalgonda, Anantapur etc. clusters).

Nutrient element	Vermicompost (%)	Farmyard manure (%)
Organic carbon	10–14	12
Nitrogen	0.51–1.6	0.8
Phosphorus	0.19–1.02	0.4
Potassium	0.15–0.73	0.5
Calcium	1.18–7.61	2.2
Magnesium	0.093–0.568	0.5
Sodium	0.06–0.16	< 0.01
Zinc	0.0042–0.011	0.0012
Copper	0.0026–0.0048	0.0017
Iron	0.2050–1.3313	1.1690
Manganese	0.0105–0.2038	0.0414

soils cultivated with vermicompost sequester more carbon and contribute to mitigating climate change. Additionally, the increased microbial activity in vermicompost-amended soils aids in breaking down pollutants and contaminants, making it a valuable tool in remediation efforts for degraded lands. The use of vermicompost aligns with the principles of regenerative agriculture by emphasizing soil regeneration, closed-loop nutrient cycles, and promoting a holistic approach to farming that benefits both the environment and the farmers.

Vermicomposting is a sustainable and efficient method of converting organic waste into nutrient-rich vermicompost using earthworms. Its straightforward process and accessible materials make it a viable option for waste management and organic farming practices. The advantages of vermicomposting, including nutrient-rich compost production, improved soil structure and support for regenerative agriculture, highlight its significance in promoting sustainable agricultural systems and contributing to a greener future.

Biodynamic compost

Biodynamic compost is a specialized type of compost created following the principles of biodynamic agriculture. It is considered essential for enhancing soil fertility and vitality in a holistic and self-sustaining farming approach. Specific organic materials, including plant residues and manure, are used, often sourced from the farm or biodynamic community to maintain a closed nutrient cycle. Biodynamic preparations, like Horn Manure (BD 500) and Horn Silica (BD 501), are applied to the soil as 'sequential sprays' to improve 'life forces' in the soil. Further, BD 502 to BD 507 preparations are incorporated into the compost to stimulate root development, enhance photosynthesis and increase plant vitality. Timing and application of the preparations are based on celestial and seasonal rhythms to leverage cosmic and earthly forces during the composting process. Building a well-balanced compost pile with carbon-rich and nitrogen-rich materials is crucial for promoting microbial activity and decomposition. The composting process can take several months to a year, and, once matured, the compost is applied to the soil on the farm or in the garden (Table 27.2).

Biodynamic farmers consider the energetic qualities of the compost when determining its use, aligning with the principles of ecological balance, biodiversity and natural rhythms in agricultural practices.

Municipal compost

Municipal compost, also known as community compost or city compost, refers to compost that is produced and managed by municipal governments or their authorized entities. It is a part of municipal waste management systems aimed at diverting organic waste from landfill and promoting sustainable practices (Alkoaik, 2010).

Here are some key aspects of municipal composting:

Table 27.2. Biodynamic preparations (BD 502 – BD 507).

BD preps	Herbs/scientific name	Planet (connected)	Nutrients	Results
BD 502	Yarrow *Achillea millefolium*	Venus	Sulphur (S), potassium (K), trace elements	Helps the plants to absorb trace elements
BD 503	Chamomile flower *Matricaria chamomilla*	Mercury	Calcium (Ca), sulphur (S)	Stabilizes nitrogen (N) within compost and increases soil life
BD 504	Stinging nettle *Urtica dioica*	Mars	Sulphur (S), potassium (K), calcium (Ca), iron (Fe)	Provides the needed nutrition compounds, enlivens the earth/soil
BD 505	Oak bark *Quercus robur*	Moon	Calcium (Ca)	Provides the healing forces
BD 506	Dandelion flower *Taraxacum officinalis*	Jupiter	Silicon (Si), potassium (K)	Attracts the cosmic forces to soil
BD 507	Valerian flower *Valeriana officinalis*	Saturn	Phosphorus (P)	Stimulates compost to have phosphorus component

Collection. Municipal composting programmes typically include a separate collection system for organic waste. This can include food scraps, yard waste and other compostable materials. Depending on the programme, residents may use separate bins or containers specifically designated for organic waste, which are then collected by municipal waste management services.

Compost facilities. Municipalities may have their own composting facilities or work with third-party operators to manage the composting process. These facilities are equipped to handle large volumes of organic waste and have the necessary infrastructure for composting operations. Composting methods employed by municipal facilities can vary, including windrow composting, aerated static pile composting, or in-vessel composting.

Process. Once the organic waste is collected, it is taken to the composting facility for processing. At the facility, the organic waste is typically mixed with bulking agents such as wood chips or yard trimmings to ensure proper airflow and balance of carbon and nitrogen. The composting process involves regular turning or agitation to provide aeration and promote decomposition. The duration of the composting process can vary, but it generally takes several months to produce mature compost.

Quality control. Municipal composting facilities monitor and manage the composting process to ensure quality and safety. Factors such as temperature, moisture content and oxygen levels are carefully controlled to promote decomposition and suppress the growth of pathogens or harmful organisms. Regular testing may be conducted to assess the maturity and quality of the compost.

Municipal composting programmes contribute to waste reduction, resource recovery and the promotion of sustainable practices within communities. By diverting organic waste from landfill and producing compost, municipalities can reduce greenhouse gas emissions, improve soil health and support local food production. These programmes align with broader waste management strategies and environmental goals aimed at achieving more sustainable and circular systems for handling organic waste.

Green Manuring

Green manuring is a fundamental technique in organic farming, aimed at enhancing soil fertility and nutrient management while promoting sustainable agricultural practices. It involves the incorporation of fresh, green plant materials into the soil as organic matter. One common form of green manure is 'green leaf manure', which primarily consists of fast-growing leguminous crops like clover, vetch or alfalfa. These plants have a remarkable ability to fix atmospheric nitrogen through symbiotic associations with nitrogen-fixing bacteria, enriching the soil with this essential nutrient. Apart from nitrogen fixation, green manure crops exhibit rapid biomass production, which contributes to organic matter build-up in the soil. The organic matter, in turn, enhances soil structure, water retention and nutrient-holding capacity, ultimately leading to improved soil health and fertility.

The process of green manuring involves carefully selecting appropriate green manure crops and timing their incorporation into the soil. Leguminous crops are often preferred for their nitrogen-fixing capabilities, while non-leguminous plants like buckwheat and mustard are used for their rapid biomass production. Farmers usually sow these crops before the main crop season or in-between crop rotations. When the green manure crop reaches its peak vegetative stage, it is cut and incorporated into the soil. The incorporation can be done through ploughing, disc harrowing, or simply by mulching the plant material. The decomposition of green manure releases nutrients gradually, ensuring a steady supply of organic nutrients to subsequent crops. Additionally, the decomposition process enhances microbial activity in the soil, further enriching its nutrient content.

Green manuring offers several advantages that make it a valuable practice in organic farming systems. First, it improves soil fertility by adding organic matter and essential nutrients like nitrogen, phosphorus and potassium. As a result, the need for synthetic fertilizers is reduced, leading to cost savings for farmers and mitigating the environmental impacts associated with their excessive use. Secondly, green manure crops act as effective weed suppressors, reducing weed competition and the need for herbicides. Thirdly, the increased soil organic matter enhances soil structure, reducing erosion and promoting water infiltration, thereby contributing to soil and water conservation (Table 27.3).

Moreover, the nitrogen-fixing ability of leguminous green manure crops can enhance the nitrogen content in the soil for subsequent crops, fostering sustainable crop rotation practices. Additionally,

Table 27.3. Nutrient content of green leaf manure[a].

Plant	Scientific name	Nutrient content (%) on air dry basis		
		N	P_2O_5	K
Gliricidia	*Gliricidia sepium*	2.76	0.28	4.60
Pongania	*Pongamia glabra*	3.31	0.44	2.39
Neem	*Azadirachta indica*	2.83	0.28	0.35
Gulmohur	*Delonix regia*	2.76	0.46	0.50
Peltophorum	*Peltophorum ferrugenum*	2.63	0.37	0.50
Weeds				
Parthenium	*Parthenium hysterophorus*	2.68	0.68	1.45
Water hyacinth	*Eichhornia crassipes*	3.01	0.90	0.15
Trianthema	*Trianthema portulacastrum*	2.64	0.43	1.30
Ipomoea	*Pomoea*	2.01	0.33	0.40
Calotrophis	*Calotropis gigantea*	2.06	0.54	0.31
Cassia	*Cassia fistula*	1.60	0.24	1.20

[a]https://agritech.tnau.ac.in/org_farm/orgfarm_green%20manure.html#II

the promotion of beneficial soil microbes through green manuring supports overall soil health and resilience to various stressors. Overall, green manuring is a versatile and eco-friendly technique that plays a vital role in nutrient management and sustainable agriculture, making it an indispensable component of modern organic farming practices.

Liquid Organic Manure

Liquid organic manures are an essential component of nutrient management in organic farming, providing a concentrated and readily available source of nutrients for plants. These nutrient-rich liquids are derived from the fermentation and decomposition of various organic materials, including crop residues, animal dung, urine and plant matter. One of the most widely used liquid organic manures is Jeevamritham, a traditional Indian preparation. Jeevamritham is made by blending cow dung, cow urine, jaggery, water and select beneficial microorganisms like *lactobacillus*. The mixture is left to ferment for several days, during which the microorganisms break down the organic matter, releasing essential nutrients and beneficial compounds. Jeevamritham not only provides vital nutrients like nitrogen, phosphorus and potassium but it also enhances soil microbial activity, promoting healthier plant growth and increased resistance to diseases and pests.

Another prominent liquid organic manure is Panchakavya, which has its roots in ancient Indian agricultural practices. Panchakavya is a concoction made from five products obtained from cows: cow dung, cow urine, cow milk, cow curd and cow ghee. These components are mixed with water and fermented for a specific period. Panchakavya is renowned for its growth-promoting substances, such as amino acids, vitamins, enzymes and growth hormones, which stimulate plant growth and development. Additionally, it contains beneficial microorganisms that contribute to soil fertility and overall plant health. Farmers often use Panchakavya as a foliar spray, applying it directly to the leaves of plants, enabling rapid absorption of nutrients and bioactive compounds. This foliar application enhances photosynthesis, nutrient uptake and enzymatic activity in plants, resulting in improved crop yields and enhanced fruit quality.

Furthermore, fish amino acid (FAA) liquid organic manure has gained popularity in recent years due to its high concentration of essential nutrients and growth regulators. Made from fish waste or fish remnants, FAA is a rich source of proteins, amino acids and micronutrients. This potent organic manure can be used in both foliar applications and soil trenching through drip irrigation. When used as a foliar spray, FAA provides a quick boost of nutrients directly to the plant's leaves, promoting vigorous growth and improving the plant's ability to withstand environmental stresses. Alternatively, incorporating FAA through soil trenching with drip irrigation ensures a steady supply of nutrients to the plant's roots, fostering long-term growth and development. Additionally, FAA aids in enhancing soil microbial diversity and activity, fostering a more balanced and nutrient-rich soil ecosystem.

ITK – Key Players in Soil Health Improvement

Indigenous Technical Knowledge (ITK) is the actual knowledge of a given population that reflects the experiences based on tradition and includes more recent experiences with modern technologies (Sundamari and Ranganathan, 2003; Vimera, 2014; Williams and Muchena, 1991). ITK offers valuable insights and practices related to soil management and applications. ITK refers to the knowledge and practices developed and passed down through generations by indigenous communities (Warren, 1991). Here are some examples of soil applications based on ITK:

Mulching. Indigenous communities have long used mulching techniques to protect and enrich the soil. Mulching involves covering the soil surface with organic materials such as leaves, straw or plant residues. This practice helps retain soil moisture, regulate soil temperature, suppress weed growth and enhance organic matter content in the soil.

Crop rotation. ITK emphasizes the importance of crop rotation to maintain soil health and productivity. Indigenous communities have traditionally employed crop rotation systems, alternating different crops in a particular field over time. This practice helps prevent nutrient depletion, manage pests and diseases, and improve soil structure by varying the root systems and nutrient requirements of different crops.

Intercropping. Indigenous agricultural practices often involve intercropping, which is the simultaneous cultivation of multiple crops in the same field. Intercropping helps optimize land use, increase biodiversity, enhance nutrient cycling, and minimize pest and disease pressure. Specific combinations of crops are selected based on their complementary characteristics and beneficial interactions.

Agroforestry. ITK promotes the integration of trees and crops in agroforestry systems. Agroforestry combines agricultural practices with the cultivation of trees, providing multiple benefits. Tree roots help stabilize the soil, improve water infiltration and enhance nutrient cycling. Additionally, trees can provide shade, windbreak, and additional sources of food, fuel, and medicinal resources.

Biochar. Biochar, a type of charcoal produced through the pyrolysis of organic materials, has been used by indigenous communities for centuries. Biochar application to the soil improves soil structure, enhances water and nutrient retention, and promotes microbial activity. It also sequesters carbon, contributing to climate change mitigation. Indigenous communities have developed specific techniques for producing and applying biochar in agricultural and horticultural practices.

Traditional amendments. Indigenous communities have traditionally used various natural amendments to enhance soil fertility. These may include locally available materials such as animal manure, fish waste, bone meal, ash or herbal extracts. These amendments provide essential nutrients, improve soil structure and enhance microbial activity, all of which contribute to healthy soil and plant growth.

Conclusion

We have covered various aspects of composting, vermicomposting, green manuring, liquid organic manures and ITKs for nutrient management in this chapter, which sheds light on vital and effective methods of regenerative agriculture. Through composting and vermicomposting, farmers can harness the power of nature to recycle organic waste and produce nutrient-rich compost, fostering healthy soil and sustainable agricultural practices. The use of green manures contributes to soil improvement by adding organic matter and essential nutrients, enhancing soil structure and fertility.

Furthermore, this chapter has explored the utilization of liquid organic manures as a valuable resource for supplying nutrients to crops in a bioavailable form, optimizing nutrient uptake and plant growth. Additionally, the incorporation of indigenous technical knowledge (ITK) provides a rich repository of traditional practices that complement modern organic farming approaches, promoting cultural and ecological diversity in agriculture.

The chapter emphasizes the importance of adopting these organic nutrient management techniques to reduce dependence on synthetic fertilizers and chemical inputs, mitigating environmental impacts and promoting a more regenerative and sustainable approach to farming. As organic farming gains prominence in addressing global agricultural challenges, the insights and practices discussed in this chapter provide valuable guidance to farmers and agricultural stakeholders seeking to embrace organic methods and contribute to a greener and healthier future for agriculture and the planet as a whole.

References

RHS (2025). Available at: https://www.rhs.org.uk/soil-composts-mulches/leaf-mould

Alkoaik, F. (2010) Some characteristics of the municipal solid waste compost: Effect of municipal solid waste compost on the growth and production of vegetable crops. Available at: https://www.researchgate.net/figure/Some-characteristics-of-the-municipal (accessed 10 February 2025).

Flannery, R. (1991) Using Leaf Compost. Available at: https://njaes.rutgers.edu/fs117/

Sundamari, M. and Ranganathan, T.T. (2003) *Indigenous agricultural practices for sustainable farming*. Agrobios (India), Jodhpur, India.

Vimera, K. (2014) Indigenous technical knowledge: Its relevance in the present-day farming with cases of your choice. Participatory Approaches in Agricultural Extension. National Institute of Agricultural Extension Management (MANAGE), Hyderabad, India.

Warren, D.M. (1991) Using IK for agricultural development. World Bank Discussion Paper 127. World Bank, Washington, DC.

Williams, D.L. and Muchena, O.N. (1991) Utilizing indigenous knowledge systems in agricultural education to promote sustainable agriculture. *Journal of Agricultural Education* 32(4), 52–57.

Zheng, X., Aborisade, M.A., Liu, S., Lu, S., Oba, B.T. *et al.* (2020) History of composting. *Journal of Cleaner Production* 276, 124232. DOI: 10.1016/j.jclepro.2020.124232.

28 Biodegradable Organic Wastes – An Eco-Friendly Approach for Nutrient Management in Sustainable Organic Farming

BHARTI SUTHAR[1*] AND BINA MISTRY[2]

[1]ICRISAT-Patancheru, Hyderabad, India; [2]Thermo Fisher Scientific, Kansas, USA

Abstract

The role of organic agriculture, whether in farming, processing, distribution or consumption, is to sustain and enhance the health of ecosystems and organisms from the smallest in the soil to human beings. Optimizing soil 'health' through composting, vermicomposting, and green and liquid manuring is the foundation of organic agriculture. Composting, vermicomposting and green manure are processes of recycling organic waste into manure as nutrient-rich soil amendment with the help of microorganisms and worms. It is a natural way of recycling organic waste into manure that is rich in humus and nutrients; and enhances soil aeration, porosity, structure, fertility, and water-holding capacity. It requires minimum effort, expertise, equipment and expense to convert organic waste into nutrient-rich manure. Indigenous technical knowledge (ITK) plays a key role in organic farming practices and sustainable agriculture because it is economically affordable, involves minimum risk and focuses on efficient utilization of eco-friendly resources.

Keywords: organic agriculture, compost, vermicompost, green manure, indigenous technical knowledge (ITK)

Introduction

Organic agriculture and Indigenous technical knowledge (ITK) are playing a major role in the changing scenario of organic farming in the modern era as increased environmental hazards associated with chemical fertilizers, biodiversity loss, natural resource conservation and sustainable agriculture become prominent concerns. 'Organic agriculture is a unique production management system which promotes and enhances agro-ecosystem health, including biodiversity, biological cycles and soil biological activity and this is accomplished by using on-farm agronomic, biological and mechanical methods in exclusion of all synthetic off-farm inputs.' (FAO, n.d)

Optimizing soil health and nutrient management are the main aims of organic agriculture with the help of reuse and recycling of plant residue and animal waste, which involves judicious use of natural resources. Optimized soil fertility may be achieved through organic matter recycling, enrichment of compost, vermicomposting, animal manures, urine, farmyard manure, litter composting, use of botanicals, green manuring, and use of biofertilizers like *Azolla*, *Azospirillium*, *Azotobacter*, *Rhizobium* culture and PSB.

Composting

Composting is a controlled, aerobic (oxygen-required) process that converts organic

*Corresponding author: bharti.suthar@yahoo.co.in

DOI: 10.1079/9781800626850.0028

Table 28.1. Liquid manure developed by progressive organic farmers and scientifically validated by various universities/ organizations.

Sr. no.	Liquid manure	Ingredients	Benefit	Method of application	Validated by
1.	Bijamrut	Cow dung – 5 kg Cow urine – 5 litres Cow milk – 1litre Lime – 250g water – 100 litres	Seed treatment	Sprinkle on seeds	TNAU, Coimbatore and CSKHPKV, Palampur
2.	Sanjivak	Cow urine – 100 litres Cow dung –100–200 kg Jaggery – 500g water – 300 litres kept for 10 days (fermentation)	Soil enrichment	Diluted 20 times before use along drip irrigation & foliar spray	University of Stellenbosch, South Africa
3.	Jivamrut	Cow dung – 10 kg Cow urine – 10 litres Jaggery – 2 kg flour of gram (tur, moong, cowpea, urad) 2 kg live soil (healthy soil) – 1 kg water – 200 litres	Promoting growth and flowering yield enhancer; soil fertility enhancer	@5–10% spray with water	TNAU, Coimbatore, CSKHPKV, Palampur and UAS, Bangalore
4.	Amritpani	Cow dung – 10 kg honey – 500g (can be replaced by 500 g jaggery) Cow desi ghee –250g or 250 ml mustard oil Water – 200 litres	Soil fertility enhancer	200 litres/ acre along with irrigation water	NEERI (CSIR Institute), Nagpur
5.	Panchagavya	Cow dung slurry – 4 kg of fresh cow dung – 1 kg of cow urine – 3 litres of cow milk – 2 litres of curd – 2 litres of cow butter oil – 1 kg	Seed and seedling treatment; soil fertility enhancer	3 litres Pachagavya diluted in 100 litres water	TNAU, Coimbatore, CSKHPKV, Palampur, UAS, Bangalore and MPUAT, Udaipur
6.	Anda-ARKH	Juice of 25 lemons (approx.) (can be collected as waste lemon from lemon field); 5 eggs Jaggery – 250g closed for 10 days and use 11th day onwards	Growth and flowering enhancer; disease control	10–50 ml/10 litres water	Akhil Bhartiya Sajiv Kheti Samaj, Mapusa, Goa
7.	Fish-ARKH	Desi fish – 1kg (depends on availability) Jaggery – 1 kg Clean cloth (to cover); clean stick (to mix); earthen pot/ plastic drum	Disease control	3 litres/100 litres water	Akhil Bharatiya Sajiv Kheti Samaj, Mapusa, Goa

materials into a nutrient-rich soil amendment or mulch through natural decomposition. The end-product is compost – a dark, crumbly, earthy-smelling material.

Microorganisms feed on the materials added to the compost pile during the composting process. They use carbon and nitrogen to grow and reproduce, water to digest materials, and oxygen to breathe (EPA, n.d.)

Benefits

- a natural way of recycling leaves, grass clippings, yard trim, and food scraps
- reduction of materials that might go to landfill or incinerators
- requires minimal effort, equipment, expense and expertise
- produces a free, high-quality soil amendment

Table 28.2. ITK was documented under the following specific categories.

Sr. no.	Categories
1.	Rainwater management
2.	Methods to check soil and water erosion
3.	Tillage and intercultural management
4.	Crops and cropping systems
5.	Pest and disease management
6.	Soil fertility management
7.	Farm implements
8.	Post-harvest technology
9.	Grain/seed storage
10.	Horticultural crops
11.	Veterinary science and animal husbandry
12.	Fisheries
13.	Ethno-botany and agrobiodiversity
14.	Methods of weather forecasting
15.	Food product development
16.	Agro-animal-based yarns/natural dyes and weaves
17.	Low-cost housing materials
18.	Unclassified
19.	Geographical indicators and other attributes
20.	Water harvesting, seepage control and ground-water recharge

Ingredients

- Carbon-rich materials (browns) provide food for the microorganisms to consume and digest – dry leaves, plant stalks and twigs.
- Nitrogen-rich materials (greens) heat up the pile to create ideal conditions for the material to break down – grass clippings and food scraps.
- Water (moisture)
- Air (oxygen)

Composting process

Aerobic-presence of O_2

1. **Collection of browns and greens.** Collect and store food garbage and vegetable and fruit scraps in a closed container on the kitchen counter or in the fridge, freezer or under the sink. Set aside an area outside to store browns as leaves, twigs or other carbon-rich material (to mix with the food scraps).
2. **Site selection and bin.** The site should be easily accessible all year round and have good drainage, water supply and sunlight. Your compost pile will break down in the sun or the shade. Choose a type of bin for your pile. Bins can be constructed from materials such as wire, wood or cinder blocks. They can also be enclosed and include barrels and tumblers.
3. **Preparation of ingredients for composting.** Before making green and brown layers in a pile, crush them into smaller pieces for fast decomposition.
4. **Building the composting pile.** Start your pile with a four- to six-inch layer of bulky browns such as twigs and wood chips. This layer will absorb extra liquids, elevate the pile and allow air to circulate at the base of the pile. Then layer the greens and browns. When adding these, add at least two to three times the volume of browns (such as dry leaves) to the volume of greens (such as food scraps). Always ensure food scraps are covered by 4–8 inches of dry leaves or other browns. Air and water are the other key ingredients for your pile. To ensure air circulation, add enough browns and turn the compost occasionally. To maintain moisture in the pile, ensure the combined materials have the consistency of a wrung-out sponge.

Table 28.3. A list of some documented ITK under different categories.

Sr. no.	Categories	Name of ITK
1.	Agronomic measures	Cultivation and sowing across the slope, wider row spacing and deep interculturing, mixed cropping, cover cropping, criss-cross ploughing, hoeing with local hoes, set-furrow cultivation, application of manure (FYM), strip cropping, green capping, green manuring (Mishra, 2002).
2.	Conservation ditches	The ditches are something like inverted contour bunds. They are constructed between fields of farms of the Vidarbha region in deep black soils in medium- to high-rainfall regions. These ditches not only collect runoff but also some sediment (1.0–1.5 tons/ha). The water stored in these ditches is primarily used by livestock and/or for spraying (*maharashtra*) (Patnaik, 1987).
3.	Tillage	Conservation furrows with traditional ploughing, deep ploughing, summer ploughing/off-season tillage, repeated tillage during monsoon season (Mishra, 2002).
4.	Vegetative barriers for soil and water conservation	The use of vegetation to moderate erosivity of rainfall both in natural stream and in sloping arable fields is an age-old practice in rainfed agriculture. Farmers in the Shekhavati region of Rajasthan, north Gujarat, semi-arid region and South Karnataka plant *Saccharum munja*, *Euphorbia caducifolia* and *Nasehallu nnisetum honekeri*, respectively, along the field boundaries to reduce runoff and soil loss (Venkateswarlu, 2000).
5.	Control of white fly in cotton	White fly (*Bemisia tabaci*) is usually found on the dorsal surface of the leaves. The nymphs are small and flat with visible beaks that suck the plant sap. To get rid of these flies, papers pasted with castor oil/grease are hung at five or six places in the cotton field. Then air is blown, using a sprayer, over the crop. The disturbed adult flies (white fly) come out of the plant and stick to the oily paper. By this method 90% of white fly can be successfully reduced (Pondicherry) (Balasubramanian, 2000).
6.	Control of blight in cumin by botanical mixture	When the crop has grown for about 30 days, a mixture of 500 ml of neem extract with 5g of washing powder, diluted in 15 litres of water is applied. Water is sprayed at 8–10 day intervals, which prevents blight (Surendra-Nagar, Gujarat) (Rathod, 2001).
7.	Control of pests and diseases in ginger	Use of mulch at 6 q/acre reduces weed infestation, restricts disease incidence, provides nutrients and increases soil moisture in the rhizosphere of ginger plants. Farmers of this area use red mud and wood ash in equal proportion as paste to enhance keeping quality of seed ginger, by providing protection against scale insects (Orissa) (Barik *et al*., 2002).
8.	Enhancement of fruit set in papaya by inserting nail in stem	Farmers insert nails into the stem at a height of about one foot to increase fruits. Nails of various materials, such as iron or wood (1–2 cm diameter) are used (Amreli, Gujarat) (Sheldiya, 1992).
9.	Disease control in bottle gourd by *hing*	A spoonful of *hing* (asafoetida) powder is put on the rotten portion and a fine-cloth bandage is tied around it. This treatment has been in existence for a long time and is mostly used in homestead gardens and produces favourable effects within a week (Jamnagar, Gujarat) (Patel, 1993)

5. **Maintain the compost pile.** As the materials begin to decompose, the temperature of the pile will initially begin to rise, especially in the centre. It can reach 130–160° F. High temperatures help reduce the presence of pathogens and weed seeds.

Fig. 28.1. Treatment steps in animal slurry processing. (Available at: https://www.survivopedia.com/wp-content/uploads/2016/07/manure-tea.jpg [accessed 11/2/25])

6. Turning and mixing the pile from time to time will help speed up the decomposition process and aerate the pile. Use a garden fork to turn the outside of the pile inwards.
7. Monitor the pile for moisture, odour and temperature, and make adjustments as needed:
   - if the pile is too dry, activity in the pile will slow or cease; moisten the pile and turn it;
   - if the pile has a bad odour, it may be too wet or need more air circulation; add more browns/dry material to the pile and turn it;
   - if the pile is not heating up, mix in greens and turn the pile.
8. **Harvest compost.** Eventually the temperature declines to ambient temperature. By the time composting is completed, the pile becomes more uniform and less active biologically, although mesophilic organisms recolonize the compost. The material becomes dark brown to black in colour. The particles reduce in size and become consistent and soil-like in texture. In the process, the amount of humus increases, the ratio of carbon to nitrogen (C:N) decreases, pH neutralizes, and the exchange capacity of the material increases (FAO, n.d.).

Precautions

- It is important to reinforce the bin with a lid and, potentially, a floor. There should be no holes or gaps in the bin that are larger than a quarter of an inch; this prevents pests or rodents.
- Maintain the ratio of materials in the pile: 2–3 parts carbon-rich material (browns) to 1 part nitrogen-rich material (greens).
- Do not add meat, dairy or greasy food to the pile (EPA, n.d.).

Vermicomposting

Vermicomposting is a process in which earthworms convert the organic waste into manure with high nutritional content. Vermicompost is nothing but the excreta of earthworms, which is rich in humus

Table 28.4. Identification of researchable issues of some selected ITK.

Name of ITK	Purpose	Researchable issues
Nadi farming system	To collect runoff during kharif for life-saving irrigation during drought spell or pre-sowing irrigation (*Palewa*) for rabi crops	• Documentation and analysis of socio-economic aspect of present nadi system for its sustainability • Evaluation of present nadi farming system
Furrow opening in standing crops	Rainwater conservation	• Modification of implementation with different serrated blades and introducing additional tines • Effectiveness in conserving soil moisture
Mixed pulses as vegetative barrier	Resource conservation	• Proportion of pulses as vegetative barrier; cost-effectiveness of the system
Application of white soil as lining material in farm pond	To work as a sealant material for lining dugout farm pond	• Standardization of application technique and economic feasibility for wider application • Study on the seepage losses at different hydraulic heads
Wider row spacing in pearl millet	Rainwater conservation and weed control	• Plant geometry and population research in different rainfall situations
Rainwater harvesting in kund/tanka	The harvested water in kund/tanka is used for drinking and establishment of trees.	• Use of stored water for arid horticulture • Design of tankas for different geo-hydrologic conditions
Crop stubbles and residue management	Improves the organic matter and water holding capacity of soil	• Quantification of soil and water conserved and yield advantage • Crop residue incorporation alternate ways of composting and application
Mulching in turmeric	To conserve rainwater	• Quantification of soil loss, improvement of soil quality and water availability • Use of alternative organic material to sal leaves as mulch
Water harvesting and recycling	Rainwater harvesting	• Recharging of water table; cost-effectiveness
Formation of gurr	Reduction of runoff and soil moisture conservation	• Effect of bullock and tractor made gurr on runoff reduction, soil water conservation and crop productivity
Green manuring practice	To conserve soil water and improve soil health	• Growing of green manure crop and its management in improving soil health and crop productivity (Kogut, n.d.)

and enhances soil and plants with vital life nutrients, air, porosity, structure, fertility and water-holding capacity.

Earthworms will consume most organic materials, including food preparation residuals and leftovers, scrap paper, animal manure, agricultural crop residues, organic by-products from industries, and yard trimmings. The most popular type of earthworm utilized worldwide for the vermiculture method is *Eisenia fetida* (Red worm). They have a fast metabolism and can break down and turn organic materials into vermicompost in 45–50 days.

Vermicompost methods

- **Bed method.** Composting is done on the pucca/kachcha floor by making a bed (dimension: 6 × 2 × 2 feet) of organic mixture. This is an easy method in which beds of organic matter are prepared.

- **Pit method.** Composting is done in cemented pits, wooden boxes, plastic buckets, silpaulin bags, baskets etc. However, this method is not prominent as it is hampered by problems of poor aeration and waterlogging.
- **Site selection.** Vermicompost production can be done in any place that is shady, high in humidity and cool. It should have protection, availability of dung or manure, access to water and be a conducive environment for earthworms.

Materials required for vermicomposting

- water, cow dung, thatched roof, soil or sand, gunny bags, earthworms, weed biomass, large bin (plastic or cemented tank), dry straw and leaves, biodegradable wastes collected from fields and the kitchen.

Procedure

1. A plastic or a concrete tank can be used to prepare the compost.
2. Collect the biomass and dry it under sunlight for about 8–12 days. Then chop into small pieces with a cutter.
3. Predigestion is the process of partial digestion of organic waste for 20–25 days by dumping the material along with cattle dung slurry that is fit for earthworm consumption.
4. Add loamy soil at the bottom of the first layer, which should be around 15–20 cm thick.
5. Prepare fine bedding by adding partially decomposed cow dung, dried leaves and other biodegradable waste collected from fields or the kitchen. Distribute them evenly on the sand layer.
6. Continue adding both the chopped biowaste and partially decomposed cow dung in layers up to a depth of 15–30 cm.
7. After adding all the biowastes, release the earthworms over the mixture and cover the compost with dry straw or gunny bags.
8. Sprinkle water on a regular basis to maintain the moisture content of the compost.
9. Cover the tank with a thatched roof to prevent the entry of ants, lizards, mice, snakes etc., and protect the compost from rainwater and direct sunlight.
10. Frequently check to ensure the compost does not overheat. Maintain proper moisture and temperature.
11. Before harvesting, stop watering to allow drying of the top part of the vermicompost. After screening to remove un-decomposed materials and worms, the final product is then dried for a few days, weighed and bagged in 50 kg amounts (Edwards and Bater, 1992).

Storing and packing of vermicompost

The harvested vermicompost should be stored in a cool, open place protected from sunlight. The compost can be stored for one year without loss of quality, if moisture is maintained at 40% level (Chanu *et al.*, 2018).

Vermiwash

Vermiwash is a liquid that is collected after the passage of water through the vermicomposting system. It is very useful as a foliar spray. Application: 1 litre of vermiwash or 0.5 litres of vermiwash +0.5 litres of cow urine are diluted in 20 litres of water and can be used as a foliar spray.

Green Manuring

Green manuring is practice of ploughing or turning into the soil un-decomposed fresh green plant tissue for the purpose of improving soil fertility.

Characteristics of green manuring crops

- capable of establishing and growing quickly, accumulating sufficient fixed N (4–6 weeks);
- tolerant to adverse climatic conditions, pests and disease;
- effective nitrogen fixer as they possess adequate *Rhizobium* nodulation;
- easy to incorporate and quickly decomposable.

Types of green manures

Green manuring in situ

Green manuring crops are grown and buried in the field either as a pure crop or as an intercrop with the main crop; e.g. sunhemp (*Crotolaria juncea*), dhaincha (*Sesbania aculeata*), cluster bean

(*Cymopsis tetragonoloba*), cowpea (*Vigna sinensis*), khesari (*Lathyrus sativus*), berseem (*Trifolium alexandrium*) green gram (*Vigna radiata*).

Green leaf manuring

Collect leaves and tender twigs from shrubs, trees and forest areas and incorporate them into cultivable fields. The common shrubs and trees are neem, mahua, wild indigo, glyricidia, karanji (*Pongamia glabra*) calotropis, avise (*Sesbania grandiflora),* subabul, ipomea, jatropha gossipifolia and other shrubs.

Benefits of green manuring

- contributes N ranging from 50 to 175 kg/ha and a huge quantity of organic matter;
- increases water, microbial population and nutrient-holding capacity of the soil;
- improves soil texture and increases the availability of various macro- and micronutrients (Warman, n.d.).

Liquid Manure

Liquid manure is a combination of animal waste and organic matter used as an agricultural fertilizer (Table 28.1). Liquid manure contains less than 5% dry matter and is produced when wash-water is added to slurry manure, or when a significant solid fraction is removed from slurry manure after solid-liquid separation (Liu and Wang, 2020).

The main purpose of liquid manure treatment is to improve the properties of manure for land application or reuse as wash-water by reducing odour and excess solids.

Benefits

- improves health, as usage of harmful chemicals is reduced;
- prevents crop, pests and diseases, and increases beneficial insects;
- provides nutrients, soil moisture, and improves soil health;
- uses local resources and knowledge, protects the environment and saves cost;
- helps us to be self-reliant.

Basic treatment steps in plant slurry processing

1. Collect the organic waste/plants, crush them in small pieces and fill the container with them, and add water.
2. Add ash and the cow dung, wrapped in a sack. Stir it to produce beneficial microorganisms.
3. Slurry starts to smell, a sign it is ready to use.
4. In cold weather, the process is completed in 2–3 weeks or in 5 days in hot weather.

Basic treatment steps in animal slurry processing (Fig. 28.1)

1. separation of colloidal and suspended particles;
2. filtration (in combination with coagulation/ flocculation);
3. centrifuging (in combination with coagulation/ flocculation);
4. sedimentation;
5. separation of urine and faeces;
6. sealing of the container;
7. leave in shade for 2 weeks.

Indigenous Technical Knowledge

Indigenous technical knowledge (ITK) is a collection of knowledge unique to a particular culture, society or geographical region and preserved over several generations. It has been efficiently used for the betterment and conservation of natural resources (Mehta *et al.*, 2010). It is a people-derived science and represents people's creativity, innovation and skills. This knowledge is thousands of years old and has contributed to sustainable food security, conservation of varieties, and arability of animals, plants and soil properties. ITK followed by farmers is seed treatment, storage methods, borne setting methods and disease treatment.

For the purpose of documentation and analysis of ITK, a simple ITK proforma was developed. This proforma contained title, purpose, location, agroecological setting, description, advantages, constraints, replicability/feasibility and researchable issues. Researchable issues opens new vistas for furthering the technologies towards development of farmer-friendly feasible technologies. A few xamples of ITK could be documented in a limited timeframe through the project centres located in various southern agricultural universities (Tables 28.2 and 28.3).

The research findings have been disseminated through the extension agencies. The research results benefit the farming community as well as the promoting agencies, i.e. the government or non-government organizations, and research institutes. Some of the researchable issues pertaining to some examples of ITK have been identified and presented in Table 28.4.

Advantages of ITK for the future

- Similar exercises can be undertaken to document in regional languages for all the agroecological regions of the country.
- Suitable modifications of the traditional practices through on-farm research would help in developing appropriate and acceptable technologies for different local environments.
- The effect of conservation measures on resource losses can be studied in detail through experimentation and use of stimulation model.
- As a policy matter, the local ITK should be built into the resource conservation programme.

Conclusion

Use of compost, vermicompost, green manure and liquid manure is an important component of integrated nutrient management and sustainable organic agriculture. It can maintain soil health and plant nutrients availability over a long period of time. Indigenous technical knowledge (ITK) is more beneficial for organic farming as it easily adaptable, less costly, economically viable, and helps in agriculture husbandry in field crops, vegetables, fruit orchards, and cattle, and even in humans. Now, ITK is becoming vital in modern-day organic farming practices. Organic agriculture is a changing scenario of agricultural farming practices and environmental concerns associated with chemical fertilizer usage.

References

Balasubramanian, K. (2000) Beyond clinical pesticides in cotton. *Honey Bee* 12(1), 21.

Barik, T., Dash, S., Samal, K.C., Dash, S.K. and Mohapatra, B.P. (2002) Traditional method of storage of pulses, grains and vegetable seeds. *International Seminar on Traditional Knowledge, Health and Environment,* Vol. 65. Bhubaneswar, India.

Chanu, L.J., Samarendra, C., Choudhury, H.B.U., Balusamy, R.T.A., Moirangthem, P. *et al.* (2018) *A guide to vermicomposting: production process and socio-economic aspects*. Umiam-793 103, Meghalaya.

Edwards, C.A. and Bater, J.E. (1992) The use of earthworms in environmental management. *Soil Biology and Biochemistry* 24, 12.

EPA (n.d.) Composting at home. Available at: https://www.epa.gov (accessed 29 June 2023).

FAO (n.d) Composting process and techniques. Available at: https://www.fao.org (accessed 29 June 2023).

Kogut, P. (n.d.) Green manure crops: Planting & management tips. Available at: https://eos.com/blog/green-manure (accessed June 2023).

Liu, Z. and Wang, X. (2020) Manure treatment and utilization in production systems. In: Bazer, F.W., Lamb, G.C. and Wu, G. (eds) *Animal Agriculture*. Academic Press, Amsterdam, pp. 455–467.

Mehta, P., Sharma, A.K. and Negi, K. (2010) Indigenous knowledge system and sustainable development with particular reference to folklores of Kumaon Himalaya, Uttarakhand. *Indian Journal of Traditional Knowledge* 9, 547–550.

Mishra, P.K. (2002) Indigenous technical knowledge (ITK) on soil & water conservation in rainfed areas. *Ecology, Environment and Conservation* 13(4), 207–215.

Patel, M.S. (1993) Disease control – rotting of bottle gourd vine. *Honey Bee* 4(2 & 3), 20.

Patnaik, U.S. (1987) Conservation ditching for soil and water conservation in low rainfall vertisol regions. Central Soil & Water Conservation Research & Training Institute. Available at: https://books.google.co.in/books?id=oSHSHAAACAAJ (accessed 11 February 2025).

Rathod, P.G. (2001) Control of blight in cumin crop. *Honey Bee* 12(2), 10.

Sheldiya, M.P. (1992) Enhancement of fruit set in papaya. *Honey Bee* 3(3 & 4), 1516.

Venkateswarlu, J. (2000) *Technical Manual on Watershed Management. I: Resource Conservation*. National Institute of Agricultural Extension Management, Hyderabad, India.

Warman, P. (n.d.) The basics of green manuring. Available at: https://www.eap.mcgill.ca/publications/EAP51.htm (accessed 30 June 2023).

29 Plant Biostimulants: Essential Tools for Sustainable Organic Farming

Gheorghe Christian Popescu[1]* and Monica Popescu[2]*

[1]Department of Applied Sciences and Environmental Engineering, National University of Science and Technology Politehnica Bucharest, University Center of Pitesti, Romania; [2]Department of Natural Sciences, National University of Science and Technology Politehnica Bucharest, Romania

Abstract

Nowadays, environmental protection in agricultural ecosystems and healthy food production are important goals for agricultural systems. Organic farming is an agricultural production system with multiple benefits for human nutrition, environment and society. Plant biostimulants (PBs) application could be an efficient tool to be integrated as sustainable practice in agriculture. PBs, a relatively new category of crop inputs, are substances or microorganisms containing different active ingredients as amino acids, humic substances, seaweed extract, fungi, bacteria, or other natural compounds to improve plant growth, nutrient uptake, abiotic and biotic stress tolerance, food production and crop quality. PBs have important potential for the global market of agricultural inputs due to the need for sustainable agricultural practices and the increasing demand for high-quality agricultural products. These products will bring new business opportunities for the bioeconomy, circular economy and agriculture sector. PBs minimize the use of chemical fertilizers and synthetic pesticides providing an environmentally friendly tool for crop management.

Keywords: plant biostimulants, organic farming, environmental sustainability, crop yields and quality, food security, plant growth

Introduction

Nowadays, environmental protection in agricultural ecosystems and healthy food production are important goals for agricultural systems. Organic farming is an agricultural production system with multiple benefits for human nutrition, environment and society. Plant biostimulants (PBs) application could be an efficient tool to be integrated as sustainable practice in agriculture. PBs, a relatively new category of crop inputs, are substances or microorganisms containing different active ingredients such as amino acids, humic substances, seaweed extract, fungi, bacteria, or other natural compounds to improve plant growth, nutrient uptake, abiotic and biotic stress tolerance, food production and crop quality. PBs have important potential for the global market of agricultural inputs due to the need for sustainable agricultural practices and the increasing demand for high-quality agricultural products. These products will bring new business opportunities for the bioeconomy, circular economy and agriculture sector. PBs minimize the use of chemical fertilizers and synthetic pesticides providing an environmentally friendly tool for crop management.

General Considerations Related to Plant Biostimulants, Policy and Legislation

Agriculture is one of the most important sectors of the global economy and must adapt to the new challenges and respond to the sustainability standards. In recent years, agricultural production systems face a multitude

*Corresponding authors: christian_popescu2000@yahoo.com and monica_26_10@yahoo.com

 Advances in Organic Farming (ed. L. P. Awasthi)
DOI: 10.1079/9781800626850.0029

of challenges related to food security, environment, resources, climate change, global population growth etc. (Popescu *et al.*, 2022). Plant biostimulants (PBs) are substances widely used in agricultural systems. These natural products could be a promising and efficient tool to improve crop productivity and sustainability in agriculture and horticulture. Therefore, sustainable agricultural practices and innovative technologies using PBs can generate efficient and sustainable management of agricultural crops by reducing the use of chemical fertilizers and synthetic pesticides.

Organic farming is part of sustainability covering all three dimensions related to environmental protection, and social and economic aspects (Fig. 29.1). The main pillars of organic farming are very well connected with sustainability and a resilience approach in agriculture (Fig. 29.2).

Agriculture covers very large terrestrial land areas, and intensive agricultural land use is one of the major sources for affecting natural resources, habitats and species diversity in the agricultural landscape. Therefore, advancements and

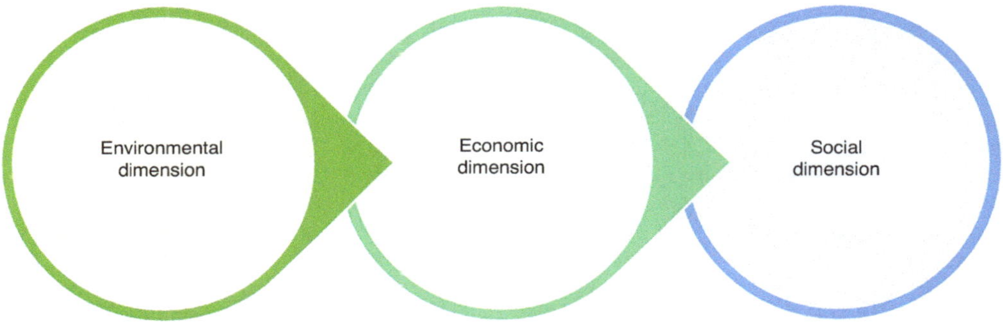

Fig. 29.1. The three dimensions of sustainability.

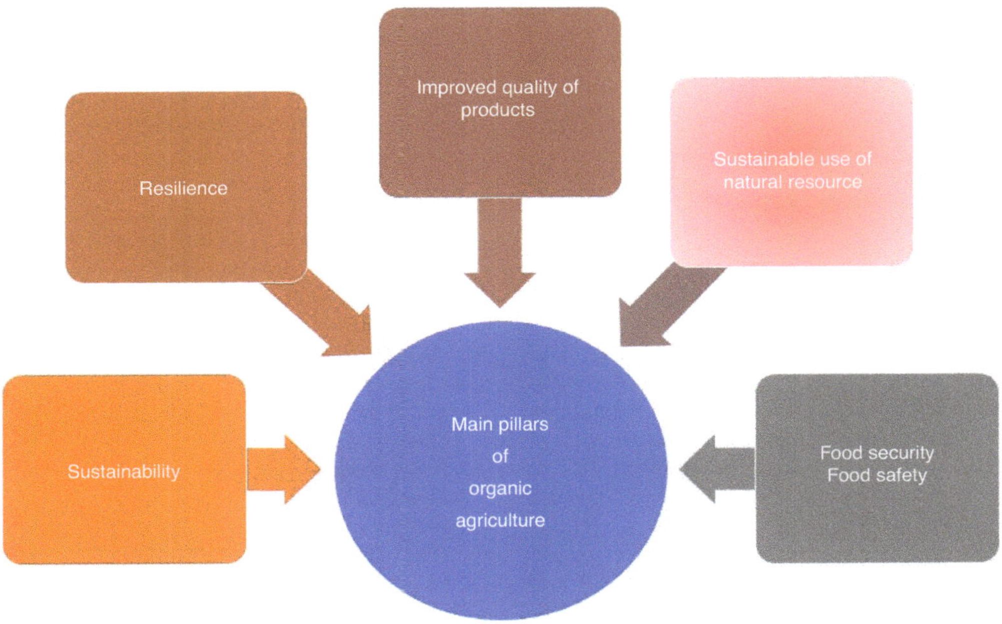

Fig. 29.2. Main pillars of organic agriculture systems.

innovations, such as the development of PBs, will lead to higher agricultural productivity, sustainability and resilience of agricultural systems.

In Europe and the USA, the legislative framework on PBs is very complex and more advanced compared to other countries or regions in the world. For example, currently the legislation framework at the level of the European Union (EU) is harmonized at the level of the member states by Regulation (EU) 2019/1009.

According to the EU, regulatory PBs are defined by Regulation (EU) 2019/1009 of the European Parliament and of the Council of 5 June 2019, laying down rules on the making available on the market of EU fertilizing products and amending Regulations (EC) No 1069/2009 (European Commission, 2016) and (EC) No 1107/2009 and repealing Regulation (EC) No 2003/2003 (https://eur-lex.europa.eu/eli/reg/2019/1009/oj [accessed 12/2/25]), 'as fertilising product with function to stimulate plant nutrition processes independently of the product's nutrient content with the aim of improving one or more of the following characteristics of the plant or the plant rhizosphere:

- nutrient use efficiency;
- tolerance to abiotic stress;
- quality traits;
- availability of confined nutrients in the soil or rhizosphere.'

The legislative framework from the USA defines PBs as:

> [A] substance, micro-organism, or mixture thereof, that, when applied to seeds, plants, the rhizosphere, soil, or other growth media, act to support a plant's natural processes independently of the biostimulant's nutrient content, including by improving nutrient availability, uptake or use efficiency, tolerance to abiotic stress, and consequent growth, development, quality, or yield.

At the European Union level, there is a strict and specific procedure for the approval of each new PBs product. Similar to the EU or USA, the regulations for PBs are present in many other countries such as India, Canada, China or South Africa. The different legislative systems of the countries define PBs and the categories of PBs differently.

Moreover, the most important markets for the sale of these commercial products are the USA and the member states of the EU. The international PBs market is expected to grow rapidly in the next few years due to the need for sustainable agricultural practices and the increasing demand for high-quality agricultural products.

However, at the global level, in many countries there are differences in approach, definition or classification regarding the production and use of biostimulating plants, and thus there is no harmonization of legislation. More harmonization of the legislation regarding plant biostimulators would be useful for the global market of PBs and sustainable organic farming (Ali *et al.*, 2022). Currently, many types of PBs are available in the global market of crops inputs in different forms and concentrations. PBs will bring new business opportunities for the bioeconomy, circular economy and agriculture sector.

Efficient regulatory framework, specific rural and agricultural policies, and representative strategies of PBs are required for efficient application of agricultural biostimulants and achieving sustainable organic farming (Fig. 29.3).

PBs are substances or microorganisms that are applied to plants, soil or seed containing different active ingredients such as amino acids, humic substances, seaweed extract, fungi, bacteria, or other natural ingredients in liquid, granular or powder form to improve plant growth, nutrient uptake, abiotic and biotic stress tolerance, and crop quality overall (Fig. 29.4). Mode of application for PBs can be direct on soil, plants or seeds. Also, PBs can be applied using irrigation systems.

PBs are natural products and alternative biosolutions containing substances or microorganisms that have a natural and positive impact on the plant's physiological mechanisms, enhance growth parameters and improve agricultural sustainable production. These types of natural compounds are a new category of crop inputs with multiple benefits for environments and agricultural technologies. At the same time, plant biostimulants are environmentally friendly solutions to secure and optimize agricultural yields in an organic way.

Biostimulators are relatively new agricultural inputs with multiple benefits for agricultural crops, especially if they are applied during plant stress periods. Through the mechanisms that promote natural defence response and the positive effects for crops, biostimulators can play a significant role in achieving agricultural productivity and crop quality (Fig. 29.5).

PBs can contribute to ensuring the sustainability of agricultural systems because they have a multitude of benefits for crops, soil and the environment such as: reducing fertilizer supply, enhancing plant tolerance to abiotic stress (salinity, low-fertility soil,

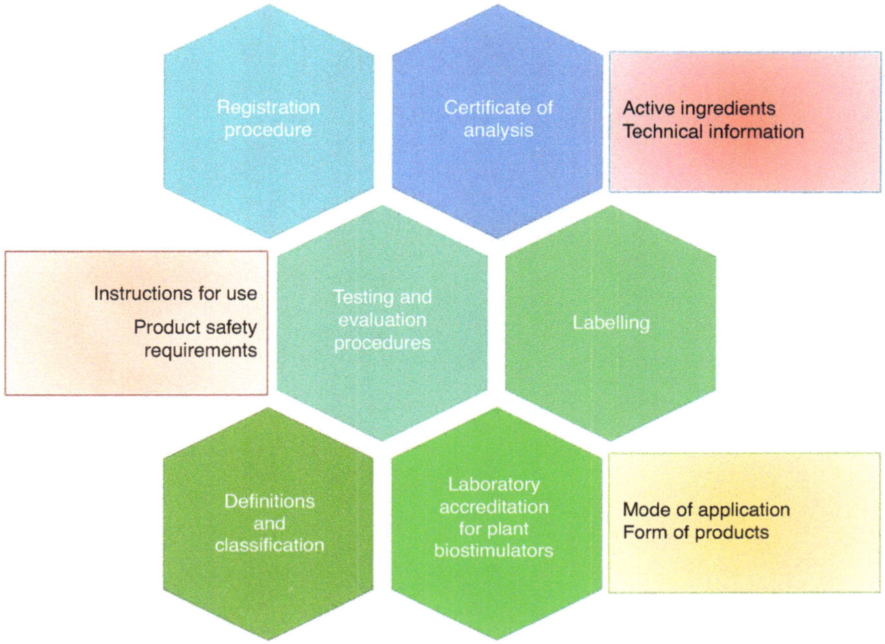

Fig. 29.3. Main components for an efficient regulatory framework and policy of PBs.

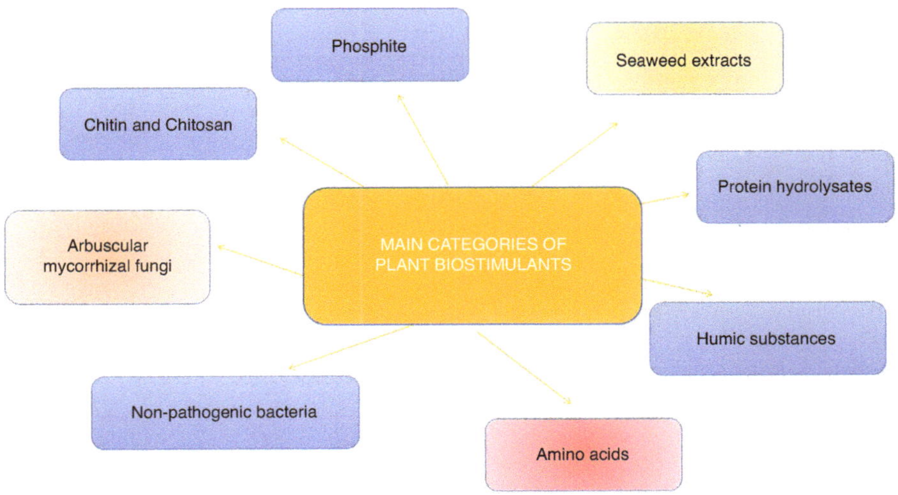

Fig. 29.4. Main categories of PBs.

heavy metal stress, drought), limiting flower abortion, stimulating plant metabolism and total antioxidant activity, improving plant vigour, increasing root macro- and micronutrient concentrations, and increasing disease resistance (Fig. 29.6).

Main sources to produce plant biostimulants include: animal protein, plant protein, agricultural waste, waste from fish processing, other agro-industrial residues, glomalin-related soil protein (GRSP), plant extracts from wild flora, microalgae,

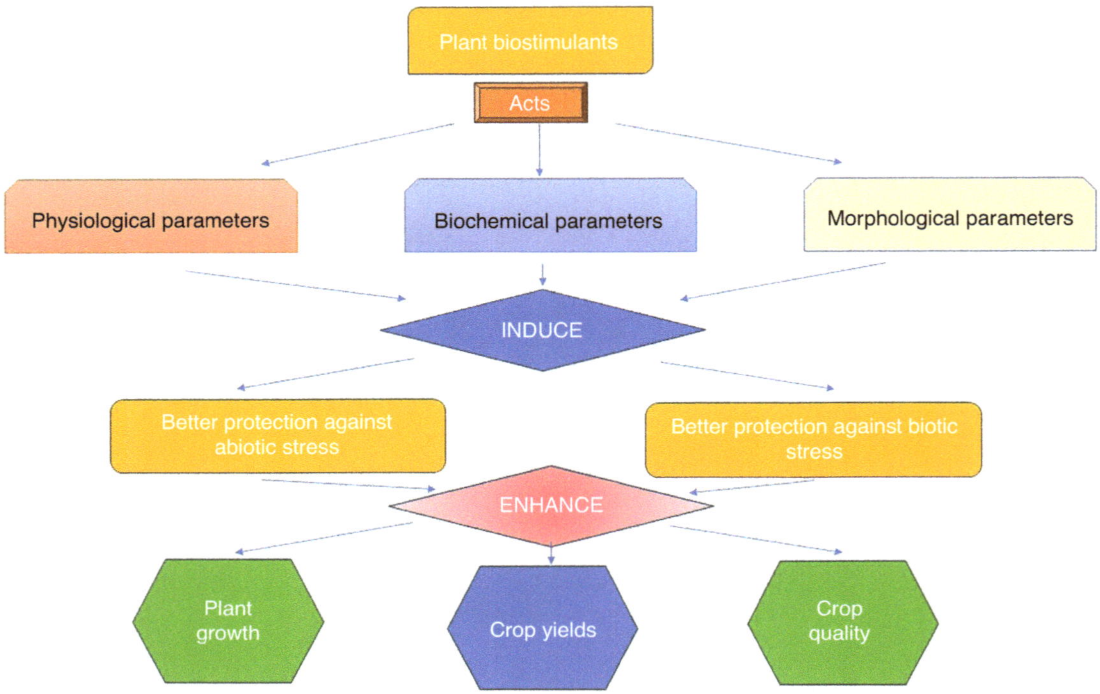

Fig. 29.5. Conceptual framework of the PBs contribution to enhanced crop yield and quality.

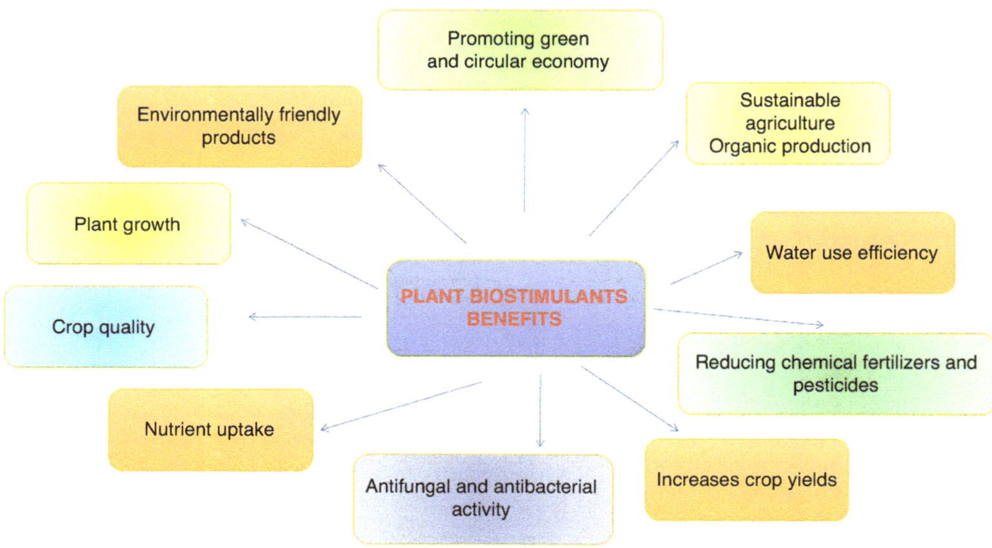

Fig. 29.6. General benefits and impact of PB application.

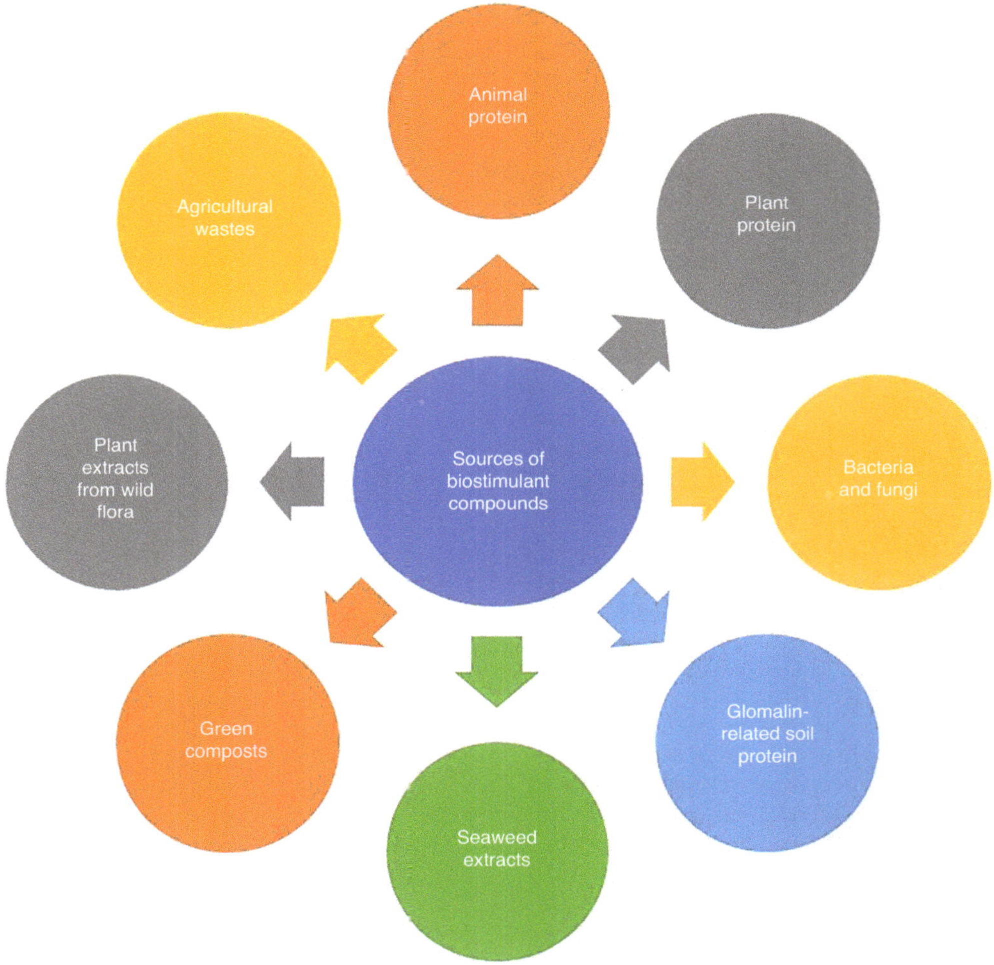

Fig. 29.7. Main sources to produce plant biostimulants.

green composts, peat, vermicompost, leonardite, alfalfa biomass, animal epithelial or connective tissues, and seaweed extract (Fig. 29.7).

Plant Biostimulant Applications in Sustainable Agriculture for Crop Yields and Quality

Many research studies were addressed to enhance the quality and productivity of agricultural production. Currently, PBs are part of agricultural technologies, especially for organic production. Enhancement of various physiological, biochemical and morphological parameters of plants using PBs has been reported for a lot of agricultural species (Table 29.1).

Many studies have reported positive effects of humic substances (HS) applications on crop performances. PBs have the ability to protect plants against abiotic stresses. Drought stress is one of the major abiotic factors that adversely affects crop growth and limits yield production. Humic acid (HA) application was used in a factorial study and completely randomized design to enhance the onion (*Allium cepa* L.) growth parameters under water-deficit conditions. Onion growth and quality

Table 29.1. PBs application in agricultural systems with implications for crop yields.

Type of plant biostimulants	Species	References
Humic acids (HA)	*Allium cepa* L. (onion)	Amiri Forotaghe *et al.*, 2022
	Cucumis sativus L. (cucumber)	De Hita *et al.*, 2020
	Zea mays L. (maize)	Monda *et al.*, 2018
	Vitis vinifera L. (grapevine)	Popescu and Popescu, 2018a
Protein hydrolysates (PH)	*Solanum lycopersicum* L. (tomato)	Wang *et al.*, 2022
Glomalin-related soil protein (GRSP)	Citrus reticulate L. (mandarin)	Liu *et al.*, 2022
	Citrus sinensis L. (orange)	
	Citrus paradisi Mac. (grapefruit)	
Seaweed extract (SE)	*Solanum tuberosum* L. (potato)	Caradonia *et al.*, 2022
Humic substances (HS)	*Solanum tuberosum* L. (potato)	Wadas and Dziugieł, 2020
Sargassum angustifolium	*Fragaria ×ananassa* Duch. (strawberry)	Tajdinian *et al.*, 2022
Arbuscular Mycorrhizal Fungi (AMF)	*Vitis vinifera* L. (grapevine)	Aguilera *et al.*, 2022
	Arachis hypogaea L. (peanut)	Liu *et al.*, 2023
	Lavandula angustifolia Mill. (lavender)	Popescu and Popescu, 2022

traits significantly increased in the experimental variant of soil with HA under water-deficit conditions, including leaf biomass, leaf dry weight, total chlorophyll, vitamin C and bulb flavonoids (Amiri Forotaghe *et al.*, 2022). In another research study, humic acids derived from green composts stimulated the plant growth parameters of maize seedling in terms of plant biomass, plant height, and root length and weight (Monda *et al.*, 2018). Foliar sprays with humic acid derived from vermicompost as PBs induced a significant increase in the total leaf area, yield and total soluble solids of grapes in treated grapevines (Popescu and Popescu, 2018a).

Potato plants were treated with three plant biostimulants: one containing seaweed extracts from *Ascophyllum nodosum*; a second was an extract from *Ecklonia maxima*; and the last was based on humic and fulvic acids leonardite. These formulations of PBs enhanced abiotic stress tolerance and increased marketable tuber yield in comparison with the control treatment (Wadas and Dziugieł, 2020).

Herbaspirillum seropedicae and humic acids stimulated shoot and root growth of tomato plants treated with *Xanthomonas euvesicatoria* (da Silva *et al.*, 2021).

Lumactud *et al.* (2022) highlighted that those humic substances extracted from different natural sources such as compost, manure, coals, peats or lignites can be used as organic agricultural amendments to improve soil health and to use less synthetic fertilizers. Humic-based products are an efficient tool to promote sustainable farming practice.

Protein hydrolysates have been used as plant biostimulants in agriculture due to their positive impacts on crop yields under abiotic stress. Foliar application of pig blood-derived protein hydrolysate increased plant growth parameters for tomato under drought stress. The plant biostimulant represented by protein hydrolysate has a positive impact for tomato crops stimulating the accumulation of total phenolic, total flavonoid, ascorbic acid, and glutathione (Wang *et al.*, 2022). Therefore, protein hydrolysate derived from animal or vegetal protein could be an environmental PB for enhanced plant performance.

Glomalin-related soil protein (GRSP) was reported in many research studies as PBs with multiple roles for crops. Liu *et al.* (2022) noted that GRSP foliar application in three citrus species (mandarin, grapefruit and orange) improved quality of fruits, including pericarp fruit weight, soluble solids, or sugar content of citrus fruits. Also, Liu *et al.* (2022) observed a positive effect at application of GRSP in trifoliate orange (*Poncirus trifoliata*) seedlings for total chlorophyll content, NPK root content or root morphology. At the same time, Meng *et al.* (2021) reported that this PB as foliar spray improved physiological parameters of fruits of late-ripening sweet oranges.

Seaweed extracts from *Ascophyllum nodosum* are widely used to produce commercial PBs.

Ascophyllum nodosum as PBs have been explored for their potential to improve plant growth, drought-stress tolerance and agricultural productivity, improving nutrient-use efficiency in treated plants. Another important application of seaweed extract based on *Ascophyllum nodosum* is to increase the heavy stress metal in plants. Popescu and Popescu (2018b) found that application of *Ascophyllum nodosum* extract increased the activity of the antioxidant enzyme catalase and relative water content in bean seedlings (*Phaseolus vulgaris* L.) and therefore drought tolerance was improved. Application of *Ascophyllum nodosum* seaweed extract improved growth, fruit-quality parameters, and water productivity of tomato under water stress (Ahmed *et al.*, 2023).

Foliar application of PBs extracted from brown alga (*Sargassum angustifolium*) increased the level of expression for vegetative, biochemical and reproductive parameters in strawberries, comparing treated plants infected with charcoal root disease compared to non-treated plants (Tajdinian *et al.*, 2022). The findings showed that the total phenol, flavonoids and total antioxidant activity were significantly increased in plants treated with *Sargassum angustifolium* extract compared with control plants.

Other important PBs are products containing arbuscular mycorrhizal fungi (AMF). Many research studies suggested that arbuscular mycorrhizal fungi are beneficial microorganisms in soil–plant interaction providing several ecosystem services to host plants, such as water and nutrient uptake efficiency (Aguilera *et al.*, 2022; Liu *et al.*, 2023). AMF form a symbiotic association with most vascular plants and therefore can be used in different formulations to create plant biostimulants. Technologies on sustainable agriculture may consider arbuscular mycorrhizal fungi as biofertilizers. For example, Liu *et al.* (2023) showed that *Rhizophagus irregularis* SA, *Rhizophagus clarus* BEG142, *Glomus lamellosum* ON393, and *Funneliformis mosseae* BEG95 could enhance salt-, drought- and cold-stress tolerance through improving plant growth in inoculated peanut plants. AMF mixture containing *Rhizophagus irregularis, Claroideoglomus claroideum, Funneliformis mosseae, Funneliformis geosporum, Claroideoglomus etunicatum,* and *Glomus microaggregatum* applied in a growing substrate with peat, coconut bark and volcanic rock enhanced plant growth, physiological processes and flowering performance in lavender plants (Popescu and Popescu, 2022). Darakeh *et al.* (2022) noted that AMF colonization is a promising alternative to chemical fertilizers for black cumin cultivation.

A lot of growth and quality parameters of different crops were evaluated under the effect of PBs application including: leaf morphology, root system, flowering performance, fruit growth and quality, physiological process, plant biomass etc. (Fig. 29.8).

Plant Biostimulant Applications in Sustainable Agriculture for Crop Protection

Agricultural crops from all over the world are severely affected by different phytopathogens that decrease the quality of products, yields, and nutritional and commercial value. Disease and pest control is an important issue in agriculture and horticulture, which requires mainly chemical treatment with synthesized products as pesticides. Plant protection for disease and pest management is an essential tool to improve crop quality and production. New plant protection alternatives are necessary in agriculture for better management of disease and pest control, especially because conventional plant protection products are no longer effective for new forms of phytopathogens that appear in culture. Also, synthetic pesticides are regarded less favourably for application in agriculture mainly due to consumers who are more attentive to product quality and environmental protection. Therefore, PBs applications provide a viable biosolution to sustain sustainable plant protection and these bioproducts can be part of integrated pest and disease management programmes. PBs have a major potential in biological control of plant diseases (Olowe *et al.*, 2020).

PBs started to be used more and more in plant protection management because they contain natural compounds, are effective for controlling biotic stress, and the toxicity of them is low or absent (Gebashe *et al.*, 2021). Recent studies point out the positive effects of PBs application in plant protection. Studies indicate that mechanisms of action for biostimulants are variable, depending on the type of phytopathogen, biostimulant and host plant (Gebashe *et al.*, 2021; Sahana *et al.*, 2022).

Many research studies were addressed to enhance crops' tolerance against diseases and pests.

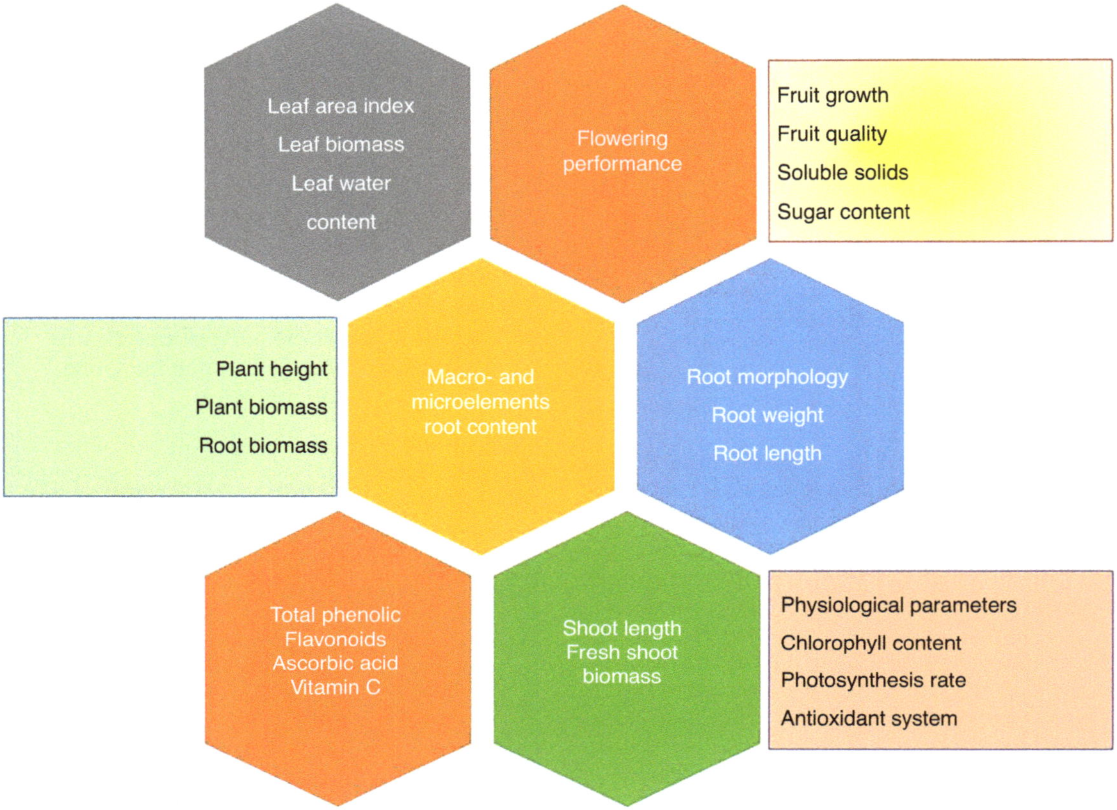

Fig. 29.8. Plant growth and quality fruit parameters studied under the effect of PBs.

PB application using plant biostimulants has been reported for many phytopatogens (Table 29.2).

Bacterial biostimulants are involved in promoting plant growth and provide tolerance against biotic stresses. Plant extract formulation from *Ascophyllum nodosum* was tested to evaluate the disease resistance in cucumber plants inoculated with *Phytophthora melonis* (Abkhoo and Sabbagh, 2016). Treated plants of cucumber with the tested pathogen had a significant reduction of disease incidence. Also, the authors noted that best treatment in reducing the incidence was found at the alternation of one fungicide with seaweed extract of *Ascophyllum nodosum*.

Natural biostimulant based on *Ascophyllum nodosum* and plant extracts (alfalfa and sugarcane) combined with a low dose of the conventional synthetic fungicide, Imazalil, reduced the incidence and severity of fruit green mould caused by *Penicillium digitatum* and also enhanced the expression level of plant defence-related encoding genes in fruit peel (β-1,3-glucanase – Glucanase, phenylalanine ammonia-lyase – PAL, and peroxidase – PEROX) (La Spada *et al.*, 2021). Another significant result of the study was that the residual concentration of synthetic fungicide in fruits treated with the biostimulant of fungicide +bistimulant mixture was significantly lower than the level found in fruits treated only with the fungicide at the same dose (La Spada *et al.*, 2021).

Aureobasidium pullulans and *Bacillus amyloliquefaciens* were used to produce PBs and then applied for biological control of *Verticillium dahliae* Kleb. in olive plantations (López-Moral *et al.*, 2021). Also, the same study evaluated the effect of different PBs based on amino acids, microorganisms or other substances of natural origin. The authors emphasized the potential of PBs for plant protection in olive orchards.

Table 29.2. PB applications for plant protection.

Biostimulant-based products	Phytopathogens	References
Ascophyllum nodosum	*Penicillium digitatum*	La Spada *et al.*, 2021
	Phytopthora melonis	Abkhoo and Sabbagh, 2016
Humic acids	*Fusarium culmorum*	Iwaniuk *et al.*, 2023
Aureobasidium pullulans and Bacillus amyloliquefaciens	*Verticillium dahliae* Kleb.	López-Moral *et al.*, 2021
Endophyte *Bacillus velezensis* KOF112	*Botrytis cinerea*, *Colletotrichum gloeosporioides*, *Phytophthora infestans*	Hamaoka *et al.*, 2021
Seaweed (*Kappaphycus* sp. and *Eucheuma* sp.)	*Magnaporthe oryzae*	Sahana *et al.*, 2022
Sargassum angustifolium	*Macrophomina phaseolina* (Tassi) Goi	Tajdinian *et al.*, 2022
Streptomyces strains	*Macrophomina phaseolina* *Rhizoctonia solani*	Díaz-Díaz *et al.*, 2022
Humic acids and *Herbaspirillum seropedicae*	*Xanthomonas euvesicatoria*	da Silva *et al.*, 2021

Plant extract is other important source for biostimulant products with application in management of plant disease. Jacquens *et al.* (2022) noted that biostimulants from plant extract (from the aerial parts of a monocotyledon plant) added to the culture medium enhanced plantlets growth of *Vitis vinifera* L. cv. Marselan and the defence responses against downy mildew agent (*Plasmopara viticola*) were induced only for treated plantlets.

PBs extracted from brown alga (*Sargassum angustifolium*) were examined for the reduction of charcoal root disease symptoms caused by *Macrophomina phaseolina* (Tassi) Goid., in strawberries cultivated under greenhouse conditions (Tajdinian *et al.*, 2022). Treated strawberries had a reduced disease incidence from *Macrophomina phaseolina* (Tassi) Goid. At the same time, the defence-related enzymes (phenylalanine ammonia-lyase and polyphenol oxidase) were significantly increased in plants treated with *Sargassum angustifolium* extract than in control plants.

Tomato plants treated with biostimulants containing *Herbaspirillum seropedicae* and humic acids, alone or combined, showed reduced disease incidence from *Xanthomonas euvesicatoria* (da Silva *et al.*, 2021).

Hamaoka *et al.* (2021) explored the potential of endophyte *Bacillus velezensis* KOF112 from shoot xylem of grapevine *Vitis* sp. cv. Koshu to inhibit the mycelial growth of *Botrytis cinerea* (grey mould), *Colletotrichum gloeosporioides* (anthracnose), and *Phytophthora infestans* (downy mildew) and demonstrated that KOF112 can be used as biocontrol product against the three fungal phytopathogens and may contribute to developing sustainable practice in viticulture for disease management.

The main contributions of PBs to achieve efficient management of plant disease include: enhance plant defence-related encoding genes; reduction of disease incidence; reduction of disease severity; better mechanisms of disease resistance; reduced dosages of the conventional synthetic fungicide; and antimicrobial activities or better defence responses (Fig. 29.9).

PBs application demonstrated multiple benefits of integration into new sustainable strategies for crop protection in order to increase plant resistance against phytopathogens and to minimize the quantities of the conventional synthetic fungicide and their residues in agricultural products. The stimulation of natural defence response of plants is key for integrated disease management in sustainable agriculture.

Conclusions

PBs have a valuable application in the agricultural sector and potential for other economic fields. It is a certainty that the use of biostimulants in agriculture will continue to increase in the agricultural sector and increase in popularity among agricultural producers. At the same time, more research and development studies are required to explore all the potential natural compounds contained in a variety

Fig. 29.9. PBs' contribution for plant disease management.

of substances, all agricultural wastes or other natural sources. Moreover, research studies are necessary to evaluate the effect of different concentrations, mode of application, time of application or new bioformulations of PBs in crop science.

The global market of plant biostimulators for agriculture has medium- and long-term potential, especially in the context where public policies and society promote the application of sustainable practices in agriculture and the demand for high-quality products is increasing. At the same time, these relatively new bioproducts bring a series of challenges and implications for companies and agricultural producers: acceptance by farmers and agricultural companies; the effectiveness of the plant biostimulators to improve crop quality and yields; the production cost or the affordable purchase price; availability of products on a large scale, etc. PBs present interest not only for the agriculture sector but also for the waste industry, circular economy, bioeconomy, etc.

PBs application promotes plant growth and development, enhancing tolerance against biotic and abiotic stresses and increasing agricultural productivity, and could be an efficient strategy for reducing chemical inputs in the agriculture sector.

References

Abkhoo, J. and Sabbagh, S.K. (2016) Control of *Phytophthora melonis* damping-off, induction of defense responses, and gene expression of cucumber treated with commercial extract from *Ascophyllum nodosum*. *Journal of Applied Phycology* 28(2), 1333–1342. DOI: 10.1007/s10811-015-0693-3.

Aguilera, P., Ortiz, N., Becerra, N., Turrini, A., Gaínza-Cortés, F. *et al.* (2022) Application of arbuscular mycorrhizal fungi in Vineyards: Water and biotic stress under a climate change scenario: New challenge for chilean grapevine crop. *Frontiers in Microbiology* 13, 826571. DOI: 10.3389/fmicb.2022.826571.

Ahmed, M., Ullah, H., Attia, A., Tisarum, R., Cha-um, S. *et al.* (2023) Interactive effects of *Ascophyllum nodosum* seaweed extract and silicon on growth, fruit yield and quality, and water productivity of tomato under water stress. *Silicon* 15(5), 2263–2278. DOI: 10.1007/s12633-022-02180-x.

Ali, S., Moon, Y.S., Hamayun, M., Khan, M.A., Bibi, K. *et al.* (2022) Pragmatic role of microbial plant biostimulants in abiotic stress relief in crop plants. *Journal of Plant Interactions* 17(1), 705–718. DOI: 10.1080/17429145.2022.2091801.

Amiri Forotaghe, Z., Souri, M.K., Ghanbari Jahromi, M. and Mohammadi Torkashvand, A. (2022) Influence of humic acid application on onion growth characteristics under water deficit conditions. *Journal of Plant Nutrition* 45(7), 1030–1040. DOI: 10.1080/01904167.2021.1994604.

Caradonia, F., Ronga, D., Tava, A. and Francia, E. (2022) Plant biostimulants in sustainable potato production: An overview. *Potato Research* 65(1), 83–104. DOI: 10.1007/s11540-021-09510-3.

Darakeh, S.A.S.S., Weisany, W., Tahir, N.A.R. and Schenk, P.M. (2022) Physiological and biochemical responses of black cumin to vermicompost and plant biostimulants: arbuscular mycorrhiza and plant growth-promoting rhizobacteria. *Industrial Crops and Products* 188, 115557. DOI: 10.1016/j.indcrop.2022.115557.

da Silva, A.P.S., Olivares, F.L., Sudré, C.P., Peres, L.E.P., Canellas, N.A. *et al.* (2021) Attenuations of bacterial spot disease *Xanthomonas euvesicatoria* on tomato plants treated with biostimulants. *Chemical and Biological Technologies in Agriculture* 8(1), 1–9. DOI: 10.1186/s40538-021-00240-9.

De Hita, D., Fuentes, M., Fernández, V., Zamarreño, A.M., Olaetxea, M. *et al.* (2020) Discriminating the short-term action of root and foliar application of humic acids on plant growth: Emerging role of jasmonic acid. *Frontiers in Plant Science* 11, 493. DOI: 10.3389/fpls.2020.00493.

Díaz-Díaz, M., Bernal-Cabrera, A., Trapero, A., Medina-Marrero, R., Sifontes-Rodríguez, S. *et al.* (2022) Characterization of actinobacterial strains as potential biocontrol agents against *Macrophomina phaseolina* and *Rhizoctonia solani*, the main soil-borne pathogens of *Phaseolus vulgaris* in Cuba. *Plants* 11(5), 645. DOI: 10.3390/plants11050645.

European Commission (2016) Proposal for a regulation laying down rules on the making available on the market of CE marked fertilising products and amending regulations (EC)1069/2009 and 157.

Gebashe, F., Gupta, S. and Staden, J. (2021) Disease management using biostimulants. In: Gupta, S. and Van Staden, J. (eds) *Biostimulants for Crops from Seed Germination to Plant Development*. Academic Press, Amsterdam, pp. 411–425. DOI: 10.1016/B978-0-12-823048-0.00005-8.

Hamaoka, K., Aoki, Y. and Suzuki, S. (2021) Isolation and characterization of endophyte *Bacillus velezensis* KOF112 from grapevine shoot xylem as biological control agent for fungal diseases. *Plants* 10(9), 1815. DOI: 10.3390/plants10091815.

Iwaniuk, P., Łuniewski, S., Kaczyński, P. and Łozowicka, B. (2023) The influence of humic acids and nitrophenols on metabolic compounds and pesticide behavior in wheat under biotic stress. *Agronomy* 13(5), 1378. DOI: 10.3390/agronomy13051378.

Jacquens, L., Trouvelot, S., Lemaitre-Guillier, C., Krzyzaniak, Y., Clément, G. *et al.* (2022) Biostimulation can prime elicitor induced resistance of grapevine leaves to downy mildew. *Frontiers in Plant Science* 13, 998273. DOI: 10.3389/fpls.2022.998273.

La Spada, F., Aloi, F., Coniglione, M., Pane, A. and Cacciola, S.O. (2021) Natural biostimulants elicit plant immune system in an integrated management strategy of the postharvest green mold of orange fruits incited by *Penicillium digitatum*. *Frontiers in Plant Science* 12, 684722. DOI: 10.3389/fpls.2021.684722.

Liu, X.-Q., Xie, Y.-C., Li, Y., Zheng, L., Srivastava, A.K. *et al.* (2022) Biostimulatory response of easily extractable glomalin-related soil protein on soil fertility mediated changes in fruit quality of citrus. *Agriculture* 12(8), 1076. DOI: 10.3390/agriculture12081076.

Liu, Y., Lu, J., Cui, L., Tang, Z., Ci, D. *et al.* (2023) The multifaceted roles of arbuscular mycorrhizal fungi in peanut responses to salt, drought, and cold stress. *BMC Plant Biology* 23(1), 1–19. DOI: 10.1186/s12870-023-04053-w.

López-Moral, A., Agustí-Brisach, C. and Trapero, A. (2021) Plant biostimulants: new insights into the biological control of verticillium wilt of olive. *Frontiers in Plant Science* 12, 662178. DOI: 10.3389/fpls.2021.662178.

Lumactud, R.A., Gorim, L.Y. and Thilakarathna, M.S. (2022) Impacts of humic-based products on the microbial community structure and functions toward sustainable agriculture. *Frontiers in Sustainable Food Systems* 6, 977121. DOI: 10.3389/fsufs.2022.977121.

Meng, L.-L., Liang, S.-M., Srivastava, A.K., Li, Y., Liu, C.-Y. *et al.* (2021) Easily extractable glomalin-related soil protein as foliar spray improves nutritional qualities of late ripening sweet oranges. *Horticulturae* 7(8), 228. DOI: 10.3390/horticulturae7080228.

Monda, H., Cozzolino, V., Vinci, G., Drosos, M., Savy, D. *et al.* (2018) Molecular composition of the humeome extracted from different green composts and their biostimulation on early growth of maize.

Plant and Soil 429(1–2), 407–424. DOI: 10.1007/s11104-018-3642-5.

Olowe, O.M., Akanmu, A.O. and Asemoloye, M.D. (2020) Exploration of microbial stimulants for induction of systemic resistance in plant disease management. *Annals of Applied Biology* 177(3), 282–293. DOI: 10.1111/aab.12631.

Popescu, G.C. and Popescu, M. (2018a) Yield, berry quality and physiological response of grapevine to foliar humic acid application. *Bragantia* 77(2), 273–282. DOI: 10.1590/1678-4499.2017030.

Popescu, G.C. and Popescu, M. (2022) Role of combined inoculation with arbuscular mycorrhizal fungi, as a sustainable tool, for stimulating the growth, physiological processes, and flowering performance of lavender. *Sustainability* 14(2), 951. DOI: 10.3390/su14020951.

Popescu, G.C., Popescu, M., Khondker, M., Clay, D.E., Pampana, S. *et al.* (2022) Agricultural sciences and the environment: Reviewing recent technologies and innovations to combat the challenges of climate change, environmental protection, and food security. *Agronomy Journal* 114(4), 1895–1901. DOI: 10.1002/agj2.21164.

Popescu, M. and Popescu, G.C. (2018b) Positive role of seaweeds extracts in bean plant responses to salt stress. In: *18th International Multidisciplinary Scientific GeoConference SGEM2018*, pp. 715–721 (Vol. 18). DOI: 10.5593/sgem2018/5.2/S20.093.

Sahana, B.N., PrasannaKumar, M.K., Mahesh, H.B., Buela Parivallal, P., Puneeth, M.E. *et al.* (2022) Biostimulants derived from red seaweed stimulate the plant defence mechanism in rice against magnaporthe oryzae. *Journal of Applied Phycology* 34(1), 659–665. DOI: 10.1007/s10811-021-02627-5.

Tajdinian, S., Rahmati-Joneidabad, M. and Ghodoum Parizipour, M.H. (2022) Macroalgal treatment to alleviate the strawberry yield loss caused by *Macrophomina phaseolina* (tassi) goid. in greenhouse cultivation system. *Frontiers in Sustainable Food Systems* 6, 1089553. DOI: 10.3389/fsufs.2022.1089553.

Wadas, W. and Dziugieł, T. (2020) Changes in assimilation area and chlorophyll content of very early potato (*Solanum tuberosum* L.) cultivars as influenced by biostimulants. *Agronomy* 10(3), 387. DOI: 10.3390/agronomy10030387.

Wang, W., Zheng, W., Lv, H., Liang, B., Jin, S. *et al.* (2022) Animal-derived plant biostimulant alleviates drought stress by regulating photosynthesis, osmotic adjustment, and antioxidant systems in tomato plants. *Scientia Horticulturae* 305, 111365. DOI: 10.1016/j.scienta.2022.111365.

30 An Overview of Organic Waste Management: Biosecurity and Arable Use of Manure and Biowaste

Dang Khoa Tran[1]*, Petr Konvalina[2] and Thi Xuan Phuong Tran[1]

[1]*Faculty of Agronomy, University of Agriculture and Forestry, Hue University, Hue, Vietnam;* [2]*Department of Agroecosystems, Faculty of Agriculture and Technology, University of South Bohemia, České Budějovice, Czech Republic*

Abstract

Municipal solid waste, general and organic, is generated continuously along with population growth and economic development. It is predicted that the world's waste will increase progressively to 3.4 billion tons by 2050. The rapid generation of wastes means the environment is paying a high price which affects human health through factors such as air pollution and water pollution. This chapter highlights organic waste – its characteristics, methods and technologies for its disposal, and ways to minimize the impact on the environment, in order to achieve the zero-waste goal and the circular economy in the near future.

Keywords: municipal solid waste, organic waste, composting, vermicomposting, biosecurity

Introduction

The global population of approximately 7.6 billion is expected to reach 8.6 billion by 2030, according to a United Nations forecast report. By 2050, the world's population will reach 9.8 billion, 70% of which will be living in urban areas (FAO, Rikolto and RUAF, 2022). The rapid population growth in urban areas, economic development, industrialization and consumerism have placed more and more pressure on natural resources and the global environment through increased solid waste generation (Yang *et al.*, 2023; Shitu *et al.*, 2015; Amrul *et al.*, 2022).

Gone are the days when humans used to live in peace with nature, releasing minimum solid waste. The amount of municipal solid waste generated is approximately 2 billion tons each year. Each person contributes 40.15 to 1657.1 kg of waste. Over a third of municipal solid waste is untreated and unmanaged, which poses potential environmental hazards. Forecasts predict that the world's waste will increase progressively to 3.4 billion tones by 2050, driven by population growth and urbanization. In general, waste generation is related to income level. Low- and middle-income countries' waste output will surge by approx. 40%, compared with high-income countries whose output is predicted to grow by only 19% by 2050. What is more, the high-income countries generate more dry waste than organic waste; the reverse is true for low- and middle-income countries. Specifically, high-income countries generate 32% organic waste, lower than middle- and lower-income countries at 21% and 24%, respectively (Kaza *et al.*, 2018; Singh, 2019).

Organic waste makes up about 45% of the total of municipal solid waste. Another source of organic

*Corresponding author: trandangkhoa@hueuni.edu.vn

DOI: 10.1079/9781800626850.0030

waste is from animals, humans (faeces), by-products of agriculture and food processing, and wastewater, which can be decomposed by microorganisms (ISWA, 2020; Contreras *et al.*, 2020; Patel *et al.*, 2021; Chernysh *et al.*, 2021).

Categorization of Organic Waste

Animal waste

Animal waste includes urine and excrement, the volume of which depends on the total liveweight of the animal, the livestock type, the size and age of animal, its feed and water intake, etc. According to Girotto and Cossu (2017), annual animal waste production was 34 billion tons from chickens,18.5 billion tons from cattle, 1.5 billion tons from pigs and 0.8 billion tons from sheep. It is projected that animal waste will increase to 55 billion tons (Girotto and Cossu, 2017).

Agro-industrial wastewater

The global amount of agro-industrial by-products being generated globally is increasing quickly. They include six types: tapioca, sugarcane, slaughterhouse, palm oil, brewery, and fruit and vegetables (Polprasert and Koottatep, 2007). The data in 2019 showed that the slaughter of animals generated a great deal of wastewater (2448 billion litres), followed by production of bioethanol (1650 billion litres). Processing of palm oil produced 256 billion litres, tapioca 85 billion litres, and milk 0.143 billion litres (Walter *et al.*, 2021).

Human waste

The combination of faeces and urine is one element of human waste. When these are diluted with flushing water or other grey water, they become domestic sewage or wastewater. Another type of human wastes is food waste, rubbish, ashes and residues, etc. The amount of human excreta, wastewater and solid wastes ranges widely from location to location. It depends on diet, socio-economic factors, weather, and water availability (Polprasert and Koottatep, 2007).

Waste Conversion Technologies

The technologies to convert waste are various, depending on type, quantity and property of feedstock, demand for energy, economic conditions, environmental standards, etc. (Singh *et al.*, 2011). Commonly, there are three wastes from conversion processes, from thermal conversions, biochemical conversions and from landfill. Thermal conversions include incineration, pyrolysis and gasification techniques. Through thermal conversion, wastes are transformed into energy. Biochemical conversions are mainly based on the action of enzymes of microorganisms. These techniques include anaerobic digestion and composting, which are much more environmentally friendly than thermal conversion. At the end of biochemical conversion of waste, compost, soil conditioner and biogas (or energy) are created. The last process of waste conversion is landfill. Although landfilling is popular around the globe, it poses a serious threat to health and pollutes the environment (Singh *et al.*, 2011; Rubagumya *et al.*, 2023).

Practically, using a single technique to dispose of municipal solid waste in general, as well as organic waste in particular, are not the optimal scenario and are relatively inefficient. Therefore, an integrated approach with a combination of methods is the best way to manage solid waste and is more eco-friendly (Asefi *et al.*, 2020; Colangelo *et al.*, 2021; Arabi *et al.*, 2021; Weihs *et al.*, 2022). For example, applying several methods such as recycling, resource recovery, and reducing the amount of final waste sent to landfill can achieve efficient and environmentally friendly waste management practices. Furthermore, the old and new technologies blend together to achieve better management efficiency and better benefits. For instance, landfill, material recovery facilities, composting, and anaerobic digestion technology combine with biometallurgy, pyrometallurgy and hydrometallurgy (Saha *et al.*, 2021; Tabelin *et al.*, 2021; Lai *et al.*, 2022; Budihardjo *et al.*, 2023).

Manure and Biowaste Management Practices for Biosecurity and Arable Use

Applying manure as a biofertilizer has long been used in agriculture. There are numerous methods to convert manure to produce liquid, semi-liquid, and solid organic fertilizer. In this chapter, we focus on eco-friendly methods such as composting and vermicomposting.

Composting

Composting is defined as a process of biological decomposition and stabilization of organic substrates under controlled conditions, using microbial populations via both mesophilic and thermophilic phases. As a result, the complex organic molecules decompose into simpler compounds such as carbon dioxide, water, minerals and stabilized organic matter (Pereira-Neta, 1987; Chia *et al.*, 2020).

Composting offers significant benefits for cultivation of crops as fertilizing with compost not only supplies N, P, K and micronutrients, but it also improves soil fertility, enhances soil structure, facilitates microbial remediation, and mitigates soil desertification (Sradnick *et al.*, 2013; Kadam *et al.*, 2022). Indeed, in the first year of applying compost, the nutrients release to the soil 5–60% N, 35–100% P, and 75–100% K. Furthermore, the soil's chemical properties positively improve, increasing pH, CEC and EC parameters (Fagnano *et al.*, 2011; Martínez-Blanco *et al.*, 2013; Bedada *et al.*, 2014). Soil's physical properties are improved, like increasing porosity and water-holding capacity, and reducing bulk density (Diacono and Montemurro, 2010; Srivastava *et al.*, 2016). Regarding material to make compost, C-poor compost, containing high nutrient value, is a good fertilizer and C-rich compost plays a role as a soil conditioner (Castán *et al.*, 2016). However, composting has brought disadvantages that can hinder the compost production process on a bigger scale because of negative effects on compost quality and harm to the environment. The first reason is loss of N during the composting process and intensification bioavailability of heavy metals. The next reason is decrease of humus content of organic matter as well as the increased residue of antibiotics (Parkinson *et al.*, 2004; Kadam *et al.*, 2022).

Vermicomposting

As well as composting, vermicomposting is a technique used in many countries, especially in urban areas, due to its simplicity and ease of implementation, and effectiveness. It is also environmentally friendly. This technique is based on the action of earthworms and microorganisms to stabilize organic matter (Adhikary, 2012; Sharma and Garg, 2019).

Earthworms play a crucial role in the vermicomposting process. They are considered a mechanical blender, breaking down organic matter. Thanks to this, organic matter is fragmented and consumed, leading to an increase of surface area where there are favourable conditions for microorganisms to operate (Domínguez, 2004; Khoshnevisan *et al.*, 2021). By using various species of earthworm, vermicomposting is an effective method to transform manure into nutrient-rich biofertilizer. This fertilizer not only supplies a balanced mineral composition, but also enhances plant nutrient availability. Moreover, the vermicompost has physical characteristics of peat such as porosity, aeration, excellent structure, drainage and moisture retention capabilities (Lee and Febrisiantosa, 2020; Khoshnevisan *et al.*, 2021). Additionally, vermicomposting is able to decrease the availability and mobility of heavy metals contained in livestock manure (Lv *et al.*, 2016).

In order to achieve cost-effective vermicompost production, physical factors such as temperature, pH and moisture need to be controlled during the vermicomposting phase. Simultaneously, minimizing the presence of toxic chemicals, high salinity, and wide C:N ratio in the material input can foster the growth and reproduction of earthworms (Maharjan *et al.*, 2022).

Conclusion

Currently, there is a worldwide focus on organic waste management. In this chapter we describe the classification of organic wastes, including animal wastes, agro-industrial wastewater and human wastes. These sources pollute air, land, rivers and oceans. Although organic waste has brought numerous drawbacks to the environment, it has also a big potential to convert waste into energy and biofertilizers. As highlighted, the technical considerations for the management of organic waste are presented in order to help readers understand the biosecurity and arable use of manure as well as biowaste and to achieve the zero-waste goal, the circular economy and to pursue sustainable development goals (Martínez-Burgos *et al.*, 2021).

Finally, the application of technologies to disposal of livestock manure must take into account the properties of the scale of organic waste, policy, the ability to invest etc. in order to minimize the impact to the environment.

References

Adhikary, S. (2012) Vermicompost, the story of organic gold: A review. *Agricultural Sciences* 3(7), 905–917.

Amrul, N.F., Kabir Ahmad, I., Ahmad Basri, N.E., Suja, F., Abdul Jalil, N.A. *et al*. (2022) A review of organic waste treatment using black soldier fly (*Hermetia illucens*). *Sustainability* 14, 4565.

Arabi, M., Ostovan, A., Li, J., Wang, X., Zhang, Z. *et al*. (2021) Molecular imprinting: Green perspectives and strategies. *Advanced Materials* 33(30), 1–33.

Asefi, H., Shahparvari, S. and Chhetri, P. (2020) Advances in sustainable integrated solid waste management systems: Lessons learned over the decade 2007–2018. *Journal of Environmental Planning and Management* 63(13), 2287–2312.

Bedada, W., Karltun, E., Lemenih, M. and Tolera, M. (2014) Long-term addition of compost and NP fertilizer increases crop yield and improves soil quality in experiments on smallholder farms. *Agriculture, Ecosystems & Environment* 195, 193–201.

Budihardjo, M.A., Priyambada, I.B., Chegenizadeh, A., Haumahu, S.A.Q. and Puspita, A.S. (2023) Environmental impact technology for life cycle assessment in municipal solid waste. *Global Journal of Environmental Science and Management* 9(s I), 145–172.

Castán, E., Satti, P., González-Polo, M., Iglesias, M.C. and Mazzarino, M.J. (2016) Managing the value of composts as organic amendments and fertilizers in sandy soils. *Agriculture, Ecosystems & Environment* 224, 29–38.

Chernysh, Y., Shtepa, V., Roy, I., Chubur, V. and Skvortsova, P. (2021) The potential of organic waste as a substrate for anaerobic digestion in Ukraine: Trend definitions and environmental safety of the practices. *Environmental Problems* 6(3), 135–144.

Chia, W.Y., Chew, K.W., Le, C.F., Lam, S.S., Chee, C.S.C. *et al*. (2020) Sustainable utilization of biowaste compost for renewable energy and soil amendments. *Environmental Pollution* 267, 115662.

Colangelo, F., Farina, I., Travaglioni, M., Salzano, C., Cioffi, R. *et al*. (2021) Eco-efficient industrial waste recycling for the manufacturing of fibre reinforced innovative geopolymer mortars: Integrated waste management and green product development through LCA. *Journal of Cleaner Production* 312(127777), 1–14.

Contreras, M., Gázquez, M.J., Romero, M. and Bolívar, J.P. (2020) Recycling of industrial wastes for value-added applications in clay-based ceramic products: A global review. In: Samui, P., D., Kim., Iyer, N.R. and Chaudary, S. (eds) *New Materials in Civil Engineering*. Butterworth-Heinemann, Oxford, UK, pp. 155–208.

Diacono, M. and Montemurro, F. (2010) Long-term effects of organic amendments on soil fertility: A review. *Agronomy for Sustainable Development* 30(2), 401–422.

Domínguez, J. (2004) State-of-the-art and new perspectives on vermicomposting research. In: Edwards, C.A. (ed.) *Earthworm Ecology*, 2nd edn. CRC Press, Boca Raton, Florida, pp. 401–424.

Fagnano, M., Adamo, P., Zampella, M. and Fiorentino, N. (2011) Environmental and agronomic impact of fertilization with composted organic fraction from municipal solid waste: a case study in the region of Naples, Italy. *Agriculture, Ecosystems & Environment* 141(1–2), 100–107.

FAO, Rikolto and RUAF (2022) *Urban and Peri-urban Agriculture Sourcebook: From Production to Food Systems*. FAO and Rikolto, Rome.

Girotto, F. and Cossu, R. (2017) Animal waste: opportunities and challenges. In: *Sustainable Agriculture Reviews*, Vol. 22. Springer, Cham, Switzerland.

ISWA (2020) Global Assessment of Municipal Organic Waste Production and Recycling. Rotterdam, The Netherlands: International Solid Waste Association.

Kadam, R., Khanthong, K., Jang, H., Lee, J. and Park, J. (2022) Occurrence, fate, and implications of heavy metals during anaerobic digestion: A review. *Energies* 15(22), 8618.

Kaza, S., Woerden, F.V., Yao, L. and Bhada-Tata, P. (2018) What a waste 2.0: a global snapshot of solid waste: management to 2050. *World Bank*. Available at: https://openknowledge.worldbank.org/entities/publication/d3f9d45e-115f-559bb14f-28552410e90a (accessed 2 February 2021).

Khoshnevisan, B., Duan, N., Tsapekos, P., Awasthi, M.K., Liu, Z. *et al*. (2021) A critical review on livestock manure biorefinery technologies: sustainability, challenges, and future perspectives. *Renewable and Sustainable Energy Reviews* 135, 110033.

Lai, X., Chen, Q., Tang, X., Zhou, Y., Gao, F. *et al*. (2022) Critical review of life cycle assessment of lithium-ion batteries for electric vehicles: A lifespan perspective. *ETransportation* 12(100169), 1–22.

Lee, J.H. and Febrisiantosa, A. (2020) Improvement of nitrogen balance (land budget) in South Korea in terms of livestock manure: A review. In: *IOP Conference Series: Earth and Environmental Science*. IOP Publishing, Bristol, UK.

Lv, B., Xing, M. and Yang, J. (2016) Speciation and transformation of heavy metals during vermicomposting of animal manure. *Bioresource Technology* 209, 397–401.

Maharjan, K.K., Noppradit, P. and Techato, K. (2022) Suitability of vermicomposting for different varieties of organic waste: a systematic literature review (2012–2021). *Organic Agriculture* 12(4), 581–602.

Martínez-Blanco, J., Lazcano, C., Christensen, T.H., Muñoz, P., Rieradevall, J. *et al*. (2013) Compost benefits for agriculture evaluated by life cycle assessment: a review. *Agronomy for Sustainable Development* 33(4), 721–732.

Martínez-Burgos, W.J., Bittencourt Sydney, E., Bianchi Pedroni Medeiros, A., Magalhães, A.I., Carvalho, J.C. *et al.* (2021) Agro-industrial wastewater in a circular economy: characteristics, impacts and applications for bioenergy and biochemicals. *Bioresource Technology* 341, 125795.

Parkinson, R., Gibbs, P., Burchett, S. and Misselbrook, T. (2004) Effect of turning regime and seasonal weather conditions on nitrogen and phosphorus losses during aerobic composting of cattle manure. *Bioresource Technology* 91(2), 171–178.

Patel, P., Modi, A., Minipara, D. and Kumar, A. (2021) Microbial biosurfactants in management of organic waste. In: Mishra, V.K. and Kumar, A. (eds) *Sustainable Environmental Clean-Up*. Elsevier, Amsterdam.

Pereira-Neta, J.T. (1987) On the treatment of municipal refuse and sewage sludge using aerated static pile composting: A low-cost technology approach. PhD dissertation, Leeds University, UK.

Polprasert, C. and Koottatep, T. (2007) *Organic Waste Recycling Technology and Management*. International Water Association, London.

Rubagumya, I., Komakech, A.J., Kabenge, I. and Kiggundu, N. (2023) Potential of organic waste to energy and bio-fertilizer production in sub-Saharan Africa: A review. *Waste Disposal & Sustainable Energy* 5(3), 259–267.

Saha, L., Kumar, V., Tiwari, J., Rawat, S., Singh, J. *et al.* (2021) Electronic waste and their leachates impact on human health and environment: global ecological threat and management. *Environmental Technology & Innovation* 24(102049), 102049.

Sharma, K. and Garg, V. (2019) Vermicomposting of waste: A zero-waste approach for waste management. In: Rousta, K. and Bolton, K. (eds) *Sustainable Resource Recovery and Zero Waste Approaches*. Elsevier, Amsterdam, The Netherlands, pp. 133–164.

Shitu, A., Izhar, S. and Tahir, T.M. (2015) Sub-critical water as a green solvent for production of valuable materials from agricultural waste biomass: a review of recent work. *Global Journal of Environmental Science and Management* 1, 255–264.

Singh, D.M. (2019) Solid waste management in urban India: imperatives for improvement. *The Journal of Contemporary Issues in Business and Government* 25(1), 87–92. Available at: https://cibgp.com/au/index.php/1323-6903/article/view/153 (accessed 13 February 2025).

Singh, R.P., Tyagi, V.V., Allen, T., Ibrahim, M.H. and Kothari, R. (2011) An overview for exploring the possibilities of energy generation from municipal solid waste (MSW) in indian scenario. *Renewable and Sustainable Energy Reviews* 15(9), 4797–4808. ISSN1363-0321.

Sradnick, A., Murugan, R., Oltmanns, M., Raupp, J. and Joergensen, R.G. (2013) Changes in functional diversity of the soil microbial community in a heterogeneous sandy soil after long-term fertilization with cattle manure and mineral fertilizer. *Applied Soil Ecology* 63, 23–28.

Srivastava, V., Araujo, A.S.F., Vaish, B., Bartelt-Hunt, S., Singh, P. *et al.* (2016) Biological response of using municipal solid waste compost in agriculture as fertilizer supplement. *Reviews in Environmental Science and Bio/Technology* 15(4), 677–696.

Tabelin, C.B., Park, I., Phengsaart, T., Jeon, S., Villacorte-Tabelin, M. *et al.* (2021) Copper and critical metals production from porphyry ores and e-wastes: a review of resource availability, processing/recycling challenges, socio-environmental aspects, and sustainability issues. *Resources, Conservation and Recycling* 170(105610), 1–35.

Walter, J.M., Eduardo, B.S., Adriane, B.P.M., Antonio, I.M., Júlio, C.D.C. *et al.* (2021) Agro-industrial wastewater in a circular economy: characteristics, impacts and applications for bioenergy and biochemicals. *Bioresource Technology* 341, 125795. DOI: 10.1016/j.biortech.2021.125795.

Weihs, G.A.F., Jones, J.S., Ho, M., Malik, R.H., Abbas, A. *et al.* (2022) Life cycle assessment of co-firing coal and wood waste for bio-energy with carbon capture and storage – New South Wales study. *Energy Conversion and Management* 273, 1–18.

Yang, M., Chen, L., Wang, J., Msigwa, G., Osman, A.I. *et al.* (2023) Circular economy strategies for combating climate change and other environmental issues. *Environmental Chemistry Letters* 21(1), 55–80.

31 Application of Marine-derived Materials as Biofertilizers in Organic Farming

Anne-Kristin Løes[1]*, Joshua Cabell[1], Sara Hansdotter[1], Violeta Alexandra Ion[2], Ingrid Olesen[3] and Tatiana Rittl

[1]*Norwegian Centre for Organic Agriculture (NORSØK), Tingvoll, Norway;* [2]*Research Center for Studies of Food Quality and Agricultural Products, University of Agronomic Sciences and Veterinary Medicine of Bucharest, Bucharest, Romania;* [3]*Department of Production Biology, Nofima, Ås, Norway*

Abstract

Significant volumes of materials from fish, seaweed and other marine organisms are poorly utilized and may be well suited for application as fertilizers and soil amendments. Brown seaweed contains significant amounts of potassium and calcium, and bone-rich residues from white fish are rich in nitrogen and phosphorus. Together, such materials are a good starting point for a well-balanced complete fertilizer, applicable for arable and feed crops. At NORSØK, several studies have been carried out in recent years with such materials as soil amendments. Whereas fish residues have a remarkably rapid and positive effect on crop growth, seaweed residues have a remarkable residual effect. Potentially toxic elements, especially arsenic and cadmium, need consideration when brown seaweeds are applied to soil, but we have not revealed increased uptake of these elements in crop plants grown in soil amended with seaweed.

Keywords: fish residues, seaweed, field trials, growth effects, potentially toxic elements

Introduction

Need for new fertilizer sources in organic farming

The proportion of organically certified farmland has increased rapidly over the last decades and comprised 1.6% of the world's total agricultural land in 2021 (Willer *et al.*, 2023). Organic farming is commonly supported by governing bodies to decrease environmental impacts from agriculture, and to satisfy consumer demands. The European Union (EU) has ambitious goals for increasing organic production, stating in the 'Farm to Fork' strategy that 25% of agricultural land should be managed organically by 2030 (EC, 2020). Concurrently, several countries aim to decrease the consumption of meat to reduce environmental impacts. Altogether, this will lead to reduced access to animal manure and other animal-derived fertilizer products. Fertilizers derived from conventional animal husbandry are widely applied in organic agriculture (Løes *et al.*, 2018). Fertilizers derived from the conventional food industry, such as vinasse from sugarbeet, may also contain pesticide residues and impact plant growth, as demonstrated in a recent study of tomatoes (McKinnon *et al.*, 2021). Hence, for several reasons, new fertilizer materials applicable for organic agriculture are required to cover the demands of crop plants, and to avoid reduced soil fertility in the long term due

*Corresponding author: anne-kristin.loes@norsok.no

DOI: 10.1079/9781800626850.0031

to the depletion of phosphorus (P) and other plant nutrients, which may be a risk, especially in organic arable farming systems (Möller *et al.*, 2018).

Significant efforts are made to transform industrial activities into a circular economy, and increased attention is given to wasted and residual materials that are currently poorly utilized. Materials such as sewage, slaughterhouse waste, and various leftovers from the marine sector contain significant amounts of P and/or other essential plant nutrients. Furthermore, using low trophic species, including plants, seaweed and mussels (i.e. carbon (C) sequestration) and increasing organic C stocks in soils and vegetation biomass (i.e. C storage) through a systemic and cross-sectorial approach, means that marine-derived materials can be utilized to contribute to climate mitigation by transfer of atmospheric carbon dioxide to the terrestrial biosphere, as suggested by Olesen *et al.* (2023).

Norway has a long coastline and a significant marine industry, and hence the utilization of marine materials for soil amendment and for the production of commercial fertilizers is of special relevance in this country. Fish, seaweed and other marine organisms have traditionally been applied both for feed purposes and for soil amendment in coastal areas. Special fertilizers from seaweed and fish (Ahuja *et al.*, 2020) are available and permitted for use in certified organic growing. However, these fertilizers are often expensive and are hence mostly applied to horticultural crops. Complete fertilizers for organic arable and feed crops need to be produced in much larger volumes at a much lower price to compete with the current alternatives. Hence, materials that are currently poorly utilized must be sought. Further, several challenges must be solved before brown seaweed and fishbones can be sold to organic farmers for fertilization purposes.

At the Norwegian Centre for Organic Agriculture (NORSØK), several studies have been conducted to develop knowledge required to produce relatively cheap fertilizers from marine materials that may replace the current extensive use of fertilizers made from conventional poultry manure in organic growing. The studies have expanded to collaborations on an international level. We believe that it should be possible to develop a complete and well-balanced marine-derived fertilizer applicable for arable and feed crops. This chapter will explain our efforts until now to realize this goal.

Sourcing of materials

Whereas the capture and wild collection of fish and other seafood show a declining trend (Naylor *et al.*, 2021), aquaculture, which includes fish farming and the cultivation of seaweeds and other marine species, is increasing. Globally, freshwater systems account for the largest proportion of cultivated aquaculture species (Naylor *et al.*, 2021), but in Norway, the production of salmon (*Salmo salar*) in sea pens (including the raising of juveniles in freshwater) constitutes the dominant type of aquaculture. From this industry, fish trimmings and intestines are utilized for food or feed purposes, whereas the sludge produced from fish faeces and uneaten feed represents a significant source of important plant nutrients such as nitrogen (N) and P and may be applicable in fertilizers. Dead fish (e.g. from disease) may also be a valuable material to produce fertilizers, preferably after undergoing digestion for biogas generation or a fermentation or composting process. However, the acceptability of materials from conventional fish farming in organic farming is not straightforward. Industrialized, large-scale fish farming may possibly be characterized as factory farming, e.g. due to high animal densities and stress caused by lack of domestication, and hence not permitted in organic farming. The current Norwegian regulation does not allow dead fish or fish sludge (= category 2 of animal by-products) as a fertilizer source in certified organic growing, only category three materials appropriate for human consumption or feed (NFSA (Norwegian Food Safety Authority), 2022). Hence, at NORSØK, we, until now, worked with materials from captured fish, not cultivated fish species.

In Norway, captured fish species are commonly divided into 'fat fish' such as herring (*Clupea harengus*) and mackerel (*Scomber scombrus*), where all trimmings etc. are used for the production of fish meal and fish oil; and 'white fish' such as cod (*Gadus morhua*) and saithe (*Pollachius virens*), where the leftover material (heads, backbones, intestines) contain less fat and a much higher proportion of fish bone. Residues from white fish were applied as feed for fur animals until these were recently banned in Norway. Due to the decrease in demand for such feed, significant amounts of bone-rich residuals from white fish are available, and the amount could be increased if all parts of captured fish were brought to shore. Currently, e.g., heads may be cut off and left at sea. Estimates have shown that close to 4000 tonnes of P could be available in bone-rich residues from white fish in Norway (Ahuja *et al.*, 2020). This amount of P corresponds to about 25% of the total annual P consumption in Norway in the form of animal and mineral fertilizers.

Fishbones contain high amounts of calcium (Ca), N and P, but less of other plant macronutrients (Ahuja *et al.*, 2020; Løes *et al.*, 2022a; Maroulis *et al.*, 2023). The amount of P that crop plants demand is at least comparable with the uptake of N, which is far too low in fishbones and needs to be complemented. Seaweeds, especially certain species of brown macroalgae, contain relatively high amounts of K. Hence, an appropriate mixture of seaweed and fishbone material could be a good starting point to develop a complete fertilizer.

For seaweeds, there is an emerging interest both to harvest wild seaweed and to cultivate macroalgae. In Norway, both collection and cultivation are done mainly with brown seaweed species. For fertilizer production, we would need to rely on residual materials from value chains designed for food, feed or technical applications to reduce the material costs. Hence, our efforts have until now been on residues from wild rockweed (*Ascophyllum nodosum*) which is collected along the Norwegian coast, dried, milled and chemically extracted to produce high-value fertilizers and biostimulants. The residual material is a sludge with about 25% dry matter (DM), which is currently incinerated and hence has a negative energy value, in addition to significant costs of transportation from the production site. Recently, we have also begun to study soil amendment with fresh and preserved residual materials of sugar kelp (*Saccharina latissima*) cultivated for feed purposes; wild bladderwrack (*Fucus serratus*) sorted out from rockweed collected for extraction of fertilizer as a 'by-product', since these species often grow together; and the fronds of cuvie (*Laminaria hyperborea*), harvested for production of alginate.

We have also worked with residues from blue mussels (*Mytilus edulis*) as an ingredient in composts, and with shells from captured crab (*Cancer pagurus*) to study if that may be applied to replace traditional lime (calcium carbonate) in peat-based growing media and if we may find any indications of a positive effect of chitin in the shells on plant health.

For all materials, we have studied their characteristics and possible applications as fertilizers or ingredients in fertilizer mixtures.

A high content of salt, especially sodium (Na) and chlorine (Cl), is a drawback of marine materials in fertilizers, particularly in climates with low precipitation. Another challenge is a relatively high content of potentially toxic elements (PTEs, also called heavy metals) in brown seaweed, and partly also in fish, especially fish species dwelling near sea-bottom sediments. Other challenges are posed by processing and logistics. In collaboration with other research institutes, universities and, not least, industry partners, several projects have been conducted at NORSØK to solve these challenges and come closer to the realization of a commercial product that may reduce the current dependency on conventional animal husbandry in organic growing, supported by various funding bodies. The aim of this chapter is to give a current overview of the research on marine-derived fertilizers conducted by NORSØK and collaborating partners, with reference to publications where results are presented in full detail for further reading.

Methods

Applied materials

Bone-rich materials from white fish may be conserved by freezing, drying or acidification to pH <4. We have worked with pure bone residues from hydrolysed fish (cod heads received from SINTEF Ocean); dried sediments from acidic hydrolysation of minced residues from various white fish species (cod, saithe etc.); freshly minced backbones from various white fish; and coarse fractions from backbones from different species of white fish that were dried, milled and sieved. Rinsed fishbones had a loss on ignition (LOI) of about 34% (Ahuja and Løes, 2019), whereas acidified fish sediment (Fig. 31.1) had 41–47% LOI, indicating some residues of more soft tissues in this material. Fish material has been tested in pot or field trials as well as in blends with seaweed material for composting.

Blue mussel residues were subject to chemical analysis and tested in compost blends in Dewar flasks (Cabell and Løes, 2023). Crab shells with different mechanical composition (2–0.25 mm vs. <0.25 mm) were studied as a liming agent for peat in growing media, with an incubation period of 47 days at room temperature. Milled calcium

Fig. 31.1. Sediments of ground and hydrolysed acidified white fish (cod, saithe etc.) being dried on a plastic sheet spread on the floor, April 2019.

carbonate from limestone, which is usually applied to increase the pH and Ca availability of peat and agricultural soil, was used as a comparison. The same amount of calcium per kg of peat was applied with all materials.

For seaweed materials, we have mostly worked with rockweed sludge (25–30% DM). This material comes in slightly different types according to the applied extraction chemicals. We have mostly worked with a sludge with about 1.5% total-N (% of DM; 'high N'), due to the application of nitric acid, and a sludge with about 0.3% N ('low N'; details given in Zikeli *et al.*, 2022).

Pot trials, field studies and soil-fertilizer-plant dynamics

An initial trial in 2018 was conducted with ryegrass (*Lolium westerwoldicum*) as a test crop in pots with low-P agricultural soil as growing media under controlled conditions indoors. In total, five cuts of plant material demonstrated a very rapid release of nutrients from fishbone material, whereas the release from rockweed sludge occurred towards the final harvest (Ahuja and Løes, 2019). The fishbone material was milled bones from hydrolysed cod heads.

In this trial, the materials were tested separately, with a fertilization rate corresponding to 300 or 600 kg total N applied per hectare. Mineral N fertilizer (calcium nitrate) was applied as a positive control.

In 2019, similar materials were tested in a pot trial located outdoors, with oats (*Avena sativa* L.) and leek (*Allium porrum* L.) as test crops (Løes *et al.*, 2022b). The rates of N application corresponded to 160 and 320 kg total N per hectare. The pot trial was located next to a plot experiment, where the same fertilizers were applied at 160 kg total N per hectare and the test crop was oats grown as a cover crop for a perennial grass-clover ley and harvested while green. A commercial poultry manure fertilizer was used as a positive control in field plots, whereas calcium nitrate and poultry manure were applied as two positive controls in the pots. Again, rockweed sludge and fishbones from white fish were compared, but this time the fishbones were derived from an industrial plant, and were sediments from minced heads, backbones, intestines and trimmings of white fish that had been hydrolysed at pH <4. For the first time, we also tested a mixed fishbone and seaweed fertilization, where 70% of the applied N came from fishbone sediments and 30% from rockweed sludge. Without any further application of fertilizers, the residual effects of the materials applied in 2019 were recorded in the ley during 2020–23, with two cuts per season to measure DM yields.

In 2020, we established a plot trial close to the plots with perennial ley, where ryegrass (*Lolium westerwoldicum*) was grown as a test crop, with four cuts (Fig. 31.2). The N rates were 300 vs. 600 kg total N per ha. In addition to rockweed sludge and hydrolysed white fish sediments (Fig. 31.3), applied individually or as a mixture with 70% of the N from fish sediment, we also tested minced backbones from various white fish.

Fig. 31.2. The experimental field with ryegrass amended with different marine-derived fertilizers ready for the third cut, 12 August 2020.

Fig. 31.3. (*top*) Application of the mixed treatment of rockweed sludge and fish sediments in a field plot; (*bottom*) close-up of the fertilizer materials. Black particles = rockweed sludge; light brown = fish sediment.

Again, the positive control was commercial poultry manure fertilizer.

In 2021, we grew potatoes (*Solanum tuberosum* L.) in the former ryegrass plots with no further application of marine-derived materials, to study if we could find any indications of higher concentrations of potentially toxic elements, especially cadmium (Cd) and arsenic (As), when the potatoes were grown in soil amended with significant amounts of seaweed material that contained more of these elements than the inherent soil (Løes *et al.*, 2022c).

In 2022, we sent marine materials to partners in Romania and Greece for further analysis and to test their performance as fertilizers with table-top cultivated strawberries (*Fragaria* ×*ananassa* Duch.) as a test crop (Moloşag *et al.*, 2023). Further, we established a perennial ley on the plots where potatoes were grown in 2021 and measured the DM yields of this crop with one cut in autumn. No significant differences were then found between the plots which had been subject to different fertilization in 2020.

In 2022 and 2023, we conducted composting trials with rockweed sludge, dried and milled rockweed, and cuvie fronds in Dewar flasks (Cabell and Løes, 2023). Dewar flasks are well insulated, open flasks with a volume of 2 litres, and are used for conducting a standard test (BS EN 16087-2:2011, n.d) for assessing compost maturity (Brinton *et al.*, 1995). They can also be applied to predict the decomposability of organic materials (Miloştean and Flori, 2021). In both applications, the temperature of the material within the flasks is recorded at least daily and is used as an indicator of biological activity. Rockweed sludge was also composted in a 200-litre insulated reactor and outdoors in a windrow. We studied the effect of increasing the N content of the material by mixing N-rich material such as dried fishbones into the substrate. We also tested the effect of using expanded clay aggregates (Leca®) as a relatively inert bulking agent to improve aeration rather than woodchips or straw which affect the carbon concentration.

In 2023, various seaweed materials were applied in late April to the experimental plots in one of the four replicate blocks where ryegrass was grown in 2020, and potatoes in 2021 (Fig. 31.4). The amounts of seaweed material were aimed at corresponding to the maximal amount a Norwegian farmer may apply to agricultural soil over ten years because of a relatively high concentration of Cd. Organic materials with concentrations of Cd between 0.8 and 2 mg per kg DM may be applied with up to 20 tons of DM per ha over ten years. For

Fig. 31.4. (*From left to right*) Seven different materials of seaweed applied to a first-year ley in late April or early May (photo taken 11 May 2023). To the right, two control plots with no application of seaweed material in 2023.

materials with 0.4–0.8 mg Cd per kg DM, the limit is up to 40 tons. Materials with <0.4 mg DC per kg DM may be applied according to crop demand. Our materials contained between 0.3 and 1 mg Cd per kg DM. The aim of this study was to see if a young ley would survive such high applications of material. DM yields were recorded in the first-year ley with two cuts.

Results

Liming effect of crab shells

The particle size of the liming materials clearly affected the pH of the peat (Table 31.1). With coarse crab shells, the peat did not reach a satisfactory pH over 47 days, whereas with limestone the increase in pH from day 1 to day 47 was very small. Fine crab shell particles (< 0.25 mm) were less reactive than limestone meal, possibly due to a coarser particle structure. For limestone, 99.8% of the particles were <0.2 mm, and this may be the main reason for the more rapid pH increase with this material. The conductivity increased from day 1 to 47 in all treatments, and, as expected, it was somewhat higher with the application of marine-derived materials, especially for the fine fraction. However, the differences were not very large.

High N concentrations in fish residues may be lost

Acidified fishbone material from minced and hydrolysed white fish species may contain significant amounts of nitrogen. For some time (weeks) after removal from the hydrolysis tank, the concentration may be close to the concentration in freshly ground white fish backbones with some soft tissue, where about 11% of DM is total N. After some months of storage in open tanks that were covered to avoid inlet of precipitation, but open to air because the tanks were only half-filled, the total N level stabilized at about 4% of DM. Concurrently, the pH in the material increased from 2.5, initially due to excess application of formic acid, to about 5 (Løes *et al.*, 2022b). This is likely due to the high buffer capacity of the bones, which are rich in Ca, but may also be due to volatilization of formic acid. Hence, during storage of fish residues, it is possible to lose N even when the material is acidified. An efficient drying procedure may conserve the N.

Growth effects

Application of high rates of fishbone material, and of rockweed sludge and other seaweed materials, may hamper the germination of seeds both from weeds and crop plants. This is likely due to the high concentration of ammonium in the fishbone material, high pH in the rockweed sludge, and a high content of salts in the seaweed materials.

In all trials, we have seen a very rapid growth effect of fishbones, and a residual effect resembling that of dried poultry manure. The rockweed sludge has not shown an immediate fertilizer effect, but a significant residual growth effect. The DM yields of the field experiment over 2019–2023 are shown in Fig. 31.5 and show significantly higher yields in treatments with rockweed sludge, especially when mixed with fishbone material. Comparable results were found in the pot studies (Løes *et al.*, 2022b), where leek had a longer period of growth than oats and hence performed somewhat better with rockweed sludge. The dry-matter content of leek was much lower with rockweed sludge (11% compared with 16–19% in other treatments and the control), and lower dry-matter content was also observed in other studies. With no fertilizer application since 2019, the yields declined over time (Fig. 31.5).

Table 31.1. Effect on pH (H_2O) and conductivity (cm/s) in peat amended with milled crab shell or limestone.

Material	pH		Conductivity	
	Day 1	Day 47	Day 1	Day 47
Peat (no lime)	4.7	4.5	13	22
Peat + crab shell 0.25–2 mm	4.7	5.4	22	39
Peat + crab shell < 0.25 mm	5.8	6.6	24	47
Limestone	6.6	6.8	17	31

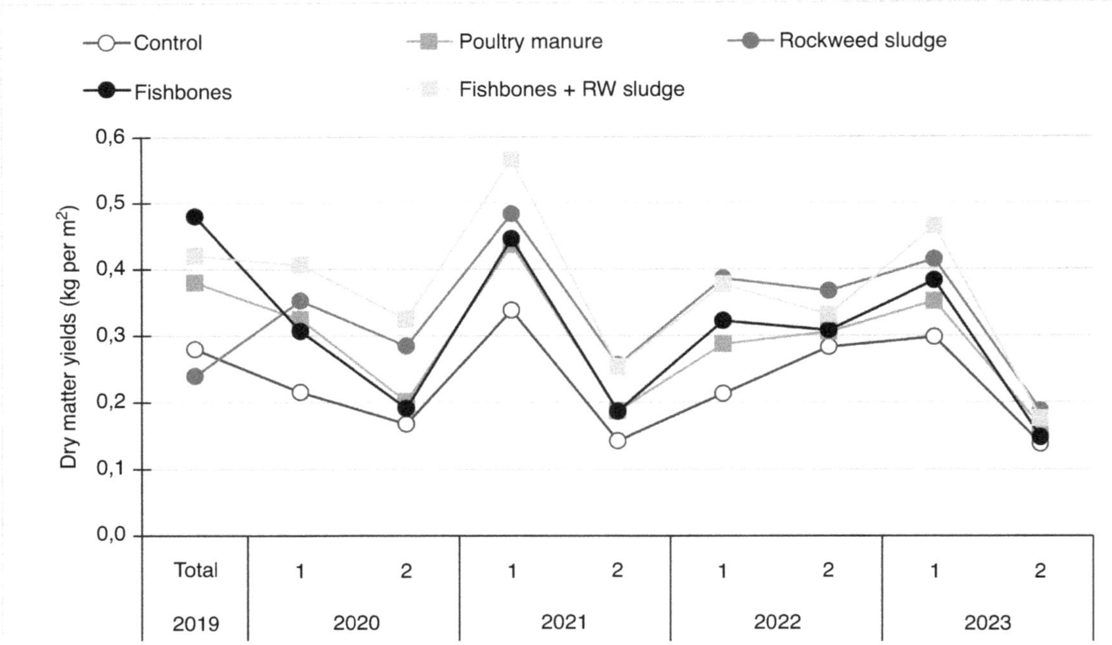

Fig. 31.5. Dry-matter yields (kg per m^2) of oat canopy (harvested while green) in 2019, followed by perennial ley with two cuts per year over 2020–2023. Fertilization was done only in 2019, with 160 kg total N per ha applied in the various fertilizers.

The differences between fertilizer treatments were statistically significantly different at each harvest date ($p < 0.05$). From visual observation, it was obvious that the red clover especially thrived in plots fertilized with rockweed sludge.

For the ryegrass in 2020 followed by potatoes in 2021, comparable results were achieved (Fig. 31.6). Again, the yields in the year when fertilizers were applied were highest with fishbones. The additional yield by doubling the N rate was rather small. Fresh fishbones were only applied at the low N rate (300 kg total N per ha) to avoid problems which could arise with a higher application, such as flies, birds or bad odour. Fresh fishbones performed somewhat poorer than the sediment from hydrolysed fish residues. Rockweed sludge was also only applied at the low rate because this material has a pH of around nine and hence may impact the germination of ryegrass seeds. Towards the end of the 2020 growing season (cut 4), some positive effects of rockweed sludge on ryegrass growth could be seen. Poultry manure performed better in 2020 than in 2019, likely because we had a better incorporation of all materials by a hand-held rotovator, whereas a rake was used in 2019. This may have caused some N loss from the poultry manure. A positive residual effect of rockweed sludge was found in potatoes in 2021 (Fig. 31.6), confirming that this material can be a valuable soil amendment. Other fertilizers also had a positive residual effect, and there was an effect of application rate. The fresh fishbones had a lower residual effect than the sediment from hydrolysed fish residues.

Application restricted by content of potentially toxic elements

To apply 160 kg of total N per ha in the form of rockweed sludge, an amount of 4 kg per m^2 was required, corresponding to 1 kg DM per m^2 (25% DM). For 300 kg total N per ha, 1.75 kg DM of rockweed sludge was required. The highest amount came close to the maximal amount permitted over ten years in the Norwegian regulation (20 tons of DM per ha = 2 kg DM per m^2) since the rockweed sludge contains about 1 mg Cd per kg DM (Løes *et al.*, 2022b). Such high amounts could be relevant for farmers located near a plant

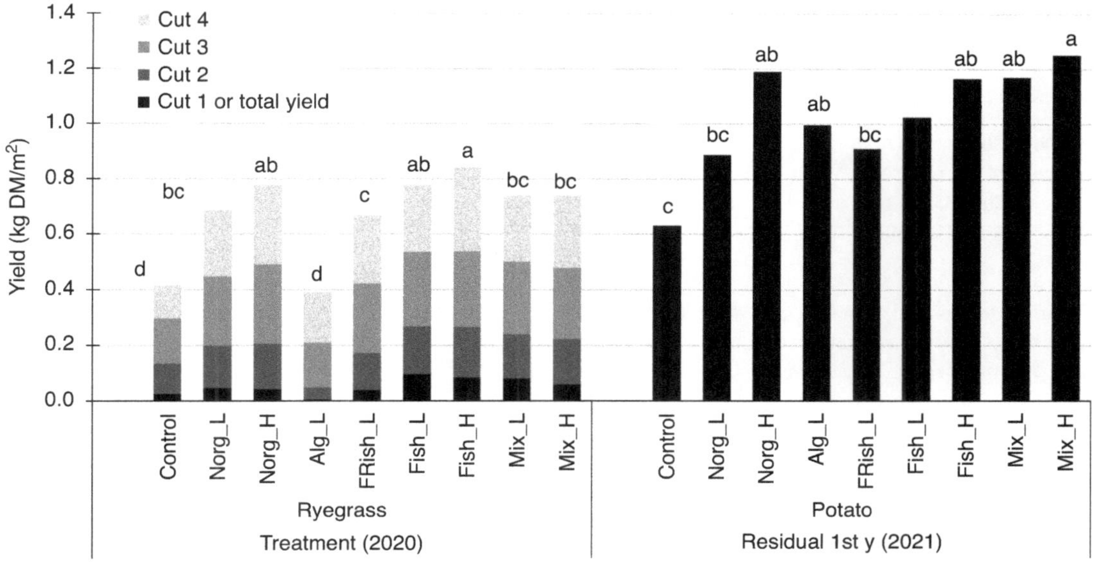

Fig. 31.6. Dry-matter yields of ryegrass (four cuts) amended with poultry manure at low and high rate (Norg_L, Norg_H), rockweed sludge at low rate (Alg_L), fresh minced fish residues at low rate (Frish_L), sediments from acidified hydrolysed fish residues at low and high rate (Fish_L, Fish_H) and a mix of FB and RW where 70% of N came from FB at low and might rate (Mix_L, Mix_H). Low rate = 300 kg total N per ha; high rate = 600 kg. Letters indicate differences (p<0.05) between mean values using the Tukey t-test.

processing seaweed giving the sludge away for free. Costs associated with transporting and spreading the sludge in the field could then be minimized by spreading larger amounts more often. However, the material is not well balanced with respect to mineral composition. Ruminants demand a certain balance between mono- and divalent cations. To assess the risk of tetany in ruminants (sudden death due to cation imbalance, commonly caused by intake of feed rich in K), the ratio between potassium over calcium and magnesium (Mg), K/(Ca +Mg) measured in milliequivalents may be computed (Manitoba Agriculture, 2004). The concentration of K is then multiplied by 256, Ca by 499 and Mg by 823, before computing the ratio. To avoid tetany, the ratio should be <2.2. For the oat plant material in 2019, the ratio was 2.5 in material fertilized with rockweed sludge, 2.0 for the mixed treatment (30% of N from rockweed sludge), 1.5 for the unfertilized control, 1.3 for poultry manure and 0.9 for material fertilized with fishbones. Hence, the first crop after application of such high amounts of rockweed sludge should likely not be applied as feed for ruminants, at least not as the main proportion of the daily feed intake. In 2020 (first-year ley), the ratio had decreased to 1.47 at the first cut and 1.23 at the second. We recommend that rockweed sludge should not be applied at a higher rate than 10 tons per ha over 5 years, to avoid excess uptake of K in crop plants.

Potatoes grown in soil amended with rockweed sludge did not have higher concentrations of As or Cd than potatoes grown in soil amended with comparable amounts of other fertilizer materials (Løes *et al.*, 2022c). The tuber yield (fresh weight) was 46.6 tons per ha when rockweed sludge was applied in the previous season, compared with 41.6 tons when poultry manure was applied (at a low rate) and 29.5 in the unfertilized control (Table 31.2). Rockweed sludge application increased the concentrations of K and Mg, but not of As or Cd.

Survival of young ley

With the application of rockweed sludge and dry materials, the ley survived and produced higher yields than the unfertilized plot in 2023. The application

Table 31.2. Concentrations of minerals (g kg^{-1} DM for K, Mg) and PTEs (μg kg^{-1} DM for As, Cd) in potatoes produced in 2021, in plots amended with various fertilizers in 2020. Tuber fresh weight (FW) yields in tons per ha. For each characteristic, different letters (a, b) indicate statistically significant differences ($p < 0.05$).

Fertilization 2020	K	Mg	As	Cd	Tuber FW
Control, no fert.	15 b	0.8 b	16 b	44 a	29.5 b
Rockweed sludge	22 a	1.2 a	15 b	37 a	46.6 a
Poultry manure	14 b	0.8 b	15 b	42 a	41.6 a

of wet materials did not function very well. With one material, only perennial weeds, especially coach grass (*Elymus repens* L.), survived (Fig. 31.4). The other material did not extinguish the crop plants, but the growth was less favourable than for the solid seaweed materials. Even if the ley survived and yields were increased, the tetany ratios of the harvested plant material will likely be very high. Further, the dry materials and the rockweed sludge were not completely decomposed during the season and especially the dry materials were present as a wet, smeary layer even four months after application. This layer would most likely have polluted the harvested material if farm-sized harvesting equipment had been applied, and possibly destroyed the sward by wheel. Hence, a single maximum application of seaweed materials every ten years may not kill the crop but may still not be recommended, although the long-term effects might be positive.

Decomposition by composting

Studies with Dewar flasks demonstrated that fronds from fresh cuvie, and dried, milled rockweed, both of which contained high concentrations of salt, initiated heat (Cabell and Løes, 2023). This shows that high salt concentrations may not be a hindrance for composting fresh seaweeds. They also demonstrated that expanded clay aggregates (Leca®) were well suited as a bulking agent and that it was possible to initiate heat and decomposition with 'high-N' rockweed sludge and Leca as the only materials in the flask substrate. However, in a 200-litre insulated reactor, the heat initiation of rockweed sludge with Leca was rather small, demonstrating the importance of physical conditions for the performance of biological processes. In the outdoor windrow, rockweed sludge was mixed with hydrolysed fish sediments and woodchips, and the heat generation was significant.

Discussion and Concluding Remarks

There is an increased interest in alternative sources of plant nutrients which may replace existing fossil resources. Most European countries are currently dependent on importing P and K, which is a challenge in a world with increasing levels of conflict. Significant amounts of nutrients are found in the sea, and utilization of minerals such as P, K, Mg and Ca in agriculture would imply a recycling of elements lost from terrestrial and aquaculture food systems. This fits well with the basic principle of ecology in organic agriculture (IFOAM, 2020). Residual materials that are poorly utilized are available, possibly at prices that may compete with current costs for alternative nutrient sources such as conventional poultry manure. Conventional poultry manure is currently often available at a cheap price or even for free because poultry farmers may have high animal densities and limited farmland on which to spread the manure. Fossil minerals, such as K salts and rock P, may be applied in organic agriculture and are still cheap as compared with what it will cost to extract these minerals from marine materials. One reason for this is that the prices for fossil resources do not include externalities such as eutrophication, or local environmental impacts caused by mineral extraction. Similar challenges exist for most recycled products. A further challenge is that fertilizers are lower on the biomass value pyramid than food and pharmaceuticals. Producers of residual materials will always search for the best price, and new applications in other sectors may offer higher prices than as raw materials in fertilizers. This increases the risk for those considering establishing a value chain for marine-derived fertilizers.

The industry producing pelleted and liquid fertilizers from various organic materials is still quite decentralized. For example, in Norway, with only 5 million inhabitants, there are two factories, each located in an area with significant production of (conventional) poultry. Hence, it may be more feasible to introduce marine materials as ingredients in existing fertilizers rather than to establish a separate industry for marine-derived fertilizers. In any case, we still need considerably more research and developmental activities to establish cross-sectorial value chains for marine-derived fertilizers which may be applied to certified organic arable and feed crops. As concluded by Olesen *et al.* (2023), innovative mixes of technologies applied to low trophic species, marine and terrestrial animal by-products, new crops, and diversified and integrated production systems can link the high primary productivity rates of the marine environment to the C storage capability of the terrestrial food sector. This can further pave the way for linking and scaling up C-neutral marine and terrestrial food production systems into a future sustainable and circular bioeconomy.

Acknowledgements

We are grateful for funding from several sources, initially from Møre and Romsdal County Council for the project 'Residual materials from marine industries as fertilisers in organic agriculture' (RESTOR, 2018–2022) and the Regional Research Council Møre og Romsdal for 'Seaweed fibre for increased soil organic matter' (FIMO, 2021–2022). Activities in these projects were linked up with the Horizon 2020 project 'Pathways to phase out contentious inputs from organic agriculture in Europe' (Organic-PLUS, 2018–2022, GA774340) funded by the EC. More recently, funding has been achieved from the Research Council of Norway and other national funding bodies over the BlueBio Cofund (agreement 817992) for 'Sustainable utilization of MARine resources to foster GREEN plant production in Europe' (MARIGREEN, 2021–2024) and 'Value creation and ecosystem services of European seaweed industry by reducing and handling potentially toxic elements from breeding to soil' (SeaSoil, 2022–2025), and 'Amplifying SeaSoil: demonstrating seaweed fertilisation effects on perennial ley' (SeaSoilDemo, 2023–2024). We also acknowledge the support from Nutrishell AS and the Regional Research Council of Trøndelag for the project 'Sea-conserved crab shells as fertiliser' (2023–2024).

References

Ahuja, I. and Løes, A.-K. (2019) Norwegian Centre for Organic Agriculture (NORSØK). NORSØK Report. Tingvoll, Norway. Available at: https://orgprints.org/id/eprint/36439/ (accessed 14 February 2025).

Ahuja, I., Dauksas, E., Remme, J.F., Richardsen, R. and Løes, A.-K. (2020) Fish and fish waste-based fertilizers in organic farming – with status in Norway: A review. *Waste Management* 115, 95–112.

Brinton, W.F., Evans, E., Droffner, M.L. and Brinton, R.B. (1995) Standardized test for evaluation of compost self-heating. *Biocycle* 36(11), 64–69.

BS EN 16087-2:2011 Soil improvers and growing media (n.d) Determination of the aerobic biological activity: Self-heating test for compost. Available at: https://www.en-standard.eu/bs-en-16087-2-2011-soil-improvers-and-growing-media-determination-of-the-aerobic-biological-activity-self-heating-test-for-compost/ (accessed 14 February 2025).

Cabell, J. and Løes, A.-K. (2023) Blue biomass composting technology. NORSØK Report. Available at: https://orgprints.org/id/eprint/50477/

EC (European Commission) (2020) Communication from the Commission to the European Parliament, the Council, the European Economic and Social Committee and the Committee of the Regions. Available at: https://eur-lex.europa.eu/legal-content/EN/TXT/?uri=CELEX:52020DC0381 (accessed 14 February 2025).

IFOAM (International Federation of Organic Agriculture Movements) (2020) The principle of ecology. Available at: https://www.ifoam.bio/why-organic/principles-organic-agriculture/principle-ecology (accessed 14 February 2025).

Løes, A.-K., Katsoulas, N., Caceres, R., De Cara, M., Cirvilleri, G. *et al.* (2018) Current use of peat, plastic and fertiliser inputs in organic horticultural and arable crops across Europe. Available at: https://orgprints.org/id/eprint/34357/ (accessed 14 February 2025).

Løes, A.-K., Ahlin, J.P., Ahuja, I., Krogstad, T., Smevoll, S. *et al.* (2022a) Effects of formic acid preservation of fishbones on the extractability of ammonium lactate–acetate soluble calcium, phosphorus, magnesium, and potassium. *Waste and Biomass Valorization* 13(8), 3547–3559.

Løes, A.-K., Ahuja, I., deBoer, A. and Rittl, T. (2022b) Fertilisation effects of marine-derived residual materials on agricultural crops. NORSØK Report 13 (7). NORSØK, Tingvoll, Norway. Available at: https://orgprints.org/id/eprint/45330/

Løes, A.-K., Grønmyr, F., Pommeresche, R., Rittl, T. and Stensrud, A.F. (2022c) Algae fibre for soil improvement (FIMO). 8. NORSØK Report (Vols. 1–7). Tingvoll: NORSØK. Available at: https://orgprints.org/id/eprint/444040/

Manitoba Agriculture (2004) Potassium and tetany ratios (Nutrition update Vol. 15). Available at: https://www.gov.mb.ca/agriculture/livestock/beef/potassium-and-tetany-ratios.html (accessed 14 February 2025).

Maroulis, M., Matsia, S., Lazopoulos, G., Pârvulescu, O.C., Ion, V.A. *et al*. (2023) Chemical and biological profiling of fish and seaweed residues to be applied for plant fertilization. *Agronomy* 13(9), 2258.

McKinnon, K., Løes, A.-K. and Almvik, M. (2021) Gjødsel med rester av herbicid: effekt av klopyralid på oppalsplanter. Fertiliser with residues of herbicide: effect of clopyralide on transplants. 6. Tingvoll, Norway: NORSØK. Available at: https://orgprints.org/id/eprint/40065/

Miloştean, D. and Flori, M. (2021) Study of biodegradable waste compostability through self-heating test. *Annals of Faculty Engineering – International Journal of Engineering* XIX(4), 107–110. Available at: https://annals.fih.upt.ro/pdf-full/2021/ANNALS-2021-4-13.pdf

Möller, K., Oberson, A., Bünemann, E.K., Cooper, J., Friedel, J.K. *et al*. (2018) Improved phosphorus recycling in organic farming: navigating between constraints. In: Sparkes, D.L. (ed.) *Advances in Agronomy 147*. Elsevier, pp. 159–237.

Moloşag, A., Pârvulescu, O.C., Ion, V.A., Asănică, A.C., Soane, R. *et al*. (2023) Effects of marine residue-derived fertilizers on strawberry growth, nutrient content, fruit yield and quality. *Agronomy* 13, 1221.

Naylor, R.L., Hardy, R.W., Buschmann, A.H., Bush, S.R., Cao, L. *et al*. (2021) A 20-year retrospective review of global aquaculture. *Nature* 591(7851), 551–563.

NFSA (Norwegian Food Safety Authority) (2022) Utfyllende informasjon om regelverket for økologisk landbruksproduksjon (Guidance on the organic regulation: Additional information on the regulation for organic agricultural production). Available at: https://www.mattilsynet.no/om_mattilsynet/gjeldende_regelverk/veiledere/veileder_for_okologisk_landbruk.2651/binary/Veileder%20for%20%C3%B8kologisk%20landbruk (accessed 14 February 2025).

Olesen, I., Bonaldo, A., Farina, R., Gonera, A., Hughesle, A.D. *et al*. (2023) Moving beyond agriculture and aquaculture to integrated sustainable food systems as part of a circular bioeconomy. *Frontiers in Marine Science* 10.

Willer, H., Schlatter, B. and Trávníček, J. (eds) (2023) The world of organic agriculture: Statistics and emerging trends. In: *Research Institute of Organic Agriculture, FiBL, Frick, and IFOAM*. Organics International, Bonn. Available at: https://orgprints.org/id/eprint/45973/ (accessed 14 February 2025).

Zikeli, S., Weiler, C.S., Løes, A.-K., Conroy, J., Rayns, F. *et al*. (2022) Technical report on alternative fertilisers (arable farming and vegetables). Deliverables from the Organic PLUS project: Pathways to phase-out contentious inputs from organic agriculture in Europe. Available at: https://orgprints.org/id/eprint/40076/ (accessed 14 February 2025).

32 Acceptability of Marine-derived Fertilizers in Organic Farming

Kanu Murmu*, Sramana Sen Sarma and Arup Sarkar

Department of Agronomy, Faculty of Agriculture, Bidhan Chandra Krishi Viswavidyalaya, Mohanpur, India

Abstract

The next century is expected to see a doubling of agricultural output. It will be difficult to maintain food security in the face of global population increase without developing solutions that reduce related environmental harm in a way that aligns with the Sustainable Development Goals of the United Nations. Using organic fertilizers derived from marine sources is one strategy that may be used to strengthen soil structure by boosting soil organism activity and replenishing the soil with vital plant nutrients. Here, we highlight potential uses of marine-derived materials such as *Ascophyllum* seaweed to create organic fertilizers. *Ascophyllum* are an annoyance in many coastal locations across the world because of their detrimental effects on the productivity of marine ecosystems. Numerous studies have shown that by lowering pollutants and enhancing soil health and quality, organic fertilizers made from *Ascophyllum* could improve the services provided by coastal ecosystems. According to recent studies, seagrass litter increases the soil's ability to retain water and the nutritional value of crops. Because it contains important macro- and micro-elements, it can also be used as a multi-purpose fertilizer. Applying *Ascophyllum* extract fertilizer improves the soil's organic matter, nitrogen, phosphorus and potassium concentrations and significantly improves seedling growth and survival. This overview includes experimental investigations on the effects of marine organic fertilizers on soil restoration and the implications for coastal management, as well as creative techniques for using seagrass as fertilizer sources.

Keywords: marine-derived fertilizer, *Ascophyllum* seaweed, biostimulants, soil fertility, organic farming

Introduction

Nutrition significantly influences agricultural output. Continuous farming operations deplete the nutritional content of the soil. Soil is the primary provider of vital nutrients to plants. If certain minerals are missing from the soil, plants will display deficiency symptoms that might impact their physiological functions, such as reproduction and growth (Sandip, 2012). To address mineral shortage and boost crop output, soil enrichment can be achieved by providing these nutrients from external sources. Compounds added to soil to enhance fertility are categorized as chemical fertilizers and organic fertilizers. Many farmers use chemical fertilizers to quickly fertilize the soil (McCauley, 2009). Chemical fertilizers contain significant levels of heavy metals such as mercury (Hg), cadmium (Cd), arsenic (As), lead (Pb), copper (Cu) and nickel (Ni), as well as natural radionuclides like uranium (238U), thorium (232Th) and polonium (210 Po). Applying chemical fertilizer causes contamination in water, soil and air. Chemical fertilizers increase crop productivity but can pose environmental risks. Hence, alternative agricultural practices need to be taken into account. Currently, organic farming shows great potential. Organic farming is an agricultural approach that involves abstaining from the use of synthetic pesticides, fertilizers, genetically modified organisms and growth hormones. Organic farming will boost agricultural production and create a pollution-free environment. Significant amounts of trash are

*Corresponding author: kanumurmu@gmail.com

DOI: 10.1079/9781800626850.0032

produced during the processing of aquatic meals derived from crustaceans and fish. These materials contain significant levels of nutrients that can be beneficial for agricultural techniques. The majority of aquatic food-processing waste was either deposited in dump sites or spread indiscriminately on land. Applying crustacean waste to land causes soil contamination. Therefore, this trash must be treated differently. Dao and Kim (2011) explored the utilization of fermented fish waste products as liquid fertilizer. The nutrients in the fermented liquid fish waste can enhance plant growth by promoting beneficial soil microbes and improving the absorption of vital nutrients. This liquid fertilizer is easily accessible, gradually distributes nutrients and helps avoid soil leaching. Nutrients can be administered through foliar sprays to both immobilize and supplement them to the plants. The research indicates that crustacean waste can be managed by fermenting it to provide a liquid fertilizer for agricultural techniques, promoting sustainable development in agriculture. Furthermore, the escalating impacts of global warming and human-induced pollution are swiftly altering marine ecosystems, causing coral-dominated reefs in several locations to transition into seaweed-dominated reefs. Ecological phase transitions described by Rohwer *et al.* (2010) decrease reef production and diminish their capacity to support communities, while also causing problems such as over-abundance, leading to concerns like unsightliness, odour and beach fouling. To address this issue, a sustainable approach would involve finding a way to utilize the excess biomass, such as converting it into agricultural composts and solid/liquid fertilizers. Historical use of chemical fertilizers and inadequate waste disposal caused significant harm to numerous coastal regions, resulting in a shift in the ecosystem towards algae dominance.

One suggested solution to address the decrease in crop production and the surplus of seaweeds is to develop techniques that redirect extra marine plant material into agricultural applications by converting the biomass into solid and liquid fertilizers. Seaweed-based fertilizers have been successfully used on crops in affluent regions like the USA and Europe. However, they are costly to import and unaffordable for most farmers in our area. Empowering local farming communities with the expertise to transform marine-derived products into sustainable agricultural fertilizers can help them create economically viable alternatives to commercial fertilizers. This approach can reduce environmental harm, curb the spread of seaweeds on coral reefs, and enhance the production of subsistence and cash crops, ultimately improving food and financial security.

Background

Marine fertilizers have been utilized in Europe since the 1920s and earlier, primarily in coastal regions, but were not as cost-efficient as chemical fertilizers. With the introduction of liquid fertilizers and foliar sprays in the 1960s, there was a resurgence of interest in utilizing seaweed-derived additions in agriculture. Algal extracts have been observed to have various positive effects, including improved germination, enhanced root development and growth, increased nitrogen and phosphorus assimilation, enhanced resistance to pests and diseases, and cleaner, longer-lasting fruits (Merigout, 2006; Kumar *et al.*, 2012).

Marine-derived Products

Marine-derived fertilizers, often produced from seaweeds or fish by-products, are widely accepted in agriculture. They supply vital nutrients such as nitrogen, phosphate and potassium that enhance plant growth. Moreover, these fertilizers can improve soil composition and increase microbial function. Nevertheless, sustainability, harvesting procedures and local rules impact their overall attractiveness. Responsible sourcing and application are essential to reduce environmental effect. Seaweed extract and fish emulsion fertilizers have become popular because of their potential benefits. They include critical nutrients such as micronutrients, amino acids and growth-promoting chemicals that enhance plant health. Seaweed fertilizers frequently function as biostimulants by enhancing plant stress tolerance, promoting root growth, and increasing overall resilience. They significantly impact soil structure and microbial activity, fostering a robust soil ecosystem. Additionally, they aid in retaining nutrients and minimizing nutrient leakage. Marine-derived fertilizers are often viewed as more environmentally benign than manufactured fertilizers. They can play a role in sustainable agriculture by helping to decrease the need for chemical inputs. They adhere to the standards of organic cultivation and are therefore utilized in many ways. Acceptance may differ depending on area restrictions and agricultural practices. Certain areas may require certain regulations or qualifications for the application of fertilizers produced from marine sources. Farmers should take into account their crops' individual requirements, local restrictions and environmental sustainability while using marine-derived fertilizers. Harvesting marine resources is essential. Uncontrolled or too exploitative harvesting can result in environmental issues and affect marine ecosystems.

Biostimulants

Definition

Biostimulants are natural or man-made compounds used on seeds, plants and soil. These compounds induce alterations in essential and structural processes to enhance plant growth by improving resistance to non-living environmental factors and boosting seed and grain production along with quality. Plant biostimulants are distinct from insecticides. They do not modify plant growth in a way that aligns with the original purpose of the federal Insecticide, Fungicide and Rodenticide Act (FIFRA) concerning plant regulators. They have a beneficial impact on the plant's physiology. They provide benefits that align with products not covered by FIFRA. The regulatory pathway is ambiguous and legal definitions are not yet established.

Biostimulants from marine sources

Seaweed extract, also known as Seaweed Liquid Fertilizer (SLF), is considered a promising new natural organic product that might enhance crop growth (Iswarya *et al.*, 2019). Seaweeds have growth-enhancing properties due to their content of carbohydrates, phenylacetic acid, macro- and micro-elements, vitamins, and plant growth regulators such as cytokinin and gibberellins. They have been found to promote plant growth and increase yield, as well as enhance plants' ability to withstand environmental stress.

Lower concentrations of liquid fertilizers derived from *Ascophyllum* and *Laminaria* promoted faster growth compared to larger concentrations in experimental crops including potatoes, corn, peppers, tomatoes, pineapples and oranges. Using *Ascophyllum* extract as a foliar spray boosted the chlorophyll levels in cucumber cotyledons and tomato plants. Applying *Ecklonia maxima* seaweed extract as a soil drench to tomato seedlings significantly increased their growth (Table 32.1). One per cent seaweed liquid fertilizer (SLF) made from *Sargassum wightii* and *Ulva lactuca* combined with 25% of the recommended chemical fertilizers resulted in the highest growth parameters and yield in *Arachis hypogaea*, including fresh weight, dry weight, root and shoot length, number of branches, leaf area, and levels of chlorophyll, protein, carbohydrate, lipid and yield (Sridhar and Rengasamy 2010, 2011). The productivity and protein levels of soybean were evaluated by applying seaweed saps derived from *Kappaphycus* and *Gracilaria* extracts at various growth stages. The study found that the greatest grain yield was achieved with the use of 15% seaweed extract, followed by 12.5% seaweed extract, which led to increases of 57% and 46%, respectively, compared to the control (Lodhi and Diwan, 2018). The seaweed liquid extract contains a mixture of red seaweed (*Kappaphycus alvarezii*) and brown showed (*Sargassum* sp.) algal extracts that significantly enhance the growth factors, yield and biochemical characteristics of greengram. The 0.5% concentration of a seaweed liquid fertilizer extracted from *S. polycystum* containing water and alcohol showed superior outcomes in seed germination, growth parameters, biochemical composition and pigment content of *C. cajan* (Erulan *et al.*, 2009). Singh *et al.* (2015) found that the liquid extract of *Ulva fasciata*, *Sargassum ilicifolium*, and *Gracilaria corticata* affected the carbohydrate, protein and free amino acids content of *Trigonella foenum* (Pise, 2010). An investigation was conducted to study the impact of SLF derived from *Gracilaria textorii* and *Hypnea musciformis* on seed germination, growth and yield metrics, including the number of leaves and fruit weight, in agricultural plants, including brinjal, tomato and chilli. The SLF was discovered to enhance growth and yield more effectively at lower doses compared to the control group. An experiment was carried out to investigate the effectiveness of liquid seaweed extracts from *Kappaphycus alvarezii* and *Gracilaria edulis* on the growth and yield of summer rice in a lowland field in Gerua, Assam. The study involved using *Kappaphycus* sap and *Gracilaria* sap at several concentrations (2.5%, 5%, 10% and 15%) together with recommended fertilizer doses. These treatments were compared to using only fertilizer with water spray, as well as a combination of 7.5% *Kappaphycus* sap and

Table 32.1. List of various seaweeds for biostimulant production.

Sl no	Seaweed name	Form of usage	Treated crops	Reference
1.	*Ascophyllum* and *Laminaria*	Liquid fertilizers	Potatoes, corns, peppers, tomatoes, pineapples, oranges	Stephenson, 1973; Whapham *et al.*, 1993
2.	*Ascophyllum*	Seaweed extract	Cucumber, tomato plants	Whapham *et al.*, 1993
3.	*Caulerpa recemosa* and *Gracilaria edulis*	Seaweed liquid fertilizer (SLF)	*V. catajung*	Anandharaj and Venkatesalu, 2001a
		Seaweed liquid fertilizer (SLF)	*Abelmoschus esculentus* and *Lycopersicon lycopericum*	Selvaraj *et al.*, 2004
4.	*Caulerpa racemosa* and *Gracilaria edulis*	Seaweed extract	*Dolichos biflorus* seeds	Anantharaj and Venkatesalu, 2001b
5.	*Gracilaria textorii* and *Hypnea musciformis*	Seaweed liquid fertilizer (SLF)	Brinjal, tomato and chilli	Rao and Chatterjee, 2014
6.	*Gracilaria textorii* and *Hypnea musciformis*	Seaweed liquid fertilizer (SLF)	Brinjal, tomato and chilli	Rao and Chatterjee, 2014
7.	*Kappaphycus* and *Gracilaria*	Foliar spray	Soybean	Lodhi and Diwan, 2018
8.	*Kappaphycus alvarezii* and *Sargassum* sp.	Foliar spray and seaweed extract	Greengram	Iswarya *et al.*, 2019
9.	*Kappaphycus alvarezii* and *Gracilaria edulis*	Seaweed sap	Maize, *Oryza sativa*	Satapathy *et al.*, 2014; Layek *et al.*, 2015; Singh *et al.*, 2015
10.	*Kappaphycus*	Seaweed sap	Potato	Pramanick *et al.*, 2017
11.	*Kappaphycus* and *Gracilaria*	Seaweed extract	Greengram	Dwivedi *et al.*, 2014
		Seaweed extract	Greengram	Gurusaravanan *et al.*, 2010
		Seaweed extracts	Pulses	Bai *et al.*, 2011
12.	*S. polycystum*	Seaweed liquid fertilizer (SLF)	*C. cajan*	Erulan *et al.*, 2009
13.	*Ulva lactuca*	Seaweed liquid fertilizer (SLF)	*Tagetes erecta*	Sridhar and Rengasamy, 2010

50% RDF. The study found that using a 5% mixture of *Kappaphycus* or *Gracilaria* sap along with the appropriate amount of fertilizer resulted in increased grain and straw yield, net income and benefit-cost ratio for summer rice. Seaweeds in China, Korea and Japan are bought in a dehydrated form. There is a market for some types of fresh seaweeds that are utilized as a salad vegetable or as garnishes for dishes like fish. Certain species of *Caulerpa*, *Eucheuma* and *Gracilaria* are utilized for this purpose, particularly in the tropical Southeast Asian nations like the Philippines, Malaysia, Thailand and Indonesia (Rao *et al.*, 2018).

Mode of action of biostimulants

Rayorath *et al.* (2008) conducted a study to explore the possible role of the plant growth hormone gibberellic acid (GA3) in seed germination responses triggered by a seaweed extract from *Ascophyllum nodosum*. α-Amylase enzyme is involved in breaking down endosperm starch into sugars during seed germination in cereals, providing energy for root and shoot growth (Fig. 32.1). Endogenous gibberellins or gibberellic acids serve as a signal to activate the α-amylase genes, which play a role in seed germination. ABA, when administered to seeds, can inhibit the induction of the α-amylase

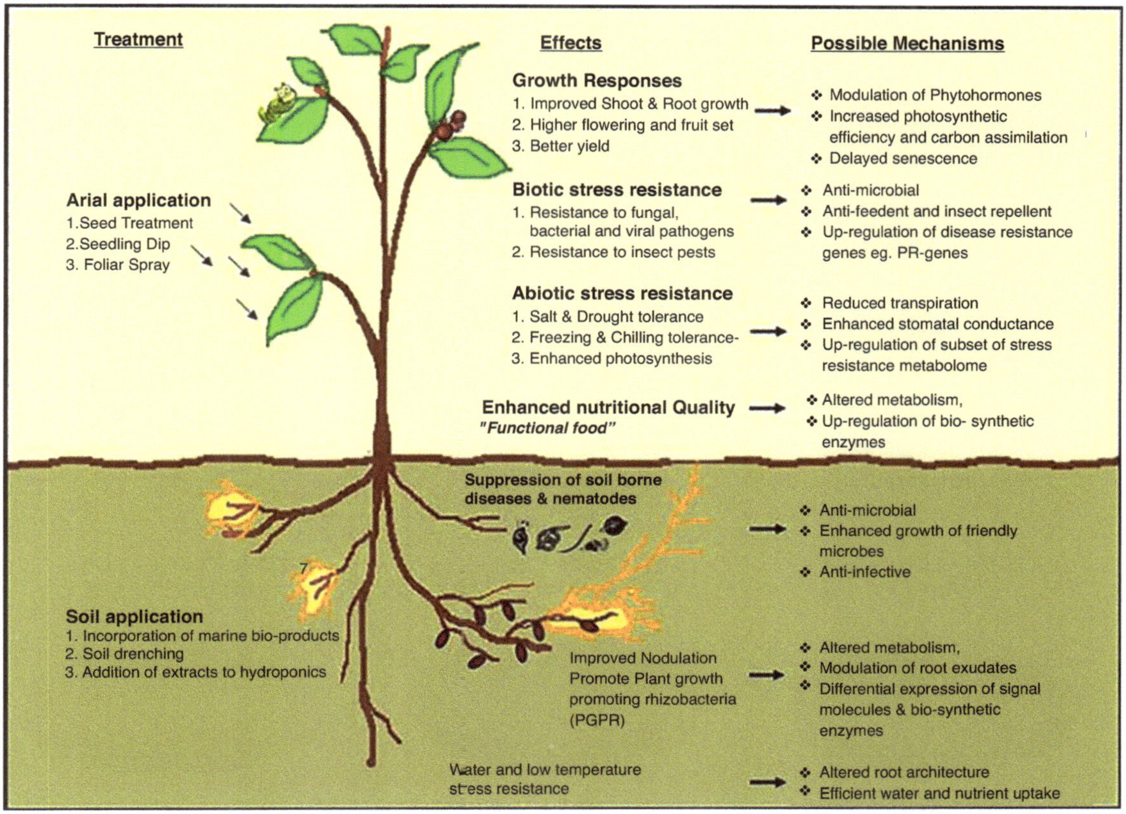

Fig. 32.1. Schematic representation of physiologic effects elicited by seaweed extracts and possible mechanism(s) of bioactivity. (From Khan *et al.*, 2009)

gene by GA3 and counteract the effects of GA3 on seed germination. Rayorath *et al.* (2008) showed that applying an alkaline extract of *Ascophyllum nodosum* (ANE), which does not contain GA3, stimulated α-amylase activity and seed germination in GA-deficient mutant barley seedlings (grd2) that cannot naturally manufacture GA. This indicates that substances in the seaweed extract, aside from GA, are accountable for the effects shown in the barley mutant. The α-amylase activity triggered by the seaweed extract in the mutant was not hindered by external ABA treatments, indicating that non-GA chemicals in ANE operate through a GA-independent mechanism for seed germination in barley. Recent research indicates that concentrated seaweed extracts containing polysaccharides can boost metabolism and impact plant development. Studying the impact of seaweed polysaccharides on plant systems is a new and developing field of inquiry. Seaweed is a diverse carbohydrate-rich food, mostly categorized into various types of brown, green and red seaweeds. Brown seaweeds, belonging to the phylum Ochrophyta, often store laminarin and mannitol as carbohydrates and have fucoidans, cellulose, sulphated xylofucoglucan, sulphated xylofucoglucuronan, β-(1-3)-glucans, and alginates in their cell walls. Green seaweeds primarily utilize starch, inulin and sucrose as store carbohydrates, and ulvans, cellulose, xyloglucan, mannans, glucuronan and β-(1-3)-glucan as cell-wall carbohydrates. Red seaweeds store carbohydrates such as floridean glycogen, starch, mannitol, floridoside and isofloridoside. Red seaweed cell walls contain agars, carrageenans, cellulose, glucomannan, other mannans and xylans. The mechanisms by which seaweed extracts work are not well comprehended. Plant priming and seaweed extract both share common attributes. Seaweed-derived carbohydrates such as laminarin, fucoidan, alginate, carrageenan, ulvan, and other non-carbohydrate compounds may act as molecular priming agents or key bioactives in seaweed extracts to trigger specific plant responses in treated plants. Variations in seaweed raw material, extraction conditions and processing factors can impact the composition of extracts and the structural integrity of polysaccharides. These differences can lead to varying effects on plants when treated. Studies at the gene level using standard microarray platforms with a restricted set of genes have detected changes in transcript levels in plants when treated with seaweed extract. In an *in vitro* experiment conducted on half-strength Murashige and Skoog basal medium, a reduced concentration of *Ascophyllum nodosum* extract enhanced plant biomass and influenced the phenylpropanoid and flavonoid pathways to promote the growth and nutritional quality of spinach (Fan *et al.*, 2013). An acidic extract obtained from freshly harvested *Ascophyllum nodosum* increased the overall dry weight of rapeseed (*Brassica napus*) when cultivated in a nutrient solution. The notable rise in shoot and root weight was linked to an improved nitrate absorption resulting from the SWE treatment. The genes that showed varying levels of expression in response to the SWE treatment in this study were linked to the regulation of internal hormones, photosynthesis and the intake and transportation of nutrients.

Biostimulant market

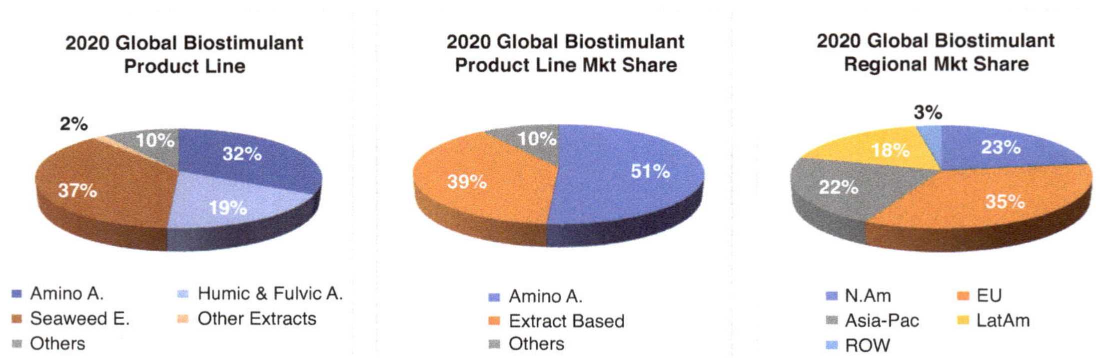

(From Dunham Trimmer International Bio Intelligence)

The growth rate of biostimulants is approximately 12–15%. The Latin American market is experiencing the highest growth rate compared to other regions. The climate and crops in Latin America are well suited to the application of biostimulants. The biostimulant market is primarily controlled by fast-expanding entrepreneurial firms. Major corporations are concentrating on microbial products, but the process of market concentration has already begun. Biostimulants are novel agricultural compounds sourced from natural origins and used as seed treatments or foliar applications to enhance crop production by boosting stress tolerance, water- and nutrient-utilization efficiency, and overall growth and yield. Natural products can enter the market faster and with reduced regulatory expenses compared to manufactured chemicals. The global biostimulants market is expected to grow at a rate of 12% annually, reaching a value over $2200 million by 2018.

Role in organic farming

Seaweed has been utilized as a form of fertilizer in agriculture since ancient times. For the past three decades, it has been extensively used as fertilizer. They are used in their entirety or finely diced as manure or liquid extract. The new liquid extract is gaining popularity as a foliar spray to enhance crop yield in vegetables, fruits and cereal crops at a faster rate. Research indicates that applying nutrients directly to plant leaves is more effective than other techniques of delivering nutrients through the soil. Seaweed extracts are biodegradable, non-toxic, non-polluting and non-hazardous, unlike chemical fertilizers. The rise of organic farming and the increased need for organic food items has allowed for the utilization of seaweed as a biofertilizer, biostimulant and soil conditioner. Abundant algal resources are found along the Pakistan coast but are not effectively utilized. Utilizing these resources sustainably can support livelihoods and enhance economic conditions, in addition to promoting sustainable agriculture practices.

Seaweeds used as manure in farming have been a long-standing practice among the Romans and in countries like France, Britain, Japan and China, and have a positive effect on crop productivity. Using marine macro-algae as fertilizer in crop production is a well-established practice in coastal regions globally. Seaweed continued to be significant for farmers, even until the early 1900s. Seaweed and beach cast are still utilized in agriculture and horticulture in numerous nations. Applying seaweed extract directly to the onion field resulted in a notable increase in crop output. A concentration of 0.5% was found to be the most effective, promoting maximum plant development, nutrient absorption and high-quality yield. Various varieties of seaweed were utilized in rice fields, resulting in a beneficial impact on crop productivity. *Ascophyllum nodosum* is acknowledged as the most effective and often used seaweed extract, functioning as a biostimulant to enhance agricultural productivity. The substance includes growth-promoting components such as vitamins, minerals, oligosaccharides and phytohormones. The application of liquid seaweed as fertilizer is expected to increase crop output by around 20–30% compared to plantations that do not receive this treatment.

Abiotic stress factors

Research shows that specific abiotic variables like drought, temperature and salinity have a permanent impact on plants and decrease significant crop yield. By 2050, it is projected that 50% of the arable land would be affected by salinization, significantly impacting global agricultural productivity.

Biotic stressors

Seaweeds are recognized as a potent reservoir of natural compounds. Seaweeds can produce a variety of secondary metabolites such as terpenes, lipids, steroids, aromatic chemicals, acerogenins, phlorotannins, amino acid derivatives and other polymeric molecules. Marine algae generate bioactive compounds in reaction to microbial processes. Seaweed extracts showed enhanced plant protection against pests and diseases. Plants protect themselves from pathogen invasion by recognizing signal molecules such as peptides, polysaccharides, oligosaccharides, lipids and proteins commonly present in the cell wall of harmful pathogens. Further research is required to thoroughly analyse and evaluate the impact of seaweed on disease control from multiple perspectives in the future. Comprehensive research is required to investigate the enhancement of systemic resistance along with the direct suppression of pathogens. For instance, biocontrol active bacilli produce antibiotic secondary metabolites that can inhibit phytopathogens and stimulate plant defence mechanisms.

Seaweed extracts have demonstrated positive impacts on plant growth and development by promoting seed germination, root growth, improved nutrient absorption and higher frost tolerance in challenging environments. Applying seaweed extract to the leaves of maize plants enhanced the growth of seedlings, leading to a notable improvement in both root and shoot development. Seaweeds enhance the soil's nutrient content by increasing levels of nitrogen, phosphorus, potassium and other minerals required for plant development. An analysis of seaweed samples collected from the Karachi shore revealed the presence of Ca, Cr, Cd, Cu, Na, Pb, K, Mg, Co, Zn and Fe. The elements with the highest concentration were sodium, potassium, calcium, iron and magnesium. Trace elements' existence suggests their potential for use in agriculture due to their role as plant growth-promoting factors.

Seaweeds are being evaluated as potential biofertilizers to improve soil condition due to the rising need for organic fertilizers in organic farming. Seaweeds, like *Ascophyllum nodosum*, are commonly utilized as soil conditioners and mineral suppliers for plants. Liquid seaweed extract application improves plant growth and soil texture. Enhanced soil structure results in increased aeration and improved nitrogen fixation. Due to higher capillary activity, increased root system growth is observed.

Remediation of contaminated soil

Heavy metal contamination and pollution are significant environmental issues currently. Various

Table 32.1. Commercial seaweed products used in the agriculture and horticulture industries. (From Khan *et al*., 2009)

Product name	*Seaweed name*	Company	Application
Acuidan®	*Ascophyllum nodosum*	Acadian Agritech	Plant growth stimulant
Acid Buf	*Lithothamnium calcareum*	Chance & Hunt Limited	Animal feed
Agri-Gro Ultra	*Ascophyllum nodosum*	Agri Gro Marketing Inc.	Plant growth stimulant
AgroKelp	*Macrocystis pyrifera*	Algas y Bioderivados Marinos, S.A. de C.V.	Plant growth stimulant
Alg-A-Mic	*Ascophyllum nodosum*	BioBizz Worldwide N.V.	Plant growth stimulant
Bio-Genesis™ High Tide™	*Ascophyllum nodosum*	Green Air Products, Inc.	Plant growth stimulant
Biotiva	*Ascophyllum nodosum*	PI Industries Ltd	Plant growth stimulant
Emerald RMA	Red marine algae	Dolphin Sea Vegetable Company	Health product
Espoma	*Ascophyllum nodosum*	The Espoma Company	Plant growth stimulant
Fartum	Unspecified	Inversiones Patagonia S.A.	Biofertilizer
Guarantee®	*Ascophyllum nodosum*	MaineStream Organics	Plant growth stimulant
Kelp Meal	*Ascophyllum nodosum*	Acadian Seaplants Ltd	Plant growth stimulant
Kalpak	*Ecklonia maxima*	BASF	Plant growth stimulant
Kelpro	*Ascophyllum nodosum*	Teinproceicos Biologicos, S.A. de C.V.	Plant growth stimulant
Kelprosoil	*Ascophyllum nodosum*	Productos del Pacifico, S.A. de C.V.	Plant growth stimulant
Maxicrop	*Ascophyllum nodosum*	Maxicrop USA, Inc.	Plant growth stimulant
Nitrozeime	*Ascophyllum nodosum*	Hydrodynamics International Inc.	Plant growth stimulant
Profert®	*Durvillaea antarctica*	BASF	Plant biostimulant
Sea Winner	Unspecified	Chaina Ocean University Product Development Co., Ltd	Plant biostimulant
Seanure	Unspecified	Farmura Ltd.	Plant growth stimulant
Seasol®	*Durvillaea potatorum*	Seasol International Pry Ltd.	Plant growth stimulant
Soluble Seaweed Extract	*Ascophyllum nodosum*	Technaflora Plant Products, LTD	Plant growth stimulant
Stimplex®	*Ascophyllum nodosum*	Acadian Agritech	Plant growth stimulant
Synergy®	*Ascophyllum nodosum*	Green Air Products. Inc.	Plant growth stimulant
Tasco®	Unspecified	Acadian Agritech	Animal feed

industries discharge untreated effluent containing pollutants, such as heavy metals, into the environment. These industries include textile, electroplating, automotive, sugar, pharmaceutical, metallurgy, iron and steel, electro-osmosis, tannery, paper, mining, beverages and food. Heavy metal does not decompose and acts as a permanent source of environmental contamination, posing a harm to human health and ecosystems. Heavy metals can be categorized into three main groups: hazardous metals (e.g. Pb, As, Hg, Ni, Cr, Cu, Sn); precious metals (e.g. Ru, Ag, Pd, Pt); and radionuclides (e.g. Am, Ra, Th, U). Seaweeds offer promising opportunities for biosorption due to their macroscopic composition, which is ideal for producing biosorbent particles suited for use as sorption material. Research shows that DDT, a significant environmental toxin, undergoes gradual biodegradation when treated with seaweed extracts. The high concentration of dissolved organic carbon in seaweed-modified soil likely enhances the biodegradation of DDT. *Kappaphycus* sp. was used as a biosorbent to effectively remove heavy metals (Pb, Cu, Fe, Zn) and yielded noteworthy outcomes. Seaweed was determined to be a promising absorbent for eliminating heavy metals from the environment. The results indicate that *G. corticata varcartecala* and *G. lithophila* are promising algae species for efficiently eliminating heavy metals such as Cr (III), Cr (VI), Pb (II), Hg (II), Pb (II) and Cd (II) from the environment. Green macroalgae (*Chaetomorpha linum*) was used to biosorb heavy metals Cu and Zn. It was viewed as a different method for handling undesired biological substances by utilizing processed algal material.

Government Initiative

The first-ever national conference on promotion of seaweed cultivation was held in Gujarat to implement seaweed cultivation on a pan-India basis and emphasise the promotion of seaweed cultivation. The Union Minister of Fisheries, Animal Husbandry & Dairying, Shri Parshottam Rupala, chaired the confernece on 27 January 2024 at Koteshwar (Kori Creek), Kutch, Gujarat. The conference put forward seaweed cultivation as an alternative for employment generation of seaweed products diversifying marine production and coastal community livelihood. The pilot project of Kori Creek can be a game-changer for seaweed cultivation. The government aims for innovations in seaweed production, and policy frameworks and regulations to facilitate networking opportunities and foster relationships. End-to-end mapping of the seaweed value chain and addressing the bottlenecks in the value chain is the need of the hour.

Conclusion

Seaweed is seen as an underutilized resource that should be maximized. Seaweeds boost plant growth and offer protection against both living and non-living stress factors. The use of chemical or synthetic fertilizers resulted in soil degradation, health risks and environmental problems. Seaweeds offer an organic alternative to fertilizer to boost crop production in order to fulfil the worldwide food requirements. The demand for seaweed extract is rising steadily because of its affordability and environmentally benign nature.

References

Anandharaj, M. and Venkatesalu, V. (2001a) Studies on the effect of seaweed extract on dolichos biflorus. *Seaweed Research and Utilisation Association* 24(1), 129–137.

Anantharaj, M. and Venkatesalu, V. (2001b) Effect of seaweed liquid fertilizer on *Vigna calajung*. *Seaweed Research and Utilisation Association* 23, 33–39.

Bai, N.R., Christi, R.M. and Kala, T.C. (2011) Effect of seaweed concentrate of *Pavonia* on the growth and yield of a pulse crop. *Plant Archive* 1, 117–120.

Dao, V.T. and Kim, J.K. (2011) Scaled-up bioconversion of fish waste to liquid fertilizer using a 5L ribbon-type reactor. *Journal of Environmental Management* 92(10), 2441–2446.

Dwivedi, S.K., Meshram, M.R., Pal, A., Pandey, N. and Ghosh, A. (2014) Impact of natural organic fertilizer (*Seaweed saps*) on productivity and nutrient status of blackgram (*Phaseolus mungo* L). *The Bioscan* 9(4), 1535–1539.

Erulan, V., Soundarapandian, P., Thirumaran, G. and Ananthan, G. (2009) Studies on the effect of *Sargassum polycystum* (C.Agardh, 1824) extract on the growth and biochemical Composition of *Cajanus cajan* L. Mill Sp. *American-Eurasian Journal of Agricultural and Environmental Science* 6(4), 392–399.

Fan, D., Hodges, D.M., Critchley, A.T. and Prithiviraj, B. (2013) A commercial extract of brown macroalga (*Ascophyllum nodosum*) affects yield and the nutritional quality of spinach in *vitro*. *Communications in Soil Science and Plant Analysis* 44(12), 1873–1884.

Gurusaravanan, P., Pandiyarajan, V. and Jayabalan, N. (2010) Effect of the seaweed liquid fertilizer on growth and productivity of *Vigna radiata* (L.) Wiliczek. *Green Farming* 1(2), 138–140.

Iswarya, S., Latha, K.R. and Srinivasan, K. (2019) Evaluation of seaweed extract on growth determinants, yield and biochemical parameters of greengram (*Vigna radiata*). *Journal of Pharmacognosy and Phytochemistry* 8(3), 1861–1864.

Khan, W., Subramanian, S., Rayorath, P., Rayirath, U.P., Jithesh, M.N. *et al.* (2009) Seaweed extracts as biostimulants of plant growth and development. *Journal of Plant Growth Regulation* 28, 386–399.

Kumar, N.A., Vanlalzarzova, B., Sridhar, S. and Baluswami, M. (2012) Effect of liquid seaweed fertilizer of *Sargassum wightii* Grev. on the growth and biochemical content of green gram (*Vigna radiata* (L.) Wiliczek. *Recent Research in Science and Technology* 4, 40–45.

Layek, J., Das, A., Ramkrushna, G.I., Trivedi, K., Yesuraj, D. *et al.* (2015) Seaweed sap: A sustainable way to improve productivity of maize in north-east India. *International Journal of Environmental Studies* 72(2), 305–315.

Lodhi, K.K. and Diwan, U.K. (2018) Effect of natural organic fertilizer (*Seaweed saps*) on productivity and protein status of soybean. *International Journal of Fauna and Biological Studies* 5(1), 157–158.

McCauley, A. (2009) Plant nutrient functions and deficiency and toxicity symptoms. Nutrient Management Module. Montana State University, Bozeman, Montana.

Merigout, P. (2006) *Étude du métabolisme de la plante en réponse à l'apport de différents fertilisants et adjuvants culturaux, Influence des phytohormones sur le métabolisme azoté*. Institut National Agronomique Grignon, Paris.

Pise, N.M. and Sabale, A.B. (2010) Effect of seaweed concentrates on the growth and biochemical constituents of *Trigonella foenum-graecum* L. *Journal of Phytology* 2, 50–56.

Pramanick, B., Brahmachari, K., Mahapatra, B.S., Ghosh, A., Ghosh, D. *et al.* (2017) Growth, yield and quality improvement of potato tubers through the application of seaweed sap derived from the marine alga *Kappaphycus alvarezii*. *Journal of Applied Phycology* 29(6), 3253–3260.

Rao, G.M.N. and Chatterjee, R. (2014) Effect of seaweed liquid fertilizer from *Gracilaria textorii* and *Hypnea musciformis* on seed germination and productivity of some vegetable crops. *Universal Journal of Plant Science* 2(7), 115–120.

Rao, P.S., Periyasamy, C., Kumar, K.S., Srinivasa Rao, A.S. and Anantharaman, P. (2018) Seaweeds: Distribution, production and uses. In: Noor, M.N., Bhalnagar, S.K. and Sinha, S.K. (eds) *Bioprospecting of Algae*. Society for Plant Research, Gaur City, India, pp. 59–78.

Rayorath, P., Khan, W., Palanisamy, R., MacKinnon, S.L., Stefanova, R. *et al.* (2008) Extracts of the brown seaweed *Ascophyllum nodosum* induce gibberellic acid (GA3)-independent amylase activity in barley. *Journal of Plant Growth Regulation* 27(4), 370–379.

Rohwer, F., Youle, M. and Vosten, D. (2010) *Coral Reefs in the Microbial Seas*, Vol. 1. Plaid Press, Granada Hills, California.

Sandip, D. (2012) *Agricultural production and food distribution to vulnerable families in India today*. Heinrich Böll Stiftung, New Delhi.

Satapathy, B.S., Pun, K.B., Singh, T. and Rautaray, S.K. (2014) Effect of liquid seaweed sap on yield and economics of summer rice. *ORYZA – An International Journal on Rice* 51(2), 131–135.

Selvaraj, R., Selvi, M. and Shakila, P. (2004) Effect of seaweed liquid fertilizers on *Abelmoschus esculentus* (L.) Moench and *Lycopersicon esculentum* Mill. Seaweed Research and Utilisation Association 26, 118–123.

Singh, S., Singh, M.K., Pal, S.K., Perween, S., Kumari, J. *et al.* (2015) Seaweed sap as productivity booster of maize. *The Bioscan* 10(3), 1303–1305.

Sridhar, S. and Rengasamy, R. (2010) Significance of seaweed liquid fertilizers for minimizing chemical fertilizers and improving yield of *Arachis hypogea* under field trial. *Recent Research in Science and Technology* 2(5), 73–80.

Sridhar, S. and Rengasamy, R. (2011) Influence of seaweed liquid fertilizer on growth and biochemical characteristics of *Arachis hypogea* L. under field trial. *Journal of Ecobiotechnology* 3(12), 18–22.

Stephenson, W.A. (1973) *Seaweed in Agriculture and Horticulture*. E.P. Publishing Ltd, Brighton, UK.

Whapham, C.A., Blunden, G., Jenkins, T. and Hankins, S.D. (1993) Significance of betaines in the increased chlorophyll content of plants treated with seaweed extract. *Journal of Applied Phycology* 5(2), 231–234.

33 Advances in Plant Disease Detection in Organic Farming Systems

Tahira Batool[1]*, Matt J. Bell[2], Wenhao Zhang[3] and Joel Allainguillaume[1]

[1]Department of Health and Applied Sciences, University of the West of England, Bristol, UK; [2]Animal and Agriculture Department, Hartpury University, Gloucester, UK; [3]Centre for Machine Vision, School of Engineering, University of the West of England, Bristol, UK

Abstract

Traditional crop monitoring relies on human observation to determine the presence of disease. As human vision can only sense a narrow segment of electromagnetic spectrum called visible light, spectrophotometers have been used to determine absorbed and reflected light from a surface in a much wider electromagnetic spectrum generating rich hyperspectral data that can support humans with plant disease detection. Furthermore, machine learning techniques to process hyperspectral data can be referred to as the frontier of innovation for non-invasive detection of asymptomatic pathogens as they provide an automatic, real-time and on-field solution for distinguishing healthy and diseased plants with high accuracy. As plants react differently in every component of electromagnetic spectrum, various studies have been conducted to use plants' hyperspectral data to measure water loss, classify asymptomatic infected samples, assess disease severity, distinguish cultivars of crops, and identify wavelength bands crucial for disease detection. Hyperspectral sensors are also popular in monitoring large fields and orchards with real-time results via unmanned aerial vehicles (UAV).

Keywords: hyperspectral data, machine learning, plant pathogens, detection

Introduction

Plant diseases are a significant cause of economic loss in crop production around the globe (Cheshkova, 2022). According to the Food and Agriculture Organization of the United States (FAO), every year 20–40% of crops are affected by plant diseases which cost $22 billion to the global economy (Sarkozi, 2009). Plant pathogens can be categorized into three classes: viruses, bacteria, and fungi (Srisakrapikoop *et al.*, 2020). Viruses are often transmitted by vectors (such as insects) depending upon hosts cells for replication (Poudel *et al.*, 2021). Bacteria and fungi are prokaryotic and eukaryotic cell organisms, respectively, where bacteria can infect the leaves through open wounds or other natural openings while fungi can penetrate the cell wall of plants and infect them (Nazarov *et al.*, 2020).

For automatic plant disease diagnosis, procedures can be typically divided into two types, direct and indirect (proxy) (Moghadam *et al.*, 2017). Molecular or serological methods are categorized as direct methods and usually take several days for sample processing and analysis. Indirect methods are mostly based on other physiological factors such as morphological or colour changes in plants. Fig. 33.1 provides an overview of multiple trending techniques currently being used for the detection of plant pathogens including non-invasive detection methods. Direct or biological methods are effective and highly sensitive in plant disease detection but are time-consuming (Cheshkova, 2022). These methods

*Corresponding author: Tahira.batool@uwe.ac.uk

DOI: 10.1079/9781800626850.0033

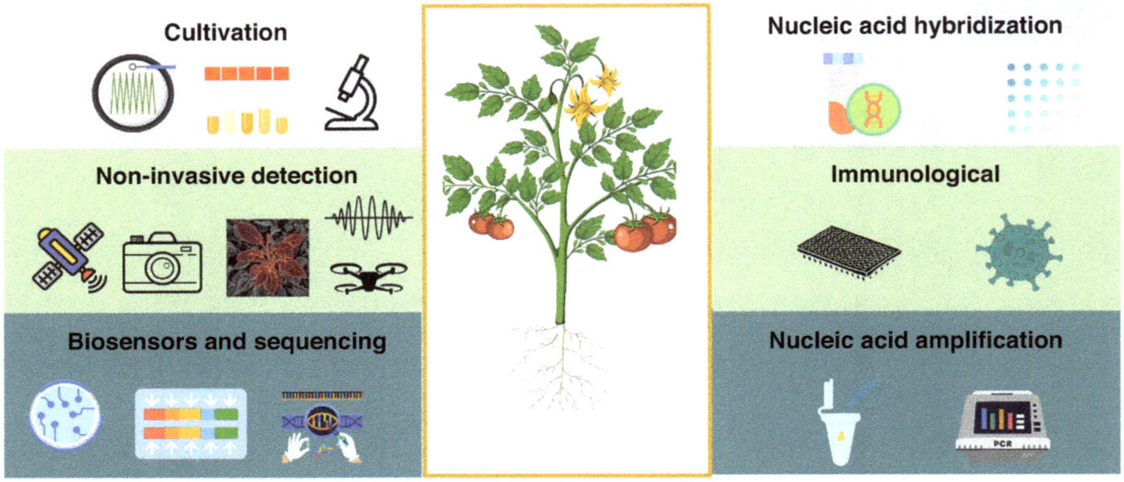

Fig. 33.1. Representation of some common biological and non-biological methods used for plant pathogen detection.

typically possess insubstantial operational robustness (Fang and Ramasamy, 2015) and they rely on meticulous and destructive leaf sampling (Golhani *et al.*, 2018). Skilled personnel with prior professional knowledge of plant pathology are an essential requirement for this procedure (Zhang *et al.*, 2020). Also, molecular methods are hard to implement in real time on large-scale farms (Wang *et al.*, 2019). Moreover, in most cases, these methods are carried out in the middle to latter phases of the infection when the visible symptoms are generally obvious (Golhani *et al.*, 2018), while to stop spreading the disease, it is crucial to adopt preventative measures at an early stage of the pathogen (Dzahini-Obiatey *et al.*, 2010). In the last decade, a large number of non-invasive techniques have been introduced that proved to be increasingly accurate, cost-effective, non-destructive and consistent (Zhang *et al.*, 2020). Applications of these non-invasive methods using sensor technologies, robotics, computer vision and machine learning are increasing rapidly for plant disease detection as these methods tend to produce real-time results with high throughput (Cheshkova, 2022).

Advancements in technology have introduced different non-invasive methods for the detection of plant pathogens, such as optical-imaging cameras, thermography, fluorescence and hyperspectral sensors (Moghadam *et al.*, 2017). It has been proved that biotic stress in plants can affect chlorophyll content and infrared radiation reflection even before the occurrence of visible disease symptoms (Martinelli *et al.*, 2015). Furthermore, progressive spectral techniques have enabled the researchers to measure changes in light emitted or reflected by plants in broad electromagnetic spectrum (Zubler and Yoon, 2020). The spectral analysis can be flexible to implement, for instance by taking high-resolution spectral images, or by collecting enriched spectral information of a single leaf spot. In addition, due to compact size, low cost, light weight and robust automation potential, these sensors are becoming more popular in agriculture to monitor large areas of fields and identify areas more vulnerable to pathogens (Venbrux *et al.*, 2023). These advanced applications of spectral analysis also include remote sensing where high-resolution satellites collect hyperspectral images for pathogen detection. For instance, in a recent study, satellite imaging proved useful for the detection of a sudden-death syndrome derived *from Fusarium virguliforme* in soybean, where the disease can actually be predicted with an accuracy of more than 75% via hyperspectral satellite images before the occurrence of visual symptoms (Raza *et al.*, 2020).

Electromagnetic Spectrum

An electromagnetic spectrum is the representation of all electromagnetic radiations with respect to their wavelengths where gamma rays possess the shortest wavelengths (Gooch, 2007). Fig. 33.2 represents a number of key components of the electromagnetic spectrum with their wavelengths.

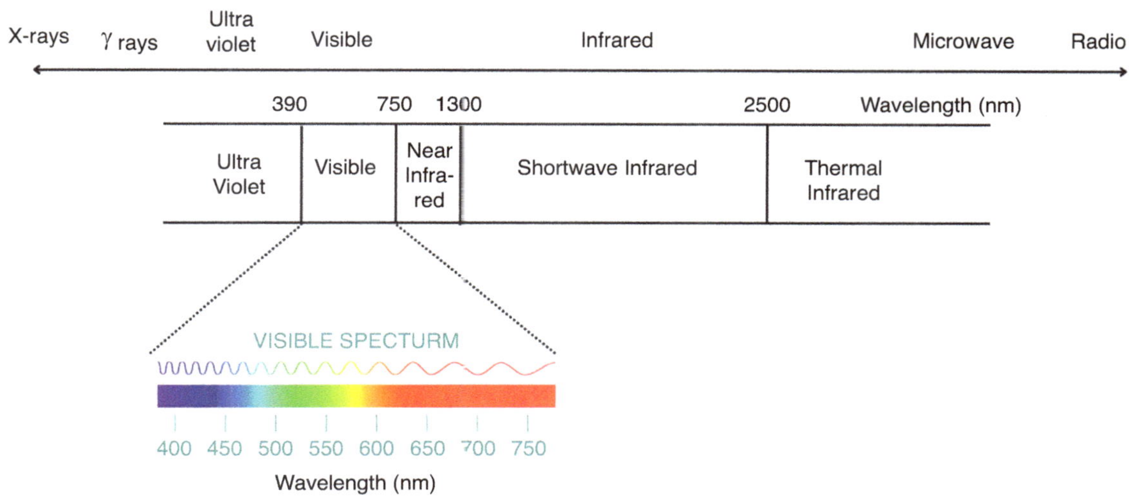

Fig. 33.2. An electromagnetic spectrum in context of reflectance, namely the visible region (390–750 nm), the near infrared (NIR, 750–1300 nm) and the shortwave infrared (SWIR, 1300–2500 nm).

Plants react differently in every component of the electromagnetic spectrum depending upon leaf anatomy and its biochemical structure (Moghadam *et al.*, 2017). Thus, various parameters, such as water absorption, biotic or abiotic stress, pathogens, photosynthesis and nitrogen content can quickly be evaluated from a leaf's hyperspectral data (Kyaw *et al.*, 2022). For instance, Zhang *et al.* (2020) conducted research to measure water loss in trees using leaves' hyperspectral data and observed that the reflectance increased in near Infrared (NIR) and shortwave infrared (SWIR) regions when water loss was increased over time. According to their study, 500, 650, 1450 and 1750 nm wavelengths were prominent in showing the water stress. Moreover, the reason for low reflectance in the visible range is chlorophyll absorption (Kyaw *et al.*, 2022). Fig. 33.3 represents the general layout of hyperspectral signatures collected from leaves, where dips in visible region are caused by chlorophyll absorption and dips in short infrared range are due to water absorption.

Spectral Data and Artificial Intelligence Approaches

The procedure of determining absorbed and reflected light from a surface in wide electromagnetic spectrum is termed as spectroscopy, carried out by specific sensors, such as spectrophotometers (Toivonen *et al.*, 2021). Therefore, spectral images contain the information of each pixel with respect to the acquired spectral bands identified by their wavelengths (Schneider and Feussner, 2017). Hyperspectral spectroscopy is the most popular technology among non-invasive techniques. The analysis of the acquired reflectance spectrum of plants has the potential to detect and identify pathogens; differentiate healthy and infected plants; evaluate the severity of disease, even at an early stage; and identify the symptoms of stressed plants, even if they are not visible to the human eye (Cheshkova, 2022). Several studies have adopted hyperspectral imaging to discriminate between healthy and infected plants. For instance, Moghadam *et al.* (2017) highlighted the importance of full-range visible to near infrared (VNIR; 400–1000 nm) hyperspectral imaging to differentiate between healthy capsicum plants and those infected with tomato spotted wilt virus (TSWV). A variety of features such as vegetation indices and probabilistic topic models were extracted to train a machine-learning model. A support vector machine (SVM) with a radial basis function (RBF) kernel yielded a classification accuracy of more than 90%. Given the difficulty of detecting any specific pathogen that is not visible to the human eye, Grisham *et al.* (2010) used hyperspectral remote sensing with a spectral range of 340–800 nm to detect sugarcane yellow leaf virus (SCYLV) in asymptomatic leaves

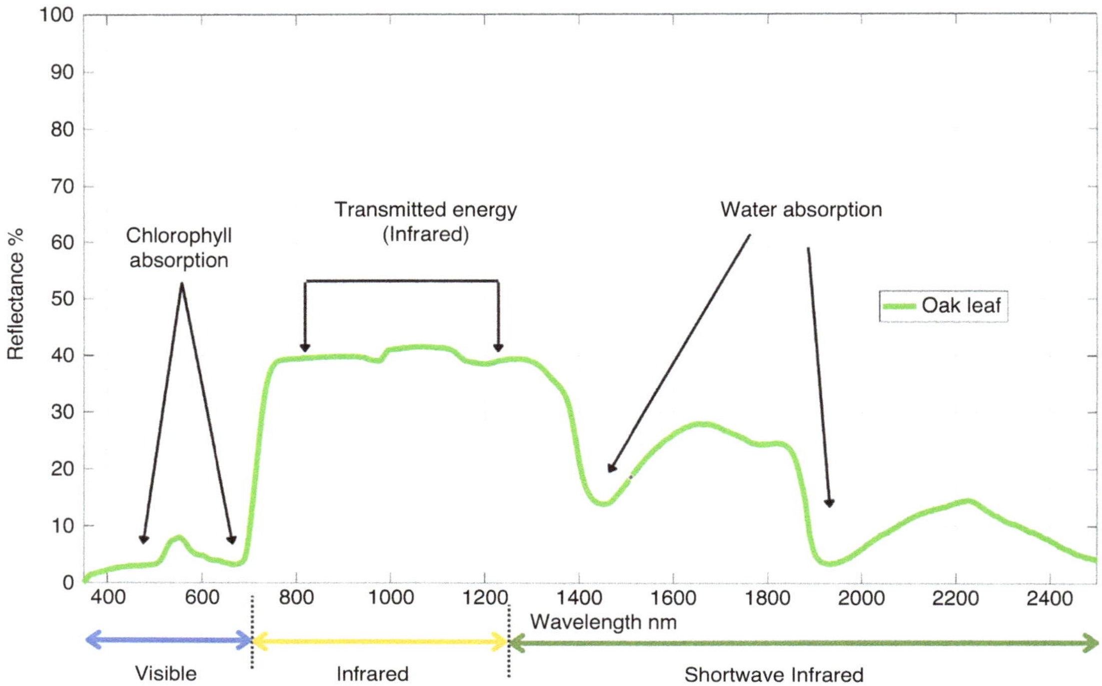

Fig. 33.3. Light gets absorbed at ~420 and ~650 nm (blue and red light) due to chlorophyll. Healthy leaves transmit the energy in infrared region (~750 nm to ~1300) and water content in plants absorbs light in SWIR region.

of two cultivars of sugarcane (LCP 85–384 and Ho 95–988). Linear discriminant analysis classified the infected and non-infected plants with an accuracy of 62%.

Hyperspectral imaging has proved to be the most efficient real-time solution compared to other non-destructive imaging techniques. A comparative analysis was presented by Mahlein *et al.* (2019) to characterize Fusarium head blight (FHB) of wheat on spikelet scale where wheat plants were analyzed with three distinct imaging approaches: infrared thermography, chlorophyll fluorescence imaging, and hyperspectral imaging at incubation periods of 3, 5 and 30 days. Hyperspectral images were collected using two hyperspectral cameras: VNIR (400–1000 nm) and SWIR (1000–2500 nm). Support vector machines classified all three forms of data samples where hyperspectral data produced an accuracy of 78% for FHB detection. When hyperspectral imaging was combined with infrared thermography and chlorophyll fluorescence imaging, the accuracy of 89% was obtained after the 30 days incubation period.

Unlike conventional biological methods, hyperspectral imaging systems can be deployed on large-scale farms and orchards with real-time results with the help of UAV. Abdulridha *et al.* (2019) proposed a study to detect citrus canker on asymptomatic, early and late symptomatic citrus plants (sugar belle leaves) by using hyperspectral imaging (400–1000 nm). The experiment was conducted with a laboratory-based hyperspectral sensor and with the same hyperon mounted on a UAV to detect the virus remotely on trees in an orchard. An artificial neural network and a K-nearest neighbour learning algorithm were used for classification, which achieved an accuracy of 100 and 96%, respectively.

Along with disease detection, hyperspectral imaging has the capacity to classify different cultivars of crops to diagnose symptom variation of disease on every cultivar. Couture *et al.* (2018) used a non-imaging hyperspectral sensor of ultraviolet to shortwave infrared range (UV-SWIR; 300–2500 nm) assembled with a leaf-clip and a standard halogen light source to detect potato virus Y, to diagnose effects of disease on asymptomatic plants and to classify different variants of the virus. Furthermore, the spectral bands of multiple potato cultivars were also analyzed to identify symptom variations with host cultivars. Partial least-squares discriminate analysis (PLS-DA) was used for the classification of hyperspectral data and accurate results were generated. For accuracy, they calculated Cohen's kappa from the resulting confusion matrix and yielded a mean validation kappa of 0.73.

Disease severity can also be assessed with help of hyperspectral imaging and Dhau *et al.* (2018) presented research to detect the severity of maize streak virus (MSV) infestation in maize crops by using hyperspectral imaging with spectral range of 350–2500 nm. The five most unique bands were selected by the guided regularized random forest (GRRF) algorithm and were given as input to a random forest classifier to distinguish healthy maize leaves and those which are at early, moderate or severe stages of MSV infection. Overall, 96% accuracy was produced by this discrimination model.

If more than one plant disease outbreak occurs in a crop, hyperspectral imaging has the capability to identify different pathogens on the same crop, for instance Zhao *et al.* (2012) presented a notable study to differentiate between powdery mildew and yellow rust on a winter wheat crop. As both pathogens usually occur simultaneously in the field, it was necessary to discriminate between them to decide control procedures using fungicide. A non-invasive and real-time solution was implemented in terms of hyperspectral remote sensing with spectral range of 350 nm–1050 nm. Spectral bands were compared in three pairs (healthy vs powdery mildew, healthy vs yellow rust, powdery mildew vs yellow rust) through an independent t test which revealed those bands (531, 570–654, 685–717 nm) that were significant for the detection of powdery mildew and yellow rust. Fishers Linear Discriminant Analysis (FDA) was used for further classification of yellow rust and powdery mildew and achieved an overall accuracy of 93%.

Plants interreact differently in every component of the electromagnetic spectrum depending upon leaf anatomy and its biochemical structure (Moghadam *et al.*, 2017). Spectral signatures of rice (*Oryza sativa* L.) panicles were analyzed by Liu *et al.* (2010) to classify four stages (healthy, light, moderate and serious) of a fungal infection caused by rice false smut disease. A portable spectrophotometer of 350 nm–2500 nm spectral range was used to extract reflectance of rice panicles. Observations showed that the healthy panicles had lower spectral responses in the visible region (400–700 nm) as compared to infected panicles. Healthy and lightly infected panicles showed similar reflectance characteristics throughout the spectrum. Moderately infected panicles showed high reflectance in the visible and SWIR (1135–2400 nm), while lower reflectance was observed in the NIR (709–1134 nm) region. However, rice panicles with severe fungal infection showed low reflectance in the 400–1297 nm range while high reflectance peaks were observed in the SWIR region. Principal components were extracted using principal component analysis and were given as input to a learning vector quantization (LVQ) neural network to classify the aforementioned four stages of fungal infection in rice panicles, which achieved an overall accuracy of 95%. Studies have shown that hyperspectral imaging using computer vision and machine learning has been superior to other real-time and non-destructive solutions (e.g. red, green, blue (RGB) imaging, multi-spectral imaging, thermal imaging, fluorescence imaging etc.) for plant disease detection, characterization and classification (Moghadam *et al.*, 2017; Zhang *et al.*, 2020).

Recently the hybridization of hyperspectral imaging and artificial neural networks has shown promising potential for plant disease diagnosis. An artificial neural network is a collection of interconnected nodes, creating a network resembling the human brain neurons. Each connection carries a weight corresponding to the input values of nodes with a given set of parameters allowing the neural networks for learning, problem solving and generating relationships within the network (Golhani *et al.*, 2018). A Convolutional Neural Network (CNN) is a deep neural network consisting of multiple layers, in particular convolutional layers are also known as building blocks of convolutional neural network architectures (Alzubaidi *et al.*, 2021) that are able to learn various complex spatial patterns. Along with classification, CNN autonomously detects the hidden patterns extracting the complex distinctive features from data (Paoletti *et al.*, 2018). For instance, Jung *et al.* (2022) proposed a study to identify gray mold disease using the hyperspectral image data of asymptomatic strawberry leaves where a three-dimensional CNN was deployed for gray mold disease classification yielding accuracy of 84% on selected wavelengths.

The architecture of a neural network can be modified according to the nature of problem. Moshou *et al.* (2004) investigated the identification of yellow rust in wheat by designing a multi-layer perceptron (MLP) to accept a number of hyperspectral bands as input. The MLP then combines the hyperspectral information in the hidden layer before outputting a binary classification result to indicate crop health. The proposed architecture generated classification accuracy of 99%.

Conclusions

Throughout this chapter, it has been emphasized how plant diseases (fungal, bacterial or viral) can severely affect crop production and cause major economic loss. By reviewing some of the screening methods (biological and non-biological), it is crucial to highlight such techniques that can provide rapid, accurate and on-field pathogen detection at an early stage of disease to take preventive measures on time. Hyperspectral spectroscopy hybridized with machine-learning algorithms can be the new frontier for non-invasive detection of asymptomatic pathogens with fast, on-field and highly efficient means of distinguishing diseased and healthy plants. Hyperspectral data are being used for several applications in agriculture including measuring water loss in crops; identifying asymptomatic plant pathogens along with their severity; distinguish crop cultivars; and monitoring large fields simultaneously.

References

Abdulridha, J.N., Batuman, O. and Ampatzidis, Y. (2019) UAV-based remote sensing technique to detect citrus canker disease utilizing hyperspectral imaging and machine learning. *Remote Sensing* 11, 1373 [online].

Alzubaidi, L., Zhang, J., Humaidi, A.J., Al-Dujaili, A., Duan, Y. *et al.* (2021) Review of deep learning: Concepts, CNN architectures, challenges, applications, future directions. *Journal of Big Data* 8(1), 53.

Cheshkova, A.F. (2022) A review of hyperspectral image analysis techniques for plant disease detection and identification. *Vavilov Journal of Genetics and Breeding* 26(2), 202–213.

Couture, J.J., Singh, A., Charkowski, A.O., Groves, R.L., Gray, S.M. *et al.* (2018) Integrating spectroscopy with potato disease management. *Plant Disease* 102(11), 2233–2240.

Dhau, I., Adam, E., Mutanga, O. and Ayisi, K.K. (2018) Detecting the severity of maize streak virus infestations in maize crop using *in situ* hyperspectral data. *Transactions of the Royal Society of South Africa* 73(1), 8–15.

Dzahini-Obiatey, H., Domfeh, O. and Amoah, F.M. (2010) Over seventy years of a viral disease of cocoa in Ghana: from researchers' perspective. *African Journal of Agricultural Research* 5(7), 476–485.

Fang, Y. and Ramasamy, R.P. (2015) Current and prospective methods for plant disease detection. *Biosensors* 5(3), 537–561.

Golhani, K., Balasundram, S.K., Vadamalai, G. and Pradhan, B. (2018) A review of neural networks in plant disease detection using hyperspectral data. *Information Processing in Agriculture* 5(3), 2214–3173.

Gooch, J.W. (2007) Electromagnetic spectrum. In: Gooch, J.W. (ed.) *Encyclopedic Dictionary of Polymers*. Springer, New York.

Grisham, M.P., Johnson, R.M. and Zimba, P.V. (2010) Detecting sugarcane yellow leaf virus infection in asymptomatic leaves with hyperspectral remote sensing and associated leaf pigment changes. *Journal of Virological Methods* 167(2), 140–145.

Jung, D.H., Kim, J.D., Kim, H.Y., Lee, T.S., Kim, H.S. *et al.* (2022) A hyperspectral data 3D convolutional neural network classification model for diagnosis of gray mold disease in strawberry leaves. *Frontiers in Plant Science* 13, 837020.

Kyaw, T.Y., Siegert, C.M., Dash, P., Poudel, K.P., Pitts, J.J. *et al.* (2022) Using hyperspectral leaf reflectance to estimate photosynthetic capacity and nitrogen content across eastern cottonwood and hybrid poplar taxa. *PLoS ONE* 17(3), e0264780.

Liu, Z.Y., Wu, H.F. and Huang, J.F. (2010) Application of neural networks to discriminate fungal infection levels in rice panicles using hyperspectral reflectance and principal components analysis. *Computers and Electronics in Agriculture* 72(2), 99–106.

Mahlein, A.-K., Alisaac, E., Al Masri, A., Behmann, J., Dehne, H.-W. *et al.* (2019) Comparison and combination of thermal, fluorescence, and hyperspectral imaging for monitoring *Fusarium* head blight of wheat on spikelet scale. *Sensors* 19(10), 2281.

Martinelli, F., Scalenghe, R., Davino, S., Panno, S., Scuderi, G. *et al.* (2015) Advanced methods of plant disease detection. *Agronomy for Sustainable Development* 35(1), 1–25.

Moghadam, P., Ward, D., Goan, E., Jayawardena, S., Sikka, P. *et al.* (2017) Plant disease detection using hyperspectral imaging. In: *International Conference on Digital Image Computing*. Sydney, Australia.

Moshou, D., Bravo, C., West, J., Wahlen, S., McCartney, A. *et al.* (2004) Automatic detection of 'yellow rust' in wheat using reflectance measurements and neural networks. *Computers and Electronics in Agriculture* 44(3), 173–188.

Nazarov, P.A., Baleev, D.N., Ivanova, M.I., Sokolova, L.M. and Karakozova, M.V. (2020) Infectious plant diseases: Etiology, current status, problems and prospects in plant protection. *Acta Naturae* 12(3), 46–59.

Paoletti, M.E., Haut, J.M., Plaza, J. and Plaza, A. (2018) A new deep convolutional neural network for fast hyperspectral image classification. [online], *ISPRS Journal of Photogrammetry and Remote Sensing* 145, 120–147.

Poudel, M., Mendes, R., Costa, L.A.S., Bueno, C.G., Meng, Y. *et al.* (2021) The role of plant-associated bacteria, fungi, and viruses in drought stress mitigation. *Frontiers in Microbiology* 12, 743512.

Raza, M.M., Harding, C., Liebman, M. and Leandro, L.F. (2020) Exploring the potential of high-resolution satellite imagery for the detection of soybean sudden death syndrome. *Remote Sensing* 12(7), 1213.

Sarkozi, A. (2009) New standards to curb the global spread of plant pests and diseases. Food and Agriculture organization of the United States. Available at: https://www.fao.org/newsroom/detail/New-standards-to-curb-the-global-spread-of-plant-pests-and-diseases/en (accessed 16 February 2025).

Schneider, A. and Feussner, H. (2017) Diagnostic procedures. In: Schneider, A. and Feussner, H. (eds) *Biomedical Engineering in Gastrointestinal Surgery*. Academic Press, Amsterdam, pp. 87–220.

Srisakrapikoop, U., Pirie, T.J. and Fellowes, M.D.E. (2020) Meta-analysis suggests differing indirect effects of viral, bacterial, and fungal plant pathogens on the natural enemies of insect herbivores. *Insects* 11(11), 765.

Toivonen, M.E., Talvitie, T., Rajani, C. and Klami, A. (2021) Visible light spectrum extraction from diffraction images by deconvolution and the Cepstrum. *Journal of Imaging* 7(9), 166.

Venbrux, M., Crauwels, S. and Rediers, H. (2023) Current and emerging trends in techniques for plant pathogen detection. *Frontiers in Plant Science* 14, 1120968.

Wang, D., Vinson, R., Holmes, M., Seibel, G., Bechar, A. *et al.* (2019) Early detection of tomato spotted wilt virus by hyperspectral imaging and outlier removal auxiliary classifier generative adversarial nets (Or-ac-gan). *Scientific Reports* 9(1), 4377.

Zhang, N., Yang, G., Pan, Y., Yang, X., Chen, L. *et al.* (2020) A review of advanced technologies and development for hyperspectral-based plant disease detection in the past three decades. *Remote Sensing* 12(19), 3188.

Zhao, J., Yuan, L., Zhang, J., Du, S., Huang, W. *et al.* (2012) Discrimination of yellow rust and powdery mildew in wheat at leaf level using spectral signatures. In: *First International Conference on Agro-Geoinformatics*. Shanghai, China, pp. 1–5.

Zubler, A.V. and Yoon, J.Y. (2020) Proximal methods for plant stress detection using optical sensors and machine learning. *Biosensors* 10(12), 193.

34 Biological Control of Insect Pests with Natural Enemies, Parasitoids, Mating Disruption, Attractants and Traps

Alessia Iacovone* and Jacopo Menghi

CBC Bioplanet, Cesena, Italy

Abstract

Organic agriculture, and an extensive adoption of it by growers, accounts for the lack of effective pest control methods. Biological control, nowadays, offers significantly more tools compared to earlier days, and they are more efficient and usable because the industry has been continuously investing in their development. Nevertheless, the use of biological control products requires a more complex approach than the adoption of chemical products, with a full knowledge of them required for their succesful implementation.

Keywords: organic agriculture, biological control, beneficial insects, semiochemicals, biopesticides

Introduction

Despite an apparent link between biological control and organic production, the relationship has often been controversial due to an old, but still rather common, bias – the assumption that an organic farm is a balanced system that almost regulates itself without the need of inputs and interventions. It is true that organic farms are more biodiverse than their conventional peers; it is, however, also true that they are artificial systems with a high degree of intensification. It is therefore sometimes hard to distinguish whether a farm is organic or conventional, because in the last couple of decades, as well as the pioneers, a new category of organic growers has been increasing and now prevailing – those not necessarily inspired by ethical values but mostly by new business opportunities offered by the double-digit growth of the organic market. These farmers are market-driven and have to fulfil the modern commercial standards of the food they produce, without compromise in spite of their organic status. It is also a matter of bottom line, as modern organic growers manage their companies like any other business, and lower yields or any lower value caused by an unprofessional growing system do not find a place in their accounts.

These organic growers are 'demand-oriented', putting the production method second while first adhering to the rules. Without biases or presumptions they are indeed more open to adopting biocontrol techniques, with the inconvenience that the organic tools are often used as a strict replacement for the chemical ones without the necessary change of paradigm when approaching integrated pest management (IPM); e.g. a common mistake is to use organic plant protection products as a curative, like they would do with chemical pesticides in a conventional context. This incorrect belief, possibly a legacy of years of pesticide industry indoctrination, leads to a poor use of organic tools and to an even poorer result in terms of

*Corresponding author: alessia.iacovone@bioplanet.it

DOI: 10.1079/9781800626850.0034

pest control – two limits that need to be worked on. Organic inputs are so different to their chemical ancestors and require a deep level of knowledge of their nature, origin, composition, formulation, application and mode of action to be used efficiently, aspects that were not so needed during the chemical pesticide age. A main role in this process of re-education may be played by the organic input industry itself, which knows its product better than any other. This process towards a better and proper use of organic tools is entrusted with their wider adoption and the sustainable growth of organic farming in the world.

Invertebrate Biological Control Agents (IBCAs)

Here we discuss the groups of IBCAs with their main features:

Entomopathogenic nematodes (EPNs)

These nematodes infect the target insects by entering via their natural openings and then releasing a symbiotic bacterium that they carry. The bacteria rapidly multiply and kill the host, generally within 24–72 hours. EPNs are highly virulent and effective BCAs once their application allows them to reach their hosts.

The most commonly used species belong to the families Heterorhabditidae (*Heterorhabditis bacteriophora*, *H. downesi* and *H. megidis*), to control the larvae of vine weevils and grubs, and Steinernematidae (*Steinernema carpocapsae*, *S. feltiae*, *S. glaseri* and *S. kraussei*), whose use is intended against a wider range of pests at their larval stage like fungus gnats, weevils, thrips, fruit moths, and mole crickets.

EPNs are soil-dwelling organisms and the soil is their natural habitat, but in recent years their use has been extended to also control pests aboveground when the conditions of use remind them of a soil-like habitat (e.g. under the bark of a tree or even directly on leaves when a proper adjuvant is added to protect them from the ultraviolet rays and fast dehydration).

EPNs have a good appeal among growers due to their usability with standard irrigation or spray equipment, and are very effective when adequate knowledge of the pest life cycle, the stage to be controlled and the substrate condition (temperature and permeability) are taken into consideration (Lee and Dreves, 2023).

Mites

Several predators have been evaluated in the Arthropoda subclass Acarina and about ten of them are commonly used in the market as predators of various pests. The most important family is represented by the Phytoseiidae with *Phytsoseiulus persimilis* at the head of the group, still the best predator of the two-spotted spider mite (TSSM) *Tetranychus urticae* after more than 60 years present in the world market. Its extreme specificity on the TSSM, which may seem like a limit, is overcompensated for by its strong demographic impact on the pest, so much so that it can also be used as a curative. The genus *Amblyseius* offers a bunch of widely adopted predatory species: *Amblyseius (Neoseiulus) cucumeris* and *A. degenerans* used against thrips; *A. (Neoseiulus) californicus* and *A. andersoni* as spider mite predators; and *A. swirskii*, *A.* (*Amblydromalus*) *limonicus* and *A.* (*Transeius*) *montdorensis*, which were introduced into the market in the last two decades to tackle whitefly species in horticultural and ornamental crops with a side action against the juvenile stages of thrips.

The *Amblyseius* mites owe their market success to their relatively low cost, which allows high unitary releases per plant, as well as to their visibility once established on the plants, which makes them friends of the growers even when their impact on the pest is not entirely satisfactory and when they need to be supported by other BCAs. More recently, the distribution by blowing of complementary foods such as pollen, *Artemia* spp., or the same mite species used as food in predator breeding (e.g. *Tyrophagus* spp. or *Carpoglyphus lactis*) on plants is suggested and adopted in order to foster their development and to improve their action. All of the above predatory mites take advantage of a high relative humidity at crop level, a factor that allows growers to achieve better control with simple watering or misting/fogging when needed.

Two other species in the Laelapidae family, *Stratiolaelaps scimitus* and *Gaeolaelaps* (*Hypoaspis*) *aculeifer*, are also commonly used. Both are soil-dwelling predators and their targets are mostly larvae of Diptera (especially sciarids) or juvenile stages (pupae) of thrips. A more recently introduced species, *Macrocheles robustulus* of the family Macrochaelidae, preys on the same pests.

Soil predatory mites are relatively polyphagous and are usually adopted with preventive releases in order to build up a population that is able to tackle the coming infestation of their targets.

Insects

Thysanoptera

The use of predators in the thrips group to control phytophagous thrips is minor but noteworthy. Only *Franklinothrips vespiformis* sees some application with its use being limited as an additional biological tool to sensitive ornamental crops, but the group may become richer in the future.

Hemiptera

This group is famous for including major plant pests, but also includes important and widely adopted pest predators which, from a phytophagous origin, have evolved a zoophagous food regimen. They maintain a bond with the plant to lay eggs and to feed when their prey is not available, a factor that determines their adoption as a biocontrol tool limiting it to a given range of vegetable species.

The family Pentatomidae offers a couple of species (*Picromerus bidens* and *Podisus maculiventris*) that find a use as biocontrol tools against Lepidopteran larvae or other pests on horticultural crops.

The Miridae family offers three very important whitefly predator species: *Macrolophus pygmaeus*, *Nesidiocoris tenuis* and *Dicyphus hesperus*. The first is the pillar of whitefly biocontrol in the modern European glasshouse tomato industry; the second replaces the former in warmer Mediterranean areas where it establishes best; and the last is the North American option due to the exotic status of the two previously cited Palearctic species.

The successful use of predatory mirids on tomatoes is based on the timely building of a population in the greenhouse capable of preventing the development of the whiteflies above a damage threshold. To achieve this they should be introduced soon after transplanting, if not in the nursery, and in certain situations supplemental feeding with *Artemia* spp. or *Ephestia*/*Sitotroga* eggs is considered. Growers may also create biodiversity areas to host and preserve the mirid population between crop cycles or even introduce specific host plants preferred for egg laying. These methods are, however, not easily replicable, nor can they replace a more systemic use of these BCAs by professional growers.

In the family Anthocoridae we find equally important predatory species. *Anthocoris nemoralis* has been known and used for decades as BCA of the pear sucker (*Psylla* spp. and *Cacopsylla* spp.); it has been largely demonstrated that even a minimal introduction in early spring can trigger a population build-up able to prevent pest outbreaks that usually occur later in the season. The *Orius* group (*O. laevigatus*, *O. majusculus* and *O. insidiosus*) includes predators that usually live in flowers where pollen is used as a staple food. During their evolution these species have developed a strong predatory attitude towards species of thrips living in flowers, such as *Frankliniella occidentalis* (Western flower thrip). To date, *Orius* species are the pillars of thrip protection in several important crops such as sweet pepper, cucumber, eggplant, strawberry, etc. Once introduced, even before the host presence occurs, they quickly develop a significant population that is able (if not disturbed) to, within two generations, keep the pest under control until the end of cultivation.

Neuroptera

This small group of insects includes many well-known generalist predators, but only a few have been considered for field application: *in primis Chrysoperla* spp. (especially *C. carnea* but also *C. lucasina* and *C. rufilabris*) used in their larval stage as an aphid predator on many vegetable crops.

Lacewing larvae are effective predators, but their distribution must be managed carefully as for any other 'non-flying' BCA. Their use at adult stage would be easier if it was not hampered by their erratic behaviour and fragility, while the use of their eggs, recently introduced, appears easy and cheap but it must be taken into account that only a minimal part of them, once blown on the crop, can reach a potential target once hatched. The Hemerobiidae family has recently joined the BCA group with the introduction of *Micromus angulatus* on the market, an aphid predator released in the adult stage whose advantage is that their adult stage is also predatory, thanks to their mouthparts.

Diptera

In this very large order of insects, only a few families include beneficial species and several of them have been adopted as BCAs.

The family Cecidomyiidae (midges) offers two commercially available species: *Aphidoletes aphidimyza* (aphid predator) and *Feltiella acarisuga* (spider mite predator). Both are used in the pupal stage for ease of handling; consequently, they must find the right conditions to be able to emerge once in the crop. After mating, the adults begin to feed on sugary sources and look for prey colonies to lay their eggs. The larvae are active and voracious predators despite their limited mobility. Once mature, they drop to the ground to pupate. Their reproduction in field conditions is not obvious, so their use, as for lacewings, is intended more as augmentative than inoculative.

Hoverfly (Fam. Syrphidae) adults are well-known pollinators, but many species act as predators when in the larval stage; only a few have been considered for field application due to difficulty in artificial rearing, resulting in a moderate to high cost to growers. This must be considered when comparing them to other biocontrol tools. *Episyrphus balteatus* has been used for many years for aphid control in cold-climate areas, while *Sphaerophoria rueppellii* and *Eupeodes corollae* have more recently been introduced on the market for warmer areas or periods. This new era for hoverflies as BCAs is due to the recurring problems caused by aphids in many crops that other more classic tools such as parasitoids or other predators are unable to fully control.

Coleoptera

This order, which represents the largest group of living species, includes several famous beneficial insects like ladybirds – the evocative lucky icons. The family Staphylinidae (rove beetles) includes predators, but only a few are used in the market: *Aleochara bilineata* and, more commonly used, *Dalota* (*Atheta*) *coriaria*. They are released into the growing substrate to balance the proliferation of various pests like small Lepidopteran larvae, fungus gnats, *Delia* spp., shore flies (Ephydridae) and thrips (less).

In the well-known family Coccinellidae we find several species, a few of them widely used in the world in important biocontrol strategies.

Among the aphid predators we mark *Adalia bipunctata*, *Propylea quatuordecimpunctata* and *Hippodamia variegata*. The last-mentioned, when coming from natural populations collected in the wild, should not be used despite its low cost.

Ladybirds, preying upon mealybugs, are represented by the notorious *Cryptolaemus montrouzieri*, widely used on citrus, grape and other important fruits or ornamental crops. Their smaller fellows of the genus *Nephus* (*N. conjunctus* and *N. bipunctatus*) are used as a complementary tool and/or in niche situations where the larger *Cryptolaemus* performs poorly.

Chilocorus spp. and *Rhyzobius lophanthae* have been in the market for many years as predators of scales in the Diaspididae group. Their use is more common on ornamental plants or in small gardens than on large citrus orchards, if not at inoculative level. The larger scales in the Coccidae family (e.g. *Pulvinaria* spp., *Saissetia* spp. and *Ceroplastes* spp.) are better prayed on by *Exochomus quadripustulatus* whose use is growing in Europe.

The small *Delphastus catalinae* is the unique ladybird species used for biocontrol of whiteflies in its natural area, North America, but also in Europe where it has now acclimatized. Its use is mostly limited to ornamental crops in glasshouses, but more recently also outdoors on citrus and other plant species affected by the newly introduced whitefly pest *Aleurocanthus spiniferus*. *Stethorus punctillum* is a rarely available but valuable tool for spider mite control on fruit crops.

The last mention in the ladybird family goes to *Rodolia cardinalis*, the pioneer of biocontrol still used worldwide against *Icerya purchasi*. Due to its natural expansion in various continents, following the classical biocontrol programmes adopted since 1888, its use is now limited to niche situations only at the level of gardens or for reintroduction programmes where its populations have been reduced by human activities.

Hymenoptera

This group includes the most evolved BCAs with sophisticated mechanisms of host detection and reproduction. It is not by chance that it is the most exploited order for biocontrol activities from the outset, first for classical then for augmentative strategies.

To date, the species used as BCAs belong to several families of this group.

The Braconidae family includes various aphid parasitoids such as the common *Aphidius colemani*, *A. ervi*, *A. matricariae* and less-important *Ephedrus cerasicola* and *Praon volucre*. They have slight differences in host specificity, mostly due to host size and habitat of use. They do not practise host feeding, but their feasible cost allows them to be used for many cash-crop strategies, which can be augmentative with weekly introductions. Furthermore, once established, they build a local population that may even eradicate aphid populations.

The 'mummy' stage, which makes the parasitized aphids more easily detectable, is another advantage of these species. Among the Braconidae, we also find three other species: *Cotesia marginiventris* and *C. flavipes*, which are adopted on a large scale in the southern hemisphere as parasitoids of various Noctuidae species on open field crops such as maize, sorghum, sugarcane etc.; and *Habrobracon hebetor*, another larval parasitoid of Lepidopteran more famous as BCA of stored products but also applied on arable crops such as sunflower and corn in eastern European countries. This family also includes *Dacnusa sibirica*, an old but still minor larval parasitoid of leafminer flies (Diptera: Agromyzidae).

The Eupelmidae family offers some *Anastatus* species that are commercially available egg parasitoids used for wide-area programmes related to fruit crops: *A. bifasciatus*, against the invasive *Halyomorpha halys* in Europe, and *Anastatus* spp., against *Amblypelta* spp. in Australia.

The Scelionidae *Trissolcus basalis* is the world-renowned egg parasitoid of the green stink bug *Nezara viridula*, a common pest of many arable crops. To date, it has played a growing role due to gradual diffusion of its host in European greenhouses.

The Platygastridae *Telenomus podisi* also acts on stink bug eggs. This parasitoid is mass-reared in South America for commercial applications on soybean.

The Encyrtidae family includes various parasitoid species used for many decades to control mealybugs, such as *Leptomastix dactylopii*. More recently, *Leptomastidea abnormis*, *Anagyrus vladimiri* and *A. fusciventris* have seen a large diffusion in various continents. In the same family a minor role player is *Coccidoxenoides perminutus*, another mealybug parasitoid.

Diglyphus isaea is the unique species representing the Eulophidae family with a commercial relevance. It is a very powerful natural enemy of leafminers in the genus *Liriomyza* (Diptera: Agromyzidae).

Another important family of BCAs is the Aphelinidae family (Hymenoptera: Chalcidoidea). Here we find the notorious *Encarsia formosa*, a classic whitefly parasitoid used against *Trialeurodes vaporariorum* in greenhouses; and in the last decade it has also been flanked with *Eretmocerus eremicus*, a more versatile species in terms of host specificity and climatic adaptability.

Two other species of Aphelinidae are noteworthy: *Aphytis melinus*, a very common parasitoid adopted in all citrus areas to control the California red scale *Aonidiella aurantii*, and *Aphelinus abdominalis*, generally applied as a complementary tool for aphid control strategies thanks to its features that differentiate it from the more common aphid parasitoids in the braconid group.

A single species of the Diapriidae family is currently on the market: *Trichopria drosophilae*, a generic Drosophilidae pupal parasitoid whose introduction is due to the recent fast diffusion of *Drosophila suzukii*, the spotted wing fly, in Europe.

The last family to mention is the Trichogrammatidae, which includes tiny wasps that parasitize eggs of various insects, but above all Lepidopteran. This group includes several species with a light host specificity whose use is often connected with local regulation and rearing traditions. The list is long but the most common species in the world market are: *Trichogramma brassicae*, *T. achaeae*, *T. dendrolimi*, *T. evanescens*, *T. minutum*, *T. pintoi*, *T. galloi* and *T. pretiosum*.

Finally, the list of species reported here does not claim to be exhaustive:

- species with only local use or niche use may not be included;
- species for use in animal breeding or in stored food are not reported; and
- species, usually exotic, reared by public or academic institutions for classical territorial biocontrol programmes are not mentioned.

Microbial Control Agents

Viruses

Viruses are ultramicroscopic, metabolically inert infectious agents that cannot grow or reproduce outside a living cell. All organisms can be infected by viruses and currently approximately 1100 viruses are known to cause disease in insects (Eberle *et al.*, 2012). These viruses are used in crop protection as insecticides. Insect pathogenic viruses belong to different virus families, but only those belonging to the family Baculoviridae have entered the market so far. These viruses have been isolated exclusively from invertebrates and are therefore considered safe for mammals and other complex life forms. The major advantages of virus applications are: (i) their high species-specificity; (ii) their ease of use (they can be applied with conventional spray equipment); and (iii) their safety for non-target organisms. Viruses have a slow action as the insects may only die days after infection, which is not always desirable for growers. Further, they are susceptible to UV degradation, and production is expensive because viruses only produce new copies of themselves in live host cells.

Among them, few have been approved as phytosanitary tools and made available in several countries. Here are the most relevant:

- *Adoxophyes orana* GV (AoGV) against summer fruit tortrix (*Adoxyphes orana*);
- *Cydia pomonella* GV (CpGV) against codling moth (*Cydia pomonella*);
- *Helicoverpa armigera* NPV (HezeNPV) against cotton bollworm (*Helicoverpa armigera*);
- *Spodoptera exigua* NPV (SpexNPV) against beet armyworm (*Spodoptera exigua*);
- *Spodoptera littoralis* NPV (SpliNPV) against cotton leafworm (*Spodoptera littoralis*);
- *Phthorimaea operculella* granulovirus (PhopGV) against South American tomato pinworm (*Tuta absoluta*).

Fungi

Fungi have been extensively studied for their effectiveness against insect pests, and over 800 species of entomopathogenic fungi have been identified worldwide. These fungi primarily act through contact with the insect cuticle, penetrating it to infect and kill the host.

Many commercial strains are available on the market, the most important belonging to the following species:

- *Akanthomyces muscarius* (*Lecanicillium muscarium*) against whiteflies, thrips and Tetranychidae mites;
- *Beauveria bassiana* against various Rhynchota, Thysanoptera, Coleoptera, Diptera and Tetranychidae mites;
- *Cordyceps* (*Isaria*) *fumosorosea* against various Rhynchota, Coleoptera and Tetranychidae mites;
- *Metarhizium anisopliae* against Coleoptera, Lepidopteran, mealybugs and thrips;
- *Paecilomyces fumosoroseus* against whiteflies and aphids.

Bacteria

Bacteria are small ubiquitous organisms that have also been exploited for biocontrol of plant pests. Records of at least 100 different species of bacteria being pathogenic to arthropods can be found in literature (Thacker, 2002), with the pioneer *Bacillus thuringiensis* (Berliner) known since the beginning of the last century as a potential BCA and, to date, the most used microbiological insecticide. The shortlist of species commercially available is:

- *Bacillus thuringiensis* subsp. *aizawai* against Lepidopteran larvae;
- *Bacillus thuringiensis* subsp. *kurstaki* against Lepidopteran larvae; and
- *Bacillus thuringiensis* subsp. *tenebrionis* against larvae of Coleoptera Chrysomelidae.

Semiochemicals

Insects rely heavily on odours as a means of communication and navigation in their environment. Chemical compounds play a crucial role in insect communication and the wide array of odour compounds produced by insects has evolved over time. These chemical messages are called semiochemicals and are further classified into allelochemicals and pheromones, depending on whether the interaction occurs between different species or within the same species, respectively.

Among the various semiochemicals, sexual pheromones are the most well-known compounds; typically emitted by female insects, they serve to facilitate encounters between males and females.

Semiochemicals are exploited as biocontrol tools to disrupt the encounter between sexes or to attract pests towards devices or baits that kill them.

Mating disruption

Mating disruption is an effective method for pest control that can be achieved in two main ways: male confusion and false trail following. Male confusion involves the use of ambient pheromone concentrations that are high enough to mask the trail of female insects. This is accomplished by deploying point-source dispensers or microcapsules emitting high concentrations of species-specific pheromones. More recently, automatic spray devices have been developed, replacing the hand application of the dispensers to facilitate the adoption of this technique by growers.

False trail following involves placing a large number of point sources of pheromone per unit area, far exceeding the anticipated number of females present in the crop (typically 2000–3000 dispensers per ha). The purpose is to significantly reduce the chances of males finding females at the end of the pheromone trail. The emission of pheromones from each dispenser, in this case, is relatively low compared to male-confusion dispensers.

To date, mating disruption systems are available against the following pests:

- *Adoxophyes orana*, *Archips* spp., *Pandemis* spp. (tortricid moths) on pome fruits
- *Anarsia lineatella* (peach twig borer) on stone fruits
- *Chilo suppressalis* (Asiatic rice borer) on rice
- *Cydia pomonella* (codling moth) on pome fruits
- *Eupocelia ambiguella* (European grape berry moth) on grapevine
- *Grapholita molesta* (oriental fruit moth) on stone and pome fruits
- *Keiferia lycopersicella* (tomato pinworm) on tomato
- *Lobesia botrana* (European grapevine moth) on grapevine
- *Ostrinia nubilalis* (European corn borer) on pepper
- *Synanthedon* spp. (clearwing) on currant
- *Tuta absoluta* (South American tomato pinworm) on tomato in protected crop
- *Zeuzera pyrina* (leopard moth) on fruit crops and olives

Attract and kill

Attract-and-kill devices (combination of an attractant with a contact insecticide; Welter *et al.*, 2005) have been commercially developed for the control of the Mediterranean fruit fly *Ceratitis capitata* and the olive fruit fly *Bactrocera oleae*. The outer surface of the attract-and-kill device is coated with an insecticide. A specific attractant is used to attract the flies to the insecticide-treated surface. According to local regulations, synthetic pesticides like deltamethrin or lambda-cyhalothrin, are allowed to be used in this way in organic productions.

Conclusion

The path towards sustainable farming via organic agriculture is both promising and challenging, particularly concerning pest control. Among these challenges, biological methods stand out as an indispensable tool. Continuous investments and innovations in industry have provided a diverse array of tools that have become increasingly effective and readily applicable. However, the successful integration of biological control products into farm management requires a nuanced approach compared to the more straightforward application of chemical solutions.

As we strive towards a future with more sustainable agricultural practices, embracing this complexity is essential to unlock the full potential of organic farming. This necessitates a complete shift in approach, moving away from obsolete concepts such as eradication or curative combat, and acknowledging their limited applicability. The focus must shift towards scientific training and continuous presence in the field. Managing pests can no longer rely only on product labels; instead it requires solid scientific knowledge and active engagement.

Regarding the products, companies with knowhow and technologies, through exploitation of economies of scale, will have the opportunity to ensure accessibility for all users.

In conclusion, we must establish a multilayered system, comprehending the selection of suitable active ingredients, understanding the historical context of every farm, biological cycles and appropriate timings to ensure a safe and sustainable future for food production.

References

Eberle, K.E., Jehle, J.A. and Huber, J. (2012) Microbial control of crop pests using insect virus. In: Abral, D.P. and Shankor, V. (eds) *Integrated Pest Management: Principles and Practice*. CAB International, Wallingford, UK, pp. 281–297. DOI: 10.1079/9781845938086.0281.

Lee, J. and Dreves, A.J. (2023) Entomopathogenic nematodes. In: *Pest Management Handbooks*. Pacific Northwest. Available at: https://pnwhandbooks.org/node/263 (accessed 17 February 2025).

Thacker, J.R.M. (2002) *An Introduction to Arthropods Pest Control*. Cambridge University Press, Cambridge, UK.

Welter, S.C., Pickel, C., Millar, J., Cave, F., Steenwyk, R.A. *et al*. (2005) Pheromone mating disruption offers selective management options for key pests. *California Agriculture* 59(1), 16–22. DOI: 10.3733/ca.v059n01p16.

35 Key Challenges and Solutions for Sustainable Pest Management in Organic Farming

PHILIPPE R. BELLIARD*

Zoology Department, University of Innsbruck, Austria

Abstract

Concerns about the sustainability of modern agriculture and the negative impacts of chemical pesticides on the environment and human health highlight the need for agroecological pest management. Integrated pest management (IPM) strategies offer a systems approach to regulate pests sustainably based on ecosystem services, farm management practices and biological control. As such, IPM is a central pillar of organic agriculture.

This chapter reviews some of the key challenges and solutions for IPM in organic agriculture, focusing on arthropod pests in crop production. Whereas conservation biological control constitutes the basis of natural pest regulation in agricultural landscapes, cultural practices on farms comprise key actionable methods that farmers can implement within cropping systems. In addition, biological inputs provide crucial correctional levers to ensure pest control with targeted interventions. While each phase of IPM has its distinct challenges and solutions, a common thread is the need to tailor control methods to specific agricultural contexts with knowledge-based approaches due to their inherent variability, specificity and complexity. Support from research, education, and policymaking is essential for promoting agroecological pest management to ensure sustainable food production.

Keywords: integrated pest management, biological control, agroecology

Introduction

Agricultural pests (weeds, arthropods, nematodes and microbial pathogens) cause significant yield losses and must be adequately managed (Oerke, 2006). However, excessive pesticide use is associated with detrimental impacts on the environment, biodiversity, human health and ecosystem services such as pollination and natural pest control (Desneux *et al.*, 2007; Geiger *et al.*, 2010; Beketov *et al.*, 2013; Brzozowski and Mazourek, 2018). Additionally, pest resurgence subsequent to pesticide application can be exacerbated due to the concurrent suppression of natural enemies, rendering agro-ecosystems more susceptible to pest infestations (Dutcher, 2007). The pressing need for alternative pest control strategies is further highlighted by pests' resistance to pesticides and genetically modified crops, the expansion of pesticide bans and pesticide reduction policies, growing retail and consumer demands, and pressure from NGOs (Buurma *et al.*, 2010; Birch *et al.*, 2011; Sparks and Nauen, 2015; Tabashnik and Carrière, 2017).

Integrated pest management (IPM) aims at reducing or eliminating pesticide use by prioritizing preventive measures and eco-friendly interventions according to a hierarchical 'pyramid' of practices in which pesticides are only used as a last resort (Naranjo, 2001; Zehnder *et al.*, 2007). In both conventional and organic systems, IPM emphasizes a systems approach that incorporates genetic crop resistance, farm-scale cultural practices, sanitation measures, mechanical intervention,

*Corresponding author: philippe.belliard@uibk.ac.at

DOI: 10.1079/9781800626850.0035

and crop-targeted biological control with natural agents. Despite leading to strong reductions in pesticide use without compromising crop yields, the success of IPM over the past decades has been limited by inconsistent levels of implementation and various adoption barriers (Deguine *et al.*, 2021). In addition, the low cost of synthetic pesticides, which does not account for the costs of socio-environmental externalities, constitutes a disincentive to phasing out their utilization (Pimentel, 2005; Bourguet and Guillemaud, 2016).

By contrast, the prohibition of synthetic pesticides in organic agriculture has led to more widespread adoption of IPM and biological control tactics (IFOAM, 2014). Biological control can be defined as the use of living organisms to suppress specific pest organisms, and is generally divided into classical biocontrol, conservation biocontrol and augmentative biocontrol (van Lenteren *et al.*, 2018). Classical biocontrol is the intentional introduction of an exotic biological control agent, usually co-evolved with a specific target pest, for permanent establishment and long-term control at regional level. Conservation biocontrol (CBC) aims at supporting natural enemies of pests present in the agro-ecosystem by managing habitat and vital resources in the immediate and broader agricultural environment. Augmentative biocontrol (ABC) is the introduction of commercially mass-reared biological control agents, either via mass release for short-term, high-impact control (inundative ABC), or via the release of smaller populations that establish and develop in a production system over several generations (inoculative ABC). Furthermore, organic IPM includes the use of so-called biorational chemical agents (e.g. biopesticides, semiochemicals) approved for organic agriculture (Isman, 2006; Saha *et al.*, 2017).

Although decades of research and technological developments have yielded significant advances in IPM solutions, major challenges persist. This chapter presents some key challenges and solutions for sustainable pest management in organic crop production, with a particular focus on biological control of arthropod pests. Following the progressive phases of IPM, the topics covered include (agro) ecosystem-based pest regulation, farm management practices, and targeted interventions using biocontrol products.

Conservation Biological Control in Agricultural Ecosystems

Landscape and conservation biocontrol

A substantial degree of pest regulation can be provided by natural enemies that naturally occur in and around agricultural environments (Naranjo *et al.*, 2015; Begg *et al.*, 2017). The delivery of this ecosystem service relies heavily on natural enemy diversity and abundance (Crowder and Jabbour, 2014; Dainese *et al.*, 2019), which is strongly linked to landscape composition, complexity and configuration (Bianchi *et al.*, 2006; Chaplin-Kramer *et al.*, 2011; Veres *et al.*, 2013). Networks of semi-natural habitat (SNH) such as woodlands, meadows and hedgerows support regional or landscape-level species pools of predators and parasitoids, from which certain species disperse and spill over into agricultural fields (Tscharntke *et al.*, 2007; Begg *et al.*, 2017). Hence, more complex landscapes with higher SNH coverage are frequently associated with higher natural enemy diversity, lower pest pressure and enhanced climate resilience of natural biological control (Veres *et al.*, 2013; Feit *et al.*, 2019; Tortosa *et al.*, 2023).

However, most intensive agricultural regions are characterized by highly simplified landscapes with low plant diversity in which SNHs are sparse and fragmented, severely limiting the prospects for CBC (Tscharntke *et al.*, 2007; Grab *et al.*, 2018). Therefore, efforts to promote CBC at landscape level typically aim to increase plant diversity, and the area, heterogeneity and connectivity of SNHs (Begg *et al.*, 2017; Tortosa *et al.*, 2023). The implementation of such conservation measures, however, is severely hampered by competing land uses, contrasting perceptions of the value of SNH, lack of agency by farmers and conservationists, and the complexity of evaluating their potential economic benefits (Tscharntke *et al.*, 2016; Shields *et al.*, 2019). Moreover, the positive effects of landscape complexity and SNHs are inconsistent and often fail to translate into higher pest control in crop fields (Chaplin-Kramer *et al.*, 2011; Karp *et al.*, 2018). Among the explanations for this failure to enhance biological control, Tscharntke *et al.* (2016) identify five hypotheses: (i) effective natural enemies for pests are lacking in the region; (ii) SNHs benefit pests to a greater extent than natural enemies; (iii) crops provide more resources than SNH for natural enemies; (iv) the proximity, composition or configuration of SNHs are inadequate to support enough natural enemies; and (v) agricultural practices counteract enemy recruitment and establishment in crop fields. Thus, despite its importance, landscape-mediated CBC presents major challenges that necessitate complementary on-farm SNH management to facilitate its effectiveness. Furthermore, boosting the potential of CBC at landscape level will require regional co-ordination and co-operation of multiple stakeholders to optimize land-use design for the provisioning of ecosystem services through targeted ecological engineering, aided by support from ecological theory and modelling of pest and natural enemy dynamics (Birch *et al.*, 2011; Naranjo *et al.*, 2015; Shields *et al.*, 2019).

Promoting conservation biological control on farms

Some of the challenges of CBC can be alleviated by managing non-crop vegetation at field level. To a certain extent, local-scale landscape elements in direct proximity to crop fields can be more important – although still dependent on the broader landscape context. Spillover of natural enemies and colonization into the crops is maximized when SNHs are adjacent to or integrated in crop fields, thereby accounting for the limited dispersal capacity of many natural enemies (Sarthou *et al.*, 2014; Tscharntke *et al.*, 2016; Holland *et al.*, 2016). Specialists, in particular, have been found to be more responsive to small-scale landscape complexity (Chaplin-Kramer *et al.*, 2011).

The establishment and maintenance of field-level SNH, including beetle banks, grassy field margins, agroforestry alleys and flower strips, promote CBC by providing natural enemies with food sources such as pollen, nectar and alternative prey or hosts, as well as providing habitat for overwintering, shelter from disturbance and predation, and improved microclimatic conditions (Collins *et al.*, 2002; Rusch *et al.*, 2010; Tschumi *et al.*, 2015; Ganser *et al.*, 2019). To be successful, these SNHs must be sufficiently large and the space between them sufficiently small in order to support large-enough natural enemy populations that can effectively colonize crop fields and exert pest control (Collins *et al.*, 2002; Tscharntke *et al.*, 2016). Although the establishment of SNH features on farms is often financially supported by agri-environmental policies (Kleijn and Sutherland, 2003; Haaland *et al.*, 2011), many farmers are still reluctant to dedicate agricultural land for this purpose, and concerns about pest and weed propagation from bordering SNHs further limits their adoption (Mante and Gerowitt, 2009). Moreover, the management of SNHs to enhance biological control may require technical knowledge with regard to the maintenance and optimal species compositions of flower strips to preferentially favour specific natural enemies while minimizing benefits to pest populations (Lavandero *et al.*, 2006; Tschumi *et al.*, 2015). Nevertheless, in general, increasing local-scale plant diversity in SNHs and within cropping systems significantly increases the abundance and diversity of natural enemies and suppresses pests (Letourneau *et al.*, 2011; Dassou and Tixier, 2016). Promoting plant diversity in and around cropping areas is therefore a primordial component of sustainable pest management that must be integrated in farmland design, as well as in the cultural practices employed in crop production operations to enhance biological control.

Cultural Practices for Sustainable Pest Management

In organic crop production in particular, cultural practices are the cornerstone of sustainable IPM, as they encompass a number of management strategies that farmers can directly implement to regulate pest populations. Two fundamental components of pest, weed and disease management are cropping system diversification and soil health management.

Cropping system diversification

Crop rotations and cover crops

The temporal diversification of crops in cropping sequences impedes the buildup of pests and pathogen populations, especially those that specialize in specific crops, by disrupting their life cycles and breaking the continuity of host-plant availability (Isbell *et al.*, 2017; Jaworski *et al.*, 2023). Crop rotations are particularly important for managing soil-borne pests such as plant-parasitic nematodes that disperse less easily to neighbouring fields (Puissant *et al.*, 2021). Yet, by increasing plant diversity at wider spatial scales, rotations also limit the spread of more mobile pests while favouring biodiversity and natural enemies at landscape level (Redlich *et al.*, 2018; Aguilera *et al.*, 2020; Beillouin *et al.*, 2021).

Cover crops play a pivotal role within crop rotations, further diversifying cropping sequences and providing the opportunity to intercede service crops between main crops that specifically target pests or natural enemies. Natural enemies benefit from the shelter and alternative food sources provided by cover crops, notably when flowering cover crops are incorporated in cover crop mixtures (Beaumelle *et al.*, 2021; Puissant *et al.*, 2021; Beillouin *et al.*, 2021). In addition, certain cover crops have the capacity to suppress pests and pathogens biochemically. Brassicaceous cover crops incorporated as green manure can reduce nematode infestations as well as suppress microbial pathogens, purportedly via the release of isothiocyanates from these glucosinolate-containing crops (Matthiessen and Kirkegaard, 2006). Marigolds (*Tagetes* spp.) produce nematicidal compounds in living roots that can be effective against several plant-parasitic nematode species (Wang *et al.*, 2007).

Crop rotations and cover crops are widely used in organic agriculture because of a broad range of benefits beyond pest control, such as soil quality conservation, carbon sequestration, nutrient loss prevention and income diversification (Letourneau *et al.*, 2011; Isbell *et al.*, 2017; Beillouin *et al.*, 2021). Farmers must therefore design their rotations according to a complex integration of multiple agronomic factors, with trade-offs often resulting in suboptimal choices with regard to pest management. Moreover, a certain crop may aid in the enhancement of a natural enemy or suppression of one pest, while concomitantly favouring the development of another pest (Poveda *et al.*, 2008; Molendijk and van Beers, 2013; Tscharntke *et al.*, 2016). Another critical bottleneck in rotation diversification is the need for specialized crop-specific machinery. The best cropping composition and sequence is ultimately contingent on local conditions of pest pressure and farming context, and necessitates specific knowledge of pest–host ranges and plant–insect interactions that farmers often lack. Advisers and decision-making tools such as rotation design software and modelling approaches can therefore be crucial to assist the development of tailored cropping systems that are effective in managing pests sustainably within multi-criteria assessment schemes (Dury *et al.*, 2012; Molendijk and van Beers, 2013; Schöning *et al.*, 2023).

Intercropping

The simultaneous mixing of crop species in a variety of intercropping configurations can have a significant impact on pest regulation (Letourneau *et al.*, 2009; Beillouin *et al.*, 2021). Spatial arrangements of intercropping range from mixed intercropping to strip cropping, combining multiple main crops or a main crop with a service crop. In addition, intercropping can take the form of agroforestry, combining tree crops with understory crops or alternating crop alleys with tree rows. Intercropping reduces the spread and growth of pest populations by diluting host plants and disrupting visual and olfactory cues that pests rely on to locate their host plants (Poveda *et al.*, 2008; Huss *et al.*, 2022). Furthermore, natural enemy diversity and abundance are boosted by the habitat complexity and alternative food sources found in intercropping systems, thereby reducing pest pressure (Rakotomalala *et al.*, 2023).

Push-pull systems are a special case of intercropping that make use of pest-attracting trap crops (pull) in combination with pest-repellent crops (push). Semiochemicals that mediate communication between organisms play a critical role in push- and pull-crop repellency or attractiveness (Cook *et al.*, 2007). Trap crops help regulate pests by luring them into vegetation where their fitness is reduced or where they can be targeted with other control measures (Shelton and Badenes-Perez, 2006). Push crops divert pests away from the main crop towards trap crops. However, push-pull systems are generally highly specific to particular crop–pest interactions, and issues of scalability may limit their adoption mainly to smallholder crop production (Khan *et al.*, 2016).

The large-scale adoption of intercropping is hindered by challenges in mechanizing farm operations such as harvesting for crops in close proximity. Moreover, determining compatible crop combinations or push-pull associations that are practicable and successfully reduce pest pressure without compromising yield is a knowledge-intensive process that necessitates trial-and-error as well as site-specific tailoring (Huss *et al.*, 2022). Crop system modelling can play a key role in optimizing intercropping configurations (Chimonyo *et al.*, 2015) and the adoption of intercropping could be stimulated through risk-mitigation schemes and special subsidies that incentivize its uptake, as well as the development of adaptable mechanization solutions.

Soil health and pest suppression

Soil management practices significantly affect pests, pathogens and their natural enemies (Zehnder *et al.*, 2007). Moreover, plant health and thus resistance to pests and pathogens is linked to physical, chemical and especially biological properties of soil health (Altieri *et al.*, 2005).

The soil microbiome

Soil-microbial communities are crucial components of biological soil health and play a pivotal role in pest regulation. Beneficial soil microbes in the rhizosphere can enhance plant resistance to pests by boosting plant vigour, promoting the release of herbivore-induced plant volatiles (HIPVs) that attract natural enemies, and activating biochemical defence pathways that confer systemic resistance against pests (Rasmann *et al.*, 2017). Indeed, several studies have demonstrated significant relationships between soil microbiomes and pest infestations in crops (e.g. Harkes *et al.*, 2020; Pineda *et al.*, 2020; Howard *et al.*, 2020). Furthermore, soil microbiomes can play a critical role in parasitic weed management (Masteling *et al.*, 2019). How to steer the soil microbiome in a way that consistently supports pest suppression, however, remains a complex and challenging question (Pineda *et al.*, 2017). While soil inoculations with specific microbes can enhance crop resistance to pests, positive effects observed in controlled environments often fail to translate to field conditions (Gadhave *et al.*, 2016). Microbially mediated pest suppression is frequently associated with enhanced microbial diversity, biomass and activity resulting from applications of organic amendments (Pineda *et al.*, 2017; Rosskopf *et al.*, 2020; Blundell *et al.*, 2020). Organic amendments can therefore be a crucial leverage point for stimulating microbial communities in a way that helps regulate pests.

Organic amendments

In addition to microbially mediated effects, organic amendments affect pest biocontrol through other mechanisms. The decomposition of various organic residues may release toxic compounds that reduce soil-borne pest infestation (Rosskopf *et al.*, 2020). Moreover, communities of detritivores enhanced by organic amendments serve as alternative prey that supports generalist predators, leading to increased pest biocontrol (Altieri *et al.*, 2005; Birkhofer *et al.*, 2008; Rowen *et al.*, 2019). Mulching, additionally, provides habitat and improved microclimate for these natural enemies, and may disrupt a pest's ability to locate host plants (Zehnder *et al.*, 2007; Alyokhin *et al.*, 2020). However, in some cases, mulching also promotes pest populations.

Another mechanism of pest suppression via organic amendments lies in soil nutrient management. Reduced pest infestations in crops grown with organic compared to synthetic fertilizer are often attributed to lower nitrogen bioavailability, affecting plant nutritional quality in ways that reduce herbivore performance or preference (Altieri and Nicholls, 2003; Zehnder *et al.*, 2007). Large fractions of nitrogen in organic fertilizers are organically bound and become available gradually during decomposition, thereby limiting excessive plant uptake (Rowen *et al.*, 2019). Managing soil fertility is also crucial for maintaining adequate balances of soil macro- and micronutrients to ensure plant health, as plant nutritional status strongly influences plant defence mechanisms and resilience to herbivory (Altieri and Nicholls, 2003).

The use of organic amendments may be limited by competing uses for crop residues and animal manure, variability of the potential benefits, and, in some cases, uneconomically high application rates required to ensure positive outcomes (Quilty and Cattle, 2011; Pannell *et al.*, 2014). The valorization of organic waste streams from various sources in circular food systems has substantial potential to augment the availability and usage of organic amendments. However, risks of pollutants and pathogens in organic waste streams must be properly mitigated (Urra *et al.*, 2019).

Tillage

Intensive soil disturbances such as tillage have major effects on soil biota, with important implications for biological pest control. Although there is considerable species-dependent variability, reduced-tillage systems generally harbour more abundant and diverse natural enemies and fewer arthropod pests compared to conventional tillage (Rowen *et al.*, 2020; Alyokhin *et al.*, 2020). This is usually attributed to better environmental stability, habitat conservation and higher abundances of alternative prey for polyphagous predators (Rusch *et al.*, 2010). On the other hand, weed infestation may be more problematic in conservation tillage systems where mechanical weed control is strongly reduced, particularly with the exclusion of herbicides in organic agriculture (Carr *et al.*, 2013). Yet the supplementary plant resources supplied by weeds that benefit pest predators also promote granivorous arthropods that, in turn, contribute to weed biocontrol (Blubaugh and Kaplan, 2015). In addition, judiciously timed cover crop management helps mitigate weed pressure in reduced tillage fields (Carr *et al.*, 2013).

Biological inputs for sustainable pest control

In many plant production systems, deficiencies in native natural enemy populations or the effectiveness of natural biological control warrant strategic interventions using biological control agents (BCAs) or biochemical substances approved for organic agriculture.

Augmentative biocontrol strategies

Recent decades have seen a strong development of ABC research and technologies. Many successful macrobial (parasitoids, predatory mites and insects) and microbial (entomopathogenic nematodes, fungi, bacteria, viruses) BCA products exist against many pests for a variety of cropping systems (van Lenteren *et al.*, 2018). However, whereas ABC is often effective in greenhouses, in the field it frequently fails to achieve satisfactory pest control (Sivinski, 2013). Factors underlying this disparity include enemy dispersal, pest immigration, and unfavourable environmental conditions, which can be controlled in closed greenhouse systems. Conducive environments are particularly important for microbial BCAs (Lacey, 2017). In addition, intraguild predation (IGP), interference and hyperparasitism in natural enemy communities may limit the suppression of target pests, notably when pest densities are low and BCA niches overlap (Liang *et al.*, 2022). On the other hand, using diverse assemblages of functionally complementary natural enemies can significantly enhance pest control, although optimizing BCA compositions requires crop- and pest-specific research (Alhadidi *et al.*, 2018).

IGP can be reduced through the provision of supplementary food such as floral resources (Liang *et al.*, 2022). Furthermore, supplementary food in the form of pollen and nectar from banker plants, pollen sprays, insect eggs or prey mites can be key in maintaining or boosting BCA populations for more effective biocontrol (Messelink *et al.*, 2014). These food sources, as well as the provision of alternative parasitoid hosts on banker plants, are crucial to ensure the establishment of a 'standing army' of natural enemies via inoculative releases in preventive ABC strategies (Pijnakker *et al.*, 2020). However, the use of supplementary food sources should be approached with caution, since they may undermine biocontrol by benefiting pests or hyperparasitoids (Messelink *et al.*, 2014).

By establishing and conserving inoculative standing armies of BCAs that control developing pest populations early on, the need for large inundative releases is reduced, which may make ABC more cost-effective (Huang *et al.*, 2011). The high cost of BCAs is a major hurdle that limits their use to mostly high-value crops in protected cropping systems (Sivinski, 2013). Cost-effectiveness can be substantially improved by regular pest monitoring, based on which targeted ABC interventions can be made according to intervention thresholds. New technologies such as automated detection tools, AI-based identification techniques, and remote sensing for pest damage detection in crop fields present promising advancements to improve the effectiveness and economic efficiency of ABC (Zhang *et al.*, 2019; Lima *et al.*, 2020). In addition, new techniques in mass-rearing could play a key role in lowering production costs and improving BCA access and availability (Sivinski, 2013).

Biopesticides and semiochemicals

When other control methods are insufficient, biopesticides can be used in organic agriculture as a last option in the IPM pyramid. Biopesticides are derivatives of living organisms (fermentation products, botanical extracts and oils) or natural non-biological substances (e.g. kaolin clay, mineral oil, copper oxide, insecticidal soaps) (Zehnder *et al.*, 2007; Weintraub *et al.*, 2017). Additionally, live microbial BCAs are sometimes considered to be biopesticides. While many biopesticides are efficacious against several pests, their effectiveness is strongly compromised by their low persistence. For instance, pyrethrins and azadirachtin, the active compound in neem oil, are rapidly degraded by sunlight, and plant essential oils volatilize shortly after application (Isman, 2006). This lack of persistence limits negative impacts on off-target organisms, but also reduces exposure to target pests. Novel formulations using microencapsulation technologies could enhance biopesticide effectiveness, with continuous delivery systems prolonging pest exposure and reducing the need for frequent application (Damalas and Koutroubas, 2018). However, as with chemical pesticides, certain biopesticides must be utilized in moderation due to their broad-spectrum activity and the possible development of resistance in pest species, especially for single-compound formulations (Isman, 2006).

Semiochemicals, such as pheromones and HIPVs, are an exception in the restricted use of synthetic compounds in organic agriculture. Pheromones can be used for monitoring, mass trapping, and mating disruption of specific insect pests (Saha *et al.*, 2017). In addition, synthetic pest-associated semiochemicals and HIPVs can lure natural enemies into crops to enhance pest control, reinforced by additional floral resources in 'attract and reward' designs (Ayelo *et al.*, 2021). Furthermore, semiochemicals could boost the efficacy of BCAs, most notably parasitoids, through olfactory conditioning (Kruidhof *et al.*, 2019). The highly species-specific aspect of semiochemicals, in particular sex pheromones, is both challenging and advantageous. On the one hand, it allows to precisely target select pests and thereby exclude off-target impacts. On the other hand, it entails that extensive research is needed for each pest or parasitoid to identify species-specific compounds that produce desired effects. Moreover, dosages, odour blends and release rates must be fine-tuned to optimize the use of semiochemicals in pest management (Ayelo *et al.*, 2021).

Conclusion

As demand for organic products and concerns about chemical pesticides grow, there is an ever-expanding need for effective ecological pest management. While not exhaustive, this chapter covered some key IPM challenges and solutions in organic agriculture, but applicable to conventional systems as well. A common thread running through the challenges across all phases of IPM is the inherent variability and specificity of many control methods, and the need for tailoring solutions to distinct agricultural contexts with integrative knowledge-based approaches. Overcoming these challenges and adoption barriers to IPM will require increased education and extension, and supportive policies such as the subsidization of cultural practices and ecosystem service valuation (Naranjo, 2001; Baker *et al.*, 2020; Deguine *et al.*, 2021). Furthermore, policy and regulatory reforms are needed to promote biocontrol research and the development of novel BCAs and biopesticides (van Lenteren *et al.*, 2018). Impending future challenges, including the effects of climate change on agroecosystems and new pest invasions, underscore the importance of resilient and ecological IPM tactics (Skendžić *et al.*, 2021). Applying principles of agroecology for crop protection will foster virtuous and sustainable changes in food systems that are essential to ensure long-term food security within planetary boundaries.

References

Aguilera, G., Roslin, T., Miller, K., Tamburini, G., Birkhofer, K. *et al.* (2020) Crop diversity benefits carabid and pollinator communities in landscapes with semi-natural habitats. *Journal of Applied Ecology* 57(11), 2170–2179. DOI: 10.1111/1365-2664.13712.

Alhadidi, S.N., Griffin, J.N. and Fowler, M.S. (2018) Natural enemy composition rather than richness determines pest suppression. *BioControl* 63(4), 575–584. DOI: 10.1007/s10526-018-9870-z. (accessed 18 February 2025).

Altieri, M.A. and Nicholls, C.I. (2003) Soil fertility management and insect pests: Harmonizing soil and plant health in agroecosystems. *Soil and Tillage Research* 72(2), 203–211. DOI: 10.1016/S0167-1987(03)00089-8.

Altieri, M.A., Nicholls, C.I. and Ponti, L. (2005) Enhanced pest management through soil health: Towards a belowground habitat management strategy. *Biodynamics* 253, 33–40. Available at: https://www.researchgate.net/publication/280132576 (accessed 18 February 2025).

Alyokhin, A., Nault, B. and Brown, B. (2020) Soil conservation practices for insect pest management in highly disturbed agroecosystems – a review. *Entomologia Experimentalis et Applicata* 168(1), 7–27. DOI: 10.1111/eea.12863.

Ayelo, P.M., Pirk, C.W.W., Yusuf, A.A., Chailleux, A., Mohamed, S.A. *et al.* (2021) Exploring the kairomone-based foraging behaviour of natural enemies to enhance biological control: A review. *Frontiers in Ecology and Evolution* 9. DOI: 10.3389/fevo.2021.641974.

Baker, B.P., Green, T.A. and Loker, A.J. (2020) Biological control and integrated pest management in organic and conventional systems. *Biological Control* 140, 104095. DOI: 10.1016/j.biocontrol.2019.104095.

Beaumelle, L., Auriol, A., Grasset, M., Pavy, A., Thiéry, D. *et al.* (2021) Benefits of increased cover crop diversity for predators and biological pest control depend on the landscape context. *Ecological Solutions and Evidence* 2(3), e12086. DOI: 10.1002/2688-8319.12086.

Begg, G.S., Cook, S.M., Dye, R., Ferrante, M., Franck, P. *et al.* (2017) A functional overview of conservation biological control. *Crop Protection* 97(July), 145–158. DOI: 10.1016/j.cropro.2016.11.008.

Beillouin, D., Ben-Ari, T., Malézieux, E., Seufert, V. and Makowski, D. (2021) Positive but variable effects of crop diversification on biodiversity and ecosystem services. *Global Change Biology* 27(19), 4697–4710. DOI: 10.1111/gcb.15747.

Beketov, M.A., Kefford, B.J., Schäfer, R.B. and Liess, M. (2013) Pesticides reduce regional biodiversity of stream invertebrates. *Proceedings of the National Academy of Sciences of the United States of America* 110(27), 11039–11043. DOI: 10.1073/pnas.1305618110. (accessed 18 February 2025).

Bianchi, F.J.J.A., Booij, C.J.H. and Tscharntke, T. (2006) Sustainable pest regulation in agricultural landscapes: A review on landscape composition, biodiversity and natural pest control. *Proceedings of the Royal Society B: Biological Sciences* 273(1595), 1715–1727. DOI: 10.1098/rspb.2006.3530.

Birch, A.N.E., Begg, G.S. and Squire, G.R. (2011) How agro-ecological research helps to address food security issues under new IPM and pesticide reduction policies for global crop production systems. *Journal of Experimental Botany* 62(10), 3251–3261 DOI: 10.1093/jxb/err064.

Birkhofer, K., Bezemer, T.M., Bloem, J., Bonkowski, M., Christensen, S. *et al.* (2008) Long-term organic farming fosters below and aboveground biota: Implications for soil quality, biological control and productivity. *Soil Biology and Biochemistry* 40(9), 2297–2308. DOI: 10.1016/j.soilbio.2008.05.007.

Blubaugh, C.K. and Kaplan, I. (2015) Tillage compromises weed seed predator activity across developmental stages. *Biological Control* 81(February), 76–82. DOI: 10.1016/j.biocontrol.2014.11.007.

Blundell, R., Schmidt, J.E., Igwe, A., Cheung, A.L., Vannette, R.L. *et al.* (2020) Organic management promotes natural pest control through altered plant resistance to insects. *Nature Plants* 6(5), 483–491. DOI: 10.1038/s41477-020-0656-9.

Bourguet, D. and Guillemaud, T. (2016) The hidden and external costs of pesticide use. In: Lichtfouse, E. (ed.) *Sustainable Agriculture Reviews*, Vol. 19. pp. 35–120. DOI: 10.1007/978-3-319-26777-7_2. (accessed 18 February 2025).

Brzozowski, L. and Mazourek, M. (2018) A sustainable agricultural future relies on the transition to organic agroecological pest management. *Sustainability* 10(6), 2023. DOI: 10.3390/su10062023.

Buurma, J.S., Lamine, C. and Haynes, I. (2010) Transition to consumer-driven value chains in the Netherlands. In: *XXVIII International Horticultural Congress on Science and Horticulture for People (IHC2010): International Symposium on 930*, pp. 69–75.

Carr, P.M., Gramig, G.G. and Liebig, M.A. (2013) Impacts of organic zero tillage systems on crops, weeds, and soil quality. *Sustainability* 5(7), 3172–3201. DOI: 10.3390/su5073172.

Chaplin-Kramer, R., O'Rourke, M.E., Blitzer, E.J. and Kremen, C. (2011) A meta-analysis of crop pest and natural enemy response to landscape complexity. *Ecology Letters* 14(9). DOI: 10.1111/j.1461-0248.2011.01642.x.

Chimonyo, V.G.P., Modi, A.T. and Mabhaudhi, T. (2015) Perspective on crop modelling in the management of intercropping systems. *Archives of Agronomy and Soil Science* 61(11), 1511–1529. DOI: 10.1080/03650340.2015.1017816.

Collins, K.L., Boatman, N.D., Wilcox, A., Holland, J.M. and Chaney, K. (2002) Influence of beetle banks on cereal aphid predation in winter wheat. *Ecosystems and Environment* 93, 1–3. DOI: 10.1016/S0167-8809(01)00340-1.

Cook, S.M., Khan, Z.R. and Pickett, J.A. (2007) The use of push-pull strategies in integrated pest management. *Annual Review of Entomology* 52, 375–400. DOI: 10.1146/annurev.ento.52.110405.091407.

Crowder, D.W. and Jabbour, R. (2014) Relationships between biodiversity and biological control in agroecosystems: Current status and future challenges. *Biological Control* 75, 8–17. DOI: 10.1016/j.biocontrol.2013.10.010.

Dainese, M., Martin, E.A., Aizen, M.A., Albrecht, M., Bartomeus, I. *et al.* (2019) A global synthesis reveals biodiversity-mediated benefits for crop production. *Science Advances* 5(10), eaax0121. DOI: 10.1126/sciadv.aax0121.

Damalas, C.A. and Koutroubas, S.D. (2018) Current status and recent developments in biopesticide use. *Agriculture* 8(1). DOI: 10.3390/agriculture8010013.

Dassou, A.G. and Tixier, P. (2016) Response of pest control by generalist predators to local-scale plant diversity: A meta-analysis. *Ecology and Evolution* 6(4), 1143–1153. DOI: 10.1002/ece3.1917.

Deguine, J.P., Aubertot, J.N., Flor, R.J., Lescourret, F., Wyckhuys, K.A.G. *et al.* (2021) Integrated pest management: Good intentions, hard realities. *Agronomy for Sustainable Development* 41(3). DOI: 10.1007/s13593-021-00689-w.

Desneux, N., Decourtye, A. and Delpuech, J.M. (2007) The sublethal effects of pesticides on beneficial arthropods. *Annual Review of Entomology*. DOI: 10.1146/annurev.ento.52.110405.091440.

Dury, J., Schaller, N., Garcia, F., Reynaud, A. and Bergez, J.E. (2012) Models to support cropping plan and crop rotation decisions. *Agronomy for Sustainable Development* 32(2), 567–580. DOI: 10.1007/s13593-011-0037-x.

Dutcher, J.D. (2007) A review of resurgence and replacement causing pest outbreaks in IPM. *General Concepts in Integrated Pest and Disease Management* 27–43. DOI: 10.1007/978-1-4020-6061-8_2.

Feit, B., Blüthgen, N., Traugott, M. and Jonsson, M. (2019) Resilience of ecosystem processes: A new approach shows that functional redundancy of biological control services is reduced by landscape simplification. *Ecology Letters* 22(10), 1568–1577. DOI: 10.1111/ele.13347.

Gadhave, K.R., Hourston, J.E. and Gange, A.C. (2016) Developing soil microbial inoculants for pest management: Can one have too much of a good thing? *Journal of Chemical Ecology* 42(4), 348–356. DOI: 10.1007/s10886-016-0689-8.

Ganser, D., Knop, E. and Albrecht, M. (2019) Sown wildflower strips as overwintering habitat for arthropods: Effective measure or ecological trap. *Agriculture, Ecosystems & Environment* 275, 123–131. DOI: 10.1016/j.agee.2019.02.010.

Geiger, F., Bengtsson, J., Berendse, F., Weisser, W.W., Emmerson, M. *et al.* (2010) Persistent negative effects of pesticides on biodiversity and biological control potential on European Farmland. *Basic and Applied Ecology* 11(2), 97–105. DOI: 10.1016/j.baae.2009.12.001.

Grab, H., Danforth, B., Poveda, K. and Loeb, G. (2018) Landscape simplification reduces classical biological control and crop yield. *Ecological Applications* 28(2), 348–355. DOI: 10.1002/eap.1651.

Haaland, C., Naisbit, R.E. and Bersier, L.F. (2011) Sown wildflower strips for insect conservation: A review. *Insect Conservation and Diversity* 4(1). DOI: 10.1111/j.1752-4598.2010.00098.x.

Harkes, P., Suleiman, A.K.A., Holterman, M.H.M. and Helder, J. (2020) Shifts in the active rhizobiome paralleling low *Meloidogyne chitwoodi* densities in fields under prolonged organic soil management. *Frontiers in Plant Science* 10. DOI: 10.3389/fpls.2019.01697.

Holland, J.M., Bianchi, F.J., Entling, M.H., Moonen, A.-C., Smith, B.M. *et al.* (2016) Structure, function and management of semi-natural habitats for conservation biological control: A review of European studies. *Pest Management Science* 72(9), 1638–1651. DOI: 10.1002/ps.4318.

Howard, M.M., Muñoz, C.A., Kao-Kniffin, J. and Kessler, A. (2020) Soil microbiomes from fallow fields have species-specific effects on crop growth and pest resistance. *Frontiers in Plant Science* 11, 1171. DOI: 10.3389/fpls.2020.01171.

Huang, N., Enkegaard, A., Osborne, L.S., Ramakers, P.M.J., Messelink, G.J. *et al.* (2011) The banker plant method in biological control. *Critical Reviews in Plant Sciences* 30(3), 259–278. DOI: 10.1080/07352689.2011.572055.

Huss, C.P., Holmes, K.D. and Blubaugh, C.K. (2022) Benefits and risks of intercropping for crop resilience and pest management. *Journal of Economic Entomology* 115(5), 1350–1362. DOI: 10.1093/jee/toac045.

IFOAM (2014) *The IFOAM Norms for Organic Production and Processing*. Bonn, Germany: IFOAM. Available at: https://www.ifoam.bio/sites/default/files/2020-09/IFOAM%20Norms%20July%202014%20Edits%202019.pdf (accessed 18 February 2025).

Isbell, F., Adler, P.R., Eisenhauer, N., Fornara, D., Kimmel, K. *et al.* (2017) Benefits of increasing plant diversity in sustainable agroecosystems. *Journal of Ecology* 105(4). DOI: 10.1111/1365-2745.12789.

Isman, M.B. (2006) Botanical insecticides, deterrents, and repellents in modern agriculture and an increasingly regulated world. *Annual Review of Entomology* 51, 45–66. DOI: 10.1146/annurev.ento.51.110104.151146.

Jaworski, C.C., Thomine, E., Rusch, A., Lavoir, A.-V., Wang, S. *et al.* (2023) Crop diversification to promote arthropod pest management: A review. *Agriculture Communications* 1(1), 100004. DOI: 10.1016/j.agrcom.2023.100004.

Karp, D.S., Chaplin-Kramer, R., Meehan, T.D., Martin, E.A., DeClerck, F. *et al.* (2018) Crop pests and predators exhibit inconsistent responses to surrounding landscape composition. *Proceedings of the National Academy of Sciences of the United States of America* 115(33), E7863–E7870. DOI: 10.1073/pnas.1800042115.

Khan, Z., Midega, C.A.O., Hooper, A. and Pickett, J. (2016) Push-pull: Chemical ecology-based integrated pest management technology. *Journal of Chemical Ecology* 42(7), 689–697. DOI: 10.1007/s10886-016-0730-y.

Kleijn, D. and Sutherland, W.J. (2003) How effective are European agri-environment schemes in conserving and promoting biodiversity? *Journal of Applied Ecology* 40(6). DOI: 10.1111/j.1365-2664.2003.00868.x.

Kruidhof, H.M., Kostenko, O., Smid, H.M. and Vet, L.E.M. (2019) Integrating parasitoid olfactory conditioning in augmentative biological control: Potential impact, possibilities, and challenges. *Frontiers in Ecology and Evolution* 7. DOI: 10.3389/fevo.2019.00084.

Lacey, L.A. (2017) Entomopathogens used as microbial control agents. In: Lacey, L. (ed.) *Microbial Control of Insect and Mite Pests*. Academic Press, pp. 3–12. DOI: 10.1016/B978-0-12-803527-6.00001-9.

Lavandero, B., Wratten, S.D., Didham, R.K. and Gurr, G. (2006) Increasing floral diversity for selective enhancement of biological control agents: A double-edged sward? *Basic and Applied Ecology* 7(3), 236–243. DOI: 10.1016/j.baae.2005.09.004.

Letourneau, D.K., Jedlicka, J.A., Bothwell, S.G. and Moreno, C.R. (2009) Effects of natural enemy biodiversity on the suppression of arthropod herbivores in terrestrial ecosystems. *Annual Review of Ecology,*

Evolution, and Systematics 40, 573–592. DOI: 10.1146/annurev.ecolsys.110308.120320.
Letourneau, D.K., Armbrecht, I., Rivera, B.S., Lerma, J.M., Carmona, E.J. *et al.* (2011) Does plant diversity benefit agroecosystems? A synthetic review. *Ecological Applications* 21(1). DOI: 10.1890/09-2026.1.
Liang, Y., Chen, X., Dai, H., Wang, J., Guo, X *et al.* (2022) Flower provision reduces intraguild predation between predators and increases aphid biocontrol in tomato. *Journal of Pest Science* 95(1), 461–472. DOI: 10.1007/s10340-021-01396-x.
Lima, M.C.F., Damascena de, M.E., Valero, C., Pereira Coronel, L.C. and Gonçalves Bazzo, C.O. (2020) Automatic detection and monitoring of insect pests. *Agriculture* 10(5), 161. DOI: 10.3390/agriculture10050161.
Mante, J. and Gerowitt, B. (2009) Learning from farmers' needs: Identifying obstacles to the successful implementation of field margin measures in intensive arable regions. *Landscape and Urban Planning* 93(3–4), 229–237. DOI: 10.1016/j.landurbplan.2009.07.010.
Masteling, R., Lombard, L., de Boer, W., Raaijmakers, J.M. and Dini-Andreote, F. (2019) Harnessing the microbiome to control plant parasitic weeds. *Current Opinion in Microbiology* 49, 26–33. DOI: 10.1016/j.mib.2019.09.006.
Matthiessen, J.N. and Kirkegaard, J.A. (2006) Biofumigation and enhanced biodegradation: Opportunity and challenge in soilborne pest and disease management. *Critical Reviews in Plant Sciences* 25(3), 235–265. DOI: 10.1080/07352680600611543.
Messelink, G.J., Bennison, J., Alomar, O., Ingegno, B.L., Tavella, L. *et al.* (2014) Approaches to conserving natural enemy populations in greenhouse crops: Current methods and future prospects. *BioControl* 59(4), 377–393. DOI: 10.1007/s10526-014-9579-6.
Molendijk, L.P.G. and van Beers, T.G. (2013) Het Aaltjesschema. Available at: http://Progemuese.Eu/Uploads/Flyer%20aaltjesschema%20PRO%20GEMÜSE.Pdf
Naranjo, S.E. (2001) Conservation and evaluation of natural enemies in IPM systems for *Bemisia tabaci*. *Crop Protection* 20(9), 835–852. DOI: 10.1016/S0261-2194(01)00115-6.
Naranjo, S.E., Ellsworth, P.C. and Frisvold, G.B. (2015) Economic value of biological control in integrated pest management of managed plant systems. *Annual Review of Entomology* 60, 621–645. DOI: 10.1146/annurev-ento-010814-021005.
Oerke, E.C. (2006) Crop losses to pests. *The Journal of Agricultural Science* 144(1), 31–43. DOI: 10.1017/S0021859605005708.
Pannell, D.J., Llewellyn, R.S. and Corbeels, M. (2014) The farm-level economics of conservation agriculture for resource-poor farmers. *Agriculture, Ecosystems & Environment* 187, 52–64. DOI: 10.1016/j.agee.2013.10.014.
Pijnakker, J., Vangansbeke, D., Duarte, M., Moerkens, R. and Wäckers, F.L. (2020) Predators and parasitoids-in-first: From inundative releases to preventative biological control in greenhouse crops. *Frontiers in Sustainable Food Systems* 4. DOI: 10.3389/fsufs.2020.595630.
Pimentel, D. (2005) Environmental and economic costs of the application of pesticides primarily in the United States. *Environment, Development and Sustainability* 7(2), 229–252. DOI: 10.1007/s10668-005-7314-2.
Pineda, A., Kaplan, I. and Bezemer, T.M. (2017) Steering soil microbiomes to suppress aboveground insect pests. *Trends in Plant Science* 22(9), 770–778. DOI: 10.1016/j.tplants.2017.07.002.
Pineda, A., Kaplan, I., Hannula, S.E., Ghanem, W. and Bezemer, T.M. (2020) Conditioning the soil microbiome through plant-soil feedbacks suppresses an aboveground insect pest. *The New Phytologist* 226(2), 595–608. DOI: 10.1111/nph.16385.
Poveda, K., Gómez, M.I. and Martínez, E. (2008) Diversification practices: Their effect on pest regulation and production. *Revista Colombiana de Entomología* 34(2), 131–144. DOI: 10.25100/socolen.v34i2.9269.
Puissant, J., Villenave, C., Chauvin, C., Plassard, C., Blanchart, E. *et al.* (2021) Quantification of the global impact of agricultural practices on soil nematodes: A meta-analysis. *Soil Biology and Biochemistry* 161(October), 108383. DOI: 10.1016/j.soilbio.2021.108383.
Quilty, J.R. and Cattle, S.R. (2011) Use and understanding of organic amendments in Australian agriculture: A review. *Soil Research* 49(1), 1. DOI: 10.1071/SR10059.
Rakotomalala, A.A.N.A., Ficiciyan, A.M. and Tscharntke, T. (2023) Intercropping enhances beneficial arthropods and controls pests: A systematic review and meta-analysis. *Agriculture, Ecosystems & Environment* 356, 108617. DOI: 10.1016/j.agee.2023.108617.
Rasmann, S., Bennett, A., Biere, A., Karley, A. and Guerrieri, E. (2017) Root symbionts: Powerful drivers of plant above- and belowground indirect defenses. *Insect Science* 24(6). DOI: 10.1111/1744-7917.12464.
Redlich, S., Martin, E.A. and Steffan-Dewenter, I. (2018) Landscape-level crop diversity benefits biological pest control. *Journal of Applied Ecology* 55(5), 2419–2428. DOI: 10.1111/1365-2664.13126.
Rosskopf, E., Hong, J.C., Pisani, C. and Kokalis-Burelle, N. (2020) Annual review of phytopathology organic amendments for pathogen and nematode control. *Annual Review of Phytopathology*. DOI: 10.1146/annurev-phyto-080516.
Rowen, E., Tooker, J.F. and Blubaugh, C.K. (2019) Managing fertility with animal waste to promote

arthropod pest suppression. *Biological Control* 134, 130–140. DOI: 10.1016/j.biocontrol.2019.04.012.

Rowen, E.K., Regan, K.H., Barbercheck, M.E. and Tooker, J.F. (2020) Is tillage beneficial or detrimental for insect and slug management? A meta-analysis. *Agriculture, Ecosystems & Environment* 294, 106849. DOI: 10.1016/j.agee.2020.106849.

Rusch, A., Valantin-Morison, M., Sarthou, J.-P. and Roger-Estrade, J. (2010) Biological control of insect pests in agroecosystems: Effects of crop management, farming systems, and seminatural habitats at the landscape scale. In: Sparks, D.L. (ed.) *Advances in Agronomy 109*. pp. 219–259. DOI: 10.1016/S0065-2113(10)09006-1.

Saha, T., Chandran, N. and Saha, C.T. (2017) Chemical ecology and pest management: A review. *International Journal of Chemical Studies* 5(6), 618–621.

Sarthou, J.P., Badoz, A., Vaissière, B., Chevallier, A. and Rusch, A. (2014) Local more than landscape parameters structure natural enemy communities during their overwintering in semi-natural habitats. *Agriculture, Ecosystems & Environment* 194, 17–28. DOI: 10.1016/j.agee.2014.04.018.

Schöning, J., Wachter, P. and Trautz, D. (2023) Crop rotation and management tools for every farmer? The current status on crop rotation and management tools for enabling sustainable agriculture worldwide. *Smart Agricultural Technology* 3, 100086. DOI: 10.1016/j.atech.2022.100086.

Shelton, A.M. and Badenes-Perez, F.R. (2006) Concepts and applications of trap cropping in pest management. *Annual Review of Entomology* 51, 285–308. DOI: 10.1146/annurev.ento.51.110104.150959.

Shields, M.W., Johnson, A.C., Pandey, S., Cullen, R., González-Chang, M. *et al.* (2019) History, current situation and challenges for conservation biological control. *Biological Control* 131. DOI: 10.1016/j.biocontrol.2018.12.010.

Sivinski, J. (2013) Augmentative biological control: Research and methods to help make it work. *CABI Reviews: Perspectives in Agriculture, Veterinary Science, Nutrition and Natural Resources* 1–11. DOI: 10.1079/PAVSNNR20138026.

Skendžić, S., Zovko, M., Pajač Živković, I., Lešić, V. and Lemić, D. (2021) Effect of climate change on introduced and native agricultural invasive insect pests in Europe. *Insects* 12(11), 985. DOI: 10.3390/insects12110985.

Sparks, T.C. and Nauen, R. (2015) IRAC: Mode of action classification and insecticide resistance management. *Pesticide Biochemistry and Physiology* 121(June), 122–128. DOI: 10.1016/j.pestbp.2014.11.014.

Tabashnik, B.E. and Carrière, Y. (2017) Surge in insect resistance to transgenic crops and prospects for sustainability. *Nature Biotechnology* 35, 10. DOI: 10.1038/nbt.3974.

Tortosa, A., Giffard, B., Sirami, C., Larrieu, L., Ladet, S. *et al.* (2023) Increasing landscape heterogeneity as a win-win solution to manage trade-offs in biological control of crop and woodland pests. *Scientific Reports* 13(1), 13573. DOI: 10.1038/s41598-023-40473-2.

Tscharntke, T., Bommarco, R., Clough, Y., Crist, T.O., Kleijn, D. *et al.* (2007) Conservation biological control and enemy diversity on a landscape scale. *Biological Control* 43(3), 294–309. DOI: 10.1016/j.biocontrol.2007.08.006.

Tscharntke, T., Karp, D.S., Chaplin-Kramer, R., Batáry, P., DeClerck, F. *et al.* (2016) When natural habitat fails to enhance biological pest control – five hypotheses. *Biological Conservation* 204, 449–458. DOI: 10.1016/j.biocon.2016.10.001.

Tschumi, M., Albrecht, M., Entling, M.H. and Jacot, K. (2015) High effectiveness of tailored flower strips in reducing pests and crop plant damage. *Proceedings of the Royal Society B* 282(1814), 20151369. DOI: 10.1098/rspb.2015.1369.

Urra, J., Alkorta, I. and Garbisu, C. (2019) Potential benefits and risks for soil health derived from the use of organic amendments in agriculture. *Agronomy* 9(9), 542. DOI: 10.3390/agronomy9090542.

van Lenteren, J.C., Bolckmans, K., Köhl, J., Ravensberg, W.J. and Urbaneja, A. (2018) Biological control using invertebrates and microorganisms: Plenty of new opportunities. *BioControl* 63(1), 39–59. DOI: 10.1007/s10526-017-9801-4.

Veres, A., Petit, S., Conord, C. and Lavigne, C. (2013) Does landscape composition affect pest abundance and their control by natural enemies? *Agriculture, Ecosystems & Environment* 166, 110–117. DOI: 10.1016/j.agee.2011.05.027.

Wang, K.-H., Hooks, C.R. and Ploeg, A. (2007) Protecting crops from nematode pests: Using marigold as an alternative to chemical nematicides. Available at: http://www.ctahr.hawaii.edu/freepubs (accessed 19 February 2025).

Weintraub, P.G., Recht, E., Mondaca, L.L., Harari, A.R., Diaz, B.M. *et al.* (2017) Arthropod pest management in organic vegetable greenhouses. *Journal of Integrated Pest Management* 8(1), 29. DOI: 10.1093/jipm/pmx021.

Zehnder, G., Gurr, G.M., Kühne, S., Wade, M.R., Wratten, S.D. *et al.* (2007) Arthropod pest management in organic crops. *Annual Review of Entomology* 52(1). DOI: 10.1146/annurev.ento.52.110405.091337.

Zhang, J., Huang, Y., Pu, R., Gonzalez-Moreno, P., Yuan, L. *et al.* (2019) Monitoring plant diseases and pests through remote sensing technology: A review. *Computers and Electronics in Agriculture* 165, 104943. DOI: 10.1016/j.compag.2019.104943.

36 Current Challenges for Conservation Biological Control with a Focus on the Olive Grove Agro-ecosystem

Alice Caselli* and Ruggero Petacchi

Center of Plant Sciences, Sant'Anna School of Advanced Studies, Piazza Martiri della Libertà 33, Pisa, Italy

Abstract

Conservation Biological Control (CBC) assumes a great importance as an alternative tool for pest control and agro-ecosystem management considering the new challenges in terms of green agriculture defined by the Institutions worldwide (e.g. the Green Deal for the European countries) and the climate change scenario. It lays the foundation on the enhancement and improvement of the intrinsic resource represented by natural enemies living in a certain agro-ecosystem, such as predators and parasitoids. Indeed, CBC requests a deep knowledge of pests' and natural enemies' ecology. In this context, understanding trophic relationships (e.g. identification of secondary hosts) and insects' habits (e.g. shelter preference) is pivotal. Insect chemical ecology studies interactions mediated by volatiles, which involve both insect–insect and insect–plant relationships; therefore it is fundamental to understand bitrophic and tritrophic relationships. Kairomones have been already tested for attracting parasitoid and predator populations in infested agro-ecosystems and have shown reliable results for a practical perspective in agriculture. However, global warming threatens insect relationships, disrupting communicative and phenological traits. The challenge lies in investigating ecological networks and linkages between the parts involved, between partners to forecast pests' phenology. Here, we report a case study concerning the olive grove agro-ecosystem, one of the oldest of the Mediterranean, and significant worldwide. Nevertheless, CBC of olive pests is nowadays lacking practical information and refers especially to *B. oleae*, while information on secondary pests is even more scarce. There is an urgent need of further studies of the relationships concerning *Olea europaea* pests' natural enemies in order to keep up with the current challenges.

Keywords: *Bactrocera oleae*, climate change, insect chemical ecology, *Olea europaea*, pest control

Introduction

General overview of conservation biological control

The human world population is estimated to reach 10.9 billion by 2100, with a major contribution from the 'demographic transition' countries, such as sub-Saharan Africa and South-east Asia (Maja and Ayano, 2021). Therefore, fulfilling future food demand is one of the main challenges of this century, considering the impact of the intensified agricultural practices on the environment (Davis *et al.*, 2016). However, on the one hand, among agricultural modern inputs, chemicals assume a great role in avoiding colossal food losses (Ngegba *et al.*, 2022), but on the other, it must be considered that their use strongly impacts human health (e.g. increasing the risk of cancer, metabolic disease, infertility, damage to immune and endocrine systems), destabilizes the environment (e.g. runoff into groundwaters and aquatic biota, eutrophication) (Ngegba *et al.*, 2022), and causes pesticide resistance (Shields *et al.*, 2019). This recent awareness has prompted scientists to investigate

*Corresponding author: alice.caselli18@gmail.com

DOI: 10.1079/9781800626850.0036

eco-friendly alternatives (Chhipa, 2017) and politicians to curb the use of some chemicals, promoting biological agriculture (e.g. EU banned the use of three neonicotinoids in 2013 on crops attractive to bees) (Butler, 2018). However, in complex agro-ecosystems, the exclusive use of both pesticides and biological methods is often thorny and not successful (Torres and Bueno, 2018).

Therefore, a combination of different techniques, including the use of biological tools and resistant varieties, habitat manipulation, modified agronomic practices, and, finally, the application of synthetic pesticides, is commonly accepted and known under the name of Integrated Pest Management (IPM) (Torres and Bueno, 2018). However, this approach has led us, over time, to think about IPM as a set of control methods rather than a technique focused on the agro-ecosystems and their ecological processes (Lescourret *et al.*, 2016). For this reason, IPM has recently made room for agroecological crop protection (ACP). ACP concentrates on restoring ecologically healthy agro-ecosystems and the regulation of ecosystem services for an efficient control of agricultural pests, and it aims to strengthen the relationships between plants and animals within the agro-ecosystem, and reduce the probability of pest infestations and outbreaks (Deguine *et al.*, 2012). In this scenario, of pivotal importance is a branch of biological control named conservation biological control (CBC) (Deguine *et al.*, 2021). CBC is made of different strategies that operate at multiple scales intended to manipulate behavioural and ecological processes aimed at the promotion of beneficial arthropods, especially predators and parasitoids (Begg *et al.*, 2017). Indeed, the adoption of CBC practices entails a deep knowledge of natural enemies' ecology and a strong understanding of the interactions in which they are involved, especially about the relationships that are the basis of trophic networks (Jonsson *et al.*, 2008). Both parasitoids and predators may depend on food sources that do not concern host prey (Jonsson *et al.*, 2008). For instance, in parasitoids, the presence of sugary sources (e.g. nectar, honeydew) is fundamental for ensuring efficient parasitization rates (Benelli *et al.*, 2017). Furthermore, diet selectivity has a great influence on several parasitoid reproductive traits, such as fecundity, fertility and sex ratio (Benelli *et al.*, 2017). In this scenario, making non-host food source available in the field became a cornerstone of CBC (Jonsson *et al.*, 2008).

In many crops the adoption of flowering strips at the field margins supports natural enemies. However, enhancing flower biodiversity *per se* does not give optimal results in attracting high diversity of predators and parasitoids, while a *tailored diversity* seems to be the key for the establishment of functional relationships (Hatt *et al.*, 2020). When natural sugary source is not available in field, CBC leans on the application of artificial food sprays that can be loaded with honey, glucose, galactose or sucrose. Among them, honey is the most studied and used since it is tasty and appreciated by insects, and it is not expensive (Benelli *et al.*, 2017). Beyond food, predators and parasitoids need also appropriate shelter for the establishment of robust relationships in agro-ecosystems. They serve as a refuge for overwintering, aestivation, reproduction and for taking cover during farming operations such as ploughing and harvesting (Griffiths *et al.*, 2008). Farmers can provide shelter habitats, either within the field or at the edge, depending on the crop. A popular example is the use of 'beetle banks' in arable fields, which consist of an earth bank sown with perennial grasses for maintaining the populations of polyphagous predators, such as ground beetles (Coleoptera: Carabidae), rove beetles (Coleoptera: Staphylinidae) and spiders (Griffiths *et al.*, 2008). In the following sections, we will debate the current situation of CBC, with a particular focus on the olive grove agro-ecosystem.

The Role of Chemical Ecology

The link between community structure and ecosystem processes is based on the dynamics among trophic interacting species (Kawatsu *et al.*, 2021). Plants, insect herbivores and natural enemies form complex multi-trophic networks influencing biodiversity, productivity and stability of ecosystems (de Araújo and Maia, 2021). In these interactions, chemical cues perform a key role, since they can potentially manipulate natural enemies in CBC programmes (Colazza *et al.*, 2017). For the host and prey location, parasitoids and predators take advantage of semiochemicals, such as kairomones and herbivore-induced plant volatiles (HIPVs), which are pivotal for directing them in the agro-ecosystem (Colazza *et al.*, 2022). In particular, HIPVs have a two-fold purpose, either as mediators between the attacked plant and the herbivore natural enemies, or as repellents against herbivores (Turlings and Erb, 2018). Indeed, the spatial scales at which HIPVs are released are generally heterogenous and the volatiles can be transported by wind resulting in a mix of complex volatile blends (Aartsma *et al.*, 2017). The ability of the natural enemies to follow the odour plume is strictly dependent on the landscape, since with increasing distance from the source, the volatile compounds have more probability of being hindered by physical barriers and to break down (Aartsma *et al.*, 2017).

Recently, synthetic HIPVs have been tested in the field to attract both predators and parasitoids, showing promising results. An example was recently reported by James (2003), explaining the capacity of a synthetic HIPV in attracting the predator *Chrysopa nigricornis* (Neuroptera: Chrysopidae) against aphid and mite populations in hops. HIPVs are used in an appealing technique in CBC named *attract and reward*. This strategy is based on the attractive capacity of synthetic HIPVs, such as methyl salicylate (MeSA) and methyl jasmonate, that are related to plant physiological pathways concerning phloem-sucking and chewing insects, respectively (Jaworski *et al.*, 2019), while the rewards are given by companion plants artificially added within cropping systems. These plants have specific functional traits that promote the maintenance and growth of herbivores' antagonists; for instance, they may be optimal providers of nectariferous and pollen sources, and they attract predators and parasitoids via visual and olfactory cues (Jaworski *et al.*, 2019).

Among the most used companion plants we cite *Calendula officinalis* (Martínez-Uña *et al.*, 2013), *Fagopyrum esculentum* (Simpson *et al.*, 2011) and *Lobularia maritima* (Brennan, 2016). A different CBC strategy is known as *push and pull* technique, based on the behavioural alteration via volatiles of a target organism, either a pest or a natural enemy, influencing its distribution in the agro-ecosystem (Cook *et al.*, 2007). When the target is represented by a herbivore pest, the 'push' action is given by a repellent that drives the insect away from the crop field, while an attractant provokes the 'pull' effect, enticing the pest towards baited traps or neighbouring trap crops (Cusumano *et al.*, 2020). This strategy has been successfully used in the field against the fall armyworm, *Spodoptera frugiperda* (Lepidoptera: Nocutidae), a polyphagous insect known as the key maize pest in tropical regions (e.g. East Africa, India) (Midega *et al.* (2018)). Furthermore, HIPVs can be used paired with botanical oil-based pesticides, and they produce a characteristic action, intended to alert plants to a specific herbivore menace and induce the emission of natural defences for attracting herbivore antagonists (Khan *et al.*, 2008).

These preliminary results show an exciting potential of HIPVs in CBC, but if, on the one hand, the use of this strategy gives a very appealing framework for pest control, on the other it must be considered that field studies are scarce and current literature is limited; therefore investigations on different crops and geographical areas are urgently needed (Khan *et al.*, 2008).

Conservation Biological Control Under the Global Climate Change Scenario

The evidence of global climate change is now overwhelming, and the forecast models for the next decades are not optimistic since temperatures are expected to increase by 4°C by 2100 (Thuiller, 2007). The major contributor is global CO_2 concentration, which, together with the other main greenhouse gases (CH_4, NO_x, halocarbons), has risen by about 25–30% during the last century, principally on account of anthropogenic activities (Selvaraj *et al.*, 2013). Among them, the agricultural sector is pivotal, but, at the same time, it is also one of the most afflicted by climate change (Clapp *et al.*, 2018). Indeed, global warming affects crops' productivity both directly and indirectly; for example, affecting insect pests and their natural enemies (Selvaraj *et al.*, 2013). Insects' phenology is strictly related to many climate factors, such as temperature and precipitation. Generally, an indirect relationship is seen between increasing temperature and insects' developmental time, resulting in more generations per year and anomalous diapause patterns (Laws, 2017). Temperature is the most influencing parameter on insect physiological traits. Therefore, like other species, natural enemies' population dynamics and distribution strictly depend on these parameters since they are heterothermic and their body temperature follows the environment's characteristics (Caselli and Petacchi, 2021).

Climate change causes pest outbreaks and breakdown cycles with negative consequences for ecosystem functioning (Harvey *et al.*, 2020). Furthermore, the consequences concerning the trophic relationships are often amplified by climate disturbance, and natural enemies are expected to show stronger effects compared to their hosts (Aguilar-Fenollosa and Jacas, 2013). However, the effect of the impact is not always the same and depends on species characteristics, such as trophic habits. Indeed, natural enemies that present monophagous traits or have an intimate relationship with the host are described as the most vulnerable to climate change, as in the case of koinobiont endoparasitoids (Hance *et al.*, 2007). They share a long co-evolutionary path and are very vulnerable to thermal changes in one or both species (Le Lann *et al.*, 2021). On the other hand, omnivorous predators and polyphagous parasitoids are considered more resistant since they can choose the food source based on its availability (Hance *et al.*, 2007). Furthermore, the alterations of food-web interactions may depend also on competitors, mutualists and antagonists, causing uncertainty about the prospective spatial distribution of species and their abundance (Jeffs and Lewis, 2013).

Parasitoids respond mainly in three ways to climate perturbances. Firstly, they can shift their distribution following suitable climate conditions, such as cooler environment (Jeffs and Lewis, 2013). This happens because insects show a maximum extent of distribution depending on precipitation and temperature tolerance. When alterations occur, we assist in a variation of insect tolerance involving space characteristics such as latitude and altitude (Walther *et al.*, 2002). Secondly, parasitoids can alter their phenology, modifying the timing of biological events (Jeffs and Lewis, 2013). These phenological shifts are not always synchronized between a parasitoid and its host. Consequently, the fitness, abundance, distribution and population dynamics of both may be affected (Voigt *et al.*, 2003), influencing food webs and ecosystem functions (Walther, 2010). Finally, parasitoids may respond to climatic alterations through *in situ* adaptation (Bellard *et al.*, 2012). This phenomenon remains poorly understood.

In this scenario, CBC needs answers about climate change that do not consider just single species but rather the relations and linkages between actors of ecological networks and food webs, considering that responses to climatic alterations are connected by interactions with others that present the same or different trophic levels (Walther, 2010).

Case Study: Conservation Biological Control in the Olive Grove

The olive tree (*Olea europaea*) is one of the oldest known cultivated species worldwide, praised by many ancient intellectuals such as Theophrastus, Sophocles and Pliny the Elder for its longevity, resistance, and usefulness of its principal product, olive oil (Rhizopoulou, 2007). This plant, hailing from a valley of the river Jordan, was successively cultivated in the whole Mediterranean basin, where it is considered endemic. Nowadays, it is spread over South Africa, Australia, China, Japan, Chile, the Caribbean, Peru, Argentina, Brazil, Mexico and California, where it arrived in the 17th century (Rhizopoulou, 2007). Recently, cultivation reached north-western India and the Caucasus (Mourtier and van der Vossen, 2001). The importance of the olive tree is principally related to oil production, which annually reaches around 3 million tons (International Olive Council, 2022), with an 81% contribution by the European countries (e.g. Spain, Italy, Greece, Portugal, France); 11% by North Africa; and 7% by the Near East, while the residual production has American origin (Uylaşer and Yildiz, 2014). The characteristic chemical composition and sensory traits make olive oil very appreciated all over the world (Malheiro *et al.*, 2015); it is a fundamental ingredient in the Mediterranean diet, and its consumption ranges between 25 and 50 mL/day per person. Olive oil is also considered pivotal in reducing the probability of cardiovascular disease and in preventing certain cancers (Reboredo-Rodríguez *et al.*, 2018).

There are several diverse factors (e.g. agricultural, technological) that compromise olive oil quality and composition; among them olive pests assume a major role (Malheiro *et al.*, 2015), causing annual crop losses of around US$ 800 million (Hoelmer *et al.*, 2011). Particularly, the olive fruit fly, *Bactrocera oleae* (Diptera: Tephritidae), is considered the key pest worldwide (Caselli *et al.*, 2021a). It is a monophagous species that attacks *Olea* spp. (e.g. *O. europaea*, *O. verrucosa*, *O. chrysophylla*, *O. ferruginea*) fruits, ovipositing inside them, thus providing a direct access to food for larvae just after emergence from the eggs (Malheiro *et al.*, 2015). It presents several generations per year, from one to five (Marchi *et al.*, 2016), and can reach extremely high population densities depending on climatic conditions, causing significant economic burden (Estes *et al.*, 2011). In addition to *B. oleae*, major secondary pests such as *Prays oleae* (Lepidoptera: Praydidae), *Saissetia oleae* (Hemiptera: Coccidae) (Haniotakis, 2005), *Dasineura oleae* (Diptera: Cecidomyiidae) (Caselli *et al.*, 2021b) and *Palpita vitrealis* (= *P. unionalis*) (Lepidoptera: Crambidae) (Chaouche *et al.*, 2019) can cause important economic losses, locally and occasionally (Haniotakis, 2005). Considering the olive orchard agro-ecosystem as a multi-functional reality (Fleskens *et al.*, 2009) that provides ecosystem services thanks to its peculiar characteristics (e.g. developing since ancient times on land in which irrigation was not feasible), and taking into account the necessity of alternative control tools, CBC has to be deemed a valid alternative for pest control. Below, we explore the main actors of CBC in the olive agro-ecosystem.

Arthropods as natural enemies

Concerning the olive grove, most studies about pests' natural enemies concerned arthropods (Martínez-Núñez *et al.*, 2021a). Very often, the biodiversity and efficacy of natural enemies are investigated together with landscape composition and supplementary non-crop vegetation (Wade *et al.*, 2008). This is because predators and parasitoids are generally omnivorous and need pollen and nectar from non-crop plants to satisfy their biological and physiological requirements (Villa *et al.*, 2021). Most investigations on predators and parasitoids of the olive grove refer to *B. oleae*, the major pest. However, we report examples of arthropod control agents of pests considered of secondary economic importance.

Predators

Arthropod predators of the olive grove can be found at different levels, such as soil and tree canopy, and are generally related to more than one prey. Those dwelling in the soil are numerous and can potentially be involved in *Bactrocera oleae* control, since during autumn the third larval stage pupates in the soil where they remain until the following spring (Dimou *et al.*, 2003). Soil predators' community is mainly composed of rove beetles (Coleoptera: Staphylinidae), ground beetles (Coleoptera: Carabidae), ants, earwigs, spiders, opiliones and myriapods (Dinis *et al.*, 2016a). Among staphylinids, *Ocypus oleae* feeds certainly on *B. oleae* pupae as demonstrated by Neuenschwander *et al.* (1983) under laboratory conditions and Albertini *et al.* (2018) in field trials. Its voracity and the higher feeding activity compared to other predaceous rove and ground beetles make it an important candidate as a *B. oleae* control agent (Albertini *et al.*, 2018). Concerning carabids, the olive agro-ecosystem supports rich and well-structured assemblages that present contemporaneity with *B. oleae* life cycle (Albertini *et al.*, 2017). In particular, lab investigations demonstrated the high predacious behaviour of *Pterostichus globosus* and *Calathus garnatensis* on *B. oleae* pupae, and, specifically, the former showed higher consumption of total biomass, probably due to its large dimensions (Dinis *et al.*, 2016b). Even if the dimensions are smaller, ants can contribute to the suppression of olive pests, in particular of *B. oleae* and *P. oleae*.

Tapinoma nigerrimum is a very abundant and dynamic ant in olive groves (Martínez-Núñez *et al.*, 2021b), having an important role as predator of the pests cited above, as reported by Morris *et al.* (2002) and Dinis *et al.* (2016a) in field tests. Additionally, *Messor barbarous*, a species normally found in olive groves, was observed carrying *B. oleae* pupae into the nest during a lab experiment (see Neuenschwander *et al.*, 1983). However, its contribution as control agent is not very clear since in the field it shows granivorous habits (Dinis *et al.*, 2016a). Among canopy predators, spiders can limit the abundance of the olive fruit fly (Picchi *et al.*, 2016). In particular, *Frontinellina frutetorum*, very abundant in olive groves, has been identified as a potential control agent. However, further studies are necessary to confirm this hypothesis (Picchi *et al.*, 2016).

Parasitoids

Parasitoids are important components of ecological trophic relationships, and they are of interest in controlling several agricultural pests (Benelli *et al.*, 2017), including those of the olive grove. Among them, many have been studied and released following classical biological programmes (Wang *et al.*, 2009). However, studies concerning the conservation of parasitoids' territory in the olive agro-ecosystem are not so numerous. Concerning *B. oleae*, significant effects of landscape composition on parasitoid fauna (*Pnigalio agraules, Eupelmus urozonus, Eurytoma martellii*) were demonstrated, highlighting that connectivity at a large scale is more important than woodland abundance. In particular, woodland fragmentation at large scale negatively influences parasitism rate (Boccaccio and Petacchi, 2009). Considering landscape studies, Picchi *et al.* (2021) reported that the parasitism rate of *D. oleae* parasitoids is influenced by a small-scale effect and leaf infestation by a large-scale effect. The parasitism rate of the polyphagous *Platygaster demades* (Hymenoptera: Platygastridae) increased in the semi-natural habitat with plants associated with alternative hosts, such as *Erica* spp. and *Pyrus communis* subsp. *pyraster* (Picchi *et al.*, 2021).

Even if vegetation cover seems to enhance the abundance of natural enemies, it seems to be not effective to control olive pests' abundance if considered alone (Paredes *et al.*, 2015). Accordingly, no top-down effects of inter-row ground covers were reported concerning some parasitoids of the olive moth *Prays oleae* and the psyllid *Euphyllura olivina*. This has to be related to the lack of synchrony of pests and the parasitoids belonging to Braconidae, Elasmidae, Encyrtidae and Scelionidae families (Paredes *et al.*, 2013). However, in a recent study, the encyrtid *Ageniaspis fuscicollis* has been reported to be efficient in controlling the anthophagous generation of *P. oleae* in an olive grove with ground vegetation cover. This may depend on the variability in plant composition that may contribute to a spillover effect of the natural enemy among plant species and seasons (Villa *et al.*, 2021).

Birds

In olive grove agro-ecosystem, the study of avian-mediated pest control has not been extensively investigated, even though birds have proved to be successful control agents and provide ecosystem services with important economic consequences in many other agro-ecosystems (Martínez-Núñez *et al.*, 2021a). Some studies refer to birds as indicators of olive agro-ecosystem health, highlighting that the communities of breeding birds are different depending on the grove management (e.g. conventional, organic, abandoned), and chemicals negatively influence species richness (Solomou and Sfougaris, 2012). Furthermore, the olive orchard is also used as a wintering quarter by frugivorous birds, particularly in the Mediterranean, since the high lipid content that characterizes olive drupe (40–66% of dry mass) makes it a highly nutritious food (Rey, 2011). Nevertheless, studies carried out so far demonstrated that insectivorous birds of the olive grove do not contribute significantly as biological control agents against *B. oleae* and *P. oleae*, as reported by Martínez-Núñez *et al.* (2021a). This may be due to the limited attractiveness of these pests and the difficulty in accessing them. For improving the abundance and efficacy of frugivorous birds as control agents in the olive grove habitat, improvement and foraging behaviour are recommended (Martínez-Núñez *et al.*, 2021a).

Conclusions

CBC lays foundations on ecological-based theories, such as food network linkages and communicative relationships. Nowadays, agriculture is facing big challenges concerning the reduction of chemical use, and CBC offers a great opportunity as green alternative. However, it is very hard to be applied in an efficient way since insects' ecology and relationships are very complex to study and time-consuming. The olive grove agro-ecosystem has been, in part, already explored in this sense, but the practical applications are still scarce. This work may shed light on the challenges of CBC with the aim of improving it in olive groves and other agro-ecosystems.

References

Aartsma, Y., Bianchi, F.J.J.A., Van der Werf, W., Poelman, E.H. and Dicke, M. (2017) Herbivore-induced plant volatiles and tritrophic interactions across spatial scales. *New Phytologist* 216(4), 1054–1063. DOI: 10.1111/nph.14475.

Aguilar-Fenollosa, E. and Jacas, J.A. (2013) Can we forecast the effects of climate change on entomophagous biological control agents. *Pest Management Science* 70(6), 853–859. DOI: 10.1002/ps.3678.

Albertini, A., Pizzolotto, R. and Petacchi, R. (2017) Carabid patterns in olive orchards and woody semi-natural habitats: First implications for conservation biological control against *Bactrocera oleae*. *BioControl* 62(1), 71–83. DOI: 10.1007/s10526-016-9780-x.

Albertini, A., Marchi, S., Ratti, C., Burgio, G., Petacchi, R. *et al.* (2018) *Bactrocera oleae* pupae predation by *Ocypus olens* detected by molecular gut content analysis. *BioControl* 63(2), 227–239. DOI: 10.1007/s10526-017-9860-6.

Begg, G.S., Cook, S.M., Dye, R., Ferrante, M., Franck, P. *et al.* (2017) A functional overview of conservation biological control. *Crop Protection* 97, 145–158. DOI: 10.1016/j.cropro.2016.11.008.

Bellard, C., Bertelsmeier, C., Leadley, P., Thuiller, W. and Courchamp, F. (2012) Impacts of climate change on the future of biodiversity. *Ecology Letters* 15(4), 365–377. DOI: 10.1111/j.1461-0248.2011.01736.x.

Benelli, G., Giunti, G., Tena, A., Desneux, N., Caselli, A. *et al.* (2017) The impact of adult diet on parasitoid reproductive performance. *Journal of Pest Science* 90(3), 807–823. DOI: 10.1007/s10340-017-0835-2.

Boccaccio, L. and Petacchi, R. (2009) Landscape effects on the complex of *Bactrocera oleae* parasitoids and implications for conservation biological control. *BioControl* 54(5), 607–616. DOI: 10.1007/s10526-009-9214-0.

Brennan, E.B. (2016) Agronomy of strip intercropping broccoli with alyssum for biological control of aphids. *Biological Control* 97, 109–119. DOI: 10.1016/j.biocontrol.2016.02.015.

Butler, D. (2018) EU pesticide review could lead to ban. *Nature* 555, 150–151. DOI: 10.1038/d41586-018-02639-1.

Caselli, A., Favaro, R., Petacchi, R. and Angeli, S. (2021a) Infestation of the gall midge *Dasineura oleae* provides first evidence of induced plant volatiles in olive leaves. *Bulletin of Entomological Research* 1–13. DOI: 10.1017/S0007485321001000.

Caselli, A. and Petacchi, R. (2021) Climate change and major pests of Mediterranean olive orchards: Are we ready to face the global heating? *Insects* 12(9), 802. DOI: 10.3390/insects12090802.

Caselli, A., Francini, A., Minnocci, A. and Petacchi, R. (2021b) *Dasineura oleae*: Morphological and physiological characterization following the midge attack on olive leaves. *Journal of Plant Diseases and Protection* 128(1), 173–182. DOI: 10.1007/s41348-020-00380-4.

Chaouche, S.T., Bengouga, K. and Fadlaoui, H. (2019) The first detection of the olive leaf moth *Palpita vitrealis* (Rossi) (Lepidoptera: Pyralidae) as a serious pest in Biskra province (Algeria). *Bulletin OEPP* 49(3), 593–596. DOI: 10.1111/epp.12607.

Chhipa, H. (2017) Nanofertilizers and nanopesticides for agriculture. *Environmental Chemistry Letters* 15(1), 15–22. DOI: 10.1007/s10311-016-0600-4.

Clapp, J., Newell, P. and Brent, ZW. (2018) The global political economy of climate change, agriculture and food systems. *The Journal of Peasant Studies* 45(1), 80–88. DOI: 10.1080/03066150.2017.1381602.

Colazza, S., Shields, M.W., Peri, E. and Cusumano, A. (2017) Ecosystem services provided by unmanaged habitats in agricultural landscapes: Challenges for agronomists, ecologists, economists and policymakers. *Environmental Pest Management. Wiley, Ch* 7. DOI: 10.1002/9781119255574.

Colazza, S., Pappas, M.L., Cortesero, A.M and Rodriguez-Saona, C. (2022) Chemical ecology and conservation biological control. *Frontiers in Ecology and Evolution* 10, 857438. DOI: 10.3389/fevo.2022.857438.

Cook, S.M., Khan, Z.R. and Pickett, J.A. (2007) The use of push-pull strategies in integrated pest management. *Annual Review of Entomology* 52, 375–400. DOI: 10.1146/annurev.ento.52.110405.091407.

Cusumano, A., Harvey, J.A., Bourne, M.E., Poelman, E.H. and de Boer, J.G. (2020) Exploiting chemical ecology to manage hyperparasitoids in biological control of arthropod pests. *Pest Management Science* 76(2), 432–443. DOI: 10.1002/ps.5679.

Davis, K.F., Gephart, J.A., Emery, K.A., Leach, A.M., Galloway, J.N. *et al.* (2016) Meeting future food demand with current agricultural resources. *Global Environmental Change* 39, 125–132. DOI: 10.1016/j.gloenvcha.2016.05.004.

de Araújo, W.S. and Maia, V.C. (2021) Topological structure of a tritrophic network composed of host plants, gall-inducing insects and parasitoids in a resting area in Brazil. *Entomological Science* 24(2), 201–216. DOI: 10.1111/ens.12468.

Deguine, J.-P., Rousse, P. and Atiama-Nurbel, T. (2012) Agroecological crop protection: Concepts and a case study from reunion. In: Larramendy, M.L. and Soloneski, S. (eds) *Integrated Pest Management and Pest Control – Current and Future Tactics*. pp. 63–76.

Deguine, J.-P., Aubertot, J.-N., Flor, R.J., Lescourret, F., Wyckhuys, K.A.G. *et al.* (2021) Integrated pest management: good intentions, hard realities. A review. *Agronomy for Sustainable Development* 41(3), 38. DOI: 10.1007/s13593-021-00689-w.

Dimou, I., Koutsikopoulos, C., Economopoulos, A.P. and Lykakis, J. (2003) Depth of pupation of the wild olive fruit fly, *Bactrocera* (*Dacus*) *oleae* (Gmel.) (Dipt., Tephritidae), as affected by soil abiotic factors. *Journal of Applied Entomology* 127(1), 12–17. DOI: 10.1046/j.1439-0418.2003.00686.x.

Dinis, A.M., Pereira, J.A., Pimenta, M.C., Oliveira, J., Benhadi-Marín, J. *et al.* (2016a) Suppression of *Bactrocera oleae* (Diptera: Tephritidae) pupae by soil arthropods in the olive grove. *Journal of Applied Entomology* 140(9), 677–687. DOI: 10.1111/jen.12291.

Dinis, A.M., Pereira, J.A., Benhadi-Marín, J. and Santos, S.A.P. (2016b) Feeding preferences and functional responses of *Calathus granatensis* and *Pterostichus globosus* (Coleoptera: Carabidae) on pupae of *Bactrocera oleae* (Diptera: Tephritidae). *Bulletin of Entomological Research* 106(6), 701–709. DOI: 10.1017/S0007485316000213.

Estes, A.M., Nestel, D., Belcari, A., Jessup, A., Rempoulakis, P. *et al.* (2011) A basis for the renewal of sterile insect technique for the olive fly, *Bactrocera oleae* (Rossi). *Journal of Applied Entomology* 136(1–2), 1–16. DOI: 10.1111/j.1439-0418.2011.01620.x.

Fleskens, L., Duarte, F. and Eicher, I. (2009) A conceptual framework for the assessment of multiple functions of agro-ecosystems: A case study of Trás-os-Montes olive groves. *Journal of Rural Studies* 25(1), 141–155. DOI: 10.1016/j.jrurstud.2008.08.003.

Griffiths, G.J.K., Holland, J.M., Bailey, A. and Thomas, M.B. (2008) Efficacy and economics of shelter habitats for conservation biological control.

Biological Control 45(2), 200–209. DOI: 10.1016/j.biocontrol.2007.09.002.

Hance, T., Baaren, J., Vernon, P. and Boivin, G. (2007) Impact of extreme temperatures on parasitoids in a climate change perspective. *Annual Review of Entomology* 52(1), 107–126. DOI: 10.1146/annurev.ento.52.110405.091333.

Haniotakis, G.E. (2005) Olive pest control: present status and prospects. In: *Proceedings of the IOBC/WPRS Conference on Integrated Protection of Olive Crops, Chania*.

Harvey, J.A., Heinen, R., Gols, R. and Thakur, M.P. (2020) Climate change-mediated temperature extremes and insects: From outbreaks to breakdowns. *Global Change Biology* 26(12), 6685–6701. DOI: 10.1111/gcb.15377.

Hatt, S., Francis, F., Xu, Q., Wang, S. and Osawa, N. (2020) Perennial flowering strips for conservation biological control of insect pests: From picking and mixing flowers to tailored functional diversity. *Integrative Biological Control* 20.

Hoelmer, K.A., Kirk, A.A., Pickett, C.H., Daane, K.M. and Johnson, M.W. (2011) Prospects for improving biological control of olive fruit fly, *Bactrocera oleae* (Diptera: Tephritidae), with introduced parasitoids (Hymenoptera). *Biocontrol Science and Technology* 21(9), 1005–1025. DOI: 10.1080/09583157.2011.594951.

International Olive Council (2022) *The world of olive oil*. Available at: https://www.internationaloliveoil.org/the-world-of-olive-oil/?lang=it (accessed November 2023).

James, D.G. (2003) Field evaluation of herbivore-induced plant volatiles as attractants for beneficial insects: methyl salicylate and the green lacewing, *Chrysopa nigricornis*. *Journal of Chemical Ecology* 29(7), 1601–1609. DOI: 10.1023/a:1024270713493.

Jaworski, C.C., Da, X., Xu, Q., Ramirez-Romero, R., Guo, X. *et al*. (2019) Varying the spatial arrangement of synthetic herbivore-induced plant volatiles and companion plants to improve conservation biological control. *Journal of Applied Ecology* 56(5), 1176–1188. DOI: 10.1111/1365-2664.13353ff.

Jeffs, C.T. and Lewis, O.T. (2013) Effects of climate warming on host–parasitoid interactions. *Ecological Entomology* 38(3), 209–218. DOI: 10.1111/een.12026.

Jonsson, M., Wratten, S.D., Landis, D.A. and Gurr, G.M. (2008) Recent advances in conservation biological control of arthropods by arthropods. *Biological Control* 45(2), 172–175. DOI: 10.1016/j.biocontrol.2008.01.006.

Kawatsu, K., Ushio, M., Veen, F.J.F. and Kondoh, M. (2021) Are networks of trophic interactions sufficient for understanding the dynamics of multi-trophic communities? Analysis of a tri-trophic insect food-web time-series. *Ecology Letters* 24(3), 543–552. DOI: 10.1111/ele.13672.

Khan, Z.R., James, D.G., Midega, C.A.O. and Pickett, J.A. (2008) Chemical ecology and conservation biological control. *Biological Control* 45(2), 210–224. DOI: 10.1016/j.biocontrol.2007.11.009.

Laws, A.N. (2017) Climate change effects on predator-prey interactions. *Current Opinion in Insect Science* 23, 28–34. DOI: 10.1016/j.cois.2017.06.010.

Le Lann, C., Baaren, J. and Visser, B. (2021) Dealing with predictable and unpredictable temperatures in a climate change context: The case of parasitoids and their hosts. *Journal of Experimental Biology* 224, jeb238626. DOI: 10.1242/jeb.238626.

Lescourret, F., Poncet, C. and Simon, S. (2016) Insights into the ecological control of pests in horticulture. *Acta Horticulturae* 1137(1137), 91–104. DOI: 10.17660/ActaHortic.2016.1137.14.

Maja, M.M. and Ayano, S.F. (2021) The impact of population growth on natural resources and farmers' capacity to adapt to climate change in low-income countries. *Earth Systems and Environment* 5(2), 271–283. DOI: 10.1007/s41748-021-00209-6.

Malheiro, R., Casal, S., Baptista, P. and Pereira, J.A. (2015) A review of *Bactrocera oleae* (Rossi): Impact in olive products: From the tree to the table. *Trends in Food Science & Technology* 44(2), 226–242. DOI: 10.1016/j.tifs.2015.04.009.

Marchi, S., Guidotti, D., Ricciolini, M. and Petacchi, R. (2016) Towards understanding temporal and spatial dynamics of *Bactrocera oleae* (Rossi) infestations using decade-long agrometeorological time series. *International Journal of Biometeorology* 60(11), 1681–1694. DOI: 10.1007/s00484-016-1159-2.

Martínez-Núñez, C., Rey, P.J., Manzaneda, A.J., García, D., Tarifa, R. *et al*. (2021a) Insectivorous birds are not effective pest control agents in olive groves. *Basic and Applied Ecology* 56, 270–280. DOI: 10.1016/j.baae.2021.08.006.

Martínez-Núñez, C., Rey, P.J., Salido, T., Manzaneda, A.J., Camacho, F.M. *et al*. (2021b) Ant community potential for pest control in olive groves: Management and landscape effects. *Agriculture, Ecosystems & Environment* 305, 107185. DOI: 10.1016/j.agee.2020.107185.

Martínez-Uña, A., Martín, J.M., Fernández-Quintanilla, C. and Dorado, J. (2013) Provisioning floral resources to attract aphidophagous hoverflies (Diptera: Syrphidae) useful for pest management in central Spain. *Journal of Economic Entomology* 106(6), 2327–2335. DOI: 10.1603/ec13180.

Midega, C.A.O., Pittchar, J.O., Pickett, J.A., Hailu, G.W. and Khan, Z.R. (2018) A climate-adapted push-pull system effectively controls fall armyworm, *Spodoptera frugiperda* (J.E. smith), in maize in East Africa. *Crop Protection* 105, 10–15. DOI: 10.1016/j.cropro.2017.11.003.

Morris, T.I., Symondson, W.O.C., Kidd, N.A.C. and Campos, M. (2002) The effect of different ant species on the olive moth, *Prays oleae* (Bern.), in Spanish olive orchard. *Journal of Applied Entomology* 126(5), 224–230. DOI: 10.1046/j.1439-0418.2002.00647.x.

Mourtier, N. and van der Vossen, H.A.M. (2001) *Olea europaea* L. vegetable oils and fats. In: van der Vossen, H.A.M. and Umali, B.E. (eds) *Plant Resources of South-East Asia*, Vol. 14. Backhuys Publishers, Leiden, pp. 107–112.

Neuenschwander, P., Bigler, F., Delucchi, V. and Michelakis, S. (1983) Natural enemies of preimaginal stages of *Dacus oleae* gmel (Dipt., Tephritidae) in western Crete. *Bionomics and Phenologies. Bollettino Del Laboratorio Di Entomologia Agraria "Filippo Silvestri"* 40, 3–32.

Ngegba, P.M., Cui, G., Khalid, M.Z. and Zhong, G. (2022) Use of botanical pesticides in agriculture as an alternative to synthetic pesticides. *Agriculture* 12(5), 600. DOI: 10.3390/agriculture12050600.

Paredes, D., Cayuela, L., Gurr, G.M. and Campos, M. (2013) Effect of non-crop vegetation types on conservation biological control of pests in olive groves. *PeerJ* 1, e116. DOI: 10.7717/peerj.116.

Paredes, D., Cayuela, L., Gurr, G.M. and Campos, M. (2015) Is ground cover vegetation an effective biological control enhancement strategy against olive pests. *PLoS ONE* 10(2), e0117265. DOI: 10.1371/journal.pone.0117265.

Picchi, M.S., Bocci, G., Petacchi, R. and Entling, M.H. (2016) Effects of local and landscape factors on spiders and olive fruit flies. *Agriculture, Ecosystems & Environment* 222, 138–147. DOI: 10.1016/j.agee.2016.01.045.

Picchi, M.S., Tondini, E., Albertarelli, N., Montefiorti, G. and Petacchi, R. (2021) Following the pest outbreak: Preliminary findings on the landscape effect on *Dasineura oleae* and its parasitoids in central Italy. *Phytoparasitica* 50(2), 375–389. DOI: 10.1007/s12600-021-00962-4.

Reboredo-Rodríguez, P., Varela-López, A., Forbes-Hernández, T.Y., Gasparrini, M., Afrin, S. *et al* (2018) Phenolic compounds isolated from olive oil as nutraceutical tools for the prevention and management of cancer and cardiovascular diseases. *International Journal of Molecular Sciences* 19(8), 2305 DOI: 10.3390/ijms19082305.

Rey, P.J. (2011) Preserving frugivorous birds in agroecosystems: Lessons from Spanish olive orchards. *Journal of Applied Ecology* 48(1), 228–237 DOI: 10.1111/j.1365-2664.2010.01902.x.

Rhizopoulou, S. (2007) *Olea europaea* L.: A botanical contribution to culture. *American-Eurasian Journal of Agriculture & Environmental Sciences* 2(4), 382–387.

Selvaraj, S., Palanisamy, G. and Pandiaraj, T. (2013) Potential impacts of recent climate change on biological control agents in agro-ecosystem: A review. *International Journal of Biodiversity and Conservation* 5(12), 845–852.

Shields, M.W., Johnson, A.C., Pandey, S., Cullen, R., González-Chang, M. *et al*. (2019) History, current situation and challenges for conservation biological control. *Biological Control* 131, 25–35. DOI: 10.1016/j.biocontrol.2018.12.010.

Simpson, M., Gurr, G.M., Simmons, A.T., Wratten, S.D., James, D.G. *et al*. (2011) Field evaluation of the 'attract and reward' biological control approach in vineyards. *Annals of Applied Biology* 159(1), 69–78. DOI: 10.1111/j.1744-7348.2011.00477.x.

Solomou, A.D. and Sfougaris, A.I. (2012) Bird community characteristics as indicators of sustainable management in olive grove ecosystems of Central Greece. *Journal of Natural History* 49(5–8), 301–325. DOI: 10.1080/00222933.2014.987839.

Thuiller, W. (2007) Climate change and the ecologist. *Nature* 448(7153), 550–552. DOI: 10.1038/448550a.

Torres, J.B. and Bueno, A. de F. (2018) Conservation biological control using selective insecticides – a valuable tool for IPM. *Biological Control* 126, 53–64. DOI: 10.1016/j.biocontrol.2018.07.012.

Turlings, T.C.J. and Erb, M. (2018) Tritrophic interactions mediated by herbivore-induced plant volatiles: mechanisms, ecological relevance, and application potential. *Annual Review of Entomology* 63, 433–452. DOI: 10.1146/annurev-ento-020117-043507.

Uylaşer, V. and Yildiz, G. (2014) The historical development and nutritional importance of olive and olive oil constituted an important part of the mediterranean diet. *Critical Reviews in Food Science and Nutrition* 54(8), 1092–1101. DOI: 10.1080/10408398.2011.626874.

Villa, M., Santos, S.A.P., Aguiar, C. and Pereira, J.A. (2021) Plants biodiversity in olive orchards and surrounding landscapes from a conservation biological control approach. *International Electronic Conference on Plant Science* 4(1), 66. DOI: 10.3390/IECPS2020-08604.

Voigt, W., Perner, J., Davis, A.J., Eggers, T., Schumacher, J. *et al*. (2003) Trophic levels are differentially sensitive to climate. *Ecology* 84(9), 2444–2453. DOI: 10.1890/02-0266.

Wade, M.R., Zalucki, M.P., Wratten, S.D. and Robinson, K.A. (2008) Conservation biological control of arthropods using artificial food sprays: Current status and future challenges. *Biological Control* 45(2), 185–199. DOI: 10.1016/j.biocontrol.2007.10.024.

Walther, G.-R. (2010) Community and ecosystem responses to recent climate change. *Philosophical Transactions of the Royal Society B* 365(1549), 2019–2024. DOI: 10.1098/rstb.2010.0021.

Walther, G.-R., Post, E., Convey, P., Menzel, A., Parmesan, C. *et al*. (2002) Ecological responses to recent climate change. *Nature* 416(6879), 389–395. DOI: 10.1038/416389a.

Wang, X.-G., Johnson, M.W., Daane, K.M. and Yokoyama, V.Y. (2009) Larger olive fruit size reduces the efficiency of *Psyttalia concolor* as a parasitoid of the olive fruit fly. *Biological Control* 49(1), 45–51. DOI: 10.1016/j.biocontrol.2009.01.004.

37 Protection of Durable Stored Products of Organic Farming

Schöller Matthias*

Department of Life Sciences, Humboldt Universität zu Berlin, Berlin, Germany

Abstract

To maintain the quality of durable organic farming products after harvesting, it is imperative to protect them from fungi, insects and mites. Stored-product protection strategies should be implemented at every stage of the production chain, from loading the grain on-farm into silos, flat-storage boxes or bags, to storage in food processing companies, and finally in retail trade and private households. To ensure the long-term storage of durable products such as grain, rice, millet, dried fruits, herbs, spices or nuts, it is crucial to provide specific storage conditions. The design of the storage rooms also plays a significant role in determining the possibilities for long-term storage. Integrated Stored Product Protection (IPM) is the strategy to address these various storage conditions. For organic farming, preventative methods, including storage room design and temperature regulation, are of utmost importance. Pest monitoring is a second component of IPM, accomplished through sampling and trap usage. In case of infestation, there are various control measures available, including biological, physical and chemical options, whereby the quantities of chemical control products should be reduced as far as possible. The regulatory standards that apply to organic agricultural field production also apply to products as they move further through the food system, including stored products. Some techniques were developed specifically for organic farming, including diatomaceous earths, pyrethrum products without synthetic synergists, and biological control applications. This chapter provides a comprehensive review of the currently applied IPM options for protecting durable products of organic farming during storage. Practical examples are provided, and future research needs are outlined.

Keywords: stored-product protection, organic products, prevention, control

Introduction

Stored-product protection is a key strategy for ensuring the quality of organic farming products and global food security. This includes not only grains, but also processed grain products such as flour or oat flakes, as well as other raw commodities like dried fruit, nuts, herbs and spices. IPM measures must be implemented in storage rooms and areas in food processing facilities to ensure the highest level of protection for these products. This section does not include perishable plant products, products of animal origin or finished food products.

According to Lipinski *et al.* (2013), cereal crop losses worldwide amount to 19% on a weight basis (Kumar and Kalita, 2017). Substantial losses can occur from farm to private households all along the chain from harvest to distribution of the final products. For example, in India, the loss due to insect stored-product pests is estimated to account for 20–25% damage (Vijayan *et al.*, 2023). Cowpea farmers in West Africa lose a substantial portion of their harvest during storage due to cowpea weevils *Callosobruchus maculatus*, ranging from 10% to as high as 100% after only a few months (Moussa *et al.*, 2014). Post-harvest losses account for direct physical losses, and quality losses due to insects (Fig. 37.1), mites, rodents, birds or mould. This chapter reviews the IPM methods currently used in organic farming, provides references for further reading, and identifies future research needs.

*Corresponding author: Matthias.Schoeller@hu-berlin.de

DOI: 10.1079/9781800626850.0037

Fig. 37.1. Common stored-product insects in Western Africa: (1) *Sitophilus zeamais* (2) *Tribolium castaneum* (3) *Cathartus quadricollis* (4) *Tenebroides mauritanicus* (5) *Cryptolestes pusilloides* (6) *Palorus subdepressus* (7) *Carpophilus dimidiatus* (8) *Anisopteromalus calandrae*.

Development of Stored-Product Protection Techniques for Organic Farming

Most of the components of Integrated Pest Management (IPM) for stored-product protection were developed already in the period 1910–1940, including aeration, heat treatment, biological control, modified atmospheres, botanicals like pyrethrum and the use of inert dusts. However, after World War II, chemical control by synthetic chemical products formulated as powders, sprays or toxic fumigants became the dominant method to protect durable stored products (Hagstrum and Philips, 2017). However, pest resistance, workers' safety and an increasing awareness of pesticide residues in food led to a renewed interest in the previously developed non-chemical stored-product methods up to the 1980s. At the same time, the land used for organic farming increased and the need for appropriate protection of products of organic farming became obvious. In detail, the definition of IPM for stored products differs from those proposed for field production. The following specific definition was proposed:

> An IPM system in stored-product protection comprises hygiene, technical, technological and biotechnical methods, physical control, biological control and chemical control. These methods should be harmonised in a way granting highest priority to the protection of the human health as well as the environment. Moreover, commercial policy is a possible further element of IPM.
>
> (Schöller *et al.*, 1997)

Methods and products for stored-product protection specifically developed for organic farming include water-based emulsions of pyrethrum products formulated without synthetic chemical synergists, amorphous diatomaceous earths, and biological control.

IPM for Protection of Organic Products

Pest management in organic systems, or organic pest management (OPM), is similar to IPM. The main difference between these two management approaches is that in OPM, products and practices must also meet the more stringent standards of national organic programmes (Grieshop *et al.*, 2012) or of organic farming associations, if the farmer or producer is a member of the respective association. Lists of pesticides compatible with organic systems are available, e.g. from the Organic Materials Review Institute (OMRI, www.omri.org) in the USA or FiBL Projekte GmbH (FiBL, www.fibl.org) in central Europe. National regulations are derived from the guidelines of the International Federation of Organic Agriculture Movements (IFOAM). In the European Union, the Regulation on Organic Production and Labelling of Organic Products applies, and in the USA, the USDA National Organic Program. The same regulatory standards that apply to organic agricultural production also apply to products as they move through the food system, including stored products.

Prevention

Sealing the building and the storage structure

Both the plant exterior and the plant interior can be modified in order to avoid pest infestation (Heaps, 2012). Once the building is optimized, protecting the organic products is a continual challenge for organic food manufacturers.

Sanitation

Appropriate sanitation programmes for stored-product facilities throughout the post-harvest supply chain have been recognized as the foundation of good IPM programmes for a long time. Decreased sanitation negatively affects the efficacy of most IPM tactics, with a mean 1.3–17-fold decrease in efficacy compared to better sanitation (Morrison *et al.*, 2019). For example, Arthur (2000) found that the presence of flour at the time of application for diatomaceous earth significantly increased the survival of the flour beetles *T. castaneum* and *T. confusum* by up to 100-fold. However, the exact impact depending on various control techniques is still under evaluation (Morrison *et al.*, 2019).

Monitoring

Monitoring insects, mites and rodents is a basic and important component of integrated pest management. Economic losses can be avoided by supressing pest populations in an early state. Monitoring provides information about species and developmental stage, age structure, density, spatial distribution, and population growth rate. Monitoring techniques differ for bulk raw commodities, and storage and processing rooms (Table 37.1). For bulk raw commodities, a sample can be obtained from the moving grain with a pelican sampler or a diverter, or with the help of an Ellis cup as grain moves along a conveyor belt.

The cost of sampling is composed of the cost of labour plus the cost of traps etc. The cost of a sampling programme must be justified by the potential economic losses. Finally, the cost of sampling plus for the pest management programme adds to the economic losses caused by stored-product pests. The sampling plan has to be uniquely suited to a particular stored-product situation. This is because few storages and buildings as well as the

Table 37.1. Sampling methods for different stored-product situations.

Stored-product situation	
Bulk commodity	Probe trap, deep-bin cup, vacuum probing, pheromone-baited traps
Moving grain	Ellis cup, pelican sampler, diverter
Bags	Triers
Processing plants/empty room	Visual inspection, sticky traps, food-baited traps, pheromone-baited traps, light traps

Fig. 37.2. Dome-trap for monitoring flour beetles *Tribolium* spp. in an organic bakery.

combination of stored products and potential pest species are identical.

Sampling

The storage of durable agricultural products is mostly characterized by large quantities, and pest insects and mites are very small, i.e. monitoring pests is like finding a needle in a haystack. This problem is approached by taking a small number of representative samples, or by using traps to monitor pest activity over time. Obviously, insect movement inside and outside grain bulks and processed products influences the monitoring results. Insect movement may result in clumped insect spatial distributions and thus makes large sample sizes necessary for monitoring (Jian, 2019).

Traps

Unbaited sticky traps are the most commonly used traps used throughout the marketing system Traps baited with pheromones or food baits (Fig. 37.2) are available for a number of key pests (Hagstrum and Subramanyam, 2000).

UV-light traps are unspecific attracting all kinds of insects, consequently the building has to be properly sealed to attract the pests only. Adhesive trapping surfaces inside the light trap prevent scattering of insect debris compared to light traps without such glueboards (Harris, 2006). These traps are very effective to monitor the biscuit beetle *Stegobium paniceum* and the tobacco beetle *Lasioderma serricorne*.

Optical systems

Recently, an integrated insect monitoring system that employs a simple RGB camera and data-driven deep-learning models to detect and identify stored-product insect species was proposed. A computer vision model from the YOLO family was trained and evaluated on an insect dataset (Badgujar *et al.*, 2023). Such systems are still under development, probably they can compensate for the lack of specialists able to identify the insects in the future and determine rapidly a high percentage of the samples.

Acoustic detection

For small grain samples (c.500 g) a hand-held device is available, which monitors the feeding sound of stored-product pests with headphones as well as indicating these by a light signal. Recently, the so-called beetle sound tube system was made available for the market, which consists of large perforated metal tubes and an acoustic device. The beetle sound tubes are available in different sizes, for monitoring big bags, flat stores and large silos. Field trials showed the detection of stored-product beetles eight weeks earlier compared to probe traps at the grain surface, and ten weeks earlier compared to temperature monitoring (Müller-Blenkle *et al.*, 2022).

Mould and mycotoxins

Moist grain will result in quality deterioration (e.g. germination loss), or grain will turn into silage. Mould growth rapidly increases if grains are not properly dried and stored, producing aflatoxin, which causes liver cancer and other health problems. Monitoring for spoilage and mycotoxins is therefore another important aspect. Grain spoiled by fungal growth has an off-odour that can be described as musty, sour, earthy, or even putrid. The odour can be easily perceived during the standard quality control, or with electronic nose technology. Mycotoxin testing is done by specialized laboratories (Martinez and Woloshuk, 2012).

Resistant grain varieties

Different physicochemical features like kernel hardness and protein content affect the susceptibility of wheat to stored-grain pests (D'Isita *et al.*, 2024). Cereals differ in their susceptibility

Fig. 37.3. Drying of teff *Eragrostis* tef in Ethiopia (*left*); detail of teff (*right*).

towards stored-product pests, the small grains of teff *Eragrostis tef* (Fig. 37.3), for example, are not suitable for the development of weevils in the genus *Sitophilus*. Resistant grain varieties play a significant role in subsistence farming. In Ethiopia, for the storability of sorghum landraces, there is a correlation between rice weevil *Sitophilus oryzae* progeny production and sorghum weight loss (Teshome *et al.*, 1999). Some old durum wheat genotypes were found to be more resistant towards *Sitophilus granarius* weevils compared to new ones (D'Isita *et al.*, 2024). However, the decision for a certain variety depends as well on many other factors, like milling quality, marketability, drought tolerance or disease resistance.

Control

The selection of an appropriate control method depends on the area of application (Table 37.2).

Biological control

Biological control is the use of living organisms to control pests. The organisms used for biological control are called beneficials, natural enemies or antagonists. In order to prevent a negative effect on the native fauna, regionally established natural enemies should be released for biological control, otherwise beneficials have to be evaluated from a nature-protection perspective (Schöller *et al.*, 2018). In central Europe, the commercialization and use of beneficial insects for stored-product environments is well-established (Hervet and Morrison III, 2021). The most frequently applied natural enemies are *Habrobracon hebetor* and *Trichogramma* spp. (Fig. 37.4) against stored-product moths, and *Ansiopteromalus calandrae* (Fig. 37.1) and *Lariophagus distinguendus* against various stored-product beetles like weevils or the drugstore beetle (Schöller *et al.*, 2018). In flat grain storage and empty rooms, the release of parasitoids every 10–20 sq. m was found to be effective in several studies, e.g. for *Anisopteromalus calandrae* against the tobacco beetle (Guo *et al.*, 2023).

Physical control

Physical control techniques are an important component of IPM, as they do not produce residues and preserve the quality of the products when properly applied.

GRAIN AND FLOUR CLEANING. Substantial levels of insect mortality can be achieved by specific grain cleaning machinery. Ninety per cent to 100% of adult insects can be removed from the grain, as well as larvae of external feeders, and mites (Armitage *et al.*, 1996). However, the immature stages of internal feeders, like the granary weevil, are not affected. A commercial laminar airflow grain cleaner was shown to remove 99.5% of mixed-age *T. castaneum* and 93.5% of mixed-age *R. dominica* with a grain weight loss of less than 1.5% (Weller *et al.*, 1998).

SIEVING. Flour is sieved prior to packaging or selling it on the market in rural areas. Sieves up to a mesh-width of 100 µm allow the elimination of

Table 37.2. Control measures for the protection of organic stored products. National registrations for the respective chemical control products have to be checked prior to application. (From Adler *et al.*, 2021)

Area of application	IPM control options
Empty rooms	Beneficials Diatomaceous earth Heat Pyrethrum
Rooms with stored products (treatment of cracks and crevices)	Beneficials Diatomaceous earth Pheromones Pyrethrum
Preventative treatment or for reduction of infestation	Beneficials Carbon dioxide under high pressure Diatomaceous earth Entoleter Freezing Sieving
Treatment of actual infestation	Carbon dioxide Diatomaceous earth Freezing Entoleter
Treatment of actual infestation by rodents	Anticoagulant baits Carbon dioxide Chloralose Mechanical traps

Fig. 37.4. Release unit for *Trichogramma parasitoid* wasps attached to a shelf with bagged stored products.

insects including the eggs of flour beetles and the Mediterranean flour moth.

MECHANICAL DISINFESTATIONS. Entoleters are used for grain prior to milling or for treating flour, and effectively destroy all developmental stages including eggs. Kernels are destroyed to a large proportion, this is why entoleters are mostly used in mills (Stratil *et al.*, 1987).

Manual disturbance methods tailored for smallholder farmers have been shown to be effective to reduce the infestation of grain by the maize weevil (Suleiman *et al.*, 2016). Recently, the effectiveness of periodic stirring of grain was shown to suppress maize weevils and other beetle populations in large-scale grain storage bins (Sserunjogi *et al.*, 2023). However, mechanical disinfestation, except for flour, affects also the quality of the product to some extent, typically by breaking seeds.

TEMPERATURE MODIFICATION. Extreme temperatures, at or above 50°C and at or below 15°C, can be used to disinfest commodities and structures (Fields *et al.*, 2012).

COOLING AND FREEZING. Cooling stored products can either be obtained with ambient or refrigerated aeration. Cooling over a longer period can lead to increased mortality in stored-product pests; however, the reliable main effect is the delay or interruption of development of insects and

Fig. 37.5. Heavy infestation of grain by *Trogoderma granarium*.

mites. Mites are generally more cold-tolerant; the threshold temperature for population growth can be as low as 7°C for *Acarus siro*. For major stored-product beetles it is considerably higher, e.g. about 14°C for *S. zeamais* at 70% r.h. and 20°C for *T. castaneum* and *Trogoderma granarium* (Fig. 37.5) (Beckett, 2010).

Most stored-product moths start development at 15°C or above, but *Ephestia kuehniella* and *Ephestia elutella* are starting development already at 12 and 10°C, respectively (Rees, 2004). The forecast of actual emergence of adult moths is further complicated by diapause of the last larval instar in *E. elutella* and *Plodia interpunctella*. On the other hand, termophilous species might be cold-tolerant as well, especially if they can acclimate. For example, 5% of an acclimated adult population of *Cryptolestes ferrugineus* was estimated to survive 75 days at –10°C (Fields *et al.*, 2012). Finally, climate, pest species, target conditions and cost determine whether cooling with ambient or refrigerated air achieves economic and effective outcomes (Beckett, 2010).

Chilled aeration is the technique usually applied for whole grain and seeds. A large high-capacity air conditioning unit is attached to the aeration system of a silo or flat storage. Typically, temperatures of 10–15°C are targeted. Organically certified dried fruits and nuts must be disinfested of field pests prior to packaging. Commercial cold-storage facilities offer –20°C to –30°C for this purpose. For certain high-value items, flash-freezing with liquid nitrogen to obtain –18°C is recommended (Fields *et al.*, 2012). Temperature data-loggers placed inside the interior of the product mass monitor the freezing process. In principle, freezing is also possible in case of infestation of processed packed products in retail facilities or warehouses; however, contrary to bulk products, these cannot be cleaned from dead insects, faeces or exuviae without unpacking.

HEAT. At moderately high temperatures, heat tolerance is species-dependent. Identification of the pest species is important here. Typically, high insect mortality can be achieved at grain temperatures as low as 43°C, which is the maximum recommended for malting and seed barley, seed wheat, canola, and grain legumes at high moisture content (Hill, 1999).

Even though stored-product pests are specialized to develop on dry plant materials and do not need to drink, most species need a certain minimum moisture content of the products. For example, in the saw-toothed grain beetle *Oryzaephilus surinamensis*, at 20°C, mortality is 46–70% r.h., but as much as 99% at 30% r.h. (approximately 9% moisture content of the grain) (Beckett and Evans, 1994). Some stored-product pests are very sensitive to drying, especially psocids and mites which cannot survive at any temperature below about 60% r.h. for psocids (Rees, 2004) and 65% r.h. for mites (Navarro *et al.*, 2002).

The use of heat to disinfest entire buildings including their equipment to 50–60°C for 24–36 h to control stored-product pests is applied in the food industry and in mills. Heaters are used to gradually heat the air in the structure. A long treatment period is necessary for the heat to penetrate wall voids and equipment to kill insects harbouring in them (Fields *et al.*, 2012).

INSECT TIGHT PACKAGES. Insect tight packaging of consumer packages is another component to prevent losses. Egg larvae of moths can invade openings down to 0.014 mm (Adler *et al.*, 2021). Packages are tested for insect tightness by exposing them to the respective pest species and identification of openings.

HERMETIC STORAGE. Hermetic storage requires airtight conditions created during storage. Respiration and metabolic activities of insects, mites or the product itself lead to the depletion of oxygen and release of carbon dioxide inside the hermetic containers (Baributsa and Conception Ignacio, 2020). Commonly used hermetic technologies include silos (metal and plastic), drums, cocoons, plastic bags, and other containers. Hermetic bags such as the Purdue Improved Crop Storage (PICS) bags are the most widely disseminated among smallholder farmers in sub-Saharan Africa and Asia. Hermetic bags consist of two basic components, an inner liner and an outer bag. The inner liner provides a gas barrier, it could be an individual or multi-layered film. The outer bag is a polypropylene bag designed to provide protection during handling. There are several types of hermetic bags, single-layer, double-layer, and triple-layer bags manufactured by several licensed plastic companies in sub-Saharan Africa, Asia and Latin America (Murdock and Baributsa, 2014). To maintain quality during storage in hermetic bags, grains must be dried to the recommended moisture content (e.g. 13–14% moisture content for maize). The introduction of hermetic bags was a major achievement in more than 35 countries in sub-Saharan Africa (mainly) and Asia, with more than 20 million bags sold and more than 65,000 villages reached with training activities (Baributsa and Conception Ignacio, 2020).

VACUUM STORAGE. Vacuum storage in flexible PVC chambers has been demonstrated as an effective means for maintaining quality and controlling insect pests in smaller volumes (approximately up to 50–60 tonnes) of peas, beans, wheat, corn and sunflowers for extended lengths of time (Finkelman *et al.*, 2002). In smaller applications, vacuum is applied, and the liner shrinks over the products (Fig. 37.6). Successful control times have been demonstrated at 55 mm of Hg. Even when a moderate infestation is already present, the pests are controlled over time (Adler *et al.*, 2021).

Fig. 37.6. Vacuum-packed nuts.

Chemical control

BOTANICALS: PYRETHRUM. Among botanical extracts used as insecticides, pyrethrum is commercially the most successful compound. In products for conventional farming, pyrethrum is mixed with the semi-synthetic synergist Piperonyl butoxide (PBO) (Marchand *et al.* (2018)). This synergist is already being phased out for the aforementioned use in most countries in organic farming (Bioland, 2022). Already in the 1930s, plant oils like sesame oil and rapeseed oil were found to be the most efficient natural synergists, and this was confirmed in recent studies (Marchand *et al.*, 2017). Various producers provide pyrethrum products with natural synergists to be applied as emulsions with water.

Essential oils are a promising alternative because of their worldwide availability and relative cost-effectiveness (Orlando *et al.*, 2018). However, except for pyrethrum, almost all the trials with botanicals were carried out under laboratory conditions, while no experiments were conducted under real operating conditions.

MODIFIED ATMOSPHERES. Carbon dioxide can be used under ambient pressure to control all developmental stages of insects and mites; the CO_2 content must be at least 60% by volume. The oxygen containing air in the storage facility is replaced by purging with the gas until the oxygen content drops to a low level at the ventilation port. After this initial purging, the necessary amount of gas to keep the oxygen content within the treated volume sufficiently low depends mainly on the gas tightness of the structure. The effective treatment period depends on temperature; at 20°C it is, e.g.,

Fig. 37.7. Pressure chamber for carbon dioxide under high pressure.

three weeks for *S. granarius*, while at 30°C it is four days for *O. surinamensis* (Adler *et al.*, 2000).

New concrete silos and welded steel silo bins may be sufficiently gas-tight without additional effort, otherwise cracks can be sealed with, e.g. polyurethane foam or silicone adhesives. Bags, pallets or other smaller quantities can be treated under properly sealed wraps.

Carbon dioxide under high pressure is lethal to insects and mites (Adler *et al.*, 2000). It is applied in an autoclave, a pressure chamber (Fig. 37.7), which has, e.g. a diameter of 2.2 m and a length of 6.4 m, and the products treated are left in their original packaging on pallets. The treatment period is typically two hours at 20 bar. Two autoclaves allow the re-use of part of the gas, which is otherwise directly exhausted into the atmosphere.

DIATOMACEOUS EARTHS. Diatomaceous earths (DEs) are a type of naturally occurring soft siliceous sedimentary rock, consisting of the fossilized exoskeleton of unicellular algae, i.e. phytoplankton, diatoms. Other types of inert dusts, such as zeolites or kaolin, and ash, have also been investigated for control of stored-product pests; however, among such different types of inert materials, DEs have a prominent position in practice (Zeni *et al.*, 2021). DEs are mainly composed of silicon dioxide. The mode of action is primarily physical, with DE particles attaching to the insect cuticle and causing death by desiccation. DEs are compatible with organic farming as they are natural substances with no toxicity to mammals or the environment. DEs are generally not compatible with biological control as natural enemies are also killed by the dust.

TREATMENT OF BULK GRAIN. Typically, DEs must be applied at relatively high concentrations to be effective. Preventative treatments work with 1000 ppm, but treatments of actual infestations often exceed 1500 ppm (Korunic *et al.*, 1996). The effectiveness is comparable with those of synthetic contact insecticides. DE is either mixed with the whole grain bulk during loading, or applied to the grain surface layer only. This is why they create a 'dusty' surface on the stored products or surfaces. Application of DE on stored products results in the reduction of the test weight (weight to volume ratio), which is a critical characteristic in the international grain market and also important for quality control in mills (Korunic *et al.*, 1996).

EMPTY ROOM TREATMENT. DEs can also be used for empty room treatment. The challenge here is the uneven distribution on walls and ceilings compared to the floor (Schöller and Reichmuth, 2004). A slurry formulation of DE, i.e. a mixture of DEs and water, can overcome this problem. It may not be as effective as dust (powder) formulations. However, a slurry formulation can be more practical in terms of application with the same technology as traditional grain protectants (Zeni *et al.*, 2021).

Conclusions and Future Research Needs

Increasing urbanization, climate change and land use for non-food crop production are factors challenging the food demand of a rapidly growing global population (Kumar and Kalita, 2017). Reducing post-harvest losses, particularly in developing countries, is a key component of increasing food availability. Global warming will pose new challenges for the protection of stored products. This issue is beginning to be addressed (Adler *et al.*, 2022). An increase in average temperature will result in an increase in the annual number of generations for most stored-product pests, as they are thermophilous and typically develop optimally at temperatures between 25 and 32°C. In temperate regions, stored-product pests of tropical and subtropical origin are likely to expand their range northwards, as has already been observed in central Europe (Müller-Blenkle *et al.*, 2022). At the same time, however, the energy required for heat treatment may decrease and some storage rooms, such as flat stores or upper floors of mills, may become too hot even for the development of stored-product insects, as has been observed in recent years. All these aspects show the need for future research to ensure the quality of organic stored products and food safety. In the short term, there is a need to train pest control professionals to meet the needs of the organic industry, as typically local pest control companies work for companies processing organic products. In tropical countries, training of smallholder farmers is needed. To establish biological control outside central Europe, more local production sites need to be established, as the range of shipment is limited for most beneficials. The large body of knowledge on the efficacy of botanicals in laboratory trials would need to be translated to a practical level for the development of new products for organic farming, particularly the potential of repellents, which is still poorly understood. However, there is limited research capacity worldwide on the protection of stored products. Government agencies, organizations and universities need to ensure the development of facilities to provide research, information and training for the adaptation and effective use of stored-product technologies.

References

Adler, C.S., Corinth, H.-G. and Reichmuth, C. (2000) Modified atmospheres. In: Subramanyam, B. and Hagstrum, D.W. (eds) *Alternatives to Pesticides in Stored-Product IPM*. Kluwer, Boston, Massachusetts, pp. 105–146.

Adler, C.S., Athanassiou, C., Carvalho, M.O., Emekci, M., Gvozdenac, S. *et al*. (2022) Changes in the distribution and pest risk of stored product insects in Europe due to global warming: Need for pan-European pest monitoring and improved food safety. *Journal of Stored Products Research* 97, 101977.

Adler, C.S., Kühne, S., Preißel-Reckling, S., Prozell, S. and Schöller, M. (2021) Vorräte richtig schützen und lagern. In: *Vorräte Richtig Schützen Und Lagern*. Landwirtschaft, Verarbeitung Und Handel. Ulmer.

Armitage, D.M., Cook, D.A. and Duckett, C. (1996) The use of an aspirated sieve to remove insects, mites and pesticides from grain. *Crop Protection* 15(8), 675–680.

Arthur, F.H. (2000) Impact of food source on survival of red flour beetles and confused flour beetles (Coleoptera: Tenebrionidae) exposed to diatomaceous earth. *Journal of Economic Entomology* 93(4), 1347–1356.

Badgujar, C.M., Armstrong, P.R., Gerken, A.R., Pordesimo, L.O. and Campbell, J.F. (2023) Real-time stored product insect detection and identification using deep learning: System integration and extensibility to mobile platforms. *Journal of Stored Products Research* 104, 102196.

Baributsa, D. and Conception Ignacio, M.C. (2020) Developments in the use of hermetic bags for grain storage. In: Maier, D. (ed.) *Advances in Postharvest Management of Cereals and Grains*. Burleigh Dodds Science Publishing, Cambridge, UK, pp. 1–28.

Beckett, S.J. (2010) Protecting and disinfesting stored products by drying and cooling, and disinfesting stored products during handling by mechanical treatments. *Julius-Kühn-Archiv* 425, 219–228.

Beckett, S.J. and Evans, D.E. (1994) The demography of *Oryzaephilus surinamensis* (L.) (Coleoptera: Silvanidae) on kibbled wheat. *Journal of Stored Products Research* 30(2), 121–137.

Bioland (2022) *Bioland Richtlinien Schädlingsbekämpfung*. Bioland e.V., Mainz.

D'Isita, I., Pistillo, O.M., Palma, A.M., Vita, P. and Germinara, G.S. (2024) Susceptibility of old and modern wheat genotypes to *Sitophilus granarius* (L.) and *Rhyzoperta dominica* (F.). *Journal of Stored Products Research* 106, 102265.

Fields, P., Subramanyam, B. and Hulasare, R. (2012) Extreme temperatures. In: Hagstrum, D., Phillips, T. and Cuperus, G. (eds) *Stored Product Protection*. Kansas State University, Manhattan, Kansas, pp. 179–190.

Finkelman, S., Navarro, S., Isikber, A., Dias, D., Azriel, A. *et al*. (2002) Application of vacuum to sealed flexible containers: A viable alternative to disinfestation of durable commodities with methyl bromide. In: Batchelor, T.A. and Bolivar, J.M. (eds) *Proceedings of the International Conference on Alternatives to Methyl Bromide*, European Commission, Brussels, Seville, Spain, pp. 145–149.

Grieshop, M.J., Rogers, D.T. and Arthur, F.H. (2012) Organic approaches and regulations for stored product pest management. In: Hagstrum, D., Phillips, T. and Cuperus, G. (eds) *Stored Product Protection*. Kansas State University, Manhattan, Kansas, pp. 233–242.

Guo, Y., Lü, J., Bai, C., Guo, C., Huang, Y. *et al*. (2023) Dispersal and parasitic ability of *Anisopteromalus calandrae* (Hymenoptera: Pteromalidae) against *Lasioderma serricorne* (Coleoptera: Anobiidae) larvae in an experimental warehouse. *Journal of Stored Products Research* 104, 102210.

Hagstrum, D.W. and Philips, T.W. (2017) Evolution of stored-product entomology: Protecting the world food supply. *Annual Review of Entomology* 62, 379–397.

Hagstrum, D.W. and Subramanyam, B. (2000) Monitoring and decision tools. In: Subramanyam, B. and Hagstrum, D.W. (eds) *Alternatives to Pesticides in Stored-Product IPM*. Kluwer Academic Publishers, Boston, Massachusetts.

Harris, J.E. (2006) Insect light traps. In: Heaps, J. (ed.) *Insect Management for Food Storage and Processing*, 2nd edn. AACC International, St. Paul, Minnesota, pp. 55–66.

Heaps, J.W. (2012) Food plant sanitation, pest exclusion, and facility design. In: Heaps, J. (ed.) *Insect Management for Food Storage and Processing*, 2nd edn. AACC International, St. Paul, Minnesota, pp. 85–93.

Hervet, V.A.D. and Morrison III, W.R. (2021) Prospects for use of biological control of insect and mites for the food industry in North America. *Agronomy* 11, 10.

Hill, M. (1999) The characteristics of grain and seeds and their reaction to climate. In: Hill, M. (ed.) *The Drying and Storage of Grain and Herbage Seeds. Foundation for Arable Research*. Lincoln, New Zealand, pp. 3–21.

Jian, F. (2019) Influences of stored product insect movements on integrated pest management decisions. *Insects* 10(4), 100.

Korunic, Z., Fields, P.G., Kovacs, M.I.P., Noll, J.S., Lukow, O.M. *et al*. (1996) The effect of diatomaceous earth on grain quality. *Postharvest Biology and Technology* 9(3), 373–387.

Kumar, D. and Kalita, P. (2017) Reducing postharvest losses during storage of grain crops to strengthen food security in developing countries. *Foods* 6(1), 8.

Lipinski, B., Hanson, C., Lomax, J., Kitinoja, L., Waite, R. *et al*. (2013) *Installment 2 of Creating a Sustainable Food Future, Reducing Food Loss and Waste*. World Resource Institute, Washington, DC.

Marchand, P.A., Dimier-Vallet, C. and Vidal, R. (2018) Biorational substitution of piperonyl butoxide in organic production: Effectiveness of vegetable oils as synergists for pyrethrums. *Environmental Science and Pollution Research International* 25(30), 29936–29942.

Martinez, E.M. and Woloshuk, C. (2012) Monitoring for spoilage and mycotoxins. In: Hagstrum, D., Phillips, T. and Cuperus, G. (eds) *Stored Product Protection*. Kansas State University, Manhattan, Kansas, pp. 283–287.

Morrison III, W.R., Bruce, A., Wilkins, R.V., Albin, C.E. and Arthur, F.H. (2019) Sanitation improves stored product insect pest management. *Insects* 10(3), 77. DOI: 10.3390/insects10030077.

Moussa, B., Abdoulaye, T., Coulibaly, O., Baributsa, D. and Lowenberg-DeBoer, J. (2014) Adoption of on-farm hermetic storage for cowpea in West and Central Africa in 2012. *Journal of Stored Products Research* 58, 77–86. DOI: 10.1016/j.jspr.2014.02.008.

Müller-Blenkle, C., Simon, U., Szallies, I., Prozell, S., Schöller, M. *et al*. (2022) Large-scale trapping and acoustic detection of beetles in grain storage using the 'beetle sound tube' system. *Integrated Protection of Stored Products IOBC-WPRS Bulletin* 159, 12–17.

Murdock, L.L. and Baributsa, D. (2014) Hermetic storage for those who need it most: Subsistence farmers. In: Arthur, F.H., Kengkanpanich, R., Chayaprasert, W. and Suthisut, D. (eds) *Proceedings of the 11th International Working Conference on Stored Product Protection*, Chiang Mai, Thailand, November 24–28, 2014, pp. 310–323.

Navarro, S., Noyes, R., Armitage, D. and Maier, D.E. (2002) Objectives of aeration. In: Navarro, S. and Noyes, R. (eds) *The Mechanics and Physics of Modern Grain Aeration Management*. CRC Press, pp. 1–34.

Orlando, C., Giunti, G., Russo, A., Plameri, V. and Zappalà, L. (2018) Essential oils in stored product insect pest control. *Journal of Food Quality* 1–18.

Rees, D.P. (2004) *Insects of Stored Products*. CSIRO Publishing, Collingwood, Australia.

Schöller, M. and Reichmuth, C. (2004) Field trials with the diatomaceous earth SilicoSec® for treatment of empty rooms and bulk grain. *Julius-Kühn-Archiv* 425, 899–905.

Schöller, M., Prozell, S., Al-Kirshi, A.-G. and Reichmuth, C. (1997) Towards biological control as a major component of integrated pest management in stored product protection. *Journal of Stored Products Research* 33(1), 81–97.

Schöller, M., Prozell, S., Suma, P. and Russo, A. (2018) Biological control of stored-product insects. In: Athanassiou, C.G. and Arthur, F.H. (eds) *Recent Advances in Stored Product Protection*. Springer-Verlag GmbH, Berlin, pp. 183–209.

Sserunjogi, M., Bern, C.J., Brumm, D.E., Maier, D.E. and Phillips, T.W. (2023) Mechanical stirring of bulk-stored maize in steel bins to suppress maize weevils and other beetle populations. *Journal of Stored Products Research* 106, 102281.

Stratil, H., Wohlgemuth, R., Bolling, H. and Zwingelberg, H. (1987) Optimierung des prallmaschinenverfahrens zum abtöten und entfernen von vorratsschädlichen insekten aus nahrungsmitteln unter besonderer berücksichtigung der qualität der mahlerzeugnisse. *Getreide, Mehl Und Brot* 41, 294–302.

Suleiman, R., Rosentrater, K. and Chove, B. (2016) Periodic physical disturbance: An alternative method for controlling *Sitophilus zeamais* (maize weevil) infestation. *Insects* 7(4), 51.

Teshome, A., Torrance, J.K., Baum, B., Fahrig, L., Lambert, J.D.H. *et al*. (1999) Traditional farmers' knowledge of sorghum (*Sorghum bicolor* [Poaceae]) landrace storability in Ethiopia. *Economic Botany* 53(1), 69–78.

Vijayan, R., Vidhya, C.S., Ramkumar, S., Neladri Sekhar, S., Maheshwari, S. *et al*. (2023) Use of botanicals plant for stored grain pest management: A critical review. *International Journal of Plant & Soil Science* 35(21), 801–809.

Weller, G.L., Cassells, J.A. and Banks, H.J. (1998) The Grainscanner® – could it play a role in the control of insects in bulk grain. In: Banks, H.J., Wright, E.J. and Damcevski, K.A. (eds) *Stored Grain in Australia. Proceedings of the Australian Postharvest Technical Conference*, Canberra, 26–29 May, Stored Grain Research Laboratory CSIRO, Collingwood, Australia.

Zeni, V., Baliota, G.V., Benelli, G., Canale, A. and Athanassiou, C.G. (2021) Diatomaceous earth for arthropod pest control: Back to the future. *Molecules* 26(24), 7487.

38 Approaches to the Control of Late Blight *Phytophthora infestans* in Organic Potato Crops

David Frost[1,2]*

[1]*Principal, Frost Consulting, formerly Senior Consultant, ADAS Pwllpeiran Research Centre, Cwmystwyth, Wales, UK;* [2]*Marketing Strategy Co-ordinator, Organic Centre Wales, Aberystwyth University, Wales, UK*

Abstract

After a brief introduction to the history of the potato and to late blight, *Phytophthora infestans*, this chapter examines methods used to prevent and/or control late blight in organic potato production. The reliance on copper-base fungicides in organic potato production is problematic and they are either restricted, banned or threatened with a ban in many countries. Alternative agronomic strategies and treatments to the use of copper-based fungicides are reviewed, with a particular focus on the development of blight-resistant potato cultivars. The chapter concludes by questioning the implications of exclusive reliance on blight-resistant varieties in organic potato supply chains.

Keywords: potato, late blight, prevention and control, resistant varieties

Introduction

Evidence suggests that potatoes were cultivated in Peru 2000 years ago (Wilson, 1993) and in the Bolivian *altiplano* even earlier (Zuckerman, 1999, citing Hawkes and Francisco-Ortega, 1992). According to local narratives, the pre-Hispanic name for the settlement where the Bolivian city of La Paz is now sited was 'the village of the potato fields' and as the invading Spanish Conquistadores discovered, it was regarded as a place of treasure. When the Spanish discovered the potato in the 1530s, however, they did not realize what tremendous wealth and power they had found: they were looking for gold and silver.

The potato took centuries after its arrival in Europe to become a staple food but by the 19th century 'it could be said that the potato was the vegetable of the industrial revolution in England and perhaps elsewhere' (Zuckerman, 1999, p. 57). But not long after its acceptance into urban and rural European diets, disaster struck: 'The potato blight that invaded Europe in August 1845 was the most vigorous, least predictable the world had ever seen…within weeks the plague had reached England, Belgium, France, Germany and Poland' (Zuckerman, 1999, p. 184).

In Ireland, the epidemic destroyed an estimated 40% of the potato harvest, and in 1846 the return of what we now know as late blight, *Phytophthora infestans*, caused an estimated loss of 90% of the harvest. In a country that had become so dependent on potato as the staple food crop, it is not surprising that by 1847 the Great Famine was causing mass emigration.

Control of Late Blight Using Copper-Based Fungicides

Late blight was first identified in 1846 by Miles Joseph Berkeley. Its origin is thought to be Mexico (most likely) or the Andes, and spread to Europe and elsewhere on infected imported tubers. It is regarded as the greatest threat to potato production worldwide (Rogozina *et al.*, 2023). It affects

*Corresponding author: davidfrost@tyn-yr-helyg.com

DOI: 10.1079/9781800626850.0038

both organic and conventional potato production and spreads rapidly with 3–4 days between infection and sporulation. The use of copper fungicides to combat plant diseases dates from the late 19th century when Alexis Millardet discovered the efficacy of neutralized copper sulphate to reduce grapevine downy mildew (Ayres, 2004; Gessler *et al.*, 2011, cited by Tamm *et al.*, 2022).

The use of copper-based fungicides has been permitted as a control for late blight in organic production systems but there are concerns about the toxicity of copper. In 2018, in the UK, the Agriculture and Horticulture Development Board (AHDB) published an Emergency Authorisation for the use of the copper oxychloride product Cuprokylt, which allowed organic control bodies to issue derogations to use this copper product but no other in the UK (Wacher and Frost, 2018).

In the European Union (EU), copper-based fungicides were due to be prohibited for use in organic farming from March 2020, but the ban has been delayed because of the increased risk of crop diseases and associated economic losses for organic producers in the medium to long term without effective alternative control methods in place (Leifert and Wilcockson, 2005). In Germany, many potato producers take advantage of a 3 kg/ha allowance for copper, and in the wet growing season of 2021, an emergency allowance increased the permitted use of copper to 4 kg/ha (Sumption, 2024).

The UK organic control body, the Soil Association, currently lists five copper compounds permitted for use in organic growing (Soil Association, 2023; Table 38.1).

Many countries have developed a system for predicting the likelihood of late-blight epidemics based on weather forecasting. In the UK, the Agriculture and Horticulture Development Board have developed BlightSpy, which potato growers and agronomists can use to monitor weather forecasts to predict the likelihood of late-blight infestation using Hutton Criteria. A Hutton Criterion occurs when two consecutive days have a minimum temperature of 10°C, and at least 6 hours of relative humidity at or above 90%. BlightSpy also provides a blight risk forecast for the next 48 hours, with up to 8 days ahead available. It also provides data on both temperature and humidity (Agriculture and Horticulture Development Board, 2023).

Aitkenhead *et al.* (2020) investigated why blight strains predominate in certain areas and their rate of spread between locations. Such information can improve the assessment of blight risk. Statistically significant patterns in time and space were found for early outbreaks, overall incidence, and the distribution of various pathogen genotypes. The authors claim that this provides valuable information on the variable risk posed by late blight for the potato production areas of UK. Practical Decision Support Systems (DSS) have also been developed for implementation into practice to reduce the use of copper products to control late blight in potatoes (Organic Research Centre, 2016).

Alternative Agronomic Strategies

The Soil Association (Blogg, 2019) recommends that potato producers should review their growing systems and look for unique improvements. An example is growing in beds rather than on ridges, the traditional method in the UK. As beds retain more moisture in the soil, the need for irrigation is reduced and thereby the associated risk of humidity and late blight is lessened. Blogg also recommends the traditional strategy of topping (defoliation) at the first sign of blight. The aim of potato haulm removal is to prevent the spread of late-blight spores and the

Table 38.1. Permitted pesticides and plant protection products (copper compounds) for UK organic potato production. (From Soil Association Organic Standards for Great Britain: Farming and Growing. Version 1.3, published 25 May 2023. For an overview of the current use of copper-based fungicide products in European organic agriculture, see Tamm *et al.*, 2022)

Copper compounds in the form of:	Guidance for use
copper hydroxide/copper oxychloride/copper oxide/ Bordeaux mixture/tribasic copper sulphate	Compliance with plant protection products (PPPs) legislation – users must check the pesticide register for the authorization of the product to be used and comply with all specifications including the application limit set by the authorization for the product.

infection of tubers. Removal of potato plant haulm means, however, that plants can no longer photosynthesize and tuber growth is restricted.

Avoiding the use of pits to dispose of potato crop outgrades etc. and ensuring the removal of groundkeepers (a.k.a. volunteer potatoes) is widely recognized as a necessary precaution to avoid the infestation of new planted crops. Careful planning of crop cultivars can also help growers avoid late blight. For example, first early varieties, even if not blight-resistant, can finish before the late-blight pathogen can take hold (Hartley, 2014). Pre-sprouting seed potatoes and planting times to minimize crop exposure to late-blight pressure are further widely deployed strategies (Leifert and Wilcockson, 2005).

Alternative treatments

The use of mesh crop covers (widely used to prevent pest damage on, e.g. brassicae crops) to control potato blight has been researched by the Biological Husbandry Unit Future Farming Center in New Zealand (Merfield, 2017) and by the Organic Research Centre (ORC) with on-farm trials in Pembrokeshire and Devon (see https://www.innovativefarmers.org/ [accessed 21/2/25]). The New Zealand research found a slight reduction in potato blight spores under mesh, and the actual occurrence of blight was slightly reduced in one trial though there have been questions as to whether the blight controlled was the less pernicious early blight, *Alternaria solani*, rather than late blight, *Phytophthora infestans* (Wacher and Frost, 2018). Other novel strategies include minimum tillage and thick layers of mulch to reduce late blight and using rows of trees as physical barriers to prevent the spread of blight (Sumption, 2018).

Nechwatal and Zellner (2015) identified potential copper-free products that could be used to control late blight in organic potato farming. They undertook laboratory, greenhouse and field tests on a selection of commercial and non-commercial copper-free preparations as potential substitutes for copper. Their findings highlighted some late-blight reduction potential for some of the alternative products, in particular chitosan, a liquorice and a horsetail test product, and a copper fertilizer. They concluded that these could be used for applications mixed or alternating with copper, which could lead to a further reduction in the use of copper fungicides in organic potato production.

In Switzerland, Dorn *et al.* (2007) evaluated 53 copper-free preparations (CFPs) based on natural compounds including plant extracts and microorganisms and five copper preparations for their potential to control late blight in potatoes. Although there were positive results in laboratory and growth-chamber trials, the rain-fastness persistence of CFPs was low compared to copper. They concluded that more-stable formulations of CFPs need to be developed.

The Soil Association (Blogg, 2019) has argued that the development of biostimulants offers potential to expand organic preventative treatments. The main routes are treating seed, inoculating compost at time of transplanting, and foliar application as the plant develops. They note an overlap with the traditional prophylactic strategies that use compost teas and solutions of horsetail, *Equisetum arvense*.

A trial that included the use of the biostimulant OptiYield Diamond among the treatments to combat late blight was undertaken in Wales in 2017 (Shaw *et al.*, 2019). The project trialled two options for blight control over two growing seasons in 2017 and 2018. These were:

1. Hederin/Alpha-hederin, a saponin occurring in common ivy, *Hedera helix*, extracted and formulated as spray for blight control; and
2. OptiYield Diamond (OptD)[1], a biostimulant product based on phosphonate that could help plants grow more vigorously and, potentially, increase their resistance to pests and diseases, and increase yield.

OptD was tested on its own and in combination with Alpha-hederin to explore its efficacy with respect to managing blight on potato crops and whether there is a synergistic relationship between the two products. Field trials were carried out at two sites in 2017 and 2018. An organic farm near Aberystwyth, which compared the efficacy of two different concentrations each of Hederin and Cuprokylt (a copper oxycholoride fungicide permitted in organic systems) and a control (tap water); and at the Henfaes Research Centre near Bangor, which compared the efficacy of Hederin, OptD, and a combination of Hederin and OptD to standard synthetic fungicides and a control (no treatment).

The results of the trials found that Hederin slowed the early progression of late blight, but it did not prevent blight infection reaching 100% of the crop and was less effective than standard fungicide. OptD was successful both in terms of slowing progression of the disease and reducing the overall level of blight by the end of the season. The results suggested that, on its own, Hederin, at the concentrations tested, does not give comparable control to standard fungicide programmes in conventional systems and to copper in organic systems. However, it may have a role to play in integrated disease management programmes, for example in combination with resistant potato varieties or when mixed with OptD or similar products.

Blight-resistant varieties

In the UK, the AHDB maintains a National List and Database on GB-certified potato varieties that have undergone independent resistance testing for key pests and diseases. For late blight, the results of the National List and Independent Variety Trials are on a 1 to 9 scale where 1 shows a variety is susceptible and 9 shows it is resistant to a disease or damage. The AHDB notes, however, that for most diseases, 'a very high resistance score should not be taken as indicating that a disease will be absent but that there is less risk of the disease developing on these varieties. In addition, pathogens may mutate and this may affect the resistance response of a variety in practice'.

Its website also reports on assessments of existing varieties scoring 5 or more for foliar late blight using the new aggressive 13 A2 genotype of the late-blight fungus. Following these assessments, 'significant changes from historic ratings were recorded' and the database has been updated for 31 varieties[2]. This underlines the importance of regular checking of a cultivar's rating for late-blight resistance.

In The Netherlands, where fungicides are completely banned in organic potato production, the whole supply chain work together through the Potato Covenant to move to 100% use of blight-resistant varieties. In 2016, organic potato producers had suffered big losses to late blight, and Bionext, the Dutch organic umbrella organization, recognized that the availability of resistant varieties was needed to prevent future disasters. Working together, 28 organic potato breeders, growers and retailers signed the covenant 'Accelerated transition to robust potato varieties'. The covenant partners agreed to give these varieties priority in breeding, in seed potato production, in potato cultivation and in sales. The transition was achieved in a very short time. In 2017, there were just three blight-resistant potato varieties on the market; by 2020, the number had risen to 22 (Sumption, 2020).

In the UK, the Soil Association (Blogg, 2019) advises growers to check blight scores on the AHDB variety database (http://varieties.ahdb.org.uk [accessed 21/2/25]) noting that 'These will lower over time'.

Breeding blight-resistant varieties

Blight-MOP was a multi-disciplinary project that aimed to develop a systems approach for the management of potato late blight in EU organic potato production without using copper-based fungicides. The acronym stands for: Development of a systems approach for the **M**anagement of potato late blight (caused by *Phytophthora infestans*) in EU **O**rganic **P**otato production in the absence of copper-based fungicides. The Blight-MOP consortium included 13 partners from seven European countries representing a range of environmental conditions and approaches to potato production in organic systems (Leifert and Wilcockson, 2005).

As part of the Blight-MOP project, assessments were made of the resistance to foliar and tuber blight and yield under 'organic growing conditions'. The assessments were made in Switzerland, France, England and Norway in 2001 and 2002. Several new blight-resistant potato varieties[3] were less susceptible to foliar and tuber blight than most of the currently grown varieties.

The project concluded that the introduction of these new varieties could be constrained by market requirements for 'particular characteristics' of the tubers and that using more blight-resistant varieties in organic systems should be encouraged, but the strategy alone is 'unlikely to eliminate the need to use copper-based fungicides'.

In 2018, the Dutch co-operative Agrico launched a range of *Phytophthora*-resistant (late-blight) varieties which they call the Next Generation potatoes (see Fig. 38.1). The company states that, 'As the fungus mutates, and the world climatic conditions continue to change, Agrico will continue its development work and expects in the years ahead to add even more sustainable, late blight-resistant varieties to its Next Generation package'.

Agrico UK is a part of the Dutch Agrico co-operative and accesses potato varieties bred by Agrico Research in The Netherlands. Bioselect UK is the Agrico brand that brings potato varieties to

Maincrop	Alouette	Beyonce	Carolus	Levante
Early	Jacky	Twinner	Twister	

Fig. 38.1. Agrico Next Generation potato varieties. (From https://www.agrico.co.uk/next-generation-varieties; accessed 4 June 2023)

the UK organic market. In a blind taste of Bioselect UK potato varieties in 2014, Alouette was rated the highest, with 'many positive comments on its flavour and appearance'. The test occurred before Alouette became available as organic seed (Hartley, 2014).

The Sárvári Research Trust (SRT) was set up in 2002 to breed late blight- and virus-resistant potatoes. The trust has made use of germplasm from the Sárvári family in Hungary. This family had been breeding potatoes for low-input agriculture for 50 years and the Sárvári KFT was registered as a breeding company in Hungary in 1996.

In the 1970s, the Soviet government wanted a hardy strain of potatoes for growing across the USSR that would withstand disease and that were not dependent on chemical inputs. Using South American and Mexican wild potato material from the Vavilov collection, genes conferring resistance to common viruses (including PVX, PVY, PLRV) were bred into existing stocks and eventually high resistance to late blight was achieved. UK contact was first made with the Sárvári family in 1994 by the Scottish seed potato grower, Adam Anderson. Anderson and David Shaw of Bangor University subsequently established the Sárvári Research Trust. Naming their new varieties Sarpo (**Sár**vari **po**tato), Sarpo Mira, a clone with high resistance to late blight, was the first SRT variety to be nationally listed in 2002, followed by Axona, another maincrop clone with similar high resistance to late-blight disease. Five varieties are currently available on the company's website: Mira, Axona, Blue Danube, Kifli and Shona (http://sarpo.co.uk/portfolio/ [accessed 21/2/25]).

In 2023, a consortium including Dyson Farming Ltd and the Sárvári Research Trust at Bangor University were awarded a three-year grant to develop ideas to allow potatoes to be grown with regenerative methods, including minimum tillage, and further breeding to develop more robust varieties. Trials are starting in 2023 at the Henfaes Research Centre, Bangor. The SRT is also currently selecting clones with Axona or Mira as one parent and another blight-resistant cultivar for the other. It is aiming to prepare varieties for national listing.

Skea Organics, a leading supplier of organic seed potatoes to farmers and market gardeners throughout the UK, announced on its website in 2023 that although only a third of its varieties were organic, by volume its organic range comprised *c.*80% of its harvest and sales (https://www.potatohouse.co.uk/about-us/ [accessed 21/2/25]).

ADAS undertook demonstrations and evaluations of blight resistant potato varieties in 2004. The main aim of the field trial was to examine the level of resistance to *Phytophthora infestans* in selected Sarpo potato clones in a high-risk blight area (West Wales, UK) and the results were presented to the Final Blight-MOP Conference at Newcastle Upon Tyne University, England in January 2005 (Frost and Clarke, 2005). The trial found blight resistance in Sarpo cultivars (Mira and Axona) was greater than in commercially available cultivars (Stirling, Lady Balfour, Cara, Maris Piper, King Edward, Desiree, Wilja, Majestic, Pentland Crown). There was a considerable range in yield between cultivars - yield differences for commercial cultivars were associated with susceptibility to foliar blight, whereas yield differences in Sarpo cultivars were associated with variations in vigour and ability to out-compete weed competition rather than susceptibility to foliar blight.

Blight-resistant potatoes – organoleptic qualities

At an 'Innovative Farmers' Field Lab in 2018 (Sumption, 2018), a group of participating farmers compared the performance of varieties of blight-resistant potatoes. There were some differences between sites, with cv. 'Cara' showing more resistance to blight than cv. 'Carolus' in Scotland (northern site), while the reverse was true in Gloucestershire, (southern site). The group then taste-tested varieties as part of their attempt to persuade consumers and retailers that 'different varieties can be good to eat'.

This led to a discussion on consumer behaviour and buying habits. Other questions raised were whether organic consumers are being denied choice in the range of varieties available to them and if varieties could be sold by their purpose (e.g. for salad, mash, roast, chipped etc.) rather than by name; or should celebrity chefs be encouraged to name blight-resistant varieties in their recipes?

This raises the issue of the taste and eating quality of new blight-resistant varieties. Noticeably, the SRT website promotes the eating quality of its potatoes in the face of some consumer resistance with, for example, quotes from 'Top River Cottage Chef, Gelf Anderson' praising the Sarpo variety Shona as making 'good mash and amazing chips'.

Similarly, the Bioselect New Generation potato varieties are each described as having good eating qualities or are praised for their flavour on the website (https://www.agrico.co.uk/uk-varieties; accessed 21/2/25).

Many taste tests simply involve varieties being given scores for flavour and eating quality by groups of growers and of consumers; for example, at on-farm events or trade open days. A more detailed project was undertaken by Food Centre Wales to determine the sensory characteristics of two blight-resistant potato varieties (Sarpo Mira and Sarpo Axona – both red-skinned varieties) and to evaluate their performance against a leading red skin potato, Desiree (Food Centre Wales, 2006). The project report concluded that it was not possible to state that one of the new varieties was significantly a match for, nor preferred to, Desiree, and Desiree was not found to be consistently better that the new varieties apart from their more regular size, weight and a lower number of defects including hollow heart defects[4] (Food Centre Wales, 2006).

Many factors can affect the organoleptic qualities of potatoes, not just varietal difference. Weather conditions in the growing period, soil conditions, growth stage at harvest and cooking method can each have an effect. Nonetheless, if, as many organic growers may wish, and as the Dutch Potato Covenant have achieved, there is movement to 100% use of blight-resistant varieties, consumers may find that varieties much cherished for their taste, texture and flavour will become unavailable to the organic consumer: varieties such as Jersey Royal Kidney, King Edward VII, Pink Fir Apple, Belle de Fontenay, and Bamberger Hörnla.

Lawrence Hills (d. 1991), in his day a leading writer on organic gardening and founder of the Henry Doubleday Research Association, became concerned at the loss of older varieties of potatoes. He published *The Good Potato Guide* (Hills, 1989) and noted that, already in 1989, there remained only eight varieties on the UK National List that had been introduced before 1900. Hills attributed the loss of varieties to the cost of maintaining a potato variety on the UK National list and the economic pressure to drop older varieties: each variety on the National List must have a Registered Maintainer. Non-maintained varieties are removed from the list and cannot legally be marketed.

In the current situation, it will be necessary for organic growers not only to support the development of new blight-resistant varieties but also to use all the other agronomic and alternative strategies discussed here – and to develop new ones – to ensure that the widest number of potato varieties remain available to organic growers and consumers.

Conclusions

Late blight, *Phytophthora infestans*, was first identified in 1846 and more than 150 years later it remains a great threat to potato production globally. The use of copper fungicides to control plant diseases also dates from the late 19th century and although copper-based fungicides are widely permitted to control late blight in organic potato production, concerns about their toxicity have led to a search for alternative control strategies. These include methods to predict the likelihood of late-blight outbreaks and the spread of epidemics, alternative agronomic strategies and treatments, and the breeding of late-blight-resistant varieties.

The development of complete supply chains for 100% blight-resistant potato varieties, as pioneered in The Netherlands, is one direction that organic production and marketing could take and there is some evidence from the organic seed market in the UK that blight-resistant varieties are increasingly accepted. It will, however, be necessary for organic growers not only to support the continued breeding of new blight-resistant varieties but also to continue with all the other agronomic and alternative strategies discussed here – and to develop new ones – to ensure that the widest number of potato varieties remain available to organic growers and consumers.

Notes

[1] OptiYield Diamond is not approved for use in organic farming in the UK.

[2] These are: Accent, Banba, Barna, Cosmos, Druid, Dundrod, Galactica, Habibi, Harlequin, Harmony, Kerr's Pink, Kikko, Kondor, Lady Balfour, Markies, Morene, Nadine, Orla, Pentland Dell, Pentland Ivory, Premiere, Remarka, Robinta, Romano, Sante, Setanta, Slaney, Spunta, Sunset ,Verity.

[3] The varieties trialled were: Appell, Derby, Innovator and Naturella in Switzerland; Derby, Eden, Escort and Naturella in France; 87418 (Axona), 874120 (Sarpo Mira), 8740133 (Tominia), Eve Balfour, Lady Balfour in England; and N89-1756 and N92-15138 in Norway.

[4] Hollow heart shows as central cavities in potato tubers. Larger cavities have brown colouring. Hollow heart is associated with senescence in large tubers and may also be caused by periods of uneven growth.

References

Agriculture and Horticulture Development Board (2023). Available at: https://ahdb.org.uk (accessed 24 May 2023).

Aitkenhead, M., Coull, M., Troldberg, M., Cooke, D. and Skelsey, P. (2020) *Spatio temporal analyses of potato late blight outbreaks in Great Britain 2017-2020*. Available at: https://ahdb.org.uk (accessed 24 May 2023).

Ayres, P.G. (2004) Alexis Millardet: France's forgotten mycologist. *Mycologist* 18, 23–26.

Blogg, H. (2019) Organic potatoes – pathways for coping without copper. Soil Association, Bristol, UK. Available at: https://organicplusnet.files.wordpress.com/2020/01/coping-without-copper-briefing-paper-2019.pdf (accessed 21 February 2025).

Dorn, B., Musa, T., Krebs, H., Fried, P.M. and Forrer, H.R. (2007) Control of late blight in organic potato production: Evaluation of copper-free preparations under field, growth chamber and laboratory conditions. *European Journal of Plant Pathology* 119(2), 217–240.

Food Centre Wales (2006) *Project Summary: Evaluating Potato Varieties Suitable for Producers, Consumers and Public Procurers in Wales*. Food Centre Wales, Horeb, UK.

Frost, D. and Clarke, A. (2005) Blight challenge – potato variety trials on organic land in a high-risk area. In: *Final Conference of the EU FP5 RTD project Blight MOP*, Newcastle University, UK.

Gessler, C., Pertot, I. and Perazzolli, M. (2011) Plasmopara viticola: a review of knowledge on downy mildew of grapevine and effective disease management. *Phytopathol. Mediterr* 50, 3–44.

Hartley, T. (2014) Bioselect UK potato variety demonstration day. *The Organic Grower* (28), 23.

Hawkes, J.G. and Francisco-Ortega, J. (1992) The potato in Spain during the late 16th century. *Economic Botany* 46(1), 86–97. Available at: http://www.jstor.org/stable/4255411 (accessed 23 February 2025).

Hills, L.D. (1989) *The Good Potato Guide*. Henry Doubleday Research Association. National Centre for Organic Gardening. Ryton-on Dunsmore, Coventry, UK.

Leifert, C. and Wilcockson, S.J. (2005) Blight-MOP: Development of a systems approach for the management of late blight (caused by *Phytophthora infestans*) in EU organic potato production. University of Newcastle, UK. Available at: https://orgprints.org/id/eprint/10650/ (accessed 21 February 2025).

Merfield, C.N. (2017) Mesh crop covers for pest control in commercial crop production. The BHU Future Farming Centre, Lincoln, New Zealand. Available at: http://www.bhu.org.nz/future-farmingcentre/ffc/information/crop-management/production/mesh-potatoes/mesh-crop-covers-forpest-control-in-commercial-crop-production-2017-ffc-merfield. pdf (accessed 26 February 2025).

Nechwatal, J. and Zellner, M. (2015) Potential suitability of various leaf treatment products as copper substitutes for the control of late blight (*Phytophthora infestans*) in organic potato farming. *European Association for Potato Research* 58(3), 261–276.

Organic Research Centre (2016) *Can We Replace Copper in Organic Fruit and Vegetable Production Systems?* No. 121.

Rogozina, E.V., Gurina, A.A., Chalaya, N.A., Zoteyeva, N.M., Kuznetsova, M.A. *et al*. (2023) Diversity of late blight resistance genes in the VIR potato collection. *Plants* 12(2), 273.

Shaw, D., Frost, D., Baird, M., Jones, O., Harris, I. *et al*. (2019) *Potato Blight Control Using Components of Indigenous Non-food Waste Plants*. Final report of the European Innovation Partnership Wales Project. Welsh Government.

Soil Association Organic Standards for Great Britain (2023) Farming and growing. Version 1.3.

Sumption, P. (2018) Managing potato blight for small growers. *The Organic Grower* 42, 12–13.

Sumption, P. (2020) The Dutch potato covenant model. *The Organic Grower* 51, 14–15.

Sumption, P. (2024) Way forward in organic farming plant health care strategies. *The Organic Grower* 66, 14–15.

Tamm, L., Thuerig, B., Apostolov, S., Blogg, H., Borgo, E. *et al*. (2022) Use of copper-based fungicides in organic agriculture in twelve European countries. *Agronomy* 12(3), 673.

Wacher, C. and Frost, D. (2018) Advances in the control of potato late blight and clarity on copper? *The Organic Grower* 44.

Wilson, A. (1993) *The Story of the Potato*. Alding and Mansell, Wisbech, UK.

Zuckerman, L. (1999) *The Potato*. Macmillan, London.

39 Biopesticides in Organic Farming

MASOUD LATIFIAN*

Agricultural Research, Education and Extension Organization, Horticultural Sciences Research Institute, Temperate Fruits Research Center, Karaj, Iran

Abstract

Demand for biopesticides is increasing globally because of growing consumer demand for safe and healthy food. Biopesticides will soon be able to replace synthetic pesticides without affecting productivity or yield. At present, there are obstacles to the full acceptance of biopesticides in agriculture, and their resolution requires research advances, including increasing the efficiency of production, formulation and packaging of formulated products as vital processes of biopesticide commercialization. In addition, knowledge of nanotechnology for formulation, genetic transfer to crops, controlled release, detection of pesticides and pathogens, and improved stability of biopesticides is needed from the perspective of social communication, understanding bioeconomy, and identifying needs and concerns. Consumers and other groups of the value chain of agricultural production, as well as the relationship between the environmental effects of non-chemical pesticides and environmental services, are the priority of society. There are different types of non-chemical pesticides that are used in the organic production of agricultural products, which are discussed in this chapter.

Keywords: microbial, toxins, botanical, pseudo-hormones, biopesticides

Introduction

Sustainable agriculture is based on the use of nature-friendly pesticides that are harmless to the environment and people. At the global level, the demand for biological pesticides is growing due to the increasing demand of consumers for safe and healthy food (Singh *et al.*, 2019). These reasonable demands can only be met by modifying the methods of production, registration, marketing and promotion of their proper use. Prioritizing non-chemical control methods is a correct and clear principle in integrated pest management (Deguine *et al.*, 2021).

But the problem is to achieve the level of 'satisfactory pest management' that should always be considered in the choice of management tactics. Of course, the satisfactory control efficiency of each method is not only reaching the highest level of control but also the stability of the control method, preventing the creation of new pest problems, the health of the environment and humans, the economy of the method, and the absence of resistance in pests. This is a standard to assess when selecting non-chemical tactics (Latifian, 2017a; b).

Environmentally compatible pesticides are mainly in harmony with the criteria of integrated pest management, and for this reason, they have been able to find their place in pest control (Santos *et al.*, 2021). There are different types of non-chemical pesticides that are used in the healthy and organic production of agricultural products, which are discussed in this chapter.

Biopesticides

Microbial pesticides

Microbial pesticides are active ingredients that are naturally or genetically modified by bacteria, fungi, algae, viruses or protozoa. The pathogenic effects of these pesticides are very specific to the target pests depending on the type of microorganism that constitutes it (Pathak *et al.*, 2017). In choosing a

*Corresponding author: masoud_latifian@yahoo.com

DOI: 10.1079/9781800626850.0039

pathogenic agent for the production of microbial pesticide, various characteristics of the microbial agent are examined in order to choose the best option for commercialization (Latifian and Rad, 2012). Although microbial pesticides differ in significant areas, the following characteristics apply to all microbial pesticides:

1. Organisms used in microbial pesticides are essentially non-toxic and non-pathogenic to wildlife, humans and other organisms that are not closely related to the target pest. The safety of microbiological pesticides is of great benefit (Essiedu *et al.*, 2020).
2. The lethality of microbiological pesticides is often unique to a group or species of living pests. This characteristic means that most microbial pesticides do not directly affect beneficial organisms (including predators/parasitoids or pathogens) in treated areas (Bahmani *et al.*, 2020).
3. If required, most microbial pesticides can be used with chemical pesticides as most microbial pesticides are not inactivated or damaged by chemical pesticide residues (Latifian *et al.*, 2014; Latifian *et al.*, 2020a).
4. Because their residues are not hazardous to humans or other animals, microbial pesticides can be used even when the crop is almost ready for harvest (Latifian *et al.*, 2020b).
5. In some cases, pathogenic microorganisms can establish in the pest population or its habitat and control the pest through generations or the seasons (Latifian and Rad, 2019).
6. Microbial pesticides also promote plant growth by stimulating the soil's beneficial microflora. In this way, they also play a role in increasing product performance (Elnahal *et al.*, 2022).

There are natural limits to microbial pesticides, but these limits do not preclude their successful use. These factors only allow users to select more effective microbial pesticides and take action to achieve greater efficiency (Essiedu *et al.*, 2020).

1. As a microbial pesticide is only toxic to a specific species or group of pests, each treatment can only control some of the pests in the field. If there are other types of pests in the field, they will survive and may continue to cause damage (Thakur *et al.*, 2020).
2. Temperature, air, soil dryness, or UV exposure reduce the effectiveness of many microbial pesticides. Consequently, timing and appropriate application methodologies are particularly important for effective management (Latifian and Mostaan, 2020).
3. Specific microbial pesticides require special formulations and storage methods. Even though the agent can complicate the production and distribution of certain products, the storage requirements for microbial pesticides are not very limited (Latifian, 2018).
4. Given that many specific microbial pesticides can often be used for a single pest, the potential market for such products may be limited. Their development, registration and production costs cannot be offset through a broad range of sales and distribution markets (Samada and Tambunan, 2020).

Many of these microbes are commercially available. In order to succeed in this direction, sustainable agriculture relies on the use of pesticides compatible with nature, which are harmless to the environment and humans. At the global level, the demand for biological pesticides is growing due to the increasing demand of consumers for safe and healthy food (Singh *et al.*, 2019). Responses to such reasonable requests may only be provided by altering the methods of production, registration, marketing and promotion of their correct use. Prioritizing non-chemical control methods is a correct and clear principle in integrated pest management (Deguine *et al.*, 2021).

But the problem is to achieve the level of 'satisfactory pest management' that should always be considered in the choice of management tactics. Of course, the satisfactory control efficiency of each method is not only reaching the highest level of control but also the stability of the control method, preventing the creation of new pest problems, the health of the environment and humans, the economy of the method, and the absence of resistance in pests. It is a standard to be evaluated in the selection of biopesticides (Latifian, 2017a; b).

Environmentally friendly pesticides primarily meet the criteria for integrated pest management, and for this reason they have found their place in pest management (Santos *et al.*, 2021). There are different types of non-chemical pesticides that are used in the healthy and organic production of agricultural products, which are discussed in this chapter. Only the experiences of international companies active in this field should be used (Moosavi and Minassian, 2021). In summary, the most important conditions

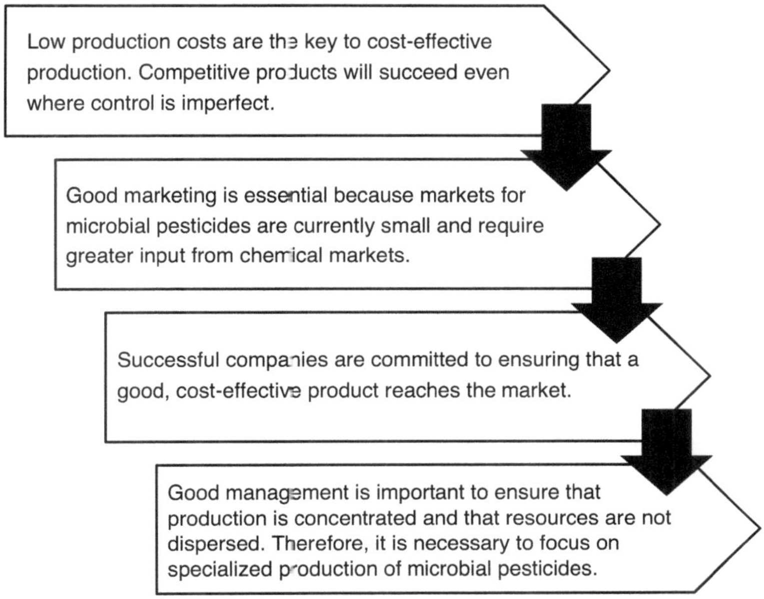

Fig. 39.1. Success conditions of the companies that produce microbial pesticides.

for the company to succeed in producing microbial pesticides are listed in Fig. 39.1.

Biopesticides derived from microbial toxins

Bacteria, actinomycetes, fungi and yeast, as well as other kinds of living organisms, produce various metabolites, some of which have pesticide properties. The most important mechanisms of effect of this group of biological insecticides (Keswani *et al.*, 2020) include the following:

1. juvenile hormone-like compounds that disrupt metamorphosis and egg-laying activity in insects (Lindeman *et al.*, 2019);
2. anti-chitin synthetic-like compounds that inhibit the formation of new exoskeleton and its hardening after moulting (Muzzarelli, 2011);
3. anti-activity of ATPase enzyme-like compounds that prevent the release of energy from respiration and thus cause the death of insects (Al-Askar *et al.*, 2022);
4. anti-RNA synthesis-like compounds such as Thuringiensis that prevent the replication of the genetic material of insect cells (Schinke *et al.*, 2017);
5. synergism of the effects of other pesticides such as Aminolevulinic acid that effect dipyridyl toxicity or increase the effects of pyrethroid pesticides (Xu *et al.*, 2015); and
6. synergism of the effects of biological control agents such as Thiolutin which intensifies the effects of Steinernema insectivorous nematodes (Dreyer, 2018).

The effect of these compounds on the insects of the orders in which the important species of farm pests are placed is remarkable. The effect of these compounds on different orders of insects is different. Some types of microbial poisons intensify each other's insecticidal properties and their combined use has a higher degree of effectiveness. Therefore, in recent years, they have been taken into consideration at the commercial and practical level, and some of them have successfully reached the stage of commercial production (Ravensberg, 2011).

In relation to the future use of these compounds, most developing countries are using old chemicals usually without the safety standards of larger companies; large international companies are making efforts; they have pushed themselves in the direction of producing compounds more compatible with integrated pest management. Such combinations have many benefits for the aforementioned companies, including more appropriate management of resistance and therefore longer life, greater satisfaction of the farmer, and support from the people and the government. Considering the existing and future problems of synthetic chemical pesticides, this cost should be minimized by the industry by providing cheaper production solutions (Mishra *et al.*, 2015).

Botanical biopesticides

Botanical biopesticides are toxic substances that have evolved in plants to protect them against herbivorous insects. Their advantages over artificial insecticides include the absence of environmental hazards being harmless to humans, the absence of new pest outbreaks, and reducing the effects of pesticides on natural enemies. However, there is evidence that the widespread use of plant-based pesticides can also adversely affect non-target organisms and the environment. Many efforts have been made in the second half of the 20th century to make plant biopesticides more stable or to make new low-risk compounds and molecules (Acheuk *et al.*, 2022).

Botanical biopesticides include several classes of molecules, terpenes, flavonoids, alkaloids, polyphenols, cyanogen glucosides, quinones, amides, aldehydes, thiophenes, amino acids, saccharides, and polypeptides. More than 2500 plant species belonging to 235 families have been reported to have biological activity against several types of pests. But even though they could be used to make commercial botanical biopesticides, not many have been found on the market. These biopesticides are only used on organic crops (Zhao *et al.*, 2022).

At the beginning of 2013, 400 active ingredients and more than 1250 botanical biopesticide products had been registered and commercialized. From that year until today, the rapid growth trends in the development of registration and commercialization of this group of pesticides can be seen. In order to be able to use native plants more effectively in the production of this group of pesticides, a wide range of special formulations of these compounds should be available and protect a wide range of agricultural products of important pests (Reddy and Chowdary, 2021).

Several factors appear to limit the commercialization of botanical biopesticides, such as: difficulties in large-scale production, unavailability of raw materials, poor shelf life, reduction of residual toxicity under field conditions, limitations related to standardization, and refinement of the final product. Furthermore, since the phytochemical profile of a plant species may vary according to its genome/transcriptome/proteome/metabolome, this variation is related to several edaphic-climatic factors (i.e. temperature, relative humidity, solar radiation level, altitude, period depending on light and type of soil) as well as ecological interactions (i.e. herbivory or mutualism), and producers must take extra care in order to maintain efficiency and ensure consistent performance of their products (standardization) (Lavoir *et al.*, 2022).

A serious drawback to the commercialization of plant materials is the high cost of processing plant materials to meet the standards of pesticide regulatory authorities. Some ideas to improve their adoption include doing on-farm demonstrations, more training and education on how the products work, and how to incorporate them into integrated pest management. Just because a compound is a natural product does not mean it is safe, as most toxins are natural products or inspired by them. In addition, if biopesticides as well as synthetics are used indiscriminately, it may lead to the development of pest resistance (Butu *et al.*, 2020).

Hormones and pseudo-hormone-based biopesticides

The act of chewing by herbivores stimulates the tendency to produce jasmonic acid (JA) in the plant, while insects feeding on phloem and biotrophic fungal pathogens induce the production of salicylic acid (SA) in plants. The production of these hormones activates a systemic cascade of physiological responses to the host plant that can lead to the production of secondary metabolites or the release of volatiles that attract natural enemies of herbivores such as predatory arthropods or parasitoids. JA plays an important role in defence against root-feeding insects, nematodes and plant pathogens. While SA defends plants against biotrophic pathogens. The application of JA and SA inhibits the absorption of chewing and feeding insects from the phloem, respectively. Also, the use of these biochemical compounds leads to the attraction of natural enemies of the treated fields (Caarls *et al.*, 2015).

Hormones are very important to different stages of growth and development of insects. Several hormonal compounds are known concerning skin change in insects. In recent years, researchers have found that juvenile hormone can be synthesized.

Compounds similar to Juvenile Hormone Mimic have been tested. The juvenile hormone prevents the insect from ageing in the moulting stages. For this reason, its amount is high in the early stages of the larva. The closer it is to the pupal stage, the smaller it will be. Therefore, in pest control, foliar spraying of juvenile hormones should be done in the larval stage (Subramanian and Shankarganesh, 2016).

Pseudo-juvenile hormones may also cause malformation or disrupt the survival of the insect. Some pseudo-hormones of juveniles are effective against the ovaries. Therefore, if they are sprayed on the insect at an inappropriate time, such as autumn, they will stimulate egg laying and the insect will lay eggs instead of going to diapause. In this way, both the female insect and the laid eggs are destroyed by adverse natural factors. The most important form of pseudo-juvenile hormones are only effective against the larval stage. Therefore, they are not applicable to pests that cause damage in the larval stage, especially in the early larval stage, and those that have only one larval stage (Loni *et al*., 2015).

Over years of research, the structural modifications that have occurred in juvenile hormones have led to the development of new pseudo-juvenile hormones. The discovery of Diflubenzuron led to the development of a series of chitin synthesis inhibitors such as Novaluron, Lefenuron, Hexaflumoron, Buprofezin and Etoxazole. The discovery of Bisacylhydrazine has led to the search for new moulting hormone agonists such as Tebofenozide, Methoxyfenozide, Halofenozide and Chromafenoside with high selectivity against certain groups of pests. All these innovations expanded the scope of application of IGRs as suitable alternatives to chemical insecticides (Sarwar, 2015).

The Future Perspective of Biopesticides

In the near future, biopesticides can replace synthetic pesticides without significant impact on productivity and performance. The limitations facing the full adoption of biopesticides are: high cost of commercial production, inability to meet the demand of the global market, difference in the standard method of preparation and application instructions, determining the concentration of active ingredients, and the sensitivity of pesticides (Latifian, 2017b).

Integrated pest management based on the application of biopesticides shows four dimensions of sustainable development: (i) environmental safety; (ii) wise use of resources; (iii) productivity; and (iv) safe/cost-effective technology. Currently, there are obstacles to the full acceptance of biopesticides in agriculture, and their resolution requires research advances, including the need to increase the efficiency of production, formulation and packaging of formulated products as vital processes of pesticide commercialization. In addition, knowledge of nanotechnology is required for formulation, transfer from genetic material into crops, controlled release, detection of pesticides and pathogens, and improved stability and (bio)encapsulation of pesticides (Mossa, 2016).

Several challenges remain unresolved regarding the outcomes of non-chemical pesticide development in relation to society. The identified social challenges are categorized into: educational; social awareness and communication; joint production of knowledge; and stakeholder participation (Latifian *et al*., 2006). Identifying the present and future skills required in the application of non-chemical pesticides is necessary for the connection between education and research within the industry. From the point of view of social communication, understanding the bioeconomy, identifying the demands and concerns of consumers and other groups in the value chain of agricultural production, as well as the communication of the environmental effects of non-chemical pesticides and environmental services are the priority of society. In this sense, many efforts have been made by the interdisciplinary scientific community to find ways for transition to a sustainable agricultural system, to create a culture of shared responsibility, and to communicate between the scientific community and civil society. Humans are effective on the development of biological pesticides in both direct and indirect ways. Correct identification of what are and are not biopesticides can help efforts to understand their benefits for humans and the environment (Moss, 2019).

Conclusion

Sustainable farming relies on the use of natural pesticides that are harmless to the environment and humans. The conducted researches have shown that non-chemical control agents have the ability to control a large number of key pests of agricultural products. Non-chemical control agents, while effectively controlling pests, maintain and preserve biodiversity. Due to their specificity, these pesticides are suitable for use in integrated pest management and have negligible effects on natural enemies. The increase in the use of non-chemical control agents depends on various factors such as control power, formulation, method of application and the level of knowledge of the farmer.

Strengthening the role of bioeconomy in sustainable agriculture production with the aim of reaching the global market requires more coherent policies and programmes. In the meantime, the Islamic Council, as the main legislative and regulatory body, is obliged to monitor the implementation of the policies and laws in this area by creating the necessary capacities and legal obligations, and regularly monitoring the performance of other responsible institutions in order to create convergence and coherence of activities, and to monitor the development of the bioeconomy.

References

Acheuk, F., Basiouni, S., Shehata, A.A., Dick, K., Hajri, H. *et al.* (2022) Status and prospects of botanical biopesticides in Europe and Mediterranean countries. *Biomolecules* 12(2), 311.

Al-Askar, A.A., Rashad, E.M., Moussa, Z., Ghoneem, K.M., Mostafa, A.A. *et al.* (2022) A novel endophytic *Trichoderma longibrachiatum* WKA55 with biologically active metabolites for promoting germination and reducing mycotoxinogenic fungi of peanut. *Frontiers in Microbiology* 769, 772417.

Bahmani, N., Latifian, M., Ostovan, H. and Hesami, S. (2020) Pathogenic effects of *Beauveria bassiana* and *Bacillus thuringiensis* on the population dynamics of *Ephestia kuehniella*. *Egyptian Journal of Biological Pest Control* 30(1), 1–9.

Butu, M., Stef, R., Grozea, I., Corneanu, M. and Butnariu, M. (2020) Biopesticides: Clean and viable technology for healthy environment. In: Hakeem, K.R., Bhar, R.A. and Qadri, H. (eds) *Bioremediation and Biotechnology: Sustainable Approaches to Pollution Degradation*. Springer, pp. 107–151.

Caarls, L., Pieterse, C.M.J. and Van Wees, S.C.M. (2015) How salicylic acid takes transcriptional control over jasmonic acid signaling. *Frontiers in Plant Science* 6, 170.

Deguine, J.-P., Aubertot, J.-N., Flor, R.J., Lescourret, F., Wyckhuys, K.A. *et al.* (2021) Integrated pest management: Good intentions, hard realities. *Agronomy for Sustainable Development* 41(3), 38.

Dreyer, J. (2018) Characterization of novel *Xenorhabdus-Steinernema* associations and identification of novel antimicrobial compounds produced by *Xenorhabdus khoisanae*. Doctoral dissertation, Stellenbosch University, Stellenbosch, South Africa.

Elnahal, A.S.M., El-Saadony, M.T., Saad, A.M., Desoky, E.-S.M., El-Tahan, A.M. *et al.* (2022) The use of microbial inoculants for biological control, plant growth promotion, and sustainable agriculture: A review. *European Journal of Plant Pathology* 162(4), 759–792.

Essiedu, J.A., Adepoju, F.O. and Ivantsova, M.N. (2020) Benefits and limitations in using biopesticides: A review. *AIP Conference Proceedings* 2313, 1.

Keswani, C., Singh, H.B., García-Estrada, C., Caradus, J., He, Y.-W. *et al.* (2020) Antimicrobial secondary metabolites from agriculturally important bacteria as next-generation pesticides. *Applied Microbiology and Biotechnology* 104(3), 1013–1034.

Latifian, M. (2017a) Integrated pest management of date palm fruit pests: A review. *Journal of Entomology* 14(3), 112–121.

Latifian, M. (2017b) A review of date palm integrated pest management (challenges and solutions). *IAU Entomological Research Journal* 8(4), 271–287.

Latifian, M. (2018) Physicochemical properties effects on different oil formulations on fungus *Metarhizium anisopliae* for control of *Oryctes elegans*. *Journal of Entomology* 15(2), 83–92.

Latifian, M. and Rad, B. (2012) Pathogenicity of the entomopathogenic fungi *Beauveria bassiana* (Balsamo) Vuillmin, *Beauveria brongniartii* Saccardo and *Metarhizium anisopliae* Metsch to adult *Oryctes elegans* Prell and effects on feeding and fecundity. *International Journal of Agriculture and Crop Sciences* 4(14), 1026–1032.

Latifian, M. and Rad, B. (2019) Efficacy evaluation of *Metarhizium anisopliae* inoculative release for biological control of date palm horned beetle *Oryctes elegans* (Coleoptera: Scarabaeidae) in garden. *Journal of Entomological Society of Iran* 39(1), 33–45.

Latifian, M. and Mostaan, A. (2020) Advanced trap design of *Metarhizium anisopliae* spore release for controlling date palm horned beetle *Oryctes elegans*. *Plant Pest Research* 10, 1.

Latifian, M., Ahmadi, A. and Pezhman, H. (2006) The effects of socio-economic factors in control of date pests and diseases in Khuzestan. *Plant Protection (Scientific Journal of Agriculture)* 29(3), 75–83.

Latifian, M., Rad, B., Rahkhodaei, I. and Shakrmi, J. (2014) Intraction between neem and *Beauveria bassiana* on survival of sawtoothed beetle (*Oryzaephilus surinaemnsis*) in laboratory conditions. *Plant Pests Research* 4(1), 35–46.

Latifian, M., Rad, B. and Ahmadizadeh, S. (2020a) Effect of inoculating of date fruit with the fungus *Beauveria* bassiana on fruit quality and shelf life of Sayer, Zahedi, and Deiri Cultivars. *Iranian Journal of Horticultural Science and Technology* 21(4), 483–498.

Latifian, M., Rad, B., Shakarami, J. and Rahkhodaei, E. (2020b) Evaluation of the lethal effect of plant extracts and their synergistic effect with *Beauveria bassiana* to control the population of *Oryzaephilus surinaemensis* fed on date palm. *BioControl in Plant Protection* 7(2), 17–30.

Lavoir, A.-V., Michel, T., Poëssel, J.L. and Siegwart, M. (2022) Challenges in developing botanical biopesticides for pest control. In: Fauverge, X., Rusch, A., Barret, M., Bardin, M., Jacquin-Joly, E. *et al* (eds) *Extended Biocontrol*. Springer, pp. 161–170.

Lindeman, L.C., Thaulow, J., Song, Y., Kamstra, J.H., Xie, L. *et al*. (2019) Epigenetic, transcriptional and phenotypic responses in two generations of *Daphnia magna* exposed to the DNA methylation inhibitor 5-azacytidine. *Environmental Epigenetics* 5(3), dvz016.

Loni, S., Soleyman-Nejadian, E., Latifian, M., Sheikhi-Garjan, A. and Vafaei-Shoshtari, R. (2015) Effect of direct contact and oral toxicity of IGR insecticides on *Oryzaephilus surinamensis* L. (Col., Silvanidae) *in vitro*. *IAU Entomological Research Journal* 7(3), 261–276.

Mishra, J., Tewari, S., Singh, S. and Arora, N.K. (2015) Biopesticides: Where we stand? In: Arora, N.K. (ed.) *Plant Microbes Symbiosis: Applied Facets*. Springer, pp. 37–75.

Moosavi, M.R. and Minassian, V. (2021) Microbial biopecticides: Opportunities and challenges. In: Karimi, J. and Madadi, H. (eds) *Biological Control of Insect and Mite Pests in Iran: A Review from Fundamental and Applied Aspects*. Springer, pp. 535–550.

Moss, S. (2019) Integrated weed management (IWM): Why are farmers reluctant to adopt non-chemical alternatives to herbicides. *Pest Management Science* 75(5), 1205–1211.

Mossa, A.-T.H. (2016) Green pesticides: Essential oils as biopesticides in insect-pest management. *Journal of Environmental Science and Technology* 9(5), 354.

Muzzarelli, R.A.A. (2011) Chitin nanostructures in living organisms. In: Gupta, N.S. (ed.) *Chitin: Formation and Diagenesis*. Springer, pp. 1–34.

Pathak, D.V., Yadav, R. and Kumar, M. (2017) Microbial pesticides: Development, prospects and popularization in India. In: Singh, D.P., Singh, H.B. and Prabha, H. (eds) *Plant-Microbe Interactions in Agro-Ecological Perspectives: Volume 2: Microbial Interactions and Agro-Ecological Impacts*. Springer, pp. 455–471.

Ravensberg, W.J. (2011) *A Roadmap to the Successful Development and Commercialization of Microbial Pest Control Products for Control of Arthropods*, Progress in Biological Control, Vol. 10. Springer Science & Business Media.

Reddy, D.S. and Chowdary, N.M. (2021) Botanical biopesticide combination concept: A viable option for pest management in organic farming. *Egyptian Journal of Biological Pest Control* 31, 1–10.

Samada, L.H. and Tambunan, U.S.F. (2020) Biopesticides as promising alternatives to chemical pesticides: A review of their current and future status. *Online Journal of Biological Sciences* 20(2), 66–76.

Santos, M.S., Rodrigues, T.F., Nogueira, M.A. and Hungria, M. (2021) The challenge of combining high yields with environmentally friendly bioproducts: A review on the compatibility of pesticides with microbial inoculants. *Agronomy* 11(5), 870.

Sarwar, M. (2015) Information on activities regarding biochemical pesticides: An ecological friendly plant protection against insects. *International Journal of Engineering and Advanced Research Technology* 1(2), 27–31.

Schinke, C., Martins, T., Queiroz, S.C.N., Melo, I.S. and Reyes, F.G.R. (2017) Antibacterial compounds from marine bacteria, 2010–2015. *Journal of Natural Products* 80(4), 1215–1228.

Singh, A., Bhardwaj, R. and Singh, I.K. (2019) Biocontrol agents: Potential of biopesticides for integrated pest management. In: Giri, B., Prasad, R., Wu, Q.-S. and Varma, A. (eds) *Biofertilizers for Sustainable Agriculture and Management*. Springer, pp. 413–433.

Subramanian, S. and Shankarganesh, K. (2016) Insect hormones (as pesticides). In: Omkar (ed.) *Ecofriendly Pest Management for Food Security*. Academic Press, London, pp. 613–650.

Thakur, N., Kaur, S., Tomar, P., Thakur, S., Yadav, A.N. *et al*. (2020) Microbial biopesticides: Current status and advancement for sustainable agriculture and environment. In: Gupta, V.K. (ed.) *New and Future Developments in Microbial Biotechnology and Bioengineering*. Elsevier, pp. 243–282.

Xu, L., Zhang, W., Ali, B., Islam, F., Zhu, J. *et al*. (2015) Synergism of herbicide toxicity by 5-aminolevulinic acid is related to physiological and ultra-structural disorders in crickweed (*Malachium aquaticum* L.). *Pesticide Biochemistry and Physiology* 125, 53–61.

Zhao, J., Liang, D., Li, W., Yan, X., Qiao, J. *et al*. (2022) Research progress on the synthetic biology of botanical biopesticides. *Bioengineering* 9(5), 207.

40 Botanical Insecticides in Crop Pest Management

Desam Nagarjuna Reddy[1]* and Abdul Jabbar Al-Rajab[2]

[1]*School of Engineering and Applied Technology, Department of Chemistry, BEST Innovation University, Sri Sathaya Sai, India;* [2]*Etcectera Publications, Chesterville, Canada*

Abstract

The increasing global population poses a challenge for food production and the need to improve and adopt synthetic pesticides as an immediate and effective option to combat pests and diseases. However, excessive dependence on synthetic pesticides or chemicals is discouraged because of deleterious effects on the environment and human health and the emergence of increasingly resistant pest and disease strains. Combined with an expanding interest in naturally delivered food sources, this is significantly driving investigations into alternative methodologies and organic pesticides. Botanical pesticides (BPs) or natural pesticides (NPs) are useful in combating various harvest issues and are economical and biodegradable. BPs and NPs have different modes of action, their sources are easily accessible, and they have low toxicity to non-target living organisms. Naturally, plants contain many chemicals with actions against crop pests. These components can be integrated into co-ordinated pest management systems and contribute to sustainable agricultural production. However, BPs have yet to be completely embraced because of difficulties in detailing and commercialization, which are attributed to the absence of phytochemical information and positive controls. There are several studies on BP phytochemicals (e.g. esters, fatty acids, alkaloids, flavonoids, essential oils and glycosides) demonstrating anti-insect activities. BPs used in eradicating insects are included in formulations, such as anti-feedants, growth retardants, attractants and repellents. BPs only target insect pests, without effects on natural enemies. Given that they leave food free of residues and are eco-friendly, they are an attractive option in insect management, in contrast to synthetic pesticides. This chapter describes BPs, their phytochemical composition, and mechanisms of action against pests to ensure sustainable agricultural production.

Keywords: botanical pesticides, pest management, insecticides, crop protection, anti-insect activity

Introduction

Generally, all crops are persistently inhabited or imperilled by insects, which influence their growth and quality. To prevent insect attacks, farmers typically use immediate pest management (PM), which is predominantly synthetic pesticides (SPs) or chemicals (Nikkhah *et al.*, 2017; FAO, 2018). Despite their efficiency, the constant use of SPs has been a major threat, such as an increase in pesticide-resistant pests (Shabana *et al.*, 2017). SPs can be overused, causing detrimental effects, such as toxicity to the environment, humans and non-target organisms (NTOs), thus adversely affecting biodiversity (Sande *et al.*, 2011). The chemical constituents of SPs have been implicated in many human illnesses due to exposure or use (Kumari *et al.*, 2014; Damalas, 2016). Generally, most SPs are not effectively biodegradable. Consequently, they accumulate in the atmosphere and cause contamination of water and soil, as well as depletion of the ozone layer (Sande *et al.*, 2011; Wimalawansa and Wimalawansa, 2014). Many of the drawbacks of SPs are associated with overuse or misuse that blends elective pest management options (Mahmood *et al.*, 2016).

Plants with bioactive chemicals are natural agents against various human infections and

*Corresponding author: alrajab@hotmail.com

 Advances in Organic Farming (ed. L. P. Awasthi)
DOI: 10.1079/9781800626850.0040

crop pests, and they have been used successfully (Thiruppathi *et al.*, 2010; Karpagam and Devaraj, 2011). Examples include *Dalmatian chrysanthemum*, *D. pyrethrum* and *Artemisia annua*, which have sources of safe chemicals for the management of insect pests and jungle or malaria fever vectors individually (Sarwar, 2015). Pest control that used plant-based chemical constituents was a long-standing tradition before technological innovations and the development of SPs. SPs were promptly embraced because of their viability and adequacy in controlling crop diseases, such as scourge and rusts (Raja, 2014; Mahmood *et al.*, 2016). Thus, the use of natural products (NPs) of plant origin gradually returned recently as the use of SPs started to undermine environmental safety and human health (Nikkhah *et al.*, 2017).

The worldwide trend is towards the production of food using safe and ideal NP protection. Identification of hazardous pesticide deposits in food sources and expanded consumer mindfulness on food safety are driving the boycotting of specific pesticides in agriculture production, and plant-based NPs are becoming ubiquitous in natural agri-business and sustainable organic agriculture (Kareru *et al.*, 2013; Karaca *et al.*, 2017). Whereas the constant use of SPs has brought about adverse consequences – for instance, environmental contamination, health hazards and loss of biodiversity – the adoption of organic pesticides (OPs) or botanical pesticides (BPs) results in a healthy climate or atmosphere as well as sustainable agriculture (Shabana *et al.*, 2017). The use of SPs has adverse effects on farmers, particularly regarding the export trade of horticulture produce (Nashwa and Abo-Elyousr, 2012). The identification of prohibited pesticides or residue traces above legislated limits results in the loss of market and income to both cultivators and exporters in emerging nations. For instance, the use of Alphadime® and Demeton® on new produce for export has been restricted (Business Daily, 2013, 2014).

The significance of BPs is attributed to their adequacy, biodegradability, various modes of activity, lower toxicity, and accessibility of source materials (Neeraj *et al.*, 2017). They have short pre-harvest and re-emergence intervals. Regularly used BPs are established in natural cultivation, where naturally produced food fetches top prices (Srijita, 2015). Hence, BPs are becoming prominent because they are safe to use on crops produced for human use, and, lately, there has been a significant market among consumers able to pay for naturally produced food (Misra, 2014). There have been many investigations of plant species with pesticidal properties (Erenso and Berhe, 2016; Jawalkar *et al.*, 2016). Plants that are sources of commercially accessible BPs include *Azadirachta indica* (neem), *Nicotiana tabacum* (tabacco), *Ryania speciosa* (ryania), *Schoenocaulon officinate* (sabadilla) and *Tanacetum cinerariifolium* (pyrethrum) (Pawar *et al.*, 2013).

Traditionally, farmers use prevention products of plant origin in post-harvest insect management, particularly in safeguarding grains during storage. BPs are the derivatives of plants that repulse, restrain growth, or kill pests (Hikal *et al.*, 2017). Most BPs are used to control insect pests, and several investigations of their use have focused significantly on insect pest management (Singh *et al.*, 2001; Ali *et al.*, 2014; Isman and Grieneisen, 2014; Mkenda *et al.*, 2015; Stevenson *et al.*, 2017; Isman, 2017). Plants with pesticidal characteristics can also have effects on plant microorganisms, such as bacteria, fungi, nematodes and viruses. The present study discusses BPs in the sustainable control of harvest bugs, aggregating scientific evidence on their phytochemical constituents, modes of action against insects, and the difficulties confronting their adoption and use.

History of Botanical Pesticides

BPs are acquired from plants of various families and are used either as plant extracts or as essential oils (Mizubuti *et al.*, 2007). Plant parts formulated into BPs include blossoms, bark, cloves, roots, stems, leaves, fruits and rhizomes due to the abundance of natural bioactive chemical constituents. Plant families that possess these bioactive chemical constituents with activity against significant crop pests include Apiaceae, Apocynaceae, Asteraceae, Caesalpinaceae, Cupressaceae, Liliaceae, Lamiaceae, Lauraceae, Myrtaceae, Piperaceae, Poaceae, Rutaceae, Sapotaceae, Solanaceae, and Zingiberaceae (Vidyasagar and Tabassum, 2013; Jnaid *et al.*, 2016; Gakuubi *et al.*, 2016; Ahmad *et al.*, 2017). The plant parts are dried under shade and ground into fine powder before extraction of the target chemical constituents with natural solvents (Chougule and Andoji, 2016). The extracts are then concentrated, formed and assessed for viability under laboratory or field conditions (Joseph and Sujatha, 2012). A few of the natural products exhibit pesticidal activity that have effectively been isolated and commercialized; for example, pyrethrin from pyrethrum (*T. cinerariifolium*), azadiractin from neem (*A. indica*), thyme (*Thymus vulgaris*), garlic (*Allium sativum*), ginger (*Zingiber officinale*), rosemary (*Rosmarinus officinalis*) and turmeric (*Curcuma longa*) (Table 40.1).

BPs in Agricultural Production

BPs are obtained naturally from plants; thus, they are obtainable from the environment, and the vast majority of them have other considerable uses, such as in flavourings, ornamentals, drugs, food and feed (Srijita, 2015). Their accessibility makes them cheap, and, consequently, they are easily integrated into agricultural production systems (Castillo-Sánchez *et al.*, 2015). Economically commercialized pesticides from plants such as neem, sabadilla and pyrethrum are much less toxic, particularly to non-target organisms, such as fish and pollinators (Dubey *et al.*, 2010). This trait makes BPs productive, authentic and adequate for sustainable yield assurance. Furthermore, they do not leave residues in crop produce or the environment, ensuring ecological conservation and food safety for customers (Dubey *et al.*, 2008). The relationship between BPs and insects is normally biochemical; thus, targeted pests will probably not develop resistance to BPs (Gaikwad *et al.*, 2012; Joseph and Sujatha, 2012). Plant-based natural products in extract and natural oils are target-specific, which ensures the well-being of non-target organisms, particularly pollinator honeybees and natural enemies (Hernandez-Moreno *et al.*, 2013; Nawaz *et al.*, 2016).

Based on the plant source and the concentrations that have been utilized, BPs have a zero or less allelopathic effect on crops (Sarkar *et al.*, 2012; Arafat *et al.*, 2015). The viability of BPs is dependent on the type of plant source; however, fresh or dried plant material and solvents are typically used in extraction methods (Gurjar *et al.*, 2012). BPs display different methods of activity on the targeted pests or insects, including toxicity, abhorrence, growth regulation and structural alteration, making them reasonable alternatives in pest management for crops (Rattan, 2010; Kushram *et al.*, 2017). They hinder insect behaviour, physiology, metabolic pathways and biochemical processes. The actions of BP metabolites in BPs are specific in their impact on insects. For instance, terpenes could block the glucose in the mouths of lepidopteran hatchling receptor cells (Rattan, 2010), although some plant natural oils have been chemosterilant (Sarwar, 2015) (Fig. 40.1). Table 40.1 shows plants with BP activity, their chemical constituents, and their target organisms. Several studies in pest management have investigated BP use, adoption and acceptance (Kareru *et al.*, 2013; Chougule and Andoji, 2016), and adequate information and evidence regarding their chemistry and viability are expected to meet established pesticide guidelines (Isman and Grieneisen, 2014). Furthermore, there are data on their administration, degradation, lifespan and harmfulness (Sarkar and Kshirsagar, 2014). The integration of BPs in agrarian production systems ensures significant advantages to farmers, including food security, food safety, decreased pest levels, enhanced quality of produce, higher income, and increased market access (Nefzi *et al.*, 2016). Customers in lucrative food production sectors are committed to bearing the expenses for naturally produced foods, consequently creating market opportunities for BPs (Misra, 2014). The potential pathways of BP and SP adoption are depicted in Fig. 40.1. SPs play a major role in crop pest management, as they add to a decrease in harm to crops and the resultant misfortunes in production and revenue (Lomborg, 2001; Damalas and Koutroubas, 2015). Nevertheless, they should be utilized sensibly and applied by educated personnel. This is particularly significant among smallholder ranchers, where a lot of care and effort may be required to ensure human and environmental safety (Lomborg, 2001). Hence, integrating BPs into co-ordinated insect management projects can prevent unnecessary use of the SPs (Cooper and Dobson, 2007; Damalas, 2016).

Natural Products of Botanical Pesticides

Recent increased interest in plants' natural substances in the agri-business, drug and food industries has led to research on the different chemical compositions of different plant families (Jnaid *et al.*, 2016; Plata-Rueda *et al.*, 2017). BPs

Table 40.1. Bioactive chemical compounds derived from plants and their antimicrobial activity on target pathogens.

Name of plant	Bioactive chemical constituents	Target pathogens	References
Aloe vera	Anthraquinones, ascorbic acid, pyrocatechol, cinnamic acid and p-coumaric acid	*Pseudomonas aeruginosa, Staphylococcus aureus, Bacillus cereus, Streptococcus pyogenes*	Rubina *et al.*, 2009; Karpagam and Devaraj, 2011
Allium cepa	2,2-diphenyl-1-picrylhydrazyl	*Escherichia coli*	Abdel-Salam *et al.*, 2014
Artemisia spp.	Borneol, camphene, camphor, 1,8-cineole, p-cymene, β-eudesmol, α-gurjunene, α-pinene, terpene-4-ol, sabinene, myrcene, chamazulene, α-thujone, β-thujone, 1,2-dehydro acenaphthylene	*Bacillus* spp., *Staphylococcus aureus, Pseudomonas aeruginosa, Candida albicans, Rhizopus japonicum*	Erel *et al.*, 2012; Juvatkar *et al.*, 2012
Azadirachta indica	Azadirachtin, nimonol, epoxyazadiradione	*Aspergillus niger, Microsporum gypseum, Aspergillus flavus*	Mahmoud *et al.*, 2011
Cinnamon spp.	Eugenol, eugenyl acetate	*Staphylococcus aureus, Escherichia coli*	Cui *et al.*, 2016
Citrus spp.	Neohesperidin, hesperetin, neoeriocitrin, eriodictyol, naringin, naringenin, limonene, tetrazin, coumarin	*Escherichia coli, Salmonella enterica, Pseudomonas putida, Staphylococcus aureus*	Mandalari *et al.*, 2007; Soković *et al.*, 2010; Dhanavade *et al.*, 2011; Plant and Stephens, 2015
Cymbopogon spp.	Geraniol	*Escherichia coli, Staphylococcus aureus*	Plant and Stephens, 2015; Sharoba *et al.*, 2015
Centella aciatica	Germacrene A, germecrene B, germacrene D	*Listeria monocytogenes*	Pitinidhipat and Yasurin, 2012
Euphorbia tirucalli	Hydroxycinnamic acid	*Staphylococcus* spp.	Araújo *et al.*, 2014
Jatropha spp.	Gallic acid, pyrogallol, rutin, myricetin, daidzein, 2-(hydroxymethyl)–2 nitro-1,3-propanediol, β-sitosterol, 2-furancarboxaldehyde	*Alternaria alternata, Aspergillus* spp. *Fusarium oxysporum, Rhizoctonia solani, Trichoderma viride*	Oskoueian *et al.*, 2011; Gaikwad *et al.*, 2012
Lantana camara	Hexadecanoic acid phytol, caryophyllene oxide, 9,12,15-octadecatrienoic acid, methyl ester, (Z,Z,Z) caryophyllene, α-caryophyllene, germacrene D, isocaryophillene, γ-muurolene, γ-elemene	*Klebsiella pneumoniae Escherichia coli*	Pawar *et al.*, 2013; Swamy *et al.*, 2015
Lavandula angustifolia	Linalool, linalyl acetate	*Micrococcus flavus*	Soković *et al.*, 2010
Mentha piperita	Menthol, menthone, isomenthone, neomenthol, acetylmenthol, pulegone, methyl acetate	*Enterococcus faecium, Salmonella choleraesuis, Staphylococcus aureus, Bacillus subtilis*	Soković *et al.*, 2010; Pramila *et al.*, 2012

Continued

Table 40.1. Continued

Name of plant	Bioactive chemical constituents	Target pathogens	References
Origanum spp.	Terpinen-4-ol, gamma-terpinen, o-cymene, cis-beta-terpineol, alpha-terpinen, beta-phellandrene, alpha-terpieol, carvacrol, p-cymene, thymol	*Bacillus* spp. *Staphylococcus saprophyticus Micrococcus luteus, Serratia marcescens*	Saeed and Tariq, 2009; Soković *et al.*, 2010; Sharoba *et al.*, 2015; Jnaid *et al.*, 2016
Ocimum basilicum	Estragole, 1,8-cineole, trans-α-bergamotene	*Escherichia coli*	Sienkiewicz *et al.*, 2013
Piper nigrum	Piperine, chavicine	*Escherichia coli, Salmonella typhi, Proteus* sp., *Staphylococcus aureus, Fusarium oxysporum*	Rani *et al.*, 2013; Ganesh *et al.*, 2014
Rosmarinus officinalis	Caffeic acid, rosmarinic acid, cineole, camphor, α-pinene, β-pinene, camphene, isocarnosol	*Listeria monocytogenes, Staphylococcus aureus Escherichia coli*	Genena *et al.*, 2008; Sienkiewicz *et al.*, 2013; Kozłowska *et al.*, 2015
Salvia officinalis	Camphor, α-thujone	*L. monocytogenes, P. aeruginosa and P. mirabilis*	Soković *et al.*, 2010
Satureja hortensis spp.	Rosmarinic acid, carvacrol, γ-terpinene	*Bacillus cereus, Candida kruzei*	Karami-Osboo *et al.*, 2010; Stanojković *et al.*, 2013
Thymus vulgaris	Thymol, β-cymene, caffeic acid, rosmarinic acid, luteolin-7-o-glucoside	*Erwinia amylovora, Escherichia coli, Staphylococcus aureus, Bacillus cereus*	Karami-Osboo *et al.*, 2010; Kozłowska *et al.*, 2015; Sharoba *et al.*, 2015
Tanacetum cinerariaefolium	Cinerin I, cinerin II, jasmolin I, jasmolin II, pyrethrin I, pyrethrin II, chrysanolide, β-cyclopyrethrosin, chrysanin, dihydro-β-cyclopyrethrosin, taraxasterol, pyrethrosin, 11, 13-dihydrotatridin-A, 11,13-dihydrotatridin-B		Sharafzadeh, 2011

contain secondary metabolites that exhibit bioactive chemical constituents and include alkaloids, phenols, flavonoids, tannins, terpenes, steroids and gums with insecticidal, anti-bacterial, anti-fungal, anti-viral, antioxidant and anti-cancer activities (Ahmad *et al.*, 2017). For example, *Mentha piperita* leaves consist of flavonoids, tannins and phenols (Pramila *et al.*, 2012), while the major substances in *Jatropha carcus* seed kernels are esters, flavonoids and phenolic substances (Oskoueian *et al.*, 2011). The particular chemical composition found in the given types of plants makes them potential agents against certain classes of pests (Table 40.1). The biologically active chemical constituents found in plants also determine their modes of activity, especially on specific target pests.

Mechanism of Action of BPs

The biologically active chemical compounds in BPs have different modes of activity against different pests, including bacteria, fungi, insects and plant host cells infected by viral microbes (Table 40.2). The modes of activity, such as restraint, repellence

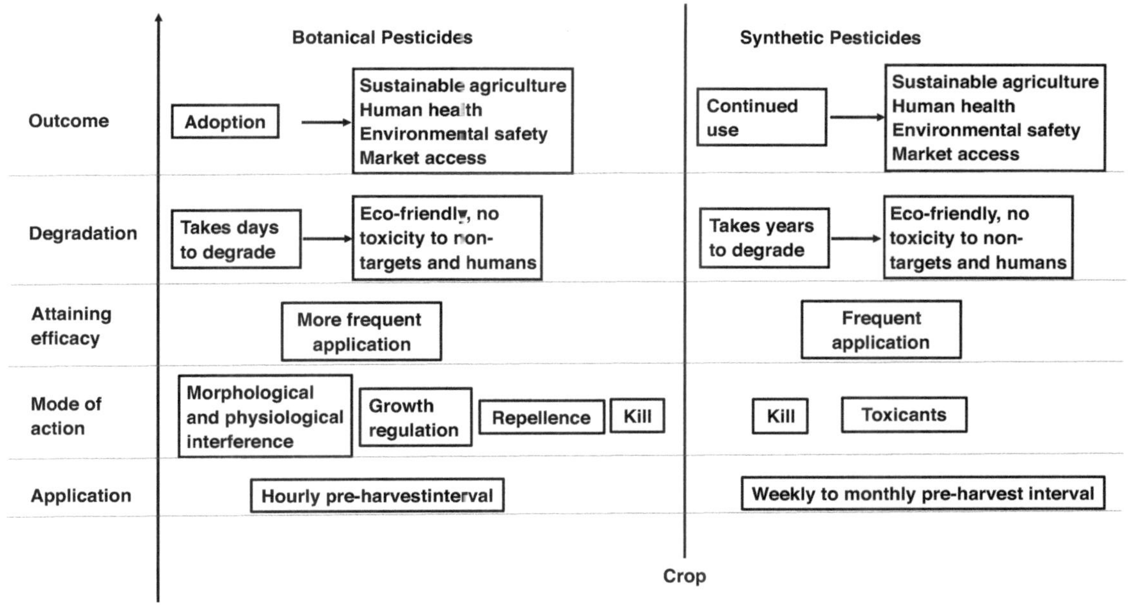

Fig. 40.1. A prototype demonstrates the contrast between botanical pesticides and synthetic pesticides with respect to utilization, method of activity, persistence and effect on ecosystem.

Table 40.2. Mode of action of selected botanical pesticides on selected crop pests.

Target pests	Plant source	Mode of action	References
Bacteria	*Aloe vera*	Inhibits cellular activities, impairs permeability of plasma membrane, denatures proteins, inhibits ATP production and glucose uptake	Karpagam and Devaraj, 2011
Fungi	Garlic (*Allium sativum*)	Delays and inhibits spore germination, inhibits protein and DNA synthesis, inhibits production of mycotoxins, disrupts cellular components and their activities, hyphal and mycelial modifications	Perelló *et al.*, 2013
Insects	Neem (*Azadirachta indica*)	Binding to acetylcholine receptors thereby disrupting the nervous system, repellence, feeding deterrence, inhibition of oviposition, egg hatching and moulting	Grdiša and Gršic, 2013
Nematodes	*Tagetes erecta*	Inhibits egg hatching, larval toxicity structural modification, mortality	Feyisa *et al.*, 2015
Viruses	*Nepeta nuda* subsp. *nuda*	Host-plant manipulation, inhibits virus replication and multiplication, prevents virus adsorption, inhibits nucleic acids liberation	Todorov *et al.*, 2015

and protein denaturation, depend on the kinds of NP and pest. For example, BPs such as pyrethrum specifically target the nerve system or nerve cells of pests or insects, causing paralysis and death, although pesticides from neem-based substances interrupt the endocrine system, upset oviposition, instigate moulting irregularities, anti-feedant and repellence properties (Grdiša and Gršic, 2013). SPs are more particular to their targets and are, for the most part, neurotoxicants with outcomes comparable to BPs (Stevenson *et al.*, 2017). Understanding the mode of action of a pesticide on a pest, which mainly depends on biological, physical and chemical interactions, is imperative in pest management to inform the implementation of the management strategy (Lee and Lee, 2015; Wink, 2015). The next subsections discuss methods of NP activity against specific populations of agricultural pests.

Modes of Action of BPs against Nematode Pests

Generally, BPs exhibit nematocidal characteristics, such as restraint of egg incubating. Some of the chemical constituents in BPs influence the populations of different microorganisms in soils, which further influences the tolerance of nematode eggs and adolescents (Khan *et al.*, 2008). A few BP chemical constituents can destroy egg masses, hinder second-stage adolescents, affect normal larval harmfulness, reduce galling, and ultimately influence the populations of nematodes in a biological ecosystem (Kepenekçi *et al.*, 2016). For instance, *Brassica napus* (*B. napus*), *Lantana camara* (*L. camara*), *Tagetes erecta* (*T. erecta*) and *A. indica* plant extracts constrain incubation of the eggs of root knot nematodes, such as *Meloidogyne incognita*, resulting in immobilization and death of second-stage adolescents. Under a greenhouse environment, utilization of *L. camara* as well as *Trichoderma harzianum* (*T. harzianum*) diminished the incidence and reproduction of root knot nematodes in tomato (Feyisa *et al.*, 2015).

Plant NPs, such as alkaloids, tannins and glycosides, mainly act against egg hatching, migration and mortality in second-stage adolescents (Akyazi, 2014; Asif *et al.*, 2017). The bioactive chemical constituents further influence root knot nematodes through loss of motion or paralysis, diminishing infectivity potential and the spread of infective adolescents (Oka *et al.*, 2012). Plant extracts and natural oils are also known to cause mortality for second-stage adolescents of root knot nematodes (*Heterodera cajani*) (Singh *et al.*, 2001). Some phytochemicals identified in plant extracts are lipophilic, enabling them to be slowly absorbed into the cytoplasmic membrane of nematodes, where they obstruct protein structures for growth and development (Pavaraj *et al.*, 2012).

Modes of Action of BPs against Bacterial Pathogens

The phytochemical constituents of BPs have different anti-bacterial properties, including growth restraint (Aljamali, 2013). Gram-negative (G-neg) microbes are more vulnerable than Gram-positive microorganisms because of the absence of a peptidoglycan cell wall (Djeussi *et al.*, 2013). BPs mostly affect microbial cell cycles; for instance, protein synthesis, enhancing penetrability of plasma membrane, prompting spillage of cell contents, and eventually death (Khan *et al.*, 2008). For example, *A. vera* methanolic extracts hinder the growth of *E. coli* and *Bacillus subtilis*, whereas its acetone extract can inhibit *Pseudomonas aeruginosa*. *A. vera* extracts deteriorate microorganism proteins, subsequently upsetting their functionality because the extracts consist of cinnamic acid, which blocks glucose uptake and ATP production (Djeussi *et al.*, 2013). Coumaric acid and ascorbic also confer *A. vera* with the ability to disrupt enzymatic activity in the microorganisms (Karpagam and Devaraj, 2011). Essential oil extracts from *Thymus vulgaris* possess excellent antimicrobial activity against *B. cereus*, *E. coli*, *Klebsiella pneumonia*, *Salmonella typhimurium* and *Staphylococcus aureus* (Sharoba *et al.*, 2015). Their activities have been attributed to the presence of thymol, which depolarizes and makes the bacterial membrane penetrable, disrupting cell components (Plant and Stephens, 2015).

Modes of Action of BPs against Insect Pests

Most BP extracts act as repellents to insects, preventing feeding, causing toxicity and death, hindering oviposition, and interfering with physiological activities (Laxmishree and Nandita, 2017). BPs play multiple active roles in insect pests, making them more popular in the market (Isman, 2014). For instance, commercially more established BPs like pyrethrum have been known to have, among others, neurotoxicant impact on insect pests causing paralysis, knockdown and thus mortality (Grdiša and Gršic, 2013). BPs also block the creation of significant proteins, such as those responsible for moulting, subsequently inhibiting development and growth (Ntalli and Menkissoglu-Spiroudi, 2011).

A. sativum and *C. longa* extracts exert toxicity, repellence and death and restrain the progeny development of red flour beetles such as *Tribolium castaneum* (Ali *et al.*, 2014). There are many reports of plant extracts blocking oviposition, egg incubation, and the general growth of insect pests. A few BPs have been associated with loss of motion and blockage of electron transport in the respiratory process of insects, as well as toxicity and immobilization (Singh *et al.*, 2001; Moreira *et al.*, 2007). Herbal insect sprays comprise the largest portion of BP formulations on the market worldwide (Isman, 2014).

Method of Action against Fungal Pathogens

The alkaloids, alcohols, phenols, tannins, terpenes and various secondary metabolites identified in BPs instigate toxicity to fungal cell walls, cell nuclei and cell membranes (Yoon *et al.*, 2013). These secondary metabolites block spore fertilization, development of mycelial and microorganism tube extension, postpone sporulation, and restrain production of significant enzymes, resulting in reduced protein and DNA synthesis (Martinez, 2012). The natural chemical constituents also cause structural alterations of the mycelia or hypha, thereby hindering the production of metabolites such as fumonisin and aflatoxin in fungi, including *Aspergillus* spp. and *Fusarium* spp. These outcomes decrease mycotoxin pathogenicity resulting in fungal pathogens (Martinez, 2012). Exposure of *Fusarium verticillioides* to various plant extracts has been shown to disrupt the cell wall, cause the loss of nuclear components, and block the production of primary chemical constituents, such as ergosterol and fumonisin (Bomfim *et al.*, 2014). For example, ginger (*Z. officinale*) extracts reduce the cytoplasmic substances of *Fusarium*, as well as actuating changes in the morphology of microconidia (Yamamoto-Ribeiro *et al.*, 2013). *Allium sativum* contains allicin, which is the most active chemical constituent that affects morphological abnormalities, such as repressed germination of spores and hyphal development, hindering hyphal development. It also hinders alcohol dehydrogenases, cysteine proteinases and thioredoxin reductases in single adaptable cells, which cause looseness of the bowels in humans (Ankri and Mirelman, 1999). *A. sativum* mainly consists of allicin, and other secondary metabolites that work in synergy with other chemicals to target sporangium development and spore impregnation (Hadi *et al.*, 2013; Wink, 2015). There are various herbal fungicides that contain significant volatile oils as active chemical constituents (Zaker, 2016).

Modes of Action of BPs against Virus Pathogens

BPs restrain viral growth primarily by controlling the production of antiviral proteins and preventing viral actions on host plants (Montanha *et al.*, 2004). Some BP chemical compounds exhibit antiviral properties, restrain propagation of the virus, and promote methodological hindrance by the host plants, as well as exert insecticidal properties against insect vectors (Waziri, 2015). General methods of activity include inhibiting virus penetration into the host cell and blocking virus replication, haemagglutination and enzymatic activity, which are essential for virus attachment to the host (Bhanuprakash *et al.*, 2008; Rajasekaran *et al.*, 2013; Kohn *et al.*, 2015). Under laboratory conditions, acetone extracts of cotton seed oil slurry potentially hindered the tobacco mosaic virus isolated from infected tobacco plants (*Nicotiana tabacum*) as well as under field conditions. The extracts also decreased the infection levels effect by the rice stripe virus and the southern rice black streaked dwarf virus in *Oryza sativa*, because the cotton seed oil slurry presence of β-sitestrol and gossypol exhibited potential antiviral activity. A derivative of gossypol can block the development of cell combination-initiated centres, while β-sitestrol, the dominant phytosterol with chemical structures similar to that of cholesterol, has been reported to have uses such as induction of apoptosis (Zhao *et al.*, 2015). *Thuja occidentalis* can block the reproduction of the water melon mosaic virus isolated from *Citrus lanatus*, water melon. A decrease in the side effects of infectious disease on the hypocotyls of the explants of the melon was observed after treatment with extracts of *T. occidentalis*, which was attributed to the blocking of nucleic acids, hence restraining reproduction of the virus (Elbeshehy *et al.*, 2015). Table 40.2 displays a few plants, their BPs and modes of action against different types of crop pests.

Natural Biodegradation of BPs

The bioactive nature of BPs enables their rapid breakdown in the environment so that they do not accumulate; for instance, in soils and water, subsequently wiping out possibilities of pollution (Soković *et al.*, 2010). The exposure of BPs to sun radiation, air, atmospheric moisture and various temperatures is sufficient to degrade them into their constituents (Gunasekara, 2005). For example, thymol can be degraded in the soil within eight days as well as in the crops within 28 hours in the presence of sunlight. Thymol is mainly found in *Piper nigrum*, *Satureja hortensis*, *Thymus vulgaris* and *Zataria multiflora* (Liu *et al.*, 2017). The degradation of azadirachtin occurs between 24 h and 28 h in crops and soil, respectively (Yan *et al.*, 2017; Yang *et al.*, 2017). These plant extracts are protected from sunlight to preserve them, as direct sunlight can effectively degrade them like pesticides; for example, neem becomes rapidly degraded in the presence of sunlight after application (El Shafie and Almahy, 2012). Under field conditions, pyrethrum-based insecticides can easily be degraded within a few hours after application (Grdiša and Gršic, 2013). In the presence of atmospheric oxygen and sunlight, 100% photodegradation of natural pyrethrins takes place in 5 h.

Microorganisms in biological systems also expedite the biodegradation process through oxidative digestion because of the abundance of detoxification enzymes (Mpumi *et al.*, 2016). A typical biodegradation pathway is mainly ester hydrolysis mediated by carboxylesterases in microorganisms, such as *Aspergillus niger* and *Bacillus cereus* (Cycon and Piotrowska-Seget, 2016). Effective biodegradation is dependent upon the complexity of the pesticide and the enzymatic capability of the microbial population in the soil. Normally, pesticides are completely biodegradable into gases, water and salts, and their by-products are either absorbed by dirt or soil or further isolated using biomagnification (Ortiz-Hernández *et al.*, 2013). Soil microorganisms possess enzymes that initially degrade the BPs into small groups that are less harmful than the parent product, before rendering the metabolites biologically inaccessible and non-poisonous (Ortiz-Hernández *et al.*, 2013; Velázquez-Fernández *et al.*, 2012). Bacteria species known to degrade carbamates, organochlorines, organophosphates and pyrethroids include genera of *Aerobacter*, *Arthrobacter*, *Bacillus*, *Brevibacillus*, *Escherichia*, *Flavobacterium*, *Ochrobactrum*, *Pseudomonas*, *Serratia*, *Sphingobacterium*, and *Streptomyces* (Porto *et al.*, 2011; Lin *et al.*, 2011; Cycon and Piotrowska-Seget, 2016). *Trichoderma* spp. is also known to degrade organochloride and DDT pesticides (Pramila *et al.*, 2012). Laboratory-synthesized pyrethroids have ester groups, which hydrolyse in the presence of microorganisms, resulting in carboxylic acids that are further broken down into carbon dioxide gas under aerobic conditions (Jones, 1995). The viability of such biotransformation depends on factors such as pH, oxygen, salinity and water supply, which affect the development and tolerance of microbial populations. Population-specific elements of target microorganisms also affect the accessibility and convergence of biotransformation enzymes (Singh, 2008). Successful microbial degradation of pesticides also depends on the state and nature of microorganisms, their interactions with the pesticides, and ecological conditions (Ortiz-Hernández *et al.*, 2013; Cycon and Piotrowska-Seget, 2016). The quick degradation of BPs is a substantial benefit that corresponds to natural conservation.

Role of BPs in Integrated Pest Management

Integrated pest management is a methodology that integrates various procedures to accomplish sustainable pest management (Mizubuti *et al.*, 2007). The goal is to economically decrease pests and accomplish high and productive yields while keeping the environment safe (Alam *et al.*, 2016). BPs are NPs that are more effective against insect pests, bacteria, nematodes, fungi, viruses and insect pests (Genena *et al.*, 2008; Karpagam and Devaraj, 2011; Perelló *et al.*, 2013; Grdiša and Gršic, 2013; Kareru *et al.*, 2013; Todorov *et al.*, 2015; Feyisa *et al.*, 2015). BPs are effectively biodegradable, have different methods of activity, are less toxic to living beings, are non-pollutant, and are easily available in the environment (Neeraj *et al.*, 2017). Accordingly, they are vital in integrated pest management, along with various yield protection methods such as host resistance or resilience, rural agricultural practices, use of natural enemies, use of parasitoids, predators, microbial pesticides, and restricted application of safe SPs (Solomon *et al.*, 2000; Onditi *et al.*, 2013;

Wegulo *et al.*, 2015; Muthomi *et al.*, 2017). These methods connected with early pest observation and identification utilizing innovative technology such as the Internet of Things and tropological information systems would accomplish convenient, successful and sustainable crop pest management (Taylor, 2015; Chougule *et al.*, 2016; Yan *et al.*, 2017). BP compounds found in plants can significantly control bacteria, insects, fungi, viruses and nematodes (Elbeshehy *et al.*, 2015; Ichim *et al.*, 2017; Ingle *et al.*, 2017; Neeraj *et al.*, 2017; Salhi *et al.*, 2017).

Some researchers have reported that plant extracts using mother tinctures exhibit potential anti-fungal activity against *Chalara paradoxa* and *Fusarium guttiforme*, which cause pineapple fusariosis (Sales *et al.*, 2016). The plant extracts hinder the development of *C. paradoxa* and *F. guttiforme* by 29% and 46%, respectively. The extracts from *A. vera*, *A. sativum* and *Glycyrrhiza glabra* show potential anti-fungal activity when contrasted with synthetic fungicide such as Tebuconazole® (Sales *et al.*, 2016). In another study, extracts of *A. indica* and *Oscimum sanctum* were most effective against *Fusarium oxysporum*, which showed inhibition by up to 100% (Chougule and Andoji, 2016). *A. indica*, *Cerbera odollam*, and *Capsicum frutescens* constrained the development of *Penicillium digitatum*, the causal agent of grey mould infection in oranges, by up to 90% (Al-Samarrai *et al.*, 2013). *Z. officinale* and *C. nardus* reduced the fungus growth of *P. digitatum* by up to 70% *in vitro*. *A. indica*, *C. frutescens*, and *C. odollam* extract treatment of orange fruits exhibited no side effects from the grey fungus infection, potentially increasing fruit yield and quality. Another study found that *Hydnocarpus anthelmithicus* fruit extracts were active against *Phytophthora palmivara*, *Pyricularia oryzae*, and *Rhizoctonia solani* and inhibited *P. oryzae* growth by 100%. This makes *H. anthelmithicus* extremely encouraging in the control of parasitic diseases of rice, which would further decrease the use of synthetic fungicides. Extracts of *Equisetum arvense*, *Glycyrrhiza glabra*, *Sambucus nigra* and *Tagetes patula* totally inhibited the mycelial growth of *Rhizoctonia solani*, a contagious microbe that causes root decay in tomatoes (Rodino *et al.*, 2012).

Even crude BPs have been shown to have insecticidal characteristics (Ali *et al.*, 2017). *Cinnamomum cassia*, *Cinnamomum zeylanicum* and *Piper nigrum* are highly repellent to thrips (*Megalurothrips sjostedti*), while formulations of extracts from *Annona squamosa*, *Aglaia odorata* and *Piper retrofractum* can reduce the population of *Crocidolomia paronana* and *Plutella xylostella* in cabbage (Abtew *et al.*, 2015). The use of these extracts does not exhibit toxicity in the natural enemies of insect pests (Dadang and Djoko, 2009). *Allium sativum* and *A. indica* extracts successfully reduced the population of *M. sjostedtion* and *Maruca vitrata* in cowpea, thereby enhancing crop yield. *Cardaria draba*, *N. tabacum* and *Sinapsis arvensis* have been tested for efficacy against *Trogoderma granarium* and have effectively decreased pest populations in stored wheat grains. *C. draba* has been shown to be extremely effective against the larval phases of the pest (Sarmamy *et al.*, 2011). *Ajuga iva*, *Aristolochia baetica*, *Pegaum harmala* and *Raphanus raphanistrum* extracts exhibit significant insecticidal activity against *Tribolium castaneum*. The extracts disturb the insects' formative cycles and repress the production of F1 progenies (Jbilou *et al.*, 2006).

A. sativum, *Citrus limon*, *Curcuma longa* and *Z. officinale* crude extracts and microbial antagonists (for instance, *Paecilomyces* and *Trichoderma*) have been shown to decrease the population of Aleyrodidae (whiteflies) and thrips. They also hinder *Phaeoisariopsis griseola* (angular leaf spot), *Colletotrichum lindemuthianum* (anthracnose), and *Uromyces appendiculatus* (intensity of rust) in *Phaseolus vulgaris* (snap beans) (Muthomi *et al.*, 2017). Extracts from *A. sativum*, *Allium cepa*, *A. indica*, *Calotropis procera*, *Curcuma zedoaria*, *Ocimum canum* and *Phyllanthus embica* combined with cow dung and mineral salts in cow urine have been shown to be effective against tomato pests, including *Helicoverpa armigera*, reducing tomato fruit destruction and enhancing yield (Sumitra *et al.*, 2014). A mixture of *A. indica*, *Metarrhiziam anisopliae* and BPs can enhance the yield of snap beans and decrease thrip populations after application (Ouma *et al.*, 2014). Further research suggested that it would be commercially more effective to utilize BPs in combination with SP formulations. The mixture of SPs and BPs would enhance crops by up to 50%, as well as reduce the thrip population by over 60%. Extracts from *A. sativum* and *C. limon* significantly impaired a polyphagous pest of cotton, as well as the larvae of *Spodoptera littoralis*. This activity was attributed to decreased lipids and proteins in the stomachs of larvae (Ali *et al.*, 2017). The extracts from *Tephrosia vogelli* are highly effective against *Bactrocera curcubitae* (melonfly) and *Diabrotica undecimpunctata* (beetle cucumber) (Alao and Adebayo, 2015). Extracts of *T. toxicaria* are significantly more effective against the second-stage juveniles of *M. javanica* and *Meloidogyne enterologbii* (Moreira *et al.*, 2018). BPs have also been successful in controlling post-harvest pests. The essential oils of *A. sativum*, *Cinnamomum zeylanicum*, *Copaifera langsdorfii* and *Eugenia caryophyllata* decreased *Colletotrichum musae* (anthracnose) by up to 90% (Cruz *et al.*, 2013). Furthermore, *Cerbera odollam*, *Swietenia macrophyllai* and *Syzygium aromaticum* alcoholic extracts at a concentration of 3000 ppm inhibited the development of post-harvest fungi, *A. niger*, *P. digitatum* and *Fusarium* spp. by 40–90% in citrus (Singh *et al.*, 2012). The use of increased concentrations of extract from *A. sativum* and *Curcuma longa* can decrease the emergence rate of adults of *Tribolium castaneum*, decrease the weight of the pests or insects, and exert anti-moulting properties on larvae, pupa and adults (Ali *et al.*, 2014). *E. terreticonis*, *lantana camara* and *Tagetes minuta* can cause impermanence in adults of *Sitophilus zeamais* at a utilization rate of 20g per 200g of maize grains (Parwada *et al.*, 2018).

The models discussed above indicate that BPs can potentially contribute to the sustainable management of crop pests in integrated pest management programmes. With their activities against a wide scope of pests, different modes of action and effectiveness in various agro-climatic zones, harvests and seasons, BPs could assume a significant role in expanding crop yields while protecting biodiversity, the environment and human well-being.

Challenges in the Adoption of BPs

Despite the accessibility and viability of BPs against an extensive variety of harvest pests, they have yet to be extensively addressed in the pesticide business (Kekuda *et al.*, 2016). Commercial exploitation of BPs mainly depends on the accessibility of the source plants in large amounts, and the plants ought to be promptly cultivated. Plant sources of BPs are either developed for different purposes such as drugs, food, shade and ornamentals, normally in forests, and others on non-cultivated land. The development of plants required for the production of BPs will require macro-scale processes, presenting significant competitiveness with food production in profoundly arable agricultural terrains. Furthermore, some plants that are natural sources of BPs are utilized as food. Selecting to use these resources for other productive undertakings can consequently jeopardize food security (Srijita, 2015). Thus, the accessibility of agricultural land for the production of the necessary amounts of BPs will be a significant limiting factor.

Additionally, the repository and refining facilities of the large amount of plant material required for producing BPs will require enormous resources in storehouses, conservation methods, and apparatus or machinery. BPs also face a great deal of competition from the SPs that are not difficult to make, with simple planning and application, a long timespan of usability, and already-laid-out production facilities (Gurjar *et al.*, 2012). Notably, the definition of BPs has been very controversial, given that one plant could have many bioactive chemical constituents in contrast to SPs (Kumar and Singh, 2015). This quality could be investigated by integrating many plant-related chemical constituents with synergistic effects against pests (Stevenson *et al.*, 2017). Despite the lower toxicity displayed by BPs, their regulatory strategies for rural use are the same as those for synthetic substances, particularly in emerging nations (Blackfor, 2013). The enlistment process of BPs is commercially expensive and has several barriers; hence, the production of BPs is relatively unavailable in the market. There is scant awareness among the few agricultural holders about the convenience of BPs in controlling crop pests (Stanojković *et al.*, 2013; Okunlola and Akinrinnola, 2014).

The use of BPs is also subject to atmospheric conditions, since they are easily degraded, particularly whenever applied in their crude form (Sales *et al.*, 2016), which shortens the half-life of BPs (Cañizares *et al.*, 2004). Despite the safety of BPs, a few plants exhibit antibacterial, anti-fungal and antiviral activity and are also associated with toxicity towards a group of non-targets. For instance, rotenone isolated from the extracts from *Derris* and *Loncho carpus* is more harmful to fish, pests and mammals (Arnason *et al.*, 2012). Although

Tephrosia spp. is a potentially powerful insecticide against a few pests, it is also intensely harmful to the cultivated *Clarias gariepinus* (clariid) (Ekpo *et al.*, 2017; Stevenson and Belmain, 2017). The stability, chemical constituents and quality of BP extracts mainly depend on the plant source, temperature range, type of extraction, solvent system, environment and storage medium (Cañizares *et al.*, 2004; Nashwa, 2011; Nashwa and Abo-Elyousr, 2012). Furthermore, the extraction of BPs requires the use of organic solvents whose disposal presents issues of polluting the environment (Nashwa, 2011). This indicates the need to investigate better extraction and waste disposal techniques. These issues are driving the reluctance of many agrochemical organizations to invest in BP production.

Conclusion

Nature is an abundant source of a broad range of plants, a few of which have been utilized to heal human, plant and animal diseases. Due to concerns about human well-being, environmental safety and severe guidelines on pesticide residues or particles in agrarian produce, the use of SPs is should be done prudently and only when necessary. However, even with careful use, dependence on these chemical compounds represents harm to the environment, non-target organisms and human well-being due to their residual impacts. Accordingly, the viability and role of BPs in controlling crop pests should be reconsidered because of their nature and the safety of human beings and the environment. Taking into account the immense amount of material from which BPs could be produced, there is an enormous scope for developing source plants in local terrains that may not be appropriate for arable farming to prevent competition with food crop production. This can support employment in semi-arid areas. Short plants, such as spices with rhizomes, can be grown in forested regions without obstructing woodland trees. The chemical constituents identified with pesticidal activities in plants could be blended using information from collaborative work between chemical engineers and agricultural researchers. The handling and extraction of BPs utilizing modest solvents ought to be investigated to lessen the expense of production and minimize the issues related to waste disposal. This would make it feasible for the smallholder ranchers to manage and embrace the protected pest control products.

The rapid biodegradability of BPs is desirable yet non-economical, as their short timespan of usability would result in more application and costs. Thus, there is a need for more exploration to foster the definition of BPs and identify an appropriate lifespan with ideal adequacy. It is recommended that future investigations focus on the dependability of BPs, particularly under field conditions, which can be supported by nanotechnology, with research on viable methods for distributing bioactive substances under field conditions. This can ensure the efficacy of BPs at the farm level.

More studies are expected to further develop the use of plants with bioactive compounds pertinent to crop protection. This might include taming and improving wild plants through breeding to develop bioactive molecules, as well as improving suitable husbandry practices, including plant sustenance and agronomic practices. The development of the plant sources might include distinguishing the gene regulating the development and accumulation of bioactive chemical compounds and informing approaches that would support a high return on investment of the target pesticidal substances. These chemical constituents should be formed into products that are accessible, particularly by smallholder farmers. Concerned partners, both researchers and policymakers, should be more mindful of the need to embrace BPs and other NPs as pest control devices. Specialists and researchers dealing with such products play an important role in providing field viability information that is predictable and reproducible. The enhanced use of natural pest control chemical constituents in co-ordinated pest management will prompt superior desirability of agrarian produce in specific markets, consequently adding to development on global exchange and food security, as well as the preservation of biodiversity and the protection of human well-being and the environment.

References

Abdel-Salam, A.F., Shaheda, M.E. and Jehan, B.A. (2014) Antimicrobial and antioxidant activities of red onion, garlic and leek in sausage. *African Journal of Microbiology Research* 27, 2574–2582.

Abtew, A., Subramanian, S., Cheseto, X., Kreiter, S., Tropea Garzia, G. *et al.* (2015) Repellency of plant extracts against the legume flower thrips *Megalurothrips sjostedti* (Thysanoptera: Thripidae). *Insects* 6, 608–625.

Ahmad, W., Shilpa, S. and Sanjay, K. (2017) Phytochemical screening and antimicrobial study of *Euphorbia hirta* extracts. *Journal of Medicinal Plants Studies* 2, 183–186.

Akyazi, F. (2014) Effect of some plant methanol extracts on egg hatching and juvenile mortality of root-knot nematode *Meloidogyne incognita*. *American Journal of Experimental Agriculture* 11, 1471–1479.

Alam, M.Z., Haque, M.M., Islam, M.S., Hossain, E., Hasan, S.B. *et al.* (2016) Comparative study of integrated pest management and farmers' practices on sustainable environment in the rice ecosystem. *International Journal of Zoology* 1–11.

Alao, F.O. and Adebayo, T.A. (2015) Comparative efficacy of *Tephrosia vogelii* and *Moringa oleifera* against insect pests of watermelon (*Citrullus lanatus* Thumb). *International Letters of Natural Sciences* 35, 71–78.

Ali, S., Sagheer, M., Hassan, M., Abbas, M., Hafeez, F. *et al.* (2014) Insecticidal activity of turmeric (*Curcuma longa*) and garlic (*Allium sativum*) extracts against red flour beetle, *Tribolium castaneum*: A safe alternative to insecticides in stored commodities. *Journal of Entomology and Zoology Studies* 3, 201–205.

Ali, S., Sagheer, M., Hassan, M., Abbas, M., Hafeez, F. *et al.* (2017) Antifeedant activity and some biochemical effects of garlic and lemon essential oils on *Spodoptera littoralis* (Boisduval) (Lepidoptera: Noctuidae). *Journal of Entomology and Zoology Studies* 3, 1476–1482.

Aljamali, M.N. (2013) Study effect of medical plant extracts in comparison with antibiotic against bacteria. *Journal of Scientific Innovative Research* 5, 843–845.

Al-Samarrai, G.F., Harbant, S. and Mohamed, S. (2013) Extracts some plants on controlling green mold of orange and on postharvest quality parameters. *World Applied Sciences Journal* 4, 564–570.

Ankri, S. and Mirelman, D. (1999) Antimicrobial properties of allicin from garlic. *Microbes Infection* 2, 125–129.

Arafat, Y., Shahida Khalid, S.K., Lin WenXiong, L.W., Fang ChangXun, F.C., Sadia, S. *et al.* (2015) Allelopathic evaluation of selected plants extract against broad and narrow leaves weeds and their associated crops. *Academic Journal of Agricultural Research* 10, 226–234.

Araújo, K.M., Lima, A., Silva, J.D.N., Rodrigues, L.L., Amorim, A.G. *et al.* (2014) Identification of phenolic compounds and evaluation of antioxidant and antimicrobial properties of *Euphorbia tirucalli* L. *Antioxidants* 3, 159–175.

Arnason, J.T., Sims, S.R. and Scott, I.M. (2012) Natural products from plants as insecticides. *Encyclopedia of Life Support Systems (EOLSS)* 1–8.

Asif, M., Tariq, M., Khan, A. and Siddiqui, M.A. (2017) Biocidal and antinemic properties of aqueous extracts of ageratum and coccinia against root-knot nematode, *Meloidogyne incognita in vitro*. *Journal of Agricultural Science* 2, 108–122.

Bhanuprakash, V., Hosamani, M., Balamurugan, V., Gandhale, P., Naresh, R. *et al.* (2008) *In vitro* antiviral activity of plant extracts on goat pox virus replication. *Indian Journal of Experimental Biology* 46, 120–127.

Blackfor, R. (2013) A guide to the development of regulatory frameworks for microbial biopesticides in sub-Saharan Africa. *African Agricultural Technology Foundation, Nairobi*. Available at: https://www.aatf-africa.org/ Regulatory-Frameworks-for-Microbial-Biopesticides-in-Sub-Saharan-Africa (accessed 20 January 2018).

Bomfim, N., Nakassugi, L.P., Oliveira, J.F.P., Kohiyama, C.Y., Mossini, S.A.G. *et al.* (2014) Antifungal activity and inhibition of fumonisin production by *Rosmarinus officinalis* L. essential oil in *Fusarium verticillioides* (Sacc.) Nirenberg. *Food Chemistry* 166, 330–336.

Business Daily (2013) Chemical ban hits vegetable exports to the EU market. 14 february.

Business Daily (2014) Regulator suspends use of pesticides on vegetables. 13 October.

Cañizares, P., Gracia, I., Gómez, L.A., García, A., Argila, C.M. *et al.* (2004) Thermal degradation of allicin in garlic extracts and its implication on the inhibition of the in-vitro growth of *Helicobacter pylori*. *Biotechnology Progress* 20, 32–37.

Castillo-Sánchez, L.E., Jiménez-Osornio, J.J., Delgado-Herrera, M.A., Candelaria-Martínez, B. and Sandoval-Gío, J.J. (2015) Effects of the hexanic extract of neem *Azadirachta indica* against adult whitefly *Bemisia tabaci*. *Journal of Entomology and Zoology Studies* 5, 95–99.

Chougule, P.M. and Andoji, Y.S. (2016) Antifungal activity of some common medicinal plant extracts against soilborne phytopathogenic fungi *Fusarium oxysporum* causing wilt of tomato. *International Journal of Development Research* 6(3), 7030–7033.

Chougule, A., Jha, V.K. and Mukhopadhyay, D. (2016) Using iot for integrated pest management. In: *Proceedings of the International Conference on Internet of Things and Applications (IoTA)*, IEEE, pp. 17–22.

Cooper, J. and Dobson, H. (2007) The benefits of pesticides to mankind and the environment. *Crop Protection* 26(9), 1337–1348.

Cruz, M.E.S., Schwan-Estrada, K.R.F., Clemente, E., Itako, A.T., Stangarlin, J.R. *et al.* (2013) Plant extracts for controlling the post-harvest anthracnose of banana fruit. *Revista Brasiliera de Plantas Medicinae* 4, 727–733.

Cui, H.Y., Zhou, H., Lin, L., Zhao, C.T., Zhang, X.J. *et al.* (2016) Antibacterial activity and mechanism of cinnamon essential oil and its application in milk. *Journal of Animal and Plant Sciences* 2, 532–541.

Cycon, M. and Piotrowska-Seget, Z. (2016) Pyrethroid-degrading microorganisms and their potential for the bioremediation of contaminated soils: A review. *Frontiers in Microbiology* 7, 1463.

Dadang, E.D.F. and Djoko, P. (2009) Effectiveness of two botanical insecticide formulations to two major cabbage insect pests on field application. *International Journal of the Southeast Asian Agricultural Sciences* 1, 42–51.

Damalas, C.A. (2016) Safe food production with minimum and judicious use of pesticides. In: Selamat J. and Iqbal, S.Z. (eds) *Food Safety*. Springer, Cham, Switzerland, pp. 43–55.

Damalas, C.A. and Koutroubas, S.D. (2015) Farmers' exposure to pesticides: Toxicity types and ways of prevention. *Toxics* 1, 1–10.

Dhanavade, D.J., Chidamber, B.J. and Jai, S.G (2011) Study of antimicrobial activity of lemon (*Citrus limon* L.) peel extract. *British Journal of Pharmacology and Toxicology* 3, 119–122.

Djeussi, D.E., Noumedem, J.A.K. and Seukep, J.A. (2013) Antibacterial activities of selected edible plants extracts against multidrug-resistant Gram-negative bacteria. *BMC Complementary and Alternative Medicine* 164, 1–8.

Dubey, N.K., Srivastava, B. and Kumar, A. (2008) Current status of plant products as botanical pesticides in storage pest management. *Journal of Biopesticides* 1(2), 182–186.

Dubey, N.K., Shukla, R., Kumar, A., Singh, P. and Prakash, B. (2010) Prospects of botanical pesticides in sustainable agriculture. *Current Science* 98(4), 479–480.

Ekpo, P.B., Uno, U.U., Effiong, E.C. and Etta, S.E. (2017) Acute toxicity of *Tephrosia vogelli* on the early life stages of farmed clariid (*Clarias gariepinus*). *Asian Journal of Advanced Agricultural Research* 3(2), 1–5.

El Shafie, H.A.M. and Almahy, A.A.M. (2012) Effect of storage conditions and duration on the potency of neem (*Azadirachta indica* A. Juss) seeds as a home-made insecticide. *Agriculture and Biology Journal of North America* 10, 385–390.

Erel, Ş.B., Reznicek, G., Şenol, S.G., Yavaşoğlu, N.Ü.K., Konyalioğlu, S. *et al.* (2012) Antimicrobial and antioxidant properties of *Artemisia* L. species from western Anatolia. *Turkish Journal of Biology* 36, 75–84

Elbeshehy, E.K.F., Metwali, E.M.R. and Almaghrabi, O.A. (2015) Antiviral activity of *Thuja orientalis* extracts against watermelon mosaic virus (WMV) on *Citrullus lanatus*. *Saudi Journal of Biological Science* 22, 211–219.

Erenso, T.F. and Berhe, D.H. (2016) Effect of neem leaf and seed powders against adult maize weevil (*Sitophilus zeamais* (Motschulsky)) mortality. *Agricultural Research* 2, 90–94.

FAO (Food and Agricultural Organization of the United Nations) (2018) Environmental and social management guidelines, environmental and social standard 5. Available at: http://fao.org/environmental-social-standard/en/ (accessed 7 August 2019).

Feyisa, B., Lencho, A., Selvaraj, T. and Getaneh, G. (2015) Evaluation of some botanicals and *Trichoderma harzianum* for the management of tomato root-knot nematode (*Meloidogyne incognita*) (Kofoid and White) Chit Wood). *Advances in Crop Science Technology* 1, 1–10.

Gaikwad, R.S., Kakde, R.B., Kulkarni, A.U., Gaikwad, D.R. and Panchal, V.H. (2012) *In vitro* antimicrobial activity of crude extracts of *Jatropha* species. *Current Botany* 3, 9–15.

Gakuubi, M.M., Wanzala, W., Wagacha, J.M. and Dossaji, S.F. (2016) Bioactive properties of *Tagetes minuta* L. (Asteraceae) essential oils: A review. *American Journal of Essential Oils and Natural Products* 2, 27–36.

Ganesh, P., Suresh, K.R. and Saranraj, P. (2014) Phytochemical analysis and antibacterial activity of pepper (*Piper nigrum* L.) against some human pathogens. *Central European Journal of Experimental Biology* 2, 36–41.

Genena, A.K., Hense, H., Smânia Junior, A. and Souza, S.M.D. (2008) Rosemary (*Rosmarinus officinalis*) – a study of the composition, antioxidant and antimicrobial activities of extracts obtained with supercritical carbon dioxide. *Ciênciae Tecnologia de Alimentos* 2, 463–469.

Grdiša, M. and Gršic, K. (2013) Botanical insecticides in plant protection. *Agriculturae Conspectus Scientificus* 2, 85–93.

Gunasekara, A.S. (2005) Environmental fate of pyrethrins. *Environmental Monitoring Branch, California Dept. of Pesticide Regulation*.

Gurjar, M.S., Ali, S., Akhtar, M. and Singh, K.S. (2012) Efficacy of plant extracts in plant disease management. *Agricultural Science* 3, 425–433.

Hadi, M., Kashefi, B., Sobhanipur, A. and Rezaarabsorkhi, M. (2013) Study on effect of some medicinal plant extracts on growth and spore germination of *Fusarium oxysporum* Schlecht. *in vitro*. *American-Eurasian Journal of Agricultural and Environmental Science* 4, 581–588.

Hernandez-Moreno, D., Casa-Resino, I., Lopez-Beceiro, A., Fidalgo, L.E., Soler, F. *et al.* (2013) Secondary poisoning of non-target animals in an Ornithological

Zoo in Galicia (NW Spain) with anticoagulant rodenticides: A case report. *Veterinary Medicine (Praha)* 10, 553–559.

Hikal, W.M., Baeshen, R.S. and Said-Al Ahl, H.A. (2017) Botanical insecticide as simple extractives for pest control. *Cogent Biology* 3(1), 1404274.

Ichim, E., Marutescu, L. and Popa, M. (2017) Antimicrobial efficacy of some plant extracts on bacterial ring rot pathogen, *Clavibacter michiganensis* ssp *sepedonicus*. *European Biotechnology Journal* 1, 93–96.

Ingle, K.P., Deshmukh, A.G., Padole, D.A., Dudhare, M.S., Moharil, M.P. *et al*. (2017) Bio-efficacy of crude extracts from *Jatropha curcas* against *Spodoptera litura*. *Journal of Entomology and Zoology Studies* 1, 36–38.

Isman, M.B. (2014) Botanical insecticides: A global perspective. In: *Biopesticides: State of the Art and Future Opportunities*. ACS Symposium Series 1172, pp. 21–30.

Isman, M.B. (2017) Bridging the gap: Moving botanical insecticides from the laboratory to the farm. *Industrial Crops Production* 110, 10–14.

Isman, M.B. and Grieneisen, M.L. (2014) Botanical insecticide research: Many publications, limited useful data. *Trends in Plant Science* 19(3), 140–145.

Jawalkar, N., Zambare, S. and Zanke, S. (2016) Insecticidal property of *Datura stramonium* L. seed extracts against *Sitophilus oryzae* L. (Coleoptera: Curculionidae) in stored wheat grains. *Journal of Entomology and Zoology Studies* 4, 92–96.

Jbilou, R., Abdeslam, E. and Fouad, S. (2006) Insecticidal activity of four medicinal plant extracts against *Tribolium castaneum* (Herbst) (Coleoptera: Tenebrionidae). *African Journal of Biotechnology* 10, 936–940.

Jnaid, Y., Yacoub, R. and Al-Biski, F. (2016) Antioxidant and antimicrobial activities of *Origanum vulgare* essential oil. *International Food Research Journal* 4, 1706–1710.

Jones, D. (1995) Environment alfate of cypermethrin. Environmental Monitoring and Pest Management, Department of Pesticide Regulation, Sacramento, California.

Joseph, B. and Sujatha, S. (2012) Insight of botanical based biopesticides against economically important pests. *International Journal of Pharmacological Life Sciences* 11, 2138–2148.

Juvatkar, P., Jalalpure, S., Waghulde, S., Kale, M., Naik, P. *et al*. (2012) Antimicrobial activity of leaves of *Artemisia vulgaris* L. PhD thesis, Department of Pharmacognosy and Phytochemistry, Konkan Gyanpeeth Rahul Dharkar College of Pharmacy, Karjat, India.

Karaca, G., Bilginturan, M. and Olgunsoy, P. (2017) Effects of some plant essential oils against fungi on wheat seeds. *Indian Journal of Pharmaceutical Education Research* 3, S385–S388.

Karami-Osboo, R., Khodaverdi, M. and Ali-Akbari, F. (2010) Antibacterial effect of effective compounds of *Satureja hortensis* and *Thymus vulgaris* essential oils against *Erwinia amylovora*. *Journal of Agricultural Science and Technology* 12, 35–45.

Kareru, P., Rotich, Z.K. and Maina, E.W. (2013) Use of botanicals and safer insecticides designed in controlling insects: The African case. In Tech Open. Available at: https://www.intechopen.com/books/insecticides-development-of-safer-and-more-effective-technologies/use-of-botanicals-and-safer-insecticides-designed-in-controlling-insects-the-african-case (accessed 29 January 2018).

Karpagam, T. and Devaraj, A. (2011) Studies on the efficacy of *Aloe vera* on antimicrobial activity. *International Journal of Research in Ayurveda and Pharmacy* 4, 1286–1289.

Kekuda, P.T.R., Akarsh, S., Nawaz, S.A.N., Ranjitha, M.C., Darshini, S.M. *et al*. (2016) *In vitro* antifungal activity of some plants against *Bipolaris sarokiniana* (Sacc.) Shoem. *International Journal of Current Microbiology and Applied Science* 6, 331–337.

Kepenekçi, I., Erdogu, D. and Erdogan, P. (2016) Effects of some plant extracts on root-knot nematodes *in vitro* and *in vivo* conditions. *Turkish Journal of Entomology* 40(1), 3–14.

Khan, A., Sayed, M., Shaukat, S.S. and Handoo, Z.A. (2008) Efficacy of four plant extracts on nematodes associated with papaya in Sindh, Pakistan. *Nematologia Mediterranea* 36, 93–98.

Kohn, L.K., Foglio, M.A., Rodrigues, R.A., Sousa, I.D.O., Martini, M.C. *et al*. (2015) *In-vitro* antiviral activities of extracts of plants of the Brazilian Cerrado against the avian Metapneumovirus (aMPV). *Brazilian Journal of Poultry Science* 3, 275–280.

Kozłowska, M., Laudy, A.E., Przybył, J., Ziarno, M. and Majewska, E. (2015) Chemical composition and antibacterial activity of some medicinal plants from *Lamiaceae* family. *Acta Poloniae Pharmaceutica – Drug Research* 4, 757–767.

Kumar, S. and Singh, A. (2015) Biopesticides: Present status and the future prospects. *Journal of Fertilizers and Pesticides* 2, 1–2.

Kumari, K.A., Kumar, K.N.R. and Rao, C.N. (2014) Adverse effects of chemical fertilizers and pesticides on human health and environment. *Journal of Chemical and Pharmaceutical Research* 3, 150–151.

Kushram, T., Yadu, Y.K., Sahu, M.K., Kulmitra, A.K. and Kumar, R. (2017) Bio efficacy of botanical insecticides against defoliators pests on soybean. *International Journal of Current Microbiology and Applied Science* 3, 2196–2204.

Laxmishree, C. and Nandita, S. (2017) Botanical pesticides – a major alternative to chemical pesticides:

A review. *International Journal of Life Science* 4, 722–729.

Lee, H. and Lee, D.G. (2015) Mode of action of bioactive phytochemicals, plant secondary metabolites, possessing antimicrobial properties. *Formatex* 185–192.

Lin, Q.S., Chen, S.H., Hu, M.Y., Haq, M.U., Yang, L. *et al*. (2011) Biodegradation of cypermethrin by a newly isolated actinomycetes HU-S-01 from wastewater sludge. *International Journal of Environmental Science and Technology* 1, 45–56.

Liu, B., Chen, B., Zhang, J., Wang, P. and Feng, G. (2017) The environmental fate of thymol, a novel botanical pesticide, in tropical agricultural soil and water. *Toxicology and Environmental Chemistry* 99(2), 223–232.

Lomborg, B. (2001) *The Skeptical Environmentalist*. Cambridge University Press, Cambridge, UK.

Mahmood, I., Imadi, S.R., Shazadi, K., Gul, A. and Hakeem, K.R. (2016) Effects of pesticides on environment. In: Hakeem, K.R., Akhtar, M.S. and Abdullah, S.N.A. (eds) *Plant, Soil and Microbes*. Springer, Cham, Switzerland, pp. 253–269.

Mahmoud, D.A., Hassanein, N.M., Youssef, K.A. and Abou Zeid, M.A. (2011) Antifungal activity of different neem leaf extracts and the nimonol against some important human pathogens. *Brazilian Journal of Microbiology* 42, 1007–1016.

Mandalari, G., Bennett, R.N., Bisignano, G., Trombetta, D., Saija, A. *et al*. (2007) Antimicrobial activity of flavonoids extracted from bergamot (*Citrus bergamia* Risso) peel, a by-product of the essential oil industry. *Journal of Applied Microbiology* 103, 2056–2064.

Martinez, J.A. (2012) Natural fungicides obtained from plants. In: Dhanasekaran, D., Thajuddin, N. and Panneerselvam, A. (eds) *Fungicides for Plant and Animal Diseases*. InTech Open, pp. 978–953.

Misra, H.P. (2014) Role of botanicals, biopesticides and bioagents in integrated pest management. *Odisha Review* 62–67.

Mizubuti, E.S., Júnior, V.L. and Forbes, G.A. (2007) Management of late blight with alternative products. *Pest Technology* 2, 106–116.

Mkenda, P.A., Stevenson, P.C., Ndakidemi, P., Farman, D.I. and Belmain, S.R. (2015) Contact and fumigant toxicity of five pesticidal plants against *Callosobruchus maculatus* (Coleoptera: Chrysomelidae) in stored cowpea (*Vigna unguiculata*). *International Journal of Tropical Insect Science* 35(4), 172–184.

Montanha, J.A., Moellerke, P., Bordignon, S.A., Schenkel, E.P. and Roehe, P.M. (2004) Antiviral activity of Brazilian plant extracts. *Acta Farm. Bonaerense* 2, 183–186.

Moreira, F.J.C., de Abreu Araújo, B., do Nascimento Lopes, F.G., de Assis Lopes de, A., Evami Cavalcante Sousa, A. *et al*. (2018) Assessment of the *Tephrosia toxicaria* essential oil on hatching and mortality of eggs and second-stage juvenile (J2) root-knot nematode (*Meloidogyne enterolobii* and *M. javanica*). *Australian Journal of Crop Science* 12(12), 1829.

Moreira, M.D., Picanço, M.C., Barbosa, L.C.D.A., Guedes, R.N.C., Campos, M.R.D. *et al*. (2007) Plant compounds insecticide activity against Coleoptera pests of stored products. *Pesquisa Agropecuária Brasileira* 7, 909–915.

Mpumi, N., Mtei, K.M., Machunda, R. and Ndakidemi, P.A. (2016) The toxicity, persistence and mode of actions of selected botanical pesticides in Africa against insect pests in common beans, *P. vulgaris*: A review. *American Journal of Plant Science* 7, 138–151.

Muthomi, J., Fulano, A.M., Wagacha, J.M. and Mwang'ombe, A.W. (2017) Management of snap bean insect pests and diseases by use of antagonistic fungi and plant extracts. *Sustainable Agriculture Research* 3, 52.

Nashwa, M.A.S. (2011) Control of tomato early blight disease by certain aqueous plant extracts. *Plant Pathology Journal* 4, 187–191.

Nashwa, S. and Abo-Elyousr, A.M.K. (2012) Evaluation of various plant extracts against the early blight disease of tomato plants under green house and field conditions. *Plant Protection Science* 2, 74–79.

Nawaz, M., Juma, M. and Hongxia, H. (2016) Current status and advancement of biopesticides: Microbial and botanical pesticides. *Journal of Entomology and Zoology Studies* 2, 241–246.

Neeraj, G.S., Kumar, A., Ram, S. and Kumar, V. (2017) Evaluation of nematicidal activity of ethanolic extracts of medicinal plants to *Meloidogyne incognita* (Kofoid and White) Chit Wood under lab conditions. *International Journal of Pure Applied Bioscience* 1, 827–831.

Nefzi, A., Abdallah, B.A.R. and Jabnoun-Khiareddine, H. (2016) Antifungal activity of aqueous and organic extracts from *Withania somnifera* L. against *Fusarium oxysporum* f.sp. *radicis-lycopersici*. *Journal of Microbial and Biochemical Technology* 3, 144–150.

Nikkhah, M., Hashemi, M., Najafi, M.B.H. and Farhoosh, R. (2017) Synergistic effects of some essential oils against fungal spoilage on pear fruit. *International Journal of Food Microbiology* 257, 285–294.

Ntalli, N.G. and Menkissoglu-Spiroudi, U. (2011) Pesticides of botanical origin: A promising tool in plant protection. In: Stoytcheva, M. (ed.) *Pesticides – Formulations, Effects, Fate*. InTech Open, pp. 1–23.

Oka, Y., Ben-Daniel, B. and Cohen, Y. (2012) Nematicidal activity of the leaf powder and extracts of *Myrtus communis* against the root-knot nematode *Meloidogyne javanica*. *Plant Pathology* 61, 1012–1020.

Okunlola, A.I. and Akinrinnola, O. (2014) Effectiveness of botanical formulations in vegetable production and bio-diversity preservation in Ondo State, Nigeria. *Journal of Horticulture and Forestry* 1, 6–13.

Onditi, J., Kiarie, N. and Solomon, S. (2013) Improving potato tuber yields using genotypes with multiple virus resistance in Kenya. *Agricultural Biology Journal of North Amercia* 4, 406–412.

Ortiz-Hernández, M.L., Sánchez-Salinas, E., Dantán-González, E. and Castrejón-Godínez, M.L. (2013) Pesticide: Mechanisms, genetics and strategies to enhance the process. In: Chamy, R. and Rosencranz, F. (eds) *Biodegradation: Life Science*. Springer, pp. 251–287.

Oskoueian, E., Abdullah, N., Ahmad, S., Saad, W.Z., Omar, A.R. *et al*. (2011) Bioactive compounds and biological activities of *Jatropha curcas* L. kernel meal extract. *International Journal of Molecular Science* 12, 5955–5970.

Ouma, B., Muthomi, J., Nderitu, J. and Toroitich, F. (2014) Management of thrips in French beans by integrating biological and synthetic pesticides in conventional spray regimes. *Journal of Renewable Agriculture* 2, 27–37.

Parwada, C., Chikuvire, T.J. and Kamota, A. (2018) Use of botanical pesticides in controlling *Sitophilus zeamais* (maize weevil) on stored *Zea mays* (maize)grain. *Modern Concepts and Developments in Agronomy* 4.

Pavaraj, M., Bakavathiappan, G. and Baskaran, S. (2012) Evaluation of some plant extracts for their nematicidal properties against root-knot nematode, *Meloidogyne incognita*. *Journal of Biopesticides* 5, 106–110.

Pawar, K., Khetmalas, S., Motkar, B., Bande, R. and Wable, H. (2013) Antimicrobial activity of *Lantana camara* (L.) *Var. aculeata* (L.) Mold (Verbanaceae). *Indo American Journal of Pharmaceutical Research* 3(4), 3284–3294.

Perelló, A., Noll, U. and Slusarenko, A.J. (2013) *In vitro* efficacy of garlic extract to control fungal pathogens of wheat. *Medicinal Research Reviews* 24, 1809–1817.

Pitinidhipat, N. and Yasurin, P. (2012) Antibacterial activity of *Chrysanthemum indicum*, *Centella asiatica* and *Andrographis paniculata* against *Bacillus cereus* and *Listeria monocytogenes* under osmotic stress. *Assumption University Journals* 4, 239–245.

Plant, J. and Stephens, B. (2015) Evaluation of the antibacterial activity of a sizable set of essential oils. *Medicinal and Aromatic Plants* 2, 1–5.

Plata-Rueda, A., Martínez, L.C., Santos, M.H.D., Fernandes, F.L., Wilcken, C.F. *et al*. (2017) Insecticidal activity of garlic essential oil and their constituents against the mealworm beetle, *Tenebrio molitor* Linnaeus (Coleoptera: Tenebrionidae). *Scientific Reports* 7, 46406.

Porto, A.L.M., Melgar, G.Z., Kasemodel, M.C. and Nitschke, M. (2011) Biodegradation of pesticides. In: Stoytcheva, M. (ed.) *Pesticides in the Modern World – Pesticides Use and Management*. *InTech Open*. pp. 407–438.

Pramila, D.M., Xavier, R., Marimuthu, K., Kathiresan, S., Khoo, M.L. *et al*. (2012) Phytochemical analysis and antimicrobial potential of methanolic leaf extract of peppermint (*Mentha piperita*: Lamiaceae). *Journal of Medicinal Plants Research* 2, 331–335.

Raja, N. (2014) Botanicals: sources for eco-friendly biopesticides. *Journal of Biofertilizers and Biopesticides* 5, e122.

Rajasekaran, D., Palombo, E.A., Chia Yeo, T., Lim Siok Ley, D., Lee Tu, C. *et al*. (2013) Identification of traditional medicinal plant extracts with novel anti-influenza activity. *PLoS ONE* 8(11), 1–15.

Rani, S.S.K., Neeti, S.U. and Udaysree, (2013) Antimicrobial activity of black pepper (*Piper nigrum* L.). *Global Journal of Pharmacology* 1, 87–90.

Rattan, R.S. (2010) Mechanism of action of insecticidal secondary metabolites of plant origin. *Crop Protection* 29(9), 913–920.

Rodino, S., Butu, A., Butu, M. and Cornea, C.P. (2012) *In vitro* efficacy of some plant extracts against damping off disease of tomatoes. *Agriculture and Food* 2, 240–244.

Rubina, L., Priyanka, T. and Ebenezer, J. (2009) Isolation, purification and evaluation of antibacterial agents from *Aloe vera*. *Brazilian Journal of Microbiology* 40, 906–915.

Saeed, S. and Tariq, P. (2009) Antibacterial activity of oregano (*Origanum vulgare* Linn.) against Gram positive bacteria. *Pakistani Journal of Pharmacological Science* 4, 421–424.

Sales, M.D.C., Costa, H.B., Fernandes, P.M.B., Ventura, J.A. and Meira, D.D. (2016) Antifungal activity of plant extracts with potential to control plant pathogens in pineapple. *Asia Pacific Journal of Tropical Biomedicine* 1, 26–31.

Salhi, N., Mohammed Saghir, S.A., Terzi, V., Brahmi, I., Ghedairi, N. *et al*. (2017) Antifungal activity of aqueous extracts of some dominant Algerian medicinal plants. *BioMedl Research International* 2017, 7526291.

Sarkar, M. and Kshirsagar, R. (2014) Botanical pesticides: Current challenges and reverse pharmacological approach for future discoveries. *Journal of Biofertilizers and Biopesticides* 5, e125.

Sarmamy, A.G., Hashim, H. and Sulayman, A. (2011) Insecticidal effects of some aqueous plant extracts on the control of Khapra *Trogoderma granarium* Evert. In: *Proceedings of the International Conference on Chemicals, Biological, and Environmental Sciences*, ICCEBS-December, pp. 55–70.

Sande, D., Mullen, J., Wetzstein, M. and Houston, J. (2011) Environmental impacts from pesticide use: A case study of soil fumigation in Florida tomato production. *International Journal of Environmental Research in Public Health* 12, 4649–4661.

Sarkar, E., Samarendra, N.C. and Chakraborty, P. (2012) Allelopathic effect of *Cassia tora* on seed germination and growth of mustard. *Turkish Journal of Botany* 36, 488–494.

Sarwar, M. (2015) Microbial insecticides: An eco-friendly effective line of attack for insect pests management. *International Journal of Engineering and Advanced Research Technology* 2, 4–9.

Shabana, Y.M., Abdalla, M.E., Shahin, A.A., El-Sawy, M.M., Draz, I.S. *et al.* (2017) Efficacy of plant extracts in controlling wheat leaf rust disease caused by *Puccinia triticina*. *Egyptian Journal of Basic Applied Science* 1, 67–73.

Sharafzadeh, S. (2011) Pyrethrum, coltsfoot and dandelion: Important medicinal plants from Asteraceae family. *Australian Journal of Basic Applied Science* 12, 1787–1791.

Sharoba, A.M., El Mansy, H.A., El Tanahy, H.H., El Waseif, K.H. and Ibrahim, M.A. (2015) Chemical composition, antioxidant and antimicrobial properties of the essential oils and extracts of some aromatic plants. *Middle East Journal of Applied Science* 2, 344–352.

Sienkiewicz, M., Monika, Ł. and Marta, P. (2013) The potential of use basil and rosemary essential oils as effective antibacterial agents. *Molecules* 18, 9334–9351.

Singh, D.K. (2008) Biodegradation and bioremediation of pesticide in soil: Concept, method and recent developments. *Indian Journal of Microbiology* 48, 35–40.

Singh, H., Alsamarai, G. and Syarhabil, M. (2012) Performance of botanical pesticides to control post-harvest fungi in citrus. *International Journal of Scientific Engineering Research* 4, 1–4.

Singh, U.P., Prithiviraj, B., Sarma, B.K., Singh, M. and Rai, A.B. (2001) Role of garlic (*Allium sativum* L.) in human and plant diseases. *Indian Journal of Experimental Biology* 39, 310–322.

Soković, M., Glamočlija, J., Marin, P.D., Brkić, D. and Griensven, L.J. (2010) Antibacterial effects of the essential oils of commonly consumed medicinal herbs using an *in vitro* model. *Molecules* 15, 7532–7546.

Solomon, M.G., Cross, J.V., Fitzgerald, J.D., Campbell, C.A.M., Jolly, R.L. *et al.* (2000) Biocontrol of pests of apples and pears in northern and central Europe – 3. predators. *Biocontrol Science and Technology* 2, 91–128.

Srijita, D. (2015) Biopesticides: An eco-friendly approach for pest control. *World Journal of Pharmaceutical Sciences* 6, 250–265.

Stanojković, T., Kolundžija, B., Ćirić, A.D., Soković, M., Nikolić, D. *et al.* (2013) Cytotoxicity and antimicrobial activity of *Satureja kitaibelii* (Wierzb. ex Heuff) (Lamiaceae). *Digest Journal of Nanomaterials and Biostructures* 2, 845–854.

Stevenson, P.C. and Belmain, S.R. (2017) *Tephrosia vogelii*: A pesticide of the future for African farming. *Boletín de La S.E.A* 19–22.

Stevenson, P.C., Isman, M.B. and Belmain, S.R. (2017) Pesticidal plants in Africa: A global vision of new biological control products from local uses. *Industrial Crops Production* 110, 2–9.

Sumitra, A., Kanojia, A.K., Kumar, A., Mogha, N. and Sahu, V. (2014) Biopesticide formulation to control tomato lepidopteran pest menace. *Current Science* 7, 1051–1057.

Swamy, M.K., Uma, R.S. and Mohd, S.A. (2015) *In vitro* pharmacological activities and GC-MS analysis of different solvent extracts of *Lantana camara* leaves collected from tropical region of Malaysia. *Complementary Alternative Medicine* 506413.

Taylor, P. (2015) *Plant Wise Diagnostic Field Guide: A Tool to Diagnose Crop Problems and Make Recommendations for Their Management*. CAB International, Wallingford, UK.

Thiruppathi, S., Ramasubramanian, V., Sivakumar, T. and Arasu, T.V. (2010) Antimicrobial activity of *Aloe vera* (L.) Burm. f. against pathogenic microorganisms. *Journal of Bioscience Research* 4, 251–258.

Todorov, D., Shishkova, K., Dragolova, D., Hinkov, A., Kapchina-Toteva, V. *et al.* (2015) Antiviral activity of medicinal plant *Nepeta nuda*. *Biotechnology and Biotechnological Equipment* 1, 39–43.

Velázquez-Fernández, J.B., Martínez-Rizo, A.B., Ramírez-Sandoval, M. and Domínguez-Ojeda, D. (2012) Biodegradation and bioremediation of organic pesticides. In: Soundararajan, R.P. (ed.) *Pesticides – Recent Trends in Pesticide Residue Assay*. InTechOpen.

Vidyasagar, G.M. and Tabassum, N. (2013) Antifungal investigations on plant essential oils: A review. *International Journal of Pharmacy and Pharmaceutical Sciences* 2, 19–28.

Waziri, H.M.A. (2015) Plants as antiviral agents. *Journal of Plant Pathology and Microbiology* 2, 1–5.

Wegulo, S.N., Baenziger, P.S., Nopsa, J.H., Bockus, W.W. and Hallen-Adams, H. (2015) Management of fusarium head blight of wheat and barley. *Crop Protection* 73, 100–107.

Wimalawansa, S.A. and Wimalawansa, S.J. (2014) Agrochemical-related environmental pollution: effects on human health. *Global Journal of Biology, Agriculture Health Sciences* 3, 72–83.

Wink, M. (2015) Modes of action of herbal medicines and plant secondary metabolites. *Medicines* 2, 251–286.

Yamamoto-Ribeiro, M.M.G., Grespan, R., Kohiyama, C.Y., Ferreira, F.D., Mossini, S.A.G. *et al.* (2013) Effect of *Zingiber officinale* essential oil on *Fusarium verticillioides* and fumonisin production. *Food Chemistry* 141, 3147–3152.

Yan, Y., Feng, C.C. and Chang, K.T.T. (2017) Towards enhancing integrated pest management based on volunteered geographic information. *ISPRS International Journal of Geo-Information* 7, 224.

Yang, X., Huang, Q., Jiang, T. and Xu, H. (2017) Degradation dynamics of Azadirachtin in cabbage and soil. *Journal of South China Agricultural University* 38(4), 37–40.

Yoon, M., Cha, B. and Kim, J. (2013) Recent trends in studies on botanical fungicides in agriculture. *Plant Pathology Journal* 1, 1–9.

Zaker, M. (2016) Natural plant products as eco-friendly fungicides for plant diseases control: A review. *Agriculturists* 14(1), 134–141.

Zhao, L., Feng, C., Hou, C., Hu, L., Wang, Q. *et al.* (2015) First discovery of acetone extract from cottonseed oil sludge as a novel antiviral agent against plant viruses. *PLoS ONE* 2, 1–13.

41 Overview of Integrated Pest Management in Organic Farming

MATTHEW RUSSELL GATES*

Cannabis Horticultural Association, Humboldt, California, USA

Abstract

Organic farming utilizes naturally occurring processes and phenomena to achieve cultivation goals that are more environmentally and economically sustainable, to which advancements in natural science has supported. Many pest management practices in organic farming are ancient or take great influence from older traditions that are honed with empirical and technological refinement while novel techniques rely on contemporary understanding of plant and pest physiology and ecology for durable salubrious results. Integrating a holistic pest management system that considers cultivation, society and the environment informs the arsenal of options available to the modern farmer that build upon historical challenges and improvements. Increasingly, emphasis is placed on many discrete natural factors such as immune-suppressing pest effectors, ultraviolet and other radiation spectra, various predatory mites and insects, and mutualistic microbes concentrated into a synergistic system.

Keywords: IPM, organic, pest, strategy, future, ecology

Introduction

Tens of thousands of years before the neolithic period, human knowledge and technical ability shaped the possibilities available for the accumulation of food and other natural resources (Mercader, 2009). It was and is a close symbiosis that strengthened between certain plants, animals, fungi and other microbes which facilitated various practices that came to be known under the banner of farming. Though the modern conception of integrated pest management matured millennia later in the 1900s, the ardent necessity to reduce crop pests persists since the inception of crop cultivation. Similarly, although many ancient practices are considered organic, organic farming as an intentional pursuit developed relatively recently and both are born from a theoretical understanding of ecology, evolution and economics.

Definitions vary (Newton *et al.*, 2020) but organic farming praxis emphasizes environmental harmony and, ideally, synergy through processes that are inspired by or explicitly utilize natural relationships such as the implementation of predatory mites or parasitic insects, fungi, bacteria and viruses for targeted pest management (Baker *et al.*, 2020). Chemical agents derived from botanical or microbial sources and even venom peptides with extremely low human and environmental toxicity and residual persistence are more common, as are simple oxidizing or pH-disrupting agents which replace noxious formulations and can be more easily integrated with biocontrols and other treatments. Various pathogenic microbes, insects and mites, and their movement through vectors, inform operational culture and techniques for mitigating their chances of encountering a crop. Simultaneously, research into the movement, aggregation and facilitation of plant-mutualistic microbes and arthropods (e.g. insects, mites) influences technologies used to attract, cultivate and sustain them in crops, resulting in pest

*Corresponding author: logikinsektgates@gmail.com

DOI: 10.1079/9781800626850.0041

management strategies which tend towards an augmented proactive organic passive support system (Mahmood *et al.*, 2018). The advent of molecular-level plant physiological and ecological understanding consolidates the concept of plant health into more practical actionable models, informing new organic options. University of California entomologist Dr Carl Huffaker opines in a 1976 *New York Times* article that, 'When we kill off the natural enemies of a pest, we inherit their work'. It may be much more economical and logistically efficient to treat the pest–plant paradigm with strategic nuance and make as much use of the ambient or commercially available mutualist organisms as is practicable.

Current and Future Prospects of Organic Integrated Pest Management

A traditional classification system for IPM methodology is to organize specific solutions into descriptive functional categories: biological, chemical, physical, cultural, environmental, among others. This simplification articulates how a particular treatment option achieves its effect and both potential incompatibilities like those between some biological and chemical treatments, as well as synergies such as the combination of crop scouting, insect mesh screen, *Trichogramma* sp. egg parasitoid wasps, and *Helicoverpa* zea Nucleopolyhedrovirus (HearNPV) to neutralize corn earworm ingress and establishment. Since the immune system and general plant physiology is affected by abiotic influences such as temperature or relative humidity, as well as reactions to certain signals acoustic (Khait *et al.*, 2023), chemical, photonic, among others, IPM considerations include selection of genetic stock for verified resistance traits, strategic immune system priming to exploit the growth–defence trade-off of plant development (Khoshru *et al.*, 2023), and selective facilitation and inoculation of the plant microbiome.

Revelations in pest physiology and behaviour have led to innovations in its disruption, such as the exploitation of insect visual, olfactory, tactile and gustatory cues which can introduce interference in the decision-making process, ultimately confusing, repelling, damaging or otherwise reducing crop establishment (Niemann *et al.*, 2021). These techniques can be lethal or sublethal, increasing net efficacy in tandem. To integrate these pest management techniques most efficiently requires an understanding of their function, compatibility and economic viability within the paradigm of organic farming by employing a strategy that minimizes negative environmental footprint.

Mechanization and sensor-driven automation is more ubiquitous in the discipline of precision agriculture as the accumulation of vast amounts of information is a culminating force that requires computerized processing to aid in analysis and interpretation for actionable decisions (Iost Filho *et al.*, 2020). Technological innovations to this end have rapidly democratized and can inform the status of plant physiological conditions, resource distribution such as nutrient application, pest presence, and environmental patterns in fine detail, sometimes augmented with interpretative heuristics based on historical data to guide the cultivator in changing environmental conditions or deciding the best time to deploy certain pest prevention measures. Pervasive advances in multispectral sensors, data computerization and wireless digital communication have facilitated various automata, particularly ground and aerial vehicles that carry out sophisticated and specialized tasks like locating, discriminating and destroying pest insects or plants in the immediate air- and groundspace using percussive force, laser instrumentation and precision chemical application. Another application is the use of automated or directed aerial vehicles for generalized detection of pests like pathogenic microbes or abiotic problems such as a malfunctioning irrigation system, through specialized imaging software using Normalized Difference Vegetation Index (NDVI) and others. More indirectly, improvements in planting and harvest automata that distribute or collect a proportion of produce or can be loaded and travel autonomously reduce many stressors of cultivation, preserving the stamina of personnel and minimizing potential wounding or hygiene complications which can result in pre-harvest, harvest and post-harvest pest damage in several crops (Shamshiri *et al.*, 2018).

Wavelengths across the electromagnetic spectrum with application for pest management include visible, ultraviolet, microwave and various radiowave spectra, either as physically damaging or otherwise disruptive emissions that reduce establishment in crop or persistence in harvested material (Yadav *et al.*, 2014). Type and exposure must be dosed precisely to reduce the potential for damage to crops and increase target effectiveness. Various visible and ultraviolet spectra are known to elicit plant physiological reactions with immunological and other developmental influences, which may be exploitable to reliably trigger desirable developmental changes such as immune signal priming or growth characteristics with consequences for pest establishment (Thomas and Puthur, 2017).

Advancements in the understanding of plant physiological mechanisms like hormone signalling pathways and immune response contextualize the plant–pest interaction at micro-scale. Structural knowledge of vasculature illuminates the physics of nutrient uptake and sequestration, as well as resilience to toxic levels of salinity or the effectors used by pests to manipulate tissues at the molecular level (Yuan *et al.*, 2021). Practical applications of this knowledge can bolster certain attributes related to vegetative or yield growth, biochemical production, immune response initiation, or microbiota composition with few or no costs to the plant's development. The application of super-oxygenated water or nanobubbles has been shown in some trials to enhance certain growth profiles and severely reduce the colonization of certain pathogens (Wang *et al.*, 2021). An interest in facilitating the molecular movement of solutes into and through the plant body like nutrients or targeted immune triggers by way of ionic stimulation has been investigated with some support for certain crops (Lee *et al.*, 2022). Applications of acoustic vibrations via thigmopriming exploits plant reactions to mechanostimulation to elicit immune responses and even disrupt certain pest insects that can be forced off the body or impair their ability to communicate percussively as some insects are known to do through stems and branches, like leafhoppers (Polajnar *et al.*, 2016). Cavitation in xylem water channels of some plants occurs under stress related to physical damage or dry substrate, producing ultrasonic sounds that may be detected by other plants and animals sensitive to them, potentially causing changes in growth and behaviour, respectively (Khait *et al.*, 2023). These discrete physiological responses can have influential immunological or signalling components that may be tactically manipulated by the cultivator to preempt and disrupt pests safely, particularly by consistently directing plant resources for optimal preventative and reactive response.

Improvements in the understanding of evolution, genetic architecture and expression greatly expanded breeding precision for trait improvements such as the volume and weight of produce, resistance to environmental and biological challenges, and changes in plant or animal microbiome function. Scientific models that synergize eco-evolutionary observations like the Modern Evolutionary Synthesis (Danchin *et al.*, 2018) and Hologenome Theory of Evolution (Carrier and Reitzel, 2017) acknowledge a fundamental unity and inherent influence between many organisms that in fine detail are outside the typical scope of human observation and experience but more simply are appreciated through art and literature, historical records, eco-conservatory organizations, and other socio-political actions.

From Control to Management: An Ecologically Conscientious Approach to Pests and Plant Stress

Integrated pest management strategy is guided by information relating to pest presence and effect on the crop. Depending on the pest and its damaging capability, some degree of presence may be tolerable, as cultivation planning can accommodate damage below a certain 'economic threshold' beyond which action must be taken before losses as costly as treatment are sustained, often described as the economic injury level (Monson *et al.*, 2022). Organic farming overlaps strongly with efforts to increase sustainability of processes and this is commonly achieved through creative fusion of low- and high-tech approaches like erecting a windbreak to deflect windborne pests, culling nearby host plants of problematic pests and leveraging scouting and other input data into a graphical interface for evaluation of operations. The practice generally avoids the use of environmentally noxious chemical pesticides and products that remain at toxic levels on or inside produce, relying instead on products derived from plants or other natural sources which may be, overall, less lethal to pests as well as ambient natural enemies that can be significantly helpful when supported through more careful treatment events. In many places this sentiment is facilitated by regulatory framework to establish definitions by which farmers and customers can evaluate such products. Many pests develop perdurable resistance not only to environmental conditions but chemical agents, obviating the over-reliance on a single, albeit destructive, chemical regimen.

Coupled with this waning efficacy, organic praxis developed with recognition of a more resilient, multi-domain approach to pest management and, more fundamentally, plant stress in general. This is because more cultivators understand plant physiological processes like nutrient uptake, sequestration and reaction to biotic and abiotic stimuli at the molecular level than before, and the fact that ultimately both detrimental and beneficial effects are mediated by processes accounted for in models like the plant–defence trade-off: the articulation of generally predictable, proportional response by plants to deploy resources to physical development or defensive reactions (Monson *et al.*, 2022).

For example, high air temperature or low substrate moisture can influence the piquancy of chili peppers (Mahmood *et al.*, 2021) and while it may be beneficial for the final product goal, a cultivation strategy must balance this with the effects these conditions will have on pest management. Some pestiferous organisms like certain fungi may be inhibited under these conditions, while others, like beetles or moth larvae, may be tolerant of the conditions, and, upon establishment, more damaging under this controlled-stress context, requiring strong preventative and reactive approaches to keep unintentional stress from herbivory or disease reduced when using such techniques. Additionally, acknowledgement of specific stress-responses to certain pests and the pathway by which signals of pest presence are detected and defences mustered allows for the strategic priming of plant defences before pests establish or expand across a crop. Controlled stress management can manifest in the application of priming factors such as Damage-Associated Molecular Patterns (DAMPs), which are often structural compounds or secreted products that trigger a specific immune response for a period of time. Receptors on and in the cells of plants bind with these molecular structures which triggers mild to dramatic changes in cell processes and often a cascade of hormonally driven systemic reactions that tap metabolic energy reserves to mobilize. Used in this way without also introducing an actually damaging effect, growers can apply these signals like chitin, which compose the structure of insect, mite and fungal bodies to intentionally induce the movement of resources towards reactive defence, crucially before the presence of the related pest, and sharply impair their establishment. In most cases, pests adapted to a plant population employ effectors that disrupt this signalling response in basic or sophisticated fashion on the onset of colonization, and it is at this liminal beginning point where the evolutionary arms race between plant detection and pest disruption is often won or at least significantly influenced. So prevalent are pest-effector agents that plants often have countermeasures in the form of receptors or other agents that detect their structure and set off additional reactions.

This first-strike context presumes that the plant is not reacting to the specific stressors of the pest yet, even if it may be reacting to environmental cues or dissimilar pest signals, a more-or-less even playing field. The main thrust of these pest adaptations is predicated on a quick and comprehensive overmatch or evasion of the plant's reactive signalling system and can be severely blunted by inducing it beforehand as the resulting physiological fortification can retard normal performance. Timing is achieved by the cultivator, potentially from using pest record history that depicts seasonal activity. Depending on the stimulus selection and documented effect, it may be viable to apply regularly throughout the active season, especially if developmental growth is not impaired, while other reactions may be more stressful under other contexts and must be plied more carefully.

Conclusion

Pest management techniques must be resource-efficient, effective and safe for the applier, environment and recipients of agricultural production (Dara, 2019). Organic farming as an eco-social *zeitgeist* has supported transition from linear and narrow treatment options to complex, layered systems that are more targeted, robust and, in many cases, less taxing and extraneously problematic to implement as integrated pest management strategy. Organic IPM makes use of foundational physiological aspects of plant growth and pest ecology to exploit natural vulnerabilities and enhance natural strengths in order to bring to bear the most advantages possible with the minimal inputs necessary to achieve a satisfactory outcome.

References

Baker, B.P., Green, T.A. and Loker, A.J. (2020) Biological control and integrated pest management in organic and conventional systems. *Biological Control* 140, 104095.

Carrier, T.J. and Reitzel, A.M. (2017) The hologenome across environments and the implications of a host-associated microbial repertoire. *Frontiers in Microbiology* 8, 802.

Danchin, E., Pocheville, A., Rey, O., Pujol, B. and Blanchet, S. (2018) Epigenetically facilitated mutational assimilation: Epigenetics as a hub within the inclusive evolutionary synthesis. *Biological Reviews* 94(1), 259–282.

Dara, S.K. (2019) The new integrated pest management paradigm for the modern age. *Journal of Integrated Pest Management* 10(1), 12.

Iost Filho, F.H., Heldens, W.B., Kong, Z. and Lange, E.S. (2020) Drones: Innovative technology for use in precision pest management. *Journal of Economic Entomology* 113(1), 1–25.

Khait, I., Lewin-Epstein, O., Sharon, R., Saban, K., Goldstein, R. *et al*. (2023) Sounds emitted by plants under stress are airborne and informative. *Cell* 186(7), 1328–1336.

Khoshru, B., Mitra, D., Joshi, K., Adhikari, P., Rion, M.S.I. *et al*. (2023) Decrypting the multi-functional biological activators and inducers of defense responses against biotic stresses in plants. *Heliyon* 9(3), e13825.

Lee, S., Song, M.-J. and Oh, M.-M. (2022) Effects of air anions on growth and economic feasibility of lettuce: A plant factory experiment approach. *Sustainability* 14(22), 15468.

Mahmood, R., Keerio, I.D., Rehman, A. and Rashid, K. (2018) Role of natural enemies field reservoir (NEFR) in farmer fields for controlling papaya mealy bug *Paracoccus marginatus* at Karachi. *Pakistan Entomologist* 40(1), 7–11.

Mahmood, T., Rana, R.M., Ahmar, S., Saeed, S., Gulzar, A. *et al*. (2021) Effect of drought stress on capsaicin and antioxidant contents in pepper genotypes at reproductive stage. *Plants (Basel)* 10(7), 1286.

Mercader, J. (2009) Mozambican grass seed consumption during the Middle Stone Age. *Science* 326(5960), 1680–1683.

Monson, R.K., Trowbridge, A.M., Lindroth, R.L. and Lerdau, M.T. (2022) Coordinated resource allocation to plant growth-defense tradeoffs. *The New Phytologist* 233(3), 1051–1066.

Newton, P., Civita, N., Frankel-Goldwater, L., Bartel, K. and Johns, C. (2020) What is regenerative agriculture? A review of scholar and practitioner definitions based on processes and outcomes. *Frontiers in Sustainable Food Systems* 4.

Niemann, J.-U., Menssen, M. and Poehling, H.-M. (2021) Manipulation of landing behaviour of two whitefly species by reflective foils. *Journal of Plant Diseases and Protection* 128(1), 97–108.

Polajnar, J., Eriksson, A., Virant-Doberlet, M. and Mazzoni, V. (2016) Mating disruption of a grapevine pest using mechanical vibrations: From laboratory to the field. *Journal of Pest Science* 89(4), 909–921.

Shamshiri, R.R., Weltzien, C., Hameed, I.A., Yule, I.J., Grift, T.E. *et al*. (2018) Research and development in agricultural robotics: A perspective of digital farming. *International Journal of Agricultural and Biological Engineering* 11(4), 1–14.

Thomas, T.T. and Puthur, J.T. (2017) UV radiation priming: A means of amplifying the inherent potential for abiotic stress tolerance in crop plants. *Environmental and Experimental Botany* 138, 57–66.

Wang, Y., Wang, S., Sun, J., Dai, H., Zhang, B. *et al*. (2021) Nanobubbles promote nutrient utilization and plant growth in rice by upregulating nutrient uptake genes and stimulating growth hormone production. *Science of the Total Environment* 800, 149627.

Yadav, D.N., Anand, T., Sharma, M. and Gupta, R.K. (2014) Microwave technology for disinfestation of cereals and pulses: An overview. *Journal of Food Science and Technology* 51(12), 3568–3576.

Yuan, M., Ngou, B.P.M., Ding, P. and Xin, X.-F. (2021) PTI-ETI crosstalk: An integrative view of plant immunity. *Current Opinion in Plant Biology* 62, 102030.

42 Weed Management in Organic Cropping Systems

Emanuele Radicetti[1]*, Ben Hassine Mortadha[1], Valentina Quintarelli[1], Daniele Borgatti[1], Taab Alireza[2], Muhammad Ahsan[3], Mohamed Allam[4] and Roberto Mancinelli[5]

[1]Department of Chemical, Pharmaceutical and Agricultural Sciences, University of Ferrara, Italy; [2]Department of Agronomy and Plant Breeding, Faculty of Agriculture, Ilam University, Iran; [3]Department of Horticultural Sciences, The Islamia University of Bahawalpur, Pakistan; [4]Department of Agronomy, Faculty of Agriculture, Assuit University, Egypt; [5]Department of Agricultural and Forestry Sciences, University of Tuscia, Viterbo, Italy

Abstract

Weeds are widely recognized as one of the main constraints that severely reduce crop yields and their quality. The efficacy of chemical compounds used for managing weeds in agro-ecosystems is gradually slowing down mainly due to the difficulty of producing new active compounds and the development of resistance issues. Nowadays, the incidence of herbicide-resistant weeds is increasing dramatically, and this requires a carefully designed approach to weed control aimed at producing safer food, respecting the environment and human health, and keeping the agro-ecosystems sustainable in the long term. Therefore, it is mandatory to have an integrated weed management strategy that combines various agronomic, ecological, natural and preventive control methods for determining sustainable food production based on a reduction of chemical means. Under organic farming systems, where the use of synthetic active compounds is avoided, weed management must be approached through the adoption of mechanical means adequately integrated with the inclusion of cover crops, thermal means, and the development of natural substances to replace herbicides.

Introduction

The increase in world population growth leads to the consequent need to ensure the supply of food by increasing agricultural production (FAO *et al.*, 2023). Although the integration of synthetic compounds, i.e. mineral fertilizers, pesticides and herbicides, supported the farmers to enhance crop productivity per unit of area, nowadays, several environmental concerns have mined the functionality of agro-ecosystems causing a gradual decrease of crop yield. It is, therefore, necessary to improve the management of global agricultural resources by minimizing the negative impact on agro-ecosystems. Weeds are considered the main cause of crop yield reduction due to interspecific competition for limited resources with cultivated plants and subsequently increased processing costs (Gharde *et al.*, 2018; Zimdahl, 2018). Indeed, they exert an action of stealing water, mineral nutrients, light and air. They also harbour insects and pathogens that attack cultivated plants (Chauhan, 2020). Unfortunately, chemical control remains the most common control practice for weed management causing serious problems, such as possible damage

*Corresponding author: emanuele.radicetti@unife.it

DOI: 10.1079/9781800626850.0042

to vegetation and non-target crops, the existence of residues in water and soil, and concerns for human health and safety (Cox, 2006). However, the over-reliance on the use of herbicides has led to the evolution of weeds that are resistant to them (Heap, 2019). To direct farmers towards more sustainable production practices, the European Union, through the Green Deal, provides for the creation of organic farming systems on 25% of cultivated area by 2030 (European Commission, 2020). Moreover, the sustainability of food production systems and the environmental health are rapidly becoming important global issues renewing the interest in no-chemical pest management. Therefore, it is necessary to implement diversified weed management strategies using combinations of a wide range of sustainable practices which often result in synergistic or additive interactions, especially in organic farming systems where the use of synthetic active compounds as herbicides is avoided (Fig. 42.1).

Nowadays, several no-chemical weed management practices have been developed and a wide range of options is becoming available to farmers as new techniques. It is important to understand that control strategies applied in organic farming aim to control weeds and not to eliminate them as is the case in conventional farming where full-action chemical herbicides are available. Furthermore, organic farming prioritizes the maintenance of biodiversity at all levels and therefore the goal is to keep biotic adversities within an acceptable threshold without annihilating them. Weed management, in the context of this chapter, is based on the application of various sustainable practices in order to reduce weed pressure, eliminate the use of chemicals and achieve the goal of sustainable and safe food production. Various indirect and direct methods can be used in weed control. The aim of this chapter is to provide a comprehensive review of the main methods available in organic cropping systems.

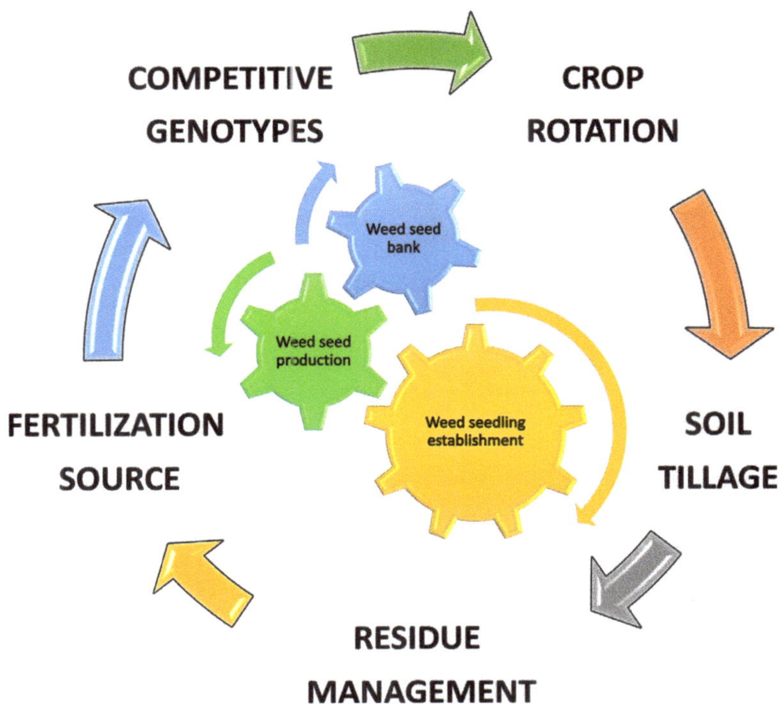

Fig. 42.1. Analysis of the main components that affect weed germination and establishment. Weed management strategy in each component affects both actual and potential weed flora, and therefore each component represents a valid field to plan the integrated approach for weed management under organic farming systems. (Adapted from Anderson, 2007)

Description of the Main No-Chemical Weed Management Methods to Use in Organic Farming Systems

From the beginnings of agriculture until the introduction of herbicides, weed management on agricultural land depended mainly on crop rotation, soil cultivation and seed cleaning. Direct or indirect control methods, preventive methods, can, therefore, be used to limit weed establishment and crop damage. The former are based on the use of mechanical and physical methods, and then by use of natural substances with herbicide effect, while the latter include genetic, ecological and agronomic methods. Mechanical methods are among the most widely used and effective for the management of the main weeds (Abbas *et al.*, 2018) as they have a greater impact on root-propagating weeds, as they bury the weeds and their seeds deep in the soil, thus preventing their germination (Hussain *et al.*, 2018). Furthermore, among many positive effects, they ensure the benefit of crop seedlings (Turner, 2000) and improve the moisture status of the soil and the accessibility of decomposers to organic residues (Hammermeister, 2016). However, mechanical weed control techniques entail side effects identifiable on the one hand in the high consumption of energy in the form of fossil-based, non-renewable products, and on the other hand, in some negative effects on soil structure. For all these reasons, mechanical weeding should be performed in rotation with other weed removal methods. Mechanical methods are effective both in the absence of the crop, by implementing techniques such as false seeding, and in the presence of the crop through weeding. False seedbed technique consists of refining the soil and irrigating it, as if one were seeding normally, but in reality, without burying the seed (Travlos *et al.*, 2020). This stimulates the germination of weed seeds in the soil, which are then removed with a harrow that works at a maximum depth of 5 cm so as not to damage the soil structure and not to bring additional weed seeds to the surface. Conversely in post-emergence, weeding can be carried out by trailed or self-propelled machines equipped with tilling devices (hoes, rotating discs etc.) in the case of multi-tillers and with flexible tines and fixed anchors in the case of weeders, which work in the inter-row of the crop.

Other physical and thermal (mulching, weeding and flaming), agroecological (intercropping, crop rotation, use of cover crops and green manure) and natural compounds-based methods (use of juglone, wood vinegar and glufosinate) are discussed below (Fig. 42.2). Mulching is a term which refers to the covering of agricultural soil with various materials (organic waste materials, straw and various materials of vegetal origin, sand, stones and plastic films etc.) with the aim of achieving a certain number of advantages (Bonciarelli, 1989). Among the many advantages, mulching allows improvement of the water balance of the soil, prevents soil compaction and reduces erosive phenomena, modifying the soil temperature and preventing the growth of weeds. Mulching or covering the soil surface can prevent weed seed germination by blocking light transmission. In organic farming, it is preferred to replace the black plastic film with a black bioplastic created from corn, wheat and potato starch, which is totally biodegradable.

Solarization is a sustainable method of management that consists of covering the soil with transparent plastic sheets in order to convert solar energy into heat with the aim of controlling weeds and pathogens before sowing (Bajwa *et al.*, 2015). Weeds can also be managed through high temperatures. It is possible to act on both developed plants and seeds. Thermal weed control is based on the use of flame, hot water, steam and freezing, which allow rapid weed control without leaving chemical residues in the soil and water. Thermal methods are selective against weeds and do not disturb the soil as with cultivation methods (Radicetti *et al.*, 2012). Flaming is the most commonly applied thermal method in organic farming systems and causes the destruction of plant membranes, resulting in death or severe weakening (Fennimore *et al.*, 2016). Flaming is most effective in controlling erect and broadleaf weeds in the early stages of growth, while it has proved less effective in controlling herbaceous and prostrate weeds (Mia *et al.*, 2020). The treatment of weed species with creeping growth habit, e.g. perennial broadleaf and protected growing points, e.g. grasses, should be repeated (Stevan, 2017).

Another method of thermal weed management is by means of soil steaming. It is a technique that can reduce germination by 50% at 62–68°C and 90% at 76–86°C. It has been reported that this method can be useful for large crabgrass (*Digitaria sanguinalis*) growth (de Oliveira *et al.*, 2023). As an alternative to herbicides, the efficacy of burning can be enhanced by its integration with tillage or mulching strategies. However, thermal weeding methods use large quantities of fossil fuels per unit

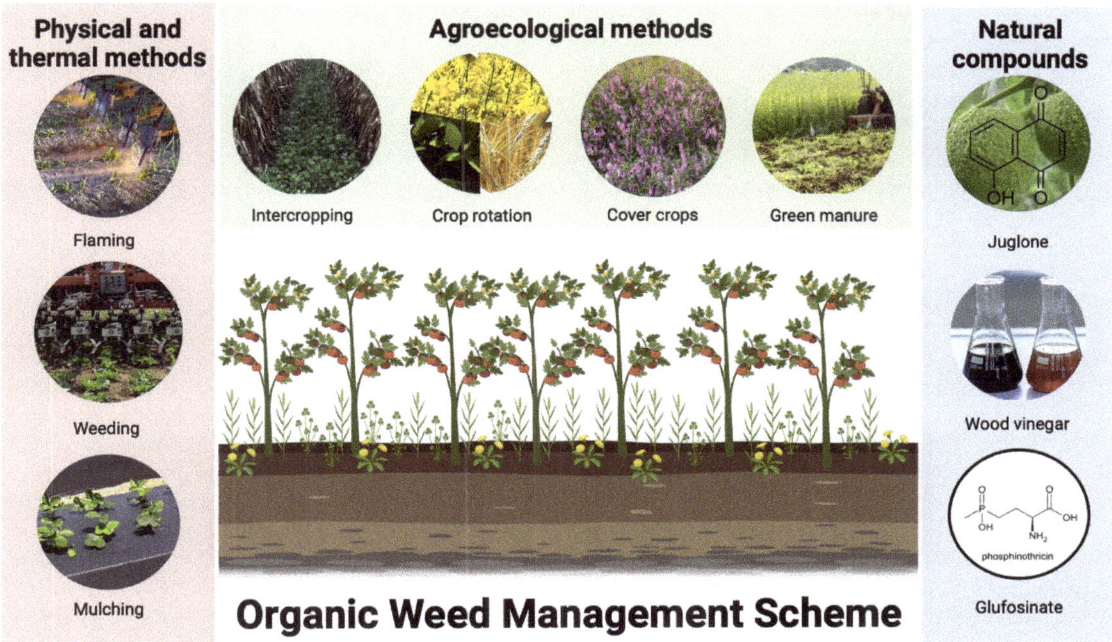

Fig. 42.2. A view of the main weed management means to be used in an integrated weed control strategy under organic farming systems.

area. The effectiveness of thermal media on weeds can be influenced by several factors, including temperature, exposure time and energy input.

It is possible to act on weeds at the ecological level by introducing other competitive but useful species (cover crop), or by promoting the multitude of species grown over time (crop rotation). To reduce the application of herbicides and to obtain multiple advantages it is possible to include, in the traditional agricultural system, the cultivation of cover crops (Quintarelli *et al.*, 2022). Among the different species of cover crops, it is interesting to note that the cultures belonging to the Brassicaceae family can produce glucosinolates, isothiocyanates and benzylamine (Ghidoli *et al.*, 2023). These molecules, which can be found in the whole plant, in root exudates, flowers or leaves (Rehman *et al.*, 2019), have an allelopathic effect; in fact it has been reported that they can reduce weed density and biomass. Hence, this is a sustainable method for weed management (Ghidoli *et al.*, 2023). Brassicaceae are among the plants best known for producing defence substances. Allelopathic potential can be found in Gramineae, in fact rye (*Secale cereale*) mulching can reduce weed germination (Tabaglio *et al.*, 2013). Allelochemical properties were found also in rice (*Oryza sativa*) and wheat (*Triticum aestivum*). Sorghum (*Sorghum bicolor*) can be incorporated into the soil, resulting in phenolic acids release (Khamare *et al.*, 2022).

Cover crops' allelopathic effect can be expressed in many ways: mulching, intercropping, aqueous extracts, soil incorporation (green manure) and including cover crops in crop rotation (Rehman *et al.*, 2019). It must be highlighted that not all weeds respond equally to various containment techniques; for example, Kruidhof *et al.* (2011) demonstrated that large-seeded weeds are more resistant to crop residues, such as oilseed rape, than small-seeded weeds. This happens because small-seeded species have larger roots which can absorb more allelochemicals, while large-seeded species contain more reserves that allow a better detoxification by cover crop molecules. Aqueous extracts of hairy vetch (*Vicia villosa*) and chickpea (*Cicer arietinum*) can affect weed growth; if applied at low concentrations (0.25 and 0.50 g/L) they can stimulate growth; if used at high concentrations (4 and 8 g/L) they inhibit growth (Hill *et al.*, 2006). These techniques influence weed seeds before the main cash crop seeding (pre-sowing); there are sustainable mechanisms which are also efficient during the development of the cash crop. In fact, cover crops can grow as a living mulch in the inter-row to control weed growth (Ngaba *et al.*, 2023). Another method is cover crops suppression and using residues as a mulch that avoids weeds growth (Allen *et al.*, 2023). Cover crops can be introduced in crop rotation; after their growth and restricting the development of other species, they can be suppressed and incorporated into soil. It has been proved that a biennial leguminous green manure crop can reduce weed seed bank problems in the subsequent organic annual crops (Melander *et al.*, 2020).

Diversified crop rotation has regained attention worldwide as a non-chemical practice for increasing the sustainability of production systems (Jastrzębska *et al.*, 2023). Monoculture has a destructive impact on agro-ecosystems (Saulic *et al.*, 2022), especially regarding weed management. Crop rotation, defined as the alternation of different crop species on the same land (Tanveer *et al.*, 2019), can avoid the application of chemical herbicides and reduce, in three to five years, the presence of weeds (Mashingaidze *et al.*, 2017). The benefits of crop rotation depend on the selection of crops and their sequence in the system. Continuous cultivation of a single crop or crops having similar management practices allows certain weed species to become dominant in the system and, over time, these weed species become hard to control (Chauhan *et al.*, 2012). Crop rotation will change the selection forces, preventing one weed from taking over and establishing itself (Sar and Padhi, 2023). The use of crop diversification could alter weed communities and increase diversity, preventing the dominance of single detrimental weed species (Hofmeijer *et al.*, 2021). Crop rotations affect weed emergence, management, composition and density of weed seed banks (Al-Hajaj, 2021). Rotating crops with different life cycles favours the natural loss of weed seeds over time because it can prevent the repeated addition of new seeds to the soil, reducing seedlings emerging in the following crops (Otto *et al.*, 2023). In a meta-analysis conducted by Weisberger *et al.* (2019), the authors demonstrated that diversification of crop rotations significantly reduced weed density by 49% compared to simple rotations. Spatial (intercropping) and temporal (rotation cropping) crop diversification had significant effects in reducing parasitic weed density and increasing crop yield (Scott and Freckleton, 2022). To manage *Convolvulus arvensis* L. Scop. (field bindweed), a difficult weed present in maize fields, Hettinger *et al.* (2023), suggested incorporating a perennial forage or an intensively tilled fallow period to prevent its expansion. However, keeping an annual cropping system can lead to increased *C. arvensis* density.

Agronomic practices are not always efficient if applied alone (Ingraffia *et al.*, 2022); in fact, no-tillage or conventional tillage techniques in organic farming must be coupled with beneficial compounds which allow weed containment. Plants, fungi and microbes producing allelochemicals are touted as promising alternatives to synthetic agrochemicals due to their reported pesticidal properties and other biochemical activities of agricultural interest with lower environmental half-lives (Islam and Widhalm, 2020). An example is represented by juglone; it is a quinoid compound that functions as an allelochemical when released from several Juglandaceae members, including *Juglans nigra* L. (black walnut) and *J. regia* L. (English walnut) in the rhizosphere. Specifically, juglone exhibits inhibitory effects on several weed species (Dayan and Duke, 2010), such as leaf wilting and yellowing or damaging the roots through induction of reactive oxygen species (ROS) together with calcium accumulation resulting in death (Massey, 1925; Chi *et al.*, 2011).

Another possible solution for the organic control of weeds is represented by using essential oils derived from a variety of aromatic plants, biomass, invasive or food crops known for their potential as non-selective natural herbicides (Vasilakoglou, 2007; Dayan *et al.*, 2009). Bioherbicides can reduce the use of chemicals and they are efficient in weed management. They contribute to crop production and quality without any damage to the agro-ecosystem (Anwar and Qureshi, 2022). An example is pelargonic acid, naturally occurring in the form of esters in the essential oil of *Pelargonium* spp. (Crmaric *et al.*, 2018). Pelargonic acid is used worldwide for weed management as a non-selective contact herbicide suitable for both garden and professional use (Muñoz *et al.*, 2020; Kanatas *et al.*, 2020). A very short time after treatment, the foliage of weeds rapidly begins to oxidize and necrotic lesions are observed on the aerial parts of plants (Lebecque *et al.*, 2019). Moreover, formulations with pelargonic acid are more effective against young plants than older ones (Dayan and Duke, 2009).

Another natural molecule with phytotoxic activity is phosphinotricin (glufosinate); it is one of the most effective non-selective herbicides. This compound is derived from bialaphos, which is produced from *Streptomyces* soil bacteria (Lebedev *et al.*, 2019). Phosphinotricin inhibits glutamine synthetase causing a decreased production of main amino acids, such as glutamate and serine. This leads to irreversible photorespiration damage (Wendler and Wild, 1990). *Streptomyces* can also release anisomycin, which is a pyrrolidine antibiotic that can be used as a herbicide but it shows toxicity for pathogenic fungi too (Abdelaal *et al.*, 2011). These main bioherbicides are directly produced from plants and microorganisms. However, new sustainable techniques allow obtaining of natural compounds from plant waste. Processes such as thermic pyrolysis of lignocellulosic materials lead to several product formations, including charcoal, flammable gases, tar and wood vinegar under inert conditions in the deficiency of oxygen (Teutscherova *et al.*, 2017). Wood vinegar, a by-product of pruning products of orchards pyrolysis mostly discarded, contains mainly water and various organic compounds, in varying amounts depending on the raw material used for the pyrolysis process. It has been seen that, at low concentrations and different application volumes, wood vinegar could be used as a foliar-applied fertilizer (Benzon and Lee, 2016). However, this substance has shown to have a phytotoxic effect that could be used as a naturally derived herbicide for weed control (Aguirre *et al.*, 2020; Liu *et al.*, 2021; Chu *et al.*, 2022).

Conclusion

Crop yield loss caused by weeds is considered one of the main constraints for achieving food security. At the farm level, the development of an efficient weed management strategy is one of the main concerns for farmers, particularly in organic farming, as weeds are the main limiting factor in vegetables and other field crops. The use of the techniques described herein can therefore help to achieve environmental sustainability and food safety objectives by reducing and eliminating the use of herbicides and agrochemicals. To eliminate the dependence of modern agriculture on herbicides, the most effective and economical weed management strategy should integrate a combination of practices into a whole-farm management approach. These strategies aim to enhance the competitive advantage of the crop and act either synergistically or additively, resulting in effective weed control and environmental safety. However, effective biological control methods have been adopted in only a fraction of countries. By expanding and diversifying the methods of weed control, it is possible to reduce the risk of new cases of evolution of herbicide resistance. The development of tailor-made weed management strategies under organic farming systems requires close interactions between researchers and all stakeholders through a continuous effort to keep the search for new innovative technologies to apply in organic farming. This creates numerous opportunities to expand the use of these techniques to help manage weeds in organic systems.

References

Abbas, T., Zahir, Z.A., Naveed, M. and Kremer, R.J. (2018) Limitations of existing weed control practices necessitate development of alternative techniques based on biological approaches. *Advances in Agronomy* 147, 239–280.

Abdelaal, T., Abdelwahed, N., Awad, G., Diwany, A.I. and Haroun, M. (2011) Improvement of anisomycin production through mutation and medium optimization for *Streptomyces griseolus*. *Australian Journal of Basic and Applied Sciences* 5, 2637–2648.

Aguirre, J.L., Baena, J., Martín, M.T., González, S., Manjón, J.L. *et al.* (2020) Herbicidal effects of wood vinegar on nitrophilous plant communities. *Food and Energy Security* 9(4), e253.

Al-Hajaj, N. (2021) Review of the influence of crop rotation and primary tillage operations on weed seed bank. *International Journal of Research – Granyhaalayah* 9(7), 407–421.

Allen, J., Pelzer, C.J., Brockmueller, B., Silva, E.M. and Ryan, M.R. (2023) No yield benefit from starter fertilizer in soybean no-till planted into rolled-crimped cereal rye. *Agrosystems, Geosciences & Environment* 6(4), e20434.

Anderson, R.L. (2007) Managing weeds with a dualistic approach of prevention and control: A review. *Agronomy for Sustainable Development* 27(1), 13–18.

Anwar, T. and Qureshi, H. (2022) Role of biocontrol agents in weed management – recent developments and trends. *Acta Scientiarum Polonorum Hortorum Cultus* 21(6), 155–162.

Bajwa, A.A., Mahajan, G. and Chauhan, B.S. (2015) Nonconventional weed management strategies for modern agriculture. *Weed Science* 63(4), 723–747.

Benzon, H.R.L. and Lee, S.C. (2016) Potential of wood vinegar in enhancing fruit yield and antioxidant capacity in tomato. *Korean Journal of Plant Resources* 29(6), 704–711.

Bonciarelli, F. (1989) *Fondamenti di Agronomia Generale*. Edagricole, Bologna, Italy.

Chauhan, B.S. (2020) Grand challenges in weed management. *Frontiers in Agronomy* 1, 3.

Chauhan, B.S., Singh, R.G. and Mahajan, G. (2012) Ecology and management of weeds under conservation agriculture: a review. *Crop Protection* 38, 57–65.

Chi, W.C., Fu, S.F., Huang, T.L., Chen, Y.A., Chen, C.C. *et al*. (2011) Identification of transcriptome profiles and signaling pathways for the allelochemical juglone in rice roots. *Plant Molecular Biology* 77(6), 591–607.

Chu, L., Liu, H., Zhang, Z., Zhan, Y., Wang, K. *et al*. (2022) Evaluation of wood vinegar as an herbicide for weed control. *Agronomy* 12(12), 3120.

Cox, C. (2006) Ten reasons not to use pesticides. *Journal of Pest Science* 26, 10–12.

Crmaric, I., Keller, M., Krauss, J. and Delabays, N. (2018) Efficacy of natural fatty acid based herbicides on mixed weed stands. In: *Deutsche Arbeitsbesprechung über Fragen der Unkrautbiologie und-bekämpfung*. Braunschweig, Germany.

Dayan, F.E. and Duke, S.O. (2010) Natural products for weed management in organic farming in the USA. *Outlooks on Pest Management* 21(4), 156–160.

Dayan, F.E. and Duke, S.O. (2009) Biological activity of allelochemicals. In: Osbourn, A.E. and Lanzotti, V. (eds) *Plant-Derived Natural Products: Synthesis, Function, and Application*. Springer, New York, pp. 361–384.

Dayan, F.E., Cantrell, C.L. and Duke, S.O. (2009) Natural products in crop protection. *Bioorganic & Medicinal Chemistry* 17(12), 4022–4034.

de Oliveira, T.R., Serafim, A.D., Breland, B., Miller, A., Beneton, K. *et al*. (2023) An integrated weed management approach in tomato using soil steaming, mulching, and winter cover crops. *Frontiers in Agronomy* 5, 1075726.

European Commission (2020) From farm to fork: Our food, our health, our planet, our future. European Commission, Brussels.

FAO, IFAD, UNICEF, WFP, and WHO (2023) *The State of Food Security and Nutrition in the World 2023: Urbanization, agrifood systems transformation and healthy diets across the rural–urban continuum*. FAO, Rome.

Fennimore, S.A., Slaughter, D.C., Siemens, M.C., Leon, R.G. and Saber, M.N. (2016) Technology for automation of weed control in specialty crops. *Weed Technology* 30(4), 823–837.

Gharde, Y., Singh, P.K., Dubey, R.P. and Gupta, P.K. (2018) Assessment of yield and economic losses in agriculture due to weeds in India. *Crop Protection* 107, 12–18.

Ghidoli, M., Pesenti, M., Colombo, F., Nocito, F.F., Pilu, R. *et al*. (2023) *Camelina sativa* (L.) Crantz as a promising cover crop species with allelopathic potential. *Agronomy* 13(8), 2187.

Hammermeister, A.M. (2016) Organic weed management in perennial fruits. *Scientia Horticulturae* 208, 28–42.

Heap, I. (2019) International survey of herbicide resistant weeds. Available at: www.weedscience.org (accessed 10 November 2023).

Hettinger, K., Miller, Z., Hubbel, K. and Seipel, T. (2023) Crop rotation and cultivation effects on *Convolvulus arvensis* population dynamics in small grain organic cropping systems. *Frontiers in Agronomy* 5, 1177461.

Hill, E.C., Ngouajio, M. and Nair, M.G. (2006) Differential response of weeds and vegetable crops to aqueous extracts of hairy vetch and cowpea. *HortScience* 41(3), 695–700.

Hofmeijer, M.A.J., Melander, B., Salonen, J., Lundkvist, A., Zarina, L. *et al*. (2021) Crop diversification affects weed communities and densities in organic spring cereal fields in Northern Europe. *Agriculture, Ecosystems & Environment* 308, 107251.

Hussain, M., Farooq, S., Merfield, C. and Jabran, K. (2018) Mechanical weed control. In: Jabran, K. and Chauhan, S. (eds) *Non-Chemical Weed Control*. Academic Press, London, pp. 133–155.

Ingraffia, R., Amato, G., Ruisi, P., Giambalvo, D. and Frenda, A.S. (2022) Early sowing can boost grain production by reducing weed infestation in organic no-till wheat. *Journal of the Science of Food and Agriculture* 102(14), 6246–6254.

Islam, A.K.M.M. and Widhalm, J.R. (2020) Agricultural uses of juglone: Opportunities and challenges. *Agronomy* 10(10), 1500.

Jastrzębska, M., Kostrzewska, M.K. and Marks, M. (2023) Is diversified crop rotation an effective non-chemical strategy for protecting triticale yield and weed diversity? *Agronomy* 13(6), 1589.

Kanatas, P., Travlos, I., Papastylianou, P., Gazoulis, I., Kakabouki, I. *et al*. (2020) Yield, quality and weed control in soybean crop as affected by several cultural and weed management practices. *Notulae Botanicae Horti Agrobotanici Cluj-Napoca* 48(1), 329–341.

Khamare, Y., Chen, J. and Marble, S.C. (2022) Allelopathy and its application as a weed management tool: A review. *Frontiers in Plant Science* 13, 1034649.

Kruidhof, H.M., Gallandt, E.R., Haramoto, E.R. and Bastiaans, L. (2011) Selective weed suppression by cover crop residues: Effects of seed mass and timing of species' sensitivity. *Weed Research* 51(2), 177–186.

Lebecque, S., Lins, L., Dayan, F.E., Fauconnier, M.L. and Deleu, M. (2019) Interactions between natural herbicides and lipid bilayers mimicking the plant plasma membrane. *Frontiers in Plant Science* 10, 329.

Lebedev, V.G., Krutovsky, K.V. and Shestibratov, K.A. (2019) Effect of phosphinothricin on transgenic

downy birch (*Betula pubescens* Ehrh.) containing bar or GS1 genes. *Forests* 10(12), 1067.
Liu, X., Zhan, Y., Li, X., Li, Y., Feng, X. *et al.* (2021) The use of wood vinegar as a non-synthetic herbicide for control of broadleaf weeds. *Industrial Crops and Products* 173, 114105.
Mashingaidze, N., Twomlow, S., Madakadze, I.C., Mupangwa, W. and Mavunganidze, Z. (2017) Weed growth and crop yield responses to tillage and mulching under different crop rotation sequences in semi-arid conditions. *Soil Use and Management* 33(2), 311–327.
Massey, A.B. (1925) Antagonism of the walnuts (*Juglans nigra* L. and *J. cinerea* L.) in certain plant associations. *Phytopathology* 15, 773–784.
Melander, B., Rasmussen, I.A. and Olesen, J.E. (2020) Legacy effects of leguminous green manure crops on the weed seed bank in organic crop rotations. *Agriculture, Ecosystems & Environment* 302, 107078.
Mia, M.J., Massetani, F., Murri, G. and Neri, D. (2020) Sustainable alternatives to chemicals for weed control in the orchard. *Horticultural Science* 47(1), 1–12.
Muñoz, M., Torres-Pagán, N., Peiró, R., Guijarro, R., Sánchez-Moreiras, A.M. *et al.* (2020) Phytotoxic effects of three natural compounds: Pelargonic acid, carvacrol, and cinnamic aldehyde, against problematic weeds in Mediterranean crops. *Agronomy* 10, 791.
Ngaba, B., Christina, M., Mansuy, A., Chetty, J., Soulé, M. *et al.* (2023) Experimental dataset of sugarcane-cover crop intercropping trials to control weeds in Reunion Island. *Data in Brief* 48, 109244.
Otto, S., Masin, R., Nikolić, N., Berti, A. and Zanin, G. (2023) Effect of 20-years crop rotation and different strategies of fertilization on weed seedbank. *Agriculture, Ecosystems & Environment* 354, 108580.
Quintarelli, V., Radicetti, E., Allevato, E., Stazi, S.R., Haider, G. *et al.* (2022) Cover crops for sustainable cropping systems. *Agriculture* 12(12), 2076.
Radicetti, E., Mancinelli, R. and Campiglia, E. (2012) Combined effect of genotype and inter-row tillage on yield and weed control of chickpea (*Cicer arietinum* L.) in a rainfed Mediterranean environment. *Field Crops Research* 127, 161.
Rehman, S., Shahzad, B., Bajwa, A.A., Hussain, S., Rehman, A. *et al.* (2019) Utilizing the allelopathic potential of brassica species for sustainable crop production. *Journal of Plant Growth Regulation* 38(1), 343–356.
Sar, K. and Padhi, I. (2023) Weed management in conservation agriculture. *Emerging Issues in Agricultural Sciences* 156.
Saulic, M., Oveisi, M., Djalovic, I., Bozic, D., Pishyar, A. *et al.* (2022) How do long term crop rotations influence weed populations: Exploring the impacts of more than 50 years of crop management in Serbia. *Agronomy* 12(8), 1772.
Scott, D. and Freckleton, R.P. (2022) Crop diversification and parasitic weed abundance: A global meta-analysis. *Scientific Reports* 12(1), 19413.
Stevan, Z.K. (2017) Flame weeding in agronomic crops. *Agricultural Research & Technology: Open Access Journal* 12(2), 555843.
Tabaglio, V., Marocco, A. and Schulz, M. (2013) Allelopathic cover crop of rye for integrated weed control in sustainable agroecosystems. *Italian Journal of Agronomy* 8(1), 5.
Tanveer, A., Ikram, R.M. and Ali, H.H. (2019) Crop rotation: Principles and practices. *Agronomic Crops* 2, 1–12.
Teutscherova, N., Vazquez, E., Santana, D., Navas, M., Masaguer, A. *et al.* (2017) Influence of pruning waste compost maturity and biochar on carbon dynamics in acid soil: Incubation study. *European Journal of Soil Biology* 78, 66–74.
Travlos, I., Gazoulis, I., Kanatas, P., Tsekoura, A., Zannopoulos, S. *et al.* (2020) Key factors affecting weed seeds' germination, weed emergence, and their possible role for the efficacy of false seedbed technique as weed management practice. *Frontiers in Agronomy* 2, 1.
Turner, B. (2000) Weeds and the organic farmer. *The Organic Way* 160, 44–45.
Vasilakoglou, I. (2007) Herbicidal potential of essential oils of oregano or marjoram (*Origanum* spp.) and basil (*Ocimum basilicum*) on *Echinochloa crus-galli* (L.) P. Beauv. and *Chenopodium album* L. weeds. *Allelopathy Journal* 20, 297–306.
Weisberger, D., Nichols, V. and Liebman, M. (2019) Does diversifying crop rotations suppress weeds? A meta-analysis. *PLoS ONE* 14(7), e0219847.
Wendler, C. and Wild, A. (1990) Effect of phosphinothricin (glufosinate) on photosynthesis and photorespiration of C3 and C4 plants. *Zeitschrift Für Naturforschung* 45(5), 535–537.
Zimdahl, R.L. (2018) *Fundamentals of Weed Science*, 5th edn. Academic Press, Cambridge, Massachusetts.

43 Recent Approaches in the Management of Weeds in Organic Cropping Systems

Arnd Verschwele*

Julius Kuehn-Institute – Federal Research Centre for Cultivated Plants, Institute for Plant Protection in Field Crops and Grassland, Braunschweig, Germany

Abstract

Weed communities in organic cropping systems are characterized by a high density and diversity. This has to do with the fact that control is often not sufficient, but also because the benefits of weeds are more appreciated than in conventional farming. However, there is evidence that diverse weed infestation leads to comparatively small yield losses. Recent studies have shown that yield losses due to weeds are lower in organic farming than in conventional farming. Nevertheless, in high-value crops, especially those with low competitive ability, a lot of effort is made to keep the fields as free of weeds as possible. Preventive measures are very important for weed management in organically grown arable crops. These include measures such as diverse crop rotations. For example, the alternation of winter and summer crops prevents the same weed species from spreading from year to year. Focusing on direct weed control, in recent years there have been promising innovations, particularly in the area of mechanical weed control. Precise GPS positioning of devices and individual tools have made it possible to detect weeds within crop rows and increase the proportion of area treated. Subsequent hand weeding is then only necessary on a small area in the crop row. The worldwide expansion of organic farming has significantly promoted such technical developments, especially in North America and Europe. In the future, robots may be the next level of technology to control weeds more efficiently and specifically. However, in many developing and emerging countries the degree of mechanization is still low, and hand weeding is the most common method of weed control. Although we can expect technical improvements, organic as well as conventional farming will be heavily affected by climate change. Weed management will be challenging by changing crop–weed interactions and by the spread of invasive neophytes.

Keywords: biodiversity, competition, crop performance, mechanical control, preventive management, seed bank, weed life cycle, yield loss

Introduction

Weeds have been viewed as harmful or useless since the dawn of agriculture. Many wild plants, on the other hand, could also be developed into cultivated plants because they had a specific direct or indirect benefit for humans. As agriculture intensified in Europe, control of weeds also improved, particularly through hand weeding and mechanical control measures, but also by better crop and land management. However, recent agricultural research and practices are focusing also on the benefits of weeds. The functions and importance of weeds for biodiversity are examined in many countries, especially in Europe and North America (Fig. 43.1).

Current basic research even questions how weeds actually compete with the crop. Direct competition for resources is generally considered the primary mechanism for weed-induced yield loss but may be misleading. Following a re-evaluation of physiological evidence, Horvath *et al.* (2023) suggest weeds initially impact crop growth and development through

*Corresponding author: arnd.verschwele@julius-kuehn.de

DOI: 10.1079/9781800626850.0043

Fig. 43.1. *Matricaria chamomilla* in organically grown oilseed rape.

resource-independent interference. Based on their studies, weed perception by crops induces a shift in crop development, before resources become limited, which ultimately reduces crop yield, even if weeds are subsequently removed. These new ideas may lead to a fundamental paradigm shift in our understanding of how weeds reduce crop yield. It can initiate new research directions and opportunities to change cropping systems and to reduce weed-induced yield losses.

Weed Specifics in Organic Cropping

The diversity of weed communities is high as far as crop rotation, soil tillage and management practices are also diverse. These are the main reasons that more weed species can be found in organic cropping systems than in conventional growing (Benaragama *et al.*, 2019). However, weed control is an essential part of crop production. Globally, crop yield losses caused by weeds average 34% (Oerke, 2006), and the costs associated with management represent a major economic burden for farmers (Atwood and Paisley-Jones, 2017). There are currently only very few specific data for organic farming to be found in the literature, although it can be assumed that weed effects on yield are similar to the conventional cropping. A meta-analysis of Seufert *et al.* (2012) shows that yield of organic and conventional systems can be similar if crops grow under best practice (including effective weed control). However, there are studies indicating that weed-induced crop yield differs between conventional and organic farming systems but the international findings are not clear on this point (Ryan *et al.*, 2009; Benaragama *et al.*, 2016).

Liebman and Davis (2009) propose three critical concerns or principal objectives for weed management in organic farming, which are equally applicable to conventional agricultural production systems. Their concerns are worthy goals for all weed management systems:

a. limiting competitive damage caused by weeds: a primary objective is to reduce weed density to tolerable levels which means accepting less than 100% weed control;
b. minimizing the size of future weed populations by decreasing seed production and survival of seeds and vegetative propagules; and
c. preventing introduction of new weed species by monitoring, cleaning of equipment, and early, targeted eradication.

Despite the high number of tools available for preventive and direct weed management, the applicability is affected by robust response, reliability and economics of different tactics (Kolb and Gallandt, 2012). Compared to conventional agriculture, weed management in organic farming differs primarily in the following point: the essential idea in management is a holistic approach to the agro-ecosystem, which aims to promote the crop and weaken the weeds. Research has been conducted on organic weed management, which means the use of methods that ultimately reduce the weed seed bank by investing in ecological interactions (Bastiaans *et al.*, 2008).

In general, systems-based weed control has not yet been fully implemented in organic farming systems (Bàrberi, 2002). This includes that the aim of management is not to completely control the weeds but rather to reduce them to an acceptable level. Positive properties of the segetal flora are valued and, if possible, will be considered in the crop management. These goals require more knowledge and experience about the biology and population dynamics of individual weed species. The primary control tools in organic farming are preventive measures. Zimdahl (2018) points out that, in contrast to conventional farming, organic farming is a farmer-dominated process over a long term, where the farmer's goals are multi-dimensional.

Direct control is carried out mainly mechanically, e.g. by hoeing and harrowing the crops or by seedbed preparation before seeding or planting, as well as stubble tillage after cropping.

The type of weed infestation is similar to that in conventional systems, at least with similar crop rotations, tillage, climate and soil conditions. Since the tools used in organic farming are generally more diverse, the risk of the spread of certain problematic species is lower. Experience in Europe has shown that weeds such as *Alopecurus myosuroides* and *Apera spica-venti* are difficult to control when winter cereals are grown repeatedly in less-diverse crop rotations. This shows that the species spectrum also depends on environmental factors, but is at least equally influenced by the type and intensity of management. Versatile management therefore creates a diverse range of weed species. One can assume that in these cases competition between weed species increases at the expense of competition against the crop. The competitive pressure and loss of yield therefore decrease with the weed diversity (if distributed evenly over the area). Storkey and Neve (2018) were able to demonstrate a negative correlation between yield loss and the number of weed species abundant on a field. Adeux *et al.* (2019) could also validate that diverse weed communities limited the negative effect of competitive and dominant species on crop productivity. This correlation will also apply to organic farming because the growing conditions are more diverse than in conventional farming. Several studies in Europe have shown that weed diversity is higher under organic crop conditions (Verschwele and Zwerger, 2005; Lundkvist *et al.*, 2008; Bàrberi *et al.*, 2017; Tyšer *et al.*, 2021).

Although the available weed management methods are used in combination, they are divided into three groups: (i) preventive; (ii) cultural; and (iii) direct methods. In principle, measures in organic farming are increasingly aimed at intervening at various points in the weed life cycle in order to reduce competition and the soil seed bank (Fig. 43.2). It mostly happens in the long term and as a preventive measure, which reduces weed density in the crop. Direct methods, on the other hand, prevent or reduce seed production or promote fatal germination through soil cultivation and seedbed preparation.

Farmers manage crop–weed competition by reducing weed density (i.e. by weeding), establishing an early-season competitive advantage to the crop and, furthermore, maximizing competitiveness by the crop using competitive species, competitive cultivars, high sowing densities, optimal spatial arrangement, intercropping complementary species, or transplanting (Gallandt and Weiner, 2015). In the long term, all successful measures cause a depletion of the weed seed bank by the following steps:

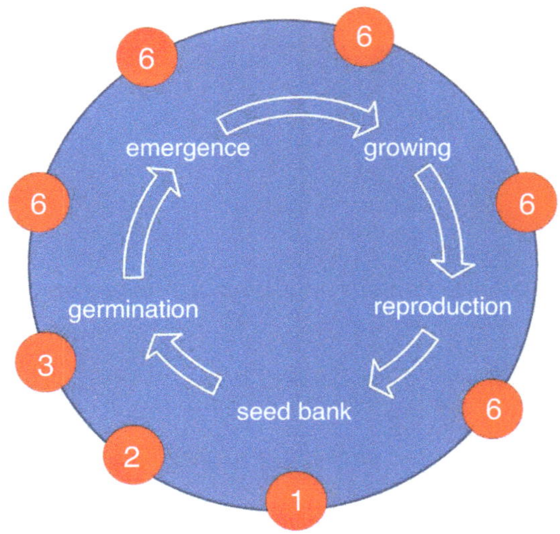

Fig. 43.2. Life cycle of annual weeds and impact points for control (for details see following text).

1. increasing seed degradation (soil tillage);
2. increasing fatal germination (soil tillage, false seedbed);
3. decreasing active germination (crop rotation, false seedbed);
4. decreasing initial weed density (false seedbed, delayed sowing, direct control);
5. decreasing weed growth and competition (crop rotation, cultivar choice, direct control); and
6. decreasing reproduction (competitive crops, direct control).

Preventive and Cultural Methods of Weed Control

Most of the preventive measures are aimed at worsening the emergence and growth conditions of weeds. In addition, other measures aim to directly promote the crop, for example by choosing a crop species or variety that is best adapted to the location. In total, all preventive methods give the crop a competitive advantage. The most important element in the cultivation of annual arable and vegetable crops is the design of the crop rotation (Bàrberi *et al.*, 2017; Mohler and Johnson, 2020).

Crop rotation is the systematic change of different crops on the same area. It is an important strategy for developing a solid, long-term weed control programme. Weeds tend to thrive with crops that have growth requirements similar to their own. Furthermore, repeating cultural practices designed to contribute to the crop can also have a positive effect on the same weed species (Kolb and Gallandt, 2012). In simplified crop rotations or monoculture, where the same crop is grown in the same field year after year, there is an accumulation of weed species that are adapted to the growing conditions of the crop (Fig. 43.3).

When different crops are used in a crop rotation, weed germination and growth cycles are disrupted by differences in the cultural practices of each crop (tillage, sowing dates, competition between crops etc.). For example, the alternation of spring and autumn sowing prevents the same weed species from continually finding favourable emergence conditions and thus being able to spread more widely. Conversely, other weed species do not find optimal conditions for germination; their seed potential is reduced by mortality, especially when the seeds reach deeper soil layers. Seed degradation can be promoted by direct mechanical intervention if weeds are stimulated to germinate and emerge through seedbed preparation, but are mechanically destroyed before or during sowing of the crop. This efficient technique is called false seedbed (Fig. 43.4) or stale seedbed (Rasmussen, 2004).

The false seedbed can be helpful in reducing initial weed infestation, especially in vegetable crops or other crops where manual weeding is common and expensive. However, the effectiveness of this method might be insufficient in dry or clayey soil. Morphological and physiological properties of the weed species such as seed weight and dormancy behaviour also influence the result (Benvenuti *et al.*, 2021). For success, it is important that all weeds are destroyed during seedbed preparation, otherwise they can lead to severe yield losses (Rasmussen, 2004).

A totally different strategy to inversion tillage systems is to avoid any soil disturbance and weed seed germination stimuli before and during crop sowing. Keeping weed seeds dormant until the crop is well-established can substantially reduce potential weed density and competition. The so-called no-tillage or zero-tillage methods have additional agronomical advantages, e.g. to reduce costs and soil erosion (Carr *et al.*, 2013). However, there are no indications that weed seed mortality is higher under organic zero-tillage compared to conventional systems (Davis *et al.*, 2005; Webster *et al.*, 2023). Reduced soil tillage requires a specific seeding technique and a very competitive crop. These crops must be well adapted to the no-tillage system, otherwise their yield will be worse than in the conventional system and they will not reliably suppress weeds (Davis *et al.*, 2005). For example, direct seeded rice is very sensitive against weeds if soil tillage is reduced (Rao *et al.*, 2017).

Once all measures have been taken to reduce the seed potential of weeds, the next aim is to increase

Fig. 43.3. Heavy infestation of loose silky-bent (*Apera spica-venti*) in winter rye.

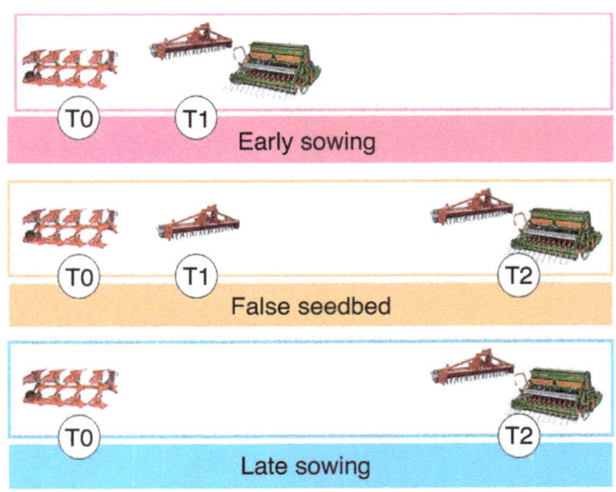

Fig. 43.4. False seedbed technique in winter wheat.

the competitiveness of the crop. With regard to weed management, the following criteria or properties of a crop are crucial: (i) weed suppression; (ii) weed tolerance; (iii) allelopathic effects; and (iv) sensitivity for direct control measures.

Strong competitors are almost all types of grain, especially rye, but also wheat. In combination with narrow row spacing and high sowing rates, dense stands are formed under good growth conditions, which shade weeds and prevent them from developing and producing seeds. If a competitive advantage quickly develops for the crop, fewer nutrients and water are available to the weeds (Kolb and Gallandt, 2012). During the critical phase, conditions can therefore change quickly to the benefit or disadvantage of the crop. Crops with low plant height, low leaf area index (LAI) and slow youth development are generally weak in competition.

An indicator of the competitive ability of a crop is the so-called critical period for weed control (CPWC). van Heemst (1985) compiled and analyzed studies conducted worldwide in different crops. Table 43.1 shows the ranking for the most relevant crops according to relative yields and the CPWC, respectively.

There are no data for conditions under organic farming but a similar ranking can be assumed. It is obvious that in organically grown crops the CPWC should be as short as possible because the possibilities and success of control through direct measures are limited. For crops that react sensitively to mechanical interventions during their juvenile phase, e.g. peas or sugarbeets, the time window for mechanical weed control is particularly small.

Table 43.1. Mean yield of weedy crops relative to weed-free crops and boundaries for the critical period for weed control (CPWC) relative to the length of the total growth cycle (van Heemst, 1985).

Crop	Relative yield of weedy crop	CPWC	
		Start	End
Wheat	0–75	n.a.	0–19
Peas	0–70	n.a.	0–21
Potato	0–65	n.a.	0–21
Soybean	0–64	0–12	0–30
Maize	0–56	0–13	0–25
Sorghum	0–52	0–15	0–21
Sweet potato	0–45	0–12	0–25
Groundnut	0–39	0–00	0–35
Cassava	0–28	0–00	0–52
Upland rice	0–25	n.a.	0–44
Sugar beet	0–23	0–10	0–42
Cotton	0–12	0–21	0–48
Onions	0–13	0–15	0–3

In addition to the differences between crops, there are also differences between varieties in terms of weed suppression. Positive varietal characteristics have been proven for a number of crop plants such as wheat, barley, rice, soy, peas, field beans and tomatoes (Jannink *et al.*, 2000). Similar to the relevant traits of a crop, varieties also differ in terms of growth habit like leaf position, crop height, or speed of growth. However, management practices such as sowing date and time also influence the competitive ability of a variety, so that it is often difficult to quantify the genotypic weed suppression (Weiner, 2023). Farmers and advisers should check on site, under their specific location and growing conditions, whether significant weed suppression can be achieved through variety selection. Variety selection can be a cost-effective tool in organic weed management but information on the specific competitive ability must be available to the farmer. So far, beside competitiveness, little attention has been given to the variety-specific weed tolerance, i.e. the ability of a variety to deliver a sufficient yield even with comparatively high weed infestation (Verschwele, 2014; Kaur *et al.*, 2021).

In addition to morphological properties, the architecture of the crop stand is also influenced by the sowing technique. Higher seed rates and/or narrower row spacing can inhibit weeds in their growth and development, especially in competitive crops like cereals (Kolb and Gallandt, 2012). This not only reduces weed-related yield losses but also reduces the weed seed return. Higher seed rates can also compensate for plant losses due to mechanical weed control, but they are directly correlated with increasing seed costs. Alba *et al.* (2020) were able to achieve a 46% higher yield by doubling the seed rate of weakly competitive lentils supported by mechanical weed control. However, the enhancing effect decreases with higher seed rates and is also impaired if competitive weed species predominate. The same applies, for example, to dry-seeded rice (Phuong *et al.*, 2005) and soybean (Liebert and Ryan, 2017). Studies show that wheat in particular reacts positively to narrower row spacing. De

Vita *et al.* (2017) demonstrated higher grain yields, higher nitrogen uptake and significantly reduced weed biomass with narrow row spacing (5 cm vs 20 cm) of durum wheat. These results have direct implications for weed and nutrient management in low-input and organic cropping systems. However, it is not valid for all crops because crop stands that are too dense promote the spread of diseases and pests and prevent optimal yield growth. In addition, narrower row spacing cannot always be achieved in certain crops such as corn or sugarbeet because they are not technically compatible with sowing and harvesting technology.

In addition to seed rate, the spatial arrangement of seeds and plants is known to affect the competitive ability of a crop stand. An even distribution of crops could lead to better utilization of resources. For example, increasing seeding and planting uniformity can increase yield and reduce weed growth in some crops (Olsen *et al.*, 2005). However, Verschwele and Herrmann (2024) were unable to find any yield effects for winter wheat under organic farming conditions. The low sowing rate in the triangular arrangement (150 vs. 300 grains x m^{-2}) could not sufficiently suppress the weeds in the early growing period. In addition, the requirements for the soil conditions and the seeding technique for an equal sowing are very high.

All measures which directly or indirectly improve the competitiveness should be implemented wherever possible. However, these tactics are much more important when the soil tillage is reduced. Such cropping systems are ambitious in organic farming, since perennial weeds cannot be controlled with adequate measures. Thus, in no-till systems, organic farmers have to establish strongly competitive crop rotations including suppressing cover crops or mulch (Anderson, 2015). Altieri *et al.* (2011) found, under Brazilian conditions, that the best cover crop mixture should include a significant proportion of rye, vetch and fodder radish. These mixtures produce large biomass and are readily killed by rolling, forming a thick mulch sufficient to provide effective weed control in the subsequent arable or vegetable crop. However, weed suppression does not necessarily increase with the number of crops in a mixture while the most important factor seems to be the total above-ground biomass (Négrier *et al.*, 2023).

Direct Control of Weeds

Direct weed control can be done using a variety of mechanical equipment. These can be hoeing devices that work row-dependently as well as harrowing devices that work independently of rows and over the entire area. There have been technical innovations in recent years that have significantly improved area performance and efficiency. Melander and McCollough (2021) provide a comprehensive overview of the status of mechanical weed control. Above all, camera and steering technology (Fig. 43.5) have led to both increasing control efficacy and area performance. This makes practical use significantly easier and at the same time reduces damage to the crop. It is to be expected that there will be further technical innovations of mechanical weed control in the near future (Machleb *et al.*, 2020).

Depending on their design, setting and travel speed, the tools work by cutting, pulling out or burying the weeds. Combinations are also possible. The effectiveness of mechanical weed control depends on many biological and technical factors:

1. proportion of treated area
2. number of treatments
3. driving speed
4. setting the device (e.g. coulter pressure, working depth)
5. timing of the treatment and size of the weeds and the crop (depending on the CPWC)
6. soil condition (loose, dry, fine-crumbly soil is beneficial)
7. dry phase after treatment (no precipitation for three days is beneficial)

Fig. 43.5. Camera-steered hoeing machine at field application.

Fig. 43.6. Weeding robot application in sugarbeets.

8. weed species and weed size (monocotyledonous and perennial species are not very susceptible to mechanical impact)

This complex structure shows that the correct timing and adjustment of the technique is of great importance (Rueda-Ayala *et al.*, 2011). The correct setting must achieve a compromise between successful control and protection of the crop, which requires a high level of experience on the part of the farmer or user. It is therefore not surprising that efficiencies can range from 0% to 90% under practical conditions. However, there are some crops, like maize and potatoes, that can be kept almost weed-free very successfully by hoeing and/or harrowing. Cereals can also be easily treated by hoeing if the row spacing is increased to at least 25 cm (Fig. 43.6).

With new tools in conjunction with very precise global positioning by RTK (Real Time Kinematic), it is now possible to specifically eliminate only really harmful weeds while moving as little soil as possible and causing little damage to the crop (Quan *et al.*, 2022).

In certain cases, hand weeding is also justified, e.g. in vegetable crops, which usually have weak competition but are more profitable than field crops. It can also be useful to prevent new colonization in the field with invasive neophytes or other problematic weed species. However, the benefits are offset by high labour costs or there are no workers available. For countries with low levels of mechanization, less technological equipment may also improve weed control. Paul *et al.* (2023) developed a manual push-pull-type conical weeder to be used in rice (Fig. 43.7). Although both field capacity (approx. 0.02 ha/h) and efficacy are rather low (approx. 82%), this hoe is significantly more efficient than hand weeding.

Beside mechanical equipment, there are also many thermal measures available like flaming, steam, hot water or solarization. Although these methods are in line with the principles of organic farming, there are some disadvantages like high energy consumption, low efficacy and weak selectivity. However, thermal weed control provides less soil disturbance and is more efficient than hand weeding (Peerzada and Chauhan, 2018). Compared to other control measures the practical relevance of thermal methods for organic farming appears low. Promising innovations can be expected in connection with sensor-based treatments and also with laser technology in the future (Bauer *et al.*, 2020).

Outlook

The expansion of organic farming in many parts of the world has significantly advanced sustainable weed management. Last, but not least, the growing problems in conventional agriculture have brought preventive weed management measures to the forefront. Mechanical weed control has also been significantly improved through the use of camera technology and GPS positioning and steering. New technical innovations can be expected in the near future resulting in more efficient weed control. Furthermore, conventional farming will learn more from organic systems because ecological approaches become more important, not only for weed management. Worldwide, there is also a strong need to replace costly hand weeding, and this is true for both organic and conventional cropping systems.

More research and development efforts are needed to specifically promote biodiversity while maintaining sufficient weed control. The actual importance of biodiversity probably depends heavily on the productivity of organic farming and food supply in the individual regions and countries. Farmers with low mechanization will probably focus on effective weed control rather than promoting biodiversity.

Attention must also be paid to climate change. It is to be expected that this will affect weed communities and crop–weed interactions, probably in terms of water competition and temperature stress. Weed species and communities will be adapted because of their diverse gene pool and greater physiological plasticity. The effect of climate change, especially on the geographic distribution of invasive weeds, will

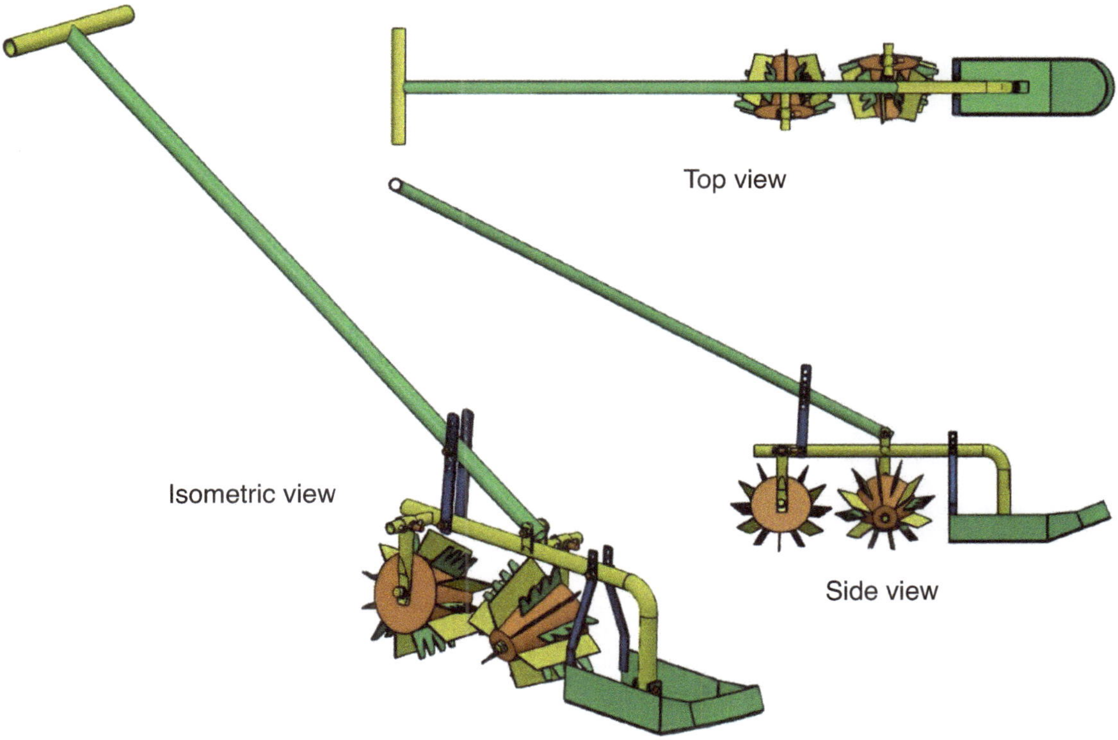

Fig. 43.7. Manual push-pull-type conical weeder. (Photo: Paul *et al.*, 2023)

be a subject of interest in the near future (Ramesh *et al.*, 2017).

Conclusions

Organic farming continues to grow strongly, although not in all parts of the world. Recently, weed management could have been significantly improved. For example, technical innovations, especially in relation to mechanical weed control, are promising. On the other hand, the range of effective preventive measures is already well described but not yet sufficiently implemented. The challenge will be to reconcile farmers' and social goals such as crop productivity, biodiversity and food supply. Weeds will still play a major role in this regard.

References

Adeux, G., Vieren, E., Carlesi, S., Bàrberi, P., Munier-Jolain, N. *et al.* (2019) Mitigating crop yield losses through weed diversity. *Nature Sustainability* 2(11), 1018–1026. DOI: 10.1038/s41893-019-0415-y.

Alba, O.S., Syrovy, L.D., Duddu, H.S.N. and Shirtliffe, S.J. (2020) Increased seeding rate and multiple methods of mechanical weed control reduce weed biomass in a poorly competitive organic crop. *Field Crops Research* 245, 107648. DOI: 10.1016/j.fcr.2019.107648.

Altieri, M.A., Lana, M.A., Bittencourt, H.V., Kieling, A.S., Comin, J.J. *et al.* (2011) Enhancing crop productivity via weed suppression in organic no-till cropping systems in Santa Catarina, Brazil. *Journal of Sustainable Agriculture* 35(8), 855–869. DOI: 10.1080/10440046.2011.588998.

Anderson, R.L. (2015) Integrating a complex rotation with no-till improves weed management in organic farming. *Agronomy for Sustainable Development* 35(3), 967–974. DOI: 10.1007/s13593-015-0292-3.

Atwood, D. and Paisley-Jones, C. (2017) Pesticides industry sales and usage: 2008–2012 market estimates. U.S. Environmental Protection Agency. Available at: https://www.epa.gov/pesticides/pesticides-industry-sales-and-usage-2008-2012-market-estimates (accessed 26 February 2025).

Bàrberi, P. (2002) Weed management in organic agriculture: Are we addressing the right issues? *Weed Research* 42(3), 177–193. DOI: 10.1046/j.1365-3180.2002.00277.x.

Bàrberi, P., Pellerin, S. and Nesme, T. (2017) Comparing crop rotations between organic and conventional farming. *Scientific Reports* 7(1), 13761. DOI: 10.1038/s41598-017-14271-6.

Bastiaans, L., Paolini, R. and Baumann, D.T. (2008) Focus on ecological weed management: What is hindering adoption? *Weed Research* 48(6), 481–491. DOI: 10.1111/j.1365-3180.2008.00662.x.

Bauer, M.V., Marx, C., Bauer, F.V., Flury, D.M., Ripken, T. *et al*. (2020) Thermal weed control technologies for conservation agriculture: A review. *Weed Research* 60(4), 241–250. DOI: 10.1111/wre.12418.

Benaragama, D., Shirtliffe, S.J., Johnson, E.N., Duddu, H.S.N. and Syrovy, L.D. (2016) Does yield loss due to weed competition differ between organic and conventional cropping systems? *Weed Research* 56(4), 274–283. DOI: 10.1111/wre.12213.

Benaragama, D., Leeson, J.L. and Shirtliffe, S.J. (2019) Understanding the long-term weed community dynamics in organic and conventional crop rotations using the principal response curve method. *Weed Science* 67(2), 195–204. DOI: 10.1017/wsc.2018.64.

Benvenuti, S., Selvi, M., Mercati, S., Cardinali, G., Mercati, V. *et al*. (2021) Stale seedbed preparation for sustainable weed seed bank management in organic cropping systems. *Scientia Horticulturae* 289, 110453. DOI: 10.1016/j.scienta.2021.110453.

Carr, P.M., Gramig, G.G. and Liebig, M.A. (2013) Impacts of organic zero tillage systems on crops, weeds, and soil quality. *Sustainability* 5(7), 3172–3201. DOI: 10.3390/su5073172.

Davis, A.S., Renner, K.A. and Gross, K.L. (2005) Weed seedbank and community shifts in a long-term cropping systems experiment. *Weed Science* 53(3), 296–306. DOI: 10.1614/WS-04-182.

De Vita, P., Colecchia, S.A., Pecorella, I. and Saia, S. (2017) Reduced inter-row distance improves yield and competition against weeds in a semi-dwarf durum wheat variety. *European Journal of Agronomy* 85, 69–77. DOI: 10.1016/j.eja.2017.02.003.

Gallandt, E.R. and Weiner, J. (2015) Crop-weed competition. In: *Encyclopedia of Life Sciences*. Wiley-Blackwell, Chichester, UK. DOI: 10.1002/047001590X.

Horvath, D.P., Clay, S.A., Swanton, C.J., Anderson, J.V. and Chao, W.S. (2023) Weed-induced crop yield loss: A new paradigm and new challenges. *Trends in Plant Science* 28, 567–582. DOI: 10.1016/j.tplants.2022.12.014.

Jannink, J.L., Orf, J.H., Jordan, N.R. and Shaw, R.G. (2000) Index selection for weed suppressive ability in soybean. *Crop Science* 40, 1087–1094. DOI: 10.2135/cropsci2000.4041087x.

Kaur, P., Sachan, S. and Sharma, A. (2021) Weed competitive ability in wheat: A peek through its functional significance, present status and future prospects. *Physiology and Molecular Biology of Plants* 27, 2165–2179. DOI: 10.1007/s12298-021-01079-y.

Kolb, L.N. and Gallandt, E.R. (2012) Weed management in organic cereals: Advances and opportunities. *Organic Agriculture* 2, 23–42. DOI: 10.1007/s13165-012-0022-y.

Liebert, J.A. and Ryan, M.R. (2017) High planting rates improve weed suppression, yield, and profitability in organically-managed, no-till-planted soybean. *Weed Technology* 31(4), 536–549. DOI: 10.1017/wet.2017.35.

Liebman, M. and Davis, A.S. (2009) Managing weed in organic farming systems: An ecological approach. In: Francis, C. (ed.) *Organic Farming: The Ecological System*. American Society of Agronomy, Madison, Wisconsin, pp. 173–196. DOI: 10.2134/agronmon-ogr54.c8.

Lundkvist, A., Salomonsson, L., Karlsson, L. and Gustavsson, A.-M.D. (2008) Effects of organic farming on weed flora composition in a long term perspective. *European Journal of Agronomy* 28(4), 570–578. DOI: 10.1016/j.eja.2008.01.001.

Machleb, J., Peteinatos, G.G., Kollenda, B.L., Andújar, D. and Gerhards, R. (2020) Sensor-based mechanical weed control: Present state and prospects. *Computers and Electronics in Agriculture* 176, 105638. DOI: 10.1016/j.compag.2020.105638.

Melander, B. and McCollough, M.R. (2021) *Advances in Mechanical Weed Control Technologies*. Burleigh Dodds Science Publishing, pp. 1–27. DOI: 10.19103/AS.2021.0098.11.

Mohler, C.I. and Johnson, S.E. (eds) (2020) Crop rotation on organic farms: A planning manual. *Sustainable Agriculture Research and Education (SARE) program*. SARE Outreach, University of Maryland. Available at: https://www.sare.org/wp-content/uploads/Crop-Rotation-on-Organic-Farms.pdf (accessed 7 March 2024).

Négrier, A., Marnotte, P., Hoareau, J., Viaud, P., Auzoux, S. *et al*. (2023) Weed control under increasing cover crop diversity in tropical summer and winter. *Biotechnologie, Agronomie, Société et Environnement* 27(2), 61–73. DOI: 10.25518/1780-4507.20220.

Oerke, E.C. (2006) Crop losses to pests. *The Journal of Agricultural Science* 144(1), 31–43. DOI: 10.1017/S0021859605005708.

Olsen, J., Kristensen, L., Weiner, J. and Griepentrog, H. (2005) Increased density and spatial uniformity increase weed suppression by spring wheat. *Weed Research* 45(4), 316–321. DOI: 10.1111/j.1365-3180.2005.00456.x.

Paul, S., Nath, B.C., Hossen, A., Pintu, K., Paul, H. *et al*. (2023) Development and fabrication of manually push-pull type conical weeder for Bangladesh condition. *Agricultural Sciences* 14(5), 685–709. DOI: 10.4236/as.2023.145046.

Peerzada, A.M. and Chauhan, B.S. (2018) Thermal weed control: History, mechanisms, and impacts. In: Jabran, K. and Chauhan, B.S. (eds) *Non-Chemical*

Weed Control. Academic Press, London, pp. 9–31. DOI: 10.1016/B978-0-12-809881-3.00002-4.

Phuong, L.T., Denich, M., Vlek, P.L.G. and Balasubramanian, V. (2005) Suppressing weeds in direct-seeded lowland rice: Effects of methods and rates of seeding. *Journal of Agronomy and Crop Science* 191(3), 185–194. DOI: 10.1111/j.1439-037X.2005.00151.x.

Quan, L., Jiang, W., Li, H., Li, H., Wang, Q. *et al.* (2022) Intelligent intra-row robotic weeding system combining deep learning technology with a targeted weeding mode. *Biosystems Engineering* 216, 13–31. DOI: 10.1016/j.biosystemseng.2022.01.019.

Ramesh, K., Matloob, A., Aslam, F., Florentine, S.K. and Chauhan, B.S. (2017) Weeds in a changing climate: Vulnerabilities, consequences, and implications for future weed management. *Frontiers in Plant Science* 8, 95. DOI: 10.3389/fpls.2017.00095.

Rao, A.N., Brainard, D.C., Kumar, V., Ladha, J.K. and Johnson, D.E. (2017) Preventive weed management in direct-seeded rice: Targeting the weed seedbank. *Advances in Agronomy* 144, 45–142. DOI: 10.1016/bs.agron.2017.02.002.

Rasmussen, I.A. (2004) The effect of sowing date, stale seedbed, row width and mechanical weed control on weeds and yields of organic winter wheat. *Weed Research* 44(1), 12–20. DOI: 10.1046/j.1365-3180.2003.00367.x.

Rueda-Ayala, V.P., Rasmussen, J., Gerhards, R. and Fournaise, N.E. (2011) The influence of post-emergence weed harrowing on selectivity, crop recovery and crop yield in different growth stages of winter wheat. *Weed Research* 51(5), 478–488. DOI: 10.1111/j.1365-3180.2011.00873.x.

Ryan, M.R., Smith, R.G., Mortensen, D.A., Teasdale, J.R., Curran, W.S. *et al.* (2009) Weed–crop competition relationships differ between organic and conventional cropping systems. *Weed Research* 49(6), 572–580. DOI: 10.1111/j.1365-3180.2009.00736.x.

Seufert, V., Ramankutty, N. and Foley, J.A. (2012) Comparing the yields of organic and conventional agriculture. *Nature* 485(7397), 229–232. DOI: 10.1038/nature11069.

Storkey, J. and Neve, P. (2018) What good is weed diversity? *Weed Research* 58(4), 239–243. DOI: 10.1111/wre.12310.

Tyšer, L., Kolářová, M., Tulačka, O. and Hamouz, P. (2021) Weed vegetation in conventional and organic farming in west Bohemia (Czech Republic). *Plant, Soil and Environment* 67(7), 376–382. DOI: 10.17221/6/2021-PSE.

van Heemst, H.D.J. (1985) The influence of weed competition on crop yield. *Agricultural Systems* 18(2), 81–93. DOI: 10.1016/0308-521X(85)90047-2.

Verschwele, A. (2014) Weed suppression and weed tolerance of wheat cultivars – relevant traits for integrated pest management. *Julius-Kuehn-Archiv* 443, 465–474.

Verschwele, A. and Zwerger, P. (2005) Effects of organic farming on weed abundance: Long-term results from a site in Northern Germany. In: *Proceedings of the 13th EWRS Symposium*, Bari, Italy. Available at: https://orgprints.org/id/eprint/5016/ (accessed 26 February 2025).

Verschwele, A. and Herrmann, D. (2024) Effect of uniform seed distribution of winter wheat on weed infestation and yield parameters. *Julius-Kuehn-Archiv* 478. DOI: 10.5073/20240109-063902-0.

Webster, T.M., Cardina, J. and White, A.D. (2023) Weed seed rain, soil seedbanks, and seedling recruitment in no-tillage crop rotations. *Weed Science* 51, 569–575. DOI: 10.1614/0043-1745.

Weiner, J. (2023) Weed suppression by cereals: Beyond 'competitive ability. *Weed Research* 63(3), 133–138. DOI: 10.1111/wre.12572.

Zimdahl, R.L. (2018) Weed management in organic farming systems. In: Zimdahl, R.L. (ed.) *Fundamentals of Weed Science*, 5th edn. Academic Press, London, pp. 337–357. DOI: 10.1016/B978-0-12-811143-7.00011-1.

44 Essential Oils and Weed Management – Bioherbicides Potential of Essential Oils

JELENA GOLIJAN-PANTOVIĆ*

Chair for Genetics, Plant Breeding and Seed Science, University of Belgrade, Faculty of Agriculture, Belgrade, Serbia

Abstract

Essential oils are products of specialized plant cells or are produced in the intercellular space and represent a mixture of different compounds such as aldehydes, ketones, alcohols and fatty acids. Due to their chemical composition, essential oils from numerous plants release allelopathic compounds of high phytotoxicity and can be used in weed control. During the last decades, the interest in using essential oils as a safe alternative to the application of herbicides in crop protection against weeds has been increasing. Essential oil-based herbicides have become a topic of interest for agriculture of the new age due to increasing concerns about the adverse effects caused by conventional pesticides. However, despite well-documented research around the world in which the effectiveness of numerous plant essential oils has been proved, little is currently known about their mechanism of action as herbicides at the molecular level. Furthermore, there are a few commercially used preparations on the world market. This should be the focus of future research and the spectrum of action should be expanded.

Keywords: essential oils, phytotoxicity, allelopathy, weeds

Introduction

The well-known problem in cultivation of food and ornamental plants is the increase of toxins and chemical residues in fresh parts of plants that are consumed by humans or animals. Herbicides used in conventional agriculture to protect crops against weeds are dangerous pollutants of crops and soil (Hazrati *et al.*, 2017). Based on studies carried out in the USA by Pimentel *et al.* (2001), crop loss caused by weeds amounted to 12%, which equals almost US$ 33 billion necessary to control weeds. Moreover, the greater resistance of plants to weeds is, the greater the use of herbicides. There are a number of methods (chemical, biological and cropping practices) used to control weeds in crop production (Singh *et al.*, 2005). Natural compounds and substances produced by medicinal plants can be applied in weed control with the aim to decrease the application of synthetic herbicides, which would result in lower levels of pollution and safer agricultural production (Sodaeizadeh *et al.*, 2010; Golijan-Pantović and Sečanski, 2022a, 2022b). There is a good possibility that essential oils (EOs) and their components, mostly terpenoids, can be substitutes for synthetic herbicides (Dudai *et al.*, 1999). Natural substances are characterized by a different structure and developed biological activities. Therefore, they can be used to produce biochemical organic herbicides that will be based on natural phytotoxin structures (Dayan and Duke, 2014). Essential oils are natural compounds with herbicidal activities, which are associated with their allochemical components (Mutlu *et al.*, 2010; Mahdavikia and Saharkhiz, 2015). Allelochemicals

*Corresponding authors: helena.ilios@gmail.com; golijan.j@agrif.bg.ac.rs

 Advances in Organic Farming (ed. L. P. Awasthi)
DOI: 10.1079/9781800626850.0044

are a divergent group of composite compounds, which encompass, among other compounds, alkaloids, phenolics, jasmonates, momilactones, flavonoids, hydroxamic acids, brassinosteroids, glucosinolates, salicylates, carbohydrates, essential oils and amino acids. The following elements regulate toxicity of allelochemicals: concentration, flux rate, age and metabolic state of plant, and predominant conditions of climate and the environment. Their quantity and quality differ over age, variety, plant organ and the season of the year (Singh *et al.*, 2003). Trials performed under laboratory and greenhouse conditions show the allelopathic competence among various essential oil plants. Essential oil-based herbicides may increase pollination because they can act as attractant agents for insect pollinators or as repellent agents for harmful insects (Romagni *et al.*, 2000; Singh *et al.*, 2002; Blázquez, 2014). Phytotoxic and allelopathic effects are particularly enhanced by monoterpenoids and sesquiterpenoids, the two most common constituents of essential oils (Martino *et al.*, 2010; Ootani *et al.*, 2017). In spite of this, it frequently appears that their herbicidal activity is selective (Vokou *et al.*, 2003).

Essential Oils and Weed Management

Herbicide applications are complex, because they have to be carefully timed to increase effectiveness of herbicides when they reach the location of their action. Herbicides need to intersect the cuticle, the 0.5–15 μm outermost skin layer consisting of cutin and waxes. Cutin components leave spaces that can be crossed, but waxes, due to their physical structure, are the main obstacle to diffusion of lipophilic compounds (Verdeguer *et al.*, 2020). Monocotyledonous and dicotyledonous plant species differ in the number and structure of xylem elements. It seems that these differences affect effectiveness of essential oils in these species (Kirkham, 2014). Although this mechanism is not well understood, it appears that the higher number of xylem elements in monocotyledonous species causes phytotoxic effects of essential oils (Kim *et al.*, 2014). Families *Lamiaceae, Myrtaceae, Asteraceae* and *Anacardiaceae* are very frequently reported to be families with the strongest phytotoxic effects of their essential oils. Limonene, α-pinene, 1,8-cineole, carvacrol, camphor and thymol are individual components of these oils with high activity (Amri *et al.*, 2013). Fig 44.1 shows the most frequently tested plant families for bioherbicidal potential.

Several authors (Saharkhiz *et al.*, 2009; Alipour and Saharkhiz, 2016; Hazrati *et al.*, 2017; Mahdavikia *et al.*, 2017) performed studies related to effects of allelochemicals on the growth and physiological traits of plants exposed to these toxic chemicals. It was confirmed that high concentrations of allelochemicals significantly intensified their phytotoxicity. Hassannejad and Ghafarbi (2013) established allelopathic effects of essential oils from the following plants in the *Lamiaceae* family: lavender, rosemary, garden sage, thyme, common balm on seed germination and seedling growth of dodder. A parasitic weed species, dodder (*Cuscuta campestris* Yunck.) reduces yields in many crops. Its suppression is very difficult. The essential oil applied to inhibit the growth of this plant species affects the post-emergence stage of dodder seedlings. The strongest effect was recorded in lavender.

According to Alipour *et al.* (2019), the inhibitory effect of rosemary essential oil varied over the plant phenological growth stage, type of plant species exposed to oil, as well as the oil concentration. Similar results gained by Saharkhiz *et al.* (2009) and Alipour and Saharkhiz (2016) showed the inhibitory effects of allelochemicals in rosemary. Based on available literature, Alipour *et al.* (2019) were the first to use starch to encapsulate rosemary essential oil and to use it as bioherbicide, because of its slow release into the soil due to the encapsulation. The increased concentration of encapsulated essential oil resulted in the significant reduction in the germination rate and percentage, leaf area, fresh and dry weights of roots and stems, root length, and chlorophyll content in redroot pigweed and radish. On the other hand, amounts of proline and relative permeability of membranes significantly rose in observed species.

According to Singh *et al.* (2005), volatile oils derived from lemon-scented gum suppressed Santa-Maria weed species. These authors considered lemon-scented gum as a possible bioherbicide to be used in weed control. The lemon-scented gum essential oil at the concentrations of 5–75 μL mL^{-1} produced fast electrolyte leakage from Santa-Maria plants. This fact indicates that this oil affected the membrane integrity. Under laboratory conditions, the increase of the oil concentration from 0.2 to 5.0 nL mL^{-1} resulted in the decrease in Santa-Maria seed germination and seedling length, chlorophyll

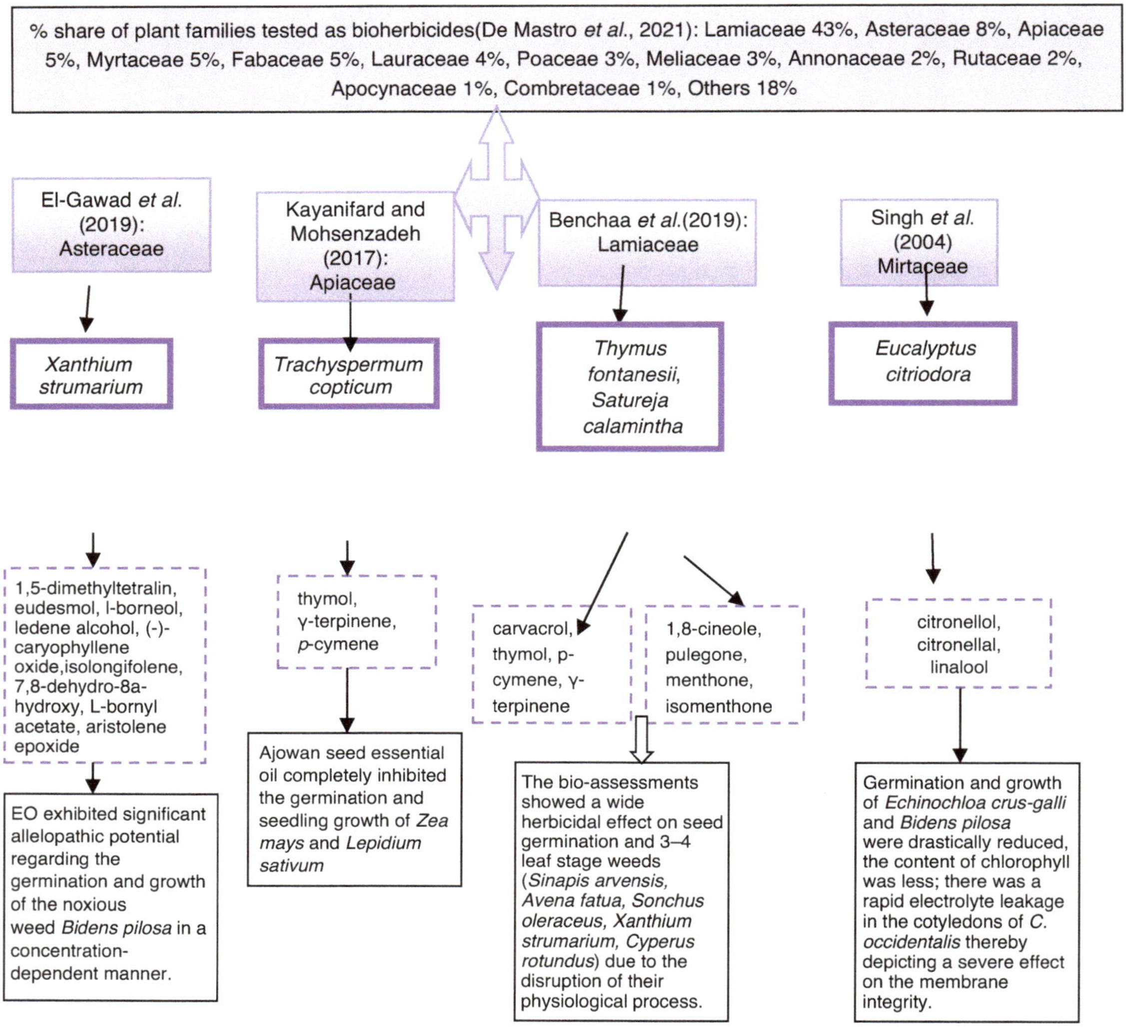

Fig. 44.1. Scheme of the most frequently tested plant families for bioherbicidal potential.

content and respiratory activity, whereas seed germination completely ceased at the oil concentration of 5.0 nL mL^{-1}. Lemon-scented gum oils consist of a mixture of a number of monoterpenes including cineole, citronellal, citronellol and linalool, which are established allelochemicals. It is possible that phytotoxicity of these oils to tested species is a result of disturbed photosynthesis and respiratory ability caused by constituent monoterpenes, which may act interactively like other allelochemicals (Einhellig, 1996).

Monoterpenes are usually water-insoluble, but their solubility is a limiting factor in passing on toxicity to other plants. The application of essential oils is considered harmless because of their fast release into environment (Beuchat *et al.*, 2001). Lemon-scented gum essential oils have no negative effects on soil and species that are not intentionally targeted for suppression by these oils. Based on *in vitro* assays, Verdeguer *et al.* (2020) stated that the conehead thyme essential oil had ceased the seedling germination and growth of Canada fleabane, annual sow thistle and lamb's quarter at the concentration of 0.125 µL/mL, hooked bristlegrass, wild oat and black nightshade at 0.5 µL/mL, redroot pigweed at 1 µL/mL and common purslane and barnyard grass at 2 μL/mL. Sodaeizadeh *et al.* (2010) performed the trial with different parts of wild rue, and found the strongest toxicity had been expressed by leaves. They seriously adversely affected the seedling length and the dry weight, leaf area and chlorophyll content of wild oat and field bindweed. According to Vaughn (1991), essential oils derived from true cinnamon tree and common thyme kill meristematic cells of potato and thus prevent the sprout growth.

Different susceptibility of various plant species to the effects of essential oils can be explained by different structures in their leaf cuticle. Diffusion, as the main pathway of transportation in plants, primarily depends on the leaf stage. During the plant growth, the accumulation of various chemical cuticular compounds increases, which results in the obstruction of the herbicide incorporation (Schreiber, 1995; Verdeguer *et al.*, 2020). According to some studies, there are no significant correlations between cuticle thickness and penetration of compounds. In contrast, other studies showed the effect of ultrastructural cuticle traits on herbicide solubility, which can to a certain degree explain the differences in cuticular permeability in various plant species (Gouret *et al.*, 1993; Verdeguer *et al.*, 2020).

Based on studies performed by Ramezani *et al.* (2008), essential oils derived from rosemary, narrow-leaved black peppermint, Lawson cypress and eastern white cedar inhibit germination of redroot pigweed, common purslane and Russian knapweed. Generally, all essential oils, particularly lemon-scented gum essential oil, powerfully impede germination of weed species. According to Hazrati *et al.* (2017) mini-emulsion of summer savoury essential oil is characterized by herbicidal traits, because it has a strong phytotoxic effect and thus affects germination, growth and physiological processes in weeds (redroot pigweed and lamb's quarter). Making mini-emulsion from summer savoury essential oil requires a low-energy input and therefore has good potential as a natural herbicide to be used in weed suppression in organic agricultural production. Cell membranes were significantly disrupted when relative electrolyte leakage was increased one and 5 days after the treatment. Due to this, the cell membrane was more permeable. Taban *et al.* (2013), reported powerful phytotoxic effects of essential oils from summer savoury and other species of *Satureja* on seed germination of weed species.

Bainard *et al.* (2006) studied effects of phytotoxicity of clove essential oil and eugenol, its main constituent, on broccoli, lamb's quarter and redroot pigweed. These authors also observed the function of leaf epicuticular wax in relation to susceptibility and retention of these essential oils. Clove essential oil inhibited the seedling growth in all species more than its constituent eugenol. Furthermore, differences in the retention of clove essential oil and eugenol solutions were not significant. It points out that their different phytotoxicity to broccoli leaves was not a result of a degree of difference in foliar retention. Ootani *et al.* (2017) studied effects of essential oils from lemon-scented gum and citronella grass and their primary constituent citronellal on the germination and growth of crabgrass and burrgrass. Seed germination was reduced by 97–99%. Essential oils applied at the concentration of 20% decreased the accumulation of dry weight of shoots and roots. Furthermore, essential oils decreased contents of chlorophyll and total protein in weeds by over 80 and 90%, respectively.

Research conducted by Ramezani *et al.* (2008) are in agreement with results of many researchers that stated that large amounts of monoterpenes and phenolic components in plants had strong inhibitory effects (Boutekedjiret *et al.*, 2003; Zrira *et al.*, 2004; Zaouali *et al.*, 2005; Baranska *et al.*, 2005; Barazandeh *et al.*, 2005; Elisabete *et al.*, 2006). According to Baum *et al.* (1998) and Romagni *et al.* (2000), volatile monoterpenes are effective inhibitors of mitosis.

Herbicidal activity of thyme essential oil could be a consequence of a high content of thymol, borneol and carvacrol (that has post-emergence effects) (Oliva *et al.*, 2015; De Mastro *et al.*, 2021). Moreover, thyme essential oil expressed pre-emergence allelopathic potential. Ali *et al.* (2015) showed that Algerian thyme inhibited the seedling growth of alfalfa and common wheat, while Zhou *et al.* (2021) established that *Thymus proximus* successfully suppressed seed germination of redroot pigweed and annual blue grass.

The allelopathic effect of species of the genus *Lavandula* (*Lavandula angustifolia, L. latifolia*, and *L. spica*), which vary in the terpene content, has been documented. A high content of both linalool and linalyl acetate was detected in lavender (Ibrahim *et al.*, 2017), whereby linalool affected pre- and post-emergence coffee senna (Singh *et al.*, 2002). They also contained 1,8-cineole and fenchone. *Lavandula* spp. were phytotoxic to seeds of redroot pigweed, *Lolium* spp., and charlock mustard (Campiglia *et al.*, 2007). For now, there are no commercial preparations based on lavender essential oil.

Phytotoxic effects of essential oils

Due to phytotoxic effects of essential oils, the interest in studying and utilizing volatile oils derived from aromatic plants has been increasing as they have potential in weed control (Singh *et al.*, 2003).

Although monoterpenoids and sesquiterpenoids pay a significant role in the phytotoxic and allelopathic effects of essential oils, their herbicidal activity is frequently selective (Vokou *et al.*, 2003; Martino *et al.*, 2010). In relation to phytotoxic effects of essential oils, α- and β-pinene, 1,8-cineole and phenylpropanoids (carvacrol and thymol) are the most analyzed compounds.

Pinene ($C_{10}H_{16}$), a collection of unsaturated bicyclic monoterpenes, includes two isomers, α- and β-pinene found in nature. Pinene is a primary component of essential oils extracted from conifers, especially pines, cypresses and pistachios, but also from rosemary and myrtle. These two pinenes are optical isomers and their interaction with polarized light differs (Salehi *et al.*, 2019). Moreover, their mirror image is not identical (Amri *et al.*, 2013). α-pinene affects energy metabolism of maize isolated mitochondria by disconnection of oxidative phosphorylation, hindering of electron transfer and mitochondrial ATP synthesis (Abrahim *et al.*, 2003). According to Singh *et al.* (2006), α-pinene adversely affects germination and seedling growth of coffee senna, slender amaranth and common wheat. The exposure of coffee senna roots to α-pinene improves leakage of solute, elevates levels of malondialdehyde, proline and hydrogen peroxide, pointing to lipid peroxidation and initiation of oxidative stress.

During the 1960s, cineole, also called eucalyptol, a monoterpenoid, was presumed to have phytotoxic effects. 1,8-cineole is well known as an allelopathic agent and an aromatic component of many plants including *Eucalyptus* sp. (Amri *et al.*, 2013). Singh *et al.* (2002) have proved that 1,8-cineole reduces germination, germination rate, seedling growth, chlorophyll content and the respiratory activity of billygoat weed. The volatile analogues, 1,4-cineole and 1,8-cineole, are components of essential oils of numerous plants, but their biological activities are still insufficiently known. According to Romagni *et al.* (2000), the inhibitory effect of 1,4-cineole on the growth of roots and shoots was strong. 1,4-cineole produced spiral-shaped morphological deformations. On the other hand, 1,8-cineole reduced the root growth and decreased germination rates. Based on chlorophyll fluorescence data (yield and F v / F m) 1,4-cineole induced considerably greater stress ($P \leq 0.001$) to photosynthesis than controls did.

Carvacrol and thymol, monoterpenoid phenols, are active ingredients of the family *Lamiaceae* that includes species such as thyme and oregano. Pure thymol and carvacrol have phytotoxic effects on redroot pigweed and lamb's quarter germination and seedling growth. These effects are stronger than effects of the commercial herbicide 2.4.D isooctyl ester (Kordali *et al.*, 2008). Azirak and Karaman (2008) tested phytotoxicity of these two phenols on the germination rate and the seedling growth of the following six weeds: charlock mustard, redroot pigweed, yellow star thistle, wild radish, Nepal dock, pale rose hollyhock and common sow thistle. These effects were strong on all weeds except pale rose hollyhock plants, which were resistant to both compounds.

Citronellol, a natural acyclic monoterpenoid, is highly phytotoxic to many mature plants and therefore is a good natural bioherbicide (Singh *et al.*, 2004). The plant species belonging to the genus *Cymbopogon* are known for their phytotoxic effects due to the presence of citronellal, geraniol and citronellol. Caryophyllene, eugenol and eugenol acetate are well-known phytotoxic compounds in clove essential oil (Almarie, 2021).

Essential Oils' Mechanisms of Action

In the available literature, there are many data on phytotoxic effects of essential oils on plants. These data also encompass the description of the mechanisms of action of essential oils. On the other hand, there are not detailed studies dealing with the mechanisms of action at the molecular level. Lins *et al.* (2019) described the mechanisms of herbicidal activities of cinnamon and *Java citronella* essential oils, including their primary chemical components such as trans-cinnamaldehyde, citronellal and citronellol on thale cress. The individual molecules of essential oils are small amphiphiles or amphipaths, which allow them to traverse the mesh of the cell wall and directly interact with plasma membrane of plants. Essential oils act on these membranes as one of their cellular targets. The modification of the lipid organization can result in vital cellular effects, such as effects on protein function (Lee, 2004; Khandelia *et al.*, 2008). Essential oils affect the lipid membrane because of their lipophilic nature and significantly alter the membrane fluidity by causing deviation of the membrane potential and/or permeability and fluidity. They can also modify the activity of ion channels. These effects may particularly provoke cell lysis or even death by means of apoptosis or necrosis (Lins *et al.*, 2019).

In relation to phytotoxic effects of essential oils, the following visible symptoms were described: reduction in the growth, acute chlorosis or leaf burn, cessation of germination and seedling development (Tworkoski, 2002; Lins *et al.*, 2019; Raveau *et al.*, 2020). Many authors indicated that the occurrence of these symptoms was related to inhibition of mitosis, reduction in cellular respiration, leakage of ions and depolarization of membranes, removal of the waxy cuticular layer, chlorophyll degradation, oxidative damages that were triggered by the accumulation of various reactive oxygen species and polymerization of microtubules (Maffei *et al.*, 2001; Zunino and Zygadlo, 2004; Amri *et al.*, 2013; Lins *et al.*, 2019).

Some herbicidal mechanisms of action of essential oils encompass the following:

1. phytotoxic effects that occur by the inhibition of DNA synthesis, interference with the mitotic activity, or disturbance of adjacent mitochondrial membranes and nuclei in certain organelles such as mitochondria (Hanana *et al.*, 2017). According to Amri *et al.* (2013), the suppression of DNA synthesis might be a mechanism of action of monoterpenes in decreasing proliferation of cells in root apical meristem. The mitotic index of the root apical meristem of annual turnip rape was significantly decreased by 1,8-cineole (Koitabashi *et al.*, 1997). Yoshimura *et al.* (2011) determined that 1,8-cineole inhibited both proliferation and elongation of BY-2 cultured tobacco cells;
2. decrease in cellular and mitochondrial respiration (Amri *et al.*, 2013);
3. inducement of oxidative damage and the loss of membrane integrity by the accumulation of reactive oxygen species that cause ion or electrolyte leakage, depolarization of membranes, disturbance of cuticular wax, obstruction of stomata and reduction in epidermal cells (Kaur *et al.*, 2011). Eugenol draws out the generation of active oxygen species in plants, which results in damages of cell membranes and inhibition of photosynthesis. The improved activity of ROS-scavenging enzymes obstructs subsequent pathways of metabolism (Acheuk *et al.*, 2022);
4. polarization of microtubules. According to Chaimovitsh *et al.* (2017), limonene and (+)-citronellal had powerful adverse effects on microtubules. These authors stated that although citral disassembled microtubules, it did not damage membrane. Effects of carvacrol on microtubules were not detectable, but it provoked membrane leakage. Limonene disturbed microtubules and caused membrane leakage. *In planta* quantification of residues showed biotransformation of citral into nerol and geraniol, and the conversion of limonene into carvacrol, which might give explanation about its dual capacity to microtubules and membrane functionality;
5. accumulation of prolines and peroxdation of lipids. Essential oils obstruct the electron flow in mitochondria. This leads to the extensive accumulation of ROS, which results in peroxidation of lipids. Since fatty acids and other lipids are components of the membrane structure, a membrane rupture could cause a lipid release within the cytoplasm of targeted cells. Lipids released in the cytoplasm may then be targeted by oxidative activity (Khare *et al.*, 2019);
6. inhibition of photosynthesis. Numerous monoterpenes inhibit photosynthesis and chlorophyll synthesis (Araniti *et al.*, 2018). Effects of citronellol and 1,8-cineole on billygoat weed were similar: the reduction in the chlorophyll content amounted to 60 and 66%, respectively (Grana *et al.*, 2012). Monoterpenoid eugenol stimulated inhibition of photosynthesis by decreasing the chlorophyll content in coffee senna and cobblers pegs (Vaid *et al.*, 2010).

Essential Oils-Based Commercial Products Used in Weed Suppression

Quite a few essential oil-based commercial products that are used for weed suppression are available on the market (Table 44.1). Nonetheless, manufacturing of new products based on essential oils and derivatives (pure compounds) is not yet developed sufficiently in the industry.

Table 44.1. Commercially formulated organic herbicides based on essential oils or plant extracts

Bioherbicide	Name of product	Target weed	References
Carum carvi, *Mentha piperita*	ATPOLAN BIO 80 EC – adjuvant (AGROMIX Niepołomice, Poland)	*Echinochloa crus-galli*	Synowiec *et al.* (2019)
d-limonene (citrus oil)	Green Match™ (USA)	Annual and perennial weeds including grasses and broadleaves	https://www2.santacruzcountyca.gov/BDS/Govstream2/Bdsvdata/non_legacy_2.0/Minutes/2008/20080226-409/PDF/056-1.pdf
d-limonene (citrus oil)	Avenger Weed Killer (USA)	All broadleaf and grassy weeds	https://www.cvear.com/wp-content/uploads/2016/05/WeedPresentationfinal3.pdf
Eugenol	Weed Slayer (USA)	All broadleaf and grassy weeds	https://www.arbico-organics.com/product/weed-slayer-herbicide/natural-organic-weed-control
Lemongrass oil 50%	GreenMatch EX (USA)	Annual and perennial weeds, grasses and broadleaves	http://legacy.picol.cahnrs.wsu.edu/~picol/pdf/WA/51459.pdf
45% clove oil + 45% cinnamon oil	WeedZap (USA)	Annual and perennial broadleaf and grassy weeds	https://www.arbico-organics.com/product/weed-zap-jh-biotech/natural-organic-weed-control
Clove oil 50%	Matratec (USA)	Annual and perennial broadleaf and grassy weeds	http://cdn.arbico-organics.com/downloads/matratec-label.pdf
Clove bud + Mint oil 5%	Torched Weed Killer (Southland Organics, USA)	All weeds	https://www.arbico-organics.com/product/torched-weed-killer/post-emergent-herbicides-foliar-sprays
50% clove oil (*Syzygium aromaticum*)	Matran	Non-selective weed and grass killer	https://www.ars.usda.gov/research/publications/publication/?seqNo115=235941
Clove oil	BurnOut Weed & Grass Killer-RTU (Bonide products, USA)	Annual, perennial and grassy weeds	https://norganics.com/index-12/pest-control/products/burnout/
Pinus radiate (pine oil)	Bioweed® (Barmac, AUS)	Annual and perennial broadleaf and grassy weeds	https://bioweed.com.au/product/bioweed-organic-herbicide/
Sunflower	Beloukha (Belchim, USA)	Annual and perennial broadleaf weeds, grasses	https://www.belchimusa.com//components/com_belchimproducts/files/assets_my_file/Beloukha06-10-19web.%7BD%7D.pdf

According to Synowiec *et al.* (2019), the oil-in-water emulsion with 2.5% of caraway essential oil will possibly be foliarly used as a botanical herbicide to control barnyard grass in maize production. These authors have chosen commercial multi-purpose adjuvant ATPOLAN BIO 80 EC (Producer: AGROMIX Niepołomice, Poland) an emulsifier of essential oils in their study. ATPOLAN BIO 80 EC primarily consists of fatty acid methyl esters of oilseed rape oil (80%). This adjuvant showed good herbicidal efficacy as an emulsifier of peppermint or caraway essential oil (Synowiec and Drozdek, 2016).

D-limonene has been detected in more than 300 crops and other plants suitable for human consumption. It is usually used as a flavouring agent, deodoriser, shampoo or mouthwash. The preparation **Avenger Weed Killer**, as a non-selective, post-emergence herbicide, effectively destroys all weeds. Avenger is now listed 'organic' by the USDA National Organics Program (https://www.cvear.com/wp-content/uploads/2016/05/FAQ_AvengerWeedKiller.pdf [accessed 27/2/25]).

GreenMatch EX is a non-selective, post-emergence weedkiller based on the lemongrass essential oil. It suppresses a wide spectrum of annual and perennial weed species, including grasses and broadleaf species. Young, immature weeds are less resistant to the product than old, mature weeds (Avila-Adame *et al.*, 2008). (http://legacy.picol.cahnrs.wsu.edu/~picol/pdf/WA/51459.pdf [accessed 27/2/25]).

WEED ZAP is an all-natural contact, non-selective herbicide that contains 45% clove essential oil and 45% cinnamon essential oil. It is also a broad-spectrum, foliar-applied herbicide, suitable for organic production. Natural ingredients of this product suppress only vigorously growing emerged greenery, including both annual and perennial broadleaf and grassy weed species. Its important characteristic is that it does not translocate or move within plants (https://www.arbico-organics.com/product/weed-zap-jh-biotech/natural-organic-weed-control [accessed 27/2/25]).

MATRATEC similar to WEED ZAP, is a contact non-selective herbicide that is applied foliarly to a broad spectrum of weeds. The degree of suppression is lower when weed plants are inactive, mature or biennial/perennial species. The herbicide does not translocate: it affects only those parts of plants that are sprayed with the herbicide. MATRATEC can cause damage in crops and therefore it is necessary to use the equipment that will prevent contact with crops (http://cdn.arbico-organics.com/downloads/matratec-label.pdf [accessed 27/2/25])

Weed Slayer consists of two products (Part A and Part B) that are mixed in order to make an exceptionally effective weedkiller. Part A is the Weed Slayer, a broad spectrum herbicide made from eugenol (extracted from clove essential oil) and molasses. On the other hand, Part B is Agro Gold, a biological adjuvant that encompass *Streptomyces* and *Bacillus* microbes. The two constituents in Part A reach roots and kill them and/or root ball of the undesirable plants. Bacteria in Part B penetrate soil and bring eugenol into roots. Due to this action, Weed Slayer rapidly destroys the entire plant, the root system as well, whereby it is more efficient than 'burndown herbicides' (https://www.arbico-organics.com/product/weed-slayer-herbicide/natural-organic-weed-control [accessed 27/2/25]).

Torched Weed Killer (USA), a blend of clove bud essential oil, mint essential oil and soaps, is a natural alternative to glyphosate herbicides. It acts fast in an environmentally friendly way. This non-selective, post-emergence weedkiller is highly effective at temperatures over 16°C and commonly just one application is sufficient (except for those most notoriously difficult weeds). As it is a contact herbicide, its activity will not cease until the entire plant including the root is destroyed (https://www.arbico-organics.com/product/torched-weed-killer/post-emergent-herbicides-foliar-sprays [accessed 27/2/25]).

Matran EC (50% clove essential oil) is applied as 5–8% solution. Matran EC is an organic post-emergence herbicide broadcast applied for control of broadleaf weeds (https://www.ars.usda.gov/research/publications/publication/?seqNo115=235941 [accessed 27/2/25])

BurnOut Weed & Grass Killer is an all-natural product based on non-selective herbicide produced from clove oil as the active ingredient and extracts from vinegar as one of the inert matters. This product is used to control broadleaf and grassy weed species. A foliar contact causes fast dehydration of plant cell tissue. The top growth of herblike perennial weeds and grasses is reduced (https://norganics.com/index-12/pest-control/products/burnout/ [accessed 27/2/25]).

Bioweed weedkiller, a herbicide designed to suppress a wide range of weeds, is made from natural hand-tapped pine oil. This herbicide also controls moss and algae (https://bioweed.com.au/product/bioweed-organic-herbicide/[accessed 27/2/25]).

Beloukha is a biodegradable, non-selective, broad-spectrum herbicide produced from sunflowers. The herbicide provides burndown of perennial and annual broadleaf weed species and grasses, in addition to the majority of mosses and other cryptogams. Beloukha acts rapidly, exclusively by contact. It attacks and breaks up cell membranes of the plant epidermis causing quick dehydration of tissues. This herbicide leaves no residue as soil organisms absorb C_9, producing CO_2 and H_2O.

Conclusion

Based on literature data, this paper reveals that essential oils have a strong inhibitory and phytotoxic effect on weed plants that can be successfully used in crop protection. The idea to study the plant-to-plant interactions by means of releasing secondary metabolites could contribute to discovering new substances with herbicidal activity. The crucial advantage of natural herbicidal products, such as the absence of 'unnatural' ring structures and the presence of hardly any heavy atoms, is their short environmental half-lives and their predisposition to influence different target sites. This latter fact is exceptionally significant since new modes of action are necessary. Due to a very small number of commercially available essential oil-based preparations on the world market, agrochemical companies keenly search for innovative mechanisms of action for which they can develop new chemistry.

References

Abrahim, D., Francischini, A.C., Pergo, E.M., Kelmer-Bracht, A.M. and Ishii-Iwamoto, E.L. (2003) Effects of α-pinene on the mitochondrial respiration of maize seedlings. *Plant Physiology and Biochemistry* 41(11–12), 985–991.

Acheuk, F., Basiouni, S., Shehata, A.A., Dick, K., Hajri, H. *et al*. (2022) Status and prospects of botanical biopesticides in Europe and Mediterranean countries. *Biomolecules* 12(2), 311.

Ali, I.B.E.H., Chaouachi, M., Bahri, R., Chaieb, I., Boussaïd, M. *et al*. (2015) Chemical composition and antioxidant, antibacterial, allelopathic and insecticidal activities of essential oil of *Thymus algeriensis* Boiss. et Reut. *Industrial Crops and Products* 77, 631–639.

Alipour, M. and Saharkhiz, M.J. (2016) Phytotoxic activity and variation in essential oil content and composition of rosemary (*Rosmarinus officinalis* L.) during different phenological growth stages. *Biocatalysis and Agricultural Biotechnology* 7, 271–278.

Alipour, M., Saharkhiz, M.J., Niakousari, M. and Seidi Damyeh, M. (2019) Phytotoxicity of encapsulated essential oil of rosemary on germination and morphophysiological features of amaranth and radish seedlings. *Scientia Horticulturae* 243, 131–139.

Almarie, A. (2021) Bioherbicidal potential of eucalyptus and clove oil and their combinations on four weedy species. *Iraqi Journal of Science* 1494–1502.

Amri, I., Hamrouni, L., Hanana, M. and Jamoussi, B. (2013) Reviews on phytotoxic effects of essential oils and their individual components: News approach for weeds management. *International Journal of Applied Biology and Pharmaceutical Technology* 4, 96–114.

Araniti, F., Landi, M., Lupini, A., Sunseri, F., Guidi, L. *et al*. (2018) *Origanum vulgare* essential oils inhibit glutamate and aspartate metabolism altering the photorespiratory pathway in *Arabidopsis thaliana* seedlings. *Journal of Plant Physiology* 231, 297–309.

Avila-Adame, C., Fernandez, L., Campbell, B., Tan, E., Koivunen, M. *et al*. (2008) Field evaluation of GreenMatch EX: A new broadspectrum organic herbicide. In: *Proceedings of California Weed Science Society* (Vol. 216).

Azirak, S. and Karaman, S. (2008) Allelopathic effect of some essential oils and components on germination of weed species. *Acta Agriculturae Scandinavica* 51, 88–92.

Bainard, L.D., Isman, M.B. and Upadhyaya, M.K. (2006) Phytotoxicity of clove oil and its primary constituent eugenol and the role of leaf epicuticular wax in the susceptibility to these essential oils. *Weed Science* 54(5), 833–837.

Baranska, M., Schulz, H., Reitzenstein, S., Uhlemann, U., Strehle, M.A. *et al*. (2005) Vibrational spectroscopic studies to acquire a quality control method of eucalyptus essential oils. *Biopolymers* 78(5), 237–248.

Barazandeh, M.M., Jaimand, K. and Asareh, M.H. (2005) Essential oil composition of *Eucalyptus porosa*. *Iranian Journal of Medicinal and Aromatic Plants Research* 20(4), 469–476.

Baum, S.F., Karanastasis, L. and Rost, T.L. (1998) Morphogenetic effect of the herbicide cinch on *Arabidopsis thaliana* root development. *Journal of Plant Growth Regulation* 17(2), 107–114.

Beuchat, L.R., Wilson, C.L. and Droby, S. (2001) Control of foodborne pathogens and spoilage microorganisms by naturally occurring antimicrobials. In: Beuchat,

L. (ed.) *Microbial Food Contamination*. CRC Press, Boca Raton, Florida, pp. 149–169.

Blázquez, M.A. (2014) Role of natural essential oils in sustainable agriculture and food preservation. *Journal of Scientific Research and Reports* 3(14), 1843–1860.

Boutekedjiret, C., Bentahar, F., Belabbes, R. and Bessiere, J.M. (2003) Extraction of rosemary essential oil by steam distillation and hydrodistillation. *Flavour and Fragrance Journal* 18(6), 481–484.

Campiglia, E., Mancinelli, R., Cavalieri, A. and Caporali, F. (2007) Use of essential oils of cinnamon, lavender and peppermint for weed control. *Italian Journal of Agronomy* 2(2), 171.

Chaimovitsh, D., Shachter, A., Abu-Abied, M., Rubin, B., Sadot, E. *et al*. (2017) Herbicidal activity of monoterpenes is associated with disruption of microtubule functionality and membrane integrity. *Weed Science* 65(1), 19–30.

Dayan, F.E. and Duke, S.O. (2014) Natural compounds as next-generation herbicides. *Plant Physiology* 166(3), 1090–1105.

De Mastro, G., El Mahdi, J. and Ruta, C. (2021) Bioherbicidal potential of the essential oils from Mediterranean Lamiaceae for weed control in organic farming. *Plants* 10(4), 818.

Dudai, N., Poljakoff-Mayber, A., Mayer, A.M., Putievsky, E. and Lerner, H.R. (1999) Essential oils as allelochemicals and their potential use as bioherbicides. *Journal of Chemical Ecology* 25(5), 1079–1089.

Einhellig, F.A. (1996) Interactions involving allelopathy in cropping systems. *Agronomy Journal* 88(6), 886–893.

El-Gawad, A.A., Elshamy, A., El Gendy, A.E.-N., Gaara, A. and Assaeed, A. (2019) Volatiles profiling, allelopathic activity, and antioxidant potentiality of *Xanthium strumarium* leaves essential oil from Egypt: Evidence from chemometrics analysis. *Molecules* 24(3), 584.

Elisabete, M., Carlos Lopes, M., Nogueira, T., Lourenco, J.A.A. and Marcelo Curto, M.J. (2006) Pilot steam distillation of rosemary (*Rosmarinus officinalis* L.) from Portugal. *Silva Lusitana* 14(2), 203–217.

Golijan-Pantović, J. and Sečanski, M. (2022a) Weed control in organic farming. *Contemporary Agriculture* 72(1–2), 43–56.

Golijan-Pantović, J. and Sečanski, M. (2022b) Biopesticides in organic agriculture. *Contemporary Agriculture* 71(1–2), 141–154.

Gouret, E., Rohr, R. and Chamel, A. (1993) Ultrastructure and chemical composition of some isolated plant cuticles in relation to their permeability to the herbicide, diuron. *New Phytologist* 124(3), 423–431.

Grana, E., Diaz-Tielas, C., Sanchez-Moreiras, A.M. and Roger, M.R. (2012) Mode of action of monoterpenes in plant-plant interactions. *Current Bioactive Compounds* 8, 80–89.

Hanana, M., Mansour, M.B., Algabr, M., Amri, I., Gargouri, S. *et al*. (2017) Potential use of essential oils from four Tunisian species of Lamiaceae: Biological alternative for fungal and weed control. *Records of Natural Products* 11, 258–269.

Hassannejad, S. and Ghafarbi, S.P. (2013) Allelopathic effects of some Lamiaceae on seed germination and seedling growth of dodder (*Cuscuta campestris* Yunck.). *International Journal of Biosciences* 3(3), 9–14.

Hazrati, H., Saharkhiz, M.J., Niakousari, M. and Moein, M. (2017) Natural herbicide activity of *Satureja hortensis* L. essential oil nanoemulsion on the seed germination and morphophysiological features of two important weed species. *Ecotoxicology and Environmental Safety* 142, 423–430.

Ibrahim, H., Salama, A.M. and Abou El-Leel, O.F. (2017) Analysis of genetic diversity of lavandula species using taxonomic, essential oil and molecular genetic markers. *Middle East Journal of Applied Sciences* 7, 141–154.

Kaur, S., Rana, S., Singh, H.P., Batish, D.R. and Kohli, R.K. (2011) Citronellol disrupts membrane integrity by inducing free radical generation. *Zeitschrift Für Naturforschung C* 66(5–6), 260–266.

Kayanifard, M. and Mohsenzadeh, S. (2017) Allelopathic analysis of four ecotypes of Ajowan. *Iranian Journal of Science and Technology, Transactions A* 41(4), 971–978.

Khandelia, H., Ipsen, J.H. and Mouritsen, O.G. (2008) The impact of peptides on lipid membranes. *Biochimica et Biophysica Acta – Biomembranes* 1778(7–8), 1528–1536.

Khare, P., Srivastava, S., Nigam, N., Singh, A.K. and Singh, S. (2019) Impact of essential oils of *E. citriodora, O. basilicum and M. arvensis* on three different weeds and soil microbial activities. *Environmental Technology & Innovation* 14, 100343.

Kim, H.K., Park, J. and Hwang, I. (2014) Investigating water transport through the xylem network in vascular plants. *Journal of Experimental Botany* 65(7), 1895–1904.

Kirkham, M.B. (2014) Stem anatomy and pressure-volume curves. In: Kirkham, M.B. (ed.) Principles of Soil and Plant Water Relations, 2nd edn. Academic Press, London, pp. 279–310.

Koitabashi, R., Suzuki, T., Kawazu, T., Sakai, A., Kuroiwa, H. *et al*. (1997) 1,8-cineole inhibits root growth and DNA synthesis in the root apical meristem *Ofbrassica campestris* L. *Journal of Plant Research* 110(1), 1–6.

Kordali, S., Cakir, A., Ozer, H., Cakmakci, R., Kesdek, M. *et al*. (2008) Antifungal, phytotoxic and insecticidal properties of essential oil isolated from turkish *Origanum acutidens* and its three components, carvacrol, thymol and p-cymene. *Bioresource Technology* 99(18), 8788–8795.

Lee, A.G. (2004) How lipids affect the activities of integral membrane proteins. *Biochimica et Biophysica Acta – Biomembranes* 1666(1–2), 62–87.

Lins, L., Dal Maso, S., Foncoux, B., Kamili, A., Laurin, Y. *et al.* (2019) Insights into the relationships between herbicide activities, molecular structure and membrane interaction of cinnamon and citronella essential oils components. *International Journal of Molecular Sciences* 20(16), 4007.

Maffei, M., Camusso, W. and Sacco, S. (2001) Effect of *Mentha x piperita* essential oil and monoterpenes on cucumber root membrane potential. *Phytochemistry* 58(5), 703–707.

Mahdavikia, F. and Saharkhiz, M.J. (2015) Phytotoxic activity of essential oil and water extract of peppermint (*Mentha* × *piperita* L. cv. Mitcham). *Journal of Applied Research on Medicinal and Aromatic Plants* 2(4), 146–153.

Mahdavikia, F., Saharkhiz, M.J. and Karami, A. (2017) Defensive response of radish seedlings to the oxidative stress arising from phenolic compounds in the extract of peppermint (*Mentha* × *piperita* L.). *Scientia Horticulturae* 214, 133–140.

Martino, L.D., Formisano, C., Mancini, E., Feo, V.D., Piozzi, F. *et al.* (2010) Chemical composition and phytotoxic effects of essential oils from four *Teucrium* species. *Natural Product Communications* 5, 12.

Mutlu, S., Atici, O. and Esim, N. (2010) Bioherbicidal effects of the essential oils of *Nepeta meyeri*. *Allelopathy Journal* 26, 291–300.

Oliva, M. de las M., Carezzano, M.E., Giuliano, M., Daghero, J., Zygadlo, J. *et al.* (2015) Antimicrobial activity of essential oils of *Thymus vulgaris* and *Origanum vulgare* on phytopathogenic strains isolated from soybean. *Plant Biology* 17(3), 758–765.

Ootani, M.A., dos Reis, M.R., Cangussu, A.S.R., Capone, A., Fidelis, R.R. *et al.* (2017) Phytotoxic effects of essential oils in controlling weed species *Digitaria horizontalis* and *Cenchrus echinatus*. *Biocatalysis and Agricultural Biotechnology* 12, 59–65.

Pimentel, D., McNair, S., Janecka, J., Wightman, J., Simmonds, C. *et al.* (2001) Economic and environmental threats of alien plant, animal, and microbe invasions. *Agriculture, Ecosystems & Environment* 84(1), 1–20.

Ramezani, S., Saharkhiz, M.J., Ramezani, F. and Fotokian, M.H. (2008) Use of essential oils as bioherbicides. *Journal of Essential Oil Bearing Plants* 11(3), 319–327.

Raveau, R., Fontaine, J. and Lounès-Hadj Sahraoui, A. (2020) Essential oils as potential alternative biocontrol products against plant pathogens and weeds: A review. *Foods* 9(3), 365.

Romagni, J.G., Allen, S.N. and Dayan, F.E. (2000) Allelopathic effects of volatile cineoles on two weedy plant species. *Journal of Chemical Ecology* 26(1), 303–313.

Saharkhiz, M.J., Ashiri, F., Salehi, M.R., Ghaemghami, J., Mohammadi, S.H. *et al.* (2009) Allelopathic potential of essential oils from *Carum copticum* L., *Cuminum cyminum* L., *Rosmarinus officinalis* L.and *Zataria multiflora* Boiss. *Medicinal and Aromatic Plant Science and Biotechnology* 3(1), 32–35.

Salehi, B., Upadhyay, S., Orhan, I.E., Jugran, A.K., Jayaweera, S.L.D. *et al.* (2019) Therapeutic potential of α- and β-pinene: A miracle gift of nature. *Biomolecules* 9(11), 738.

Schreiber, L. (1995) A mechanistic approach towards surfactant/wax interactions: Effects of octaethyleneglycolmonododecylether on sorption and diffusion of organic chemicals in reconstituted cuticular wax of barley leaves. *Pesticide Science* 45(1), 1–11.

Singh, H.P., Batish, D.R., Kaur, S., Ramezani, H. and Kohli, R.K. (2002) Comparative phytotoxicity of four monoterpenes against *Cassia occidentalis*. *Annals of Applied Biology* 141(2), 111–116.

Singh, H.P., Batish, D.R. and Kohli, R.K. (2003) Allelopathic interactions and allelochemicals: New possibilities for sustainable weed management. *Critical Reviews in Plant Sciences* 22(3–4), 239–311.

Singh, H.P., Batish, D.R., Kaur, S., Vaid, S. and Kohli, R.K. (2004) Weed suppressing ability of some monoterpenes. *Zeitschrift Fur Pflannzenkrnakheiten Und Pflanzenschutz-Sonderheft* 19, 821–830.

Singh, H.P., Batish, D.R., Setia, N. and Kohli, R.K. (2005) Herbicidal activity of volatile oils from *Eucalyptus citriodora* against *Parthenium hysterophorus*. *Annals of Applied Biology* 146(1), 89–94.

Singh, H.P., Batish, D.R., Kaur, S., Arora, K. and Kohli, R.K. (2006) Alpha-pinene inhibits growth and induces oxidative stress in roots. *Annals of Botany* 98(6), 1261–1269.

Sodaeizadeh, H., Rafieiolhossaini, M. and Van Damme, P. (2010) Herbicidal activity of a medicinal plant, *Peganum harmala* L., and decomposition dynamics of its phytotoxins in the soil. *Industrial Crops and Products* 31(2), 385–394.

Synowiec, A. and Drozdek, E. (2016) Physicochemical and herbicidal properties of emulsions of essential oils against *Avena fatua* L. and *Chenopodium album* L. *Journal of Plant Diseases and Protection* 123(2), 65–74.

Synowiec, A., Możdżeń, K., Krajewska, A., Landi, M. and Araniti, F. (2019) *Carum carvi* L. essential oil: a promising candidate for botanical herbicide against *Echinochloa crus-galli* (L.) P. Beauv. in maize cultivation. *Industrial Crops and Products* 140, 111652.

Taban, A., Saharkhiz, M.J. and Hadian, J. (2013) Allelopathic potential of essential oils from four *Satureja* spp. *Biological Agriculture & Horticulture* 29(4), 244–257.

Tworkoski, T. (2002) Herbicide effects of essential oils. *Weed Science* 50(4), 425–431.

Vaid, S., Batish, D., Singh, D.H. and Kohli, R. (2010) Phytotoxic effect of eugenol towards two weedy species. *Bioscan* 5, 339–341.

Vaughn, S.F. (1991) Natural compounds from spices could replace potato sprouting inhibitors. *Industrial Bioprocessing* 13, 5.

Verdeguer, M., Torres-Pagan, N., Muñoz, M., Jouini, A., García-Plasencia, S. *et al*. (2020) Herbicidal activity of *Thymbra capitata* (L.) cav. essential oil. *Molecules* 25(12), 2832.

Vokou, D., Douvli, P., Blionis, G.J. and Halley, J.M. (2003) Effects of monoterpenoids, acting alone or in pairs, on seed germination and subsequent seedling growth. *Journal of Chemical Ecology* 29(10), 2281–2301.

Yoshimura, H., Sawai, Y., Tamotsu, S. and Sakai, A. (2011) 1,8-cineole inhibits both proliferation and elongation of BY-2 cultured tobacco cells. *Journal of Chemical Ecology* 37(3), 320–328.

Zaouali, Y., Messaoud, C., Salah, A.B. and Boussaïd, M. (2005) Oil composition variability among populations in relationship with their ecological areas in Tunisian *Rosmarinus officinalis* L. *Flavour and Fragrance Journal* 20(5), 512–520.

Zhou, S., Han, C., Zhang, C., Kuchkarova, N., Wei, C. *et al*. (2021) Allelopathic, phytotoxic, and insecticidal effects of *Thymus proximus* Serg. essential oil and its major constituents. *Frontiers in Plant Science* 12, 689875.

Zrira, S., Bessiere, J.M., Menut, C., Elamrani, A. and Benjilali, B. (2004) Chemical composition of the essential oil of nine Eucalyptus species growing in morocco. *Flavour and Fragrance Journal* 19(2), 172–175.

Zunino, M.P. and Zygadlo, J.A. (2004) Effect of monoterpenes on lipid oxidation in maize. *Planta* 219(2), 303–309.

Benchaa, S., Hazzit, M., Zermane, N. and Abdelkrim, H. (2019) Chemical composition and herbicidal activity of essential oils from two *Labiatae* species from Algeria. *Journal of Essential Oil Research* 31(4), 335–346.

45 Climatic Impact on Organic Farming

Ben Hassine Mortadha[1], Emanuele Radicetti[1]*, Daniele Borgatti[1], Roberto Mancinelli[2], Mohamed Allam[3], Taab Alireza[4], Muhammad Ahsan[5] and Valentina Quintarelli[1]

[1]Department of Chemical, Pharmaceutical and Agricultural Sciences, University of Ferrara, Ferrara, Italy; [2]Department of Agricultural and Forestry Sciences, University of Tuscia, Viterbo, Italy; [3]Department of Agronomy, Faculty of Agriculture, Assuit University, Asyut, Egypt; [4]Department of Agronomy and Plant Breeding, Faculty of Agriculture, Ilam University, Ilam, Iran; [5]Department of Horticultural Sciences, The Islamia University of Bahawalpur, Bahawalpur, Pakistan

Abstract

Climate change represents one of the major threats to modern society related to food safety and security. In this context, the agricultural sector and related chains have a dual role: (i) to contribute to climate change; and (ii) to represent the first sector affected by climate change. To provide healthy and safe food, it is necessary to transform the agri-food system and move towards sustainable practices of agricultural and food production that can mitigate and/or adapt to inevitable climate changes. Climate change, which is increasingly evident, is challenging agriculture, specifically organic farming. Agricultural production under organic practices is characterized by the adoption of sustainable practices in an integrated approach, even if sometimes the integrated pest management may show weak points that limit control and/or management techniques against new threats brought by global warming. Therefore, organic farming is facing new stresses brought by increasing temperatures, rainfall reduction, allochthonous weeds and pathogens in the agro-ecosystems. Therefore, in order to achieve a sustainable and resilient agricultural production system against climate change, it is strongly necessary to follow regenerative agricultural practices that replace intensive approaches of conventional agriculture, which is a major cause of greenhouse gas emissions and alteration of agro-ecosystems.

Introduction

Nowadays, agro-ecosystems are subjected to different biotic and abiotic stresses that highlight the need to adopt sustainable agricultural practices able to develop resilience aimed at improving soil quality, reducing synthetic inputs, increasing biodiversity, decreasing soil tillage, maintaining profitability and taking advantage of the circular economy (Reidsma *et al.*, 2023). According to the FAO (1999):

> Organic agriculture promotes and enhances agro-ecosystem health, through a holistic production management system by emphasizing the use of management practices in preference to using off-farm inputs, considering the use of locally adapted systems at regional level. This is carried out by means of agronomic, biological, and mechanical methods, as opposed to using synthetic materials, to fulfil any specific function within the system.

Agriculture is an economic activity that depends heavily on climate, and certain weather conditions may threaten food production to sustain and satisfy human needs. Climate change has a clear impact on the agricultural production systems (Shah *et al.*, 2024). In particular, climate change is estimated to have more negative than positive impacts on agricultural systems globally (Müller, 2013); however, its impact depends on latitude, altitude and crop

*Corresponding author: emanuele.radicetti@unife.it

DOI: 10.1079/9781800626850.0045

type. Actually, there is scientific evidence that climate change may have significant influence on crop production; insects, diseases and weed dynamics; soil properties; and microbial composition of agricultural systems (Skendžić *et al.*, 2021). Accordingly, temperature increases, altered precipitation patterns, and increased CO_2 concentrations have significant feedback on ecosystems, ranging from species to ecosystem levels (Stevens *et al.*, 2004).

Climate change affects agriculture and, conversely, the overuse of synthetic fertilizers has led, over time, to the emission of greenhouse gases (GHG), eutrophication, alteration of biogeochemical cycles and release of carbon into the atmosphere (Mahankale, 2024). To mitigate the negative effects of climate change, it has been reported that pollution coming from the agricultural sector must be reduced by 39% by 2050 (Ume *et al.*, 2023). It is important to understand the relationships between the causes and effects of climate change on agriculture, specifically in organic farming. It is crucial not only to support the productivity of the agricultural sector to ensure food security worldwide but also to contribute positively to the environmental management regime. In this chapter, the interaction of climate change and organic agriculture, and how they affect each other, is analyzed; some organic solutions to maintain balance in the agro-ecosystem are also reported.

Impact of Climate Change on Crop Physiology and Production

Global warming is able to put crops under severe stress, and consequently, this leads to a significant decrease in crop yield. In 2015, the Paris Agreement stipulated that the global average temperature should not increase by 1.5–2.0°C. If it does, many harvests are at risk; in fact, a study conducted in the USA demonstrated that crop yields may decrease up to 14% for corn (*Zea mays*) and 20% for soybean (*Glycine max*) (Tijjani *et al.*, 2024). Ume *et al.* (2023) suggest that agroecological practices, in an organic farming context, are able to reduce climate change impact and, moreover, they can lead to enhanced crop productivity.

The main impacts of climate change on agricultural production systems are summarized in Fig. 45.1. Indeed, it is possible to simulate the impact of climate change on agricultural production systems through forecasting models. Thanks to this software, winter crops such as wheat (*Triticum*

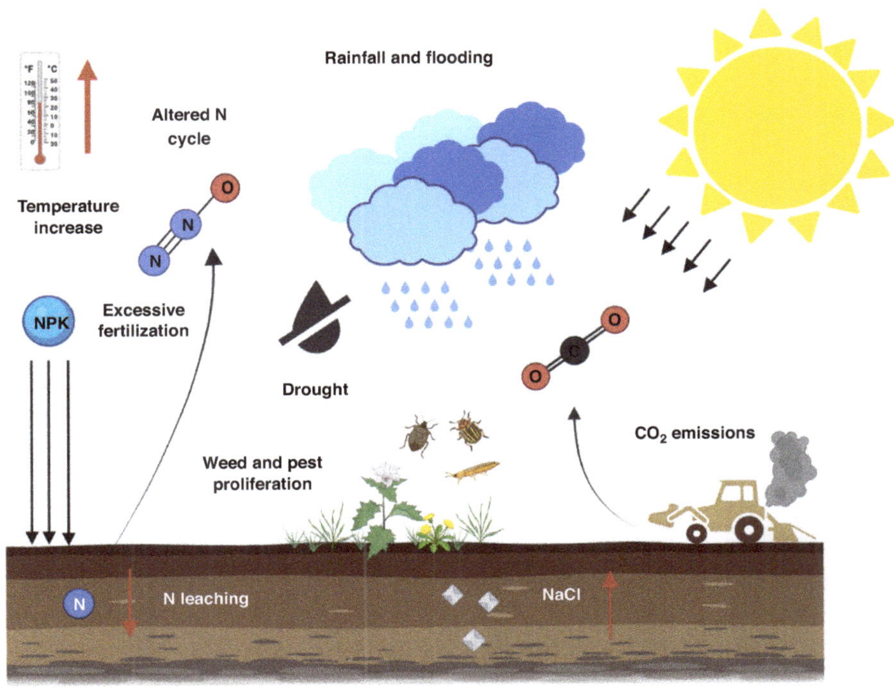

Fig. 45.1. The main impacts of climate change on the agricultural production system.

aestivum) have been found to be better adapted to climate change. On the other hand, summer crops, like sunflower (*Helianthus annuus*), could be more affected by heat stress (Moriondo *et al.*, 2011).

Temperature is a key factor that limits the geographical areas where different crops can be grown, as well as affecting the rate of development, growth and crop yield (Rehman *et al.*, 2015). Each crop has different temperature requirements for each phase of growth. In fact, higher or lower temperature levels can disturb plant development, such as the anthesis phase, and damage yield (Luo, 2011). According to the IPCC (2013), temperature change in tropical areas has generally had a negative impact on food production. As climate change continues, the spring-summer season is expected to be characterized by an increase of maximum temperatures, which would be beneficial for crop production in northern locations where the length of the growing season is currently a limiting factor (Cui, 2020; Zhang *et al.*, 2021). However, the negative effects of elevated temperature are associated with and aggravated by other environmental factors, such as the availability of water, the presence of strong winds, and the intensity and duration of sunlight. For example, increased temperature enhances the water demand from the atmosphere, which may lead to further water stress due to increased water pressure deficits, resulting in reduced soil moisture and, consequently, lower yields (Zhao *et al.*, 2017). Other indirect effects of increased temperature are the increased frequency of heatwaves and the impact on pests, weeds and plant diseases (Field *et al.*, 2014).

Climate change, specifically temperature increase, may lead to the introduction of new invasive species (weeds and pathogens) inside the agro-ecosystem; for example, tropical species are commonly found in temperate areas (Aluja *et al.*, 2014). However, a large number of chemicals are available in conventional agriculture, and they can slow down the development of these biological threats. In fact, herbicides and pesticides are able to neutralize numerous species that can harm crops (Monteiro and Santos, 2022). Organic farming may be disadvantaged because it does not use artificial herbicides and pesticides but only substances of natural origin in an integrated approach with other cultural, mechanical and agronomical means (Regulation EU, 2018/848) that are well-known to be less efficient against specific biotic stresses. The excessive damage that these invasive species bring in organic farming and the absence of agrochemicals can negatively affect crop yield (Vaglia *et al.*, 2022). In fact, organic farming has a mean production which is 18% lower compared to conventional farming; in particular, crop production decreases according to climate, geographic site, crop species, soil pH and textural classes (de la Cruz *et al.*, 2023).

Since the second point of the 2030 agenda calls for 'zero hunger' and the world population will reach 10 billion by 2050, some aspects of organic farming appear to be at odds with these goals (Nti *et al.*, 2023). Despite the advantages, organic farming has several weaknesses and threats at the production unit level; both energy use and carbon footprint are higher in growing organic vegetables, prices increase due to reduced production, exports decrease, and unattractive carbon contracts arise (Avasiloaiei *et al.*, 2023).

Another factor that strongly impacts agricultural production is water cycle and its availability at agro-ecosystem level. Indeed, climate change affects rainfall patterns, soil moisture storage, evaporation and runoff (Hakimovich and Alishovich, 2023). Since more than 80% of the world's total crop production is water-supplied by rainfall, changes in total seasonal precipitation or its patterns are very important. Due to climate change, drought is increasingly common, and in sites affected by dryness and desertification risk it is possible to reduce tillage and thus encourage soil moisture increase (O'Rourke and Petersen, 2016). In fact, in this condition, organic reduced tillage systems are able to limit crop yield loss compared to conventional ploughed systems; this is due to a better soil quality, structure and water content. Crops whose losses are limited under this condition are corn (*Zea mays*), pea barley (*Lathyrus oleraceus*) and wheat (*Triticum aestivum*) (Wittwer *et al.*, 2023).

Heat stress and drought stress are the main consequences of global warming. These factors can affect plant tolerance and defence (Faiz *et al.*, 2024). Since global mean temperature is constantly increasing, organic systems adaptation seems one of the few solutions to limit crop loss and keep the agro-ecosystem intact. Considering that organic agriculture bans synthesized chemical compounds, new strategies must be adopted. For example, chitosan, which is obtained from arthropods or insects, and microbe-based biofertilizers are able to improve thermal tolerance by increasing plants' immunity, antioxidants and hormone production, boosting nitrogen fixation, phosphorus dissolution in soil, crop biomass and photosynthesis rate. All of these considerations lead to better crop productivity and greater yield (Dastogeer *et al.*, 2022; Toaima *et al.*, 2023).

Impact of GHG Emissions and Carbon Sequestration as a Strategy to Mitigate Climate Change

Agriculture is a significant contributor to GHG, and organic farming practices can potentially offset some of these emissions (Squalli and Adamkiewicz, 2023). When considering the many GHGs, CO_2 is the essential chemical compound for photosynthesis, so an increase in CO_2 concentration allows for more rapid growth due to the rapid assimilation of carbon (Maschler *et al.*, 2022). The main effects of increased CO_2 concentrations on plants include a reduction in transpiration and stomatal conductance, improved water and light utilization efficiency, and thus an increase in photosynthetic rate. Consequently, high atmospheric CO_2 concentrations could have a direct impact on ecosystems by stimulating plant development and growth. This can be considered a positive effect for organic farming because the use of artificial fertilizers could be reduced. Organic agriculture can sequester up to one-third of the current emissions from the use of ecologically sound processes to manage the land, soil and crops (Jordan *et al.*, 2009).

Another very important GHG is nitrogen oxide (N_2O), the emissions of which are directly related to the concentration of available mineral N (ammonium and nitrate) in soils resulting from the nitrification and denitrification of available soil and added fertilizer (Bouwman *et al.*, 2013). High emission rates of this GHG were detected subsequently, with the addition of mineral fertilizers. The use of organic fertilizers is sustainable because environmental impact on the agro-ecosystem is reduced, and in addition, soil health is preserved while still producing a good yield (Mahankale, 2024). The obligation to use organic fertilizers in comparison to the use of N mineral fertilizers on organic farms should lead to a reduction in the concentration of easily available mineral N in soils, resulting in a reduction in N_2O emissions. Furthermore, organic fertilizers reduce nitrate leaching because they need to be mineralized before being absorbed by plants. Consequently, this process limits nitrogen losses and allows nutrients availability over time (Diacono *et al.*, 2018). Moreover, due to the improvement of soil organic matter levels resulting in a consequent improvement in soil structure, the soils in a biological way are better aerated than those of conventionally managed soils. These effects on the physical properties of the soil, reducing the denitrification process, cause the decrease of N_2O emissions (Skinner *et al.*, 2019). Tuomisto *et al.* (2012) demonstrated, in a meta-analysis, that nitrous oxide emissions were about 31% lower per unit of land but 8.5% higher per unit of output in European organic farms.

Since CO_2 is one of the main GHGs, a sustainable solution to limit its presence in the atmosphere must be found (Lal, 2023). Innovative techniques must be used given the limitations of organic farming. Resilient factors related to organic agricultural systems are described in the Fig. 45.2. Nowadays, carbon

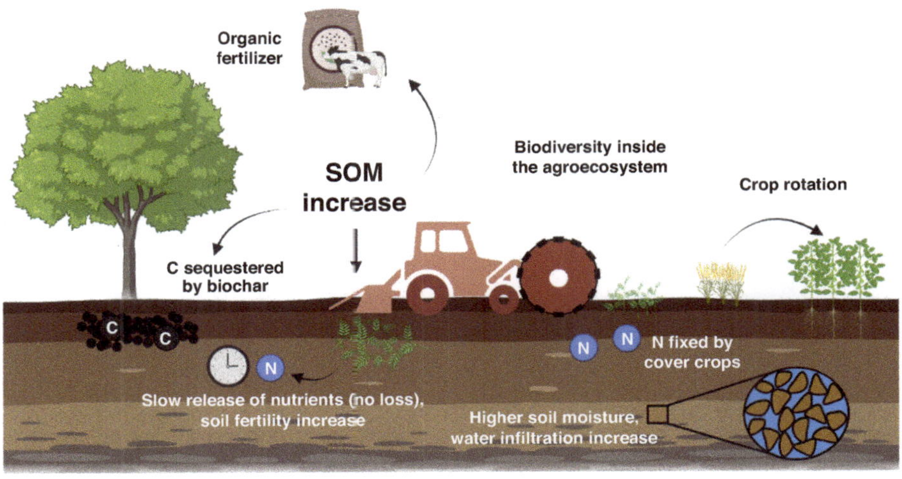

Fig. 45.2. Resilient factors related to organic agricultural systems.

sequestration is fundamental because it contributes to soil health, food security, agro-ecosystem equilibrium and climate change mitigation (Siddique *et al.*, 2024).

In organic farming, it is possible to store carbon in different ways. Moreover, it is possible to identify long-term and short-term carbon sequestration. Another important concept where it is possible to combine near carbon sequestration is the circular economy. It is defined, by the EU, as 'a system where products are reused, repaired, remanufactured or recycled' (COM, 2020/98). This concept can be used for recycling materials that in agriculture are considered wastes, for example pruning material, uprooted trees, urban green; all these materials can be used for innovative soil conditioner production. In fact, biochar is a natural product, applied as a soil amendment, that is based on a circular economy process; it is derived from a pyrolysis process (which takes place in the absence of oxygen at high temperatures) (Karim, 2018). Since it is a stable carbon-rich material, its use represents a long-term carbon pool over decades (Zhang *et al.*, 2022). Furthermore, carbon can be stocked in living biomass as a short-term carbon sequestration (Siddique *et al.*, 2024).

Organic farming, through the use of cover crops, crop plan diversification and organic soil amendments, contributes to maintaining a healthy soil system. These practices, when applied over the long term, lead, through increased soil organic matter (SOM), to greater carbon sequestration per unit area than conventional agricultural soils. In addition to storing more carbon in the soil, the use of organic farming practices directly and indirectly emit less carbon from fossil fuels to power machinery by not using synthetic fertilizers or pesticides to manage soil fertility and pests (Meemken and Qaim, 2018).

Towards a Resilient Organic Agriculture: An Essay of Bibliometric Analysis (2014–2024)

Nowadays, climate change and its impact on organic farming has gained interest in the agricultural research sector. Analyzing the papers and reviews (*c.*1880) published from 2014 to 2024, present in the database of Scopus, it is possible to make some reflections. For Cluster analysis (Fig. 45.3) 'VOS Viewer' software is used. VOS Viewer is a software tool which generates the maps based on link data

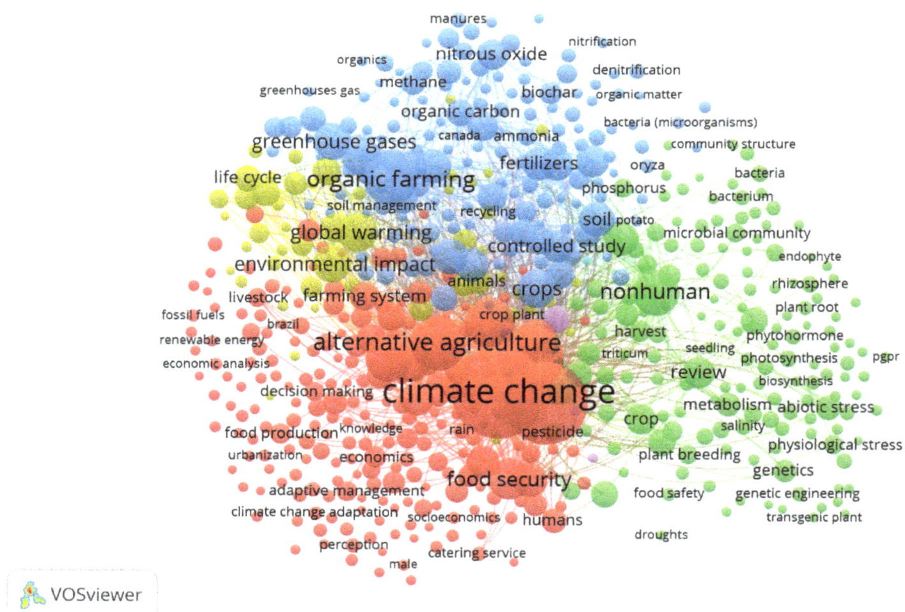

Fig. 45.3. Network of papers and reviews published from 2014 to 2024 stored in the database of Scopus in clusters of major keywords (data accessed 21 November 2023).

and for predicting and grasping these maps (Kwasi Bannor *et al.*, 2023). Cluster analysis shows the relationship between different keywords. In this case, the green cluster represents the relationship between fertilizer and plant growth, abiotic stress, crop and soil fertility. The blue cluster represents the relationship between organic farming and the N and C cycle (stock and emission). The yellow cluster represents the relationship between global warming and life cycle assessment, GHG emissions and environmental impact on crops. The red cluster represents the relationship between climate change and agricultural aspects (policy, pollution, production and management), water use and sustainable development.

China, the USA and India, which represent more than 40% of the global population in 2023, have the larger number of publications (296, 277 and 253, respectively) in the Scopus database (2014–2024) related to organic farming (Fig. 45.4). These countries are the leaders of climate change mitigation research by applying organic agriculture practices. Not only these countries but also European countries (where Italy is the leader – 168 publications) are strongly involved in reducing global warming effects by converting the actual conventional agricultural system into a more resilient agriculture applying the principles of organic agriculture.

Figure. 45.5 demonstrates that climate change is the most occurring keyword (922 times) in the publications on the Scopus database, followed by 'sustainable agriculture' (719 times), 'agriculture' (451 times), and 'alternative agriculture' (311 times).

Conclusion

One of the main effects of climate change is increased uncertainty, both in terms of weather pattern and the cultivation aspect of crops. It has been reported that much of the damage caused by climate change is brought by agriculture itself (emissions, overuse of pesticides and fertilizers), specifically by conventional agriculture. So new sustainable strategies such as organic farming should be adopted to mitigate the negative impact of climate change. Indeed, organically managed systems have the potential to make an important contribution in reducing GHGs emanating from

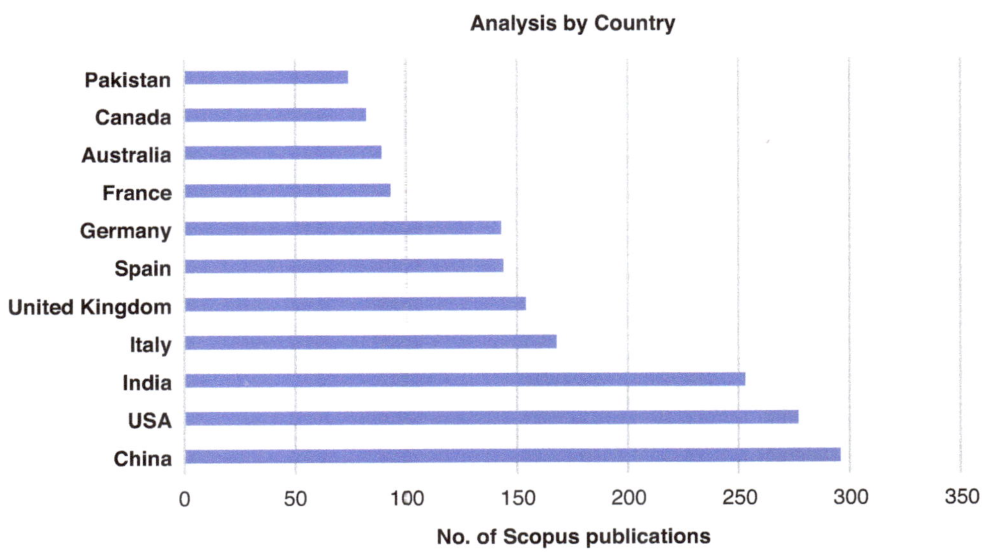

Fig. 45.4. Classification of papers and reviews (*c.* 1880) published from 2014 to 2024 stored in the database of Scopus by country (data accessed 21 November 2023).

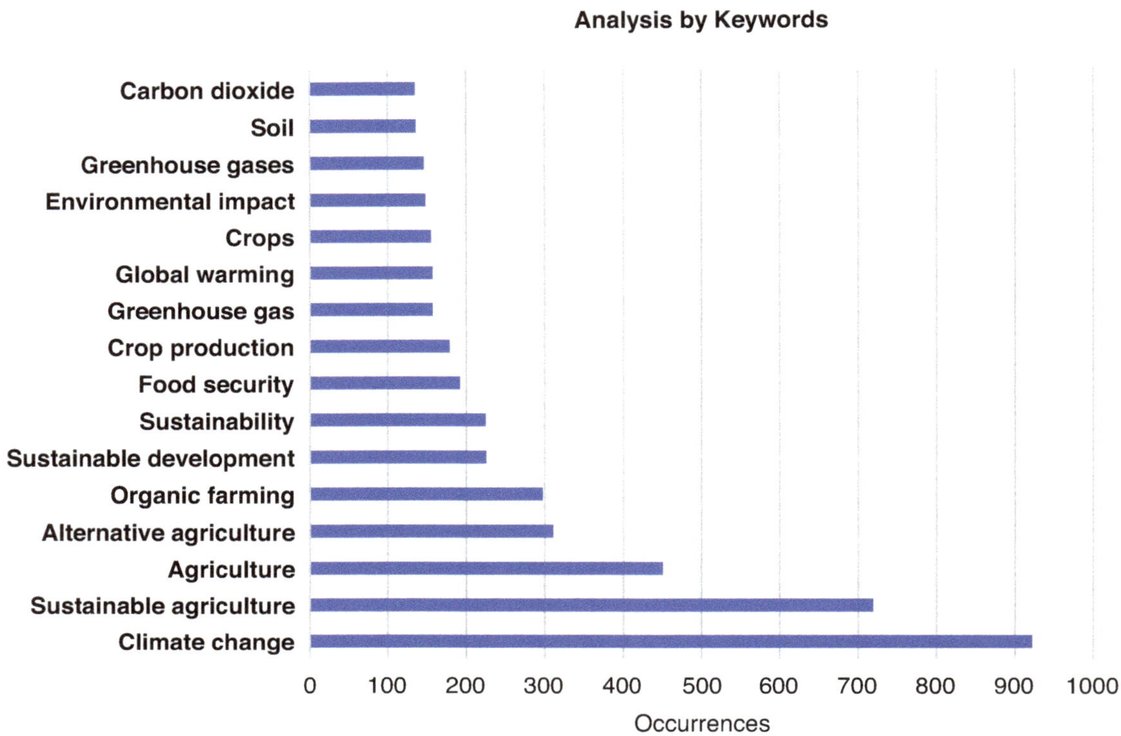

Fig. 45.5. Classification of papers and reviews (*c*.1.880) published from 2014 to 2024 stored in the database of Scopus by keyword occurrence (data accessed 21 November 2023).

agriculture. Therefore, organic farming has a strong potential to make the global agricultural sector resilient to face climate variability. Hence, a possible solution could be more land intended for organic farming. This would reduce the use of inputs that are the main cause of climate change. As a resilient system, organic farming is able to build more human capital than conventional systems, this leads to an improvement of household well-being, economy, technologies and infrastructure.

References

Aluja, M., Birke, A., Ceymann, M., Guillén, L., Arrigoni, E. *et al.* (2014) Agroecosystem resilience to an invasive insect species that could expand its geographical range in response to global climate change. *Agriculture, Ecosystems & Environment* 186, 54–63.

Avasiloaiei, D.I., Calara, M., Brezeanu, P.M., Gruda, N.S. and Brezeanu, C. (2023) The evaluation of carbon farming strategies in organic vegetable cultivation. *Agronomy* 13(9), 2406.

Bouwman, A.F., Beusen, A.H.W., Griffioen, J., Van Groenigen, J.W., Hefting, M.M. *et al.* (2013) Global trends and uncertainties in terrestrial denitrification and N_2O emissions. *Philosophical Transactions of the Royal Society of London. Series B, Biological Sciences* 368(1621), 20130112.

COM (2020/98) Communication from the commission to the European parliament, the council, the European economic and social committee and the committee of the regions. A new circular economy action plan: for a cleaner and more competitive Europe.

Cui, X. (2020) Climate change and adaptation in agriculture: Evidence from US cropping patterns. *Journal of Environmental Economics and Management* 101, 102306.

Dastogeer, K.M.G., Zahan, M.I., Rhaman, M.S., Sarker, M.S.A. and Chakraborty, A. (2022) Microbe-mediated thermotolerance in plants and pertinent mechanisms: A meta-analysis and review. *Frontiers in Microbiology* 13, 833566.

de la Cruz, V.Y.V., Cheng, W. and Tawaraya, K. (2023) Yield gap between organic and conventional farming systems across climate types and sub-types: a meta-analysis. *Agricultural Systems* 211, 103732.

Diacono, M., Persiani, A., Canali, S. and Montemurro, F. (2018) Agronomic performance and sustainability indicators in organic tomato combining different agro-ecological practices. *Nutrient Cycling in Agroecosystems* 112(1), 101–117.

Faiz, H., Khan, O., Ali, I., Hussain, T., Haider, S.T. *et al.* (2024) Foliar application of triacontanol ameliorates heat stress through regulation of the antioxidant defense system and improves yield of eggplant. *Brazilian Journal of Biology* 84, e253696.

FAO (1999) Report of fifteenth session of committee on agriculture. *Food and Agriculture Organization of the United Nations*.

Field, C.B., Barros, V.R., Dokken, D.J., Mach, K.J., Mastrandrea, M.D. *et al.* (2014) Climate change 2014: Impacts, adaptation, and vulnerability, part A: Global and sectoral aspects: contribution of working group II to the fifth assessment report of the Intergovernmental Panel on Climate Change [IPCC Summary for Policymakers. pp. 1–32.

Hakimovich, H.H. and Alishovich, K.B. (2023) Assessing the role of climate change in desertification processes. *Web of Technology: Multidimensional Research Journal* 1(4), 3–10.

IPCC (2013) *Climate Change: The Physical Science Basis. Contribution of Working Group I to the Fifth Assessment Report of the Intergovernmental Panel on Climate Change*. Cambridge University Press, Cambridge, UK and New York.

Jordan, R., Müller, A. and Oudes, A. (2009) *High Sequestration, Low Emission, Food Secure Farming. Organic Agriculture: A Guide to Climate Change & Food Security*. Available at: https://orgprints.org/id/eprint/16769/ (accessed 28 February 2025).

Karim, M.R. (2018) Prospects of biochar for clean energy, organic farming and climate change mitigation in Northern Bangladesh. In: Paredes, R. (ed.) *A Closer Look at Climate Change*. Nova Science Publishers, Inc, Hauppauge, New York, pp. 19–35.

Kwasi Bannor, R., Arthur, K.K., Oppong, D. and Oppong-Kyeremeh, H. (2023) A comprehensive systematic review and bibliometric analysis of food fraud from a global perspective. *Journal of Agriculture and Food Research* 14, 100686.

Lal, R. (2023) Carbon farming by recarbonization of agroecosystems. *Pedosphere* 33(5), 676–679.

Luo, Q. (2011) Temperature thresholds and crop production: A review. *Climatic Change* 109(3–4), 583–598.

Mahankale, N.R. (2024) Global influence of synthetic fertilizers on climate change. Applied Geomatics 16 [related article]. Available at: https://link.springer.com/10.1007/s12518-023-00511-0

Maschler, J., Bialic-Murphy, L., Wan, J., Andresen, L.C., Zohner, C.M. *et al.* (2022) Links across ecological scales: Plant biomass responses to elevated CO_2. *Global Change Biology* 28(21), 6115–6134.

Meemken, E.M. and Qaim, M. (2018) Organic agriculture, food security, and the environment. *Annual Review of Resource Economics* 10(1), 39–63.

Monteiro, A. and Santos, S. (2022) Sustainable approach to weed management: The role of precision weed management. *Agronomy* 12(1), 118.

Moriondo, M., Giannakopoulos, C. and Bindi, M. (2011) Climate change impact assessment: The role of climate extremes in crop yield simulation. *Climatic Change* 104(3–4), 679–701.

Müller, C. (2013) African lessons on climate change risks for agriculture. *Annual Review of Nutrition* 33(1), 395–411.

Nti, I.K., Zaman, A., Nyarko-Boateng, O., Adekoya, A.F. and Keyeremeh, F. (2023) A predictive analytics model for crop suitability and productivity with tree-based ensemble learning. *Decision Analytics Journal* 8, 100311.

O'Rourke, M.E. and Petersen, J. (2016) Reduced tillage impacts on pumpkin yield, weed pressure, soil moisture, and soil erosion. *HortScience* 51(12), 1524–1528.

Regulation (EU) (2018/848) European Parliament and of the Council of 30 May 2018 on organic production and labelling of organic products and repealing Council Regulation (EC) No 834/2007. Available at: https://agriculture.ec.europa.eu/farming/organic-farming/organic-production-and-products_en (accessed 28 February 2025).

Rehman, M.U., Rather, G.H., Gull, Y., Mir, M.R., Mir, M.M. *et al.* (2015) Effect of climate change on horticultural crops. In: Hakeen, K.R. (ed.) *Crop Production and Global Environmental Issues*. Springer International Publishing, Cham, Switzerland, pp. 211–239.

Reidsma, P., Accatino, F., Appel, F., Gavrilescu, C., Krupin, V. *et al.* (2023) Alternative systems and strategies to improve future sustainability and resilience of farming systems across Europe: From adaptation to transformation. *Land Use Policy* 134, 106881.

Shah, W.U.H., Lu, Y., Liu, J., Rehman, A. and Yasmeen, R. (2024) The impact of climate change and production technology heterogeneity on China's agricultural total factor productivity and production efficiency. *The Science of the Total Environment* 907, 168027.

Siddique, K.H.M., Bolan, N., Rehman, A. and Farooq, M. (2024) Enhancing crop productivity for recarbonizing soil. *Soil and Tillage Research* 235, 105863.

Skendžić, S., Zovko, M., Živković, I.P., Lešić, V. and Lemić, D. (2021) The impact of climate change on agricultural insect pests. *Insects* 12(5), 440.

Skinner, C., Gattinger, A., Krauss, M., Krause, H.-M., Mayer, J. *et al.* (2019) The impact of long-term organic farming on soil-derived greenhouse gas emissions. *Scientific Reports* 9(1), 1702.

Squalli, J. and Adamkiewicz, G. (2023) The spatial distribution of agricultural emissions in the United States: The role of organic farming in mitigating climate change. *Journal of Cleaner Production* 414, 137678.

Stevens, C.J., Dise, N.B., Mountford, J.O. and Gowing, D.J. (2004) Impact of nitrogen deposition on the species richness of grasslands. *Science (New York, N.Y.)* 303, 1876–1879.

Tijjani, S.B., Qi, J., Giri, S. and Lathrop, R. (2024) Crop production and water quality under 1.5°C and 2°C warming: Plant responses and management options in the mid-atlantic region. *The Science of the Total Environment* 907, 167874.

Toaima, W.I.M., Hamed, E.S. and Abd El-Aleem, W.H. (2023) Evaluation of Turkish oregano (*Origanum onites* L.) under organic farming system. *Plant Science Today* 10(2), 272–280.

Tuomisto, H.L., Hodge, I.D., Riordan, P. and Macdonald, D.W. (2012) Does organic farming reduce environmental impacts? A meta-analysis of European research. *Journal of Environmental Management* 112, 309–320.

Ume, C.O., Onah, T.L., Ezeibe, A.B., Onah, O.G., Opata, P.I. *et al*. (2023) Agroecology as a response to sustainable development under climate change in southeast Nigeria. In: Egbueri, J.C., Ighalo, J.O. and Pande, C.E. (eds) *Climate Change Impacts on Nigeria: Environment and Sustainable Development*. Springer, Cham, Switzerland, pp. 279–302.

Vaglia, V., Bacenetti, J., Orlando, F., Alali, S., Bosso, E. *et al*. (2022) The environmental impacts of different organic rice management in Italy considering different productive scenarios. *The Science of the Total Environment* 853, 158365.

Wittwer, R.A., Klaus, V.H., Miranda Oliveira, E., Sun, Q., Liu, Y. *et al*. (2023) Limited capability of organic farming and conservation tillage to enhance agroecosystem resilience to severe drought. *Agricultural Systems* 211, 103721.

Zhang, Y., Niu, H. and Yu, Q. (2021) Impacts of climate change and increasing carbon dioxide levels on yield changes of major crops in suitable planting areas in China by the 2050s. *Ecological Indicators* 125, 107588.

Zhang, Y.L., Xie, H.T., Wang, F.P., Sun, C. and Zhang, X.D. (2022) Effects of biochar incorporation on soil viable and necromass carbon in the luvisol soil. *Soil Use and Management* 38(1), 318–330.

Zhao, C., Liu, B., Piao, S., Wang, X., Lobell, D.B. *et al*. (2017) Temperature increase reduces global yields of major crops in four independent estimates. *Proceedings of the National Academy of Sciences of the United States of America* 114(35), 9326–9331.

46 Transformation of the Agrifood System as an Urgent Need in the Face of Climate Change

Maria Claudia Dussi*

Department of Agricultural Sciences, National University of Comahue, Rio Negro, Patagonia, Argentina

Abstract

Depletion of natural resources, pollution of watercourses, land-use change, destruction of fertile soils, loss of biodiversity, loss of agrobiodiversity, and contamination by agrochemicals are some of the impacts directly caused by industrial agriculture. Due to the amount of greenhouse gas emissions generated on farms and agricultural land, it contributes significantly to climate change. Global emissions from agriculture and associated land use currently account for around one-fifth of total emissions from all economic activities. The new paradigm for feeding the world requires an agroecological approach based on the right of all people to food and the recognition of the social function of the soil. Agroecology is conceived as a holistic model of global change that encompasses technical, social, organizational and political dimensions.

The transformative process of the entire agri-food system is a complex and multi-faceted challenge, including its perspectives on equity, justice, and access, and implies the redesign of the food system and the integration of both horizontal and vertical diversification of production systems within sustainable food systems. This implies that the ways of studying, designing and evaluating agro-ecosystems will need to change considering aspects such as footprint/biocapacity, carbon and water footprinting, agro-ecosystem diversity and technology, energy efficiency, multi-functionality of agro-ecosystems, ecological economics, human values (commitment, ethics, dignity and respect), social organization, environmental costs, food sovereignty, among others. In summary, rethink the sense of development. This requires the collective will to change.

Keywords: agroecology, energy efficiency, food sovereignty, sustainability

Introduction

Agriculture is a significant contributor to climate change due to the amount of greenhouse gas (GHG) emissions generated within the farm and on agricultural land. Global emissions from agriculture and associated land use currently represent about one-fifth of the total from all economic activities.

However, the impact of the entire food system on climate change is much greater after incorporating pre- and post-production activities along the supply chain, retail, consumption and waste disposal phases. After accounting for all food system activities, food system emissions may be as large as 25–30% of total anthropogenic emissions (10–12% from crop and livestock activities on farms, 8–10% from land use and land-use change, and 5–10% from the supply chain) (Willett *et al.*, 2019; Tubiello *et al.*, 2021).

Total GHG emissions from the food system were about 16 CO_2eq. yr^{-1} in 2018, or one-third of the global anthropogenic total. Three-quarters of these emissions, 13 Gt CO_2eq. yr^{-1}, were generated either within the farm gate or in pre- and post-production activities, such as manufacturing, transport, processing and waste disposal. The

*Corresponding author: mariaclaudiadussi@gmail.com

 Advances in Organic Farming (ed. L. P. Awasthi)
DOI: 10.1079/9781800626850.0046

remainder was generated through land-use change at the conversion boundaries of natural ecosystems to agricultural land (Tubiello *et al.* (2021).

Land-use patterns in general reveal the importance of agriculture as a major land management system transforming and making use of natural ecosystems. More than half of the Earth's land surface is intensively used for agricultural purposes such as cultivation, grazing, plantation forestry and aquaculture; and since 1950, one-third of the soil has been profoundly altered from its natural ecosystem state because of moderate to severe soil degradation (IAASTD, 2009). Between 2000 and 2010, agriculture drove 80% of deforestation worldwide. Agriculture uses 86% of anthropogenic nitrogen and half the nitrogen applied to soils is not taken up by plants.

The adverse effects of climate change are increasingly visible and dangerous around the world. Experts point out that in the coming decades, climate change will increase in all regions and that global warming of 1.5°C between 2030 and 2052 will be related to an increase in heatwaves, longer hot seasons and shorter cold seasons (IPCC, 2022). Published results indicate that the impacts of climate change are significant, however, with a wide projected range (between 5 million and 170 million additional people at risk of hunger by 2080), disrupting the production and distribution, access, and stability of the food supply (Schmidhuber and Tubiello, 2007).

Zhao *et al.* (2017), compiling extensive published results from four analytical methods, found that each degree-Celsius increase in global mean temperature would, on average, reduce global yields of wheat by 6.0%, rice by 3.2%, maize by 7.4% and soybean by 3.1%, four crops that provide two-thirds of human caloric intake.

The Intergovernmental Science-Policy Platform on Biodiversity and Ecosystem Services (IPBES, 2019) global assessment warned that biodiversity is declining faster than at any time in human history. Food systems are responsible for around 60% of global terrestrial biodiversity loss. At the same time, one-third of all food goes to waste between the points of production and consumption, while around 11% of the world's population is undernourished and 39% are overweight or obese (FAO, IFAD, UNICEF, WFP and WHO, 2019). There is a close link between human health, plant, soil, animal and ecological health.

Local, national and global agri-food systems have, and will have in the future, increasingly complex challenges such as adapting to and mitigating climate change, conserving natural resources, reducing food losses, providing enough nutritious food for a continuously growing world population, increasing healthy diets, and ending social injustice and cultural erosion.

The challenges related to hunger eradication will vary in different countries in 2030, the average global availability of food per person will grow by 4%, reaching little more than 3025 kcal/day (OECD/FAO, 2021). Although this global average will be highly inequitable, consumers in middle-income countries will increase their food intake significantly, while the diet in low-income countries will remain about the same. In sub-Saharan Africa, where 224.3 million people were undernourished in 2017–2019, per capita daily caloric availability is projected to increase by only 2.5% over the next decade, to 2500 kcal in 2030 (OECD/FAO, 2021). Over the next decade, world agricultural production is projected to increase by 1.4% per year, with the additional amount coming mostly from emerging economies and low-income countries.

The OECD/FAO report (2021) forecasts that global GHG emissions from agriculture will increase by 4% over the next ten years, with livestock accounting for more than 80% of this increase. Therefore, the agricultural sector will be required to adopt additional policy measures to contribute effectively to the global reduction of GHG emissions.

Land-use changes, destruction of the fertile layer, depletion of natural resources, contamination of water courses, loss of agrobiodiversity, loss of biodiversity and contamination by agrochemicals are some of the effects generated directly by industrial agriculture (IPBES, 2018), to which we must add the serious problems generated towards the health of the population derived from the agrifood system that combines products from the agriculture that does not feed, with an intense use of salts, sugars and fats (Pengue, 2021). This system in many cases has not nourished people, and its health consequences are as serious or more serious than the environmental, cultural and even economic damage caused by this intensive production model.

The world faces a polycrisis of a systemic nature: energy scarcity, water scarcity, environmental degradation, biodiversity loss, climate change, economic inequality, food insecurity, military conflicts, and other problems affecting the planet. These problems cannot be addressed in isolation because they are interconnected and interdependent. When one of the problems worsens, the effects spread throughout the system, and exacerbate the other problems. The causes of pandemics are the same global economic forces and environmental changes that are driving biodiversity loss and climate change. This accentuates the importance of thinking from a holistic and systemic approach when we refer to the agrifood system.

Steiner *et al.* (2020) set out four key action areas necessary for food system reconfiguration: (i) reroute farming and rural livelihoods to new trajectories that conserve ecosystems and the natural resource base, reduce social inequality, and provide food and nutrition security for all; (ii) de-risk livelihoods, farms and value chains, allowing food system actors to anticipate, respond to and recover from increasingly frequent and extreme weather events; (iii) reduce emissions from diets and value chains, targeting health and climate outcomes; produce, process, distribute and consume food in ways that lower GHGs and reduce food loss and waste; (4) realign policies, finance, support to social movements, and innovation systems to facilitate action in the areas mentioned. This action area cuts across the other three, as policy, finance, innovation and social change are integral parts of those action areas. Loboguerrero *et al.* (2020) highlight food system levers that, along with activities within these four action areas, may shift food systems towards more sustainable, inclusive, healthy and climate-resilient futures.

Within these four action areas are 11 transformative actions that were identified through consultations with over 100 partners, individuals and organizations and through a series of background science papers.

Thornton *et al.* (2023) estimated the annual cost of implementing these 11 actions stated by Steiner *et al.* (2020) to be US\$ 1.3 ± 0.1 trillion. Half of this is needed to halt conversion of forests and peatlands for agriculture, with the remainder used to reduce producer risk, lower emissions, and strengthen the policy, finance and innovation enablers of change. The authors stated that this cost is equivalent to less than 7% of the negative externalities generated annually by current food systems and that the costs of inaction will far outweigh the benefits.

The environmental crisis is determined as a serious problem resulting from human activities and can be analyzed according to Ecological Economics (EE). This current of thought is constructed as a critique of neoclassical-Keynesian environmental economics, incorporates the laws of thermodynamics into the analysis of the economic process, and gives nature a value *per se* (Foladori, 2005). EE has the characteristic of being transdisciplinary. This is because, when addressing the relationship between natural ecosystems and the economic system, it requires the intervention not only of economists but also of professionals who study natural sciences, sociologists, and other disciplines. This multidisciplinary characteristic is especially essential when you want to study environmental issues and understand the 'finitude of planet Earth' (Dussi and Flores, 2018).

EE thinks of the economy as an open process within a larger system that is planet Earth where biogeochemical cycles must be considered. Ecosystems are not only a source of resources for economic activity, but they also fulfil different functions for society and in turn are affected by human activities. In addition, there are non-renewable natural resources, and the Earth is a system closed in materials and open in solar energy. Therefore, economic growth cannot be unlimited since it would at some point be made impossible by the physical limits of the ecosystem rather than by economic reasons.

For example, the very high extraction of nutrients that some soils have is an environmental liability not included in the global transaction accounts of world agricultural materials and this leaves a cost in degradation, pollution and decrease in soil quality. In general, agricultural monoproduction systems suffer from a selective extraction of nutrients from the soil that depletes it and forces replacement via mineral fertilizers that on the one hand recover the current fertility but on the other hand lead to increasing levels of pollution and eutrophication, and generate greater external dependence for many countries by having to import increasing amounts of mineral fertilizers at dollar value, increasing the indebtedness of nations. Added to this is the leaching process and erosive processes due to poor soil management (Pengue and Feinstein, 2013).

Georgescu–Roegen (1971), in his work *The Law of Entropy and the Economic Process*, states that the economic process is neither circular nor mechanical but unidirectional. His view of the economy is as a low- to high-entropy thermodynamic system. The presence of life increases entropy.

In summary, EE broadens the focus of the different modes of valuation so that not only economic conditions but also issues such as social metabolism, biophysical indicators (nutrients, virtual soil, virtual water, net primary appropriation of biomass), energy consumption trends, natural degradation and pollution are included in the analysis. There are then clear links between the study of ecological distribution, consumerism, climate change and the study of the carrying capacity of humans on Earth.

The systemic approach adopted by TEEB (2018), analyses food value chains in their entirety and demonstrates that there are important, although economically invisible (i.e. unrelated to the market) reserves and flows that must also be considered. Although these reserves and flows may not have a specific price and have not been incorporated into macroeconomic models or the calculation of gross domestic product (GDP), they are undoubtedly real reserves and flows that can be observed, described and measured, and are, in fact, important factors driving the success (or failure) of many of the Sustainable Development Goals (SDGs), since the agrifood value chain significantly affects the climate (SDG 13), freshwater (SDG 6), biodiversity and ecosystems (SDG 14 and 15), human health (SDG 3), social equity (SDG 5 and 10) and livelihoods (SDG 1 and 8).

The TEEB AgriFood Evaluation Framework is based on three guiding principles: universality, comprehensiveness and inclusiveness. These principles translate into the design and approach of a framework that can truly represent a comprehensive perspective of any food system. They strengthen the framework by recognizing and valuing the functions of the four forms of capital reserves (produced, natural, human, and social capital) used in the agri-food systems (TEEB, 2018).

The characteristics of the agriculture of the future should therefore consider oil-independent agricultural models, agro-ecosystems with low environmental impact that are resilient to climate change, multi-functional agriculture (economic, social and environmental services) and local food systems, among other aspects.

Demand of consumers for organic products is increasing in the world. More farmers grow organically, more land is certified organic, and 186 countries report organic farming activities. Globally, 1.5% of the farmland is organic. The countries with the most organic agricultural land are Australia (35.7 million ha), Argentina (3.6 million ha) and China (3.1 million ha) (Willer *et al.*, 2020). The ten countries with the largest organic agricultural areas represent 74% of the world's organic agricultural land. World organic food market reached more than €95 billion in 2018. The USA is the leading market with €40.6 billion, followed by Germany (€10.9 billion) and France (€9.1 billion) (Willer *et al.*, 2020).

Organic food production allows rural workers to work in a healthy environment and consumers to eat food with no chemical residues. It can also be observed that organic or ecological agriculture that maintains monocultures depends on external biological and/or botanical inputs, which substitute for chemical inputs. 'Input substitution' is not based on agroecological principles; it essentially follows the same paradigm as conventional agriculture, i.e. overcoming the limiting factor, but this time with biological or organic inputs. Many of these 'alternative inputs' have become commodified; therefore, farmers continue to depend on suppliers, co-operatives or companies. According to FiBL, the world's smallholder organic producers are in low- and middle-income countries, for whom individual organic certification would be unaffordable and administratively too complex to manage (Willer *et al.*, 2020).

Therefore, in organic agriculture the process remains halfway by not promoting with the same commitment the development of fairer trade, the massiveness of production to reach more people and markets, especially the poorest, and a permanent demand of farmers in the world on access to natural resources such as soil or water.

In summary, organic and agroecological agriculture are differentiated not only by their production process and certification system, but also by the fact that the former is an agriculture of input substitution, while agroecology responds strongly to process agriculture and the construction of a strong productive diversity of both plants and animals, with a socio-ecological approach (Pengue, 2021).

The ETC Group (2017) calculates that smallholder farmers in the world (peasant food network), which make up 80% of the total farm numbers, produce, using agroecological methods, 70% of the food available for human consumption (measured in calories and weight) and use less than 25% of the world's agricultural land. In turn, they use approximately 10% of the fossil energy and no more than 20% of the water that all agricultural production demands.

In contrast, the agri-business chain uses more than 75% of the world's agricultural land, provides food for only 30% of the world's population, and is responsible for the consumption of almost 90% of the fossil fuels used in agriculture (and consequently the corresponding emissions of greenhouse gases), as well as at least 80% of the fresh water (ETC, 2017). This type of industrial agriculture produces vulnerable, genetically homogeneous and ecologically narrow monocultures, which are dependent on large quantities of agrochemical inputs that are ultimately injected into the biosphere annually.

The new paradigm to feed the world requires an agroecological approach based on the right to food of all people and the vindication of the social function of the land. The conversion of a specialized agricultural system to an agroecological system follows three principles: diversification (by including different species of crops, trees and animals), integration (by the dynamic exchange and recycling of energy and nutrients between the components of the system), and the achievement of food self-sufficiency. It is not just about replacing industrial inputs with others with low environmental impact, it is about reducing the number of inputs used per product obtained (Fig. 46.1).

In its second international symposium on agroecology 'Scaling Up Agroecology to Achieve the Sustainable Development Goals', in 2018, the FAO reinforced the interest of agroecology to enhance smallholder' and family farmers' adaptation and resilience to the impacts of climate change, to improve food security and nutrition through healthy food and diversified diets, to protect and enhance agrobiodiversity in support of ecosystem services such as pollination, soil health and recovery of degraded lands and forests, to improve livelihoods in rural areas, and to achieve a transformative change in agricultural practices towards sustainable development. These statements are reinforced by the contribution to this debate of the High-Level Panel of Experts (HLPE) on food security and nutrition (FSN) that produce a report on 'Agroecological approaches and other

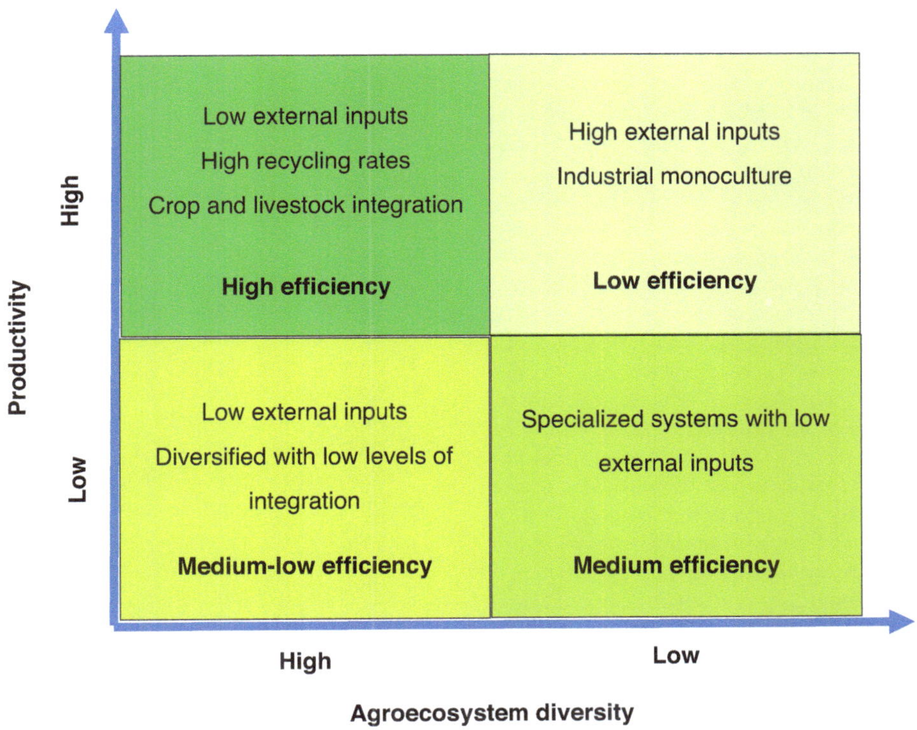

Fig. 46.1. Characteristics of the agro-ecosystems of the future: productivity, diversity, integration and efficiency. (From Dussi and Flores (2018), modified from Altieri *et al.*, 2012)

innovations for sustainable agriculture and food systems that enhance food security and nutrition' (https://www.fao.org/agroecology/database/detail/en/c/1242141/).

Thousands of agroecological initiatives around the world that revitalize traditional and peasant farming systems that have stood the test of time enhance food sovereignty while contributing to the conservation of biodiversity at farm and landscape scales. Increased plant species diversity and genetic diversity increase the overall resilience of food systems to new environmental changes and climate fluctuations (Dussi and Lauri, 2023). These farmers opt for intercropping and/or silvopastoral systems, agroforestry practices as well as other diversified agricultural approaches that incorporate genetic diversity in their crops. This choice not only improves the biodiversity of their land but also revitalizes the health of the soils and strengthens the resilience of their agricultural systems in the face of extreme weather events.

Agroecology as a Model for AgriFood Transformation

Agroecological transformation refers to multidimensional and fundamental transformational processes through which the established sociotechnical systems shift to sustainable production and consumption systems, which also involves changes in consumption practices, policies, cultural meanings, infrastructure and business models.

The concept of agroecology was proposed over a century ago as a framework for more sustainable farming systems. Agroecology seeks to reduce our dependence on fossil fuels (gasoline, gas), reduce pollution, and mitigate the effects of climate change. It is based on principles that, collectively, aim to increase the autonomy and the resilience of the agro-ecosystem: recycling biomass, enhancing soil biotic activity through better management of organic matter, minimizing nutrient losses from the agro-ecosystem, increasing energy efficiency, increasing species and within-species diversity, and enhancing beneficial interactions and synergisms within agro-ecosystems (Altieri, 2015).

Agroecology is a scientific discipline that involves the holistic study of agro-ecosystems and food systems; a set of principles and practices that improves the resilience and durability of food and agricultural systems while preserving social integrity; and a socio-political movement, which focuses on the practical application of agroecology, pursues new ways of considering agriculture, food processing, distribution and consumption, and its relationships with society and nature (Wezel *et al.*, 2009; CIDSE, 2018) (Fig. 46.2).

There is an interdependence of agroecology and food sovereignty. The declaration of food sovereignty made in the international forum for agroecology at Nyéléni, Mali, in 2015 defines agroecology as a movement led by people, and a practice that needs to be supported, rather than directed, by science and policy.

Agroecology is conceived as a holistic model of global change that includes technical, social, organizational and political dimensions. Its principles are a set of general guidelines that constitute the fundamental pillars of agroecology, its practice and implementation. They are encompassed in four dimensions (CIDSE, 2018; Dussi, 2022).

The first pillar is the *environmental* dimension of agroecology with the following principles:

- Agroecology enhances positive interaction, synergy, integration and complementarities between the components of agro-ecosystems (plants, animals, trees, soil, water etc.) and food systems.
- Agroecology increases biomass recycling, optimizing organic matter decomposition and nutrient cycling over time and promotes the biological functioning of the soil.
- Agroecology optimizes and maintains biodiversity above- and below-ground (a wide range of species and varieties, genetic resources, locally adapted varieties/breeds etc.) over time and space (at plot, farm and landscape level).
- Agroecology eliminates the use of and dependency on external synthetic inputs by enabling farmers to control pests and weeds, and improve fertility through ecological management.
- Agroecology supports climate adaptation and resilience while contributing to GHG mitigation (reduction and sequestration) through lower use of fossil fuels and higher carbon sequestration in soils.

The second pillar is the *social and cultural* dimension of agroecology with the following principles:

- Agroecology is rooted in the culture, identity, tradition, innovation and knowledge of local communities.

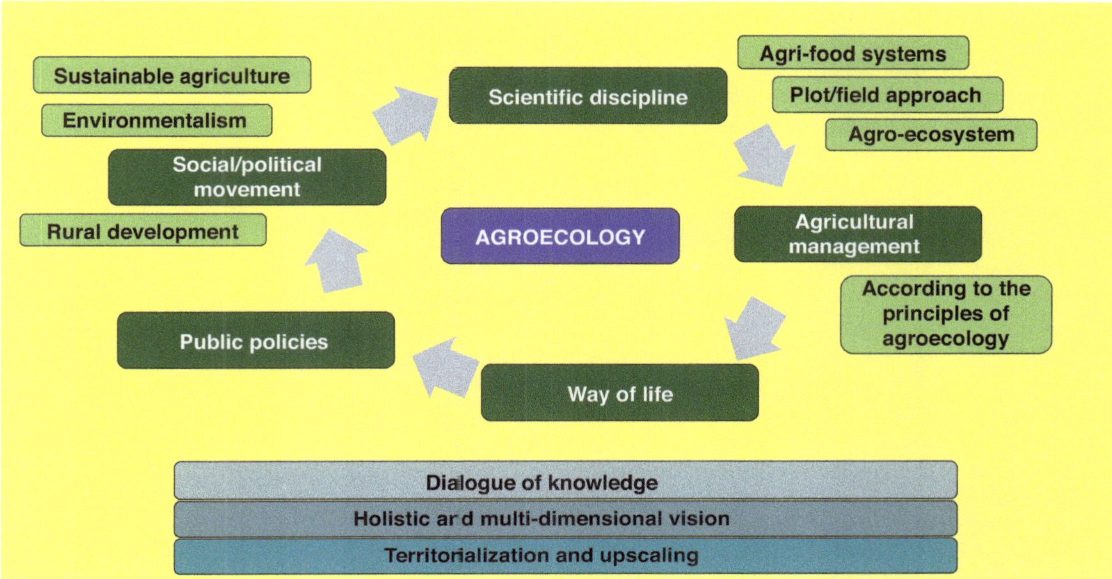

Fig. 46.2. Agroecology definitions. (Adaped from Wezel *et al.*, 2009; Dussi and Flores, 2018)

- Agroecology contributes to healthy, diversified, seasonally and culturally appropriate diets.
- Agroecology is knowledge-intensive and promotes horizontal (farmer-to-farmer) contacts for sharing of knowledge, skills and innovations, together with alliances giving equal weight to farmer and researcher.
- Agroecology creates opportunities for and promotion of solidarity and discussion between and among culturally diverse peoples (e.g. different ethnic groups that share the same values yet have different practices) and between rural and urban populations.
- Agroecology respects diversity between people in terms of gender, race, sexual orientation and religion, creates opportunities for young people and women and encourages women's leadership and gender equality.
- Agroecology supports peoples and communities in maintaining their spiritual and material relationship with their land and environment.
- Agroecology does not necessarily require expensive external certification as it often relies on producer–consumer relations and transactions based on trust, promoting alternatives to certification such as PGS (participatory guarantee system) and CSA (community-supported agriculture).

The third pillar is the *economic* dimension of agroecology with the following principles:

- Agroecology promotes fair, short distribution networks rather than linear distribution chains and builds a transparent network of relationships (often invisible in formal economy) between producers and consumers.
- Agroecology helps provide livelihoods for peasant families and contributes to creating stronger local markets, economies and employment.
- Agroecology is built on a vision of a social and solidarity economy.
- Agroecology promotes diversification of on-farm incomes giving farmers greater financial independence, increases resilience by multiplying sources of production and livelihood, promoting independence from external inputs and reducing crop failure through its diversified system.
- Agroecology encourages food producers to sell their product at fair prices and actively respond to the demand of nearby markets.
- Agroecology increases community autonomy by enhancing livelihoods and dignity.

The fourth pillar is the *political* dimension of agroecology with the following principles:

- Agroecology prioritizes the needs and interests of small-scale food producers who supply most of the world's food (ETC, 2017).
- Agroecology establishes control of seed, biodiversity, land and territories, water, knowledge and the commons into the hands of the people who are part of the food system and so achieves better-integrated resource management.
- Agroecology requires a set of supportive, complementary public policies, supportive policymakers and institutions, and public investment to achieve its full potential.
- Agroecology encourages forms of social organization needed for local adaptive management of food and agricultural systems and decision making. It also incentivizes the self-organization and collective management of groups and networks at different levels, from local to global (farmers' organizations, consumers, research organizations, academic institutions, etc.).

Scales and dimensions of agroecological research range from plot and field scale to agro-ecosystem and farm scales and expand into the full dimensions of the food system. 'Massification', 'scaling', 'expansion', 'amplification' or 'territorialization' of agroecology are defined as the process that leads an increasing number of families to practise agroecology in increasingly larger territories, and that involves more people in the processing, distribution and consumption of agroecologically produced food. Scaling means that a greater fraction of the population, both urban and rural, can produce and access healthy, nutritious, diverse food that is environmentally compatible and culturally appropriate (Mier y Terán Giménez Cacho *et al.*, 2018).

A study group on the massification of agroecology, analyzing emblematic cases of its expansion worldwide, identified eight key drivers of the process of taking agroecology to scale: (i) recognition of a crisis that motivates the search for alternatives; (ii) social organization; (iii) constructivist learning processes; (iv) effective agroecological practices; (v) mobilizing discourses; (vi) external allies; (vii) favourable markets; and (viii) favourable policies (Mier y Terán Giménez Cacho *et al.*, 2018). A more detailed understanding is needed on how these multiple dimensions interact with, reinforce and generate positive feedback with each other to make the territorial expansion of agroecology possible.

The transformative process of the entire food system includes its perspectives on equity, justice and access, and implies the redesign of the food system and the integration of both horizontal and vertical diversification of production systems within sustainable food systems (Gliessman, 2016).

Agro-ecosystems are considered open systems, which means that they receive inputs of energy, materials, capital and information, and produce results of the same nature in the form of products, services and externalities. They are influenced by external factors, which may be biophysical (such as rainfall) or socio-economic (such as market prices), and operate within a specific context, which may include political, demographic and other relevant factors. Agro-ecosystems often consist of nested subsystems, such as agricultural, livestock and/or forestry subsystems, which have panarchic effects of functioning as a whole (Tittonell, 2023).

Climate change is pushing the transformation of agro-ecosystems towards multi-functional systems. This implies that the ways of studying and evaluating agro-ecosystems will need to have another form of assessment that includes indicators like footprint/biocapacity, carbon and water footprinting, agro-ecosystem diversity and technology, energy efficiency, multi-functionality of agro-ecosystems, ecological economics, human values (commitment, ethics, dignity and respect), social organization, environmental costs, food sovereignty, among others, and, in summary, to rethink the sense of development (Dussi *et al.*, 2019).

For example, Dussi *et al.* (2020, 2022) studied energy efficiency in fruit agro-ecosystems in the production area of north Patagonia, Argentina. When comparing an organic-biodynamic (OB) agro-ecosystem with a conventional (C) one, typical of this region, the energy input was 21,578.49 and 53,517.83 $MJ.ha^{-1}$ in the OB and C orchards, respectively. Regarding the energy indicators, energy efficiency in the OB system was 2.85 and 1.44 in C; and specific energy was 0.94 and 1.86, respectively (Dussi, 2022). These results imply that the orchard under conventional agriculture (C) consumed more energy to obtain the same unit of product (Table 46.1).

Direct GHGs obtained from OB production were 1659.56 kg CO_2eq. per ha, while those generated in system C were 4080 kg CO_2eq. (Dussi *et al.*, 2022). Optimizing energy use through appropriate management of fertilization based on the actual needs of the crop, pest control based on the biological cycles of pests and predators, and a well-organized

Table 46.1. Energy efficiency and specific energy for the yield obtained from one hectare of Red Delicious apple trees cv. 'Top Red' in an organic-biodynamic (OB) and a conventional (C) agro-ecosystem characteristic of the Patagonia region of Argentina. (Adapted from Dussi *et al.*, 2022)

Agro-ecosystem	Energy efficiency energy output ($MJ.kg^{-1}$)/ energy input ($MJ.kg^{-1}$)	Specific energy ($MJ.ha^{-1}$)/ ($kg.ha^{-1}$)	Apple yields ($kg.ha^{-1}$)	Energy input ($MJ.ha^{-1}$)
OB	2.85	0.94	22,800	21,578.49
C	1.44	1.86	30,000	53,517.83

Table 46.2. Energy efficiency, specific energy and net energy values for the production of 1 ha of pear trees. (Adapted from Dussi *et al.*, 2021)

Energy efficiency	Specific energy	Net energy
$\frac{2.44\ \text{MJ kg}^{-1}\ \text{(energy output)}}{0.79\ \text{MJ kg}^{-1}\ \text{(energy input)}}$	$\frac{16769.79\ (\text{MJ ha}^{-1})}{21000\ (\text{kg ha}^{-1})}$	51240 ($MJ\ ha^{-1}$) – 16701.67 ($MJ\ ha^{-1}$)
3.08	0.79	34,538.33

harvest are key aspects of reducing the consumption of energy subsidies and thus GHG emissions.

The carbon footprint of the OB orchard was 0.073 kg CO_2eq. kg^{-1} apple while that of the C was 0.136 kg CO_2eq. kg^{-1} apple. This implies that OB fruit production has half the carbon footprint of typical C production in the region under study (Dussi *et al.*, 2018, 2022). These authors also found that of the total GHG emissions generated in the orchards, 41% corresponded to cultural practices carried out during the spring, 36% during the summer and the remaining 23% during the winter. When the highest GHG emissions occurred, the capture of CO_2 by the orchard increased. In this sense, it is important to reassess the ecosystem functions provided by the orchards by capturing atmospheric CO_2. This becomes more important if the total area planted in the different regions of the world is also considered. To evaluate an agro-ecosystem performance, not only the contamination generated by this agricultural activity should be considered, but also the capture of CO_2 that they provide.

The photosynthetic activity carried out by fruit trees grown in temperate climate regions functioning as a 'deciduous fruit forest' is an important regional potential in terms of atmospheric carbon sequestration. Each fruit tree has potential as a carbon sink in the medium term, which allows for part of the emissions to be used by the plants themselves through the process of photosynthesis since the photosynthetic peaks coincide with the emission peaks (Dussi *et al.*, 2022).

Another example of energy flow in fruit tree agro-ecosystems was presented by Dussi *et al.* (2021). These authors, working in a 40-year-old pear agro-ecosystem, analyzed all the cultural labours carried out and input used to represent the flow of income and the output flow constituted by the energy contained in the harvested product with commercial quality. Table 46.2 shows the values of energy efficiency, specific energy and net energy obtained in the study done.

The energy indicators show a favourable relationship energy input vs energy output. However, the high percentage of industrial energy used in the fruit system under analysis reflects a high dependence on energy subsidies. To reduce this dependence, it is essential to regulate the different activities that are carried out in the fruit-productive systems through the application of appropriate technologies; that is, reduce the use of industrial energy, especially that which comes from non-renewable or polluting sources, such as fuel; use tillage systems that require less use of machinery; irrigate efficiently; and use alternative sources of cultural-industrial energy, such as photovoltaic systems and wind turbines (Dussi *et al.*, 2021).

Energy flow analysis allows the evaluation of agricultural practices, identifying those aspects to be improved. Agricultural sustainability is achieved, among other factors, through adequate knowledge of the ecological processes that occur at farm level and in their context. With these bases, socio-economic changes that promote sustainability in all sectors of the agri-food system can be made.

Energy efficiency and carbon footprint are key indicators to identify cultural practices with excessive energy demand, therefore with high GHG emissions, and link them with the energy subsidy requirements for each one. The identification of these cultural practices allows for establishing the bases to elaborate mitigation strategies and obtain a system capable of producing better and impacting less on the environment, towards sustainable agriculture.

These ways to study and analyze agro-ecosystems should enable future professionals in the agricultural sciences to integrate the different dimensions of sustainability and to apply new solutions to address the climatic, economic and environmental conflicts that are present in current agro-industrial farming practices (Dussi, 2019a).

Agroecological transformation is a gradual process of co-innovation for biological input integration, adjustments and technological changes in the design and management of agro-ecosystems, which leads to leaving behind the conventional focus on agricultural production. At farm scale this transformation can be diagrammed as seen in Fig. 46.3. The spiral in the centre of the figure indicates that the process is continuously increasing, leading to greater complexity of the farm, and thus increasing its stability and resilience to climatic or economic changes. These types of farms are called 'agroecological lighthouses'; they are diversified farms that serve as models on how to 'design and manage farms based on agroecological principles' (Nicholls and Altieri, 2018; Flores *et al.*, 2022; Dussi *et al.*, 2023).

Although agroecological transformation can be initiated for different reasons, which are particular to different contexts, the degradation of natural resources and food sovereignty are driving factors.

They are the farmers of the lighthouse farms that play a key role in promoting agroecological principles (Nicholls *et al.*, 2020) in the surrounding community and to farmers in other regions. These lighthouse farms succeed in establishing and maintaining networks for knowledge and experience exchange that bring about change in the territories. They also promote agroecological principles

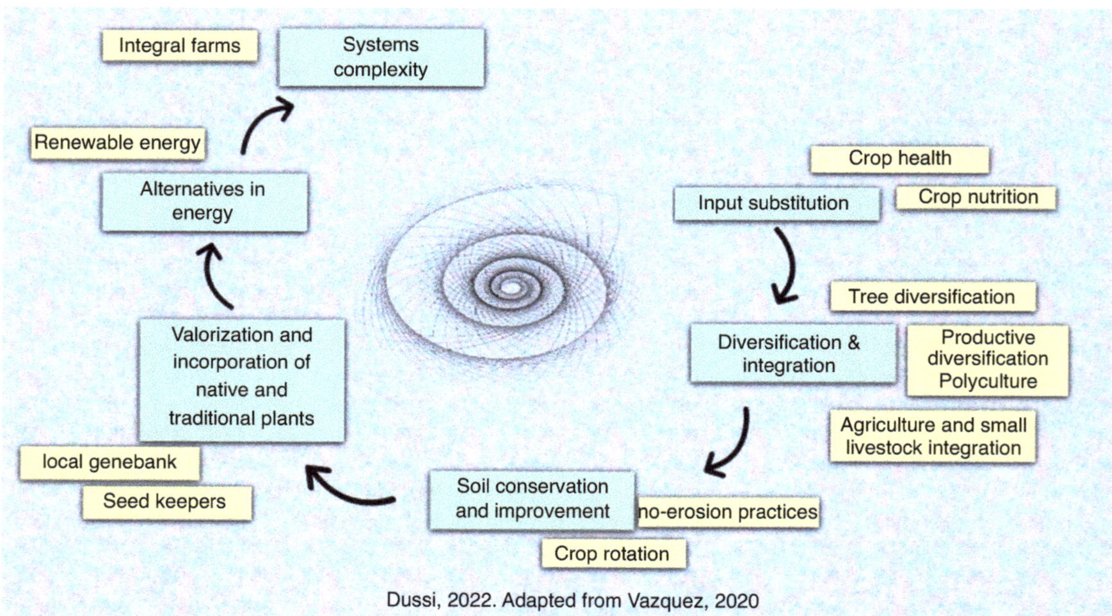

Fig. 46.3. Example of a gradual agroecological transformation at the farm or productive unit level. (Adapted from Vázquez, 2020)

through the demonstration and dissemination of production and management practices at farm level (McGreevy *et al.*, 2021; Flores *et al.*, 2022).

Through agroecology, productivity, sustainability and resilience are achieved by increasing the diversity and complexity of agricultural systems in which ecological interactions and synergisms between their biophysical components replace external inputs to provide the necessary mechanisms that increase fertility, productivity and crop protection (Nicholls and Altieri, 2016). By enhancing functional biodiversity in farming systems, one of the main goals of agroecology is to strengthen weak ecological functions in the agro-ecosystem, allowing farmers to gradually eliminate their dependence on external inputs.

At the farm level, a farming system may be considered 'agroecologically based' if it implements established agroecological principles in the design and management of the agro-ecosystem. Agroecological principles guide the spatial and temporal design of a farm taking the form of different practices (i.e. intercropping, cover cropping etc.), which, in turn, set in motion key ecological processes (nutrient cycling, pest regulation etc.). For example, variety mixture is a known strategy to reduce disease incidence through the buffering effects of genetic diversity. Intercropping is a commonly applied management practice to enhance functional biodiversity, contributing to all principles in agroecology with multiple benefits. Cover cropping is also a practice that reflects the recycling principle and organic matter accumulation, and promotes processes such as nutrient cycling, soil biological activation, weed suppression and water conservation, which are key for crop productivity and soil health. Through indicators, it is possible to evaluate how effectively practices aimed at improving the landscape and crop diversity, soil quality, plant health etc. pursue specific principles or a combination of principles, thus promoting vital processes for agro-ecosystem performance (Nicholls *et al.*, 2020).

The agroecological transformation must, from the outset, have industrial and artisanal technologies for the local production of bioproducts for the nutrition and health of plants, because the substitution of inputs contributes first to the appropriation of these new types of bioproducts by the people who cultivate the land, and at the same time helps to reduce the toxic load on agro-ecosystems. Although formal systems for the production and marketing of seeds and progeny are necessary, a profound transformation is required to rescue traditional species, varieties and breeds, as well as co-innovation systems for their adoption by farmers themselves, strengthening the capacities of resilience.

Carrying out soil conservation and improvement programmes, as well as tree integration in agro-ecosystems, even if they were financed as an environmental service, constitutes levers that promote agroecological transformation, mainly because these natural resources have been degraded by conventional agriculture. Likewise, the gradual integration of renewable energy sources makes the energy sovereignty and resilience of production systems viable.

The redesign of cropping systems (polyculture, polyfruit) and livestock (silvopastoral), as well as production systems, must go hand-in-hand with the substitution of inputs, because otherwise the facilitation of interactions that guarantee the ecological self-regulation capacity of the systems is delayed. Creating capacities for the local benefit and processing of agricultural and livestock products, in addition to adding value increases the resilience capacity of production systems (Vázquez, 2020).

To trigger the agroecological transformation, public policies are needed that facilitate the process at different scales and that are consistent with the sustainability of food systems. The transformation of conventional production systems towards agroecological-based systems includes not only technical, productive and ecological elements but also socio-cultural and economic aspects of farmers, their families and their community, which goes much further beyond the transformation of the production system from conventional to agroecological: it goes through achieving internal capacities, recovering and conserving natural resources, improving the quality of habitat for productive species and workers, and being efficient in the productive, economic, ecological and social order (Vázquez and Martínez, 2015). It also involves changes in consumption practices, policies, cultural meanings, infrastructures and business models. The process through which agroecological practices are initiated and promoted at the agricultural, community and regional levels is called amplification. This requires facilitating synergies that guarantee the success of the process at a territorial scale.

According to Gliessman (2016), the transformational process involves five levels: level 1: increase the efficiency of industrial and conventional practices to reduce the use and consumption of costly, scarce or environmentally damaging inputs; level 2: substitute conventional inputs and practices with agroecological alternatives; level 3: redesign agro-ecosystems so that they function based on a new set of ecological processes; level 4; reconnect producers' and consumers' through the development of alternative food networks; and level 5: construct a new global food system based on equity, participation, democracy and justice. Whereas levels 1 and 2 are incremental, levels 3–5 are transformational (Fig. 46.4).

The transformation of the food system requires a major change in the market economy, creating alternative food networks that reduce the distance between producers and consumers, guaranteeing healthy and accessible food for all people where wealth and employment are created and retained in local economies. There are examples of this, such as participatory guarantee systems (PGS) and community-supported agriculture (CSA) (Dussi, 2019b).

PGSs are quality-assurance systems for a product or process applied at the local level based on the active collaboration of actors. They are built on trust, networks and knowledge sharing. These systems are a counterproposal to certifications, input substitution models and the hegemonic marketing system. The increase in complexity and costs of certification systems, among other reasons, have encouraged many farmers to market considering a relationship with consumers where equity prevails (Dussi, 2019b).

On the other hand, CSA consists of a community of individuals that is committed to supporting the trajectory of a productive unit so that producers and consumers support each other, assuming both the risks and benefits of

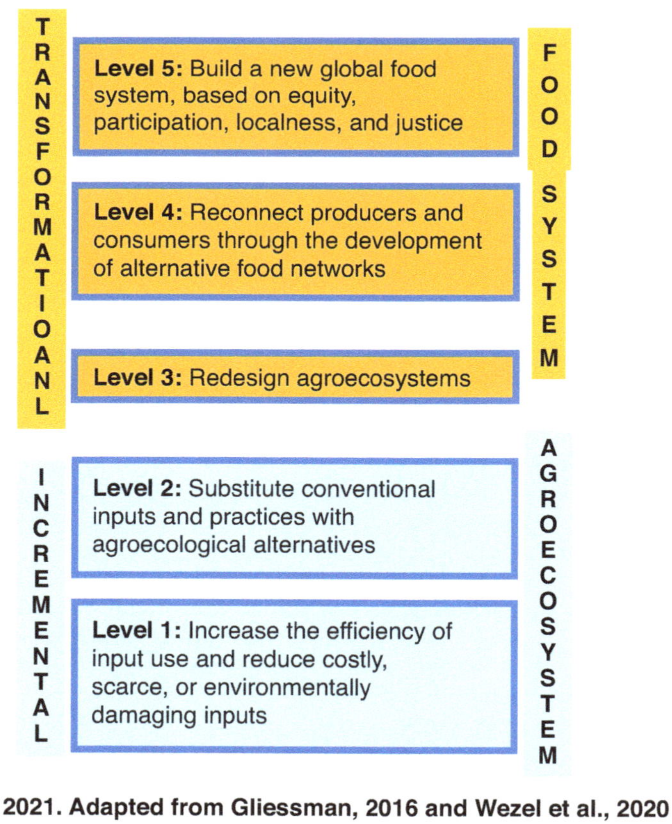

Fig. 46.4. Transformational process from the conventional form of agriculture towards agroecological food systems. (Adapted from Dussi *et al.*, 2021, Gliessman, 2016 and Wezel *et al.*, 2020)

food production. Generally, the members of the productive unit commit in advance to cover the costs of the operation of the farm and the farmer's salary; in exchange, they receive products from the farm as well as the satisfaction obtained by reconnecting to the land and participating directly in food production. Members also share the risks of agriculture, including crop failures due to unfavourable weather conditions or pests, among others (Dussi, 2019b).

These examples create alternative food networks for local markets and autonomous territories, which invites us to think about an agriculture that makes it possible to make adequate levels of production compatible with nature conservation. Building 'food citizenship' through participation in these alternative food networks is the fourth level (Fig. 46.4) of agroecological conversion. Reciprocal arrangements such as solidarity networks have often been critical to the advancement of alternative markets and the socio-economic viability of agroecology.

In summary, strategies that incorporate the amplification of agroecology must be developed considering social, spatial and temporal inequality in the human use of resources to change the agri-food system as an urgent need in the face of climate change.

For the construction of a sustainable eco-agri-food model (Fig. 46.5), analysis can be carried out starting at the farm level towards sustainable food systems where agroecology is the discipline that generates knowledge, validates and applies adequate strategies to design, manage and evaluate sustainable agro-ecosystems and pursue this construction over time (TEEB, 2018; Dussi and Flores, 2018; Dussi, 2019a).

Resilience and resistance strategies must be based on agroecological principles, solidarity and innovation. This can be a new alternative to help farmers achieve more equitable and sustainable opportunities, rather than systems that compromise social well-being, land, water and biodiversity. Resistance to climate disasters is closely related to the biodiversity present in production systems. Greater diversity increases resilience and homeostatic capacity in the face of climate change, so consideration should be given to designing agro-ecosystems surrounded by a complex landscape, with diversified production systems and soils rich in organic matter.The agroecological transformation focuses on moving from a technological model of conventional agriculture

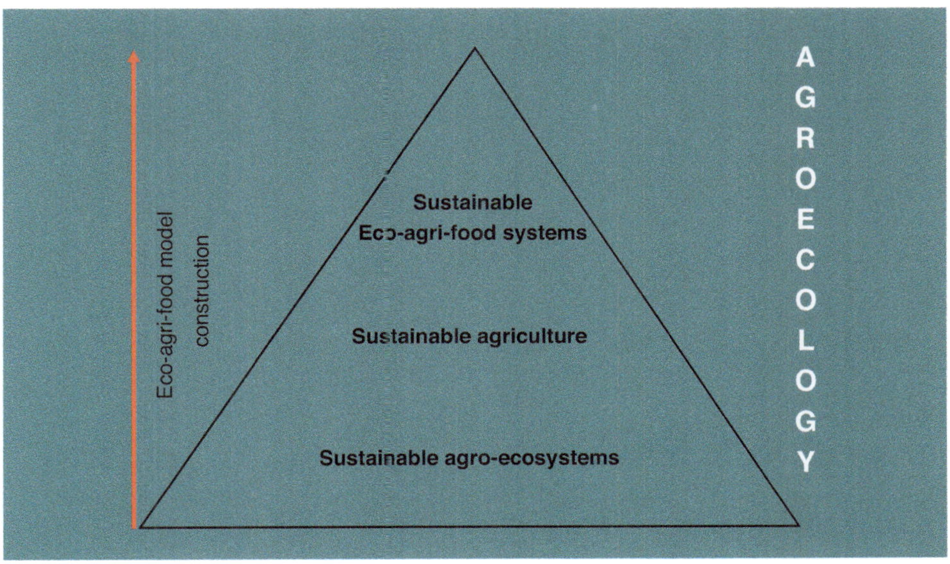

Fig. 46.5. Complexity of the eco-agri-food model construction. (Adapted from Dussi and Flores, 2018; TEEB, 2018; Dussi, 2019a)

to a sustainable one, which scales from the fields and productive units, the production system and the agricultural landscape to the territory, through multi-sector transformations that consolidate territorial food systems.

Conclusion

In the current environmental crisis we face, agrifood systems are one of the main pressures on the environment and resources, and, in turn, food consumption is an essential force to preserve a healthy environment. If there are no changes in global food systems by 2050, increases in GHG emissions, cropland use, freshwater use, and nitrogen and phosphorus applications would exceed tolerable limits. Diet should be based on the production of appropriate food in sufficient quantity and quality to sustain a healthy life with full human dignity, and that social equity, justice and ethical considerations should be fundamental values of our food system and the Sustainable Development Goals.

The future of food sovereignty will likely be characterized by intense dialogue and better information, increasingly complex alliances between producers and consumers, and greater integration with agroecology and with new paradigms of social and cultural change.

The transformation of the food system is a complex and multi-faceted challenge, but it is essential for addressing the impacts of climate change and the well-being of communities around the world. This requires the collective will to change.

References

Altieri, M.A. (2015) Agroecology, key concepts, principles and practices (Third World Network and Sociedad Cientifica Latinoamericana de Agroecologia (SOCLA)). Available at: https://www.fao.org/agroecology/database/detail/en/c/443630/ (accessed 1 March 2025).

Altieri, M.A., Funes-Monzote, F.R. and Petersen, P. (2012) Agroecologically efficient agricultural systems for smallholder farmers: contributions to food sovereignty. *Agronomy for Sustainable Development* 32(1), 1–13. DOI: 10.1007/s13593-011-0065-6.

CIDSE (2018) *The Principles of Agroecology: Towards Just, Resilient and Sustainable Food Systems*. CIDSE, Brussels.

Dussi, M.C. (2019b) Producir y consumir en la agroecología. Dossier Bioeconomía. Herramientas Para El Desarrollo Comarcal, pp. 85–88.

Dussi, M.C. (2022) Is it possible to construct a sustainable agrifood system as a resilience strategy to climate change? Research for organic agriculture to tackle future challenges. In: *ISOFAR International Scientific Workshops. Proceedings*. International Society of Organic Agriculture Research (ISOFAR), Goesan, South Korea, pp. 234–240.

Dussi, M.C. and Flores, L.B. (2018) Visión multidimensional de la agroecología como estrategia ante el cambio climático. In: *INTERdisciplina. Revista Del Centro de Investigaciones Interdisciplinarias En Ciencias y Humanidades Universidad Nacional Autónoma de México*. pp. 129–153. DOI: 10.22201/ceiich.

Dussi, M.C. (2019a) Agroecology and education: Socio-ecological resilience to climate change. *Chronica Horticulturae* 59(1), 20–22.

Dussi, M.C. and Lauri, P.E. (2023) Agroecology and its role within ISHS. *Chronica Horticulturae* 63(4), 12–14.

Dussi, M.C., Fernández, C. and Flores, L. (2018) Huella de carbono en agroecosistemas de hoja caduca. In: *ISSN 2236-7934 – Anais Do VI CLAA, X CBA e V SEMDF*. Cadernos de Agroecologia, pp. 8–15.

Dussi, M.C., Flores, L.B., Barrionuevo, M.E. and Dussi, S.E. (2019) Agroecology in higher education: a multidimensional vision as a resilience strategy to climate change. *Acta Horticulturae* 1258(1258), 79–86.

Dussi, M.C., Flores, L.B., Gómez, E., Dussi, S., Barrionuevo, M. *et al.* (2023) Reconversión agroecológica en una finca. patagonia, argentina. In: IX Congreso Latinoamericano de Agroecología: Diversidad Biocultural Para La Salud de Las Comunidades y Los Ecosistemas, 2022 Costa Rica, Memoria. pp. 242–247.

Dussi, M.C., Fernández, C. and Flores, L.B. (2020) Energy flux analysis in fruit agroecosystems. *Acta Horticulturae* 1286, 171–178. DOI: 10.17660/ActaHortic.2020.1286.24.

Dussi, M.C., Fernández, C., Machuca, Y. and Flores, L. (2021) Energy efficiency in a pear agroecosystem. *Acta Horticulturae* 1303(1303), 419–426. DOI: 10.17660/ActaHortic.2021.1303.58.

Dussi, M.C., Fernández, C. and Flores, L.B. (2022) Energy efficiency and carbon emissions in fruit orchards. *Acta Horticulturae* 1355, 431–440. DOI: 10.17660/ActaHortic.2022.1355.55.

ETC Group (2017) *Quién nos alimentará? 'La red campesina alimentaria o la cadena agroindustrial*, 3rd edn.

FAO, IFAD, UNICEF, WFP and WHO (2019) Review of the state of food security and nutrition in the world, 2018. In: *Building Climate Resilience for Food Security and Nutrition*. FAO, IFAD, UNICEF, WFP and WHO.

Flores, L.B., Dussi, M.C., Barrionuevo, M. and Zon, K. (2022) Evaluation of fruit orchards as potential

agroecological lighthouses. *Acta Horticulturae* 1355, 335–344. DOI: 10.17660/ActaHortic.2022.1355.42.
Foladori, G. (2005) La economía ecológica. capítulo 7. In: Pierri, N. and En Foladori, G. y (eds) *Sustentabilidad: Desacuerdos Sobre El Desarrollo Sustentable*. Miguel Ángel Porrúa.
Georgescu–Roegen, N. (1971) *The Entropy Law and the Economic Process*. Harvard University, Cambridge, Massachusetts.
Gliessman, S. (2016) Transforming food systems with agroecology. *Agroecology and Sustainable Food Systems* 40(3), 187–189. DOI: 10.1080/21683565.2015.1130765.
IAASTD (2009) Agriculture at a crossroads. In: *International Assessment of Agricultural Knowledge. Science and Technology for Development Global Report*. Island Press, Washington, DC.
Intergovernmental Panel on Climate Change-IPCC (2022) *Climate Change 2022: Impacts, Adaptation and Vulnerability*. Geneva, Switzerland.
IPBES (2018) Current and future interactions between nature and society. In: Rice, J., Seixas, C.S., Zaccagnini, M.E., Bedoya-Gaitan, M. and Valderrama, N. (eds) *PBES 2018: The IPBES Regional Assessment Report on Biodiversity and Ecosystem Services for the Americas*. Secretariat of the Intergovernmental Science-Policy Platform on Biodiversity and Ecosystem Services, Bonn, Germany, pp. 437–521.
IPBES Brondízio, E.S., Settele, J. and Díaz, S. (eds). (2019) *Global Assessment Report of the Intergovernmental Science-policy Platform on Biodiversity and Ecosystem Services*. Bonn, Germany.
Loboguerrero, A.M., Thornton, P.K., Wadsworth, J., Campbell, B., Herrero, M. *et al*. (2020) Actions to reconfigure food systems. *Global Food Security* 26. DOI: 10.1016/j.gfs.2020.100432.
McGreevy, S.R., Tamura, N., Kobayashi, M., Zollet, S., Hitaka, K. *et al*. (2021) Amplifying agroecological farmer lighthouses in contested territories: Navigating historical conditions and forming new clusters in Japan. *Frontiers in Sustainable Food Systems* 5, 271. DOI: 10.3389/fsufs.2021.699694.
Mier y Terán Giménez Cacho, M., Giraldo, O.F., Aldasoro, M., Morales, H., Ferguson, B.G. *et al*. (2018) Bringing agroecology to scale: Key drivers and emblematic cases. *Agroecology and Sustainable Food Systems* 42(6), 637–665. DOI: 10.1080/21683565.2018.1443313.
Nicholls, C.I. and Altieri, M.A. (2016) Agroecology: Principles for the conversion and redesign of farming systems. *Journal of Ecosystem & Ecography* 1(5). DOI: 10.4172/2157-7625.S5-010.
Nicholls, C.I., Altieri, M.A., Kobayashi, M., Tamura, N., McGreevy, S.R. *et al*. (2020) Assessing the agroecological status of a farm: A principle-based assessment tool for farmers. *Agro Sur* 48(2), 29–41. DOI: 10.4206/agrosur.2020.v48n2-04.
Nicholls, C.I. and Altieri, M.A. (2018) Pathways for the amplification of agroecology. *Agroecology and Sustainable Food Systems* 42(10), 1170–1193. DOI: DOI: 10.1080/21683565.2018.1499578.
OECD/FAO (2021) *OCDE-FAO perspectivas agrícolas 2021-2030*. OECD Publishing, Paris.
Pengue, W. and Feinstein, H.A. (2013) Nuevos enfoques de la economía ecológica: Unaperspectiva latinoamericana sobe el desarrollo. Lugar Editorial.
Pengue, W.A. (2021) *Principios agroecológicos y sistemas participativos de garantías: una guía hacia la agricultura, la alimentación y el desarrollo rural sostenible*. Orientación Gráfica Editora, Buenos Aires.
Schmidhuber, J. and Tubiello, F.N. (2007) Global food security under climate change. *Proceedings of the National Academy of Sciences of the United States of America* 104(50), 19703–19708. DOI: 10.1073/pnas.0701976104.
Steiner, A., Aguilar, G., Bomba, K., Bonilla, J.P., Campbell, A. *et al*. (2020) Actions to transform food systems under climate change. CGIAR Research Program on Climate Change, Agriculture and Food Security (CCAFS), Wageningen, The Netherlands.
TEEB (The Economics of Ecosystems and Biodiversity) (2018) *TEEB for Agriculture Food: Scientific and Economic Foundations*. UN Environment, Geneva.
Thornton, P., Chang, Y., Loboguerrero, A.M. and Campbell, B. (2023) What might it cost to reconfigure food systems? *Global Food Security* 36.
Tittonell, P. (2023) *A Systems Approach to Agroecology*. DOI: 10.1007/978-3-031-42939-2.
Tubiello, F.N., Rosenzweig, C., Conchedda, G., Karl, K., Gütschow, J. *et al*. (2021) Greenhouse gas emissions from food systems: building the evidence base. *Environment Research Letters* 16(6), 065007. DOI: 10.1088/1748-9326/ac018e.
Vázquez, L.L. (2020) Transformación agroecológica hacia la sostenibilidadalimentaria. In: *Lecturas Agroecológicas No. 2*. Contribución ante la emergencia porla Covid-19, Proyecto Red Ar, La Habana.
Vázquez, L.L. and Martínez, H. (2015) Propuesta metodológica para la evaluación del proceso de reconversión agroecológica. *Agroecología, España* 10(1), 33–47.
Wezel, A., Bellon, S., Doré, T., Francis, C., Vallod, D. *et al*. (2009) Agroecology as a science, a movement and a practice. *Agronomy for Sustainable Development* 29(4), 503–515. DOI: 10.1051/agro/2009004.
Willer, H., Schlatter, B., Trávnícek, J., Kemper, L. and Lernoud, J. (2020) *The World of Organic Agriculture: Statistics and Emerging Trends 2020*. Research Institute of Organic Agriculture (FiBL), Frick, and IFOAM, Organics International, Bonn.
Willett, W., Rockström, J., Loken, B., Springmann, M., Lang, T. *et al*. (2019) Food in the anthropocene: the EAT–lancet

commission on healthy diets from sustainable food systems. *The Lancet* 393(10170), 447–492.
Zhao, C., Liu, B., Piao, S., Wang, X., Lobell, D.B. *et al.* (2017) Temperature increase reduces global yields of major crops in four independent estimates. *Proceedings of the National Academy of Sciences of the United States of America* 114(35), 9326–9331. DOI: 10.1073/pnas.1701762114.

47 Organic Rabbit Farming in Europe: Technical and Economic Performance in French Organic Rabbit Systems

Thierry Gidenne*, Laurence Fortun-Lamothe and Davi Savietto

GenPhySE, Université de Toulouse, INRAE, ENVT, Toulouse France

Abstract

In the Europe Union (EU), organic rabbit farming (ORF) is still marginal (≈ 100 farms) and mainly developed in France (≈ 50 farms). As the rabbit is a herbivorous animal, ORF favours grazing, and well managed, it can pay a half-time salary with 40 does on 4 ha (pastures and crops). Access to land, management of prophylaxis, and feeding are the main constraints for ORF systems. ORF is under EU regulation, which recommends the maximum use of pasture. However, a herd of 40 reproducing does on only 200 m² of pasture is allowed. Knowledge on the functioning and performance of pastured rabbit farming systems are reviewed. Recent studies have quantified grass intake (30–80 g DM/day/rabbit) and growth of rabbits on pasture (15–25 g/day). The production cycle of an ORF system is extensive with an average of 2.7 parturitions/doe/year.

Keywords: rabbit, grazing, organic farming, system, performances

Introduction

The development of organic agriculture has grown since 2009 in the Europe Union (EU). However, organic rabbit farming (ORF) remains a niche market (about 100 farms in the EU and 50 in France), and consumer demand exceeds supply (Gidenne *et al.*, 2020b). The first regulations for ORF were developed in 2000 in France, with several rules, such as: a whole-year grazing, natural mating, slaughter after 100 days old, housing in movable cages (Fig. 47.1) or in paddocks (Fig. 47.2). Since 2022, an EU regulation is applicable for organic rabbit farming (JOUE, 2020). Contrary to indoor conventional rabbit farming (≈ 5000 farms in the EU), the performances of ORF systems are not referenced in national databases. This lack of technical references is an obstacle to the development of the 'alternative' rabbit sector, since such information allows farmers and supervisors to establish, organize and size an installation project.

Studies on ORF started in France in 2010 and now the constraints and production potential of this type of farming are better-known, and are the subject of this chapter. We present the context and components of an organic rabbit farming system (feed, reproduction, prophylaxis) in Europe, mainly in France; the current state of knowledge about production capacity; and possible developments in the design and management of these systems in the light of recent changes in EC regulations on organic farming.

*Corresponding author: thierry.gidenne@inrae.fr

DOI: 10.1079/9781800626850.0047

Fig. 47.1. Movable cages on pasture for organic rabbit farming. (© INRAE, T. Gidenne)

The Context of Organic Rabbit Farming in Europe and France

China ranks first in the world for conventional rabbit farming with over 500,000 tons of carcass produced. Europe ranks second with 130,000 tons (Spain 48,000; France 35,000; Italy 28,000; Wu and Lukefahr, 2021). While rabbit production is increasing worldwide, it has been falling steadily in Europe over the last 30 years due to a reduction in the profitability of rabbit farms and rising set-up costs, while farmers increase the size of their herd. Since ORF is only just emerging in the EU (100 farms), France is leading the organic rabbit production, with probably over 90% of the whole production (15,000 rabbits/year produced in France by about 50 farmers). This small supply of organic rabbit meat does not meet consumer demand.

Fig. 47.2. Fixed enclosure with grazing rabbits. (© INRAE, T. Gidenne)

Conventional rabbit housing is based on indoor caging and feeding with pelleted compound diets. In contrast, ORF is outdoors with maximum use of grazing.

Pastured Rabbit Farming Systems and Certification for Organic Agriculture

Main grazing systems used in France

Past French regulations for ORF impose the principle of maximum use of pasture, and minimum 60% roughage in the diet, as for ruminants. Two systems based on grazing were developed: (i) fixed-fence pens; and (ii) movable cages (or paddocks) (see Figs 47.1 and 47.2). In practice, grazing in movable cages is often intended for breeding adults. Small, individual, fixed enclosures (10 m^2) are also used. Fixed enclosures (50 m^2) are mainly used for rabbit fattening. A third system, in a fixed building with an outdoor exercise area, is also within the regulations. It requires intensive feeding to comply with EU regulations. This system may be useful for coping with cold climates, or for better control of reproduction (parturitions and protection of litters), as shown by Legendre *et al.* (2019).

Rabbit housing is designed by breeders and includes a grazing area and a 'shelter' area made from hard-wearing wood to protect the animals from the weather and predators. Protection against predators (dogs, foxes, weasels, birds of prey) is provided either by wire fencing (movable pens) or electrified fencing (fixed enclosure).

The production cycle in organic rabbit farming is fairly extensive (Fig. 47.3) compared with conventional indoor rabbit farming. The EU regulation stipulates for ORF a weaning from 42 days old, although the weaning often occurs later when rabbits weigh about 1 kg. Since the parturition interval is over four months (Table 47.3), a doe weans around 17 rabbits a year, which equates to the sale of 14 rabbits weighing 2.5 kg. So a herd of 40 rabbits produces around 560 rabbits for an estimated turnover of €14,000, corresponding to a half-salary for a year. Thus, ORF is mostly combined with other livestock or crop production on the same farm.

In agreement with the EU regulations, and to maximize the use of pasture, it is advisable to have an area of pasture and hay production of around 2 ha for a herd of 40 does and their offspring (Table 47.1). For the ORF system most commonly

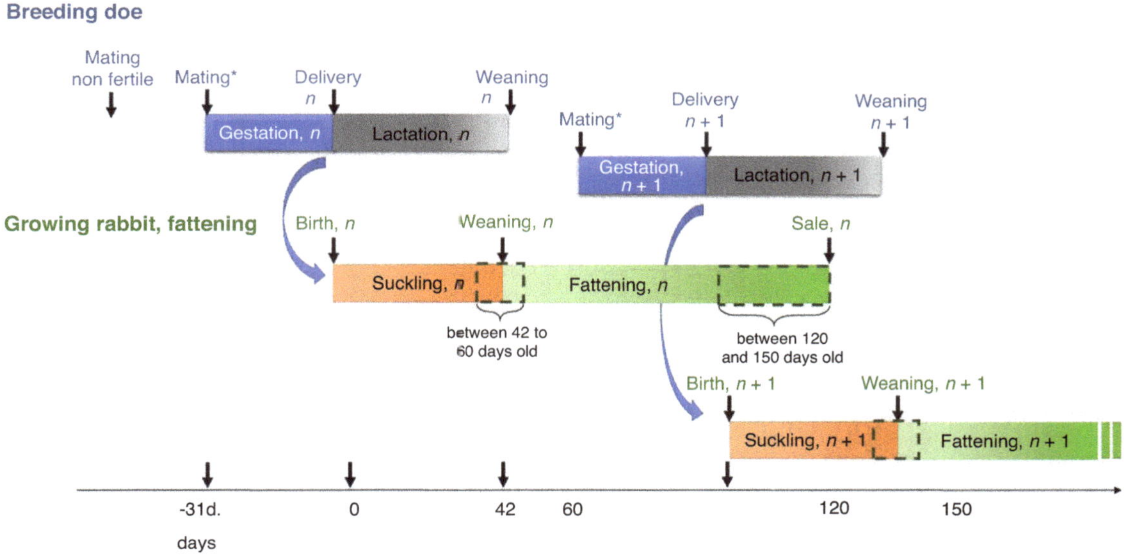

Fig. 47.3. Average production cycle in organic rabbit farming. Average production per cycle (see Table 47.3): 5 weanlings per delivery; 140-day interval between farrowings (on average); annual production: 17 weanlings per rabbit/year, i.e. for 40 female rabbits a sales potential of 580 rabbits/year (with a survival rate of 85% between weaning and sale).

Table 47.1. Pasture surfaces according to EU regulation for organic rabbit farming (for a farm with 40 does and their offspring + 5 males + 5 renewal does, i.e. a half-time occupation).

	Outside the grazing period, fixed building, with a planted outdoor run, preferably grassland	Outdoor space Movable cage
40 does (gestating or not, or lactating), with the litter till weaning	[40 × 2.5 m²/doe with kits] = 100 m² minimum	2.5 m²/doe with kits, per day if daily move: on 73 days (31 + 72 d): [73 d × 2.5 m²] × 40 = 7300 m²
five renewal does	[5 × 2.5 m²/doe] = 12.5 m² minimum	50 d × 2.5 m²] × 5 = 625 m²
five bucks	5 × 2.5 m² = 12.5 m² minimum	73 d × 2.5 × 5 m² = 912 m²
150 fattening rabbits[a], from weaning (42 d) to slaughter (110 d)	150 × 0.5 m² = 75 m² minimum	150 × 0.4 m² = 600 m²
TOTAL	200 m² minimum	8425 m² minimum

[a]25 gestating rabbits (out of 40) and 6 weaned rabbits per litter = 150 fattening rabbits per 73-day breeding cycle.

used in France, 1 ha will be allocated to the fixed fattening enclosures, and 1 ha to the movable pens (for does). Two hectares for cereal and protein crops must be added to supply concentrated feed to complement grazing or hay. In total, 4 ha are recommended to set up an ORF rabbit farm with 40 does and their offspring, corresponding to a half-time salary.

Rabbit Farming Under Organic Certification – Consistency and Inconsistency, Impact on the Development of Farming Systems

EU regulation for ORF states a maximum use of pasture, a diet containing 60% roughage, and types of housing. On the other hand, it does not impose a minimum age for slaughter, or pasture rotation. Additionally, the notion of vegetated runs or fixed pens is no longer mentioned, and the daily movement of a mobile pen is not compulsory. Apart from the fact that these last points increase the risk of parasites, the new regulations do not require the outside surface to be grassed, but they do prohibit the use of concrete. The minimum area of land required to set up an organic rabbit farm with 40 rabbits (half-time occupation) is calculated in Table 47.1 for the two possible systems: fixed building and movable cages.

With a fixed-building system, an outdoor area of 225 m^2 is all that is needed today to keep a herd of 40 rabbits; and 8425 m^2 for a system with mobile pens, assuming daily movement (not compulsory).

Therfore, in a fixed-building system, the minimum surface area of outdoor space indicated in the specifications is far from sufficient to cover the rabbit's grazing needs, and does not comply with the minimum threshold of 60% roughage in the diet. A growing rabbit weighing 2 kg grazes between 0.5 and 1.0 m^2/day. Therefore, the 0.5 m^2 area proposed by the regulation is fully grazed in one day. To reach 60% of 120g of dry feed intake (i.e. 72g of dry matter (DM)), the area to be grazed would have to be around 1 m^2/rabbit/day. The minimum area proposed for a fixed building, therefore, imposes intensive daily feeding, which does not correspond to the basic EC regulation, where it is specified that (i) rabbits have access to pasture for grazing whenever conditions allow; and (ii) breeding systems are based on maximum use of pastures.

For the movable-pen system, the previous regulation in France required that the pens be moved daily. However, the minimum grazing area for a herd of 40 does is estimated at 1.2 ha. In a season of good fodder production, this area can produce enough to cover the 60% of roughage, which is no longer the case in the intermediate or winter season, when hay will be necessary. Besides, the threshold of 60% roughage in the diet is often achieved in specific periods, such asat the end of the fattening period and with an abundant supply of grass of good nutritional quality (Martin *et al.*, 2016; Legendre *et al.*, 2019). In the case of a lactating doe, it is recommended to give a complementary concentrated feed, i.e. rich in protein and energy (cereal/protein mixes, or commercial pelleted feed) in addition to grazing and hay. Under these conditions, roughage intake represents around 40% of feed intake. Therefore, in view of the actual measurements of grass intake (see below 'Grass intake by pastured rabbits') and of the nutritional recommendations, it would be appropriate to set a regulatory threshold for the proportion of forage in the diet adapted to the digestive physiology of the rabbit (monogastric and herbivorous), estimated at 40% for suckling rabbits and 50% for growing rabbits.

Technical and Economic Performance in French Organic Rabbit Farms

The lack of a performance benchmark for organic or independent rabbit farming is holding back the development of this sector, particularly in terms of enabling project owners to organize and size their facilities. INRAE started a performance benchmark for ORF in 2017 (Gidenne *et al.*, 2020b) and is now extending it, via a smartphone application 'GAELA' (Huang *et al.*, 2021). GAELA is used by farmers and enables daily management of the rabbit farm tasks (breeding, births, weaning etc.) with direct (in the farm) and unique (no transcription from a farm notebook) entry of technical data (Fig. 47.4), secured on a national database via automated synchronization.

Over 30 ORFs are registered to GAELA since 2018. The reproductive performances of 4 AB rabbit farms that regularly recorded their performances are shown in Table 47.2 and Table 47.3. The size of the herd varies greatly depending on the farm (Table 47.2), ranging from 15 females (new farmers) to more than 70 (farmers who have been in business for more than 15 years). The rate of reproduction (natural breeding) is extensive, with 104 farrowings per year for an average herd of 37 does, i.e. 2.7 parturitions/doe/year (Table 47.3).

The parturition interval averaged 138 days (compared with 56 days on conventional indoor systems). The fertility rate averaged 71%, but varied greatly from farm to farm. Litter size (6.9 live births) varied considerably between farms (from 3.9 to 9.0). For these four farms, the weaning averaged 52 days, but varied from 42 to 65 days, which illustrates the diversity of practices. The average survival rate of young rabbits to weaning was 72%, but with great variations, with the oldest farms (A–C) recording an average survival rate of 80%. This suggests several improvements in managing of the maternity unit, such as the survival rate before weaning, or a shorter parturition interval. For example, if the

Fig. 47.4. GAELA screenshots: smartphone application to assist in the management of a rabbit farm.

Table 47.2. Descriptive data of the herds for four organic rabbit farms. Data collected from 2018 to 2020 with GAELA application.

Farm	A	B	C	D
Does (mean number/year)	29.1	70.8	32.0	15.5
Bucks (mean number/year)	4.9	10.4	3.8	6.1
Ratio doe/bucks	5.9	6.8	8.5	2.6
Mating/year	108	405	87	47
Parturitions/year	68	241	69	38
Weaning/year	64	204	45	26

survival rate at weaning could be increased from 72% to 85%, turnover could increase by €2508/ year for a herd of 40 does producing three litters/ year.

Production Techniques for Pastured Rabbit Systems

Feeding systems

The feeding system must comply with two regulatory criteria:

- Maximum use of pasture, whenever weather conditions allow. Fibrous feed must be provided if direct grazing is insufficient, by feeding with dry roughage (hay) or green roughage (grass cuttings, branches of fodder trees, grass preserved by various techniques such as silage or wrapping).
- Minimum 60% roughage in the diet (in DM). Depending on the season, hay may also be necessary, bearing in mind that coarse hay (poor in leaves) will be poorly ingested (15–30g DM/day; Duprat *et al.*, 2016).

Rabbits are monogastric herbivores and their diets should be supplemented with feeds that are richer in energy and protein than forage, such as cereal/

Table 47.3. Productive performance of four organic rabbit farms.

Farm	A	B	C	D
Mating/doe/year	3.7	5.7	3.4	3.0
Mating/buck/year	21.8	38.9	23.0	7.8
Fertility rate (%)	63.0	59.5	79.3	80.9
Parturition/doe/year	2.3	3.4	2.7	2.4
Parturition interval (days)	159	107	135	152
Born alive/parturition	7.3	9.0	6.6	3.9
Age at weaning (days)	42.5	65.0	45.1	55.3
Weaned number/parturition	6.1	6.4	5.9	1.8
Weaned number/female/year	14.4	21.8	9.6	4.3
Survival rate at weaning (%)	81.9	70.2	89.0	46.2

protein mixtures (oats, barley, peas, broad beans etc.), roots (beetroot etc.) or balanced complete pelleted feeds. This supplementation is particularly important for lactating rabbits, to cover their energy and protein requirements.

Grass intake by pastured rabbits

Grass intake is only documented for growing rabbits. In a movable pen with 0.4 m^2/day of grazing and 60g of pelleted complete feed per rabbit, Martin *et al.* (2016) showed that grass intake represented 46–50% of total dry-matter intake (120 g/d/rabbit). The level of grass intake depended on the liveweight of the rabbit (+7g DM/day for +100g liveweight$^{0.75}$) and the biomass available. Grass intake was higher (79g DM/day/rabbit) in a natural grassland (spring season) where the available biomass was more abundant than in a tall fescue (44g DM/day/rabbit) or sainfoin (51g DM/day/rabbit) pasture. The greater the biomass available, the higher the grass intake: on average +6.8g DM/day per 10g DM increase in biomass available.

Grass intake by growing rabbits is, on average, 51g DM/rabbit/day for a quantity of grass offered of 69g DM/day/rabbit, i.e. a fresh grass intake capacity (at 20% DM) of 200–250 g/day/rabbit, and it can reach 750 g/day/rabbit, i.e. 35% of liveweight (2 kg). Behaviour of pastured rabbits was recently studied (Fetiveau *et al.*, 2021, 2023). When grazing, and in a situation where there is a choice between grazing and a 'complete pelleted' feed distributed *ad libitum*, the rabbit consumes at least half of its diet (in DM) with pellets, and half by grazing.

When grazing, and without the choice of a more concentrated feed, rabbits can ingest large quantities of grass provided it is palatable. Fresh alfalfa intake can exceed 400 g/day (i.e. around 80g DM) for a 1.5 kg rabbit, and can reach 600 g/day (120g DM) for fresh, palatable plants (whole carrots, cabbage; Goby *et al.*, 2017, 2021). Rabbits therefore have a high grass ingestion capacity of up to 150g DM/day/rabbit, which could cover their growth requirements.

The intake of dry hay is very poorly documented, although it is frequently distributed in current systems. As with grazing, the quality of the hay is a determining factor: good quality legume hay, distributed on its own, can be consumed at a fairly high level: 50–80g DM/day for a 2 kg rabbit (Goby *et al.*, 2017). However, hay consumption is often moderate or even low (15–30g DM/day) as it is generally distributed as a supplement to a more concentrated feed, such as a cereal mix or a pelleted feed.

Now, methods are proposed for estimating the grass biomass consumed in movable cages (Plagnet *et al.* 2023) that show the great intake capacity of the rabbit for grass. A grazing intake simulator (Joly *et al.*, 2018) has been made available to rabbit farmers and agricultural advisers. It can be used to compare various types of management for growing rabbits. The simulator provides technical (growth and grass intake) and economic (gross margin on feed costs) results.

Rabbit growth on pasture

Growing rabbits start grazing at 3 to 4 weeks of age. Grazing lasts about 10 weeks, which allows the animals to reach a selling weight of about 2.4 kg when the feeding system is mixed: grazing, supplementary 'concentrated' feed (meslin) and hay. The growth rate of rabbits depends largely on the quality of the pasture, and in particular its protein content (grasses vs legumes): a grass rich in protein is ingested less by the rabbit (on average, –4g DM for each additional point of protein) but better covers its protein requirements for growth. The growth rate of rabbits in grazing systems range between 15 and 25 g/d, enough to reach a live sale weight of 2.4 kg at 100 days of age. Joly *et al.* (2018) estimate that 90 g/day of concentrate feed (meslin grown on the farm) and a grazing area of 0.4 m²/rabbit/day enable a growth rate of around 22 g/day. Without supplementary feed, the grazing area would have to be tripled (Duprat *et al.*, 2016), and the use of legumes (sainfoin, alfalfa, clovers etc.) would have to be encouraged, to achieve similar growth.

Practical feeding in organic rabbit farming

To cover the grazing needs of a growing rabbit, from weaning (minimum regulatory age of 42 days) to sale (2.4 kg at 100 days of age), it would be reasonable to allocate: (i) in a fixed enclosure, an area of 25 m² per rabbit to cover growth needs with 100% grazing (for a grass production of 3 t/ha, and if the proportion of legumes is at least 50%). If concentrated feed (cereals etc.) is added at 40%, this area can be equal to or greater than 15 m² per rabbit; (ii) in a movable cage, the daily grazing area would have to be increased from 2.4 m² to around 5 m², which would mean moving the mobile pen (around 2.5 m²) twice each day or enlarging the pen.

So for a herd of 40 breeding rabbits that will wean around 150 young rabbits per cycle (2–3 weanings per year), you need a pen of 2250 m² (if 40% concentrated feed) or even 3750 m² without concentrate feeds. To rotate enclosure between weaning groups, you need at least two enclosures (i.e. 4500–7500 m² of land).

It is also necessary to supplement the diet with cereals. In most cases, cereals are produced on the farm. For example, the quantity of cereals for a herd of 40 does and their offspring is estimated at 2.8 tonnes/year, which corresponds to an area of 1 ha. In addition to pasture and hay, other fodder resources can be used: fodder trees, roots, cabbage etc. Rabbits are selective in their feeding behaviour (Gidenne *et al.*, 2020a). Like other herbivores, they prefers the leaves rather than the stems of a plant, and the soft green parts rather than the dry ones. In a choice situation, rabbits will prefer to graze rather than eat dry hay. It will take longer to adapt to eating dry hay than green forage, especially in young rabbits.

In return, excessive intake of green fodder, particularly in spring and if it is a young legume, is possible in rabbits (120g DM, i.e. >500g fresh/day for a 1.5 kg rabbit). It can lead to excessive consumption of digestible proteins and an increased incidence of digestive disorders. For example, over-ingestion of young clover, protein concentration of which may exceed 20% of the total ration, increases the risk of diarrhoea due to a probable imbalance between protein and fibre intake. In this case, grazing access should be limited (less moving of the movable pen), so that the rabbit consumes more dry forage (and/or a fibrous supplementary feed). It may be advisable to cultivate pastures combining legumes and grasses and to distribute dry fodder (hay or straw) permanently and freely. Fodder or fruit trees (leaves and twigs from ash, mulberry, lime, apple trees etc.) can be used as a source of fodder, especially in late summer and autumn when grass growth is lower.

At present, we do not have nutritive values for forages measured on pasture, as the first measurements of digestibility on pasture were only carried out very recently (Goby *et al.*, 2021). However, a comparison of the nutritional values of feedstuffs for rabbits and horses (see www.feedipedia.org; equipedia.ifce.fr; www.feedtables.com [accessed 2/3/25]), an animal that is physiologically close to the rabbit from a digestive point of view (caecal fermenter) shows similar values (cereals, protein seeds) or 5–10% lower values (forage) for the rabbit. Where data are lacking, an estimate can be made based on results obtained for horses. Some data on the nutritional value of forages are available for rabbits (Table 47.4), such as alfalfa or ryegrass hay.

In the future, it will therefore be necessary to document better the capacity of growing rabbits to ingest grass, depending on the season and the quality of the pasture. But it also seems necessary to document the grass intake of adult breeding rabbits. A suckling rabbit's capacity to ingest a low-concentration feed (rich in forage, for example) is undoubtedly insufficient to meet its nutritional requirements, particularly if the minimum requirement of 60% roughage in the ration has to be met.

Table 47.4. Nutritional value of hay and pasture for rabbits.

Feeds	Digestible energy (kcal/kg DM)	Crude proteins (% DM)	Crude fibre (% DM)
Hays			
Alfalfa type A	1800	15	27
Alfalfa type B	1670	14	29
White clover	2400	18	24
British ray-grass	2460	18	19
Pasture			
Sainfoin	2300	18	26
Tall fescue	1300	13	30
White clover	2400	21	25

Reproduction and genetics

In organic rabbit farming, reproduction is carried out exclusively by natural mating, whereas artificial insemination is the main method used in conventional rabbit farming. ORF regulation prohibits the use of hormones. However, ovulation in rabbits is not spontaneous, but is induced by mating. Consequently, artificial insemination requires the injection of a GnRH analogue to trigger ovulation. In rabbits, there is no lactation anoestrus and the female can be fertilized again as soon as she gives birth and throughout lactation. This is the main practice in conventional rabbit farming, since the interval between two inseminations is generally 42 days (insemination 10–11 days after farrowing) and weaning takes place between 31 and 35 days of lactation.

Within EU regulations, a doe could have four breeding cycles a year. But, in practice, weaning is carried out between 42 and 60 days of age, or even later, and matings are often carried out after weaning. What is more, during the winter months (when the days are short), female rabbits are less receptive and several matings are required to achieve a pregnancy. The average number of births per female rabbit per year is three for ORF (Table 47.3).

Organic rabbit farmers essentialy use traditional breeds (Fauves de Bourgogne, Papillon etc.), with fertility rates varying between 35% and 68%, a high and variable nest mortality rate (10–38%) and an average number of weanlings per litter of 4.8 (Table 47.3). ORF could benefit from more productive rabbit strains, or from crossing them with heritage breeds (Savietto *et al.*, 2021). This aspect remains to be studied, bearing in mind that no genetics programme is currently being developed specifically for organic rabbit farming or grazing. Participatory selection approaches should also be considered.

Prophylaxis in grazing rabbits

In conventional rabbit farming, respiratory problems (pasteurellosis) and pododermatitis (staphylococcosis) dominate the pathology of adults, while diarrhoea (colibacillosis, epizootic rabbit enteropathy, coccidiosis) dominates the pathology of growing rabbits. In ORF, adults are mainly affected by viral diseases (myxomatosis and viral haemorrhagic disease, VHD), and digestive diseases are still predominant in young rabbits: peri-weaning diarrhoea and parasitism (helminthosis and coccidiosis).

Prophylaxis against myxomatosis and VDH is essentially based on vaccination. In practice, and in areas with a high presence of wild rabbits and hares, early vaccination of young rabbits (before weaning) is recommended. Finally, farmers should pay close attention to the vaccination schedule for their livestock, especially breeding stock. Booster doses are essential for both myxomatosis and VHD.

The risk of parasites, dominated by coccidiosis, remains a major problem in rabbit farming (whether organic or not). Moreover, the EU regulation for ORF no longer imposes a minimum period between two visits by rabbits to a pasture, whereas previous regulation imposed a two-month delay to reduce the risk of infestation. In the case of movable pen, the EU regulations no longer require a daily move on the pasture. These two factors are therefore additional risk factors for parasitic infestation, particularly for coccidiosis. We therefore need to develop our knowledge of how to control the risk of parasites in rabbit farming on pasture, as this remains a major factor in the loss of growing animals.

Research and Perspectives on Pastured Rabbit Farming – What Are the Alternatives to Indoor Conventional Systems?

There is little research on pastured rabbit or on ORF, often limited to describe performances, meat quality and sustainability of the systems. Although ORF is very marginal in Italy, a specific work on this type of farming was published in 2011 (Finzi and Mariani, 2011), and in 2020 in France (CAB, 2020). Most studies on ORF were performed in France, with a focus on nutrition (grass intake and grazing) and parasitism (see above 'Grass intake by pastured rabbits').

Nevertheless, several constraints still limit rabbit farming on pasture and would deserve specific studies: the low fertility of the doe in autumn and winter, management of parasitism, nutritional value of pastures, metabolic diseases in young rabbits in relation to a high intake of fresh grass.

In terms of alternatives to indoor rabbit farming, an experimental model with access to pasture has been developed in France by INRAE, in line with societal concerns about animal welfare. Fetiveau *et al.* (2021) studied two group sizes (25 or 50 rabbits) with or without access to a 23 m² grazing area. With a growth rate of 26.7 g/day and an overall mortality rate of 7.3%, the results were close to those obtained without access to a grass run (D'Agata *et al.*, 2009), and similar to those reported in French ORF (Legendre *et al.*, 2019). Other studies on the integration of arboriculture and pastured rabbit are in progress (Fetiveau *et al.*, 2023). Preliminary results are encouraging and indicate that rabbits are a potential alternative for developing new grass management strategies for various fruit crops. Rabbit farming associated with other types of production is the subject of various research projects, such as the association between rabbit farming and a farm.

Conclusions

Although France has the largest number of organic rabbit farms in Europe, this type of production is still very marginal. But the situation could change, given that consumer demand outstrips supply and that it enjoys a good public image. The creation of a technical reference system combined with a new digital tool, and the development of a social network of professionals, could support this development and help to meet several challenges: contributing to the technical rationalization of production, better management of geographical dispersion with collaborative working tools, increasing the size of the sector to structure it better, and maintaining or reducing production costs.

The European regulations, which are less demanding in terms of surface area and access to pasture, could change the way current systems operate, leading to a higher risk of parasitism due to the higher density of animals and the absence of rotation on grazing land, and consequently lowering the image of this profession.

References

CAB (2020) Guide éleveur.se.s: Elever des lapinsBio. Coordination Agricole Biologique (CAB) des Pays de la Loire Editions.

D'Agata, M., Preziuso, G., Russo, C., Zotte, A.D., Mourvaki, E. *et al.* (2009) Effect of an outdoor rearing system on the welfare, growth performance, carcass and meat quality of a slow-growing rabbit population. *Meat Science* 83(4), 691–696. DOI: 10.1016/j.meatsci.2009.08.005.

Duprat, A., Goby, J.P., Roinsard, A., Horst, V.D.J., Le Stum, J. *et al.* (2016) Pasture finishing of organic rabbit: Grass intake and growth – first results. In: *11th World Rabbit Congress,Qingdao, China*. Qingdao, China, pp. 931–934.

Fetiveau, M., Savietto, D., Gidenne, T., Pujol, S., Aymard, P. *et al.* (2021) Effect of access to outdoor grazing and stocking density on space and pasture use, behaviour, reactivity, and growth traits of weaned rabbits. *Animal* 15(9), 1–11. DOI: 10.1016/j.animal.2021.100334.

Fetiveau, M., Savietto, D., Bannelier, C., Fillon, V., Despeyroux, M. *et al.* (2023) Effect of outdoor grazing area size and genotype on space and pasture use, behaviour, health, and growth traits of weaned rabbits. *Animal – Open Space* 2, 100038. DOI: 10.1016/j.anopes.2023.100038.

Finzi, A. and Mariani, G. (2011) *L'allevamento Ecologico Del Coniglio*. Edagricole, Milan, Italy.

Gidenne, T., Lebas, F. and Fortun-Lamothe, L. (2020a) Feeding behaviour of rabbits. In: De Blas, C. and Wiseman, J. (eds) *Nutrition of the Rabbit*, 3rd edn. CAB International, Wallingford, UK, pp. 254–274.

Gidenne, T., Savietto, D., Goby, J.P., Fortun-Lamothe, L. and Roinsard, A. (2020b) A referencing system to analyse performances of French organic rabbit farms. *Organic Agriculture* 10(S1), 125–129.

Goby, J.P., Coullet, S., Huck, C., Bannelier, C. and Gidenne, T. (2017) Ingestion et digestion de fourrages secs et verts pour le lapin – première approche pour la luzerne et la carotte entière. In: INRA (ed.) *17 Èmes Journ. Rech. Cunicoles*. ITAVI publications, Le Mans, France, pp. 189–192.

Goby, J.P., Bannelier, C. and Gidenne, T. (2021) Nutritive value of cabbage and pelleted complete feed in free-ranged organic growing rabbit. In: *Proceeding of 72th Annual Conference of the European Federation of Animal Production*, Davos Swiss.

Huang, Y., Gigou, M., Goby, J.P., Roinsard, A., Savietto, D. *et al*. (2021) Digital breeding and assisted management in organic rabbit farming: The first results. In: Gidenne, T. (ed.) *12th World Rabbit Congress*. INRAE-ASFC publisher, Nantes, France.

Joly, L., Goby, J.P., Duprat, A., Legendre, H., Savietto, D. *et al*. (2018) PASTRAB: A model for simulating intake regulation and growth of rabbits raised on pastures. *Animal* 12(8), 1642–1651.

JOUE (Official Journal of European Union) (2020) Implementing Regulation (EU) 2020/464. Laying down certain detailed rules for the implementation of Regulation (EU) 2018/848, 26 March (L98/2).

Legendre, H., Goby, J.-P., Duprat, A., Gidenne, T. and Martin, G. (2019) Herbage intake and growth of rabbits under different pasture type, herbage allowance and quality conditions in organic production. *Animal* 13(3), 495–501.

Martin, G., Duprat, A., Goby, J.-P., Theau, J.P., Roinsard, A. *et al*. (2016) Herbage intake regulation and growth of rabbits raised on grasslands: Back to basics and looking forward. *Animal* 10(10), 1609–1618.

Plagnet, A.S., Bannelier, C., Fillon, V. and Savietto, D. (2023) Estimation of grass biomass consumed by rabbits housed in movable paddocks. *World Rabbit Science* 31(1), 21–34.

Savietto, D., Debrusse, A.M., Bonnemere, J.M., Labatut, D., Aymard, P. *et al*. (2021) Reproductive performance of a maternal rabbit cross: Fauve-de-Bourgogne X INRA1777 [3–5 Nov]. In: Gidenne, T. (ed.) *12th World Rabbit Congress*. INRAE Occitanie Toulouse, Nantes, France.

Wu, L.P. and Lukefahr, S.D. (2021) Rabbit meat trade of major countries: Regional patterns and influencing factors [3–5 Nov]. In: Gidenne, T. (ed.) *12th World Rabbit Congress*. Nantes, France, INRAE Occitanie Toulouse.

48 Selective Breeding for Sustainability in Organic Cattle Farming Systems

Matt J. Bell*

Hartpury University, Gloucester, UK

Abstract

Selective breeding on a wide range of biological traits across livestock populations, breeds and production systems is now possible. Ruminant livestock such as cattle are able to convert non-human edible plant material into human edible foods such as fresh milk and meat. This chapter uses dairy cattle as an example of developments in milk production through selective breeding. Cattle are the main source of fresh milk in the world, which has steadily increased over the last 60 years to meet global demand. In developing countries, the increase in milk production has largely been due to increasing animal numbers; whereas in developed countries, improved breeding, nutrition and management has led to higher milk yields per animal. To balance productivity with animal welfare and resource use there has been a shift towards a more balanced approach to selecting animals on a range of production (e.g. milk yield and composition), fitness (e.g. health and fertility plus overall survival) and resource efficiency (feed intake and methane emissions) traits. More balanced multi-trait selection has reduced the carbon footprint of production per unit product over several decades, and the sustainability footprint of cattle is expected to reduce with future selection on efficiency traits or emissions per cow.

Keywords: cattle, milk production, biological traits, genetics, sustainability

Introduction

There are around 3.6 billion domestic ruminant animals worldwide of which 39% are cattle (*Bos taurus*, *Bos indicus*), 31% sheep (*Ovis aries*), 24% goats (*Capra hircus*) and 6% buffalo (*Bubalus bubalis*) (Hackmann and Spain, 2010). Domestic ruminant species, such as cattle, produce human edible foods and associated products from non-human edible plant material. For example, cattle are the largest source of fresh milk globally, with 81% of production (753 million tonnes in 2022), followed by buffalo (15%), goat (2%), sheep (1%) and camel (0.4%) (Fig. 48.1; FAO, 2024a). The amount of milk produced across different animal species appears to have peaked in 2021, which is attributed to a stabilizing of animal numbers.

Along with milk, cattle are also a useful source of meat, offal, fats and hides (Fig. 48.2).

Countries in the developing (e.g. India) and developed (e.g. USA) worlds dominate milk production. While the growth in milk production in developed countries can be attributed to improved breeding, nutrition and management in dairy cows, leading to higher milk yields per animal, in developing countries the growth is largely the result of increased animal numbers and often across different animal species. The consumer demand for milk products across both developing and developed countries is projected to continue growing (FAO and GDP, 2018). Therefore, in developing countries, there are great opportunities for improvements in productivity per animal in areas such as breeding, nutrition, disease management and husbandry.

*Corresponding author: matt.bell@hartpury.ac.uk

DOI: 10.1079/9781800626850.0048

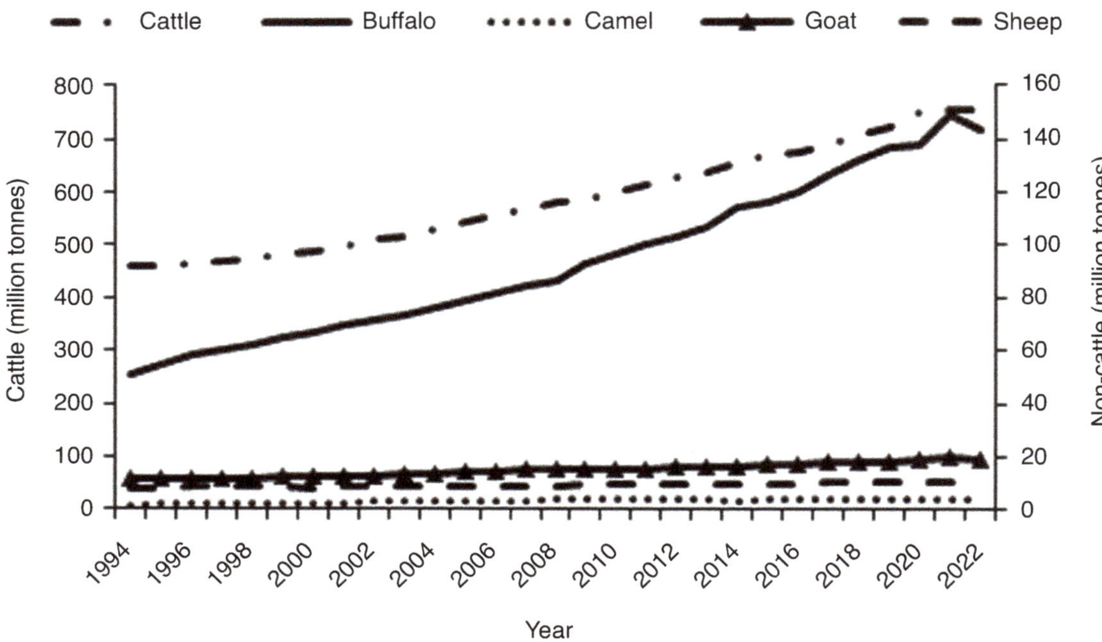

Fig. 48.1. Global fresh milk production from cattle, buffalo, goat, sheep and camel between the years 1994 and 2022. (From FAO, 2024a)

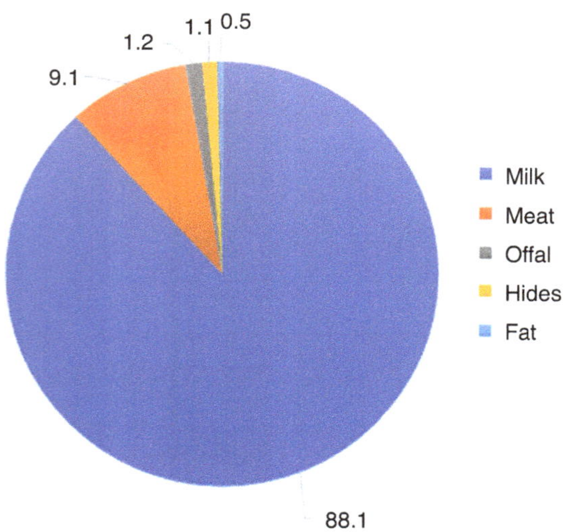

Fig. 48.2. Percentage of total global milk, meat, hides, offal and unrendered fat produced by cattle. (From FAO, 2024a)

Milk from cattle is not only an important source of nutrition for the newborn and pre-weaned calf, but also for humans in the form of fresh or dried milk, or other derived products (e.g. cheese, yoghurt, butter, cream etc.). The solid content of milk contains beneficial nutrients such as fat, protein, lactose, calcium, selenium, iodine and vitamins A-, E- and B-complex (Haug *et al*., 2007). The nutrient composition of milk can vary due to breed, diet and production system changes such as forage content (e.g. hay, haylage, grass, maize or whole-crop silages).

Different Production Systems

About 37% of terrestrial land area in the world is grassland (Suttie *et al*., 2005) and provides an affordable source of nutrients for ruminant livestock. The more productive permanent and temporary ley grasslands are often used for milk-producing animals due to their higher growth rate and often enhanced nutrient composition to meet the needs of the animal. Permanent pasture swards are classified as more than five years old, and often host a diverse range of native plant species compared to temporary leys, which are replaced within five years of sowing to maintain the productivity of the sward and its plant composition such as perennial ryegrass (*Lolium perenne*) and white clover (*Trifolium repens*) in more temperate climates (Frame and Laidlaw, 2011). In recent years, varieties of ryegrass with high growth vigour and high sugar content, and legumes rich in protein, are often used in temporary grassland swards to enhance forage productivity for feed. Legumes have the added benefit of helping to fix atmospheric nitrogen into soils to aid plant growth for forage production and reduce reliance on synthetic fertilizers (Moorby *et al.*, 2006; Hopkins and Wilkins, 2006).

Milk production at the herd level can be characterized by either:

1. seasonal calving: milk production and calving pattern (i.e. spring and/or autumn) timed to utilize grazing and available forage supplies; or
2. all-year-round calving: even profile of milk production and calving pattern that often has a higher reliance on conserved forages used in a partial or total mixed ration with cereal-based feeds.

Housed systems and all-year-round calving rely more on conserved forages (e.g. grass, maize and/or whole-crop silages) and blended cereal-based feeds, whereas seasonal calving often utilizes grazed pasture when possible. Forage is important for ruminant diets and provides a natural low-cost feed at 50–100% of the diet compared to high-energy, dense, cereal-based diets (Eastridge, 2006). Compared to forages, cereal-based feeds are often more consistent in nutrient content and can be blended to meet nutritional requirements of the animal at a given stage of production. Inefficiencies in the total output of milk produced can be caused by the metabolic requirements of the animal not being met and poor animal health and well-being resulting (Eastridge, 2006), which may also be linked to the genetic background of the individual animal (i.e. genotype × environment) (Dillon *et al.*, 2006).

Selective Breeding

There are, it is estimated, about 1300 different breeds and over 1 billion cattle in the world (FAO, 2024b). Holstein genetics are present in a large proportion of milking cows globally due to their potential for high milk yields. A genotype by feeding system interaction appears to exist whereby high-volume milk-yielding cows of predominantly North American Holstein genes tend to grow faster, be larger, have a shorter calving interval, have higher milk yield per kg dry matter and have reduced carbon footprint per unit milk when fed a high-energy dense diet (Capper *et al.*, 2009; Bell *et al.*, 2011), whereas smaller cows of predominantly New Zealand Holstein genes had similar performance on a high-forage or grazing diet (Macdonald *et al.*, 2007). Therefore, selecting animals for an environment is important. High genetic-potential cows for milk production mobilize their body energy reserves for production, which has been found to be detrimental for health and fertility, and overall survival (Pryce *et al.*, 1999; Dillon *et al.*, 2006). For this reason, we have seen a reduction in selection emphasis on milk production traits over the last 20 years (Fig. 48.3) and an increasing emphasis on health and fertility to improve the lifespan of the cow and its offspring (Egger-Danner *et al.*, 2015; Miglior *et al.*, 2017).

The average weighting on milk production traits in multi-trait selection indices has decreased by about 20% over the last two decades (Fig. 48.3), from 60% in the early 2000s (Miglior *et al.*, 2005). The selection emphasis has shifted away from traits related to animal productivity towards those related to efficient resource use including health, fertility, survival and feed intake (Cole *et al.*, 2020). The Nordic total merit index is currently the only multi-trait selection index for dairy cattle to maintain a balanced weighting across production, fitness and functional traits over the years (Fig. 48.3). It appears that other countries are adopting a similar approach as the general trend is to have a greater range of traits, which are associated with improved sustainability.

The carbon footprint of production can be minimized by improvements in herd health and fertility (improving longevity and productivity), and by reducing the number of replacement animals retained on the farm (Garnsworthy, 2004). Bell and Wilson (2018) found that regional differences in longevity of cows exist within UK dairy herds, with cows having a shorter life (averaging 2.6 lactations) in the regions with the highest milk yields and longest interval between calvings (associated with poor fertility), compared to other regions studied (about 2.8 lactations on average) with lower milk yields and calving intervals; the average number

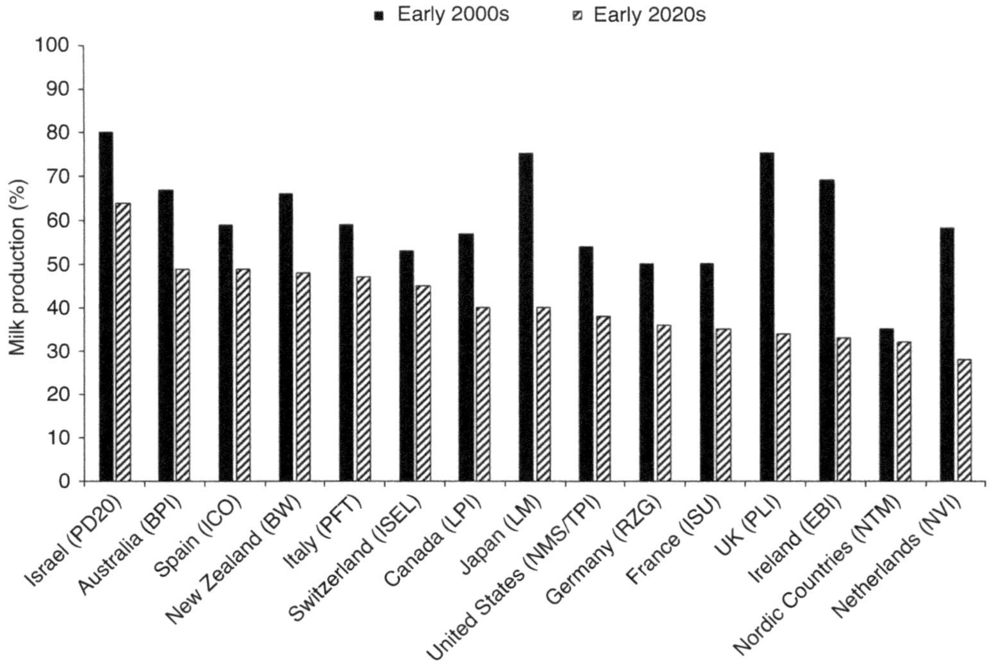

Fig. 48.3. Change in percentage weighting on milk production in national total merit selection indices from the early 2000s to early 2020s. (From Miglior *et al.*, 2005; Cole *et al.*, 2020).

of lactations across the UK was still below three lactations when cows are expected to reach their mature and optimum level of productivity.

The main benefits of selection to improve production efficiencies are increased productivity and gross efficiency (i.e. the ratio of yield of milk to resource input), by (i) diluting the maintenance cost of animals in the system, and (ii) fewer animals are needed. The permanent and cumulative improvements gained through a multi-trait selection index have typically been aimed at the objective of lifetime profitability, but this is now broadening to environmental and health objectives. Historically, the genetic gains over time have been evenly divided between increased genetic potential and improvements in nutrition and management (environment) (Cole *et al.*, 2020). Performance gains in more-heritable production traits (e.g. yield and composition) are more likely through selective breeding, whereas gains in less-heritable fitness and functional traits are more likely through improved environment (e.g. health, fertility and lifespan) (Egger-Danner *et al.*, 2015; Cole and VanRaden, 2018; Peñagaricano, 2020) (Fig. 48.4).

Hazel and Lush (1942) proposed the use of an economic selection index as a means of ranking animals on multiple traits of interest, where the aggregate genetic value (H) of the genotype is the trait (*1 to m*) genetic values (G) multiplied by their economic value or relative value (a) as $H = a_1 \times G_1 + a_2 \times G_2 + \ldots + a_m \times {}_{Gm}$. The economic value of a trait is calculated based on a unit change effect to inputs (e.g. feed and veterinary costs) and outputs (e.g. milk prices, calf prices) for the average animal over its lifetime within the herd. To determine the genetic background and breeding value of an animal involves collating phenotypic, genotypic and pedigree information. The genetic value is often the predicted transmitting ability (PTA) for each trait under selection and allows animals to be ranked based on the most profitable or desirable combination of genetic traits to suit your farming system. The PTA (youngstock, dam and bull) represents the estimated increase or reduction in a single trait that an animal will pass to its offspring. Therefore,

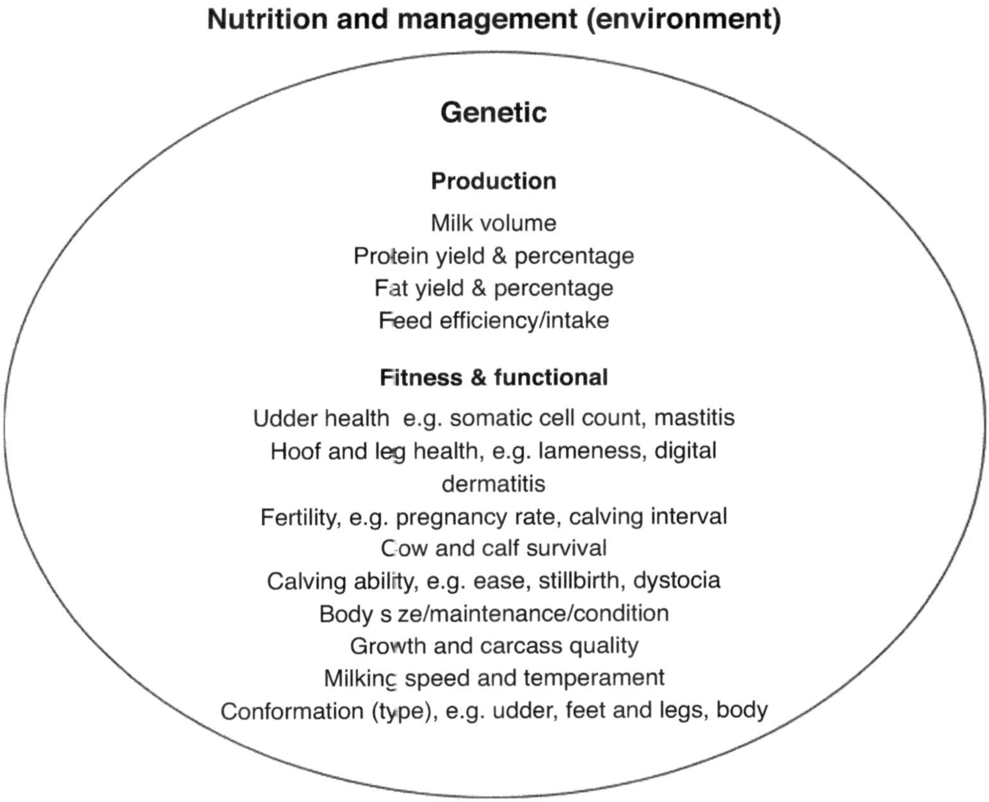

Fig. 48.4. Typical dairy cow traits used in national total merit indices for genetic selection within a given environment.

the dam and sire PTA have great importance for breeding future herd replacements and the genetic progress of the herd.

Sustainability

A challenge for modern food production is balancing the increasing need for food products from agricultural land whilst minimizing environmental impact. Sustainable development is commonly described as 'meeting the needs of the present without compromising the ability of future generations to meet their own needs' (World Commission on Environment and Development, 1987). With a continued demand for livestock products such as milk over several decades, selective breeding and economic selection indices have helped countries around the world meet the need for products and maintain marginal gains in financial performance for producers. With a high emphasis now on sustainable food production and reducing carbon emissions, the need for a more balanced approach across economic, social and environmental objectives has never been greater (Fig. 48.5).

For this reason, various national economic, environmental and health-breeding indices now exist (e.g. Profitable Lifetime, Envirocow, and Healthy Cow indices for dairy cows in the UK, and Balanced Performance Index, Health Weighted Index and Sustainability Index in Australia) including different traits and a more equal weighting placed on milk production, survival and feed-efficiency traits. Historically, economic indices have helped increase average milk yields and reduce the carbon emissions per unit of milk by about 1% per annum with improved efficiencies (Capper *et al.*, 2009; Bell *et al.*, 2015; Pryce and Bell, 2017). However, the emissions per cow are estimated to be increasing by 1.0% per annum due to increased production (Bell

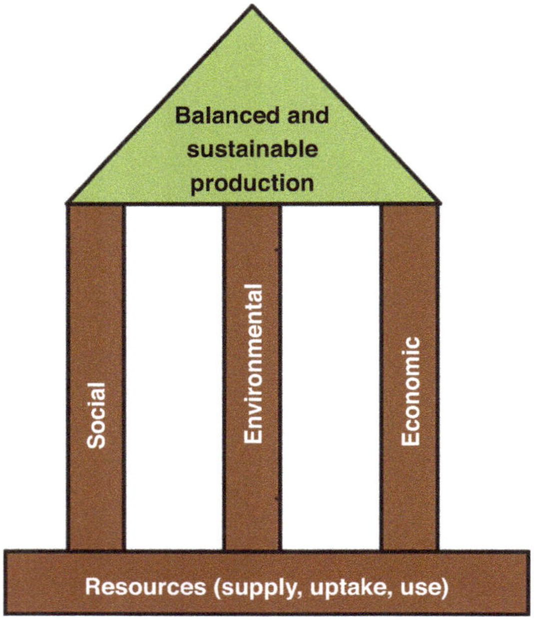

Fig. 48.5. Sustainable and balanced production across economic, social and environmental objectives.

et al., 2015). A higher weighting on fitness traits (e.g. health, fertility and survival) in a selection index should put more emphasis on environmental and social priorities; hence, we are seeing the reduction in weighting on production traits in national breeding indices (Fig. 48.3).

Developments in fitness and efficiency traits

While genetic selection in dairy cows has helped bring a remarkable reduction in emissions per unit milk over several decades (Capper *et al.*, 2009; Bell *et al.*, 2015; Pryce and Bell, 2017), there are still inefficiencies in production. Genetic improvements in health and fertility traits can be slower to achieve and more challenging to record than more heritable and measurable production traits. However, with the increase in monitoring technologies (activity, imaging, GPS, spectral, gas and temperature sensors), phenotypic measures are increasingly becoming more available for direct or indirect traits (Bell and Tzimiropoulos, 2018; Halachmi *et al.*, 2019; Cole *et al.*, 2020) (Fig. 48.6).

Accelerometers and motion sensors are already standard technology for monitoring oestrus

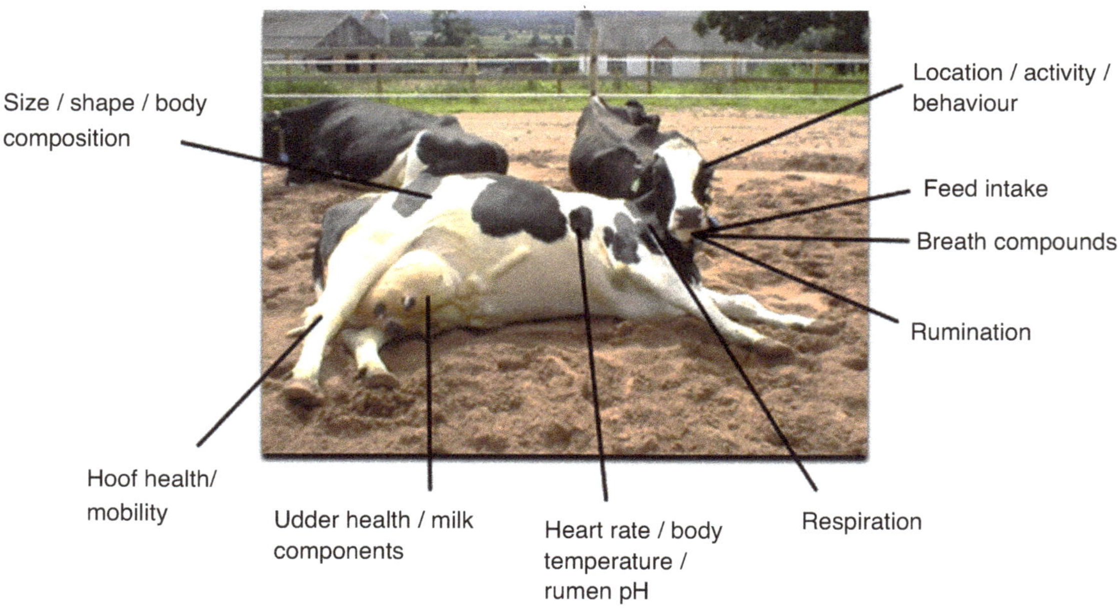

Fig. 48.6. Phenotypic measures becoming available for health, fertility, well-being and efficiency traits. (After Bell and Tzimiropoulos, 2018; Halachmi *et al.*, 2019; Cole *et al.*, 2020)

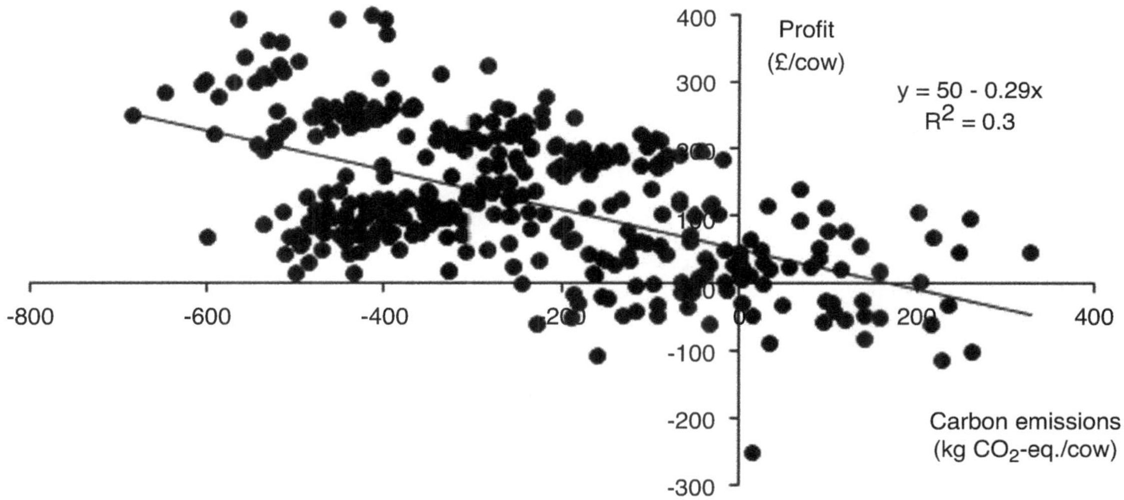

Fig. 48.7. Profit (£/cow) and carbon (kg CO_2-eq./cow) index values for a commercial dairy herd based on a multi-trait selection index including production (milk volume, milk fat and protein yield) and fitness (somatic cell count, fertility, lifespan) traits (Bell and Jauernik, 2023).

activity and provide help to farmers on when to breed animals. In-line milk sensors are not only being used to measure milk yields but also to monitor milk components (e.g. fat, protein, lactose, water, somatic cell count, urea and electric conductivity) and the potential energy status of the animal and its health (e.g. mastitis) (Halachmi *et al.*, 2019).

Ruminant livestock and, in particular, cattle have come under increasing scrutiny for their methane emissions from enteric and manure sources (Gerber *et al.*, 2013). About two-thirds of the carbon footprint of fresh milk is associated with the animal in the form of enteric and manure methane (FAO and GDP, 2018). Bell and Jauernik (2023) found that for a total merit index focused on carbon per cow (rather than carbon emissions per unit milk, such as Envirocow in the UK or Sustainability Index in Australia) coupled with an index value for profit per cow could bring more sustainable animals. Applying this approach of selecting animals on carbon and profit to a commercial herd, Bell and Jauernik (2023) found 76% of cows were classified as sustainable with a desirable increase in profit and reduction in carbon emissions (Fig. 48.7).

Carbon weightings to genetic values, rather than economic weightings, in a multi-trait selection index should bring about a permanent and cumulative reduction in emissions per cow as well as per unit product. In cattle, large inefficiencies exist in the use of the food they consume, with about 35% of energy in the diet being lost in the form of enteric methane, faeces or urine and 77% of nitrogen consumed lost in excreted faeces or urine (Bell *et al.*, 2015). Genomic breeding values are now available for feed saved or feed utilization efficiency (the difference between an animal's actual feed intake and its expected feed intake based on its size and growth over a defined period) and enteric methane emission traits (Richardson *et al.*, 2021; Manzanilla-Pech *et al.*, 2022), which are not routinely recorded on farms. It is well known that feed intake has a high and positive correlation with animal enteric methane emissions (Bell and Eckard, 2012), and due to wasted feed both traits are economically valuable for reducing the carbon footprint of milk (Bell *et al.*, 2016). Garnsworthy *et al.* (2012) found that spot measurements of enteric methane emissions during milking were correlated with total daily methane emissions by the same cows when housed subsequently in respiration chambers. Phenotypic measurement of enteric methane emissions on commercial farms using mobile gas analyzer equipment may allow direct selection (Lassen *et al.*, 2012; Garnsworthy *et al.*, 2012; Manafiazar *et al.*, 2017) and the ability to monitor changes in emissions due to dietary changes.

Conclusions

Cattle will continue to be an important source of food products and nutrition for humans. The ability of ruminants to turn forage into milk and meat is important for many countries around the world. Milk production from dairy cows has continued to evolve and help meet growing demand, with selective breeding and required nutrition developing to support this growth. The emphasis on milk production is now balanced with the need for more sustainable production practices and the increasing availability of more genetic traits being monitored across livestock populations, breeds and production systems. For environmental and societal reasons, fitness (e.g. health and fertility plus overall survival) and resource efficiency (feed intake and methane emissions) traits are equally as important to milk production systems in genetic selection. More balanced multi-trait selection has reduced the carbon footprint of production per unit product over several decades, and the sustainability footprint of cattle is expected to reduce with future selection on efficiency traits and emissions per cow.

References

Bell, M.J. and Tzimiropoulos, G. (2018) Novel monitoring systems to obtain dairy cattle phenotypes associated with sustainable production. *Frontiers in Sustainable Food Systems* 2, 31.

Bell, M.J. and Wilson, P. (2018) Estimated differences in economic and environmental performance of forage-based dairy herds across the UK. *Food and Energy Security* 7, e00127. DOI: 10.1002/fes3.127.

Bell, M.J. and Jauernik, G.-M. (2023) Selecting the 'sustainable' cow using a customized breeding index: Case study on a commercial UK dairy herd. *Agriculture* 13(2), 423.

Bell, M., Pryce, J. and Wilson, P. (2016) A comparison of the economic value for enteric methane emissions with other biological traits associated with dairy cows. *American Research Journal of Agriculture* 2, 1–17.

Bell, M.J., Wall, E., Russell, G., Simm, G. and Stott, A.W. (2011) The effect of improving cow productivity, fertility, and longevity on the global warming potential of dairy systems. *Journal of Dairy Science* 94(7), 3662–3678.

Bell, M.J. and Eckard, R.J. (2012) Reducing enteric methane losses from ruminant livestock – its measurement, prediction and the influence of diet. In: Javed, K. (ed.) *Livestock Production*. InTech Open, pp. 135–150.

Bell, M.J., Garnsworthy, P.C., Stott, A.W. and Pryce, J.E. (2015) Effects of changing cow production and fitness traits on profit and greenhouse gas emissions of UK dairy systems. *The Journal of Agricultural Science* 153(1), 138–151.

Capper, J.L., Cady, R.A. and Bauman, D.E. (2009) The environmental impact of dairy production: 1944 compared with 2007. *Journal of Animal Science* 87(6), 2160–2167.

Cole, J.B. and VanRaden, P.M. (2018) Symposium review: Possibilities in an age of genomics: The future of selection indices. *Journal of Dairy Science* 101(4), 3686–3701.

Cole, J.B., Eaglen, S.A.E., Maltecca, C., Mulder, H.A. and Pryce, J.E. (2020) The future of phenomics in dairy cattle breeding. *Animal Frontiers* 10(2), 37–44.

Dillon, P., Berry, D.P., Evans, R.D., Buckley, F. and Horan, B. (2006) Consequences of genetic selection for increased milk production in European seasonal pasture based systems of milk production. *Livestock Science* 99(2–3), 141–158.

Eastridge, M.L. (2006) Major advances in applied dairy cattle nutrition. *Journal of Dairy Science* 89(4), 1311–1323.

Egger-Danner, C., Cole, J.B., Pryce, J.E., Gengler, N., Heringstad, B. *et al.* (2015) Invited review: Overview of new traits and phenotyping strategies in dairy cattle with a focus on functional traits. *Animal* 9(2), 191–207.

FAO (2024a) FAOSTAT Statistical Database. Food and Agriculture Organization, Rome.

FAO (2024b) Domestic Animal Diversity Information System (DAD-IS). Food and Agriculture Organization, Rome.

FAO and GDP (2018) Climate Change and the Global Dairy Cattle Sector: The Role of the Dairy Sector in a Low-carbon Future. Food and Agriculture Organization, Rome.

Frame, J. and Laidlaw, A.S. (2011) *Improved Grassland Management*. The Crowood Press, Marlborough, UK.

Garnsworthy, P.C. (2004) The environmental impact of fertility in dairy cows: A modelling approach to predict methane and ammonia emissions. *Animal Feed Science and Technology* 112(1–4), 211–223.

Garnsworthy, P.C., Craigon, J., Hernandez-Medrano, J.H. and Saunders, N. (2012) On-farm methane measurements during milking correlate with total methane production by individual dairy cows. *Journal of Dairy Science* 95(6), 3166–3180.

Gerber, P.J., Hristov, A.N., Henderson, B., Makkar, H., Oh, J. *et al.* (2013) Technical options for the mitigation of direct methane and nitrous oxide emissions from livestock: A review. *Animal* 7(Suppl 2), 220–234.

Hackmann, T.J. and Spain, J.N. (2010) Invited review: Ruminant ecology and evolution: Perspectives useful to ruminant livestock research and production. *Journal of Dairy Science* 93(4), 1320–1334.

Halachmi, I., Guarino, M., Bewley, J. and Pastell, M. (2019) Smart animal agriculture: Application of real-time sensors to improve animal well-being and production. *Annual Review of Animal Biosciences* 7, 403–425.

Haug, A., Høstmark, A.T. and Harstad, O.M. (2007) Bovine milk in human nutrition: A review. *Lipids in Health and Disease* 6, 25.

Hazel, L.N. and Lush, J.L. (1942) The efficiency of three methods of selection. *Journal of Heredity* 33(11), 393–399.

Hopkins, A. and Wilkins, R.J. (2006) Temperate grassland: Key developments in the last century and future perspectives. *The Journal of Agricultural Science* 144(6), 503–523.

Lassen, J., Løvendahl, P. and Madsen, J. (2012) Accuracy of noninvasive breath methane measurements using Fourier transform infrared methods on individual cows. *Journal of Dairy Science* 95(2), 890–898.

Macdonald, K.A., McNaughton, L.R., Verkerk, G.A., Penno, J.W., Burton, L.J. *et al.* (2007) A comparison of three strains of Holstein-Friesian cows grazed on pasture: Growth, development, and puberty. *Journal of Dairy Science* 90(8), 3993–4003.

Manafiazar, G., Zimmerman, S. and Basarab, J.A. (2017) Repeatability and variability of short-term spot measurement of methane and carbon dioxide emissions from beef cattle using greenfeed emissions monitoring system. *Canadian Journal of Animal Science* 97, 118–126.

Manzanilla-Pech, C.I.V., Stephansen, R.B., Difford, G.F., Løvendahl, P. and Lassen, J. (2022) Selecting for feed efficient cows will help to reduce methane gas emissions. *Frontiers in Genetics* 13, 885932.

Miglior, F., Muir, B.L. and Doormaal, B.J. (2005) Selection indices in Holstein cattle of various countries. *Journal of Dairy Science* 88(3), 1255–1263.

Miglior, F., Fleming, A., Malchiodi, F., Brito, L.F., Martin, P. *et al.* (2017) A 100-year review: Identification and genetic selection of economically important traits in dairy cattle. *Journal of Dairy Science* 100(12), 10251–10271.

Moorby, J.M., Evans, R.T., Scollan, N.D., MacRae, J.C. and Theodorou, M.K. (2006) Increased concentration of water-soluble carbohydrate in perennial ryegrass (*Lolium perenne* L.). evaluation in dairy cows in early lactation. *Grass and Forage Science* 61(1), 52–59.

Peñagaricano, F. (2020) Genetic and genomics of dairy cattle. In: Bazer, F.W., Lamb, G.C. and Wu, G. (eds) *Animal Agriculture*. Academic Press, Cambridge, Massachusetts, pp. 101–119.

Pryce, J.E. and Bell, M.J. (2017) The impact of genetic selection on greenhouse-gas emissions in Australian dairy cattle. *Animal Production Science* 57(7), 1451.

Pryce, J.E., Nielsen, B.L., Veerkamp, R.F. and Simm, G. (1999) Genotype and feeding system effects and interactions for health and fertility traits in dairy cattle. *Livestock Production Science* 57(3), 193–201.

Richardson, C.M., Sunduimijid, B., Amer, P., Berg, I. and Pryce, J.E. (2021) A method for implementing methane breeding values in Australian dairy cattle. *Animal Production Science* 61(18), 1781–1787.

Suttie, J.M., Reynolds, S.G. and Batello, C. (2005) *Grassland of the World*. Food and Agriculture Organization, Rome.

World Commission on Environment and Development (1987) *Our Common Future*. Oxford University Press, Oxford, UK.

49 Animal Health and Nutrition in Organic Farming

Luciana Bignardi Brisola da Costa*

College of Veterinary Medicine, The Ohio State University, Columbus, Ohio, USA

Abstract

Organic dairy farming consists of a holistic approach to producing milk while prioritizing the health and welfare of the animals and the environment. This farming system is guided by strict regulations that promote natural behaviours, sustainable practices and the use of organic feed. Organic dairy cows are provided with proper living conditions, including access to pasture and open-air exercise areas. While organic dairy farming has many benefits, it also faces challenges such as weed control, maintaining yields, managing soil fertility and controlling insect pests. Despite these challenges, organic dairy farming continues to grow in popularity due to its numerous benefits to animal health, the environment and human health (Hughner *et al.*, 2007). However, further research is necessary since there is currently less scientific investigation into organic production compared to conventional methods (Manuelian *et al.*, 2020).

Introduction

The health of the animals in (organic) dairy farming is of paramount importance. Thus, a preventive management plan, good observation and early intervention are key to maintaining the health of the herd by supporting the organic standards that prohibit the routine use of synthetics and antibiotics (NOP Standard, n.d.; European Union, 2018).

Nowadays, animal farming faces several concerns and criticisms regarding its impact on environmental degradation, air pollution, greenhouse gas emissions and biodiversity, as well as on animal welfare and global health. To maintain the sector's value, farming practices must evolve to address these concerns. In this context, animal health and welfare must be seen as key contributors to the sustainability of the industry. Animal husbandry and animal health stakeholders must work together to improve their ability to anticipate, prevent and control the spread of animal diseases to ensure the required level of animal protein production. This is a challenge in the context of climate change, which is driving seasonal variations and changes in the epidemiology and distribution of many diseases, particularly vector-borne diseases such as bluetongue, lumpy skin disease, and others that are influenced by seasonal migration. Avian influenza is a prime example of this. In aquatic environments, even small changes, such as rising water temperatures, can influence disease development (Lavaine *et al.*, 2020; Espinosa-Marrón *et al.*, 2022; Eisen and Brown, 2022).

Organic products are viewed as a desirable choice for niche markets targeting consumers who prefer safe and nutritious products and those produced with environmental sensitivity (Rana and Paul, 2017). The industry of organic or natural food has developed exponentially in recent years with a growth rate of approximately 10–30% (Rana and Paul, 2017). This trend is demonstrated by consumers' readiness to pay for meat, milk or eggs that are free from antibiotics, and by import regulations enforced by large international bodies like the European Union. The growing domestic and international interest in reducing antibiotic use in livestock production implies that research to identify viable alternatives to antibiotics is of utmost importance. Results from

*Corresponding author: dacosta.28@osu.edu

 Advances in Organic Farming (ed. L. P. Awasthi)
DOI: 10.1079/9781800626850.0049

a study in Pakistan, indicate a positive correlation between an individual's awareness of health and food safety, and their intention to buy organic food products, facilitated by consumer engagement (Iqbal *et al.*, 2021).

In Europe, organic farming falls under the regulation of the European Commission. It has implemented various rules governing the production, distribution and marketing of organic products. Since 1991, the EU Organic Regulation has overseen organic farming at the EU level. Currently, Council Regulation (EC) No. 834/2007 outlines the European organic production requirements, including its goals and principles. The European Commission ensures that any substances used in organic agriculture adhere to EU rules and undergo rigorous assessment and approval before being employed in organic practices (Organic farming - European Commission (europa.eu) [accessed 5/3/25].

In 1990, the Organic Foods Production Act was enacted by the U.S. Congress. Two years later, in 1992, the National Organic Program (NOP; www.ams.usda.gov/nop [accessed 5/3/25]) was set up. This is a federal regulatory programme under the USDA that formulates and implements national standards for organic agricultural products sold in the USA. An advisory board, the National Organic Standards Board (NOSB), was also established to advise the NOP on new regulations and standards. By 2002, the NOP began accrediting certifiers. The USDA gives accreditation to certification agencies to ensure that farms and processing facilities adhere to organic regulations on its behalf (www.ams.usda.gov/nop [accessed 5/3/25]). These agencies have the authority to either grant or revoke organic certification.

Farmers have the liberty to select a certification agency based on various considerations. Besides confirming the certification status, the agency also informs the farmer or processor about any modifications to the regulations. Animal products from outside the USA can be marketed there as certified organic if they comply with USDA organic regulations or a recognized international standard and have trade agreements with the USA. At present, the USA maintains organic trade agreements with Canada, the European Union, Japan, the Republic of Korea and Switzerland.

As per the NOP, organic food should be cultivated without the use of traditional pesticides, fertilizers derived from petroleum or sewage sludge, herbicides, biotechnology (genetic engineering), antibiotics, growth hormones, or irradiation (Code of Federal Regulations or CFR 205.105). Animals that are reared in an organic environment must adhere to standards of health and welfare, should not be given antibiotics or growth hormones, must be fed entirely organic feed, and should have outdoor access (CFR 205.236 through 205.240 eCFR :: 7 CFR Part 205 -- National Organic Program [accessed 5/3/25]) (eCFR, n.d.).

Organic agriculture merges tradition, innovation and science to benefit the shared environment and foster fair relationships and a good quality of life for everyone involved. The International Federation of Organic Agricultural Movements (IFOAM) defines organic agriculture as a production system that '... should sustain and enhance the health of the soil, plant, animal, human, and planet as one and indivisible', and espouses the Principle of Care, asserting that 'Practitioners of organic agriculture can enhance efficiency and increase productivity, but this should not be at the risk of jeopardizing health and well-being' (IFOAM – Organics International, n.d.). However, this fundamental definition does not directly address an important aspect of production farming regarding the health or welfare of farmed animals.

The definitions of health and welfare can vary, and these variations can have different implications (Gunnarsson, 2006). The World Organisation for Animal Health (OIE, 2019) offers the most widely accepted definition of animal health and welfare, which is 'the physical and mental state of an animal in relation to the conditions in which it lives and dies'. This definition encompasses the 'five freedoms': (i) freedom from hunger, malnutrition, and thirst; (ii) freedom from fear and anxiety; (iii) freedom from heat stress or physical discomfort; (iv) freedom from pain, injury, and disease; and (v) freedom to express normal patterns of behavior (Wathes, 2010; Farm Animal Welfare Council, 2013). There is no need to define animal health and welfare differently in organic animal husbandry than in conventional methods. Fraser *et al.* (1997) acknowledged that there are several overlapping ethical issues concerning animal welfare based on the relative importance that individuals assign to different aspects of an animal's life. However, to properly evaluate all these concerns, a multi-disciplinary approach is needed that can assess biological function (e.g. health, production, malnutrition), affective states (e.g. fear, pain, emotions), and natural living (e.g. presence of natural behaviours).

In organic systems, the use of synthetic substances is not allowed, hence the rules for using antimicrobials vary by country. In non-organic systems, these are used to prevent or treat health issues in animals and minimize their suffering. Consumers who buy organic food often expect that the standards of animal welfare are met and are superior to those in non-organic production systems. A study conducted in the USA, which assessed the welfare status of organic dairy farms, revealed a comparable status to conventional dairy farms, albeit with a significant degree of variation (Bergman *et al.*, 2014).

A study from Brock *et al.* (2022) is among the few that investigate the current involvement of local and industry veterinarians in decision making on organic dairy farms. Strengthening collaborative initiatives among producers, local veterinarians, organic industry professionals and organic certifiers is important to maintain animal health. These efforts can facilitate stronger connections between parties and highlight the advantages for producers in working closely with local veterinarians. Providing additional resources for organic dairy herd health management, such as scientific research and continuing education opportunities, is crucial to supporting local veterinarians' efforts and enhancing producers' confidence in their professional abilities (Brock *et al.* 2021). A recent survey attracting 426 responses from across European countries provides insights and concerns of organic livestock producers. This study suggested that organic farmers rely on veterinarians as the main source of information, but the internet is gaining popularity for finding information on alternative treatments, and although farmers still largely rely on conventional treatments, phytotherapy is gaining popularity. Also, farmers identified 'feeding/nutrition', 'animal health' and 'welfare' as the most relevant issues in their farms (Manuelian *et al.*, 2023).

In both organic and conventional (or non-organic) dairy farms, the most prevalent and significant health problems are related to mastitis, lameness and reproductive issues, with mastitis being the most common disease.

Mastitis

Udder health or, more precisely, an indirect indicator of udder health called somatic cell count (SCC) is the most described health trait in the mastitis dairy literature. Even though much research and effort has been dedicated to mastitis control, it remains a persistent problem and is the most expensive disease of dairy cows (Schepers and Dijkhuizen, 1991; Barkema *et al.*, 1999). Mastitis is usually caused by a bacterial intramammary infection (IMI) and the financial implications of a mastitis case include reduced milk yield and quality, decreased reproductive efficiency, early culling, increased veterinary costs and, nowadays, the scrutiny on antibiotic use. Mastitis not only has economic consequences for the producer but also significantly affects the welfare of dairy cattle due to the negative states experienced by the animal. The adverse effects of mastitis are multi-faceted and impact the herd's longevity and success.

Mastitis, defined as inflammation of the mammary gland primarily caused by bacterial infection, is the most common animal disease reported on dairy farms. SCCs, typically used as an indirect measure of udder health and mastitis, are obtained from either individual animals or milk collected in the bulk tank. Cows with clinical mastitis (CM) show clear disease symptoms, while those with subclinical mastitis (SCM), show no apparent signs of disease and are more likely to recover with proper treatment. Detecting SCM is crucial to mitigate the negative effects of mastitis on animal production and welfare and prevent its progression to CM. Currently, diagnosing SCM is a significant challenge for dairy farmers. Failure to identify cows with SCM leads to prolonged infection durations and, consequently, prolonged impairments in animal well-being. Krieger *et al.* (2017) reported that the median prevalence of subclinical mastitis on 192 organic dairy farms in Germany, Spain, France and Sweden was 51%.

A significant caveat when assessing the prevalence of mastitis in dairy cattle, especially in Europe, is that only infections treated by a veterinarian are routinely recorded and reported. On organic farms, where alternative management strategies or remedies for mastitis are typically administered by the farmer rather than a veterinarian, fewer treatment records are kept and lower reporting rates may occur. This potential discrepancy between actual and reported cases of mastitis on organic farms was highlighted in a study from Sweden where the incidence of clinical mastitis treated and recorded by veterinarians was 9.1% in organic and 14.7% in non-organic herds; this contrasted with an incidence of 11.7% recorded by organic farmers themselves (Sutherland *et al.*, 2013).

The prevalence of antimicrobial resistance among mastitis-causing agents is a significant public health issue with studies showing that more than 60% of these agents exhibit resistance to at least one antimicrobial, and some of them even have zoonotic potential (Karns *et al.*, 2007; Boss *et al.*, 2016; Holko *et al.*, 2019). A study to measure antimicrobial usage done on adult dairy cows and pre-weaned calves in large U.S. non-organic farms showed large variability among farms, indicating that further studies are needed to identify risk factors associated with the consumption of antimicrobials (de Campos *et al.*, 2021). According to the United States Department of Agriculture's organic regulations, any animal treated with antibiotics will lose its organic status permanently and cannot return to organic production, even if the animal is managed organically. The use of antibiotics or other prohibited substances on organic animals is strictly regulated. There are no FDA-approved products specifically for mastitis treatment on organic dairy farms, and using unapproved products is contrary to FDA guidelines (USDA, 2015). Thus, organic farmers treat clinical mastitis using a variety of alternative therapies including essential oils, whey-based products, botanicals, vitamin supplements and homeopathy (Pol and Ruegg, 2007; Brock *et al.*, 2021). Effective prevention is crucial for managing mastitis, especially on organic dairy farms, where the options for treating infections in organic cattle are limited (Ruegg, 2009).

Although past studies have indicated lower levels of antimicrobial use on organic farms when compared to conventional dairy systems in the USA (Zwald *et al.*, 2004; Pol and Ruegg, 2007) and Europe (Bennedsgaard *et al.*, 2010), little research has explored the decision-making process surrounding necessary antimicrobial treatment.

Lameness

Lameness is a multi-factorial disease and is considered a welfare concern since it presents as a painful condition that can lead to behavioural changes in dairy animals (Galindo and Broom, 2002). Lameness is commonly considered to be one of the three most costly diseases in a dairy herd (after reproductive failure and mastitis), with financial losses attributable to decreased milk production and fertility as well as treatment and culling (Bruijnis *et al.*, 2010).

The prevalence of lameness varies significantly among dairy herds across different regions and countries due to diverse management practices that influence risk factors. In western Thailand, the cow-level prevalence of lameness was around 22% (Rahman *et al.*, 2014), slightly higher than a study on Australian dairy herds, where the prevalence was 19% (Ranjbar *et al.*, 2016). In India, a study found an incidence of lameness at 17%, which contrasts with the 9.4% recorded previously (Asit and Pankaj, 2016). Another study by von Keyserlingk *et al.* (2017) revealed that lameness was less common on organic farms compared to non-organic ones. In the latter, fewer than 5% of lactating cows had access to pasture during the grazing season. In the USA, researchers have highlighted significant differences in management practices between organic and conventional dairy farms. These differences can impact the occurrence of lameness in cows. For instance, a study involving 292 herds found that lameness prevalence ranged from 0% to 54%, with an average of 8%. Interestingly, this prevalence did not vary significantly between grazing systems, whether conventional non-grazing herds or organic and conventional herds with grazing practices (Richert *et al.*, 2013).

In a comprehensive study conducted by Browne *et al.* in 2016 (Browne *et al.*, 2022), they examined 85 U.S. organic dairy farms and focused on the associations between risk factors and lameness throughout the entire year, considering both grazing and housing periods. Age and genetic predicted transmitting ability for lameness were recognized as cow-level risk factors while factors such as farm, herd size, presence of other diseases (digital dermatitis), and the farmer's perception of lameness as a problem were identified as herd-level risk factors. In a meta-analysis systematic review studying dairy cows housed in free stall barns or tie-stall facilities, a total of 128 factors were identified with body condition score ≤ 2.5/5 linked to lameness, claw overgrowth, days in milk, herd size, parity for being lame (Oehm *et al.*, 2019). Flooring or housing systems can also result in long-term chronic pain for cows. Eicher *et al.* (2013) studied the effects of acute pain on chronic inflammation for cows in free stall housed on concrete with diamond-groove flooring compared with those housed on rubber. Their conclusion was that cows housed on concrete tended to have greater cortisol concentrations or be in 'more pain and (or) be more stressed'. These findings emphasize the importance of considering management practices when addressing lameness issues in dairy farming. Organic practices, such as providing access to pasture, appear to play a role in reducing lameness incidence. In a study comparing farm factors and prevalence of lameness on 40 organic and 40 non-organic dairy farms in the UK over a 2.5 year period, the average herd lameness prevalence was lower for organic farms compared to non-organic farms and similar between seasons, with autumn and winter around 16% and 19% during spring. This study provides evidence that organic management reduces herd lameness and emphasizes the multi-factorial etiology when numerous specific factors were found to significantly influence the prevalence of lameness (Rutherford *et al.*, 2009).

Metabolic Diseases

Metabolic diseases in cows can significantly impact their health and productivity.According to USDA organic regulations, organic dairy cows should have access to pastures during grazing season for a minimum of 120 days and 30% of the dry-matter intake should come from a certified organic pasture (USDA, 2011). As known, metabolic diseases in dairy cows can be influenced by nutrition and management practices.

Animal nutrition is highly dependent on local geographical conditions, and mineral deficiencies may occur in certain areas due to low mineral content or bioavailability of some trace elements in the soil. However, a major issue is regarding the risk of severe negative energy balance in cows in early lactation. However, according to Blanco-Penedo *et al.* (2012) and Richert *et al.* (2013), no evidence was found to support that organic cows were metabolically prone or had a severe negative energy balance.

Undoubtedly, organic dairy cows benefit from well-managed pastures, which contribute to their overall health and help mitigate metabolic diseases. Proper nutrition, access to pasture, and adherence to organic guidelines are essential for maintaining a healthy herd. Additionally, organic certification indicates that animal diets have a higher proportion of roughage and lower use of feed supplements and vitamins compared to non-organic operations (OEFFA, 2016). This could put those organically managed animals at an increased risk of metabolic diseases like milk fever and ketosis due to a relatively lower quality and/or restricted diet (Marley *et al.*, 2010).

A product recently approved for use in US-certified organic dairies was evaluated in a study that aimed to assess the impact of organic rumen-protected fat (RPF, Organilac®, Organic Animal Nutrition, Boulder, Colorado, USA) added to the diet of organic certified Holstein cows on milk yield, milk components, glucose and non-esterified fatty acid levels, reproductive performance, and health indicators (disease occurrence, culling and mortality). Cows from the organic group (n=100) were supplemented with organic RPF while the control group had pellets (n=102). Findings suggest that adding rumen-protected fat to the diet led to higher daily milk yield (by 1.6 kg/d during the supplementation period) and prevented body condition loss at 90 days in milk. However, RPF supplementation did not impact health, serum metabolite levels, milk components or reproductive outcomes (Manriquez *et al.*, 2019). On the other hand, Pappritz *et al.* (2011) did not find statistical differences when evaluating body condition scores in 30 cows supplemented with RPF-conjugated linoleic acid. Other studies have not detected improvements in milk yield when RPF was supplemented as well (Lohrenz *et al.*, 2010), or results differ based on the product used (McNamara *et al.*, 2003). The fact that it increases daily milk yield and affects body condition might be of interest to farmers, but it is important to note that most commercial rumen-protected fats are not allowed on U.S.-certified organic dairy farms (OEFFA, 2016).

Hardie *et al.* (2023) evaluated individual cow health data for treatment cost and proportion of disease in 16 certified U.S. organic farms. Among those farms, 11 provided treatment costs. The proportion of cases attributed to each disease presented in percentage and mean cost (shown between brackets) are the following: 53% mastitis ($46.10), 26% lameness ($66.36), 9% died ($64.98), 3% retained placenta ($45.59), 3% metritis ($28.66), 2% milk fever ($39.05), 2% respiratory ($48.35), 1% ketosis ($29.81), 1% displaced abomasum ($439.71). Krieger *et al.* (2017) assessed the prevalence of production disease from a convenient sample of 192 organic dairy farms in Germany, Spain, France and Sweden and found the median prevalence to be 51.3% for subclinical mastitis defined as somatic cell count > 1,00,000 cells/ml, 10% for risk of ketosis, 3.2% for risk of acidosis, 42% for prolonged calving interval, and 14.2% for clinical lameness.

A study mentioned that the risk of milk fever rises by 5% for each kilogram increase in milk production. The relatively lower level of milk production from organic herds is an explanation suggested for the reduced incidence of milk fever in those farms (Hardeng and Edge, 2001).

It is common knowledge that ketosis occurs following the accumulation of large quantities of ketone bodies in blood and tissue, leading to metabolic acidosis. When the incidence of ketosis was compared between organic and non-organic farms, no difference was observed (Hamilton *et al.*, 2002). The incidence of ketosis was higher in non-organic than long-time converted organic herds, but it was similar between non-organic herds and newly converted organic herds (Bennedsgaard *et al.*, 2003). Another comparative study between organic and non-organic dairy farms found a tendency for more cows on organic farms to have subclinical ketosis (Rutherford *et al.*, 2009; Sutherland *et al.*, 2013). However, cows were treated by a veterinarian for ketosis less regularly on organic than on non-organic farms in Norway (Valle *et al.*, 2007; Marley *et al.*, 2010.) As is the case with milk fever, a possible explanation for this lower incidence of ketosis on organic farms is relatively lower milk production.

Conclusion

Although organic farms share similarities with non-organic herds in terms of disease prevalence, organic farms rely primarily on preventive practices and alternative treatments to safeguard animal health. However, this equivalence should be approached with caution, as many of these diseases have multi-factorial causes, primarily correlate with management practices, and may be underestimated on organic farms. Notably, mastitis and lameness are relevant health issues in organic dairy as well, which require prompt attention to ensure animal welfare, longevity and economic viability. It is important to note that in organic dairy farming the use of antibiotics is strictly regulated to maintain the organic status of the animals. This approach can raise concerns from an animal welfare perspective. When an animal falls ill, conventional medicines, such as antibiotics, may be withheld until alternative treatments have been exhausted. Therefore, it is essential to scientifically evaluate organic alternative treatments to empower organic farmers with effective tools for maintaining animal health and welfare.

References

Asit, C. and Pankaj, K. (2016) Incidences of foot diseases of cattle in Bihar. *Indian International Journal of Agricultural Sciences and Research* 6, 267–272.

Barkema, H.W., Schukken, Y.H., Lam, T.J., Beiboer, M.L., Benedictus, G. *et al.* (1999) Management practices associated with the incidence rate of clinical mastitis. *Journal of Dairy Science* 82(8), 1643–1654.

Bennedsgaard, T.W., Thamsborg, S.M., Vaarst, M. and Enevoldsen, C. (2003) Eleven years of organic dairy production in Denmark: Herd health and production related to time of conversion and compared to conventional production. *Livestock Production Science* 80(1–2), 121–131.

Bennedsgaard, T.W., Klaas, I.C. and Vaarst, M. (2010) Reducing use of antimicrobials: Experiences from an intervention study in organic dairy herds in Denmark. *Livestock Science* 131(2–3), 183–192.

Bergman, M.A., Richert, R.M., Cicconi-Hogan, K.M., Gamroth, M.J., Schukken, Y.H. *et al.* (2014) Comparison of selected animal observations and management practices used to assess the welfare of calves and adult dairy cows on organic and conventional dairy farms. *Journal of Dairy Science* 97(7), 4269–4280.

Blanco-Penedo, I., Fall, N. and Emanuelson, U. (2012) Effects of turning to 100% organic feed on metabolic status of Swedish organic dairy cows. *Livestock Science* 143(2–3), 242–248.

Boss, R., Cosandey, A., Luini, M., Artursson, K., Bardiau, M. *et al.* (2016) Bovine *Staphylococcus aureus*: Subtyping, evolution, and zoonotic transfer. *Journal of Dairy Science* 99, 515–528.

Brock, C.C., Pempek, J.A., Jackson-Smith, D., Weaver, K., da Costa, L. *et al.* (2021) Organic dairy producer experiences and decisions related to disease prevention and treatment. *Journal of Dairy Science* 104(5), 5867–5880.

Brock, C.C., Pempek, J.A., Jackson-Smith, D., Habing, G.G., da Costa, L. *et al.* (2022) Managing organic dairy herd health: current roles and possible future roles for veterinarians with organic dairy clientele. *Journal of Dairy Science* 105(10), 8328–8341.

Browne, N., Hudson, C.D., Crossley, R.E., Sugrue, K., Kennedy, E. *et al.* (2022) Cow- and herd-level risk factors for lameness in partly housed pasture-based dairy cows. *Journal of Dairy Science 105*(2), 1418–1431.

Bruijnis, M.R.N., Hogeveen, H. and Stassen, E.N. (2010) Assessing economic consequences of foot disorders in dairy cattle using a dynamic stochastic simulation model. *Journal of Dairy Science* 93(6), 2419–2432.

de Campos, J.L., Kates, A., Steinberger, A., Sethi, A., Suen, G. *et al.* (2021) Quantification of antimicrobial usage in adult cows and preweaned calves on 40 large Wisconsin dairy farms using dose-based and mass-based metrics. *Journal of Dairy Science* 104(4), 4727–4745.

eCFR (n.d.) 7 CFR Part 205. Available at: https://www.ecfr.gov/current/title-7/subtitle-B/chapter-I/subchapter-M/part-205?toc=1 (accessed 5 March 2025).

Eicher, S.D., Lay, D.C., Arthington, J.D. and Schutz, M.M. (2013) Effects of rubber flooring during the first 2 lactations on production, locomotion, hoof health, immune functions, and stress. *Journal of Dairy Science* 96(6), 3639–3651.

Eisen, M.B. and Brown, P.O. (2022) Rapid global phaseout of animal agriculture has the potential to stabilize greenhouse gas levels for 30 years and offset 68 percent of CO_2 emissions this century. *PLoS Clim* 1(2), e00000102.

Espinosa-Marrón, A., Adams, K., Sinno, L., Cantu-Aldana, A., Tamez, M. *et al.* (2022) Environmental impact of animal-based food production and the feasibility of a shift toward sustainable plant-based diets in the United States. *Frontiers in Sustainability* 3, 8411063.

European Union (2018) Regulation (EU) 2018/848 of the European Parliament and of the Council of 30 May 2018 on organic production and labelling of organic products and repealing Council Regulation (EC) no 834/2007. *Official Journal of the European Union* 61, 1150.

Farm Animal Welfare Council (2013) *The Five Freedoms*. Farm Animal Welfare Council, London.

Fraser, D., Weary, D.M., Pajor, E.A. and Milligan, B.N. (1997) A scientific conception of animal welfare that reflects ethical concerns. *Animal Welfare* 6(3), 187–205.

Galindo, F. and Broom, D.M. (2002) The effects of lameness on social and individual behavior of dairy cows. *Journal Applied Animal Welfare Science* 5, 193–201.

Gunnarsson, S. (2006) The conceptualisation of health and disease in veterinary medicine. *Acta Vet Scand* 48(20), 1–6.

Hamilton, C., Hansson, I., Ekman, T., Emanuelson, U. and Forslund, K. (2002) Health of cows, calves and young stock on 26 organic dairy herds in Sweden. *Veterinary Record* 150, 503–508.

Hardeng, F. and Edge, V.L. (2001) Mastitis, ketosis, and milk fever in 31 organic and 93 conventional Norwegian dairy herds. *Journal of Dairy Science* 84, 2673–2679.

Hardie, L.C., Haagen, I.W., Heins, B.J. and Dechow, C.D. (2023) Genetic evaluation of health costs in US organic Holstein calves and cows. PMC10692295. *JDS Communication* 4(6), 464–468.

Holko, I., Tančin, V., Vršková, M. and Tvarožková, K. (2019) Prevalence and antimicrobial susceptibility of udder pathogens isolated from dairy cows in Slovakia. *The Journal of Dairy Research* 86(4), 436–439. DOI: 10.1017/S0022029919000694.

Hughner, R.S., McDonagh, P., Prothero, A., Shultz II, C.J. and Stanton, J. (2007) Who are organic food consumers? A compilation and review of why people purchase organic food. *Journal of Consumer Behaviour* 6(2–3), 94–110. DOI: 10.1002/cb.210.

IFOAM – Organics International (n.d.) Homepage. Available at: https://www.ifoam.bio/ (accessed 6 March 2025).

Iqbal, J., Yu, D., Zubair, M., Rasheed, M.I., Khizar, H.M.U. *et al.* (2021) Health consciousness, food safety concern, and consumer purchase intentions toward organic food: the role of consumer involvement and ecological motives. *SAGE Open* 11(2).

Karns, J.S., Van Kessel, J.S., McClusky, B.J. and Perdue, M.L. (2007) Incidence of *Escherichia* coli O157:H7 and *E. coli* virulence factors in US bulk tank milk as determined by polymerase chain reaction. *Journal of Dairy Science* 90(7), 3212–3219.

Krieger, M., Blanco-Penedo, I., Madouasse, A., Duval, J.E., Bareille, N. *et al.* (2017) Prevalence of production disease related indicators in organic dairy herds in four European countries. *Livestock Science* 198, 104–108.

Lavaine, E., Majerus, P. and Treich, N. (2020) Health, air pollution, and animal agriculture. *Review of Agricultural, Food and Environmental Study* 101(4), 517–528.

Lohrenz, A.K., Duske, K., Schneider, F., Nürnberg, K., Losand, B. *et al.* (2010) Milk performance and glucose metabolism in dairy cows fed rumen-protected fat during mid lactation. *Journal of Dairy Science* 93, 5867–5876.

Manriquez, D., Chen, L., Melendez, P. and Pinedo, P. (2019) The effect of an organic rumen-protected fat supplement on performance, metabolic status, and health of dairy cows. *BMC Veterinary Research* 15(1), 450.

Manuelian, C.L., Penasa, M., da Costa, L., Burbi, S., Righi, F. *et al.* (2020) Organic livestock production: A bibliometric review. *Animals* 10(4), 618.

Manuelian, C.L., Valleix, S., Bugaut, H., Fuerst-Waltl, B., da Costa, L. *et al.* (2023) Farmers concerns in relation to organic livestock production. *Italian Journal of Animal Science* 22(1), 1268–1282.

Marley, C.L., Weller, R.F., Neale, M., Main, D.C.J., Roderick, S. *et al.* (2010) Aligning health and welfare principles and practices in organic dairy systems: A review. *Animal* 4, 259–271.

McNamara, S., O'Mara, F.P., Rath, M. and Murphy, J.J. (2003) Effects of different transition diets on dry matter intake, milk production, and milk composition in dairy cows. *Journal of Dairy Science* 86, 2397–2408.

NOP Standard (n.d.) National Organic Program. Available at: https://www.ams.usda.gov/about-ams/programs-offices/national-organic-program (accessed 6 March 2025).

Oehm, A.W., Knubben-Schweizer, G., Rieger, A., Stoll, A., Hartnack, S. *et al*. (2019) A systematic review and meta-analyses of risk factors associated with lameness in dairy cows. *BMC Veterinary Research* 15, 346.

OEFFA (2016) *OEFFA Approved Products List for Producers*. OEFFA Certified Organic, Columbus, Ohio.

OIE (2019) Introduction to the recommendations for animal welfare. In: *Terrestrial Animal Health Code*. World Organisation for Animal Health. Available at: https://www.woah.org/fileadmin/Home/eng/Health_standards/tahc/2018/en_chapitre_aw_introduction.htm (accessed 6 March 2025).

Pappritz, J., Meyer, U., Kramer, R., Jahreis, G. and Rehage, J. (2011) Effects of long-term supplementation of dairy cow diets with rumen-protected conjugated linoleic acids (CLA) on performance, metabolic parameters and fatty acid profile in milk fat. *Archive Animal Nutrition* 65, 89–107.

Pol, M. and Ruegg, P.L. (2007) Treatment practices and quantification of antimicrobial drug usage in conventional and organic dairy farms in Wisconsin. *Journal of Dairy Science* 90(17183093), 249–261.

Rahman, M.A., Imtiaz, M.A., Ahaduzzaman, M., Ghosh, K.K., Masud, A.A. *et al*. (2014) Effects of flooring and rearing system on hoof health of dairy cows in some selected areas of Bangladesh. *Bangladesh Journal of Animal Science* 43, 132–137.

Rana, J. and Paul, J. (2017) Consumer behavior and purchase intention for organic food: A review and research agenda. *Journal of Retailing and Consumer Services* 38, 157–165.

Ranjbar, S., Rabiee, A.R., Gunn, A. and House, J.K. (2016) Identifying risk factors associated with lameness in pasture-based dairy herds. *Journal of Dairy Science* 99, 7495–7505.

Richert, R.M., Cicconi, K.M., Gamroth, M.J., Schukken, Y.H., Stiglbauer, K.E. *et al*. (2013) Perceptions and risk factors for lameness on organic and small conventional dairy farms. *Journal of Dairy Science* 96(8), 5018–5026.

Ruegg, P.L. (2009) Management of mastitis on organic and conventional dairy farms. *Journal of Animal Science* 87(13), 43–55.

Rutherford, K.M.D., Langford, F.M., Jack, M.C., Sherwood, L., Lawrence, A.B. *et al*. (2009) Lameness prevalence and risk factors in organic and non-organic dairy herds in the United Kingdom. *Veterinary Journal* 180(1), 95–105.

Schepers, J.A. and Dijkhuizen, A.A. (1991) The economics of mastitis and mastitis control in dairy cattle: A critical analysis of estimates published since 1970. *Preventive Veterinary Medicine* 10, 213–224.

Sutherland, M.A., Webster, J. and Sutherland, I. (2013) Animal health and welfare issues facing organic production systems. *Animals* 3(4), 1021–1035.

USDA (2011) Pasture for organic ruminant livestock: understanding and implementing the National Organic Program (NOP) pasture rule. Available at: https://www.ams.usda.gov/sites/default/files/media/NOP-UnderstandingOrganicPastureRule.pdf (accessed 6 March 2025).

USDA (2015) Treatment of sick or injured organic livestock. Available at: https://www.ams.usda.gov/sites/default/files/media/Treatment%20of%20Sick%20or%20Injured%20Organic%20Livestock_FINAL.pdf (accessed 6 March 2025).

Valle, P.S., Lien, G., Flaten, O., Koesling, M. and Ebbesvik, M. (2007) Herd health and health management in organic versus conventional dairy herds in Norway. *Livestock Science* 112, 123–132.

von Keyserlingk, M.A.G., Amorim Cestari, A. and Franks, B. (2017) Dairy cows value access to pasture as highly as fresh feed. *Scientific Reports* 7, 44953.

Wathes, C. (2010) Guarding the welfare of farm animals. *Veterinary Record* 167, 583–584.

Zwald, A.G., Ruegg, P.L., Kaneene, J.B., Warnick, L.D., Wells, S.J. *et al*. (2004) Management practices and reported antimicrobial usage on conventional and organic dairy farms. *Journal of Dairy Science* 87(14765827), 191–201.

50 Animal Health and Welfare in Organic Livestock Production: Current Status and Future Challenges

Georgios I. Papakonstantinou[1*], Georgia Terzidou[2], Dimitrios A. Gougoulis[1], Lampros Fotos[1] and Vasileios G. Papatsiros[1]

[1]*Clinic of Medicine, Faculty of Veterinary Medicine, School of Health Sciences, University of Thessaly, Karditsa, Greece;* [2]*Food Law Experts, Themistokleous, Athens, Greece*

Abstract

This chapter provides up-to-date information on the production, development, health and welfare of organic livestock as well as the potential prospects for sustainability and future development of its production. In particular, animal health and welfare are increasingly important in organic livestock farming due to rising consumer demand for ethically produced food. Current practices emphasize natural living conditions, access to the outdoors and the use of organic feed. However, there are other challenges, such as treating diseases without synthetic drugs and ensuring sufficient space and exercise. Future developments are likely to focus on expanded breeding programmes aimed at improving disease resistance, as well as welfare. There is an increasing focus on developing sustainable, natural treatments and preventative measures for common animal diseases. It is expected that policymakers and certification bodies will refine standards to ensure continuous improvements in animal welfare in line with evolving scientific knowledge and society's expectations.

Introduction

Over the last century, the production of animal-based foods has surged due to increasing demand, resulting in around 70 billion domestic animals being bred worldwide each year. Livestock farming will keep more animals to meet consumer demand, leading to significant expansion and development. Moreover, extensive livestock farming is the main or only source of livelihood for many people, which underlines the importance of protecting and promoting this sector (Vanhonacker *et al.*, 2009). The sector plays a crucial role in creating jobs, sustaining the rural economy and stabilizing rural population, all of which are essential for sustainable development in rural areas worldwide (Aertsens *et al.*, 2009). Despite its important role in the rural economy and ecosystems, the livestock sector significantly impacts the environment. Livestock farming contributes to water pollution and biodiversity loss. In addition, intensive livestock farming often leads to deforestation, soil degradation and contamination of water bodies with nutrients and pathogens (Brčić-Stipčević *et al.*, 2011). Consequently, the livestock sector represents a major challenge for European sustainability strategies. While it is an important pillar of the global food industry and can contribute to tackling the food crisis and achieving the Sustainable Development Goals, it also leads to environmental degradation. Organic farming is proposed as an alternative to intensive farming as it

*Corresponding author: geopapak@vet.uth.gr

DOI: 10.1079/9781800626850.0050

can minimize these negative impacts. In particular, organic livestock farming emphasizes pasture-based feeding, reducing dependence on intensive fodder cultivation, leading to deforestation and soil depletion (Vanhonacker *et al.*,2009). By avoiding the use of synthetic fertilizers and pesticides, organic farming reduces water pollution and helps maintain soil health. Through these methods, organic farming offers a more sustainable approach to livestock farming that reduces environmental impact and promotes ecological balance (United Nations).

Organic Livestock Production

Current status

The organic livestock farming sector varies widely by region and species, with notable geographical distribution and species trends. The diversity of farm types can be shown according to what is grown or reared, based on whether there is a single dominant activity or not. A farm is considered to be specialized when a particular activity provides at least two-thirds of the production or the business size of an agricultural holding. Others have a mix of activities in which no single activity dominates. In particular, some farms specialize in crop production, whether that be where field crop activities dominate, where permanent crops (like apples, grapes and olives) dominate, or indeed horticultural activities. Some farms specialize in animal production and animal products, whether that be where grazing livestock or granivores (such as pigs and poultry) dominate. Other farms have a mix of crops, livestock, or crops and livestock.

A majority of the EU's 245,000 fully organic farms in 2020 were specialized in crop production (53.8%). This was less than the corresponding share of all EU farms that specialized in crop production (58.3 %) (Fig. 50.1). Within crop specialists, there was a much lower proportion of cereals, oilseed and protein crop specialists in organic farming than all across all farms (8.3% vs. 15.9%) but a far higher proportion of organic fruit and citrus fruit farms (9.1% vs. 5.5% across all farms) and vineyards (7.8% vs. 4.7% across all farms). One-third of organic farms in the EU (33.0%) were livestock specialists in 2020. This was much higher than the corresponding share of all EU farms that were livestock specialists (21.7%). This higher share principally reflected much higher shares of specialist organic cattle-rearing and -fattening farms (12.6% vs. 4.3% across all farms) and specialist organic dairy farms (8.8% vs. 5.2%) (Fig. 50.2). Organic poultry, including chickens for meat and eggs, is one of the fastest-growing segments, especially in North America and Europe. Organic pork production is smaller but growing, especially in Europe. In addition, organic sheep and goat farming is strong in regions such as Europe and Australia, often for dairy and meat products (Boyazoglu *et al.*, 2005).

Production systems

Organic livestock production is based on five different production systems which are designed to ensure animal welfare, environmental sustainability and compliance with ecological principles. These systems can be categorized based on the types of animals and management methods (den Herder *et al.*, 2017):

- Grazing systems, where the animals are mainly kept on pasture and have access to natural feed. This system is common for ruminants such as cattle, sheep and goats. Key features include rotational grazing, minimal use of external inputs and maintaining pasture health so that animals can graze on organic pastures, which reduces feed costs and improves their health and welfare.
- Indoor housing with outdoor runs, where the animals are housed indoors but have regular access to outdoor areas. This system is typical for keeping poultry and pigs. It is crucial that the animals have sufficient space, access to outdoor runs or pastures and that organic feed is used. The housing in the barn is tailored to the animals' needs, while the outdoor run allows for natural behaviours such as foraging and exercise.
- In year-round free-range systems, the animals are outdoors all-year-round and have access to shelters that protect them from bad weather. This system is often used for hardy livestock breeds. Permanent pastures or runs, natural shelters and minimal confinement are the key features, with an emphasis on maintaining pasture quality and creating natural living conditions. Mixed systems combine indoor and outdoor housing depending on the season or production phase and are used for a variety of livestock, including pigs and poultry. The flexibility to move animals between indoor and outdoor housing, adaptation to climatic conditions and optimization of animal health and productivity are key components of this concept.

Fully organic farm specializations
(% share of fully organic farms, EU, 2020)

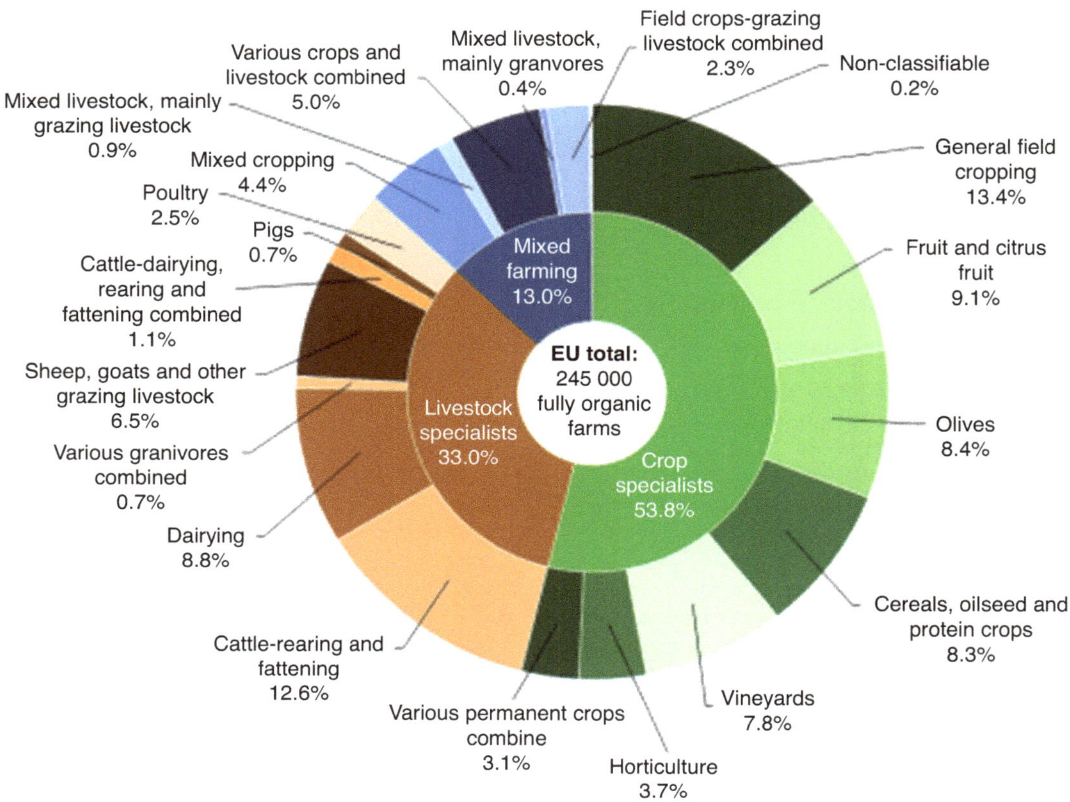

Fig. 50.1. EU fully organic farms in 2020. (From Eurostat)

- Free-range systems allow animals to roam freely within a specific area, which is typically used for poultry. Access to pasture, natural foraging behaviour and organic feed are key features that promote animal welfare and natural living conditions.

Each of these systems is tailored to the specific needs of the different animal species and follows the principles of organic farming, striving for a balance between animal welfare, environmental sustainability and economic profitability (den Herder *et al.*, 2017).

Feed requirements

The EU Council Regulations (EC-1804/1999 (Council Regulation, 1999) and EC-834/2007 (Council Regulation, 2007)) describe the rules for the appropriate risk assessment measures and contain the concept of the production of processed feed and criteria for the products and substances used in feed production (e.g. non-organic feed materials and feed additives). In particular:

- The feed requirements for organic livestock production are designed to ensure that animals receive nutritious, safe and ethically produced feed. These requirements comply with specific guidelines that promote animal health, environmental sustainability and ecological integrity. This means that it is produced without synthetic fertilizers, pesticides, herbicides or genetically modified organisms (GMOs). Farmers must keep records proving the organic status of the

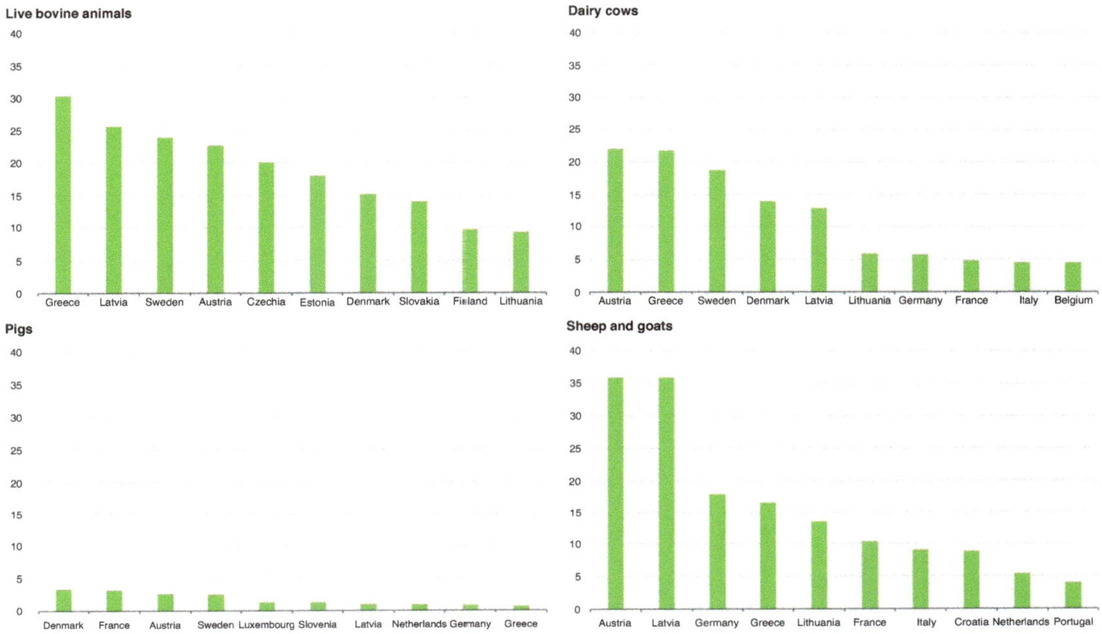

Fig. 50.2. Share of organic livestock in all livestock sectors, by countries with the highest shares, 2020 (% of the number of heads) (https://ec.europa.eu/eurostat/statistics-explained/index.php?title=File:Fig6_Share_of_organic_livestock_in_all_livestock,_by_countries_with_the_highest_shares,_2020_(%25_of_number_of_heads).png#filelinks [accessed 6/3/25])

feed, including certificates from suppliers and detailed documentation of feed purchases.

- Feed should consist primarily of natural ingredients such as organic grains, forages and pastures. Animal by-products, antibiotics and synthetic additives are prohibited. In addition, genetically modified feed ingredients are not permitted. All feed ingredients must be GM-free. The feed must provide a balanced diet that meets the nutritional requirements of the animals in the various phases of their life cycle, including proteins, vitamins and minerals.
- Ruminants, such as cattle, sheep and goats, must have access to pasture during the grazing season to promote natural behaviour and nutrition. Farmers should rotate grazing to prevent overgrazing, maintain soil health and promote biodiversity on pastures. The quality of pasture forage should be high to ensure adequate nutrition by maintaining healthy soils and plant diversity in grazing areas.
- The use of synthetic pesticides, herbicides and fertilizers for fodder plants is strictly prohibited. Antibiotics and hormones are not allowed in organic livestock feed; instead, preventative health measures and natural treatments are used. The feeding of animal by-products such as blood meal or feather meal is also prohibited in order to prevent the transmission of diseases and ensure the integrity of the organic feed.
- Organic livestock farms are encouraged to produce their own feed whenever possible to ensure better control over feed quality and organic certification. In addition to grazing, organic supplementary feed can be provided to meet nutritional needs, especially during periods when grazing is not possible, such as in the winter months. Livestock must have access to clean drinking water at all times, which is an essential component of overall nutrition and health.
- Organic feed must be stored separately from non-organic feed to avoid contamination, which

requires special storage facilities and careful handling procedures. Pest control in feed storage facilities must meet organic standards and rely on physical, mechanical and biological controls rather than synthetic pesticides. Detailed records of feed sources, quantities and storage conditions are essential, including purchase receipts, organic certificates and feed formulation details. This documentation should provide a clear audit trail for inspectors to verify during organic certification.

In summary, the requirements for organic feed are strict and aim to ensure the highest standards of animal welfare, environmental compatibility and food safety. By adhering to these guidelines, organic farmers can produce high-quality animal products that meet consumer expectations for ethical and sustainable farming (Aertsens *et al.*, 2009; Akaichi *et al.*, 2022) . The emphasis on organic feed, natural grazing and the ban on synthetic substances underline organic livestock farming's commitment to holistic and environmentally sound practices (Hölker *et al.*, 2019a; Hölker *et al.*, 2019b) .

Challenges in organic livestock farming

Animal health issues

Organic livestock farming, a method based on the principles of sustainability and natural processes, has many benefits, such as improved animal welfare and reduced environmental impact (Spooner *et al.*, 2014). However, it presents unique animal health challenges that require careful management (Fig. 50.3).

DISEASE MANAGEMENT. One of the most important health issues in organic livestock farming is disease management (Eijck *et al.*, 2005). Organic farming often restricts the use of synthetic medicines, including antibiotics and antiparasitics, which are commonly used in conventional farming to prevent and treat infections. This restriction can make it difficult to control disease outbreaks.

Coccidiosis, which is caused by protozoan parasites of the genus *Eimeria*, is a major problem in organic poultry farming. Symptoms include diarrhoea, weight loss, poor growth and a high mortality rate in severe cases. Avian influenza, caused by the avian influenza virus (AIV), is also a threat and manifests itself in respiratory problems, swelling of the head, reduced egg production and sudden death. Infectious bronchitis virus (IBV), leads to respiratory problems, reduced egg production and malformed eggs. Another problem is Marek's disease virus (MDV), which has symptoms such as tumours, paralysis and high mortality. In addition, *Mycoplasma gallisepticum* and *Mycoplasma synoviae* are also common health issues in organic poultry farming leading to respiratory symptoms, reduced growth and problems with egg production.

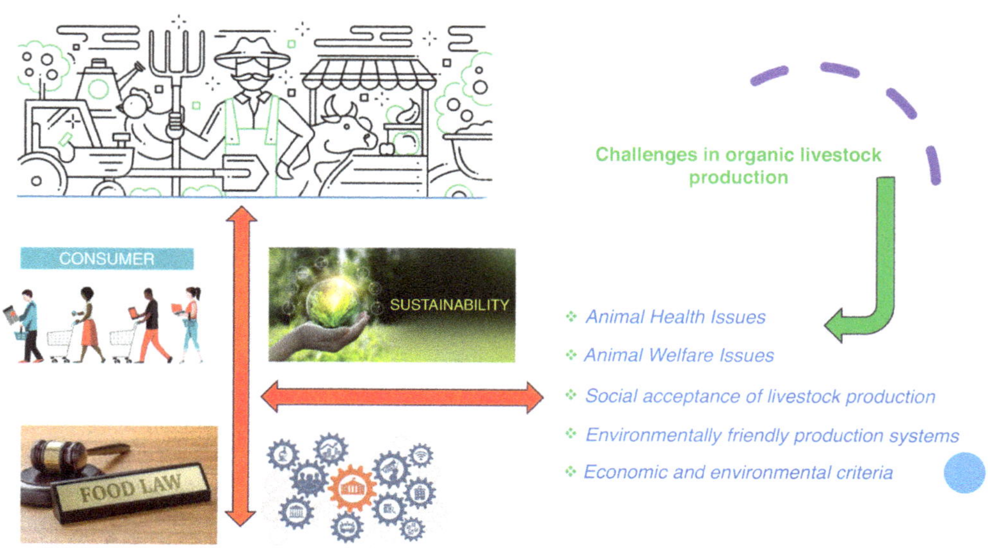

Fig. 50.3. Challenges in modern organic livestock production.

Salmonellosis leads to diarrhoea, dehydration, high mortality in chicks and chronic infections in adults (Eijick *et al.*, 2005). Parasite infestation is also a major health problem in organic poultry farming. Organic poultry can be infested by both internal and external parasites. Internal parasites, especially those that colonize the gastrointestinal tract, such as *Ascaridia galli* and *Heterakis gallinarum*, are the most common and can cause weight loss, reduced growth rates and digestive disorders. External parasites (lice, mites, fleas) live on the poultry's skin and feathers, cause irritation and can transmit diseases (Simoneit *et al.*, 2012).

Organic ruminant husbandry, including cattle, sheep and goats, is confronted with various diseases that can affect animal health and farm productivity. A major problem is mastitis, which is caused by bacterial infection of the udder, usually by *Streptococcus* and *Staphylococcus* species. Symptoms include a swollen, hard and painful udder and reduced milk production. Furthermore, common health problems are caused in organic ruminant farms due to Foot and Mouth Disease Virus (FMDV), Bovine Respiratory Disease (BRD) complex, Johne's disease (*Mycobacterium avium* subsp. *paratuberculosis*, MAP) and *Clostridial* diseases. Organic ruminants can also be infested by several internal and external parasites. Internal parasites, especially gastrointestinal nematodes (worms), are the most common and problematic. These include species such as *Haemonchus contortus*, *Ostertagia ostertagi* and *Trichostrongylus* spp., which can lead to symptoms such as diarrhoea, weight loss, anaemia or even death. External parasites such as lice, mites and ticks also pose a significant health risk as they cause skin irritation and itching and can transmit diseases (Eijck *et al.*, 2005).

The most common health problems in the Greek organic pig sector are respiratory infections (e.g. *Mycoplasma hyopneumoniae*). Due to their location (mountainous or semi-mountainous regions), organic pig farms have problems with wild boars that frequently enter their farms, either to access feed or to come into contact with sexually mature females. As a result, various seropositivities have been reported in farmed feral pigs, e.g. *Actinobacillus pleuropneumoniae*, Aujeszky's Disease Virus (ADV), Porcine Reproductive and Respiratory Syndrome Virus (PRRSV), *Mycoplasma hyopneumoniae* and *Erysipelothrix rhusiopathiae*. In view of the increase in wild boar populations and the frequent entry of them into organic pig farms, there is also an increased risk of the spread of African swine fever (ASF) (Eijck *et al.*,2005). Parasite infestation is also a common health problem in organic pig farming. Internal parasites (*Trichuris suis*, *Ascaris suum*, *Metastrongylus* spp., *Strongyloides ransomi*) and external parasites (*Sarcoptes scabei*) affect the health and productivity of pigs and lead to increased piglet mortality and poor body condition of sows after weaning. In addition, several studies have shown that *Trichinella britovi* poses a significant risk to organic pig farms (Simoneit *et al.*, 2012).

The control of animal diseases on organic farms requires a combination of preventative measures, natural treatments and strict adherence to the principles of organic farming. Preventative measures are crucial to maintaining animal health. Implementing strict biosecurity measures to prevent the introduction and spread of disease is just as important as controlling the movement of animals, people and equipment on the farm. Good nutrition plays a crucial role in strengthening the animal's immune system. A balanced and organic diet and access to clean water and minerals are essential. Effective pasture management, including rotating pastures to prevent the build-up of parasites and pathogens and avoiding overgrazing, helps to reduce the risk of infection from contaminated soils. Clean and well-ventilated stables, and regular cleaning and disinfection of feeding and watering facilities are important hygiene measures. The selection of breeds that are resistant to local diseases and appropriate breeding management to avoid genetic disorders also contribute to disease prevention (Simoneit *et al.*,2012). Regular health checks to detect early signs of disease and keeping detailed records of the animals' health status and treatments are important monitoring measures. Daily observation of the animals' behaviour, appetite and physical condition as well as immediate isolation and treatment of sick animals can prevent the spread of disease. Vaccination is also important for disease prevention, using vaccines approved for organic farming and following a vaccination schedule recommended by a veterinarian. Awareness campaigns are crucial to prevent the spread of disease. They include a variety of preventative information measures within the country and at border crossings for passengers, vehicles, river ports and airports (Eijck *et al.*, 2005). To raise awareness, it is essential to organize a series of lectures for all stakeholders, including veterinarians, farmers and hunters, especially in regions where backyard and free-range farming is prevalent. The distribution of simple information posters and leaflets with the most important basic information will also help to inform or remind all free-range farmers and their employees. Public media, such as local radio stations, television and newspapers, should also be part of the awareness campaign. Popular radio and television programmes dealing with agriculture and livestock production are among the most effective and quickest methods of disseminating biosecurity recommendations for this type of livestock production.

NUTRITIONAL DEFICIENCIES. Organic animal husbandry aims to maintain animal health through natural practices and without the use of synthetic products. However, this approach can sometimes lead to nutritional deficiencies and associated health problems. The most common deficiencies include vitamin and mineral deficiencies, protein deficiencies and energy deficiencies, each of which causes various health problems.

Vitamin E and selenium deficiencies can lead to white muscle disease, muscle weakness and reproductive problems as they are important antioxidants. A lack of vitamin A can lead to night blindness, reproductive problems, poor growth and respiratory problems, and can affect vision, immune function and reproduction. Calcium and phosphorus are essential for bone development and their deficiency can lead to rickets in young animals, and osteomalacia and poor milk production in adults. Iodine is important for thyroid function and its deficiency can lead to goiter, stillbirths, weak offspring and reduced fertility.

Protein deficiency manifests itself in poor growth, weight loss, reduced milk production and lower fertility. Protein provides essential amino acids for growth, tissue repair and metabolic functions, so inadequate protein intake can stunt growth and reduce productivity. Energy deficiency is characterized by weight loss, poor growth, lower milk yield and lethargy, as energy is crucial for all bodily functions.

Particular challenges in organic farming include limited access to concentrated feed and the lower mineral content of the soil. In organic farming, emphasis is often placed on feeding green fodder, which leads to lower energy and protein intake, especially during peak periods such as lactation. Soil quality also has an impact on pasture nutrition. Organic systems rely on natural fertilizers and crop rotations, which do not always ensure a balanced diet, especially when forage quality is poor or the climate is harsh.

To mitigate these problems, organic farmers can use various strategies. Ensuring diverse feed sources can help meet nutrient requirements, while natural mineral and vitamin supplements approved for organic farming can address specific deficiencies. Improving soil health and forage quality through crop rotation, composting and other organic soil amendments can improve pasture management. Regular health and nutritional testing can help to identify and correct deficiencies at an early stage. Finally, selecting breeds that are better adapted to organic farming and local conditions can ensure that livestock remain healthy and productive. In summary, by being aware of these potential deficiencies and applying appropriate management practices, organic farmers can ensure that their livestock remain healthy and productive and maintain the integrity and success of organic farming.

REPRODUCTIVE HEALTH ISSUES. Reproductive health is another area where organic animal husbandry faces challenges. The emphasis on natural mating and the reduced use of hormone treatments can lead to lower reproductive performance compared to conventional systems. Problems such as irregular estrus cycles, anestrus and poor conception rates may occur more frequently. In addition, dystocia or difficult labour is a common reproductive problem in livestock, which can lead to maternal and neonatal mortality and reduced fertility in affected animals. In ecological systems where interventions such as assisted deliveries and caesarean sections are limited, the treatment of dystocia requires careful monitoring of pregnant animals and adequate support during labour.

Addressing reproductive health issues in organic livestock production requires the selection of robust, reproductively fit animals and excellent nutrition and care to support reproductive health. In addition, natural breeding practices, such as ensuring good body condition and minimizing stress, are essential for maintaining high reproductive performance. Collaboration between farmers, veterinarians, researchers and agricultural extension services is also crucial for developing effective solutions and promoting sustainable reproductive health management in organic farming systems.

Animal welfare issues

Animal welfare is one of the most sensitive and controversial issues in today's society. Significant progress has been made, in particular with the recognition of the 1997 Amsterdam Treaty (Protocol on Protection and Welfare of Animals, OJ 1997, C 340/110), which recognizes that animals are 'beings with the capacity to feel'. This Protocol makes an important contribution to animal welfare by obliging the EU institutions and member states to take animal welfare fully into account when developing new policies in the fields of agriculture, transport, research and the internal market. With this in mind, the Farm Animal Welfare Council (FAWC) has developed the Five Freedoms framework, a widely recognized standard for assessing animal welfare. These are: (i) freedom from hunger, malnutrition and thirst; (ii) freedom from fear and anxiety; (iii) freedom from heat stress or physical discomfort; (iv) freedom from pain, injury and disease; and (v) freedom to exhibit normal behaviour patterns (Fig. 50.4). Therefore, organically reared animals should have the opportunity to live a natural life, appropriate to their physiological and behavioural conditions and well-being, in an environment that most closely resembles that to which the species is evolutionarily adapted. The ability of animals to lead a natural life is thus seen as a prerequisite for good animal welfare.

Organic animal husbandry aims to improve animal welfare by promoting natural living conditions and humane treatment. However, various animal welfare problems can arise in organic systems. In addition to the health and veterinary aspects already mentioned, access to exercise and pasture is an important welfare issue. Organic standards stipulate that animals

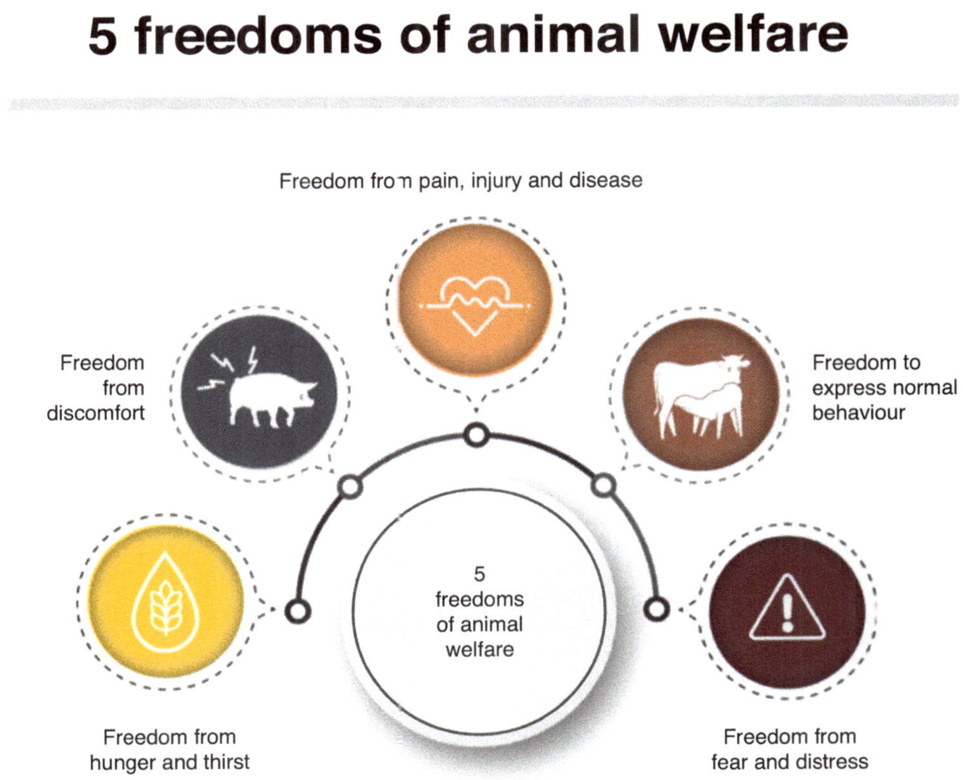

Fig. 50.4. The 'five freedoms" of the EU's legal framework on animal welfare. (From https://www.euractiv.com/section/agriculture-food/infographic/animal-welfare-in-the-eu/ [accessed 6/3/25])

must have access to exercise and pasture, but this can be a challenge in regions with extreme weather conditions or inadequate infrastructure. For example, depending on the geographical location, animals can be exposed to both heat and cold stress throughout the year. However, limited access to outdoor areas can lead to stress, reduced physical activity and health problems due to poor ventilation and hygiene. Therefore, it is crucial to develop housing and pasture management systems that protect animals from harsh weather while ensuring regular access to the outdoors. Furthermore, while housing and space requirements in organic farming are designed with animal welfare in mind, they can be difficult to maintain. Overcrowding and inadequate housing can lead to increased aggression, injuries and disease transmission. Therefore, the premises of these farms must meet or exceed organic standards and provide sufficient space, ventilation and exercise to encourage natural behaviours and social interactions to improve animal welfare. In addition, breeding on organic farms sometimes prioritizes high productivity over animal welfare, leading to health problems such as lameness in poultry or reproductive problems in dairy cows. Selecting breeds and animals that are well suited to organic systems and that prioritize health, longevity and welfare traits over maximum productivity can help mitigate these problems. Transportation and slaughter are also critical aspects of animal welfare. In particular, long transportation times and stressful slaughter processes can have a negative impact on animal welfare as they cause stress and poor treatment during transportation and slaughter, leading to suffering and poorer meat quality. Minimizing transport times, using low-stress handling techniques and ensuring humane slaughter methods are crucial to maintaining high animal welfare standards (Spooner *et al.*, 2014).

While organic livestock farming provides a framework for improved animal welfare, it also presents particular challenges. By understanding these issues and addressing them through thoughtful management practices, farmers can ensure that their animals lead healthy, comfortable and fulfilling lives. Continuous improvement and adherence to animal welfare standards are essential to the sustainability and integrity of organic livestock farming.

Future Directions and Prospects for Sustainability

The EU actively engages in legislative measures and initiatives to achieve climate neutrality by 2050. Specifically, the Sustainable Development Goals (SDGs) include: protecting biodiversity by 2030 (European Commission, n.d.); transitioning to circular systems in production and consumption; and safeguarding and optimizing the health of humans, animals and ecosystems. The EU aims for a fair green transition and, to this end, takes sustainable financing initiatives, establishing the Just Transition Mechanism (European Commission, n.d.) to provide funding. The livestock sector is the greatest challenge for the implementation of European sustainability strategies because, on the one hand, as a key pillar of the global food industry, it can support addressing the food crisis and actively contribute to achieving Sustainable Development Goals (Janseen *et al.*, 2016) . On the other hand, it impacts climate change significantly, contributing to environmental degradation through greenhouse gas emissions, water pollution and biodiversity loss. Addressing these challenges requires a holistic approach, necessitating the collaboration of public and private entities. Specifically, taking measures with environmental, social and governance (ESG) dimensions is essential for transitioning to a more sustainable, fairer and less risky livestock sector for both animal and human health. Organic livestock farming meets the economic and ecological criteria, as it could be a profitable industry and an environmentally friendly production system (United Nations).

Driven by the sustainable development of the organic livestock sector, the principles of the ESG agenda have emerged, focusing on regulating environmental, social and governance issues. These principles contribute to monitoring and evaluating organizations and companies that adopt best practices for the green transition.

Environmental aspects of sustainable livestock production

Although all three ESG pillars are important, the organic livestock sector tends to focus on actions for sustainable free-range farming that primarily involve environmental initiatives. The environmental aspects of ESG relate to reducing negative environmental impacts from activities in livestock production.

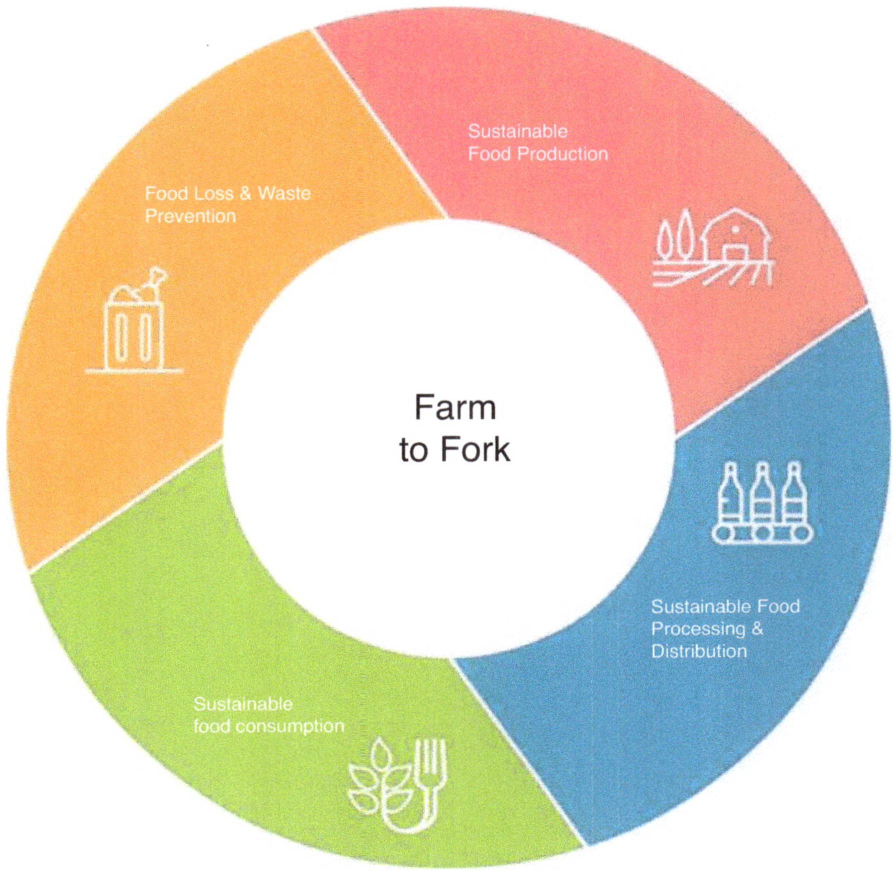

Fig. 50.5. EU's 'Farm to Fork' strategy for a fair, healthy and environmentally friendly food system. (https://food.ec.europa.eu/horizontal-topics/farm-fork-strategy_en [accessed 6/3/25])

In the context of the European Green Deal and the 'Farm to Fork' strategy (Fig. 50.5), the Common Agricultural Policy (CAP) 2023–2027 (European Commission, n.d.), which came into effect on 1 January 2023, aims to achieve environmental sustainability by focusing on three environmental objectives: combating climate change; protecting natural resources; and enhancing biodiversity. Specifically, the environmental measures under CAP 2023–2027 include, firstly, cross-compliance with the EU's high standards regarding public health and animal welfare; and, secondly, the green direct payments (eco-schemes) that support agricultural practices related to climate, environment, animal welfare and antimicrobial resistance. These measures contribute to achieving the Green Deal targets, such as maintaining permanent pastures, crop diversification, and ecological focus areas (United Nations).

Social aspects of sustainable livestock production

The social aspects of sustainable organic livestock farming are intertwined with working conditions, food safety and quality, and animal welfare at every stage from breeding to slaughter. Therefore, the social responsibility of livestock businesses is closely linked to best practices that respect the rights of both workers and farm animals. Consequently, this aspect of the ESG principles includes initiatives related, on the one hand, to participants in the production processes (workers, suppliers, investors), and, on the other, to the prevention of animal diseases, to limiting the transmission of diseases to humans and the occurrence of fatal diseases in the animal population, and to promoting the overall One Health strategy (World HealthOrganization).

Governance aspects of sustainable livestock production

The third pillar of ESG pertains to initiatives related to the corporate governance of an organization or entity. In the context of the organic livestock sector, this involves actions related to development issues, transparency, ethics and compliance with regulations, standards and national and European legislation.

In this framework, the EU encourages transparency and good governance in the supply chain by establishing a traceability system for animal-derived foods. Through monitoring systems applied at every stage of the production process, information on the animal population and producers is gathered, reinforcing the European 'Farm to Fork' strategy (European Commission, n.d.).

Critical assessment of the legal framework for sustainable organic livestock production in the light of ESG aspects

The ESG aspects that can be applied to organic livestock farming to achieve the goals set by the European Green Deal (European Commission, n.d.) were developed above. However, the sustainable and green transition of the livestock sector will not be achieved merely through arbitrary private initiatives and typical incorporation of environmental, social and governance solutions. Achieving climate neutrality by 2050 requires targeted actions harmonized with a robust European legislative framework (Zoiopoulos *et al.*, 2013).

In general, the EU has legislated a few individual issues, the assembly of which aids its efforts towards green transition. Specifically, Regulation (EU) 2023/1115 (EUR-LEX, n.d.), prohibiting the placing on the market of products (including bovine animals and animal feed) produced on 'deforested' or 'degraded lands'; Regulation (EU) 2016/429 (EUR-Lex, n.d.) ('The EU Animal Health Law'); Regulation (EU) 2017/625 (EUR-Lex, n.d.) ensuring the welfare of animals; and Regulation (EU) 2021/2115 (EUR-Lex, n.d.) ('Rules and financial support for CAP Strategic Plans') aimed at improving animal welfare; combating microbial resistance; and modernizing agriculture through digitalization, innovation and access to research; are some of the legislative texts indirectly related to the environmental aspects of sustainable livestock farming.

Furthermore, the social aspect of sustainable livestock farming is intertwined with addressing the changing needs of workers arising from digital and green transition. The current EU legislative framework provides for the protection of workers from traditional risks related to safety and health at work (e.g. risks of occupational accidents or exposure to hazardous chemicals). However, legislative initiatives are required for re-regulation and re-evaluation of existing legislation to ensure safe and healthy workspaces during the digital, green transition of livestock farming and to provide a comprehensive plan for preventing future crises, such as pandemics.

Lastly, within the governance framework, there is clear legislative provision, particularly regarding the traceability of animal populations, their derivatives, and animal feed (e.g. Regulation (EC) No 183/2005 (EUR-Lex, n.d.); Regulation (EC) No 1830/2003) (EUR-Lex, n.d.). However, the broader legislative approach is hindered because of the lack of a unified legislative text or even a package of updated legislation, which complicates comprehensive harmonization aimed at green transition and achieving more sustainable livestock farming.

However, it is worth noting at this point that the EU, recognizing the legislative gaps, regulatory failures and the need to institutionalize the principles and goals of the European Green Deal, issued Regulation (EU) 2020/852 (EUR-Lex, n.d.) on the establishment of a framework to facilitate sustainable investment. Essentially, the 'Taxonomy' Regulation, by setting common EU criteria, aims to inform investors about the environmental sustainability of economic activity. Specifically, the Regulation establishes the criteria for determining whether an economic activity is environmentally sustainable. For an economic activity to be considered sustainable, it must significantly contribute to achieving at least one environmental objective and simultaneously not harm others. The environmental objectives include climate change mitigation, climate change adaptation, the transition to a circular economy, pollution prevention and control, and the protection and restoration of biodiversity and ecosystems.

Thus, for a livestock activity to be classified as environmentally sustainable according to the EU taxonomy, it must meet all the technical criteria that ensure the activity has a positive environmental impact. To this end, delegated acts are issued, establishing lists of environmentally sustainable activities and specifying the technical screening criteria for each environmental objective.

Therefore, it is of great importance for organic livestock farming to be better organized, advised and supported by livestock experts. Financial and scientific support as well as the implementation of training programmes (especially for young farmers) could improve the current conditions in organic farming. In addition, awareness-raising campaigns such as lectures and/or the distribution of brochures with basic information should be organized. In addition, the European Union and the European Commission must contribute holistically to the gradual sustainable development and green adaptation through initiatives, the development of new strategies, and through a unified and comprehensive legislative framework.

Conclusions

Organic livestock farming has remained stable in recent years, but has potential in the European context as demand for more environmentally friendly production increases. The current status of animal health and welfare in organic livestock production is characterized by a strong commitment to natural and ethical practices, even though it faces constant challenges, such as treating diseases without synthetic drugs and ensuring sufficient space and exercise. The future of this sector looks promising, with advances expected in health-monitoring technologies and breeding programmes focused on improving disease resistance and welfare traits. In addition, the development of sustainable, natural treatments for animal diseases will be crucial. Ongoing improvements in policy and certification standards will play a crucial role in bringing organic farming practices in line with the latest scientific knowledge and public expectations, ensuring a progressive improvement in animal welfare in organic farming systems.

References

Aertsens, J., Mondelaers, K. and Huylenbroeck, G. (2009) Differences in retail strategies on the emerging organic market. *British Food Journal* 111(2), 138–154.

Akaichi, F., Glenk, K. and Revoredo-Giha, C. (2022) Bundling food labels: What role could the labels 'organic', 'local' and 'low fat' play in fostering the demand for animal-friendly meat. *Agribusiness* 38(2), 349–370.

Boyazoglu, J., Hatziminaoglou, I. and Morand-Fehr, P. (2005) The role of the goat in society: Past, present and perspectives for the future. *Small Ruminant Research* 60(1–2), 13–23.

Brčić-Stipčević, V. and Petljak, K. (2011) Research on organic food purchase in Croatia. *Tržište* 23(2), 189–207.

Council Regulation (1999) Council Regulation No. 1804/1999 of supplementing Regulation No. 2092/1991/EEC on organic production of agricultural products and indications referring thereto on agriculturalproducts and foodstuffs to include livestock production. Available at: https://eur-lex.europa.eu/legal-content/EN/TXT/PDF/?uri=CELEX:01999R1804-19990824 (accessed 8 August 2023).

Council Regulation (2007) Council Regulation No. 834/2007/EC of 28 June 2007 for organic production labelling of organic products, repealing Regulation No. 2092/1991/EEC. Available at: http://eur-lex.europa.eu/LexUriServ/LexUriServ.do?uri=OJ:L:2007:189:0001:0023:EN:PDF (accessed 8 August 2023).

den Herder, M., Moreno, G., Mosquera-Losada, R.M., Palma, J.H.N., Sidiropoulou, A. *et al.* (2017) Current extent and stratification of agroforestry in the European Union. *Agriculture, Ecosystems & Environment* 241, 121–132.

Eijck, I.A.J.M. and Borgsteede, F.H.M. (2005) A survey of gastrointestinal pig parasites on free-range, organic and conventional pig farms in The Netherlands. *Veterinary Research Communications* 29(5), 407–414.

EUR-Lex (n.d.) Regulation (EC) No 183/2005 of the European Parliament and of the Council of 12 January 2005 laying down requirements for feed hygiene (Text with EEA relevance).

EUR-Lex (n.d.) Regulation (EC) No 1830/2003 of the European Parliament and of the Council of 22 September 2003 concerning the traceability and labelling of genetically modified organisms and the traceability of food and feed products produced from genetically modified organisms and amending Directive 2001/18/EC.

EUR-Lex (n.d.) Regulation (EU) 2016/429 of the European Parliament and of the Council of 9 March 2016 on transmissible animal diseasesand amending and repealing certain acts in the area of animal health ('Animal Health Law') (Text with EEA relevance).

EUR-Lex (n.d.) Regulation (EU) 2017/625 of the European Parliament and of the Council of 15 March 2017 on official controls and other official activities performed to ensure the application of food and feed law, rules on animal health and welfare, plant health and plant protection products, amending Regulations (EC) No 999/2001, (EC) No 396/2005, (EC) No 1069/2009,(EC) No 1107/2009, (EU) No 1151/2012, (EU) No 652/2014, (EU) 2016/429 and (EU) 2016/2031 of the European Parliament and of the Council, Council Regulations (EC) No 1/2005 and (EC) No 1099/2009 and Council Directives 98/58/EC,1999/74/EC, 2007/43/EC, 2008/119/EC and 2008/120/EC, and repealing Regulations (EC) No 854/2004 and (EC) No 882/2004 of the European Parliament and of the Council, Council Directives 89/608/EEC, 89/662/EEC, 90/425/EEC, 91/496/EEC, 96/23/EC, 96/93/EC and 97/78/EC and Council Decision 92/438/EEC (Official Controls Regulation) Text with EEA relevance.

EUR-Lex (n.d.) Regulation (EU) 2020/852 of the European Parliament and of the Council of 18 June 2020 on the establishment of a framework to facilitate sustainable investment, and amending Regulation (EU) 2019/2088 (Text with EEA relevance).

EUR-Lex (n.d.) Regulation (EU) 2021/2115 of the European Parliament and of the Council of 2 December 2021 establishing rules on supportfor strategic plans to be drawn up by Member States under the common agricultural policy (CAP Strategic Plans) and financed by the European Agricultural Guarantee Fund (EAGF) and by the European Agricultural Fund for Rural Development (EAFRD) and repealing Regulations (EU) No 1305/2013 and (EU) No 1307/2013.

EUR-LEX (n.d.) Regulation (EU) 2023/1115 of the European Parliament and of the Council of 31 May 2023 on the making available on the Union market and the export from the Union of certain commodities and products associated with deforestation and forest degradation and repealing Regulation (EU) No 995/2010 (Text with EEA relevance).

European Commission (n.d.) Biodiversity strategy for 2030. Available at: https://environment.ec.europa.eu/strategy/biodiversity-strategy-2030_en (accessed 7 March 2025).

European Commission (n.d.) Delivering the European Green Deal. Available at: https://commission.europa.eu/strategy-and-policy/priorities-2019-2024/european-green-deal/delivering-european-green-deal_en (accessed 7 March 2025).

European Commission (n.d.) Farm to Fork strategy. Available at: https://food.ec.europa.eu/horizontal-topics/farm-fork-strategy_en (accessed 7 March 2025).

European Commission (n.d.) The common agricultural policy: 2023–27. Available at: https://agriculture.ec.europa.eu/common-agricultural-policy/cap-overview/cap-2023-27_en (accessed 7 March 2025).

European Commission (n.d.) The Just Transition Mechanism: Making sure no one is left behind. Available at: https://commission.europa.eu/strategy-and-policy/priorities-2019-2024/european-green-deal/finance-and-green-deal/just-transition-mechanism_en (accessed 7 March 2025).

Hölker, S., Meyer-Höfer, M. and Spiller, A. (2019a) Animal ethics and eating animals: Consumer segmentation based on domain-specific values. *Sustainability* 11(14), 3907.

Hölker, S., Meyer-Höfer, M. and Spiller, A. (2019b) Inclusion of animal ethics into the consumer value-attitude system using the example of game meat consumption. *Food Ethics* 3(1–2), 53–75. DOI: 10.1007/s41055-019-00036-6.

Janssen, M., Busch, C., Rödiger, M. and Hamm, U. (2016) Motives of consumers following a vegan diet and their attitudes towards animal agriculture. *Appetite* 105, 643–651.

Simoneit, C., Bender, S. and Koopmann, R. (2012) Quantitative and qualitative overview and assessment of literature on animal health in organic farming between 1991 and 2011 – Part II: Pigs, poultry, others. *vTI Agriculture and Forestry Research* 62(105), 110–115.

Smith, L.G., Kirk, G.J.D., Jones, P.J. and Williams, A.G. (2019) The greenhouse gas impacts of converting food production in England and Wales to organic methods. *Nature Communications* 10(1), 4641.

Spooner, J.M., Schuppli, C.A. and Fraser, D. (2014) Attitudes of Canadian citizens toward farm animal welfare: A qualitative study. *Livestock Science* 163, 150–158.

United Nations (n.d.) Department of Economic and Social Affairs –Sustainable Development, 'The 17 Goals'. Available at: https://sdgs.un.org/goals (accessed 7 March 2025).

United Nations (n.d.) Climate Change, 'Sustainable livestock management'. Available at: https://unfccc.int/technology/sustainable-livestock-management (accessed 7 March 2025).

Vanhonacker, F., Verbeke, W., Poucke, E., Buijs, S. and Tuyttens, F.A.M. (2009) Societal concern related to stocking density, pen size and group size in farm animal production. *Livestock Science* 123(1), 16–22.

World Health Organization (WHO) (n.d.) One Health. Available at: https://www.who.int/news-room/questions-and-answers/item/one-health (accessed 7 March 2025).

Zoiopoulos, P. and Hadjigeorgiou, I. (2013) Critical overview on organic legislation for animal production: Towards conventionalization of the system. *Sustainability* 5(7), 3077–3094.

51 Assessing the Impact of Organic Farming on Productivity among Norwegian Dairy Farmers: Evidence from a Semi-parametric Production Model

Fikru K. Alemayehu[1]*, Habtamu Alem[2] and Gudbrand Lien[1]

[1]*Inland Business School, Inland Norway University of Applied Sciences, Lillehammer, Norway;* [2]*Department of Economics Society, The Norwegian Institute of Bioeconomy Research (NIBIO), Ås, Norway*

Abstract

The literature shows that organic farming has become the centre of policies aiming to achieve sustainable agriculture due to its environmental benefits, such as increased biodiversity, reduced greenhouse emissions, etc. However, there is a gap in the literature on the productivity effects of organic farming over and above the conventional method to understand whether widely converting conventional farms pays off. The current study estimated the productivity function using a semi-parametric smooth-coefficient (SPSC) approach based on unbalanced panel data set from Norwegian dairy and crop farms during 1991 to 2020. The results show that organic farming, compared to conventional farming, increase productivity for most of the dairy farms, while for crop farms the effect is mixed. This finding suggests that organic farming for many farms can yield a productivity higher than or equal to conventional farming. However, the results depend on the farm under consideration, and there exists a large degree of heterogeneity among the farms. Likewise, the technical change is heterogeneous, indicating that some farms underwent technical progress (regress) or a neutral change during the study period. Finally, the returns to scale (RTS) are at the mean about 0.89 and 1.05 for dairy and crop farms, respectively, implying that these farms operate at a decreasing (increasing) returns to scale and can improve their productivity by decreasing (increasing) the current scale of operation.

Keywords: organic vs. conventional farming, semi-parametric smooth-coefficient regression, farm productivity, sustainable agriculture

Introduction

Sustainable agriculture has emerged as a major issue and topic of debate among researchers, policymakers, and entrepreneurs (Alem, 2021). Organic farming is one method of achieving sustainable production while adhering to environmental and animal welfare standards. Thus, promoting organic farming has become a crucial component of international and national food policy in Europe and other continents to encourage the production of safe and environmentally friendly foods (Sauer *et al.*, 2008). Organic agriculture increases biodiversity and reduces nitrogen losses to the environment

*Corresponding author: fikru.alemayehu@inn.no

DOI: 10.1079/9781800626850.0051

while increasing soil carbon sequestration that offsets 60–92% of current agricultural greenhouse gas emissions if organic farming practices were applied to all farmlands (Leifeld, 2012). As a result, evaluating the performance of organic farming and transferring knowledge from organic farming to conventional farming is critical.

This chapter's main research question is how organic farms perform compared to conventional ones, with evidence from Norwegian crop and dairy farms. Policy insight drawn from such analyses is important to decide whether to support a more widespread conversion of Norwegian crop and dairy farms. Indeed, given that organic farming is a complex innovation that calls for a strategic or systemic change on the part of the farmer, understanding and promoting its adoption at the farm level requires a thorough understanding of the role of the various driving forces (Yadava and Komaraiah, 2021).

Several studies were carried out to learn more about how farmers decide between organic and conventional production (e.g. Sipilainen and Oude Lansink, 2005; Sauer *et al.*, 2008; Koesling *et al.*, 2008; Yadava and Komaraiah, 2021). Also, Koesling *et al.* (2008) examined the determinants of the decision to convert to organic farming methods in a survey of 1018 Norwegian crop and dairy farmers using bivariate and multi-nomial logit models. Alonso and Guzmán (2010) investigated 78 different types of organic and conventional crops in the Spanish agricultural system. They compared the energy-use efficiency of these farming practices and found that organic farming uses less non-renewable energy inputs than conventional. Auerswald *et al.* (2003) measured the soil erosion potential of these farms and found that organic farming yields 15% less soil erosion than conventional. Yadava and Komaraiah (2021) assessed the performance of organic farming in India using the data of 21 states from 2016 to 2017. Their findings conclude that organic farming practices benefit India's land-degraded states. Sipilainen and Oude Lansink (2005) applied a stochastic distance frontier in a translog specification based on a sample of conventional and organic dairy farms in Finland from 1995 to 2002. Their findings show a lower average farm efficiency of organic farming. However, the marginal effects initially decline during the transition to organic farming and rise later in about 6 years after conversion. Lansink *et al.* (2002) compared the efficiency of conventional and organic crop and dairy farms in Finland. The study concluded a higher efficiency of organic farms relative to the conventional ones, but a lower efficiency compared to the overall frontier when conventional farms were considered. Sauer *et al.* (2008) estimated the change in the productivity of Danish organic farming using data from 56 organic milk farms from 2002 to 2004 and found evidence for a positive relationship between subsidy payments and increased farm efficiency, technological improvements, and a lower probability of organic market exit.

This chapter contributes to the literature by estimating the impact of organic farming on productivity among Norwegian dairy farmers using a semi-parametric smooth-coefficient (SPSC) production model. The advantage of the SPSC approach is its point estimates for each observation, which also helps identify the effects of these farming methods from farm heterogeneity. Alem (2023) emphasizes accounting for unobserved heterogeneity in these analyses as previous studies on Norwegian agriculture are criticized on this ground. The methodology also enables us to thoroughly understand the effects of organic farming by extending it into direct and indirect sources.

Case Analysis and Data Source

In this study, we used the annual farm-level panel data of crop and dairy farms in Norway from 1991 to 2020. These data consist of total observations of 8720, of which 6506 belong to dairy farms and 2214 to the crop. We received these data from the Norwegian Farm Accountancy database of the Norwegian Institute of Bioeconomy Research (NIBIO), which conducts surveys on approximately 1000 farms annually. The survey includes information on farm production and economic data collected from various regions, farm-size groups, and farm types. Participants were chosen randomly, but participation in the survey was completely voluntary. Also, there are no other restrictions or selection criteria to the number of years a farm is included in the survey except that approximately 10% of the surveyed farms are replaced by others yearly, which is why there is unbalanced data. That said, the survey is a representative sample of agricultural farms, particularly of those families that generate a significant portion of their income from dairy and crop farm activities.

Table 51.1 shows the summary of the case study analysis. The table shows that only about 9 and 42%, respectively, of the dairy and crop farm samples practised the organic farming methods, while the dominant rest used the conventional farming systems. The production function is constructed using sales revenue (Y_i), where i refers to dairy and crop farms and four inputs such as labor (X_{1i}), land (X_{2i}), materials (X_{3i}) and capital (X_{4i}). Further, the model includes the organic farming and the year dummy as the environmental or Z variables. More details are provided later in the chapter on this model construction. The revenue is generated from dairy and crop sales. Labour is the total number of hours spent working on the farm, including hired help, owner help and family help.

The land is the portion of farmland used for these activities. Materials is the cost of feed, oil and fuel, power, crop, animal protection expenditures, building supplies and other expenses. Capital includes expenses on fixed-cost items and depreciation and maintenance associated with and incurred by animals, buildings and machinery. All monetary values are measured in NOK1 and have been CPI-adjusted for 2020. We analyzed both farming systems separately.

Modelling Framework

Conceptual model

In measuring farm-level productivity, the production function is typically one of the main 'workhorses' for econometric analysis. The standard production function is a technical relationship between inputs and output, written as $Y = f(X)$, where Y is the output, and X is the vector of inputs. However, the amount of outputs produced given a certain level of inputs often depends on environmental factors, Z. These variables do not directly enter the production function like the inputs but externally influence the constant term, i.e. the productivity change and the input production technology. In our model, the Z variables denote dummy variables of organic farming and the year (i.e. technical change). Thus, we use a framework in which the Z variables change the production function, i.e. $Y = f|_Z\,(X)$, where $f|_Z\,(\cdot)$ indicates that the production function changes (shifts) with Z; thus, there is no single production function for all levels of Z.

The standard production function can sometimes be extended to include these Z variables as control variables as in $Y = f\,(X, Z)$ and estimated using the linear parametric functions like the ordinary least-squares method. However, a drawback of this approach is that it assumes the Z variables as additional inputs rather than as factors externally influencing the production function. The above specification $Y = f(X, Z)$ also assumes the same technology for all producers, but in our specification, $Y = f|_Z\,(X)$, the parameters are specific to the Z variables. For example, in a Cobb–Douglas specification of $Y = f|_Z\,(X)$, i.e., $Y_i = X_i'\beta$, the β parameters (the input elasticities) that define the technology are functions of Z; that is, $Y_i = X_i'\beta\,(Z)$. To make our specification more flexible, we assume that $\beta\,(Z)$ is non-parametric. This modelling framework is referred to as the semi-parametric smooth-coefficient (SPSC) regression model, in which the parameters (both the intercept and the slope parameters) are non-parametric functions of Z (Li *et al.*, 2002; Li and Racine, 2010).

Model specification

The SPSC Cobb-Douglas regression model can be written as:

$$Y_i = \beta_0\ (Z_i) + X_{1i}\,\beta_1\ (Z_i) + X_{2i}\,\beta_2\ (Z_i) + X_{3i}\,\beta_3\ (Z_i) + X_{4i}\,\beta_i\ (Z_i) + u_i\,,\ \ i = 1,\cdots,n \qquad (1)$$

where, Y_i is the logarithm (log) of output (measured using total revenue) for farm i, X_{1i} is the log of labour, X_{2i} is the log of land, X_{3i} is the log of materials, X_{4i} is the log of capital, and Z_i is a vector of environmental variables. We include two Z variables in our empirical model: the dummy variable of organic farming Z_1; and the year Z_2. Finally, the β parameters are smooth non-parametric functions to be estimated, and u_i is a random disturbance term.

Note that the model has a parametric structure regarding the input variables. The SPSC model is flexible because it allows the parameters to change with Z. Thus, the parameters are observation-specific if the Z variables are observation-specific. The overall marginal effect of the Z variables on output can be decomposed into direct and indirect marginal effects. If the β_j parameters, as well as the intercept, are linear functions of the Z variables, the model reduces to a restricted translog function.

Table 51.1. Data summary statistics (n=8720).

	Diary							
	Organic (n=531)				Conventional (n=5975)			
	Mean	First quartile	Median	Third quartile	Mean	First quartile	Median	Third quartile
Total revenue (in NOK 1000)	1293.82	681.92	993.36	1525.30	1427.60	593.84	1050.00	1890.00
Labour (100 hours)	35.79	29.12	34.40	41.05	35.17	28.34	34.10	40.45
Land (in hectares)	36.34	21.90	30.90	42.55	31.70	18.20	26.70	39.60
Materials (in NOK 1000)	368.66	165.33	257.94	449.84	471.44	166.23	312.50	613.07
Capital (in NOK 1000)	377,047	165,449	236,166	424613	391,897	136,422	242,126	482,105
Time in years (1991–2020)	18.62	15.00	18.00	23.00	18.50	15.00	19.00	25.00
	Crop							
	Organic (n=659)				Conventional (n=1555)			
Total revenue (in NOK 1000)	334.95	168.66	266.70	440.92	501.12	211.98	345.33	596.71
Labour (100 hours)	10.95	6.56	9.00	13.99	9.97	4.68	7.40	12.00
Land (in hectares)	33.35	18.75	29.00	42.45	35.46	18.90	28.90	42.20
Materials (in NOK 1000)	65.96	33.11	52.23	83.67	123.60	48.65	83.16	152.34
Capital (in NOK 1000)	139,989	73,973	119,545	186,297	225,403	93,212	158,341	281,261
Time in years (1991–2020)	11.58	6.00	13.00	15.00	19.28	12.00	21.00	26.00

Table 51.2. Elasticity estimates.

	Diary				Crop			
Variables	Mean	First quartile	Median	Third quartile	Mean	First quartile	Median	Third quartile
Intercept	1.82	1.46	1.77	2.07	1.06	0.48	0.97	1.28
Log (labour)	0.09	0.07	0.09	0.11	0.04	0.02	0.06	0.08
Log (farm land)	0.22	0.19	0.20	0.24	0.70	0.62	0.72	0.77
Log (materials)	0.47	0.45	0.51	0.54	0.17	0.09	0.20	0.26
Log (capital)	0.11	0.07	0.12	0.14	0.13	0.10	0.13	0.16
RTS	0.89	0.88	0.92	0.93	1.05	1.05	1.09	1.12

(See Appendix A on the matrix form and further derivation.)

Constrained estimation

According to neoclassical production theory, marginal products should be non-negative. However, in some cases in empirical work, and especially when applying the flexible SPSC model, marginal products (elasticities) can be negative. Because this is counterintuitive, it is desirable to impose constraints to ensure they are non-negative. One way to mitigate this potential problem is to consider constrained estimation. There is extensive literature on constrained semi-parametric estimation (see Chapter 12 in Henderson and Parmeter, 2015 for an overview). See Appendix B for more details on the restricted estimation.

Marginal effects of the Z variables

The change in the expected value of Y_i (which is in log form) with respect to a change in a particular element of $X_i = X_j$ $(i = 1,2,\ldots j = 1,2,\ldots)$ (which are also in log form), is $\beta_j(Z_i)$, based on equation (1). That is, the input elasticities $\beta_j(Z_i)$ are defined as:

$$\frac{\partial \hat{Y}}{\partial X_j} = \beta_j(Z_i) \quad (2)$$

Note that as long as the Z variables are observation-specific, and these input elasticities are functions of Z_i, we can obtain observation-specific input elasticities. Based on equation (1), the change in the expected value of Y_i with respect to a change in a particular element of Z_i, say, Z_{ki} $(= 1,2,\ldots k = 1,2\ldots)$ is defined as:

$$\frac{\partial \hat{Y}}{\partial Z_{ki}} = \frac{\partial \beta_0(Z_i)}{\partial Z_{ki}} + \sum_{j=i}^{J} X_j \frac{\partial \beta_j(Z_i)}{\partial Z_{ki}} \quad (3)$$

Then, the marginal impacts of the Z variables depend on each Z and X variable included in the analysis. The first component of (3) is the direct marginal impact of Z_{ki}. The second component is the indirect marginal impact of Z_{ki} – the sum of the direct and indirect effects results in the overall marginal effects.

Results and Discussion

Elasticity estimates and returns to scale

Table 51.2 shows the point elasticity estimates and returns to scale, which is the sum of these elasticity estimates for dairy and crop production summarized at the mean, and at the first, second and third quartiles. The elasticity estimates show a percentage change in the respective outputs, i.e. dairy and crops, from a percentage change in these inputs. At the mean, for example, the elasticity of labour, land, materials and capital are 0.09%, 0.22%, 0.47 and 0.11%, respectively, for dairy farms, but the elasticity of these inputs become 0.04%, 0.70%, 0.17%, and 0.13%, respectively, for crop farms. (See Figs. 51.1 and 51.2 in Appendix C for the distributions of these elasticity estimates.)

Table 51.3 provides the classification of these estimates based on the sign and statistical significance, i.e. the shares of observations with negative or positive (and significant) point estimates and/or insignificant ones. Our results show that the input elasticity of dairy farms is either positive or insignificant, and there are no negative values. However, in the case of crop farms, the estimates include some outlier observations, i.e. observations with negative values, which should be excluded according to the rules for restricted estimation outlined in 'Constrained estimation' below. Except for these outliers, our results are consistent with the microeconomic theory that conjectures a positive relationship between the inputs and outputs, i.e. the higher (lower) farm inputs, the higher (lower) outputs will be. Once we ensure this assumption holds, the interpretation of the elasticity estimates is straightforward.

The results also show a difference between dairy and crop farms regarding the elasticity of labour, land and materials. Labour and materials are more elastic for dairy production than crops, but this becomes the opposite in the case of land, which is more elastic for crop relative to dairy. In both cases, capital is more elastic than labour as these farms are capital-intensive and heavily replace labour with machines and robots in Norway. Thus, these findings corroborate the natural phenomenon in the Norwegian agricultural system.

The returns to scale (RTS) are about 0.89 for dairy and 1.05 for crop farms, at the mean. The interpretation of RTS follows like this: when the RTS is equal to 1, it means that a 1% change in inputs yields a proportionate 1% change in outputs, indicating the firm is operating at the optimal (desired) scale of operation. Nevertheless, when the RTS is less (greater) than 1, these findings indicate decreasing (increasing) returns to scale, i.e. the farm's operations lie below (above) this desired level. Based on these criteria, the results show that dairy farms operate close to but below the optimal scale, i.e. decreasing returns to scale. In contrast, crop farms operate at a scale above this level, i.e. increasing returns to scale, implying that crop farms can improve their productivity by increasing their current scale. While we found decreasing returns of dairy farms, Lien *et al.* (2017) found increasing returns to scale for an earlier time period.

Marginal effects

Organic farming

To recap, the main focus of this study is to find out the effects of organic farming on dairy and crop productivity compared to the conventional method. Note that the direct marginal effect (DE) refers to the change in output/productivity via the constant term, β0, in our model. In contrast, the indirect marginal effect (i.e.) refers to the effects via technology parameters, such as labour, land, materials and capital. Thus, overall marginal effects (OE) are the effect of these two sources.

Table 51.4 presents the direct, indirect and overall marginal effects of the Z variables *vis-à-vis* organic farming and technical change. The table shows that the direct (indirect) marginal effects of organic farming are negative (positive), and the overall effect becomes positive at the mean for dairy farms. These results are also similar in sign for the remaining quartiles except for the first quartile of indirect effects. For crop farms, however, the results show negative direct and indirect marginal effects and hence neutralizing, i.e. close to zero, overall effects. The exception to these findings is the positive direct and indirect marginal effects at the third quartile and the overall effects at the median. (For more details, see the density distribution of these point estimates in Figs 51.3 and 51.4 in Appendix D.)

Table 51.5 also provides that the majority of observations for direct (indirect) marginal effects are negative (positive), but overall, slightly above one-third and one-half of the observations are negative and positive for dairy farms. Similarly, the direct and indirect marginal effects dominate in opposite directions, while the overall effects balance these counteracting effects for the crop farms. Thus, these results show heterogeneous

Table 51.3. Significance of coefficient elasticities.

	Diary			Crop		
	Negative	Not sign.	Positive	Negative	Not sign.	Positive
Intercept	0.00	0.00	1.00	0.02	0.04	0.94
Log (labour)	0.00	0.00	1.00	0.14	0.10	0.76
Log (farm land)	0.00	0.00	1.00	0.00	0.00	1.00
Log(materials)	0.00	0.00	1.00	0.07	0.03	0.90
Log (capital)	0.00	0.03	97	0.04	0.01	0.95
RTS	0.00	0.00	1.00	0.00	0.00	1.00

Table 51.4. Direct, indirect and overall marginal effects

	Dairy (*n*=6504)				Crop (*n*=2214)			
Variables	Mean	First quartile	Median	Third quartile	Mean	First quartile	Median	Third quartile
				Organic farming				
DE	−0.04	−0.06	−0.03	−0.01	−0.00	−0.16	−0.00	0.03
IE	0.04	−0.10	0.06	0.21	−0.01	−0.40	−0.02	0.44
OE	0.05	0.01	0.03	0.06	0.00	−0.03	0.01	0.14
				Technical change				
DE	−0.02	−0.19	−0.02	0.12	0.02	−0.40	0.06	0.40
IE	0.00	−0.00	0.00	0.00	0.00	−0.01	−0.00	0.01
OE	0.01	−0.00	0.01	0.03	0.01	−0.03	0.01	0.04

Table 51.5. Significance of direct and indirect marginal effects.

	Dairy (*n*=6504)			Crop (*n*=2214)		
Variables	Negative	Not sign.	Positive	Negative	Not sign.	Positive
Organic farming						
DE	0.68	0.22	0.10	0.50	0.09	0.40
IE	0.10	0.16	0.73	0.38	0.12	0.50
OE	0.35	0.12	0.53	0.46	0.14	0.40
Technical change						
DE	0.35	0.15	0.51	0.49	0.12	0.39
IE	0.49	0.10	0.42	0.40	0.07	0.53
OE	0.24	0.10	0.66	0.38	0.14	0.49

marginal effects of organic farming, i.e. negative, zero or positive, depending on the farm under consideration. These findings suggest that organic farming can boost (dampen) or yield similar farm productivity to conventional farming. These are in line with the findings of Sipilainen and Oude Lansink (2005), which explained the heterogeneity in terms of the time lags from the adoption of organic farming. Applying the same logic to our findings, farms adopt the organic farming system at different periods so that those adopting it earlier can improve their productivity, while farms that joined later face no change or declining productivity, as the benefit can take some years to reap. The finding can also make sense as the productivity of organic farming varies depending on other farm characteristics. That means the findings highlight the need to identify the farms that can improve their productivity and those that do not, and the reason behind the failure to improve it after adopting organic farming and sustainable agriculture effectively. Nevertheless, we leave this to future research, as the relationship between farm characteristics and productivity impacts of organic farming lies beyond the scope of this chapter.

Technical change

The results also report the effects of technical change (TC). The direct (indirect) marginal effects are negative (positive), and the overall marginal effects are positive at the mean for dairy farms. However, it shows positive direct, indirect, and overall marginal effects for crop farms. Figs 51.3 and 51.4 in Appendix D further provide the density distribution of TC. The classification table further shows that the TC is negative for some observations, but it becomes neutral or positive for others. This suggests the Norwegian dairy and crop farms underwent mixed technical change regimes, i.e. technical regress (progress) or neither of the two depending on the specific farm.

Conclusion

The study estimates the productivity effects of organic farming compared to the conventional method of farming. The empirical analysis is based on unbalanced panel data from Norwegian dairy and crop farms from 1991 to 2020. We used a semi-parametric smooth-coefficient (SPSC) approach and estimated separately for each farm type. The analysis revealed heterogeneous direct, indirect and overall marginal effects of organic farming and technical change. Furthermore, the findings show that returns to scale are decreasing (increasing) for dairy and crop farms, respectively. These results indicate that not all organic farms respond similarly in terms of productivity. As a result, policy tools should segment these farms to provide effective farm management and contribute to sustainable agriculture.

References

Alem, H. (2021) The role of technical efficiency achieving sustainable development: A dynamic analysis of Norwegian dairy farms. *Sustainability* 13(4), 1841.

Alem, H. (2023) Accounting for heterogeneity in performance evaluation of Norwegian dairy and crop-producing farms. *Economies* 11(1), 9.

Alonso, A.M. and Guzmán, G.J. (2010) Comparison of the efficiency and use of energy in organic and conventional farming in Spanish agricultural systems. *Journal of Sustainable Agriculture* 34(3), 312–338.

Auerswald, K., Kainz, M. and Fiener, P. (2003) Soil erosion potential of organic versus conventional farming evaluated by USLE modelling of cropping statistics for agricultural districts in Bavaria. *Soil Use and Management* 19(4), 305–311.

Henderson, D.J. and Parmeter, C.F. (2015) *Applied Nonparametric Econometrics*. Cambridge University Press, Cambridge, UK.

Koesling, M., Flaten, O. and Lien, G. (2008) Factors influencing the conversion to organic farming in Norway. *International Journal of Agricultural Resources, Governance and Ecology* 7(1/2), 78.

Lansink, A.O., Pietola, K. and Bäckman, S. (2002) Effciency and productivity of conventional and organic farms in Finland 1994–1997. *European Review of Agriculture Economics* 29(1), 51–65.

Leifeld, J. (2012) How sustainable is organic farming. *Agriculture, Ecosystems & Environment* 150, 121–122.

Li, Q. and Racine, J.S. (2010) Smooth varying-coefficient estimation and inference for qualitative and quantitative data. *Econometric Theory* 26(6), 1607–1637.

Li, Q., Huang, C.J., Li, D. and Fu, T.-T. (2002) Semiparametric smooth coefficient models. *Journal of Business & Economic Statistics* 20(3), 412–422.

Lien, G., Kumbhakar, S.C. and Hardaker, J.B. (2017) Accounting for risk in productivity analysis: an application to Norwegian dairy farming. *Journal of Productivity Analysis* 47(3), 247–257.

Sauer, J., Park, T. and Graversen, J. (2008) Organic farming in Denmark – productivity, technical change and market exit. *Proceedings of Schriften Der Gesellschaft Für Wirtschafts-Und Sozialwissenschaften Des Landbaues eV* 43(874–2017–1023), 379–391.

Sipilainen, T. and Oude Lansink, A.G. (2005) Learning in organic farming: An application on Finnish dairy farms. [Technical report] [Technical report].

Yadava, A.K. and Komaraiah, J.B. (2021) Benchmarking the performance of organic farming in India. *Journal of Public Affairs* 21(2), e2208.

Appendices

A. Matrix form and further derivations

In matrix form, (1) can be written as:

$$Y_i = X_i{}'\delta\,(Z_i) \;\; + u_i \tag{4}$$

where $X_i' = \;(1, X_{1i},\, X_{2i}, X_{3i})$ and $\delta(Z_i) = (\alpha(Z_i), \beta_1(Z_i), \beta_2(Z_i), \beta_3(Z_i))'$

This model can be estimated by employing the kernel method (Li *et al.*, 2002). Pre-multiplying (4) by X_i and taking the conditional expectation $\mathrm{E}\,(\cdot|Z_i)$yields the following:

$$(X_i Y_i|Z_i) \;=\; \mathrm{E}\;(X_i X_i{}'|Z_i)\;\delta\;(Z_i) \tag{5}$$

because $\mathrm{E}\;(X_i u_i|Z_i) \;=\; 0$. Rearranging (5), we obtain:

$$\delta\,(Z_i) \;=\; [\mathrm{E}\;(X_i X_i{}'|Z_i)]^{\;-1}\mathrm{E}\;(X_i Y_i|Z_i) \tag{6}$$

Then, the parameter function IV (Z_i) can be estimated by:

$$\hat{\delta}\,(Z_i) = \left[\;\sum_{i=1}^{n} X_i'\mathrm{K}\;(Z_i,\;\mathrm{z})\right]^{-1} X_i X_i'\mathrm{K}\;(Z_i,\;\mathrm{z})^{-1}\, X_i Y_i \mathrm{K}\;(Z_i,\;\mathrm{z}), \tag{7}$$

where K $(\mathrm{Zi},\;\mathrm{z})$ is the kernel function. As discussed in the literature (for example Henderson and Parmeter, 2015), there are several ways to specify the kernel function (for example, Gaussian, uniform or Epanechnikov) and several ways to specify the bandwidth (for instance, least-squares cross-validation or likelihood cross-validation). We choose Gaussian kernel functions and least-squares cross-validation for this study to select the bandwidths.

B. More on constrained estimation

To impose the required constraints, we rewrite (7) as:

$$\hat{\delta}(Z_i) = \sum_{i=1}^{n} \mathrm{A}(X_i, Z_i, \mathrm{z})Y_i, \tag{8}$$

where $\mathrm{A}\;(X_i,\;Z_i,\;\mathrm{z}) \;=\; \left[\sum_{i=1}^{n} X_i X_i{}'\mathrm{K}\;(Z_i,\;\mathrm{z})\right]^{-1} \sum_{i=1}^{n} X_i \mathrm{K}\;(Z_i,\;\mathrm{z})$. This equation can be written as:

$$\hat{\delta}\;(Z_i) \;=\; n.\sum_{i=1}^{n} \mathrm{A}\;(X_i,\;Z_i,\;\mathrm{z})\,.\,p_u.\;Y_i, \tag{9}$$

where $p_u = n^{-1}$ denotes uniform weights. Then, we introduced a re-weighting (from the uniform weight) scheme that changes the dependent variable as little as possible but simultaneously avoided negative input elasticities. We can specify the constrained SPSC estimator as:

$$\hat{\delta}^* \left(Z_i\right) = n \cdot \sum_{i=1}^{n} \mathrm{A} \left(X_i, Z_i, z\right) \cdot p_i \cdot Y_i, \tag{10}$$

where p_i denotes the observation-specific weights and $\sum_i p_i = 1$. To select the optimal p_i, we minimize $\sum_i \left(p_i - p_u\right)^2$ subject to β (Zi) ≥ 0 using quadratic programming (the quadprog package in R).

C. Plots of input elasticity and returns to scale

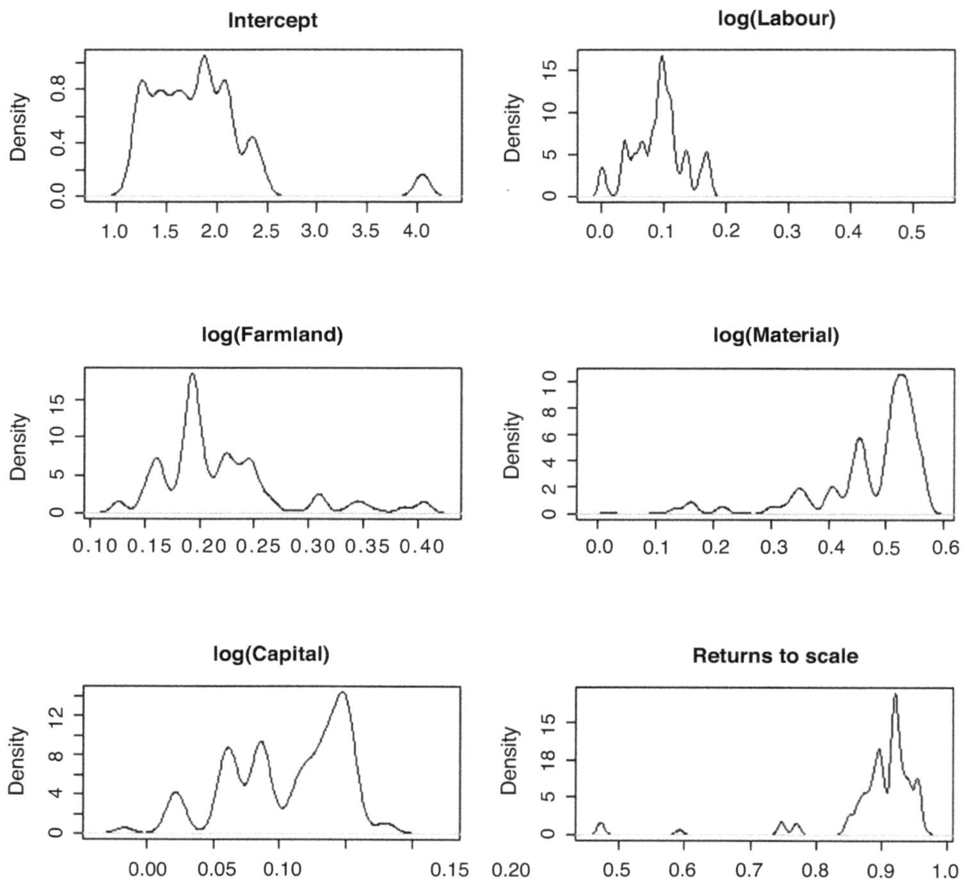

Fig. 51.1. Density distribution of input elasticity and returns to scale of dairy farms.

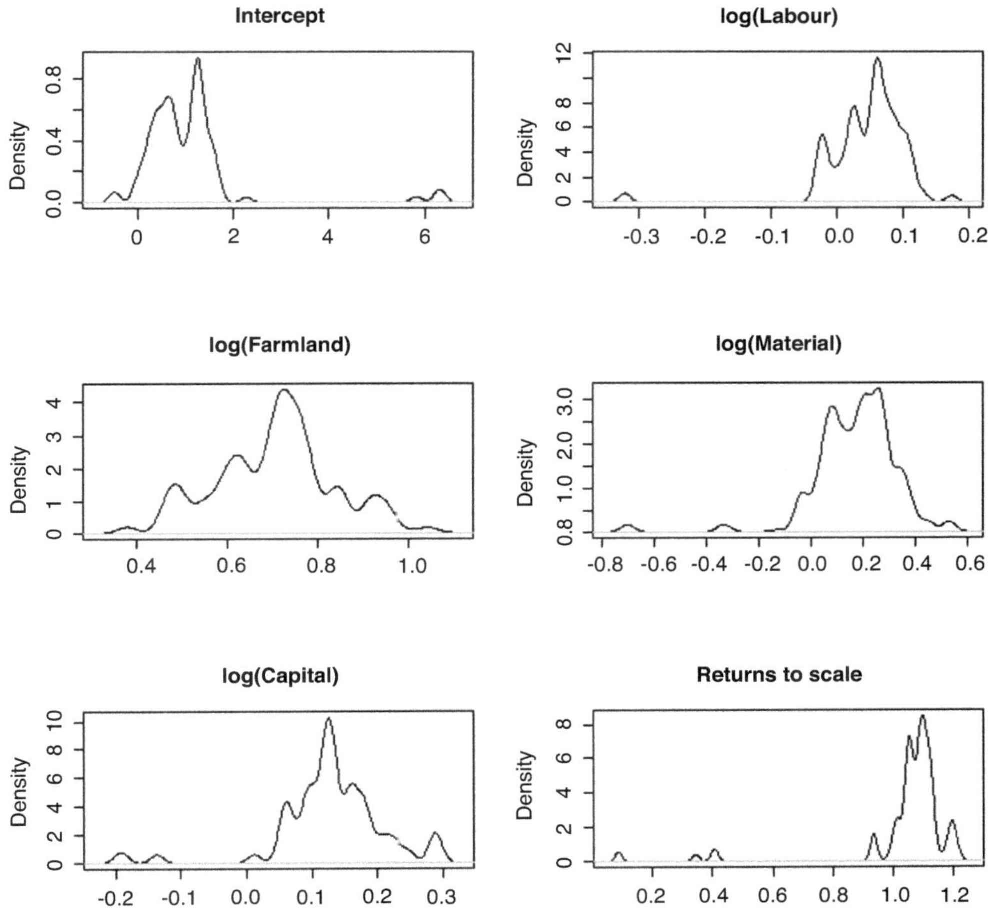

Fig. 51.2. Density distribution of input elasticity and returns to scale of crop farms.

D. Plots of marginal effects

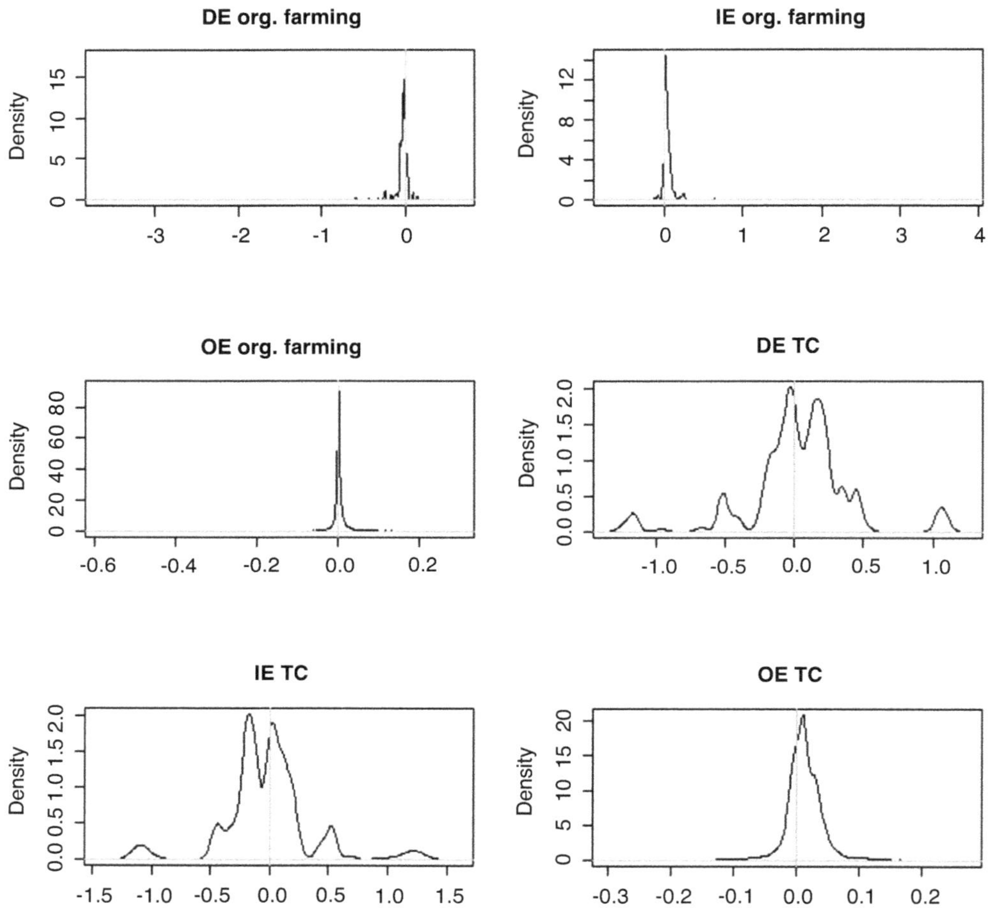

Fig. 51.3. Density plots for marginal effects of organic farming and technical change on the productivity of dairy farms.

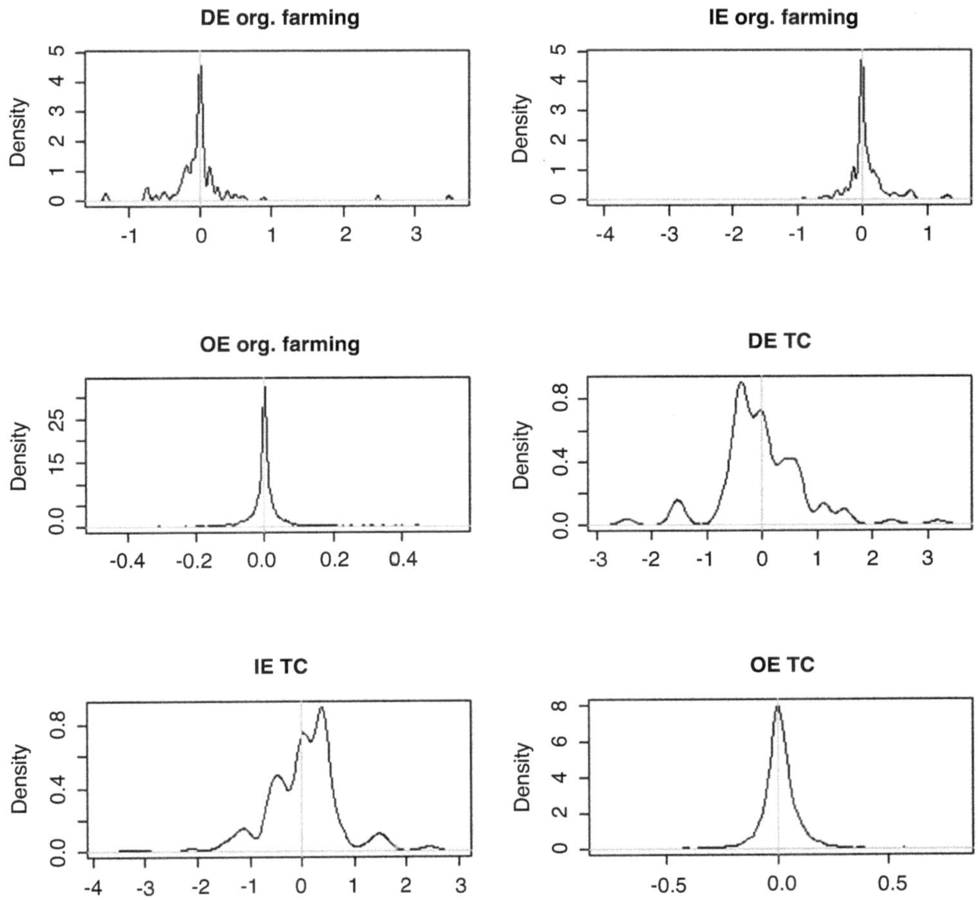

Fig. 51.4. Density plots for marginal effects of organic farming and technical change on the productivity of crop farms.

52 Socio-economic Impact of Organic Vegetable Farming on Smallholder Farmers' Economic Well-Being

Linda C. Familusi[1]*, Abdi-Khalil Edriss[2], Mthakati A. R. Phiri[2], John A. Kazembe[3] and Anthony O. Onoja[4]

[1]Department of Agricultural Economics and Agribusiness Management, University of Port Harcourt, Rivers State, Nigeria; [2]Department of Agricultural and Applied Economics and Africa Centre of Excellence for Agriculture Policy Analysis, Lilongwe University of Agriculture and Natural Resource, Lilongwe, Malawi; [3]Department of Environmental Science, Lilongwe University of Agriculture and Natural Resource, Lilongwe, Malawi; [4]Department of Agriculture and Animal Health, University of South Africa, Johannesburg, South Africa

Abstract
With the increased awareness of the benefits of organic farming and its potential to alleviate poverty among smallholder farmers, providing empirical evidence on its impact on household economic well-being becomes crucial. This study examined the socio-economic impact of organic vegetable farming on household consumption expenditures and household deprivation (poverty) scores in the Oyo and Ekiti states of Nigeria. The study employed a multi-stage sampling technique, and a total of 384 vegetable farming households were sampled. Primary data collection was through a structured questionnaire. For impact analysis, the study adopted the Endogenous Treatment Effect Model (ETEM), which considered the possibility of unobservable factors affecting the treatment effect in the model. The results showed that adopting organic green leafy vegetables (GLV) decreased poverty significantly and had the potential to increase farmers' consumption expenditures. Hence, organic GLV farming can be adopted as a strategy to improve the economic well-being of smallholder farmers.

Keywords: economic well-being, endogenous treatment effect model, multi-dimensional poverty, organic farming

Introduction

Agriculture plays a crucial role in Nigeria's economy, contributing significantly to the country's gross domestic product (26%), with crop production (92%) contributing the most (National Bureau of Statistics (NBS, 2023). However, smallholder farmers, who make up a substantial portion of the agricultural workforce with limited acreage (< 1 ha), often face challenges in adopting effective production techniques, resulting in poor economic well-being (Oluwasusi, 2014; Ibeawuchi *et al.*, 2015 and Aderemi, 2017). Conventional production methods use inorganic fertilizers and agrochemicals, thereby contributing to environmental degradation, soil nutrient depletion, and health-related issues (Mayowa and Meludu, 2014; Oyedele *et al.*, 2018). These practices also incur high input costs, low production and distribution inefficiencies, leading to low income for farmers (FAO, 2019).

Conversely, organic vegetable farming is a sustainable farming method with a beneficial

*Corresponding author: linda.familusi@uniport.edu.ng

DOI: 10.1079/9781800626850.0052

impact on soil health and the environment and can enhance the economic well-being of households (Kutama *et al.*, 2013; Alawode and Abegunde, 2016). Moreover, organic farming is described as an efficient production method to combat the negative effects of conventional methods, such as poor edaphic factors that lead to poor economic well-being for farmers (Oyedele *et al.*, 2018; Familusi *et al.*, 2022). Organic vegetable farming has gained significant attention in recent years due to its potential to contribute to sustainable agriculture, enhance food security, combat the effects of climate change, improve farmers' income, and alleviate poverty (Ayuya *et al.*, 2015; Gama and Millinga, 2019; Anderberg, 2020; Familusi *et al.*, 2023). In addition, it is highly profitable and could be utilized as a measure to alleviate poverty and improve food security in Nigeria (Ibeawuchi *et al.*, 2015). According to Ayuya *et al.* (2015), Gama and Millinga (2019) and Anderberg (2020), adopting organic farming can lift millions of households out of poverty.

Furthermore, organic agriculture is a sustainable agricultural practice that is critical for achieving the United Nations Sustainable Development Goals (SDGs) 1 and 2, which are to eradicate severe poverty and achieve food security by 2030 (Willer and Lernoud, 2019). While Nigeria has aligned with the Sustainable Development Goals, the role of horticulture, specifically organic vegetable production, to enhance farmers' economic well-being is poorly researched. Furthermore, literature claims that organic farming can potentially increase farmers' income, productivity, resilience and profitability (Ayuya *et al.*, 2015; Gama and Millinga, 2019; Anderberg, 2020), but limited empirical evidence exists to support this statement, specifically regarding its impact on farmers' economic well-being, including consumption expenditure (used as a proxy for income) and poverty levels.

Therefore, the study bridges the knowledge gap by assessing the socio-economic impact of organic vegetable farming on farmers' economic well-being in south-west Nigeria. Specifically, the study aims to investigate the socio-economic impact of organic vegetable farming on household consumption expenditure and poverty deprivation scores in the Ekiti and Oyo states of Nigeria. This study provides empirical evidence that can inform policy formulation, decision making and programme development at the local, regional and international levels. The research findings contribute to the existing literature on the role of organic farming practices in improving the income and poverty levels of farmers.

Literature Review

Concept of well-being

According to Clark (2014), well-being is the state of living well, being happy, being satisfied and having one's needs met. Well-being can be viewed from different perspectives, including economic well-being, psychological well-being and social well-being (United Nations Development Programme (UNDP, 2017). This study focuses on economic well-being. The Organization for Economic Cooperation and Development (OECD, 2013) asserts that the economic well-being of households is multi-dimensional. However, household income, consumption expenditure and wealth are major determinants of economic well-being. Therefore, the study measures the economic well-being of vegetable-producing households through their consumption expenditures and poverty. Poverty in the study is measured using weighted deprivation scores from the multi-dimensional poverty analysis methodology.

Analytical framework

The socio-economic impact of organic vegetable farming on household consumption expenditures and deprivation scores is analyzed using the Endogenous Treatment Effect Model (ETEM). This is used to determine the outcome of a specific variable due to the impact of a treatment (Nyaaba *et al.*, 2019). Propensity score matching (PSM) is commonly used for impact analysis. The PSM is used to estimate the impact of a treatment based on only observable factors, while the ETEM uses both observable and unobservable characteristics for impact analysis. Hence, ETEM was adopted for the study to overcome the shortcomings of the PSM model. The ETEM model estimates the selection/treatment equation and the outcome equation. The treatment model is represented in Equation 1:

$$P_i = \alpha + \pi_1 x_1 + \pi_2 x_2 + \ldots + \pi_n x_n + \varepsilon_i \qquad (1)$$

where $x's$represents the independent variables and $\pi's$ represents the parameter estimates. The dependent variable P_i is the participating variable which takes the value of 1 for participants/treated and zero otherwise. The error term is assumed to be normally distributed, have a zero mean and a constant variance: $\varepsilon \sim N(0, \sigma^2)$. The continuous outcome equation is represented as Equation 2:

$$y_i = \alpha + \gamma P + \beta_1 x_1 + \beta_2 x_2 + \ldots + \beta_n x_n + u_i \quad (2)$$

where $x's$ represent the independent variables, u_i represents the error term, α represents the constants, $\beta's$ represents the parameter estimates and represents the coefficient of the participation dummy variable. The dependent variable y_i is the observed household parameter. The error term is assumed to be normally distributed and has zero mean and a constant variance: $u_i \sim N(0, \sigma^2)$. If the error terms on both equations are correlated, then there is sample selection bias which brings about an endogeneity problem (Equation 3):

$$Cov(\varepsilon_i, \mu_i) \neq 0 \quad (3)$$

Therefore, the Endogenous Treatment Effects Model (ETEM) solves the endogeneity and sample selection problems. ETEM estimates the average treatment effect (ATE) and the average treatment effect on the treated (ATT) (Nyaaba *et al.*, 2019), even in the presence of endogeneity. ATE is the average effect of the treatment on the whole sample while ATT is the difference between the mean of the outcome variable of the treated and the non-treated units (Edriss, 2019).

Methodology

The study area consists of Ekiti and Oyo states, which are geographically located in the western part of Nigeria. The study area was selected because of the high proportion of organic farmers and their associations. The study adopts the multi-stage sampling technique, and primary data collection was done using a structured questionnaire. Hence, the sample size (384) consists of vegetable farmers who are registered and certified under the Justice Development and Peace Initiative (JDPI) Ekiti State; Farmers Development Union (FADU) Ibadan; the Association of Organic Agriculture Practitioners of Nigeria (NOAN); and Agricultural Development Programme (ADP) Ekiti State. Data analysis was based on the three commonly cultivated vegetable crops: green leafy vegetables (GLV), tomatoes and pepper. Therefore, the results presented were disaggregated based on these vegetable crops.

Model specification

The study estimates the Endogenous Treatment Effect Model (ETEM) instead of the popularly used PSM model because one of the weaknesses of the PSM model is that it only estimates observable factors, while the ETEM considers both observable and unobservable factors; therefore having an added advantage over the PSM model. The impact that adopting organic vegetable farming has on consumption expenditure and deprivation scores of farmers is estimated. The household multi-dimensional deprivation scores are used as a poverty measure. The ETEM is a two-stage estimation process and comprises the selection/treatment/participation equation and the continuous outcome equation, which are presented in Equations 4 and 5, as adopted in Kapesa (2018):

$$P_i = \alpha + \pi_1 x_1 + \pi_2 x_2 + \ldots + \pi_{13} x_{13} + \varepsilon_i \quad (4)$$

where the $x's$ represent the independent variables – sex of household head (representing 1 = male, 0 = female); age of household head in years; marital status of household (representing 1 = married, 0 = otherwise); years of education of household head; household size; member of association (representing 1 = yes, 0 = no); member of co-operative (representing 1 = yes, 0 = no); number of extension visits per year; size of land owned in hectares; whether the household has a formal job (1 = yes, 0 = no); joint decision (representing one when decision making on farming is jointly done by head and spouse, 0 otherwise); unit price per kg for organic product (in US$); whether household has a radio set (1 = yes, 0 = no); $\varepsilon_i$represents the error term, α represents the constant, and $\pi's$ represents the parameter estimates. The dependent variable P_i is the participating variable which takes the value of 1 for organic participants and zero otherwise.

The continuous outcome equation is specified in Equation 5 as adopted from Kapesa (2018):

$$y_i = \alpha + \gamma P + \beta_1 x_1 + \beta_2 x_2 + \ldots + \beta_{13} x_{13} + u_i \quad (5)$$

where the $x's$ represents the independent variables which are described above, u_i represents the error term, α represents the constant γ and $\beta's$ represents the parameter estimates. The dependent variable y_i is the observed total consumption expenditure (in US$) and/or weighted deprivation scores. P represents the endogenous variable in the equation. The endogeneity of the equation is analyzed using the Wald test, which is used to test for the independence (or endogeneity) of the two equations.

Furthermore, poverty is measured using weighted deprivation scores from a multi-dimensional poverty analysis. The multi-dimensional poverty analysis methodology is as shown in Familusi *et al.* (2022) study. On the other hand, total consumption expenditures are calculated as the sum of households' monthly food and non-food expenditures measured in US$. *A priori* expectation for the analysis is that participating in organic farming should positively impact the farmers' consumption expenditure and negatively impact on deprivation score of households. This would imply that organic farmers have better consumption expenditures than non-organic farmers and thus better economic well-being. The treatment/participation model is estimated using Probit, and the outcome model is estimated using a linear model. However, the model is estimated using the control function, thereby enabling a direct ATT and ATE comparison between organic farmers (OFs) and non-organic farmers (NOFs) without showing the actual regression results.

Results and Discussion

Descriptive statistics

The results imply that most of the households are male-headed (64%), married (92%), members of an association (79%), and receive extension visits (80%). On average, households are 48 years old and have ten years of schooling, with an average household size of six who cultivate on less than one hectare of land (0.9 ha). In addition, the summary result of the multi-dimensional poverty analysis and the mean consumption expenditures for organic and non-organic vegetable farmers are shown in Table 52.1.

Table 52.1. Multi-dimensional deprivation score and consumption expenditures for NOFs and OFs. (From field survey, 2021)

Variables	NOFs	OFs	P-value
Mean consumption expenditures (in US$)	363	337	0.391
Deprivation scores	0.172	0.148	0.075[a]

[a]represents 10%

The study found no significant difference in monthly consumption expenditures between organic (US$ 337) and non-organic vegetable farmers (US$ 363). This is in line with the findings of Uematsu and Mishra (2012) and Familusi *et al.* (2023). However, non-organic farming households had higher weighted deprivation scores (17.2%) than organic farming households (14 8%), suggesting they are multi-dimensionally poorer. This agrees with the findings of Ayuya *et al.* (2015) and Familusi *et al.* (2022).

Table 52.2. Endogeneity test result for total consumption expenditures. (From field survey, 2021)

Model	Chi-square	P-value	Remarks
GLV	50.78***	0.0000	Endogeneity present
Tomato	38.34***	0.0000	Endogeneity present
Pepper	3.26*	0.0710	Endogeneity present

H_0: Error term of the treatment and the outcome model are uncorrelated (rho = 0)

Impact of organic vegetable farming on farmers' consumption expenditures

To check whether there is endogeneity in the model, an endogeneity test is run using the Wald test of independent equations before checking for the ATT and ATE estimations using the *eteffects* command in STATA. The result of the Wald test of independence of equations is shown n Table 52.2.

The endogeneity test results show the presence of endogeneity for GLV ($p < 0.01$), tomato ($p < 0.01$) and pepper ($p < 0.1$) models. Hence, the estimation of the ETEM is justified. Therefore, the ETEM results for the impact of organic farming on household consumption expenditures are presented in Table 52.3.

The results show that the average treatment effect on the treated (ATT) for the adoption of organic GLV, tomato and pepper farming is not statistically significant. However, if none of the

Table 52.3. Impact of organic vegetable farming on household consumption expenditure. (From field survey, 2021)

Total consumption expenditures	ATT	P-value	Mean potential outcome for ATT	P-value	ATE	P-value	Mean potential outcome for ATE	P-value
GLV	−1152	0.129	1506**	0.046	−1248	0.109	823*	0.007
Tomato	877	0.206	−500	0.468	−8	0.984	−12	0.968
Pepper	134	0.772	235	0.609	−448*	0.070	314	0.138

Table 52.4. Endogeneity test result for deprivation score. (From field survey, 2021)

Model	Chi-square	P-value	Remarks
GLV	0.87	0.3509	No endogeneity
Tomato	0.18	0.6729	No endogeneity
Pepper	3.41*	0.0648	Endogeneity present

H_0: Error term of the treatment and the outcome model are uncorrelated (rho = 0)

Table 52.5. Impact of organic vegetable farming on household deprivation score. (From field survey, 2021)

Deprivation score	ATT	P-value	Mean potential outcome for ATT	P-value	ATE	P-value	Mean potential outcome for ATE	P-value
GLV	0.228	0.302	−0.086	0.697	−0.515**	0.031	0.062	0.484
Tomato	0.057	0.848	0.129	0.666	0.049	0.757	0.162	0.230
Pepper	0.741*	0.079	−0.551	0.186	0.393*	0.052	−0.162	0.411

organic GLV households participate in organic farming, their potential consumption expenditure would probably be higher (US$ 1506) on average than that of their non-organic counterparts ($p = 0.046 < 0.05$). This implies that adopting organic GLV, tomato, and pepper farming has no significant impact on farmers' expenditures. Furthermore, the ATE results show that, on average, the adoption of organic pepper farming has a negative (US$ 448) effect on the consumption expenditure for all sampled households ($p = 0.070 < 0.10$). The implication is that participating in organic pepper farming decreases consumption expenditure if all sampled households adopt organic farming. However, if none of the GLV households participated in organic farming, their potential consumption expenditure would be higher (US$ 823) on average. Hence, organic vegetable farming has a negative to no impact on farmers' consumption expenditures. This is in agreement with the findings of Adebiyi *et al.* (2020) and the results further suggest that households would probably grow their products using non-organic methods.

Impact of organic vegetable farming participation on farmers' deprivation score

This subsection examines the impact of organic vegetable farming on farmers' deprivation scores. Table 52.4 shows the Wald test.

The endogeneity test results indicate no endogeneity for the GLV ($p > 0.1$), and tomato ($p > 0.1$) models, while endogeneity is present for the pepper ($p < 0.1$) model. Hence, the ETEM results for the impact of organic farming on household deprivation scores are shown in Table 52.5.

The results show that the ATT of organic GLV and tomato farming is not significant. This implies that there are no differences in the poverty levels for organic and non-organic GLV and tomato farmers. On the other hand, organic pepper farmers are more likely (74.1%) to be multi-dimensionally poorer than NOFs on average (Table 52.5). This implies that organic pepper farming positively impacts farmers' poverty levels, thus influencing poorly their well-being. This is in line with the findings of Adebiyi *et al.* (2020), who reported that the adoption of organic farming is perceived to increase farmers' vulnerability by making them poorer. In addition, the ATE results show that, on average, participating in organic GLV farming leads to a decrease (51.5%) in the multi-dimensional poverty levels for all sampled households if they adopt organic farming. This agrees with the findings of Ayuya *et al.* (2015). However, participating in organic pepper (39.3%) farming increases in multi-dimensional poverty levels for all sampled households if they adopt organic farming. This is in line with Adebiyi *et al.* (2020) findings.

Conclusion

The study analyzed the socio-economic impact of organic GLV, tomato and pepper farming on the economic well-being of households. The study concludes that there is no significant impact on consumption expenditures or poverty levels in tomato households brought about by their participation in organic vegetable farming. However, adopting organic pepper farming positively impacted on farmers' poverty levels, as adopters were multi-dimensionally poorer. In addition, adopting organic GLV farming significantly reduced farmers' deprivation scores, thus improving their economic well-being. Therefore, organic vegetable farming is a means to alleviate poverty and improve farmers' well-being for certain crops. However, this potential has yet to be fully maximized by all vegetable farmers in the study area. Hence, the government should set and implement policies that would improve farmers' economic well-being through improved accessibility and affordability to food and non-food items given that such expenses augment vegetable farmers' multi-dimensional poverty and consumption expenditures. In addition, since adopting organic vegetable farming has both positive and negative impacts on farmers' economic well-being, organic certification bodies (such as NOAN) and organic farmer associations should ensure that organic principles and methods are strictly adhered to by setting up a continuous monitoring system for organic farmers in the study area.

Acknowledgements

We wish to thank the Carnegie Corporation of New York, RUFORUM, the University of Port Harcourt, UNIPORT World Bank Centre of Excellence, the Lilongwe University of Agriculture and Natural Resources and AQUAFISH for their financial and research support.

References

Adebiyi, J.A., Olabisi, L.S., Richardson, R., Liverpool-Tasie, L.S.O. and Delate, K. (2020) Drivers and constraints to the adoption of organic leafy vegetable production in Nigeria: A livelihood approach. *Sustainability* 12(1), 96.

Aderemi, K. (2017) Agriculture: Southwest Nigeria & investment opportunities. *Journal of Agricultural Extension and Rural Development* Available at. Available at: https://www.linkedin.com/pulse/agriculture-southwest-nigeriainvestment-kunle-aderemi/ (accessed January 2022).

Alawode, O.O. and Abegunde, V.O. (2016) Economic analysis of pepper marketing in Oyo State, Nigeria. *Applied Tropical Agriculture* 21(3), 116–121.

Anderberg, S. (2020) The contribution of organic agriculture to poverty reduction. In: Breiling, M. and Anbumozhi, V. (eds) *Vulnerability of Agricultural Production Networks and Global Food Value Chains Due to Natural Disasters*. Economic Research Institute for ASEAN and East Asia, Jakarta, pp. 42–72. Available at: https://www.eria.org/uploads/media/Books/2020-Jan/06_Vulnerability-of-Agricultural-Production-Networks_Chapter-3.pdf (accessed 8 March 2025).

Ayuya, O.I., Gido, E.O., Bett, H.K., Lagat, J.K., Kahi, A.K. *et al.* (2015) Effect of certified organic production systems on poverty among smallholder farmers: Empirical evidence from Kenya. *World Development* 67, 27–37.

Clark, D.A. (2014) Defining and measuring human well-being. In: Freedman, B. (ed.) *Global Environmental Change*. Springer Science, Dordrecht, The Netherlands, pp. 833–855.

Edriss, A. (2019) *Distilled Econometrics: Using African Data with Stata*. International i-Publishers, Beecsi series publications, Canada.

Familusi, L.C., Edriss, A., Phiri, M.A.R., Kazembe, J. and Onoja, A.O. (2022) Determinants of multidimensional poverty index for smallholder organic and non-organic vegetable farming households in southwest Nigeria.

Agricultural Economics and Extension Research Studies (AGEERS) 10(1), 1–8.

Familusi, L.C., Edriss, A., Phiri, M.A.R., Kazembe, J. and Onoja, A.O. (2023) A triple hurdle model of organic vegetable awareness, adoption, and production among smallholder farmers in Ekiti and Oyo states of nigeria. *Organic Agriculture* 13(3), 351–366.

FAO (2019) *Nigeria at a Glance*. Available at: http://www.fao.org/nigeria/fao-in-nigeria/nigeria-at-a-glance/en/ (accessed November 2019).

Gama, J. and Millinga, M.L. (2019) Latest developments in organic agriculture in Africa. In: Willer, H. and Lernoud, J. (eds) *The World of Organic Agriculture: Statistics & Emerging Trends 2019*. Research Institute of Organic Agriculture and FiBL IFOAM – Organics International, pp. 174–178. Available at: https://ciaorganico.net/documypublic/486_2020-organic-world-2019.pdf (accessed 24 April 2023).

Ibeawuchi, I.I., Obiefuna, J.C., Tom, C.T., Ihejirika, G.O. and Omobvude, S.O. (2015) Indigenous and current practices in organic agriculture in Nigeria: A review. *Agriculture, Forestry and Fisheries* 4(6), 257–262.

Kapesa, L.L. (2018) The impact of beekeeping on livelihoods in Mzenga and Mpamba extension planning areas of Nkhata-bay district Malawi. PhD thesis, Lilongwe University of Agriculture and Natural Resources.

Kutama, A.S., Abdullahi, M.A., Umar, S., Binta, U.B. and Ahmad, M.K. (2013) Organic farming in Nigeria: Problems and future prospects. *Global Advanced Research Journal of Agricultural Science* 2(10), 256–262.

Mayowa, O.O. and Meludu, N. (2014) Need assessment of women vegetable farmers on ecological organic agriculture (EOA) in Nigeria. In: *Fourth ISOFAR Scientific Conference*, Building Organic Bridges, 13–15 October, Istanbul, pp. 871–874.

National Bureau of Statistics (NBS) (2023) *Nigerian Gross Domestic Product Report Q4 2022*. NBS, Abuja. Available at: https://nigerianstat.gov.ng/elibrary/read/1241288 (accessed 8 March 2025).

Nyaaba, N.J, Nkrumah-Ennin, K and Anang, B.T. (2019) Willingness to pay for crop insurance in Tolon district of Ghana: Application of an endogenous treatment effect model. *AIMS Agriculture and Food* 4(2), 362–375.

OECD (2013) Economic well-being. In: *OECD Framework for Statistics on the Distribution of Household Income, Consumption and Wealth*. Organization for Economic Cooperation and Development, Paris, pp. 25–38.

Oluwasusi, J.O. (2014) Vegetable farmers' attitude towards organic agriculture practices in selected states of south west Nigeria. *Journal of Agricultural Extension and Rural Development* 6(7), 223–230.

Oyedele, G.T., Wole-Alo, F.I., Owolabi, K.E. and Okunlola, J.O. (2018) Small-scale farmers perception on organic farming status in Ondo state, Nigeria. *American Journal of Agriculture and Forestry* 6(6), 186–190.

Uematsu, H. and Mishra, A.K. (2012) Organic farmers or conventional farmers: Where's the money. *Ecological Economics* 78, 55–62.

United Nations Development Programme (UNDP) (2017) 2016 UNDP human development report background paper: Happiness, well-being and human development: The case for subjective measures. UNDP, New York. Available at: https://core.ac.uk/download/pdf/84310326.pdf (accessed 8 March 2025).

Willer, H. and Lernoud, J. (2019) *The world of organicagriculture: Statistics and emerging trends 2019*. Research Institute of Organic Agriculture (FiBL) and IFOAM, Bonn. Available at: https://ciaorganico.net/documypublic/486_2020-organic-world-2019.pdf (accessed 24 April 2023).

53 Socio-economic Impacts of Organic Farming: Opportunities and Challenges

Habtamu Alem[1]*, Fikru K. Alemayehu[2], Timo Sipiläinen[3] and Gudbrand Lien[1,2]

[1]Department of Economics and Society, Norwegian Institute of Bioeconomy Research (NIBIO), Ås, Norway; [2]Department of Economics, Inland Norway University of Applied Sciences, Lillehammer, Norway; [3]Department of Economics and Management, University of Helsinki, Finland

Abstract

Organic farming has sparked discussions about environmental sustainability, conservation, health and community involvement. In essence, it represents a broader shift in societal values, paving the way for a more sustainable and socially responsible future that includes all segments of society. This chapter thoroughly investigates the multi-faceted socio-economic impacts of organic agriculture, including economic, social and cultural dimensions. It also addresses the difficulties that farmers face when transitioning to organic farming methods and offers solutions to these difficulties. Its primary goal is to provide readers with the necessary knowledge to capitalize on the socio-economic benefits of organic farming while effectively addressing the associated challenges. The chapter focuses on how organic agriculture can promote sustainable livelihoods, boost economic well-being, and increase social and environmental resilience. It also highlights the potential of organic farming to drive long-term growth and revitalize rural landscapes.

Keywords: organic farming, socio-economic benefits, sustainability, challenges, policy

Introduction

The realms of sustainable agriculture, environmental conservation, and the production of healthy foods have merged into global movements, with organic agriculture at the forefront (Eyhorn *et al.*, 2019). Organic agriculture accounted for 1.6% of total agricultural land in 2021 (Trávníček *et al.*, 2023). Furthermore, the organic farming area has grown over the past two decades (see Fig. 53.1). According to the Research Institute for Organic Agriculture (FiBL), the global organic agricultural landscape spanned more than 76 million hectares in 2021, including certified organic regions and those undergoing conversion. Oceania, with a sprawling 36 million hectares, and Europe, boasting 18 million hectares, stood as the epicentres of organic agricultural expanses. Latin America followed with 10 million hectares, followed by Asia with 7 million hectares, Northern America with 4 million and Africa with 3 million (Schlatter *et al.*, 2021). Africa currently has the smallest proportion in the world of organic agriculture. Taking into account individual countries, Oceania, which includes Australia, New Zealand and subregions such as Micronesia, Melanesia and Polynesia, emerged as a global leader in organic agricultural production, with Australia alone accounting for 35 million

*Corresponding author: habtamu.alem@nibio.no

DOI: 10.1079/9781800626850.0053

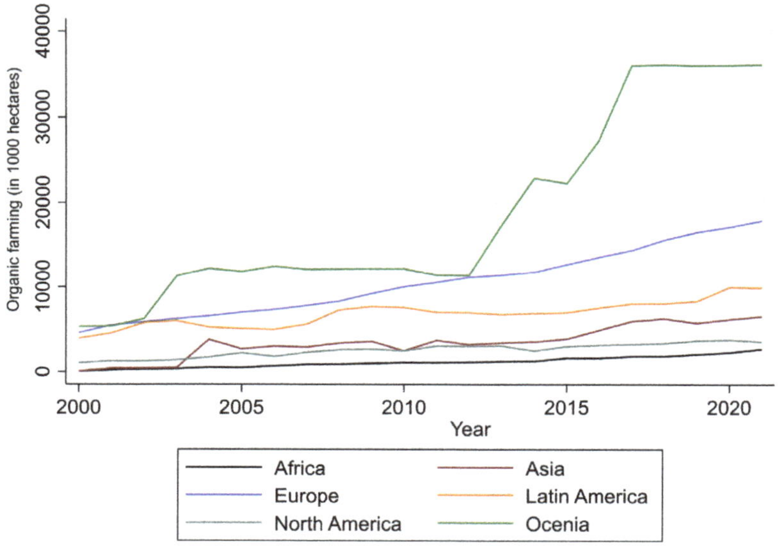

Fig. 53.1. World organic farming in hectares in the years 2000–2021. (Authors' formulation based on FiBL Statistics (2023) data)

hectares. Argentina followed with 4 million hectares and France with 3 million (Yussefi and Willer, 2007; Trávnıček *et al.*, 2023).

Many countries recognize the importance of organic agriculture and are actively formulating strategies to promote its expansion. For example, the 'Farm to Fork' strategy of the EU Commission has set an ambitious goal of having at least 25% of agricultural land in the EU under organic cultivation by 2030 (European Commission, 2020b). The motivation for this endeavour is the potential positive effects that organic farming can have on a variety of aspects of society, including improved socio-economic conditions and improved environmental sustainability. The overarching vision is that these collective efforts will propel community development, protect ecosystem integrity, and foster responsible stewardship of invaluable resources (Debuschewitz and Sanders, 2022). Organic farming is becoming more popular due to the growing concern of society for health and the environment. Its economic and social benefits are also becoming more and more significant. The surge in demand for organic products is related to increasing concerns about the sustainability of the food system. The organic farming paradigm shift in agriculture, as epitomized by the adoption of organic practices, has started a domino effect that resonates through various socio-economic dimensions. These spillover effects extend to traditional agricultural practices but are also reflected in local economies, consumer behaviour, environmental protection, and social dynamics. Furthermore, this movement, rooted in environmental and health-conscious principles, has influenced consumer choices, moulding purchasing patterns and consumption habits. Consequently, supply chains, retail strategies and economic opportunities for myriad stakeholders are undergoing a profound change. Beyond economic considerations, the social impact of the expansion of organic farming is clear. This movement, which extends beyond economic considerations, has spurred discussions on sustainable and conscientious agriculture, sparking conversations on environmental conservation, health awareness and community participation. Collectively, these factors contribute to a more comprehensive understanding of socio-economic well-being.

This chapter embarks on a comprehensive exploration of the socio-economic implications of organic farming, including its economic, social and cultural dimensions. Our aim is not only to scrutinize how organic farming is changing agricultural practices, but also to dissect its impact on livelihoods, societies and global sustainability endeavours. Armed with empirical research findings and practical case studies, we navigate the intricate web of socio-economic consequences arising from the increasing popularity of organic farming practices, along with its associated challenges. From the dynamics of organic markets to the revitalization of rural areas, from the well-being of organic farmers to evolving consumer preferences, the narrative of the chapter unfolds by initially delving into the manifold benefits of, before addressing the hurdles confronted by, organic farming. Ultimately, it culminates in a discussion of the policy implications based on the insights gleaned from this study.

Economic Benefits

Price premiums and market access

The global organic food market has experienced rapid growth, making organic farming one of the fastest-expanding segments of the global food industry (Seufert *et al.*, 2017). The exponential increase in its market value, which soared from a modest $15.2 billion in 1999 to $72 billion by 2013, underscores this significant development (Willer *et al.*, 2019). The USA and the European Union jointly command a 90% share of the global organic market. In contrast, developing countries have only a fractional presence in this vibrant market (Jouzi *et al.*, 2017; Willer *et al.*, 2019). This transformation attests to the increasing interest of consumers in organic products and a growing commitment to healthier, environmentally responsible food choices in developed countries. Thus, organic farming has significant economic implications and transformative opportunities in the global realm of agriculture and food systems (Crowder and Reganold, 2015).

The first of the socio-economic effects of organic farming is the access to the market that it offers. Organic produce commands higher prices than conventional products in both local and international markets (Jouzi *et al.*, 2017). Locally, the increasing concern of consumers for health and the environment has increased the demand for organic products. This growing demand often results in price premiums for organic produce. Farmers can explore various markets, from local co-operatives and farmers' markets to international export opportunities, diversifying their revenue streams and improving the financial resilience of farming communities. A global meta-analysis by Crowder and Reganold (2015) found that despite lower yields, organic agriculture showed significantly better, surpassing other methods by 22–35%. The price premiums associated with organic products play a pivotal role in this increased profitability. Oelofse *et al.* (2010), in their study of certified organic agriculture in China and Brazil, discovered that organic farmers perceived positive impacts on prices, incomes and market opportunities. The improvement in farmers' income perceptions is probably due to improved market accessibility coupled with increased returns or reduced costs, and diversification (Jouzi *et al.*, 2017). According to Natural resources institute of Finland – LUKE (2023), organic crop and dairy farms have been more profitable than conventional farms in most years since 2000. This is related to both higher output prices, lower input costs, and higher public subsidies for organic farming. According to these same statistics, the average yields of cereals and grass silage are typically lower on organic farms than on conventional farms.

Price stability

Organic farming contributes to price stability in agricultural markets. This stability arises from a reduced reliance on volatile external synthetic inputs, such as chemical fertilizers and pesticides. Organic farming methods prioritize natural and sustainable practices, including crop rotation, biological pest control, and soil enrichment through compost and organic matter. By minimizing the need for external inputs and focusing on soil health and biodiversity, organic farmers can often maintain stable yields even in the face of changing environmental conditions (Underwood *et al.*, 2011). The consistent production associated with organic farming is instrumental in establishing a stable supply of organic goods, thereby mitigating the volatility of prices in the organic food market. Su *et al.* (2013), in their assessment of organic milk prices in the USA, found that the prices exhibited greater stability and consistent increases over the years, in contrast to the seasonal fluctuations and annual variability seen in conventional milk prices. Price stability benefits both consumers and producers by ensuring the availability of products at consistent prices, fostering trust in the organic industry and underscoring its commitment to sustainable agriculture.

Income generation and reducing indebtedness

Organic farming offers economic advantages, with a prominent increase in income generation for farmers. Farmers adopting organic practices typically enjoy higher profit margins due to the elevated market prices commanded by organic products. The perceived quality of organic produce, along with its adherence to environmental and health standards, results in higher prices. Furthermore, organic farmers benefit from lower input costs because they eschew costly synthetic pesticides and fertilizers. Sudheer (2013) conducted an economic analysis in Andhra Pradesh, India, comparing organic and conventional farming methods for various crops. The results showed that organic farmers outperformed their conventional counterparts in terms of gross income, enjoying an advantage of 5%, 10% and 7% in paddy, red gram and groundnut, respectively. Furthermore, organic farmers realized significantly higher profits due to lower input costs, with margins surpassing those of conventional farmers by 37%, 33% and 59%, respectively, for the corresponding crops.

Reganold and Wachter (2016) conducted a comprehensive study that spanned four decades and included research on 55 different crops grown on five continents within the realm of organic agriculture. Their findings illuminated notably higher profitability in organic farming, with net present values exceeding conventional farming by 22–35%. Organic agriculture also exhibited superior benefit-to-cost ratios, ranging from 20% to 24%, particularly when considering the actual price premiums associated with organic produce. When comparing risk in organic, integrated and conventional cropping systems in Norway, Lien *et al.* (2005) found that the organic farm system had the greatest variability in net income, but both the payment system and the organic price premiums make it the most economically viable alternative. Lien *et al.* (2007) found that the organic farming system is somewhat less economically sustainable than the conventional system, especially if organic price premiums and organic area payments were to be phased out. A study of conventional and organic dairy farms in Finland found that conventional technology is more productive and organic farms are, on average, less technically efficient than conventional farms (Kumbhakar *et al.*, 2009). However, the findings in the literature are mixed, and multiple studies underscore the financial benefits of organic farming (Twarog, 2006; Parrott *et al.*, 2006; Kilcher, 2007; Seufert *et al.*, 2017).

Furthermore, organic farming plays an essential role in reducing farmers' debt burdens by reducing input costs or by improving profitability and promoting financial stability. Unlike conventional farming, which is heavily based on synthetic pesticides and fertilizers, organic farming emphasizes natural and sustainable techniques. This results in reduced production costs. The subsequent decrease in debt levels empowers farmers with improved financial stability, equipping them to effectively manage their farms, invest in sustainable agricultural practices, and secure their livelihoods while mitigating the risks associated with agricultural debt. Consequently, organic farming emerges as a financially viable option for farmers, increasing their income and financial security.

Rural employment

Organic farming contributes to rural employment due to its labour-intensive practices and the opportunities it creates in rural areas. With a focus on sustainable techniques such as manual weeding, composting, and organic pest management, organic farming frequently demands more manual labour than conventional agriculture. This labour requirement generates employment opportunities for local communities, particularly in rural areas where economic prospects are limited. Organic farming operations can expand, creating jobs ranging from farm labourers to positions in processing and marketing. A study by Green and Maynard (2006) found that organic farming in the UK generates 32% more employment opportunities per farm than comparable non-organic farming operations. This increase in job creation in the organic agriculture sector supports local economies and enhances employment prospects in rural areas, fostering socio-economic sustainability. Consequently, organic farming not only addresses unemployment issues but also helps to retain rural populations, contributing to rural development and stability.

Rural development

The promotion of rural development is a major concern in the agricultural policies of most countries. Various measures have been implemented to promote economic growth in rural areas, with the primary objective of increasing rural household incomes and reducing unemployment rates (Fuller *et al.*, 2005). Studies by Privitera (2010) and Khanal *et al.* (2019) of agri-tourism and organic farming find that they are complementary. It implies, then, from a sustainable rural development perspective, a potential for increased rural organic tourism and agri-tourism by integrating organic agriculture and agri-tourism strategies.

Organic farming can also play an essential role in rural development through the production of biogas, a mechanization approach aligned with sustainable practices. Biogas systems efficiently convert organic waste, such as crop residues and livestock manure, into renewable energy sources, particularly methane. This energy can be used for cooking, heating and electricity generation. Biogas systems provide a sustainable energy source that benefits both the farm and the surrounding community. Excess energy can even be sold back to the grid, offering farmers an additional income stream. This environmentally friendly mechanization method reduces greenhouse gas emissions, improves access to energy and diversifies income, thereby contributing to rural development.

Siegmeier *et al.* (2015) performed a system analysis, revealing that agricultural biogas production has the potential to improve overall nitrogen supply and nitrogen-use efficiency. These agronomic effects can, in turn, enhance farm productivity, stability and resilience. Organic farms that incorporate biogas production contribute to the renewable energy supply without requiring additional land, while also increasing food production and reducing greenhouse gas emissions from livestock manure. This offers opportunities for eco-functional intensification within organic farming systems, addressing interconnected challenges related to food production, energy and climate change.

Social Benefits

Empowerment and community engagement

Organic farming is unique in its ability to empower people and foster community involvement. Organic farming practices empower farmers by giving them greater control over their production methods, leading to increased self-sufficiency. Unlike conventional farmers, organic farmers have the flexibility to adopt alternative techniques that are aligned with their values and beliefs. This agency enables them to experiment with innovative approaches such as permaculture and agroforestry, promoting sustainable farming practices while improving their livelihoods. Furthermore, organic farming extends its empowerment effects to the broader community through various initiatives. Farmers' markets serve as vibrant local hubs where consumers can interact directly with growers, learning about environmentally friendly farming practices. Community-supported agriculture programmes establish direct relationships between farmers and consumers, nurturing a sense of connection and mutual support. In addition, farm tours and workshops facilitate knowledge sharing and create forums for dialogue between farmers and the community at large.

Research, such as that conducted by Nath and Athinuwat (2021), has highlighted the empowerment of women in the organic agriculture sector. Moderate empowerment among female farmers is associated with positive outcomes, since organic farming provides more opportunities for empowerment and cash-crop cultivation compared to conventional agriculture, due to its low-input approach (van Elzakker and Eyhorn, 2010; Jouzi *et al.*, 2017). This empowerment extends beyond individual circumstances, with broader implications for society and national living standards.

Strengthening local networks and co-operation

Organic farming promotes the development of local networks and co-operation, strengthening the social fabric of communities. Emphasizing community participation and collaboration, organic farming practices encourage interactions such as co-operatives, local farm markets, and community-supported agriculture (CSA) systems. These interactions facilitate knowledge exchange, resource sharing, increased use of local resources, and collective marketing efforts among farmers (Milford *et al.*, 2021). Furthermore, such local networks and co-operation not only enhance the socio-economic resilience of farming communities but also nurture a sense of belonging and shared purpose.

In addition, the focus on sustainability in organic agriculture often prompts community members to collaborate to address environmental concerns and advocate for policies that support sustainable agriculture. These joint efforts strengthen rural vitality and reinforce the social bonds crucial for thriving communities. For example, Favilli *et al.*

(2015) studied an innovative local food network in Italy, the Crisoperla Society, which focuses on the production and supply of organic foods. Organic farming serves as a unifying element in this network, allowing participants to forge shared visions, terminology and goals, ultimately enhancing their co-ordination and strengthening their organizational and political participation. Organic farming also has the potential to strengthen social bonds and empower small-scale farmers through co-operative organizations (Rice, 2001).

Preserving traditional knowledge and facilitating transfer

Incorporating stakeholders in policy development has become essential, as it improves the legitimacy of policies and the allocation of research funding. Indigenous farming practices and organic farming align, showcasing the rich cultural heritage associated with sustainable agriculture (Seufert, 2012). Indigenous communities often have deep local knowledge of the land, its ecosystems, and traditional farming methods that promote biodiversity and soil health. Organic farming principles such as minimal chemical use and crop rotation harmonize with these indigenous practices, promoting a more respectful relationship with the land and preserving cultural heritage (Guo *et al.*, 2021).

Organic farming serves as a platform for knowledge transfer, advocacy for sustainable agriculture, and preservation of local wisdom. Organic farmers often champion sustainable agricultural systems, sharing their knowledge and experiences within their communities. This knowledge transfer helps spread organic farming and strengthens local economies. Furthermore, organic farming highlights the value of local knowledge in adapting to changing environmental conditions (Gambelli *et al.*, 2023). Traditional farming practices are recognized and appreciated within the organic framework, preserving vital aspects of local culture and wisdom while promoting sustainable and climate-resilient agricultural systems by combining these with accumulating new knowledge.

Building Climate Resilience

Sustainable farming practices

Organic farming practices are closely aligned with sustainability, particularly in terms of climate resilience (Holleman *et al.*, 2020). In particular, crop rotation is a vital sustainable technique within organic farming systems, known for its positive impacts on soil health and mitigation of climate change (Watson *et al.*, 2002; Yu *et al.*, 2022). Integrating agroecological principles into soil management practices improves plant resistance to pests and diseases (Altieri, 2002). By diversifying crops grown on a specific plot of land, organic farmers can improve soil fertility, mitigate erosion, and reduce susceptibility to crop-specific diseases and pests. In addition, the emphasis on compost and natural organic materials enriches the soil with essential nutrients and organic carbon, enhancing both crop productivity and the soil's carbon-capture capacity. Consequently, sustainable and organic farming practices improve climate resilience by increasing the adaptability and robustness of agricultural systems (Fließbach *et al.*, 2007). As highlighted by Kareem *et al.* (2022), organic farming stands out as a novel and sustainable agricultural approach known for its eco-friendly practices. A meta-analysis conducted by Bengtsson *et al.* (2005) revealed that organic farming systems, on average, exhibit a 30% increase in species diversity and a 50% increase in organism abundance compared to conventional agricultural systems (Jouzi *et al.*, 2017). In this context, organic farming emerges as a pivotal strategy to achieve the goal of sustainable agriculture (Soni *et al.*, 2022).

Mitigating vulnerabilities

Agriculture's sensitivity to climate variability poses significant challenges to food security, with increasing temperatures, changing rainfall patterns, and increased extreme weather events intensifying these vulnerabilities (WHO and FAO, 2003). However, organic farming systems exhibit resilience in the face of extreme weather events, positioning them as a crucial component of climate adaptation strategies. Organic farms often demonstrate superior resistance to climate-related issues such as droughts, floods and heatwaves. This adaptability can be attributed to the holistic approach of organic agriculture, which includes practices that improve soil health and structure, preserve biodiversity, and minimize chemical inputs (Jouzi *et al.*, 2017). For example, organic soils are better equipped to retain moisture during droughts while efficiently absorbing excess water during heavy rains, thus reducing the risk of soil erosion and flooding. Furthermore, the emphasis on biodiversity in organic agriculture, including the use of cover crops and various crop rotations, acts as a buffer against climate-related crop failures.

Scialabba and Müller-Lindenlauf (2010) conducted a study that focuses on three key aspects of the design of the agricultural system: cropland, grassland and livestock management. Their findings underscored the significant potential of organic agricultural systems to fortify food systems against uncertainties, thanks to strategies such as farm diversification and using organic materials to enhance soil fertility. Organic farming also provides alternatives to energy-intensive inputs, reducing vulnerability, especially for rural communities facing limited financial resources as they transition from synthetic fertilizers. Similarly, Brzezina *et al.* (2016) used system dynamics and structural thinking tools to examine vulnerabilities in the conventional food system, both internal structural issues and external influences. Their research concluded that organic farming has the potential to mitigate these vulnerabilities and improve the resilience of the European food system. Additionally, organic farming significantly reduces risks for small farmers by diversifying their product offerings through practices such as agroforestry, intercropping and crop rotation. These strategies reduce vulnerability to major crop failures, significantly contributing to the resilience of agricultural communities and ensuring food security amid a changing climate landscape by mitigating the adverse effects of climate change.

Navigating Socio-Economic Challenges in Organic Farming

Although organic farming offers a multitude of socio-economic benefits, it also presents its fair share of challenges. Below are some of the key socio-economic challenges associated with organic farming.

Farm labour dynamics and training

Organic farming, unlike conventional agriculture, places a significant emphasis on manual labour and sustainable solutions based on nature for tasks such as weed control and pest control (Goewie, 2003; Casagrande *et al.*, 2016). This change in labour practices has several implications. Relying on manual labour in organic farming can create more employment opportunities, particularly in rural areas where job opportunities can be limited. Tasks such as weeding, mulching and biological pest control require a large workforce, which can be especially beneficial in regions with high unemployment rates, helping to reduce poverty and improve livelihoods. However, the labour-intensive nature of organic farming can influence the demographics of the agricultural workforce. Physical demands and the attractiveness of non-farming opportunities in urban areas may discourage young people from pursuing careers in organic agriculture. This shift in labour dynamics could affect the age distribution of farmers and the generational succession in agricultural communities.

In addition, successful organic farming requires specialized knowledge and training in organic practices, soil health management, pest and disease control, and certification requirements (Goewie, 2003; Jouzi *et al.*, 2017). Farmers who lack access to relevant training programmes and extension services may struggle to transition to organic methods. Accessible and comprehensive training initiatives play a crucial role in ensuring the adoption and success of organic farming practices. Although organic agriculture requires significant labour, it also provides opportunities for job creation, skill development and sustainable agriculture. Efforts to offer accessible training, particularly to aspiring young farmers, can help bridge the knowledge gap and ensure the long-term viability of organic farming systems.

Consumer confusion and low disposable income

Consumer uncertainty often arises from the availability of substitute products, such as low-pesticide foods, which leads to questions about organic products and their certifications. As highlighted by Yiridoe *et al.* (2005), consumer scepticism regarding the actual qualities of organic products and organic labels can discourage them from purchasing organic products. This scepticism and uncertainty can be attributed to reported cases of unintentional mislabelling, misrepresentation of products, and variations in organic standards and certification processes. However, the significant price premium associated with organic products remains the main constraint that affects consumer demand. Many individuals, especially in developing countries, face financial constraints that make it difficult to pay the additional costs associated with organic food products (Jouzi *et al.*, 2017). The phenomenon is also visible in many developed countries today, where rising inflation rates are influencing shifts in consumption patterns toward more cost-effective categories.

Limited access to organic markets for small farms

Gaining access to organic markets, particularly international markets, can pose significant challenges to small-scale farmers due to certification requirements, quality standards and the dominance of large-scale producers. In some regions, only a quarter of the population can access markets within two hours of travel by motorized transportation (Smale *et al.*, 2013; Jouzi *et al.*, 2017). As organic farms expand to meet market demand, maintaining organic principles and practices becomes increasingly challenging. Large-scale organic farms may struggle to meet strict organic standards, while small farmers may struggle to adopt organic practices due to limited resources, access to organic inputs, and knowledge gaps. These challenges may limit small-scale farmers' participation in the organic sector.

Quality control and certification costs

Organic farmers must ensure consistent product quality because quality control issues and compliance with organic standards can affect market acceptance and profitability. Quality assurance and certification systems can be susceptible to fraudulent activity (Jahn *et al.*, 2005). Despite established control measures, there have been instances of fraudulent organic labelling. For example, in 2000, 10% of the organic corn sold in Germany originated from 'conventional' agriculture (Baummann, 2001). Estimates suggest that fraud in organic labelling ranges from 15% to 40% in the southern states of the EU (Giannakas, 2002). The incidence of food fraud cases has increased significantly over the years, highlighting the importance of combating food fraud, which can harm consumers, producers and the entire food supply chain (European Commission, 2020a). Organic certification is costly and time-consuming (Jahn *et al.*, 2005), which can pose challenges to small-scale farmers in covering costs and navigating the certification process (Barrett *et al.*, 2001).

Higher initial investment and yield variability

Transitioning from conventional to organic agriculture often requires substantial upfront investments in organic certification, soil improvement and organic input. These costs can be prohibitive for small-scale farmers with limited resources. The transition can also lead to yield variability (Smith *et al.*, 2019), particularly during the transition period, which can affect farmers' income stability and food security.

Doubts about ensuring food security

A recurring debate revolves around whether organic agriculture is an effective strategy to ensure food security and whether it will remain relevant over time. Various scholars have presented arguments and evidence both in favour and against this perspective (see, for example, Setboonsarng, 2006; Kirchmann *et al.*, 2008; Seufert *et al.*, 2012; Aune, 2012; Pickett, 2013; Abdalla and Chen, 2021). Given the multi-faceted nature of this debate, a comprehensive analysis is beyond the scope of this chapter. Instead, our objective is to provide a nuanced examination of the socio-economic dimensions of organic agriculture, highlighting its impact and challenges while acknowledging ongoing debates and diverse perspectives surrounding this complex subject.

To promote sustainability in organic farming, a comprehensive approach is essential, considering a wide range of factors that include political, social, economic and environmental dimensions. These dimensions range from local to international in scope and involve individuals, societies, governments and organizations. To develop effective practices, a thorough examination of the inherent challenges and opportunities within the agricultural landscape and its broader ecosystem is required (Abdalla and Chen, 2021). Promoting sustainability and growth of organic agriculture often requires the implementation of supportive policies, gaining access to resources and training, developing markets, and raising consumer awareness.

Conclusion

This chapter examined the various impacts and challenges of organic farming, highlighting its potential to create sustainable livelihoods and economic advantages. It has provided a comprehensive analysis of the social and economic effects of organic farming. The remarkable growth of organic agriculture on a global scale, with millions of hectares dedicated to it, and the emphasis on its importance in the agricultural strategies of several countries, reflect the widespread belief in the positive results that organic agriculture can provide for society and the environment.

Organic agriculture has had a significant socio-economic impact that extends beyond traditional agricultural practices. It has changed consumer behaviour, reshaped supply chains, and altered economic opportunities. This movement has also sparked discussions about sustainability, health and community involvement, which has contributed to a more comprehensive understanding of social well-being. Organic farmers face a variety of challenges in the face of these opportunities, ranging from market dynamics to the need for rural revitalization, from ensuring farmer well-being to understanding changing consumer preferences.

This chapter provided valuable information tailored to the readers to address the challenges ahead and harness the potential of organic agriculture. These findings will help direct efforts to maximize the socio-economic benefits of organic farming while mitigating its drawbacks. The organic farming movement has the potential to play an important role in fostering sustainable livelihoods, environmental stewardship, and healthier communities, ultimately contributing to a more sustainable and resilient future. In summary, the organic farming movement represents a shift in social values and priorities, as well as a change in agricultural practices. It provides a path towards a more sustainable and socially responsible future, and its continued growth and success will depend on the combined efforts of all stakeholders.

References

Abdalla, T.A.O. and Chen, Y. (2021) A case study on the opportunity and challenges faced by organic farming in Jiangsu province. *Agroecology and Sustainable Food Systems* 45(9), 1327–1374.

Altieri, M.A. (2002) Agroecology: The science of natural resource management for poor farmers in marginal environments. *Agriculture, Ecosystems & Environment* 93(1–3), 1–24.

Aune, J.B. (2012) Conventional, organic and conservation agriculture: Production and environmental impact. *Agroecology and Strategies for Climate Change* 149–165.

Barrett, H.R., Browne, A.W., Harris, P.J.C. and Cadoret, K. (2001) Smallholder farmers and organic certification: Accessing the EU market from the developing world. *Biological Agriculture & Horticulture* 19(2), 183–199.

Baummann, P. (2001) Securing against fraud. In: *Proceedings of the European Conference Organic Food and Farming*, Copenhagen, Denmark, 10-11 May.

Bengtsson, J., Ahnström, J. and Weibull, A.-C. (2005) The effects of organic agriculture on biodiversity and abundance: A meta-analysis. *Journal of Applied Ecology* 42(2), 261–269.

Brzezina, N., Kopainsky, B. and Mathijs, E. (2016) Can organic farming reduce vulnerabilities and enhance the resilience of the European Food System? A critical assessment using system dynamics structural thinking tools. *Sustainability* 8(10), 971.

Casagrande, M., Peigné, J., Payet, V., Mäder, P., Sans, F.X. *et al.* (2016) Organic farmers' motivations and challenges for adopting conservation agriculture in Europe. *Organic Agriculture* 6(4), 281–295.

Crowder, D.W. and Reganold, J.P. (2015) Financial competitiveness of organic agriculture on a global scale. *Proceedings of the National Academy of Sciences* 112(24), 7611–7616.

Debuschewitz, E. and Sanders, J. (2022) Environmental impacts of organic agriculture and the controversial scientific debates. *Organic Agriculture* 12(1), 1–15.

European Commission (2020a) *The EU Agri-food Fraud Network and the Administrative Assistance and Cooperation System*. European Commission Directorate-General for Health and Food Safety.

European Commission (2020b) Farm-to-fork strategy: For a fair, healthy, and environmentally-friendly food system. *Communication from the Commission to the European Parliament, the Council, the European Economic and Social Committee, and the Committee of the Regions* 381, 1–9.

Eyhorn, F., Muller, A., Reganold, J.P., Frison, E., Herren, H.R. *et al.* (2019) Sustainability in global agriculture driven by organic farming. *Nature Sustainability* 2(4), 253–255.

Favilli, E., Rossi, A. and Brunori, G. (2015) Food networks: Collective action and local development. The role of organic farming as boundary object. *Organic Agriculture* 5(3), 235–243.

Fließbach, A., Oberholzer, H.-R., Gunst, L. and Mäder, P. (2007) Soil organic matter and biological soil quality indicators after 21 years of organic and conventional farming. *Agriculture, Ecosystems & Environment* 118(1–4), 273–284.

Fuller, R.J., Norton, L.R., Feber, R.E., Johnson, P.J., Fuller, R. *et al.* (2005) Benefits of organic farming to biodiversity vary among taxa. *Biology Letters* 1(4), 431–434.

Gambelli, D., Vairo, D., Cuoco, E. and Zanoli, R. (2023) The role of stakeholder involvement in EU research and innovation policy: A case study of technology platform organics. *Organic Agriculture* 13(2), 293–308.

Giannakas, K. (2002) Information asymmetries and consumption decisions in organic food product markets. *Canadian Journal of Agricultural Economics/Revue Canadienne d'agroeconomie* 50(1), 35–50.

Goewie, E.A. (2003) Organic agriculture in the Netherlands: Developments and challenges. *NJAS* 50(2), 153–169.

Green, M. and Maynard, R. (2006) The employment benefits of organic farming. In: Atkinson, C., Ball, B., Davies, D.H.K., Rees, R., Russell, G. *et al*. (eds) *Aspects of Applied Biology 79: What Will Organic Farming Deliver? COR 2006*. Association of Applied Biologists, pp. 51–55.

Guo, T., García-Martín, M. and Plieninger, T. (2021) Recognizing indigenous farming practices for sustainability: A narrative analysis of key elements and drivers in a Chinese dryland terrace system. *Ecosystems and People* 17(1), 279–291.

Holleman, C., Rembold, F., Crespo, O. and Conti, V. (2020) *The Impact of Climate Variability and Extremes on Agriculture and Food Security: An Analysis of the Evidence and Case Studies*. Food and Agriculture Organization of the United Nations, Rome.

Jahn, G., Schramm, M. and Spiller, A. (2005) The reliability of certification: Quality labels as a consumer policy tool. *Journal of Consumer Policy* 28(1), 53–73.

Jouzi, Z., Azadi, H., Taheri, F., Zarafshani, K., Gebrehiwot, K. *et al*. (2017) Organic farming and small-scale farmers: main opportunities and challenges. *Ecological Economics* 132, 144–154.

Kareem, A., Farooqi, Z.U.R., Kalsom, A., Mohy-Ud-Din, W., Hussain, M.M. *et al*. (2022) Organic farming for sustainable soil use, management, food production, and climate change mitigation. In: Bandh, S.A. (ed.) *Sustainable Agriculture: Technical Progressions and Transitions*. Springer, pp. 39–59.

Khanal, A.R., Mishra, A.K. and Omobitan, O. (2019) Examining organic, agritourism, and agri-environmental diversification decisions of American farms: Are these decisions interlinked? *Review of Agricultural, Food and Environmental Studies* 100(1–4), 27–45.

Kilcher, L. (2007) How organic agriculture contributes to sustainable development. *Journal of Agricultural Research in the Tropics and Subtropics* 89(1), 31–49.

Kirchmann, H., Bergström, L., Kätterer, T., Andrén, O. and Andersson, R. (2008) Can organic crop production feed the world? In: Kirchmann, H. and Bergström, L. (eds) *Organic Crop Production: Ambitions and Limitations*. Springer, pp. 39–72.

Kumbhakar, S.C., Tsionas, E.G. and Sipiläinen, T. (2009) Joint estimation of technology choice and technical efficiency: An application to organic and conventional dairy farming. *Journal of Productivity Analysis* 31(3), 151–161.

Lien, G., Flaten, O., Korsaeth, A., Schumann, K., Richardson, J. *et al*. (2005) Comparison of risk in organic, integrated, and conventional cropping systems in Eastern Norway. *Journal of Farm Management* 12, 384–401.

Lien, G., Hardaker, J.B. and Flaten, O. (2007) Risk and economic sustainability of crop farming systems. *Agricultural Systems* 94(2), 541–552.

Milford, A.B., Lien, G. and Reed, M. (2021) Different sales channels for different farmers: Local and mainstream marketing of organic fruits and vegetables in Norway. *Journal of Rural Studies* 88, 279–288.

Nath, T.D. and Athinuwat, D. (2021) Key factors of women empowerment in organic farming. *GeoJournal* 86(6), 2501–2520.

Natural resources institute of Finland – LUKE (2023) Profitability factor (organic/regular production).

Oelofse, M., Høgh-Jensen, H., Abreu, L.S., Almeida, G.F., Hui, Q.Y. *et al*. (2010) Certified organic agriculture in China and Brazil: Market accessibility and outcomes following adoption. *Ecological Economics* 69(9), 1785–1793.

Parrott, N., Olesen, J.E. and Høgh-Jensen, H. (2006) Certified and non-certified organic farming in the developing world. In: Halberg, N., Alrøe, H.F., Knudsen, M.T. and Kristensen, E.S. (eds) *Global Development of Organic Agriculture: Challenges and Prospects*. CAB International, Wallingford UK, pp. 153–179.

Pickett, J.A. (2013) Food security: Intensification of agriculture is essential, for which current tools must be defended and new sustainable technologies invented. *Food and Energy Security* 2(3), 167–173.

Privitera, D. (2010) The importance of organic agriculture in tourism rural. *Applied Studies in Agribusiness and Commerce* 4(1–2), 59–64.

Reganold, J.P. and Wachter, J.M. (2016) Organic agriculture in the twenty-first century. *Nature Plants* 2(2), 15221.

Rice, R.A. (2001) Noble goals and challenging terrain: Organic and fair trade coffee movements in the global marketplace. *Journal of Agricultural and Environmental Ethics* 14(1), 39–66.

Schlatter, B., Travnicek, J., Lernoud, J., Willer, H., Willer, H. *et al*. (2021) Current statistics on organic agriculture worldwide: Area, operators, and market. In: Willer, H., Trávníček, J. and Schlatter, B. (eds) *The World of Organic Agriculture. Statistics and Emerging Trends*. FiBL and IFOAM - Organics International, pp. 32–129.

Scialabba, N.E.-H. and Müller-Lindenlauf, M. (2010) Organic agriculture and climate change. *Renewable Agriculture and Food Systems* 25(2), 158–169.

Setboonsarng, S. (2006) Organic agriculture, poverty reduction, and the Millennium Development Goals (tech. rep). ADBI Discussion Paper.

Seufert, V. (2012) Organic agriculture as an opportunity for sustainable agricultural development. *Research to Practice Policy Briefs, Policy Brief* 13.

Seufert, V., Ramankutty, N. and Foley, J.A. (2012) Comparing the yields of organic and conventional agriculture. *Nature* 485(7397), 229–232.

Seufert, V., Ramankutty, N. and Mayerhofer, T. (2017) What is this thing called organic? How organic farming is codified in regulations. *Food Policy* 68, 10–20.

Siegmeier, T., Blumenstein, B. and Möller, D. (2015) Farm biogas production in organic agriculture: System implications. *Agricultural Systems* 139, 196–209.

Smale, M., Byerlee, D. and Jayne, T. (2013) *Maize Revolutions in Sub-Saharan Africa*. Springer.

Smith, O.M., Cohen, A.L., Rieser, C.J., Davis A.G., Taylor, J.M. *et al.* (2019) Organic farming provides reliable environmental benefits but increases variability in crop yields: A global meta-analysis. *Frontiers in Sustainable Food Systems* 3.

Soni, R., Gupta, R., Agarwal, P. and Mishra, R. (2022) Organic farming: A sustainable agricultural practice. *Vantage* 3(1), 21–44.

Su, Y., Brown, S. and Cook, M.L. (2013) Stability in organic milk farm prices: A comparative study. In: *2013 Annual Meeting*, Agricultural and Applied Economics Association, 4-6 August, Washington, DC.

Sudheer, P. (2013) Economics of organic versus chemical farming for three crops in Andhra Pradesh, India. *Journal of Organic Systems* 8(2), 36–49.

Trávníček, J., Schlatter, B. and Willer, H. (2023) Key results from the FiBL survey on organic agriculture worldwide 2013.

Twarog, S. (2006) Organic agriculture: A trade and sustainable development opportunity for developing countries. *Trade and Environment Review* 141, 223.

Underwood, T., McCullum-Gomez, C., Harmon, A. and Roberts, S. (2011) Organic agriculture supports biodiversity and sustainable food production. *Journal of Hunger & Environmental Nutrition* 6(4), 398–423.

van Elzakker, B. and Eyhorn, F. (2010) *The Organic Business Guide: Developing Sustainable Value Chains with Smallholders*. IFOAM.

Watson, C.A., Atkinson, D., Gosling, P., Jackson, L.R. and Rayns, F.W. (2002) Managing soil fertility in organic farming systems. *Soil Use and Management* 18(s1), 239–247.

WHO (World Health Organization) and FAO (Food and Agriculture Organization of the United Nations) (2003) Diet, Nutrition, and the Prevention of Chronic Disease. WHO, Geneva, Switzerland.

Willer, H., Schaack, D. and Lernoud, J. (2019) Organic farming and market development in Europe and the European Union. In: Willer, H. and Lernoud, J. (eds) *The World of Organic Agriculture. Statistics and Emerging Trends 2019*. FiBL, IFOAM – Organics International, pp. 217–253.

Yiridoe, E.K., Bonti-Ankomah, S. and Martin, R.C. (2005) Comparison of consumer perceptions and preference toward organic versus conventionally produced foods: A review and update of the literature. *Renewable Agriculture and Food Systems* 20(4), 193–205.

Yu, T., Mahe, L., Li, Y., Wei, X., Deng, X. *et al.* (2022) Benefits of crop rotation on climate resilience and its prospects in China. *Agronomy* 12(2), 436.

Yussefi, M. and Willer, H. (2007) Organic farming worldwide 2007: Overview & main statistics. In: Willer, H. and Yussefi, M. (eds) *The World of Organic Agriculture: Statistics and Emerging Trends 2007*. IFOAM. Germany & Research Institute of Organic, Bonn, pp. 9–16.

54 Organic Food Quality and Safety

Jagjeet Singh Gill*

Scientist and Quality Assurance, Green Mountain Health Alliance, Kaleden, BC, Canada

Abstract

The quality and safety of organic food were the main topics of the current study, which compared organic and conventional food to accurately interpret the data. According to the study, there is significantly reduced maximum residual limit in organic food as compared to conventional food, meaning that organic food is safe to consume. Nitrogen and traces of synthetic pesticides showed similar kinds of outcomes as MRL. When it comes to heavy metals, organic food is thought to be safe, while conventional food has been found to contain higher levels of heavy metals and fungal toxins. In certain circumstances, organic food has more protein than other foods; however, products like pasta and milk have distinct effects. Overall, this study found that organic food is safer and healthier for daily consumption since it has more dietary fibre, tastes better, has more vitamins and minerals, more phosphorus, low residual levels, low pesticide content, low nitrates, less heavy metals, and less fungal toxins.

Keywords: organic food, quality, safety, conventional food

Introduction

In 20th century after the World Wars, many powers consolidated to their areas and many countries became independent. But the major problem faced by most of the countries was food security. The 1959–1961 famine in China was the single largest famine in history. Along with this, the 1950 Canadian caribou famine, the 1957 Bihar famine, the 1958–1975 Wollo–Tigray famine, the 1983–1985 famine in Ethiopia, the 1992 famine in Somalia, and the 1993 famine in Sudan were other notorious examples. Many famines had environmental, demographic, political and economical causes but all were related to food or grain scarcity. During the 20th century, famine claimed the lives of more than 70 million people. The world saw 30 major famines, some of which were caused by natural disasters and others by political upheaval (Brown, 1996).

At the same time, some south-Asian countries and Mexico emphasized growing high-yield varieties which led to increased food grain production. Transfer of technology, good-quality seeds and use of fertilizers made many countries self-sufficient in grain. High rates of investment in crop research, infrastructure, market growth, and proper policy support during the first Green Revolution (GR) were major contributors to this success. The post-GR period can be defined as the following two decades and the first GR period as 1966–1985. Large-scale public investment in crop genetic advancement relied on scientific advancements made for the three main staple crops (wheat, rice and maize) in the industrialized world, and tailored those advancements to the needs of developing nations Pingali (2012).

A study on plant protection practices used by vegetative producers in Punjab state, India, was carried out by Gill *et al.* in 2020. Based on their analysis, they found that 26.58% of farmers use Indofil, a polymeric complex made of manganese ethylenebis (di thio carbanate), and zinc salt to treat early potato blight, and 39.24% use it to treat late potato blight; 45.59% of tomato farmers treat early tomato blight using Indofil and zinc salt (Sharma *et al.*, n.d.). By the

*Corresponding author: Jagjeetsinghgill@gmail.com

 Advances in Organic Farming (ed. L. P. Awasthi)
DOI: 10.1079/9781800626850.0054

end of the 20th century, people realized the impact of fertilizers, chemicals, pesticides and insecticides on their health; these increase yield and make countries self-sufficient but they have also deteriorated health, the environment and the ecosystem during the last decades. The widespread use of pesticides contributes significantly to pollution by contaminating the air, soil and water, causing a decline in biodiversity and fostering pest resistance. Chemical pesticide exposure in humans has been related to neurological, cardiovascular and respiratory conditions as well as cancer (Abubakar *et al.*, 2020).

After understanding the impact of chemical traces in food, the discussion was started between members of families, communities and even governments of developing countries as to how to cope with these problems at grassroots level.

The global market for organic goods (including food, apparel, cosmetics, personal care items, household goods and fabrics) has expanded quickly in the 2000s. More nations are developing official, legally binding organic certification. Keeping tabs on and contesting certification guidelines and decisions has developed into a regular, prominent activity for activists in the organic movement. Seventy-five million hectares (190,000,000 acres) or 1.6% of all agricultural area in the world was being cultivated organically as of 2020 (FiBL, 2022).

Organic Food Quality

Many discussions about the quality of food products have led to various definitions of this concept. Food quality is a concept that is always evolving. At first, it was symbolized by the numerical/measurable parameters. The holistic strategy for solving the quality issue is becoming more and more well-liked today.

Maximum residue level

The number of pesticides that can be present in food is subject to regulatory restrictions set by governments known as the maximum residue level (MRL). Typically, the MRL is calculated through rat testing of specific pesticides. Governments insist that there is no health danger from consuming pesticides below the MRL. Pesticides are known to cause or are suspected of causing a wide range of illnesses and health issues, including cancer, at lower concentrations (BMA, 1992). According to Aubert, consuming organic foods causes the amounts of pesticides in human milk and bodily tissues to decrease. There is some evidence that supports this theory. French researchers discovered that when the percentage of organic food in the daily diet increased (from 25% to 80%), the amount of pesticide residues in the milk of nursing mothers considerably dropped (Aubert, 1987). The findings of an *ad hoc* data extraction and comparison of the monitoring results on food samples produced conventionally and organically are included in the EFSA report. The data extraction process concentrated on samples collected during the reference years 2013, 2014 and 2015 as part of the EU-co-ordinated control programmes (for a total of 28,912 conventional and 1940 organic food samples). In all, 4.4% of the food samples produced using conventional methods had one or more quantifiable residues; in contrast, only 6.5% of the food samples generated using organic methods had detectable pesticide residues (European Food Safety Authority, 2018).

Vitamins and minerals

According to Sheila Zahm, researchers at the University of Missouri found that smaller, organically grown oranges contained 30% more vitamin C than large, conventionally cultivated oranges. They discovered that whereas conventionally farmed vegetables did not give the recommended daily dose of vitamin C, an organic serving of lettuce, spinach, carrots, potatoes and cabbage did (Zahm and Blair, 1992). Virginia Worthington, the certified nutritionist (2001), studies discovered that an organic fruit or vegetable has an average of 27% more vitamin C than a comparable fruit or vegetable grown conventionally. Put another way, a comparable organic fruit or vegetable would have 127 mg of vitamin C if an average conventional fruit or vegetable had 100 mg (Woese *et al.*, 1997). A thorough assessment of research comparing organic and conventional production methods for vitamin and phenolic components in fruits and vegetables was carried out by Brandt *et al.* (2011). They discovered that organic foods had much higher levels of vitamin C and 'secondary plant metabolites', which is a term that includes polyphenols, total phenolics, and other non-nutritive bioactive dietary components. Minerals such as sodium, potassium, calcium, phosphorus, iron and vitamins were determined by Kim *et al.* (2004).

Compared to regular vegetables, organic vegetables have increased vitamin and mineral content. The total mineral content of ordinary and organic kale were 688.1 and 811.8 mg/l00g, respectively. Celery, lettuce, *Angelica keiskei* and *Allium fistulosum* had total mineral content of 447.9, 486.5, 368.6 and 320.9 mg/100g, respectively (Kim *et al.*, 2004).

Proteins

Brandt studied the variables affecting the concentrations of chemicals in plants. According to the carbon/nitrogen (C/N) balance theory, plants will first produce compounds with a high nitrogen content, such as proteins for growth and nitrogen-containing secondary metabolites like alkaloids, glucosinolates and non-protein amino acids, like the precursors to the *Allium* flavour. The metabolism shifts more towards carbon-containing molecules, such as starch and cellulose, and non-nitrogen-containing secondary metabolites, such as phenolics and terpenoids, when nitrogen availability is a growth constraint (Brandt and Mølgaard, 2001).

Worthington contends that nitrogen from all types of fertilizers has an impact on the quantity and calibre of protein that plants generate. When a plant receives a lot of nitrogen, protein production rises but carbohydrates production falls. In addition, the increased protein created in response to high nitrogen levels has a lower quality in terms of human nutrition since it contains fewer of some key amino acids, such as lysine (Worthington, 2001). Dall'Asta *et al.* (2020) found significant variations were seen in the category of 'jams, chocolate spreads, and honey', wherein organic items displayed a reduced ($p < 0.001$) median energy, total carbohydrate, and sugar content in comparison to their conventional counterparts. Conversely, in this category, organic products had a substantially greater median protein content than conventional items ($p = 0.002$) (Dall'Asta *et al.*, 2020). Pasta products that are not organic have a much higher protein content than those that are ($p < 0.00$) (Dall'Asta *et al.*, 2020). The amount of whey protein in cows' milk from conventional and organic farms in Poland that was collected in the late pasture and early indoor feeding seasons was examined; the percentage of protein was higher in conventional farming than in organic farming ($p < 0.01$) (Kuczyńska *et al.*, 2012).

Dietary fibre

Fifty-three organic products out of 77 pairs of products with nutrition claims related to fibre content were found by assessing the kind of nutrition claim for the 'pasta' subcategory; just two items had regular counterparts. The only way that organic flours varied from conventional ones within the same product category was by having a higher sugar content ($p = 0.012$) (Dall'Asta *et al.*, 2020). Considerable nutrient changes in the organic 'jams, chocolate spreads and honey' subcategory may be the cause of the lower energy, carbohydrate and sugar amounts seen in these products. In this instance, disparities cannot be accounted for by the existence of multiple nutrition claims, as none was included in either organic or conventional products (Dall'Asta *et al.*, 2020).

Sensory quality of organic foods

Rembiałkowska stated numerous studies have conclusively shown that produce from organic farms tastes and smells better. This was discovered for apples, cherries and redcurrants as well as for celery, red beets, head cabbage and tomatoes. More total sugars were present in organic fruits, which likely contributed to customers' perceptions of a better flavour (Rembiałkowska, 2000).

Bjon analysis of bread made from organic grains also demonstrated improved crumb elasticity and had better taste and scent. Animals were given either organic or conventional feed in interesting experiments on animal dietary preferences. Most investigations, which were conducted on rats, mice, chickens and rabbits, have shown that animals clearly prefer organically produced diets (Bjørn and Fruekilde, 2003).

Additionally, Woese noted that organic fodder was preferred when both organic and conventional fodder met all of the physiological requirements of the test animals, as determined by chemical analysis. This was most likely caused by the flavour difference between conventional and organic feed (Woese *et al.*, 1997).

Safety of Organic Foods

Pesticides

Synthetic pesticides are used extensively in conventional agriculture for plant protection. Contrarily, organic farming typically focuses on biological preventative measures and plant protection methods, including crop rotation, intercropping, resistant cultivars, biological management using natural enemies, cleanliness practices, and other measures (Gamliel *et al.*, 2016; Gurr *et al.*, 2016).

The use of some pesticides is permitted in organic farming. Pesticides are approved in the EU after thorough evaluation, which includes a variety of toxicological tests in animal studies. In this context, pesticides are more specifically chemical plant-protection products; micro- and macrobiological agents are excluded from this discussion due to their low relevance for human health (European Commission, 2009).

The majority of pesticides authorized for use in organic farming are of relatively low toxicological concern for consumers because they are not linked to any known toxicity (e.g. spearmint oil, quartz sand); are components of a typical diet or are nutrients for humans (e.g. iron, potassium bicarbonate, rapeseed oil); or are only permitted for use in insect traps and therefore pose a negligible risk of getting into the food chain (e.g. synthetic pyr copper and pyrethrins are two significant exceptions). Though less stable than synthetic pyrethroid insecticides, pyrethrins, a plant extract from *Chrysanthemum cinerariaefolium*, work in a similar manner Even though it is poisonous at high intakes and of ecotoxicological concern due to its toxicity to aquatic life, copper is a necessary nutrient for plants, animals and humans (European Commission, 2008, 2017).

The entire agricultural system may gain from plant protection techniques created for and in organic agriculture. This is especially important for the EU's transition to sustainable pesticide usage, which places a significant focus on non-chemical plant protection techniques including biological agents (Gurr *et al.*, 2016; Garibaldi *et al.*, 2016).

The limited use of synthetic pesticides in organic food production results in low residual levels in foods and thus lesser pesticide exposure for consumers. Additionally, it lessens the occupational exposure of farm workers to pesticides and the exposure of rural residents to chemical drift. EFSA reported pesticide residues below maximum residue levels (MRL) in 43.7% of all food samples and 13.8% of organic food samples, on average, over the previous three years that data are available (Council of the European Union, 2007; European Food Safety Authority, 2015; European Food Safety Agency, 2017; European Commission, 2017).

Nitrogen and phosphorous

In comparison to conventional crops, organic crops consistently exhibit greater phosphorus levels (7–8%) and lower total nitrogen levels (7–10%), according to previous systematic reviews. These results do not directly relate to human health. However, taking into account the various fertilization methods covered above and the fundamental significance of N, P and the N:P ratio for plant development, this may lend some credence to further documented production system effects on crop composition (Güsewell, 2004; Dangour *et al.*, 2009; Huber *et al.*, 2012; Smith-Spangler *et al.*, 2012; Barański *et al.*, 2014; Bindraban *et al.*, 2015).

Nitrates

Szponar and Kierzkowska demonstrated that conventionally grown foods contain significantly more nitrates and nitrites than organically grown crops. This is crucial because nitrates are easily turned into nitrites, which can harm infants, children and the elderly by causing methemoglobinemia (Szponar and Kierzkowska, n.d.). According to Mirvich, nitrosamines, which are mutagenic and carcinogenic compounds that cause leukemia and cancer of the digestive tract, can be produced when nitrites and amines react (Mirvich, 1993). According to research by Rembiałkowska, adults of all ages as well as young children are at risk from this process. The nitrate content of organic and conventional crops has been examined using a significant amount of data. The nitrate level of organic crops was, on average, 49% higher than that of conventional ones (Rembiałkowska, 2000).

These findings give support to the claim that using organic farming techniques reduces nitrate and nitrite intake by people by about 50%.

Heavy metals

Cadmium (Cd) can demineralize bones, is carcinogenic, and is harmful to the kidneys. In addition to being naturally present in soils, P fertilizers, air deposition and other sources also add Cd to them. The soil's chemistry, humus content and pH all have an impact on how readily available Cd is to plants. Cd concentrations in the crops rise as a result of the use of Cd-containing fertilizers. Organically managed farms typically have higher soil organic matter than conventionally managed farms, while low soil organic matter generally increases the availability of Cd for crops (de Meeûs *et al.*, 2002; Åkesson *et al.*, 2014; Grant, 2015; Reganold and Wachter, 2016).

The raw material phosphate rock is where the Cd in mineral fertilizers comes from. According to reports, mineral fertilizers in Europe have an average Cd concentration of 68 mg/kg P or 83 mg/kg P. Although it varies, farmyard manure generally has lower levels of Cd: between 14 and 37 mg Cd/kg P were present on average across many types of animal manure in a German collection (Directorate-General for Enterprise and Industry (European Commission), Environmental Resources Management, 2001; Kratz and Schnug, 2005; Nziguheba and Smolders, 2008).

Therefore, a lower Cd concentration in organic crops is likely as a result of using fertilizers with lower Cd content and maybe as a result of organic fields having higher soil organic matter. The general population should have less exposure to Cd because it is near to, and in some cases over, the acceptable intake. Food is the main source of exposure for non-smokers, with cereals and vegetables being the most significant sources (European Food Safety Authority, 2009).

There have not been any recorded concentration variations between conventional and organic crops for other harmful metals like lead, mercury or arsenic. Manure-based cropping systems have a lower uranium load than mineral-fertilized systems at equal P loads. Uranium is also present as a contaminant of concern in mineral P fertilizers, but less so in organic fertilizers. In soils fertilized with minerals, uranium appears to accumulate, and agricultural activity may raise the uranium levels in surface and groundwater. No data comparing the uranium concentrations of organic and conventional goods, however, could be discovered (Kratz *et al.*, 2008; Birke *et al.*, 2009; Smith-Spangler *et al.*, 2012; Barański *et al.*, 2014; Schnug and Haneklaus, 2015; Liesch *et al.*, 2015; Bigalke *et al.*, 2017).

Various sources, including industry, transportation, municipal garbage and agriculture, introduce heavy metals into the food chain, including cadmium, lead, mercury and zinc. For instance, cadmium can be introduced into crops through the use of mineral phosphoric fertilizers in conventional agriculture, but it can also contaminate soil and crops through the metal industry and transportation. As a result, research contrasting the amounts of heavy metals in conventional vs organic crops have shown mixed results. While some statistics indicate that conventional crops have higher amounts, other data indicate the contrary (Rembiałkowska, 2000). Głodowska and Krawczyk (2017) examined soil and vegetable samples that were often utilized and obtained from both conventional and organic farms. The plant samples were examined and analyzed for the concentration of heavy metals (Cd, Co, Cr, Cu, Fe, Mn, Ni, Pb and Zn) using atomic absorption spectroscopy. The vegetables grown in the two growing systems differ significantly from one another, according to a chemical analysis of the samples that were obtained. Vegetables cultivated conventionally typically have higher quantities of certain components (Głodowska and Krawczyk, 2017). Głodowska and Krawczyk (2017) show that significant variations in the concentrations of some heavy metals between the vegetables grown conventionally and organically were found by statistical analysis, while 64% of vegetables produced conventionally had higher levels of heavy metals.

Fungal toxins

Regarding fungal toxins in crops, one meta-analysis found that organic cereal crops were less likely to be contaminated with deoxynivalenol (DON), which is a toxin generated by several fusarium species. The effects of fungicide sprays on cereal leaves are unknown, however they may change the microbial populations there and diminish disease-suppressive species. Additionally, non-cereal crops in crop rotations may help reduce fusarium infestation, because N availability is favourably correlated with the DON level of cereals. They found decreased DON contamination in organic cereals is plausible given these variables. In the EU, grains and grain-based products are the main sources of total exposure, which is above the tolerated daily intake (TDI) for toddlers, babies and children to DON. The TDI is based on mice with reduced bodyweight increase. Another important fungal toxin in the development of cereal, ochratoxin A (OTA), is not observed to be affected by the production system (Kabak *et al.*, 2006; Bernhoft *et al.*, 2012; Barański *et al.*, 2014;

Karlsson *et al.*, 2014, 2017; European Food Safety Authority, 2017).

Conclusion

According to the study's findings, organic foods are higher in dietary fibre, minerals and vitamin C. In certain circumstances, organic food has more protein than other foods; however, products like pasta and milk have distinct effects. Compared to other food types, organic food is thought to be safer due to its lower pesticide content and heavier nutrients. Further studies on the safety of organic food are required, considering other common foods, to paint a complete picture of the advantages and disadvantages of organic food in comparison to conventional food. Overall, all things considered, this study found that organic food is safer and healthier for daily use since it contains more dietary fibre, tastes better, has more vitamins and minerals, more phosphorus, low residual levels, low pesticide content, low nitrates, less heavy metals and less fungal toxins.

References

Abubakar, Y., Tijjani, H., Egbuna, C., Adetunji, C.O., Kala, S. *et al.* (2020) Pesticides, history, and classification. In: Egbuna, C. and Sawicka, B. (eds) *Natural Remedies for Pest, Disease and Weed Control*. Academic Press, Amsterdam, pp. 29–42.

Åkesson, A., Barregard, L., Bergdahl, I.A., Nordberg, G.F., Nordberg, M. *et al.* (2014) Non-renal effects and the risk assessment of environmental cadmium exposure. *Environmental Health Perspectives* 122(5), 431–438.

Aubert, C. (1987) Pollution du lait maternel, une enquete de terre vivante. *Quatre Saisons Jardinage* 42(33), 39.

Barański, M., Srednicka-Tober, D., Volakakis, N., Seal, C., Sanderson, R. *et al.* (2014) Higher antioxidant and lower cadmium concentrations and lower incidence of pesticide residues in organically grown crops: A systematic literature review and meta-analyses. *The British Journal of Nutrition* 112(5), 794–811.

Bernhoft, A., Torp, M., Clasen, P.-E., Løes, A.-K. and Kristoffersen, A.B. (2012) Influence of agronomic and climatic factors on Fusarium infestation and mycotoxin contamination of cereals in Norway. *Food Additives & Contaminants. Part A: Chemistry, Analysis, Control, Exposure & Risk Assessment* 29(7), 1129–1140.

Bigalke, M., Ulrich, A., Rehmus, A. and Keller, A. (2017) Accumulation of cadmium and uranium in arable soils in Switzerland. *Environmental Pollution* 221, 85–93.

Bindraban, P.S., Dimkpa, C., Nagarajan, L., Roy, A. and Rabbinge, R. (2015) Revisiting fertilisers and fertilisation strategies for improved nutrient uptake by plants. *Biology and Fertility of Soils* 51(8), 897–911.

Birke, M., Rauch, U. and Lorenz, H. (2009) Uranium in stream and mineral water of the Federal Republic of Germany. *Environmental Geochemistry and Health* 31(6), 693–706.

Bjørn, G. and Fruekilde, A.M. (2003) Cepa onions (*Allium cepa* L.) grown conventionally and organically – similarities and differences. *Grøn Viden* 153(1), 6.

BMA (1992) *The BMA Guide to Pesticides, Chemicals and Health. Report of Science and Education*. British Medical Association. London.

Brandt, K. and Mølgaard, J.P. (2001) Organic agriculture: Does it enhance or reduce the nutritional value of plant foods? *Journal of the Science of Food and Agriculture* 81(9), 924–931.

Brandt, K., Leifert, C., Sanderson, R. and Seal, C.J. (2011) Agroecosystem management and nutritional quality of plant foods: The case of organic fruits and vegetables. *Critical Reviews in Plant Sciences* 30(1–2), 177–197.

Brown, L. (1996) *Tough Choices: Facing the Challenge of Food Scarcity*. W.W. Norton, New York.

Council of the European Union (2007) Council regulation no 834/2007 of 28 June 2007 on organic production and labelling of organic products and repealing regulation (EEC) no 2092/91. *Official Journal of the European Union*.

Dall'Asta, M., Angelino, D., Pellegrini, N. and Martini, D. (2020) The nutritional quality of organic and conventional food products sold in Italy: Results from the food labelling of Italian products (FLIP) study. *Nutrients* 12(5), 1273.

Dangour, A.D., Dodhia, S.K., Hayter, A., Allen, E., Lock, K. *et al.* (2009) Nutritional quality of organic foods: A systematic review. *The American Journal of Clinical Nutrition* 90(3), 680–685.

de Meeûs, C., Eduljee, G.H. and Hutton, M. (2002) Assessment and management of risks arising from exposure to cadmium in fertilisers. *The Science of the Total Environment* 291(1–3), 167–187.

Directorate-General for Enterprise and Industry (European Commission), Environmental Resources Management (2001) European Commission: Analysis and conclusions from member States' assessment of the risk to health and the environment from cadmium in fertilisers. Available at: https://publications.europa.eu/en/publication-detail/-/publication/6cfa95d3-2346-4c9f-8a06-35a5a0bef0ff/language-en (accessed 15 March 2025).

European Commission (2008) Commission regulation (EC) no 889/2008 of 5 September 2008 laying down detailed rules for the implementation of council regulation (EC) no 834/2007 on organic production and labelling of organic products with regard to organic production, labelling and control. *Official Journal of the European Union*.

European Commission (2009) Commission regulation (EU) no 1107/2009 of 21 October 2009 concerning the placing of plant protection products on the market and repealing council directives 79/117/EEC and 91/414/EEC. *Official Journal of the European Union*.

European Commission (2017) EU Pesticides Database. Available at: https://Ec.europa.eu/food/plant/pesticides/eu-pesticides-database

European Food Safety Agency (2017) The 2015 European Union report on pesticide residues in food. *EFSA Journal* 15(4), 4791.

European Food Safety Authority (2009) EFSA panel on contaminants in the food chain cadmium in food: Scientific opinion of the panel on contaminants in the food chain on a request from the European Commission on cadmium in food. *EFSA Journal* 980, 1–139.

European Food Safety Authority (2015) The 2013 European Union report on pesticide residues in food. *EFSA Journal* 13, 3.

European Food Safety Authority (2017) Risks to human and animal health related to the presence of deoxynivalenol and its acetylated and modified forms in food and feed. *EFSA Journal* 15(9), 4718.

European Food Safety Authority (2018) Technical Report requested by European Parliament. Available at: https://efsa.onlinelibrary.wiley.com/doi/epdf/10.2903/sp.efsa.2018.EN-1397 (accessed 15 March 2025).

FiBL (2022) *The World of Organic Agriculture: Statistics & Emerging Trends*. FiBL and IFOAM.

Gamliel, A., van Bruggen, A.H. and Finckh, M.R. (2016) Plant disease management in organic farming systems. *Pest Management Science* 72(1), 30–44.

Garibaldi, L.A., Carvalheiro, L.G., Vaissière, B.E., Gemmill-Herren, B., Hipólito, J. *et al*. (2016) Mutually beneficial pollinator diversity and crop yield outcomes in small and large farms. *Science* 351(6271), 388–391.

Głodowska, M. and Krawczyk, J. (2017) Heavy metals concentration in conventionally and organically grown vegetables. *Quality Assurance and Safety of Crops & Foods* 9(4), 497–503.

Grant, C.A. (2015) Influence of phosphate fertilizer on cadmium in agricultural soils and crops. In: Selim, H.M. (ed.) *Phosphate in Soils: Interaction with Micronutrients, Radionuclides and Heavy Metals*. CRC Press, Boca Raton, Florida, Chapter 5.

Gurr, G.M., Lu, Z., Zheng, X., Xu, H., Zhu, P. *et al*. (2016) Multi-country evidence that crop diversification promotes ecological intensification of agriculture. *Nature Plants* 2, 16014.

Güsewell, S. (2004) N:P ratios in terrestrial plants: Variation and functional significance. *The New Phytologist* 164(2), 243–266.

Huber, D., Römheld, V. and Weinmann, M. (2012) Relationship between nutrition, plant diseases and pests. In: Marschner, P. (ed.) *Marschners Mineral Nutrition of Higher Plants*, 3rd edn. Elsevier, pp. 283–298.

Kabak, B., Dobson, A.D.W. and Var, I. (2006) Strategies to prevent mycotoxin contamination of food and animal feed: A review. *Critical Reviews in Food Science and Nutrition* 46(8), 593–619.

Karlsson, I., Friberg, H., Steinberg, C. and Persson, P. (2014) Fungicide effects on fungal community composition in the wheat phyllosphere. *PLoS ONE* 9(11), e111786.

Karlsson, I., Friberg, H., Kolseth, A.-K., Steinberg, C. and Persson, P. (2017) Organic farming increases richness of fungal taxa in the wheat phyllosphere. *Molecular Ecology* 26(13), 3424–3436.

Kim, H.Y, Lee, K.B and Lim, H.Y. (2004) Contents of minerals and vitamins in organic vegetables. *Korean Journal of Food Preservation* 11.

Kratz, S. and Schnug, E. (2005) Schwermetalle in P-Düngern (Heavy metals in P fertilisers). *Landbauforschung Völkenrode Spec* [in German], 37–45.

Kratz, S., Knappe, F. and Schnug, E. (2008) Uranium balances in agroecosystems. In: de Kok, L. and Schnug, E. (eds) *Loads and Fate of Fertilizer-Derived Uranium*. Backhuys, Leiden, The Netherlands, pp. 179–190.

Kuczyńska, B., Puppel, K., Gołebiewski, M., Metera, E., Sakowski, T. *et al*. (2012) Differences in whey protein content between cow's milk collected in late pasture and early indoor feeding season from conventional and organic farms in Poland. *Journal of the Science of Food and Agriculture* 92(14), 2899–2904.

Liesch, T., Hinrichsen, S. and Goldscheider, N. (2015) Uranium in groundwater: Fertilizers versus geogenic sources. *The Science of the Total Environment* 536, 981–995.

Mirvich, S.S. (1993) Vitamin C inhibition of *N*-nitroso compound formation. *The American Journal of Clinical Nutrition* 57(4), 598–599.

Nziguheba, G. and Smolders, E. (2008) Inputs of trace elements in agricultural soils via phosphate fertilizers in European countries. *The Science of the Total Environment* 390(1), 53–57.

Pingali, P.L. (2012) Green revolution: Impacts, limits and the path ahead. *Proceedings of the National Academy of Sciences of the United States of America* 109(31), 12302–12308.

Reganold, J.P. and Wachter, J.M. (2016) Organic agriculture in the twenty-first century. *Nature Plants* 2(2), 15221.

Rembiałkowska, E. (2000) *Wholesomeness and Sensory Quality of Potatoes and Selected Vegetables from the Organic Farms*. Fundacja Rozwo´j SGGW, Warszawa.

Schnug, E. and Haneklaus, N. (2015) Uranium in phosphate fertilizers – review and outlook. In: *Uranium - Past and Future Challenges: Proceedings of the 7th International Conference on Uranium Mining and Hydrogeology*, Springer, Cham, Switzerland, pp. 123–130.

Sharma, R., Arora, D., Singh, R.P. and Pareek, N.K. (n.d.) Impact of Krishi Vigyan Kendra on pearlmillet growers. *Indian Journal of Extension Education* 56(4), 151–154.

Smith-Spangler, C., Brandeau, M.L., Hunter, G.E., Bavinger, J.C., Pearson, M. *et al.* (2012) Are organic foods safer or healthier than conventional alternatives? A systematic review. *Annals of Internal Medicine* 157(5), 348–366.

Szponar, L. and Kierzkowska, E. (n.d.) Azotany i azotyny w s´rodowisku oraz ich wpływ na zdrowie człowieka (1990). *Postępy Higieny i Medycyny Doświadczalnej* 44(327), 350.

Woese, K., Lange, D., Boess, C. and Bögl, K.W. (1997) A comparison of organically and conventionally grown foods: Results of a review of the relevant literature. *Journal of the Science of Food and Agriculture* 74(3), 281–293.

Worthington, V. (2001) Nutritional quality of organic versus conventional fruits, vegetables, and grains. *The Journal of Alternative and Complementary Medicine* 7(2), 161–173.

Zahm, S.H. and Blair, A. (1992) Pesticides and non-Hodgkin's lymphoma. *Cancer Research* 52(19 Suppl), 5485s–5488s.

55 Organic Farming Systems and Food Security: Identifying Values and Investigating Perspectives

RACHID MRABET*

Institut National de la Recherche Agronomique (INRA), Rabat, Morocco

Abstract

This chapter examines the question of whether organic farming (OF), which occupies 2% of agricultural lands, can produce enough food to meet future demand in a nutritious way without impacting the environment. The available scientific literature proved that OF was more nutrient-dense than conventional agriculture (CA) and allowed higher environmental benefits. OF offered health benefits by reducing exposure to harmful chemicals and reducing disease risks. On average, OF had 20% lower crop yields than CA. Low external inputs and higher compensation through premium prices for quality products can help offset the economic disadvantages due yield reduction. Thus, OF economic benefits vary depending on farm size and assets, market access, scale of production, and local economic conditions. It was also found that the gaps can be reduced when soils regenerate their fertility and integrity. However, if soils are not appropriately and timely managed, yield gaps can translate into economic losses, posing challenges to the long-term viability of organic farming, especially among smallholders. To realize OF's full potential as a critical player for food and nutrition security, it is necessary to address the challenges through continued research, innovation, capacity and skill development, motivation schemes and supportive policies. OF adoption and its bundling with other regenerative agricultural systems lays the foundation for lowering yield gaps and acceleration of environmental sustainability.

Keywords: food security, organic farming and foods, crop production, nutritional quality, environmental sustainability

Introduction

Organic farming is getting higher importance from consumers and farmers and its growth has been accelerating worldwide. OF is practised in 188 countries, with 4.5 million farmers and producers operating over 96 million ha of farmland. OF is revolutionary as it has moved from local and niche initiatives to being an industry (Willer *et al.*, 2024; Panday *et al.*, 2024). According to Willer *et al.* (2024), from 2013 to 2022, the change in area of arable and permanent crops was 82.2%. Bouttes *et al.* (2019) found that farmers regarded OF to enhance their adaptive capacity through four mechanisms: reducing risks, stimulating learning, increasing professional satisfaction, and enabling maintenance of the family farm.

On the other hand, the OF sector has not reached its maturity and several questions are at the centre of lively debate. In this sense, Giampieri *et al.* (2022) and Verdi *et al.* (2022) affirmed that OF is not a sustainable strategy to optimize land-use efficiency confirming the observations of Tuomisto *et al.* (2012), reporting that approximately 84% more land is needed for OF compared to CA to satisfy a growing population's food demand. These authors showed that due to lower yields and the requirement to build the fertility of land, lower environmental impact not always guarantees per product unit. Hence, the key question repeatedly

*Corresponding author: rachid.mrabet@inra.ma

 Advances in Organic Farming (ed. L. P. Awasthi)
DOI: 10.1079/9781800626850.0055

raised was 'To what extent can OF systems enhance simultaneously food and nutrition security and environmental sustainability?' (Xu *et al.*, 2018; Aihounton and Henningsen, 2023). Hence, this chapter, through an analysis of the literature, aims to offer a sound understanding of the benefits and effects of OF systems in terms of crop productivity and nutrition quality as pillars of food security but also environmental sustainability indicators.

Crop Production and Food Security

Literature is brimming with studies that attempt to link OF to food security, although the feasibility of OF systems for achieving food and nutrition security is still contested, with mixed evidence.

Olle and Williams (2012) reported about 59% of studies observed positive effects on vegetable yields from organic cultivation, while 29% reported a negative impact, and 12% found no significant influence. Such diversity in results can be explained by a narrower spectrum of choices in organic nutrient sourcing and applications under organic agriculture (Panday *et al.*, 2024).

Badgley *et al.* (2007) mark the launch of the discussion as to whether organic agriculture can feed the global population. They used results from various studies to conclude that at the global level, organic agriculture had 33% higher average crop yields than CA. Based on 46 paired organic-conventional systems, they reported that in developed countries, organic yields were 9% lower than conventional yields, but in developing countries, organic practices would increase crop yields by 74%. Clark and Tilman (2017) found comparable conclusions.

de Ponti *et al.* (2012), Seufert and Ramankutty (2017) and Ponisio *et al.* (2015) showed that average yields from organic farms are, globally, 20% lower than from conventional farms. Seufert *et al.* (2012) reported that yields of organic fruits and oilseed crops show a small yield gap (–3% and –11%, respectively) compared to conventional crops, whereas organic cereals and vegetables have significantly lower yields than conventional crops (–26% and –33%, respectively). The same study found that organic performance is 35% less than CA under irrigated conditions, but only 17% less under rainfed conditions. The authors concluded that yield differences are highly contextual, depending on system and site characteristics, and range from 5% lower organic yields for rainfed legumes and perennials on weak-acidic to weak-alkaline soils; 13% lower yields when best organic practices are used; and 34% lower yields when the conventional and organic systems are most comparable. Seufert *et al.* (2012) suggested that perennial crops and legumes tended to produce organic crop yields more like conventional crop yields compared to other organic crops, which is supported by the superior performance of the organic perennial hay crops compared to the annual silage crop in Kniss *et al.*'s (2016) analysis in USA.

A global analysis by Meier *et al.* (2015) found that crop yield differences between OF and conventional systems ranged between –68% to +32% for arable crops and between –65–76% for fruits and vegetables. Kniss *et al.* (2016) found the following: (i) averaged across all crops, organic yield averaged 80% of conventional yield; (ii) for cereals, yields on organic farms were approximately 20% lower than on conventional farms, which is like the yield gaps reported in the meta-analyses; (iii) for certain vegetables (tomato, potato, bell pepper, and onion) and fruits (watermelon, grape, blueberry and apple), reported yield gaps were 50% and more, which is significantly higher than the yield reported by meta-analyses; (iv) organic yields surpassed conventional yields for some hay crops; (v) organic squash, snap bean (*Phaseolus vulgaris*), sweet maize and peach (*Prunus persica*) yields were not statistically different from conventional.

According to Meemken and Qaim's (2018) analyses, organic farming reduces crop yields by an average of 19–25%. In their analysis of multiple studies, the authors found significant variations among crop species, with fruits and legumes exhibiting smaller yield gaps than cereals and root and tuber crops.

Smith *et al.* (2019) concluded that efforts to reliably curb environmental impacts through organic certification may come at a cost of lower and more variable yields. They also pointed out that despite the lower yields and greater yield variability on organic farms, organic farms were more profitable and had similar production costs compared to conventional farms. They reported that while organic farms provided a 'win-win' (high means and low variability) for environmental sustainability, conventional farms provided a 'win-win' for production by promoting high crop yields with low variability. This raises two crucial and important issues: (i) OF crop high yield variability does not allow producers to achieve consistent production and avoid unprofitable years, and does not ensure that consumers have reliable access to nutritious and sufficient food; (ii) OF systems, due their low variability in ecosystem services, are more resilient to changing climatic conditions, including an increased frequency of drought and extreme weather events.

A 40-year long-term Farming Systems Trial (FST) conducted by Rodale Institute has shown how regenerative organic agriculture outperforms conventional grain cropping systems during times of extreme weather, where corn yield was 31% higher in organic than in conventional during droughts (Rodale Institute, 2022).

According to Reganold and Wachter (2016), even though crop yields were higher under conventional systems, organic agricultural systems deliver healthy products, greater ecosystem services and social benefits. They reported that OF permitted better water balance due to higher moisture storage and infiltration in soils. Such soil improvements can protect farmers from the consequences of lower crop yields if shifting from CA. In other words, yield gaps may diminish over time as improvements in soil biotic and abiotic properties occur (Sharma *et al.*, 2018). Heavy reliance on ecosystem service providers, such as biological control of pests, helps to produce adequate yields and reduce gaps under unpredictable and variable climate (Karp *et al.*, 2018).

Several authors consider organic agriculture inefficient, especially in terms of land use (Kirchmann and Andersson, 2008; Connor and Mínguez, 2012; Röös *et al.*, 2018). Manna *et al.* (2021) raised the question whether the availability of organic feed materials (animal manures, off-farm organic wastes, crop residues, green manures, biofertilizers) is sufficient to support widespread OF and to meet the global food needs. Muller *et al.* (2017) revealed that organic production on a global scale can be feasible in terms of land availability if coupled with demand-side mitigation options, including dietary change, food losses, and waste reduction. They reported that food availability could be met if these options are considered, and acceptable N-supply can be possible if integrated crop-livestock is promoted. Clark and Tilman (2017) found that since over 30% of food production is wasted, reducing food waste would offer environmental benefits without requiring shifts in production practices or diets. These authors also recommended to shift to low-input/low-impact food systems.

Organic Food Healthiness and Nutrition Security

Demand for organic foods is mostly driven by consumers' perceptions that they are more nutritious and healthier than non-organic ones. In other words, by reducing the application of synthetic chemicals, OF minimizes the potential health risks posed by pesticides, nitrates and metal residues (Crinnion, 2010; Matt *et al.*, 2011; Barański *et al.*, 2014; Hurtado-Barroso *et al.*, 2019; Gomiero, 2021; Giampieri *et al.*, 2022). Other studies showed that OF provides improved nutrition and health outcomes for consumers (Lairon, 2010; Brandt *et al.*, 2011; Barański *et al.*, 2014; Yu *et al.*, 2018; Rani *et al.*, 2023; Munné-Bosch and Bermejo, 2024; Rahman *et al.*, 2024; Panday *et al.*, 2024).

A meta-analysis by Barański *et al.*, 2014 revealed that organic crops contained on average 48% lower Cd concentrations than conventional crops. Gomiero (2021) confirmed the absence of pesticide residues and mycotoxins in organic foods. The author also reported that animal organic products are much healthier due to restriction or strict regulation of the use of hormones and antibiotics under OF.

Olle and Williams (2012) found that organically grown vegetables have, in most studies (65%), better nutritional values than conventionally grown ones; 20% were not significantly different and only in 15% was there a reduction in nutritional value. Worthington (2001) revealed 27% higher vitamin C levels in organically grown lettuce, spinach, potatoes and cabbage. Lairon (2010) showed higher vitamin C contents in organic potatoes, tomatoes, kale and celeriac as well as higher vitamin E content in organic olive oils.

Rani *et al.* (2023) showed that organic apples, tomatoes, eggplants, cauliflowers, carrots and cereals contain higher nutrition like vitamin C, carotenoids, amino acid and polyphenols. Compared to the conventional tomato, soluble solids, soluble sugar, lycopene, vitamin C and nitrate were raised by 11.86%, 42.18%, 23.95%, 18.97%, and 8.36%, respectively, for organically grown tomato (Gao *et al.*, 2023). In their review, Rahman *et al.* (2024) confirmed that organic food contained higher levels of iron, magnesium and vitamin C. However, a recent study by Reddy *et al.* (2023) did not find significant differences in water-soluble vitamins between organic and conventional crops.

Reddy *et al.* (2023) and Czech *et al.* (2022) reported a decrease in the protein content among the analyzed conventional crops in comparison to organic ones. Dos Santos *et al.* (2019) from Brazil did not find differences between the two types of sweet potato in terms of centesimal concentrations (protein, lipids, carbohydrates).

Yu *et al.* (2018) found 20% higher magnesium (Mg) content and 30% higher phosphorus (P) and potassium (K) contents in organic compared to conventionally grown summer corn. These findings were confirmed by Rembiałkowska (2016), where the results of many experiments demonstrated a higher level of iron (Fe), P and Mg contents in organically grown compared to non-organically grown products. Reddy *et al.* (2023) concluded there was a tendency to higher mineral content (Fe, Zn, K, Na, Mg and Mn) in organic than in conventional Indian foods. Other studies agreed with these findings of high mineral composition under organic foods (de Souza *et al.*, 2014; Dos Santos *et al.*, 2019).

Barański *et al.* (2017) observed that organic crops have between 18% and 69% increased antioxidant activity than do conventional crops. In a ten-year experiment, Mitchell *et al.* (2007) found that the levels of flavonoids increased over time in samples from organic treatments, whereas the levels of flavonoids did not vary significantly in conventional treatments. Ten-year mean levels of quercetin and kaempferol in organic tomatoes were 79% and 97% higher, respectively, than in conventional tomatoes. Such changes were linked to increased organic matter in soils under OF.

Rembiałkowska (2016) reported the concentration of phenolic compounds to be 20% higher in organic as compared to conventional products. According to Das *et al.* (2020), dry matter, minerals and antioxidants like polyphenols and salicylic acids are all higher in organic products. Reeve *et al.* (2016) linked these health-promoting molecules or phytochemicals to greater plant stress, rhizosphere microbial communities, and/or lower available nitrogen under OF.

Because organic produce has more sugar and a higher metabolic integrity as well as a superior cellular structure, it tastes better and keeps longer (Bourn and Prescott, 2002). According to Pawar *et al.* (2022), organic foods are tastier with characteristic fragrance. From the review by Gopalakrishnan (2019), organic fruits and vegetables are much richer in flavour than non-organic ones. This will reduce the use of quality and flavour enhancers (preservatives, synthetic flavours) commonly added in the food industry.

The high levels of vitamins and minerals found in organic foods and products aid in the strengthening of the human immune system while antioxidants have a good effect on heart disease, cancer, immune deficiency, visual problems, premature ageing, and cognitive dysfunction (Talhaoui *et al.*, 2015; Pawar *et al.*, 2022). According to Gopalakrishnan (2019), organic crops have greater antioxidant activity, ranging from 18% to 69% higher concentrations of a variety of distinct antioxidants, which is related to a lower chance of illness including cardiovascular and neurodegenerative disease, as well as specific malignancies. From the review by Rahman *et al.* (2024), health benefits for organic consumers include a reduction in obesity and body mass index, improvements in blood nutrient composition as well as reductions in maternal obesity and pregnancy-associated preeclampsia risks. All such benefits are directly linked to improved food security and hunger alleviation.

Environmental Sustainability and SDGs

As OF continues to gain momentum, it becomes essential to understand its potential environmental benefits, main sustainability challenges, and implications for sustainable agriculture and SDGs (Eyhorn *et al.*, 2019; Manna *et al.*, 2021). In this regard, the scientific community has assessed the environmental impacts of organic and conventional farming systems (Tuomisto *et al.*, 2012; Meier *et al.*, 2015; Chiriacò *et al.*, 2017; Smith *et al.*, 2019) and proved the superiority of organic systems per unit of area.

Organic agriculture is a sustainable and environmentally friendly production system that offers a wide range of economic, environmental, social and cultural benefits at various timescales (Parizad and Bera, 2023; Panday *et al.*, 2024). Eyhorn *et al.* (2019) showed that policies that specifically support OF systems and their upscaling can contribute importantly to advancing the SDGs. As shown earlier, OF induces lower crop yields compared to CA and exhibits lower yield stability, thus impeding efforts to combat hunger. However, certification of organic systems for the export market can positively impact both SDG 1 and 2. Premium prices of organic products can undoubtedly reduce poverty, which is the major factor of food insecurity (UNCAD and UNEP, 2008). From the study by Crowder and Reganold (2015), the profits of organic farms are 22–35% higher compared to conventional farms due to organic premium prices. They compensate for the lower yields of organic crops (Panday *et al.*, 2024). The price premium leads to improved financial stability for farmers and producers (Crowder and Reganold, 2015; Reganold and Wachter, 2016; Seufert and Ramankutty, 2017; Panday *et al.*, 2024).

It was found that consumption of organic foods or products lessen micronutrient deficiencies in several countries (Reddy *et al.*, 2023) and contribute to achievement of SDG 2. Such a situation is more challenging for the least developed countries striving to reduce poverty and hunger. Hence, it is crucial to exercise attentiveness, especially in sub-Saharan African countries that have not yet attained SDG one and SDG 2, when prompting OF policies (Zhang *et al.*, 2024).

SDG 12 is about ensuring sustainable consumption and production patterns, which is key to sustaining the livelihoods of current and future generations. This can involve consuming less, choosing products with lower environmental impact, and reducing the carbon footprint of day-to-day activities, consumptions and products. In fact, certification of OF systems could also reduce levels of contaminants in food, feed and the environment, especially in developed countries where regulations are stronger and stricter.

There is a wealth of research indicating that organic systems could mitigate the impact of agriculture on climate change (Aguilera *et al.*, 2015; Biernat *et al.*, 2020; Niemiec *et al.*, 2020; Goel *et al.*, 2021; Bellassen *et al.*, 2021; Nabuurs *et al.*, 2022; Singh *et al.*, 2024). From these authors, OF has been shown to emit 45% less greenhouse gas compared to CA, thereby enhancing environmental sustainability as well. In addition, OF practices can improve carbon sequestration in soils and permanent crops, and hence strengthen climate resilience (Skinner *et al.*, 2014; Hansen *et al.*, 2019; Madhavi *et al.*, 2021; Kishore *et al.*, 2023; Gamage *et al.*, 2023). Numerous investigations have shown that soil organic matter was improved in fields under organic management compared to conventional management.

OF systems are practised in all major climate zones from the warm, humid tropics to the cool, temperate and arid areas, and in all the world's main farming systems (i.e. rainfed and irrigated systems, short-term and perennial crops, mixed crop-animal farms, and organic systems). In addition, when combined with frontier technologies (precision agriculture, plant breeding and biotechnology, microbial biotechnology, smart fertilizers, biochar additions etc.), OF systems can help to soak up even more carbon in the soil, create soil resilience to achieve food security, mitigate climate change, and allow higher and stable yields in drought conditions (Venkat, 2012; Yang *et al.*, 2024).

Ume *et al.* (2022) investigated motivation to adopt agroecological practices including OF under varying disciplinary perspectives and in the global south, and showed that values are closely linked to environmentalism. Pimentel *et al.* (2005) and Seitz *et al.* (2019) estimated that soil erosion potential can be reduced by over 61% with organic practices. Jouzi *et al.* (2017) concluded that even though OF might present some significant challenges to small-scale farmers, it could/should still be considered as a part of the solution for climate change and food security and a means of improving their livelihoods. Their study showed that the most significant advantages of OF are environmental protection and a higher resilience to environmental change, increasing farmers' income, reducing external input costs, enhancing social capacity, and increasing employment opportunities. OF tends to be more labour-intensive than CA, potentially creating more employment opportunities.

Because organic foods frequently bring higher prices in the marketplace, the net economic return per ha is often equal to or higher than that of conventionally produced crops (Pimentel *et al.*, 2005). It combines tradition, innovation and science to benefit the shared environment and promote fair relationships and good quality of life for all involved. Techniques like planting cover crops, rotating crops with incorporation of legumes and spreading manure and compost instead of relying heavily on synthetic fertilizers are all hallmarks of organic farming (Sharma *et al.*, 2018).

The role of organic agriculture in maintaining the natural resource base is evident. A study by Tscharntke *et al.* (2021) indicated that local species richness and abundance can increase by approximately 34% and 50%, respectively, across various crops worldwide compared to conventional farming practices. Organic certification schemes effectively limit negative environmental effects of agriculture on beneficial natural enemy populations and pollinators. This beneficial biodiversity is highly needed for enhanced crop production, which is the opposite under agricultural systems based on pesticides (Kennedy *et al.*, 2013; Porcel *et al.*, 2018).

Under conventional systems, the decoupling of crop and livestock production is an important cause of pollution due to high nitrogen loss and erosion, especially in countries with rapid specialized livestock production (Gu, 2022; Yang *et al.*, 2022). However, OF aims for stronger integration of tree, livestock and crops for lower environmental impacts. In addition, organic principles stipulate good care of animals. Such integration improves nutrient cycling and energy efficiency. Several authors reported that animal productivity is much higher and of better quality, in terms of either meat or milk under OF (van Wagenberg *et al.*, 2017; Röös *et al.*, 2018).

Tuck *et al.* (2014) summarized a meta-analysis by stating that, on average, OF increased species richness by about 30%. Schneider *et al.* (2014), comparing 12 European and African countries, found that species richness is, on average, 10.5% higher in organic than non-organic production fields, with highest gains in intensive arable fields (around 45%). These results were confirmed by Crowder *et al.* (2010, 2012); Gattinger *et al.* (2012); Kennedy *et al.* (2013); Crowder and Reganold (2015); Reganold and Wachter (2016) and Lichtenberg *et al.* (2017).

Shrivas *et al.* (2019) reported the great importance of organic amendments and fertilizers on the development of microbial communities and their function in building healthy soils resilient to climatic and anthropic shocks. Under OF systems, soil-enhancing practices such as mulching, green manures, nitrogen-fixing crops and trees, animal manures, recoupling animals with crop-nutrient cycles enhances below-ground diversity, e.g. valuable mycorrhizal fungi and other soil microorganisms important to many ecosystem and soil fertility functions are of extreme importance for mitigating drought effects and improving water-use efficiency by crops. Bacteria, protozoa, nematodes and arthropod densities were higher in soils under OF than in soils under traditional farming (Sow *et al.*, 2023). OF improves the overall microclimate in the rhizosphere (Khadse and Rosset, 2019). In other terms, a balanced application of these practices help crops to improve their productivity and be more resilient to drought and water deficits.

Enabling Organic Farming Drivers for Enhancing of Food Security

The challenges and opportunities for improving global food security and the environment through adopting or shifting to OF systems are not evenly distributed across the globe and the movement towards organic agriculture are driven by many factors that play together. Hence, promoting and increasing the OF uptake should be enabled through both country and regional strong policies and incentives. According to Eyhorn *et al.* (2019), several European and Asian countries have defined comprehensive strategies which include push measures (e.g. support to research and extension, area-based payments), pull measures (e.g. consumer information campaigns) and enabling measures (e.g. data collection, institutional development) to increase share of organic foods in the population consumption.

The interests of both farmers, including their organizations, and the public authorities should have synergistic effects when it comes to providing public goods such as climate and environmental protection. Such synergistic effects require incentives and motivations and help avoid political compromises and enforce evidence-based investigations for scaling up OF. Effective and durable implementation of OF requires several managerial, financial and technological measures in addition to institutional support for increasing productivity of crops and livestock.

Due to the high investment cost and delayed effect of organic fertilizers and biopesticides (Cafer and Rikoon, 2018), small-scale farmers can hardly afford the higher input costs, particularly when subject to credit constraints. Consequently, credit constraints within traditional financial services emerge as a critical inhibiting factor for farmers' OF adoption (Sieber *et al.*, 2018). Hence, one of the most important policy enablers can be setting up multi-annual incentive schemes designed to increase the adoption of OF practices or systems by farmers and across value chains.

Information asymmetry has imposed unnecessary restrictions on farmers' production input decisions and agricultural productivity (Ullah *et al.*, 2020), which is the fundamental reason why farmers do not adopt OF systems or practices. The digital technology can significantly alleviate this asymmetry in agricultural production and marketing and enable farmers to optimize fertilizer application and technology adoption as well as resource allocation and precision management (Harou *et al.*, 2022). Hence, vital enablers for OF enactment and functioning are access to up-to-date information on innovative technologies and organic quality standards; simplified certification and labelling of products; investment opportunities; public-private partnerships; communication strategy to raise awareness of farmers; and continuous improvement of stimulating innovation and research. In addition, developing a robust supply chain and infrastructure for organic farming is crucial to support farmers and make their products more accessible to consumers, both domestically and abroad.

Bundling Organic Farming with Regenerative Systems

There is consensus among scholars that OF is associated with positive environmental, economic and social impacts and is an ethical approach for humanity (Sumberg, 2022). However, if OF systems maintain the use of tillage for seedbed preparation and weed control, soil erosion remains a great concern (Cooper *et al.*, 2016; Moos *et al.*, 2017), which reduces crop yields further. In addition, inefficient planning of OF and rapid transition can also create food insecurity.

Low organic crop yields were associated with limited and unpredictable nutrient availability, mainly N and P under OF. Discarding synthetic N-fertilizers could lead to nutrient undersupply, even with increased legume in organic cropping systems (Barbieri *et al.*, 2021). Consequently, OF suitability, and production and environmental impacts, are highly context-dependent as shown by Seufert and Ramankutty (2017) and Riar *et al.* (2024). In order to address such context-specific issues, several essays tried to integrate other transformative systems also referred to as regenerative or resilient systems (i.e. OF, conservation agriculture and agroforestry) to uphold the sustainability pillars and ensure sustainable and healthy food systems (Peigné *et al.*, 2016; Cooper *et al.*, 2016; Ali and Dahlhaus, 2022; Puyu *et al.*, 2023; van Balen *et al.*, 2023; Riar *et al.*, 2024). Migliorini and Wezel (2017), while comparing similarities and divergences in principles of OF and agroecology concluded that beyond regulations, both approaches offer promising contributions for the future development of sustainable agricultural production and food systems because they are based on holistic approaches, put forward sustainable use of natural resources and inputs, and consider biodiversity conservation and increased ecosystem service providers. Riar *et al.* (2024) showed that combination of OF and agro-forestry can be considered a viable alternative to conventional practices that can help alleviate the negative impacts on planetary boundaries, specifically in tropical regions.

Using modelling and scenario analysis, Muller *et al.* (2017) showed that OF can contribute to providing sufficient food and improving environmental impacts, only if adequately high proportions of legumes are produced and with significant reductions of food-competing feed use, livestock product quantities, and food wastage. The development of organic agriculture in the future should take up these challenges on the consumption side, and not only focus on sustainable production. This scenario will also reduce the need for increased lands under OF and yield gaps over time.

Conclusion

OF systems can be seen as transformative systems with proven sustainability benefits (i.e. enhanced ecosystem services including better soil quality, improved biodiversity, reduced pollution, and increased farm incomes), but in many contexts they result in yield gaps so that OF sustainability per unit product is sometimes queried. Due to lower crop yields than conventional agriculture, OF is criticized for its higher food costs, since it is usually more expensive than conventionally grown food, due to costly expenditure, certification, and use of labour-intensive methods. Hence, improvements in management techniques that address factors limiting yields in organic systems should be further studied and investigated. If not, OF adoption will be limited since expansion is needed to fill the yield gaps and satisfy food demand by a growing consumer population.

According to Mondejar *et al.* (2021), Mesías-Ruiz *et al.* (2023) and Yang *et al.* (2024), big data and artificial intelligence should be integrated in the rural economy to enhance global food security by lowering crop losses and increasing resource-use efficiency and crop yields, influencing the entire food supply chain. Such AI-based precision tools are emerging driving forces for sustainable development and should increasingly be used in organic agriculture to further enhance production and effectiveness by improving economic growth and reducing costs of energy and water uses and recycling wastes (Dong *et al.*, 2020; Timsina *et al.*, 2021; Panday *et al.*, 2024; Yang *et al.*, 2024).

The shift from competing ecology-based narratives to a collaborative green strategy for reaching the SDGs is needed for nature-positive systems, including OF, to reach a 'tipping point' to be globally transformative. In other terms, there is great momentum in merging principles of OF with those of agroecology, regenerative agriculture, agroforestry, and conservation agriculture, especially those related to reducing trade-offs due to tillage, synthetic fertilizers and pesticide use (Newton *et al.*, 2021). Potentially additive benefits are high and should be assessed and understood (Cooper *et al.*, 2016). By adopting nature-positive farming techniques, such as drought-resistant crops, efficient irrigation methods, no-till-based OF and other soil conservation practices, farmers can enhance both yields and quality of crops but also reduce the impact of drought and degraded soils on their crops and incomes. However, issues related to GMOs are still largely debated within the OF stream. Crop breeders and biotechnologists need to identify plant traits that can be targeted to improve nutritional quality of organic crops in relation to increasing atmospheric CO_2 concentration to insert them in the value chains as strong and mobilizing alternatives for healthy and durable food systems.

As the global community continues to grapple with environmental degradation and the need for sustainable development, organic farming stands as a crucial component in the transition towards more sustainable and resilient agricultural systems. The continued support for research, innovation, and policy development in OF will be key to maximizing its potential and addressing its challenges. OF can be a pillar for agricultural sustainability by combining scientific research, community-based innovation and traditional food consumption. However, more inclusive research (at all value-chain levels plus the ecosystem) is required to completely value and capture the extent of benefits of organic foods, and this will help to provide more evidence-based information and data for consumers, farmers, researchers, industries and policymakers to make informed decisions about food choices and diets. Coupling knowledge construction and social learning is needed for durable OF development while bringing together actors with heterogeneous backgrounds and foci, thus accommodating knowledge from different sources and enhancing the exchange of practices, ideas and assets (both material and intangible) (Charatsari *et al.*, 2022). As a conclusion, the organic movement and broad uptake need a new strategic course to become more relevant and impactful in view of the challenges that global agriculture is facing, as presented by IFOAM's Organic 3.0 (www.ifoam.bio/organic3zero [accessed 4/3/25]) (Arbenz *et al.*, 2017).

References

Aguilera, E., Guzmán, G. and Alonso, A. (2015) Greenhouse gas emissions from conventional and organic cropping systems in Spain. II. Fruit tree orchards. *Agronomy for Sustainable Development* 35(2), 725–737.

Aihounton, G.B.D. and Henningsen, A. (2023) *Does Organic Farming Jeopardize Food and Nutrition Security?* IFRO Working Paper, No.2023/02. University of Copenhagen, Department of Food and Resource Economics (IFRO).

Ali, B. and Dahlhaus, P. (2022) Roles of selective agriculture practices in sustainable agricultural performance: A systematic review. *Sustainability* 14(6), 3185.

Arbenz, M., Gould, D. and Stopes, C. (2017) Organic 3.0 – the vision of the global organic movement and the need for scientific support. *Organic Agriculture* 7(3), 199–207.

Badgley, C., Moghtader, J., Quintero, E., Zakem, E., Chappell, M.J. *et al.* (2007) Organic agriculture and the global food supply. *Renewable Agriculture and Food Systems* 22(2), 86–108.

Barański, M., Srednicka-Tober, D., Volakakis, N., Seal, C., Sanderson, R. *et al.* (2014) Higher antioxidant and lower cadmium concentrations and lower incidence of pesticide residues in organically grown crops: a systematic literature review and meta-analyses. *British Journal of Nutrition* 112(5), 794–811. DOI: 10.1017/S0007114514001366.

Barański, M., Srednicka-Tober, D., Volakakis, N., Seal, C., Sanderson, R. *et al.* (2014) Higher antioxidant and lower cadmium concentrations and lower incidence of pesticide residues in organically grown crops: A systematic literature review and meta-analysis. *The British Journal of Nutrition* 112(5), 794–811.

Barański, M., Rempelos, L., Iversen, P.O. and Leifert, C. (2017) Effects of organic food consumption on human health: The jury is still out!. *Food & Nutrition Research* 61(1), 1287333.

Barbieri, P., Pellerin, S., Seufert, V., Smith, L., Ramankutty, N. *et al*. (2021) Global option space for organic agriculture is delimited by nitrogen availability. *Nature Food* 2(5), 363–372.

Bellassen, V., Drut, M., Antonioli, F., Brečić, R., Donati, M. *et al*. (2021) The carbon and land footprint of certified food products. *Journal of Agricultural & Food Industrial Organization* 19(2), 113–126.

Biernat, L., Taube, F., Loges, R., Kluß, C. and Reinsch, T. (2020) Nitrous oxide emissions and methane uptake from organic and conventionally managed arable crop rotations on farms in Northwest Germany. *Sustainability* 12(8), 3240.

Bourn, D. and Prescott, J. (2002) A comparison of the nutritional value, sensory qualities, and food safety of organically and conventionally produced foods. *Critical Reviews in Food Science and Nutrition* 42(1), 1–34.

Bouttes, M., Darnhofer, I. and Martin, G. (2019) Converting to organic farming as a way to enhance adaptive capacity. *Organic Agriculture* 9(2), 235–247.

Brandt, K., Leifert, C., Sanderson, R. and Seal, C.J. (2011) Agroecosystem management and nutritional quality of plant foods: The case of organic fruits and vegetables. *Critical Reviews in Plant Sciences* 30(1–2), 177–197.

Cafer, A.M. and Rikoon, J.S. (2018) Adoption of new technologies by smallholder farmers: The contributions of extension, research institutes, cooperatives, and access to cash for improving tef production in Ethiopia. *Agriculture and Human Values* 35(3), 685–699.

Charatsari, C., Lioutas, E.D., Papadaki-Klavdianou, A., Koutsouris, A. and Michailidis, A. (2022) Experiential, social, connectivist, or transformative learning? Farm advisors and the construction of agroecological knowledge. *Sustainability* 14, 2426.

Chiriacò, M.V., Grossi, G., Castaldi, S. and Valentini, R. (2017) The contribution to climate change of the organic versus conventional wheat farming: A case study on the carbon footprint of wholemeal bread production in Italy. *Journal of Cleaner Production* 153, 309–319. DOI: 10.1016/j.jclepro.2017.03.111.

Clark, M. and Tilman, D. (2017) Comparative analysis of environmental impacts of agricultural production systems, agricultural input efficiency, and food choice. *Environmental Research Letters* 12(6), 064016.

Connor, D.J. and Mínguez, M.I. (2012) Evolution not revolution of farming systems will best feed and green the world. *Global Food Security* 1(2), 106–113.

Cooper, J., Baranski, M., Stewart, G., Nobel-de Lange, M., Bàrberi, P. *et al*. (2016) Shallow non-inversion tillage in organic farming maintains crop yields and increases soil C stocks: A meta-analysis. *Agronomy for Sustainable Development* 36(1), s13593–016.

Crinnion, W.J. (2010) Organic foods contain higher levels of certain nutrients, lower levels of pesticides, and may provide health benefits for the consumer. *Alternative Medicine Review Journal of Clinical Therapy* 15, 4–12.

Crowder, D.W. and Reganold, J.P. (2015) Financial competitiveness of organic agriculture on a global scale. *Proceedings of the National Academy of Sciences of the United States of America* 112(24), 7611–7616.

Crowder, D.W., Northfield, T.D., Strand, M.R. and Snyder, W.E. (2010) Organic agriculture promotes evenness and natural pest control. *Nature* 466(7302), 109–112.

Crowder, D.W., Northfield, T.D., Gomulkiewicz, R. and Snyderi, W.E. (2012) Conserving and promoting evenness: Organic farming and fire-based wildland management as case studies. *Ecology* 93(9), 2001–2007.

Czech, A., Szmigielski, M. and Sembratowicz, I. (2022) Nutritional value and antioxidant capacity of organic and conventional vegetables of the genus allium. *Scientific Reports* 12(1), 18713.

Das, S., Chatterjee, A. and Pal, T.K. (2020) Organic farming in India: A vision towards a healthy nation. *Food Quality and Safety* 4(2), 69–76.

de Ponti, T., Rijk, B. and van Ittersum, M.K. (2012) The crop yield gap between organic and conventional agriculture. *Agricultural System* 108, 1–9.

de Souza, Araujo, D.F., da Silva, A.M.R.B., de Andrade Lima, L.L., da Silva Vasconcelos, M.A. *et al*. (2014) The concentration of minerals and physicochemical contaminants in conventional and organic vegetables. *Food Control* 44, 242–248.

Dong, Y., Xu, F., Liu, L., Du, X., Ren, B. *et al*. (2020) Automatic system for crop pest and disease dynamic monitoring and early forecasting. *IEEE Journal of Selected Topics in Applied Earth Observations and Remote Sensing* 13, 4410–4418.

Dos Santos, A.M., Lima, J.S., Dos Santos, I.F., Silva, E.F. and de Santana, F.A. (2019) Mineral and centesimal composition evaluation of conventional and organic cultivars sweet potato (*Ipomoea batatas* (L.) Lam) using chemometric tools. *Food Chemistry* 273, 166–171.

Eyhorn, F., Muller, A., Reganold, J., Frison, E., Herren, H.R. *et al*. (2019) Sustainability in global agriculture driven by organic farming. *Nature Sustainability* 2, 253–255.

Gamage, A., Gangahagedara, R., Gamage, J., Jayasinghe, N., Kodikara, N. *et al*. (2023) Role of organic farming for achieving sustainability in agriculture. *Farming System* 1(1), 100005.

Gao, F., Li, H., Mu, X., Gao, H., Zhang, Y. *et al.* (2023) Effects of organic fertilizer application on tomato yield and quality: A meta-analysis. *Applied Sciences* 13(4), 2184.

Gattinger, A., Muller, A., Haeni, M. and Niggli, U. (2012) Enhanced topsoil carbon stocks under organic farming. *Proceedings of the National Academy of Science* 109, 18226–18231.

Giampieri, F., Mazzoni, L., Cianciosi, D., Alvarez-Suarez, J.M., Regolo, L. *et al.* (2022) Organic vs conventional plant-based foods: A review. *Food Chemistry* 383, 132352.

Goel, R., Debbarma, P., Kumari, P., Suyal, D.C., Kumar, S. *et al.* (2021) Assessment of soil chemical quality, soil microbial population and plant growth parameters under organic and conventional rice–wheat cropping system. *Agricultural Research* 10(2), 193–204

Gomiero, T. (2021) Organic agriculture: Impact on the environment and food quality. In: Galanakis, C.M. (ed.) *Environmental Impact of Agro-Food Industry and Food Consumption*. Academic Press, London, pp. 31–58.

Gopalakrishnan, R. (2019) Advantages and nutritional value of organic food on human health. *International Journal of Trend in Scientific Research and Development* 3(4), 242–245.

Gu, B. (2022) Recoupling livestock and crops. *Nature Food* 3(2), 102–103.

Hansen, S., Froseth, R.B., Stenberg, M., Stalenga, J., Olesen, J.E. *et al.* (2019) Reviews and syntheses: Review of causes and sources of N_2O emissions and NO_3 leaching from organic arable crop rotations. *Biogeosciences (Online)* 16(14), 5194.

Harou, A.P., Madajewicz, M., Michelson, H., Palm, C.A., Amuri, N. *et al.* (2022) The joint effects of information and financing constraints on technology adoption: Evidence from a field experiment in rural Tanzania. *Journal of Development Econonics* 155, 102707.

Hurtado-Barroso, S., Tresserra-Rimbau, A., Vallverdú-Queralt, A. and Lamuela-Raventós, R.M. (2019) Organic food and the impact on human health. *Critical Reviews in Food Science and Nutrition* 59(4), 704–714.

Jouzi, Z., Azadi, H., Taheri, F., Zarafshani, K., Gebrehiwot, K. *et al.* (2017) Organic farming and small-scale farmers: Main opportunities and challenges. *Ecological Economics* 132, 144–154.

Karp, D.S., Chaplin-Kramer, R., Meehan, T.D., Martin, E.A., DeClerck, F. *et al.* (2018) Crop pests and predators exhibit inconsistent responses to surrounding landscape composition. *Proceedings of the National Academy of Sciences of the United States of America* 115(33), E7863–E7870.

Kennedy, C.M., Lonsdorf, E., Neel, M.C., Williams, N.M., Ricketts, T.H. *et al.* (2013) A global quantitative synthesis of local and landscape effects on wild bee pollinators in agroecosystems. *Ecology Letters* 16(5), 584–599.

Khadse, A. and Rosset, P.M. (2019) Zero budget natural farming in India – from inception to institutionalization. *Agroecology and Sustainable Food Systems* 43(7–8), 848–871.

Kirchmann, H. and Andersson, R. (2008) Can organic crop production feed the world? In: Kirchmann, H. and Bergström, L. (eds) *Organic Crop Production – Ambitions and Limitations*. Springer, Dordrecht, The Netherlands, pp. 39–72.

Kishore, A.R., Niveditha, K., Uriti, A., Anilkumar, C., Sarabu, A. *et al.* (2023) An in-depth analysis of the elements shaping organic farmers: A systematic review. In: *International Conference on Energy, Materials and Communication Engineerin (ICEMCE),* Madurai, India, pp. 1–5. DOI: 10.1109/ICEMCE57940.2023.10434263.

Kniss, A.R., Savage, S.D. and Jabbour, R. (2016) Commercial crop yields reveal strengths and weaknesses for organic agriculture in the United States. *PLoS ONE* 11(8), e0161673.

Lairon, D. (2010) Nutritional quality and safety of organic food: A review. *Agronomy for Sustainable Development* 30(1), 33–41.

Lichtenberg, E.M., Kennedy, C.M., Kremen, C., Batáry, P., Berendse, F. *et al.* (2017) A global synthesis of the effects of diversified farming systems on arthropod diversity within fields and across agricultural landscapes. *Global Change Biology* 23, 4946–4957.

Madhavi, R., Vijaya, G.S. and Malathi, H. (2021) Sustainable organic farming in India. *International Journal of Modern Agriculture* 10(2), 1080–1088.

Manna, M.C., Rahman, M.M., Naidu, R., Bari, A.F., Singh, A.B. *et al.* (2021) Organic farming: A prospect for food, environment and livelihood security in Indian agriculture. *Advances in Agronomy* 170, 101–153.

Matt, D., Rembialkowska, E., Luik, A., Peetsmann, E. and Pehme, S. (2011) Quality of Organic vs. Conventional Food and Effects on Health: Report. Estonian University of Life Sciences.

Meemken, E.-M. and Qaim, M. (2018) Organic agriculture, food security, and the environment. *Annual Review of Resource Economics* 10(1), 39–63.

Meier, M.S., Stoessel, F., Jungbluth, N., Juraske, R., Schader, C. *et al.* (2015) Environmental impacts of organic and conventional agricultural products – are the differences captured by life cycle assessment? *Journal of Environmental Management* 149, 193–208.

Mesías-Ruiz, G.A., Pérez-Ortiz, M., Dorado, J., Castro, A.I. and Peña, J.M. (2023) Boosting precision crop protection towards agriculture 5.0 *via* machine learning and emerging technologies: A contextual review. *Frontiers in Plant Science* 14, 1143326.

Migliorini, P. and Wezel, A. (2017) Converging and diverging principles and practices of organic agriculture regulations and agroecology. *Agronomy for Sustainable Development* 37(6), 1–18.

Mitchell, A.E., Hong, Y.-J., Koh, E., Barrett, D.M., Bryant, D.E. *et al.* (2007) Ten-year comparison of the influence of organic and conventional crop management practices on the content of flavonoids in tomatoes. *Journal of Agricultural and Food Chemistry* 55(15), 6154–6159.

Mondejar, M.E., Avtar, R., Diaz, H.L.B., Dubey, R.K., Esteban, J. *et al.* (2021) Digitalization to achieve sustainable development goals: Steps towards a smart green planet. *Science of the Total Environment* 794, 148539.

Moos, J.H., Schrader, S. and Paulsen, H.M. (2017) Reduced tillage enhances earthworm abundance and biomass in organic farming: A meta-analysis. *Landbauforschung* 67, 3–4. DOI: 10.3220/LBF1512114926000.

Muller, A., Schader, C. and El-Hage Scialabba, N. (2017) Can organic agriculture feed the world in 2050? A critical assessment of the challenges and opportunities. *Nature Communication* 8, 1290.

Munné-Bosch, S. and Bermejo, N.F. (2024) Fruit quality in organic and conventional farming: Advantages and limitations. *Trends in Plant Science* 29(8), 878–894.

Nabuurs, G.-J., Mrabet, R., Abu Hatab, A., Bustamante, M., Clark, H. *et al.* (2022) Agriculture, forestry and other land uses (AFOLU). In: Shukla, P.R., Skea, J., Slade, R., Al Khourdajie, A. and Diemem, R. (eds) *Climate Change 2022: Mitigation of Climate Change*. Cambridge University Press, Cambridge, UK and New York.

Newton, A.C., Hawes, C. and Hackett, C.A. (2021) Adaptation of winter barley cultivars to inversion and non-inversion tillage for yield and rhynchosporium symptoms. *Agronomy* 11, 30.

Niemiec, M., Chowaniak, M., Sikora, J., Szeląg-Sikora, A., Gródek-Szostak, Z. *et al.* (2020) Selected properties of soils for long-term use in organic farming. *Sustainability* 12(6), 2509.

Olle, M. and Williams, I.H. (2012) Organic farming of vegetables. In: Lichtfouse, E. (ed.) *Sustainable Agriculture Reviews*. Springer, Dordrecht, The Netherlands, pp. 63–76.

Panday, D., Bhusal, N., Das, S. and Ghalehgolabbehbahani, A. (2024) Rooted in nature: The rise, challenges, and potential of organic farming and fertilizers in agroecosystems. *Sustainability* 16(4), 1530.

Parizad, S. and Bera, S. (2023) The effect of organic farming on water reusability, sustainable ecosystem, and food toxicity. *Environmental Science and Pollution Research International* 30(28), 71665–71676.

Pawar, J., Choudhari, V., Choudhari, G., Wagh, P., More, G. *et al.* (2022) Organic food: The importance from public health perspective. *Journal of Pharmaceutical Research International* 34(29B), 28–37. DOI: 10.9734/JPRI/2022/v34i29B36055.

Peigné, J., Casagrande, M., Payet, V., David, C., Sans, F.X. *et al.* (2016) How organic farmers practice conservation agriculture in Europe. *Renewable Agriculture and Food Systems* 31(1), 72–85.

Pimentel, D., Hepperly, P., Hanson, J., Douds, D. and Seidel, R. (2005) Environmental, energetic, and economic comparisons of organic and conventional farming systems. *BioScience* 55(7), 573.

Ponisio, L.C., M'Gonigle, L.K., Mace, K.C., Palomino, J., de Valpine, P. *et al.* (2015) Diversification practices reduce organic to conventional yield gap. *Proceedings of the Royal Society B* 282(1799), 20141396.

Porcel, M., Andersson, G.K.S., Pålsson, J. and Tasin, M. (2018) Organic management in apple orchards: Higher impacts on biological control than on pollination. *Journal of Applied Ecology* 55(6), 2779–2789.

Puyu, V., Vakhnyak, V., Lapchynskyi, V., Koberniuk, O. and Tarasyuk, V. (2023) Comparative study of organic farming systems in different geographical regions. *Scientific Horizons* 26(12), 124–134.

Rahman, A., Baharlouei, P., Koh, E.H.Y., Pirvu, D.G., Rehmani, R. *et al.* (2024) Comprehensive analysis of organic food. *Evaluating Nutritional Value and Impact on Human Health. Foods* 13, 208.

Rani, M., Kaushik, P., Bhayana, S. and Kapoor, S. (2023) Impact of organic farming on soil health and nutritional quality of crops. *Journal of the Saudi Society of Agricultural Sciences* 22(8), 560–569.

Reddy, A.A., Vimala, Y., Goudar, G., Mergu, N., Rao, J.S. *et al.* (2023) Nutritional compilation of commonly consumed organic and conventional fruits and vegetables from India. *Food and Humanity* 1, 1652–1658. DOI: 10.1016/j.foohum.2023.11.006.

Reeve, J.R., Hoagland, L.A., Villalba, J.J., Carr, P.M., Atucha, A. *et al.* (2016) Organic farming, soil health, and food quality: Considering possible links. *Advances in Agronomy* 137, 319–367.

Reganold, J.P. and Wachter, J.M. (2016) Organic agriculture in the twenty-first century. *Nature Plants* 2, 15221.

Rembiałkowska, E. (2016) Production system and composition of plant foods. *Human Health Implications of Organic Food and Organic Agriculture* 33–40.

Riar, A., Goldmann, E., Bautze, D., Rüegg, J., Bhullar, G.S. *et al.* (2024) Farm gate profitability of organic and conventional farming systems in the tropics. *International Journal of Agricultural Sustainability* 22(1), 2318933.

Rodale Institute (2022) *Farming Systems Trial: 40-Year Report*. Rodale Institute, Kutztown, Pennsylvania.

Röös, E., Mie, A., Wivstad, M., Salomon, E., Johansson, B. *et al.* (2018) Risks and opportunities of increasing yields in organic farming: A review. *Agronomy for Sustainable Development* 38(2).

Schneider, M.K., Lüscher, G., Jeanneret, P., Arndorfer, M., Ammari, Y. *et al.* (2014) Gains to species diversity in organically farmed fields are not propagated at the farm level. *Nature Communications* 5(1), 4151.

Seitz, S., Goebes, P., Puerta, V.L., Pereira, E.I.P., Wittwer, R. *et al.* (2019) Conservation tillage and organic farming reduce soil erosion. *Agronomy for Sustainable Development* 39(1).

Seufert, V. and Ramankutty, N. (2017) Many shades of gray: The context-dependent performance of organic agriculture. *Science Advances* 3(3), e1602638.

Seufert, V., Ramankutty, N. and Foley, J.A. (2012) Comparing the yields of organic and conventional agriculture. *Nature* 485(7397), 229–232.

Sharma, S.K., Choudhary, R., Jat, G. and Ravishankar, N. (2018) Strategies for achieving sustainable food systems through organic agriculture. Lead paper: Symposium on Doubling Farmers' Income through agronomic Interventions under Changing Scenario, organized by Indian Society of Agronomy, New Delhi and Maharana Pratap University of Agriculture and Technologym 24–26 october. Udaipur.

Shrivas, V.L., Singh, U., Weisskopf, L., Hariprasad, P. and Sharma, S. (2019) Effect of organic farming on structural and functional diversity of soil microbiome: Benefits and risks. In: Varma, A., Tripathi, S. and Prasad, R. (eds) *Plant Biotic Interactions*. Springer, Cham, Switzerland. DOI: 10.1007/978-3-030-26657-8_9.

Sieber, S., Graef, F., Amjath-Babu, T.S., Mutabazi, K.D., Tumbo, S.D. *et al.* (2018) Trans-SEC's food security research in Tanzania: From constraints to adoption for out- and upscaling of agricultural innovations. *Food Security* 10(4), 775–783.

Singh, M., Jaswal, Y., Manavi, M., Mishra, S., Kushwaha, A. *et al.* (2024) Organic farming in agricultural practices: A review. *Asian Journal of Biology* 20(3), 34–39.

Skinner, C., Gattinger, A., Muller, A., Mäder, P., Fließbach, A. *et al.* (2014) Greenhouse gas fluxes from agricultural soils under organic and non-organic management: A global meta-analysis. *Science of the Total Environment* 468–469, 553–563.

Smith, O.M., Cohen, A.L., Rieser, C.J., Davis A.G., Taylor, J.M. *et al.* (2019) Organic farming provides reliable environmental benefits but increases variability in crop yields: A global meta-analysis. *Frontiers in Sustainable Food Systems* 3, 82.

Sow, S., Ranjan, S., Kumari, J., Kumar, A., Kumar, N. *et al.* (2023) Organic farming: A step towards sustainable agriculture and its challenges. In: Kumar, S., Kumar, V., Singh, D.V. and Patel, J.N. (eds) *Novel Technologies in Modern Agriculture*. Vital Biotech Publication Kota, Rajasthan, India, pp. 256–268.

Sumberg, J. (2022) Future agricultures: The promise and pitfalls of a (re)turn to nature. *Outlook on Agriculture* 51(1), 3–10.

Talhaoui, N., Taamalli, A., Gómez-Caravaca, A.M., Fernández-Gutiérrez, A. and Segura-Carretero, A. (2015) Phenolic compounds in olive leaves: Analytical determination, biotic and abiotic influence, and health benefits. *Food Research International* 77, 92–108.

Timsina, J., Dutta, S., Devkota, K.P., Chakraborty, S., Neupane, R.K. *et al.* (2021) Improved nutrient management in cereals using nutrient expert and machine learning tools: Productivity, profitability and nutrient use efficiency. *Agricultural Systems* 192, 103181.

Tscharntke, T., Grass, I., Wanger, T.C., Westphal, C. and Batáry, P. (2021) Beyond organic farming – harnessing biodiversity-friendly landscapes. *Trends in Ecology & Evolution* 36(10), 919–930.

Tuck, S.L., Winqvist, C., Mota, F., Ahnström, J., Turnbull, L.A. *et al.* (2014) Land-use intensity and the effects of organic farming on biodiversity: A hierarchical meta-analysis. *The Journal of Applied Ecology* 51(3), 746–755.

Tuomisto, H.L., Hodge, I.D., Riordan, P. and Macdonald, D.W. (2012) Does organic farming reduce environmental impacts? A meta-analysis of european research. *Journal of Environmental Management* 112, 309–320.

Ullah, A., Arshad, M., Kächele, H., Khan, A., Mahmood, N. *et al.* (2020) Information asymmetry, input markets, adoption of innovations and agricultural land use in Khyber Pakhtunkhwa, Pakistan. *Land Use Policy* 90, 104261.

Ume, C., Nuppenau, E.-A. and Domptail, S.E. (2022) A feminist economics perspective on the agroecology–food and nutrition security nexus. *Environmental and Sustainability Indicators* 16, 100212.

UNCAD and UNEP (2008) Organic agriculture and food security in Africa. Available at: https://unctad.org/system/files/official-document/ditcted200715_en.pdf (accessed 16 March 2025).

van Balen, D., Cuperus, F., Haagsma, W., de Haan, J., van den Berg, W. *et al.* (2023) Crop yield response to long-term reduced tillage in a conventional and organic farming system on a sandy loam soil. *Soil and Tillage Research* 225, 105553.

van Wagenberg, C.P.A., de Haas, Y., Hogeveen, H., van Krimpen, M.M., Meuwissen, M.P.M. *et al.* (2017) Animal board invited review: Comparing conventional and organic livestock production systems on different aspects of sustainability. *Animal* 11(10), 1839–1851.

Venkat, K. (2012) Comparison of twelve organic and conventional farming systems: A life cycle greenhouse gas emissions perspective. *Journal of Sustainable Agriculture* 36(6), 620–649.

Verdi, L., Marta, A.D., Falconi, F., Orlandini, S. and Mancini, M. (2022) Comparison between organic and conventional farming systems using life cycle assessment (LCA): A case study with an ancient wheat variety. *European Journal of Agronomy* 141, 126638.

Willer, H., Trávníček, J. and Schlatter, B. (eds) (2024) *The World of Organic Agriculture: Statistics and Emerging Trends*. Research Institute of Organic Agriculture

FiBL, Frick, and IFOAM - Organics International, Bonn, Germany.
Worthington, V. (2001) Nutritional quality of organic versus conventional fruits, vegetables, and grains. *Journal of Alternative and Complementary Medicine* 7(2), 161–173.
Xu, Q., Huet, S., Poix, C., Boisdon, I. and Deffuant, G. (2018) Why do farmers not convert to organic farming? Modeling conversion to organic farming as a major change. *Natural Resource Modeling* 31(3).
Yang, C., Ji, X., Cheng, C., Liao, S., Obuobi, B. *et al.* (2024) Digital economy empowers sustainable agriculture: Implications for farmers' adoption of ecological agricultural technologies. *Ecological Indicators* 159, 111723.
Yang, G., Li, J., Liu, Z., Zhang, Y., Xu, X. *et al.* (2022) Research trends in crop-livestock systems: A bibliometric review. *International Journal of Environmental Research and Public Health* 19(14), 14.
Yu, X., Guo, L., Jiang, G., Song, Y. and Muminov, M.A. (2018) Advances of organic products over conventional productions with respect to nutritional quality and food security. *Acta Ecologica Sinica* 38(1), 53–60.
Zhang, C., Gu, B., Liang, X., Lam, S.K., Zhou, Y. *et al.* (2024) The role of nitrogen management in achieving global sustainable development goals. *Resources, Conservation and Recycling* 201, 107304.

Index

www.ingramcontent.com/pod-product-compliance
Lightning Source LLC
LaVergne TN
LVHW070903300926
849162LV00011B/1844